COMPREHENSIVE RENEWABLE ENERGY

COMPREHENSIVE RENEWABLE ENERGY

EDITOR-IN-CHIEF
Ali Sayigh
Chairman of WREC, Director General of WREN, and Chairman of IEI, Brighton, UK

VOLUME 6
HYDRO POWER

VOLUME EDITOR
André G.H. Lejeune
University of Liège, Liège, Belgium

ELSEVIER

AMSTERDAM BOSTON HEIDELBERG LONDON NEW YORK OXFORD
PARIS SAN DIEGO SAN FRANCISCO SINGAPORE SYDNEY TOKYO

Elsevier
Radarweg 29, PO Box 211, 1000 AE Amsterdam, The Netherlands
The Boulevard, Langford Lane, Kidlington, Oxford OX5 1GB, UK
225 Wyman Street, Waltham, MA 02451, USA

Copyright © 2012 Elsevier Ltd. All rights reserved.

4.04   Hydrogen Safety Engineering: The State-of-the-Art and Future Progress
Copyright © 2012 V Molkov

5.16   Renewable Fuels: An Automotive Perspective
Copyright © 2012 Lotus Cars Limited

The following articles are US Government works in the public domain and not subject to copyright:
1.19   Cadmium Telluride Photovoltaic Thin Film: CdTe
1.37   Solar Power Satellites
4.02   Current Perspective on Hydrogen and Fuel Cells
5.02   Historical Perspectives on Biofuels

No part of this publication may be reproduced, stored in a retrieval system or transmitted in any form or by any means electronic, mechanical, photocopying, recording or otherwise without the prior written permission of the publisher

Permissions may be sought directly from Elsevier's Science & Technology Rights Department in Oxford, UK: phone (+44) (0) 1865 843830; fax (+44) (0) 1865 853333; email: permissions@elsevier.com. Alternatively you can submit your request online by visiting the Elsevier web site at http://elsevier.com/locate/permissions, and selecting *Obtaining permission to use Elsevier material*

Notice
No responsibility is assumed by the publisher for any injury and/or damage to persons or property as a matter of products liability, negligence or otherwise, or from any use or operation of any methods, products, instructions or ideas contained in the material herein. Because of rapid advances in the medical sciences, in particular, independent verfication of diagnoses and drug dosages should be made.

British Library Cataloguing in Publication Data
A catalogue record for this book is available from the British Library

The Library of Congress Control Number: 2012934547

ISBN: 978-0-08-087872-0

For information on all Elsevier publications
visit our website at books.elsevier.com

Printed and bound in Italy

11 12 13 14   10 9 8 7 6 5 4 3 2 1

Working together to grow
libraries in developing countries

www.elsevier.com | www.bookaid.org | www.sabre.org

ELSEVIER   BOOK AID International   Sabre Foundation

Editorial: Gemma Mattingley, Joanne Williams
Production: Edward Taylor, Maggie Johnson

EDITOR-IN-CHIEF

Professor Ali Sayigh, BSc, DIC, PhD, CEng, a British citizen, graduated from Imperial College London and the University of London in 1966. He is a fellow of the Institute of Energy, a fellow of the Institution of Electrical Engineers, and is a chartered engineer.

From 1966 to 1985, Prof. Sayigh taught in the College of Engineering at the University of Baghdad and at King Saud University, Saudi Arabia, as a full-time professor, and also at Kuwait University as a part-time professor. From 1981 to 1985, he was Head of the Energy Department at the Kuwait Institute for Scientific Research (KISR) and expert in renewable energy at the Arab Organization of Petroleum Exporting Countries (AOPEC), Kuwait.

He started working in solar energy in September 1969. In 1984, he established links with Pergamon Press and became Editor-in-Chief of his first international journal, *Solar & Wind Technology*. Since 1990 he has been Editor-in-Chief of *Comprehensive Renewable Energy* incorporating *Solar & Wind Technology*, published by Elsevier Science Ltd., Oxford, UK. He is the editor of several international journals published in Morocco, Iran, Bangladesh, and Nigeria.

He has been a member of the International Society for Equitation Science (ISES) since 1973, founder and chairman of the ARAB Section of ISES since 1979, chairman of the UK Solar Energy Society for 3 years, and consultant to many national and international organizations, among them, the British Council, the Islamic Educational, Scientific and Cultural Organization (ISESCO), the United Nations Educational, Scientific and Cultural Organization (UNESCO), the United Nations Development Programme (UNDP), the Economic and Social Commission for Western Asia (ESCWA), and the United Nations Industrial Development Organization (UNIDO).

Since 1977 Prof. Sayigh has founded and directed several renewable energy conferences and workshops in the International Centre for Theoretical Physics (ICTP) – Trieste, Italy, Canada, Colombia, Algeria, Kuwait, Bahrain, Malaysia, Zambia, Malawi, India, the West Indies, Tunisia, Indonesia, Libya, Taiwan, UAE, Oman, the Czech Republic, Germany, Australia, Poland, the Netherlands, Thailand, Korea, Iran, Syria, Saudi Arabia, Singapore, China, the United States, and the United Kingdom.

In 1990 he established the World Renewable Energy Congress (WREC) and, in 1992, the World Renewable Energy Network (WREN), which hold their Congresses every 2 years, attracting more than 100 countries each time. In 2000, he and others in UAE, Sharjah, founded the Arab Science and Technology Foundation (ASTF) and regional conferences have been held in Sweden, Malaysia, Korea, Indonesia, Australia, UAE, and Libya, to name but a few. Prof. Sayigh has been running an annual international seminar on all aspects of renewable energy since 1990 in the United Kingdom and abroad. In total, 85 seminars have been held.

Prof. Sayigh supervised and graduated more than 34 PhD students and 64 MSc students at Reading University and the University of Hertfordshire when he was a professor from 1986 to 2004.

He has edited, contributed, and written more than 32 books and published more than 500 papers in various international journals and conferences.

In 2000–09, he initiated and worked closely with Sovereign Publication Company to produce the most popular magazine at annual bases called *Renewable Energy*, which was distributed freely to more than 6000 readers around the world. Presently, he is the editor-in-chief of *Comprehensive Renewable Energy*, coordinating 154 top scientists', engineers', and researchers' contributions in eight volumes published by Elsevier Publishing Company, Oxford, UK.

VOLUME EDITORS

Dr. Wilfried G. J. H. M. van Sark graduated from Utrecht University, the Netherlands, with an MSc in experimental physics in 1985, and with an MSc thesis on measurement and analysis of *I–V* characteristics of c-Si cells. He received his PhD from Nijmegen University, the Netherlands; the topic of his PhD thesis was III–V solar cell development, modeling, and processing. He then spent 7 years as a postdoc/senior researcher at Utrecht University and specialized in a-Si:H cell deposition and analysis. He is an expert in plasma chemical vapor deposition, both radio frequency and very high frequency. After an assistant professor position at Nijmegen University, where he worked on III–V solar cells, he returned to Utrecht University, with a focus on (single-molecule) confocal fluorescence microscopy of nanocrystals. In 2002, he moved to his present position as assistant professor at the research group Science, Technology and Society of the Copernicus Institute at Utrecht University, the Netherlands, where he performed and coordinated research on next-generation photovoltaic devices incorporating nanocrystals; for example, luminescent solar concentrators, as well as photovoltaic performance, life cycle analysis, socioeconomics, and policy development. He is member of the editorial board of Elsevier's scientific journal *Renewable Energy*, and member of various organizing committees of the European Union, the Institute of Electrical and Electronics Engineers (IEEE), and the SPIE PV conferences. He is author or coauthor of over 200 peer-reviewed journal and conference paper publications and book chapters. He has (co-)edited three books, including the present one.

Professor John K. Kaldellis holds a mechanical engineering degree from the National Technical University of Athens (NTUA) and a business administration diploma from the University of Piraeus. He obtained his PhD from NTUA (Fluid Sector) sponsored by Snecma–Dassault, France, and Bodossakis Foundation, Greece. He is currently the head of the Mechanical Engineering Department and since 1991 the director of the Soft Energy Applications and Environmental Protection Laboratory of the Technological Education Institute (TEI) of Piraeus. Prof. Kaldellis is also the scientific director (for TEI of Piraeus) of the MSc in Energy program organized by Heriot-Watt University and TEI of Piraeus. His scientific expertise is in the fields of energy and the environment. His research interests include feasibility analysis of energy sector applications; technological progress in wind, hydro, and solar energy markets; hybrid energy systems; energy storage issues; social attitudes toward renewable energy applications; and environmental technology–atmospheric pollution. He has participated in numerous research projects, funded by the European Union, European/Greek Industries, and the Greek State. Prof. Kaldellis has published six books concerning renewable energy applications and environmental protection. He is also the author of more than 100 scientific/research papers in international peer-reviewed journals and more than 300 papers for international scientific conferences. During the last decade, he was also a member of the Scientific Committee of the Hellenic Society of Mechanical–Electrical Engineers as well as a member of the organizing and scientific committee of several national and international conferences. He is currently a member of the editorial board of the *Renewable Energy International* journal and reviewer in more than 40 international journals in the energy and environment sector. He is the editor of the book *Stand-Alone and Hybrid Wind Energy Systems: Technology, Energy Storage and Applications* that has recently been published.

Dr. Soteris A. Kalogirou is a senior lecturer at the Department of Mechanical Engineering and Materials Science and Engineering at the Cyprus University of Technology, Limassol, Cyprus. He received his Higher Technical Institute (HTI) degree in mechanical engineering in 1982, his MPhil in mechanical engineering from the Polytechnic of Wales in 1991, and his PhD in mechanical engineering from the University of Glamorgan in 1995. In June 2011, he received the title of DSc from the University of Glamorgan.

For more than 25 years, he has been actively involved in research in the area of solar energy and particularly in flat-plate and concentrating collectors, solar water heating, solar steam generating systems, desalination, and absorption cooling. Additionally, since 1995, he has been involved in pioneering research dealing with the use of artificial intelligence methods, such as artificial neural networks, genetic algorithms, and fuzzy logic, for the modeling and performance prediction of energy and solar energy systems.

He has 29 books and book contributions and published 225 papers, 97 in international scientific journals and 128 in refereed conference proceedings. To date he has received more than 2550 citations on this work. He is Executive Editor of *Energy*, Associate Editor of *Renewable Energy*, and Editorial Board Member of another 11 journals. He is the editor of the book *Artificial Intelligence in Energy and Renewable Energy Systems*, published by Nova Science Inc.; coeditor of the book *Soft Computing in Green and Renewable Energy Systems*, published by Springer; and author of the book *Solar Energy Engineering: Processes and Systems*, published by Academic Press of Elsevier.

He has been a member of the World Renewable Energy Network (WREN) since 1992 and is a member of the Chartered Institution of Building Services Engineers (CIBSE), the American Society of Heating Refrigeration and Air-Conditioning Engineers (ASHRAE), the Institute of Refrigeration (IoR), and the International Solar Energy Society (ISES).

Dr. Andrew Cruden, a British citizen, was born in 1968. He obtained his BEng, MSc, and PhD in electrical engineering from the University of Strathclyde and CEng, MIEE Dr. Cruden is a past member of BSI GEL/105 Committee on Fuel Cells and Committee member of the IET Scotland Power Section. He is Director of the Scottish Hydrogen and Fuel Cell Association (SHFCA; www.shfca.org.uk) and Director of Argyll, Lomond and the Islands Energy Agency (www.alienergy.org.uk).

Dr. Cruden has been active in the field of hydrogen and fuel cells since 1995, when he acted as a consultant for Zevco Ltd., providing assistance with power electronic interfaces for early fuel cell systems. Later in 1998, he helped found the Scottish Fuel Cell Consortium (SFCC), supported by the Scottish Enterprise Energy Team, which ultimately developed a battery/fuel cell hybrid electric vehicle based on an AC Cobra kit car. The experience and contacts from the SFCC eventually gave rise to the formation of the Scottish Hydrogen and Fuel Cell Association (SHFCA), a trade body for the industry to promote and commercialize Scottish expertise in this field. Dr. Cruden was the founding chairman of the SHFCA.

Dr. Cruden is currently investigating alkaline electrolyzers in terms of improving their part load efficiency and lifetime when powered by variable renewable power sources, for example, wind turbines, as part of a £5 million EPSRC Supergen project on the 'Delivery of Sustainable Hydrogen' (EP/G01244X/1). He is also working with a colleague within Electronic and Electrical Engineering (EEE) at Strathclyde, studying the concept of vehicle-to-grid energy storage, as a mechanism not only to allow controlled load leveling on the power system, but also to potentially 'firm' up renewable energy generation. This work is supported by two research grants, an international E.On Research Initiative 2007 award and an ESPRC grant (EP/F062133/1).

Dr. Cruden is a senior lecturer within the Department of Electronic and Electrical Engineering at the University of Strathclyde. His current fields of research are modeling fuel cell and electrolyzer systems, fuel cell combined heat and power (CHP) systems, power electronic devices for interfacing both vehicular and stationary fuel cell systems, condition monitoring systems for renewable energy sources (i.e., wind turbines as part of EPSRC Supergen on Wind Energy Technologies, EP/D034566/1), and energy management systems for hybrid electric vehicles.

His areas of expertise include hydrogen-powered fuel cells and electrolyzers, energy storage for electric vehicles, and renewable energy generation.

Professor Dermot J. Roddy, BSc, PhD, CEng, FIET, joined Newcastle University as Science City Professor of Energy in 2008 after a period of some 20 years in the energy industry and petrochemical sectors. He is also Director of the Sir Joseph Swan Centre for Energy Research, which integrates energy research across Newcastle University and links with a powerful external industrial base in the energy sector. Outside of the university he is Chairman of Northeast Biofuels, Finance Director of the UK Hydrogen Association, and Vice-President of the Northern England Electricity Supply Companies Association. Prior to coming to Newcastle University, he was Chief Executive of Renew Tees Valley Ltd. – a company which he set up in 2003 to create a viable and vibrant economy in the Tees Valley based on renewable energy and recycling – where he was instrumental in a wide range of major renewable energy and low-carbon projects relating to biomass, biofuels, hydrogen, carbon capture and storage, wind, and advanced waste processing technologies. From 1998 to 2002, he ran the crude oil refinery on Teesside as a site director for a $5 billion turnover facility before moving to the Netherlands to work on Petroplus' international growth plans. Roddy's experience in the petrochemical industry began in 1985, involving a variety of UK and international roles in operations, engineering, and technology with ICI and others. Prior to that he developed leading-edge technology at Queen's University, Belfast, for optimization and control in aerospace applications.

André G. H. Lejeune was born on 2 August 1942 in Belgium. He was graduated in 1967 as a civil engineer, in 1972 as doctor in applied sciences (PhD), and in 1973 as master in oceanography in the University of Liège in Belgium. He was appointed full-time professor in the same university in 1976, and was visitor professor at the UNESCO–IHE Institute for Water Education in the Netherlands and Ecole Polytechnique Fédérale de Lausanne (EPFL) in Switzerland. Within the framework of his activities of professor, director of the Hydraulic Constructions and Hydraulic Research Laboratory, and expert, he took part in studies of dams and hydraulic structures and went on site in more than 90 countries of the world. In particular, he was for the last 6 years the chairman of the Technical Committee on Hydraulics for Dams in ICOLD (International Commission of Large Dams). He is a member of the Belgian Royal Academy of Sciences. He made his PhD thesis in hydraulic numerical modelization. This thesis received the Lorenz G. Straub Award in Minneapolis, USA (H. Einstein Jr. was a member of the Jury), and was used in particular by Chinese colleagues in the Three Gorges Project. Due to his practice and experience, he has a very complete knowledge of the hydraulic phenomena modelizations through both numerical and physical means.

With his wife, he has 3 children and 11 grandchildren. He likes books, tennis, and diving.

Thorsteinn I. Sigfusson is an internationally recognised physicist, educated in Copenhagen, Denmark, and Cambridge, UK. He is Director-General of the Innovation Center, Iceland and Professor of physics at the University of Iceland. He has been a visiting professor at Columbia University, New York, and he is currently the lead scientist in a prize-winning energy technology project performed at Tomsk Polytechnic University in Tomsk, Russia.

He has been a key figure in the introduction of new ideas and opportunities in the further greening of Icelandic society through the energy industry, and instrumental in the challenge of saving imported hydrocarbons by focusing on hydrogen from renewable energy.

He has started over a dozen start-up companies from research in Iceland and chaired various international societies in alternative energy. Among his achievements in geothermal energy is the construction of the world's largest solid-state thermoelectric generator powered with geothermal steam in southern Iceland. At the Innovation Center, Iceland, efforts are made to develop materials to withstand erosion in geothermal environments.

AbuBakr S. Bahaj is Professor of Sustainable Energy at the University of Southampton. After completing his PhD, he was employed by the University, progressing from a researcher to a personnel chair of Sustainable Energy. Over the past 20 years, Prof. Bahaj has established the energy theme within the University and directed his Sustainable Energy Research Group (SERG, www.energy.soton.ac.uk), which is now considered to be one of the United Kingdoms's leading university-based research groups in renewable energy and energy in buildings. He initiated and managed research in ocean energy conversion (resources, technologies, and impacts), photovoltaics, energy in buildings, and impacts of climate change on the built environment in the University. This work has resulted in over 230 articles published in academic refereed journals and conference series of international standing (see www.energy.soton.ac.uk).

Prof. Bahaj is the head of the Energy and Climate Change Division (ECCD) within the highly rated Faculty of Engineering and the Environment – Civil Engineering and the Environment – (www.civil.soton.ac.uk/research/divisions/divlist.asp?ResearchGroupID=1) (second in the United Kingdom, Research Assessment Exercise in 2008, with 80% of research judged to be either 'World Leading' or 'Internationally Excellent'). The aims of the Division and SERG are to promote and execute fundamental and applied research and preindustrial development in the areas of energy resources, technologies, energy efficiency, and the impact of climate change.

Prof. Bahaj is an experienced research team director and has many internationally focused research projects including collaborative projects in China, the European Union, the Middle East, and Africa. He also coordinated (2006–10) the United Kingdom's Engineering and Physical Sciences Research Council (EPSRC), Ecoregion Research Networks that aim to develop research themes and projects to study eco-city development encompassing resource assessment, technology pathways for the production and conservation of energy, planning, and social and economic studies required in establishing eco-regions in China and elsewhere (http://www.eco-networks.org). He is a founding member of the Sino-UK Low Carbon City Development Cooperation (LCCD) which aims to promote and undertake research into pathways for low-carbon development in Chinese cities. His work also encompasses an ongoing multimillion pound program in Africa, 'Energy for Development' for promoting and implementing village electrification systems, addressing villager's needs, and establishing coherent approaches to the commercial sustainability of the projects. This program is funded by the Research Councils and the UK Department for International Development (DFID; www.energyfordevelopment.net).

Prof. Bahaj is the editor-in-chief of the *International Journal of Sustainable Energy* and associate Editor of the *Renewable & Sustainable Energy Review*. He was on the editorial boards of the journals *Sustainable Cities and Society* and *Renewable Energy* (2005–11), and the United Kingdom's Institute of Civil Engineering journal *Energy* (2006–09). He was a member of the Tyndall Centre for Climate Change Research Supervisory Board (2005–10), and from 2001 to 2007 he was a member of the UK Government Department of Business, Enterprise and Regulatory Reform (now Department for Business Innovations and Skills, BIS), Technology Programmes Panels on Water (including ocean energy) and Solar Energy, now being administered by the Technology Strategy Board (TSB). Prof. Bahaj was the chair of the Technical Committees of the World Renewable Energy Congress – held in Glasgow (July 2008) and in Abu Dhabi (September 2010). He was a member of the Technical Committee of the 27th International Conference on Offshore Mechanics and Arctic Engineering (OMAE, 2008), a member of the management and technical committees of the European Wave and Tidal Energy Conferences (EWTEC, Porto, Portugal, September 2007; and Uppsala, Sweden, September 2009). He is also a member of the British Standards Institution (BSI) Committee GEL/82 on PV Energy Systems. Recently, at the invitation of the International Energy Agency, he has completed the 2008 status report on tidal stream energy conversion and in September 2009 was elected to chair the next EWTEC conference in the series – EWTEC2011 which was held in Southampton, 5–9 September 2011, and attended by around 500 participants.

To address training in the areas of energy and climate change Prof. Bahaj has coordinated and developed a set of MSc programs under the banner 'Energy and Sustainability' that address Energy Resources and Climate Change and Energy, Environment and Buildings.

CONTRIBUTORS FOR ALL VOLUMES

P Agnolucci
Imperial College London, London, UK

EO Ahlgren
Chalmers University of Technology, Gothenburg, Sweden

D Aklil
Pure Energy Center, Unst, Shetland Isles, UK

D-C Alarcón Padilla
Centro de Investigaciones Energéticas Medioambientales y Tecnológicas (CIEMAT), Plataforma Solar de Almeria, Almeria, Spain

K Alexander
University of Canterbury, Christchurch, New Zealand

S Alexopoulos
Aachen University of Applied Sciences, Jülich, Germany

A Altieri
UNICA – Brazilian Sugarcane Industry Association, São Paulo, Brazil

A Anthrakidis
Aachen University of Applied Sciences, Jülich, Germany

E Antolín
Universidad Politécnica de Madrid, Madrid, Spain

P Archambeau
University of Liège, Liège, Belgium

H Ármannsson
Iceland GeoSurvey (ISOR), Reykjavík, Iceland

MF Askew
Wolverhampton, UK

A Athienitis
Concordia University, Montreal, QC, Canada

G Axelsson
University of Iceland, Reykjavik, Iceland

V Badescu
Polytechnic University of Bucharest, Bucharest, Romania

AS Bahaj
The University of Southampton, Southampton, UK

P Banda
Instituto de Sistema Fotovoltaicos de Concentración (ISFOC), Puertollano, Spain

VG Belessiotis
'DEMOKRITOS' National Center for Scientific Research, Athens, Greece

P Berry
ADAS High Mowthorpe, Malton, UK

F Bidault
Imperial College London, London, UK

D Biro
Fraunhofer Institute for Solar Energy Systems, Freiburg, Germany

G Boschloo
Uppsala University, Uppsala, Sweden

C Boura
Aachen University of Applied Sciences, Jülich, Germany

E Bozorgzadeh
Iran Water and Power Resources Development Company (IWPCO), Tehran, Iran

CE Brewer
Iowa State University, Ames, IA, USA

M Börjesson
Chalmers University of Technology, Gothenburg, Sweden

RC Brown
Iowa State University, Ames, IA, USA

F Bueno
University of Burgos, Burgos, Spain

K Burke
NASA Glenn Research Center, Cleveland, OH, USA

LF Cabeza
GREA Innovació Concurrent, Universitat de Lleida, Lleida, Spain

L Candanedo
Dublin Institute of Technology, Dublin, Ireland

YG Caouris
University of Patras, Patras, Greece

UB Cappel
Uppsala University, Uppsala, Sweden

JA Carta
Universidad de Las Palmas de Gran Canaria, Las Palmas de Gran Canaria, Spain

P Chen
Dalian Institute of Chemical Physics, Dalian, China

DG Christakis
Wind Energy Laboratory, Technological Educational Institute of Crete, Crete, Greece

DA Chwieduk
Warsaw University of Technology, Warsaw, Poland

J Clark
University of York, York, UK

G Conibeer
University of New South Wales, Sydney, NSW, Australia

AJ Cruden
University of Strathclyde, Glasgow, UK

MC da Silva

B Davidsdottir
University of Iceland, Reykjavík, Iceland

O de la Rubia
Instituto de Sistema Fotovoltaicos de Concentración (ISFOC), Puertollano, Spain

E Despotou
Formerly of the European Photovoltaic Industry Association, Brussels, Belgium

BJ Dewals
University of Liège, Liège, Belgium

AL Dicks
The University of Queensland, Brisbane, QLD, Australia

R DiPippo
University of Massachusetts Dartmouth, Dartmouth, MA, USA

E Dunlop
European Commission DG Joint Research Centre, Ispra, Italy

NM Duteanu
Newcastle University, Newcastle upon Tyne, UK; University 'POLITEHNICA' Timisoara, Timisoara, Romania

LM Eaton
Oak Ridge National Laboratory, Oak Ridge, TN, USA

H-J Egelhaaf
Konarka Technologies GmbH, Nürnberg, Germany

T Ehara
Mizuho Information & Research Institute, Tokyo, Japan

B Erable
Newcastle University, Newcastle upon Tyne, UK; CNRS-Université de Toulouse, Toulouse, France

S Erpicum
University of Liège, Liège, Belgium

G Evans
NNFCC, Biocentre, Innovation Way, Heslington, York, UK

AFO Falcão
Instituto Superior Técnico, Technical University of Lisbon, Lisbon, Portugal

G Faninger
University of Klagenfurt, Klagenfurt, Austria; Vienna University of Technology, Vienna, Austria

GA Florides
Cyprus University of Technology, Limassol, Cyprus

ÓG Flóvenz
Iceland GeoSurvey (ISOR), Reykjavík, Iceland

RN Frese
VU University Amsterdam, Amsterdam, The Netherlands

Þ Friðriksson
Iceland GeoSurvey (ISOR), Reykjavík, Iceland

VM Fthenakis
Columbia University, New York, NY, USA; Brookhaven National Laboratory, Upton, NY, USA

M Fuamba
École Polytechnique de Montréal, Montreal, QC, Canada

A Fuller
University of Canterbury, Christchurch, New Zealand

LMC Gato
Instituto Superior Técnico, Technical University of Lisbon, Lisbon, Portugal

R Gazey
Pure Energy Center, Unst, Shetland Isles, UK

TA Gessert
National Renewable Energy Laboratory (NREL), Golden, CO, USA

MM Ghangrekar
*Newcastle University, Newcastle upon Tyne, UK;
Indian Institute of Technology, Kharagpur, India*

M Giannouli
University of Patras, Patras, Greece

EA Gibson
University of Nottingham, Nottingham UK

A Gil
Hydropower Generation Division of Iberdrola, Salamanca, Spain

SW Glunz
Fraunhofer Institute for Solar Energy Systems, Freiburg, Germany

JC Goldschmidt
Fraunhofer Institute for Solar Energy Systems ISE, Freiburg, Germany

R Gottschalg
Loughborough University, Leicestershire, UK

MA Green
The University of New South Wales, Sydney, NSW, Australia

J Göttsche
Aachen University of Applied Sciences, Jülich, Germany

J Guo
China Institute of Water Resources and Hydropower Research (IWHR), Beijing, China

A Hagfeldt
Uppsala University, Uppsala, Sweden

B Hagin
Ingénieur-Conseil, Lutry, Switzerland

K Hall
Technology Transition Corporation, Ltd., Tyne and Wear, UK

O Hamandjoda
University of Yaounde, Yaounde, Republic of Cameroon

AP Harvey
Newcastle University, Newcastle upon Tyne, UK

JA Hauch
Konarka Technologies GmbH, Nürnberg, Germany

D Heinemann
University of Oldenburg, Oldenburg, Germany

V Heller
Imperial College London, London, UK

GP Hersir
Iceland GeoSurvey (ISOR), Reykjavík, Iceland

T Heyer
Technical University of Dresden, Dresden, Germany

P Hilger
Aachen University of Applied Sciences, Jülich, Germany

B Hillring
Swedish University of Agricultural Sciences, Skinnskatteberg, Sweden

T Hino
CTI Engineering International Co., Ltd., Chu-o-Ku, Japan

LC Hirst
Imperial College London, London, UK

B Hoffschmidt
Aachen University of Applied Sciences, Jülich, Germany

H Horlacher
Technical University of Dresden, Dresden, Germany

N Hughes
Imperial College London, London, UK

SL Hui
Bechtel Civil Company, San Francisco, CA, USA

D Husmann
University of Wisconsin–Madison, Madison, WI, USA

JTS Irvine
University of St Andrews, St Andrews, UK

D Jacobs
Freie Universität Berlin, Berlin, Germany

Y Jestin
Advanced Photonics and Photovoltaics Group, Bruno Kessler Foundation, Trento, Italy

A Jäger-Waldau
Institution for Energy Transport, Ispra, Italy

S Jianxia
Design and Research Institute, Yangzhou City, Jiangsu Province, China

E Johnson
Pure Energy Center, Unst, Shetland Isles, UK

HF Kaan
TNO Energy, Comfort and Indoor Quality, Delft, The Netherlands

JK Kaldellis
Technological Education Institute of Piraeus, Athens, Greece

SA Kalogirou
Cyprus University of Technology, Limassol, Cyprus

HD Kambezidis
Institute of Environmental Research and Sustainable Development, Athens, Greece

M Kapsali
Technological Education Institute of Piraeus, Athens, Greece

M Karimirad
Norwegian University of Science and Technology, Trondheim, Norway

T Karlessi
National and Kapodistrian University of Athens, Athens, Greece

SN Karlsdóttir
Innovation Center Iceland, Iceland

D Al Katsaprakakis
Wind Energy Laboratory, Technological Educational Institute of Crete, Crete, Greece

O Kaufhold
Aachen University of Applied Sciences, Jülich, Germany

CA Kaufmann
Helmholtz Zentrum für Materialien und Energie GmbH, Berlin, Germany

KA Kavadias
Technological Education Institute of Piraeus, Athens, Greece

LL Kazmerski
National Renewable Energy Laboratory, Golden, CO, USA

A Kazmi
University of York, York, UK

K Kendall
University of Birmingham, Birmingham, UK

J Kenfack
University of Yaounde, Yaounde, Republic of Cameroon

R Kenny
European Commission DG Joint Research Centre, Ispra, Italy

HC Kim
Brookhaven National Laboratory, Upton, NY, USA

L Kloo
KTH—Royal Institute of Technology, Stockholm, Sweden

G Knothe
USDA Agricultural Research Service, Peoria, IL, USA

FR Kogler
Konarka Technologies GmbH, Nürnberg, Germany

D Kolokotsa
Technical University of Crete, Crete, Greece

K Komoto
Mizuho Information & Research Institute, Tokyo, Japan

E Kondili
Technological Education Institute of Piraeus, Athens, Greece

H Kristjánsdóttir
University of Iceland, Reykjavík, Iceland

LA Lamont
Petroleum Institute, Abu Dhabi, UAE

GA Landis
NASA Glenn Research Center, Cleveland, OH, USA

JGM Lee
Newcastle University, Newcastle upon Tyne, UK

G Leftheriotis
University of Patras, Patras, Greece

A Lejeune
University of Liège, Liège, Belgium

T Leo
FuelCell Energy Inc., Danbury, CT, USA

E Lester
The University of Nottingham, Nottingham, UK

E Lorenz
University of Oldenburg, Oldenburg, Germany

JW Lund
Geo-Heat Center, Oregon Institute of Technology, Klamath Falls, OR, USA

A Luque
Universidad Politécnica de Madrid, Madrid, Spain

BP Machado
Intertechne, Curitiba, PR, Brazil

EBL Mackay
GL Garrad Hassan, Bristol, UK

T-F Mahdi
École Polytechnique de Montréal, Montreal, QC, Canada

GG Maidment
London South Bank University, London, UK

A Malmgren
BioC Ltd, Cirencester, UK

C Manson-Whitton
Progressive Energy Ltd., Stonehouse, UK

Á Margeirsson
Magma Energy Iceland, Reykjanesbaer, Iceland

A Martí
Universidad Politécnica de Madrid, Madrid, Spain

M Martinez
Instituto de Sistema Fotovoltaicos de Concentración (ISFOC), Puertollano, Spain

S Mathew
University of Brunei Darussalam, Gadong, Brunei Darussalam

PH Middleton
University of Agder, Grimstad, Norway

R Mikalsen
Newcastle University, Newcastle upon Tyne, UK

D Milborrow
Lewes, East Sussex, UK

H Müllejans
European Commission DG Joint Research Centre, Ispra, Italy

V Molkov
University of Ulster, Newtownabbey, Northern Ireland, UK

M Moner-Girona
Joint Research Centre, European Commission, Institute for Energy and Transport, Ispra, Italy

PE Morthorst
Technical University of Denmark, Roskilde, Denmark

N Mortimer
North Energy Associates Ltd, Sheffield, UK

E Mullins
Teagasc, Oak Park Crops Research Centre, Carlow, Republic of Ireland

P Mulvihill
Pioneer Generation Ltd., Alexandra, New Zealand

DR Myers
National Renewable Energy Laboratory, USA

D Nash
University of Strathclyde, Glasgow, UK

GF Nemet
University of Wisconsin–Madison, Madison, WI, USA

H Nfaoui
Mohammed V University, Rabat, Morocco

T Nikolakakis
Columbia University, New York, NY, USA

X Niu
Changjiang Institute of Survey, Planning, Design and Research, Wuhan, China

B Norton
Dublin Institute of Technology, Dublin, Ireland

A Nuamah
The University of Nottingham, Nottingham, UK; RWE npower, Swindon, UK

B O'Connor
Aachen University of Applied Sciences, Jülich, Germany

O Olsson
Swedish University of Agricultural Sciences, Skinnskatteberg, Sweden

V Ortisi
Pure Energy Center, Unst, Shetland Isles, UK

H Ossenbrink
European Commission DG Joint Research Centre, Ispra, Italy

AG Paliatsos
Technological Education Institute of Piraeus, Athens, Greece

A Pandit
VU University Amsterdam, Amsterdam, The Netherlands

E Papanicolaou
'DEMOKRITOS' National Center for Scientific Research, Athens, Greece

A Paurine
London South Bank University, London, UK

N Pearsall
Northumbria University, Newcastle, UK

RJ Pearson
Lotus Engineering, Norwich, UK

RD Perlack
Oak Ridge National Laboratory, Oak Ridge, TN, USA

H Pettersson
Swerea IVF AB, Mölndal, Sweden

GS Philip
KCAET, Malapuram, Kerala, India

S Pillai
The University of New South Wales, Sydney, NSW, Australia

M Pirotton
University of Liège, Liège, Belgium

BG Pollet
University of Birmingham, Birmingham, UK

D Porter
Association of Electricity Producers, London, UK

A Pouliezos
Technical University of Crete, Hania, Greece

R Preu
Fraunhofer Institute for Solar Energy Systems, Freiburg, Germany

CM Ramos

C Rau
Aachen University of Applied Sciences, Jülich, Germany

AA Refaat
Cairo University, Giza, Egypt

TH Reijenga
BEARiD Architecten, Rotterdam, The Netherlands

AHME Reinders
*Delft University of Technology, Delft, The Netherlands;
University of Twente, Enschede, The Netherlands*

G Riley
RWE npower, Swindon, UK

DJ Roddy
Newcastle University, Newcastle upon Tyne, UK

S Rolland
Alliance for Rural Electrification, Brussels, Belgium

A Roskilly
Newcastle University, Newcastle upon Tyne, UK

F Rubio
Instituto de Sistema Fotovoltaicos de Concentración (ISFOC), Puertollano, Spain

F Rulot
University of Liège, Liège, Belgium

L Rybach
GEOWATT AG, Zurich, Switzerland

M Santamouris
National and Kapodistrian University of Athens, Athens, Greece

J Sattler
Aachen University of Applied Sciences, Jülich, Germany

M Sauerborn
Aachen University of Applied Sciences, Jülich, Germany

TW Schmidt
The University of Sydney, Sydney, NSW, Australia

N Schofield
University of Manchester, Manchester, UK

REI Schropp
Utrecht University, Utrecht, The Netherlands

K Scott
Newcastle University, Newcastle upon Tyne, UK

SP Sen
NHPC Ltd., New Delhi, India

TI Sigfusson
Innovation Center, Reykjavik, Iceland

L Sims
*Konarka Technologies GmbH, Nürnberg, Germany;
Universität Augsburg, Augsburg, Germany*

C Smith
NNFCC, Biocentre, Innovation Way, Heslington, York, UK

K Sæmundsson
Iceland GeoSurvey (ISOR), Reykjavík, Iceland

BK Sovacool
Vermont Law School, South Royalton, VT, USA

J Spink
Teagasc, Oak Park Crops Research Centre, Carlow, Republic of Ireland

JN Sørensen
Technical University of Denmark, Lyngby, Denmark

T Stallard
The University of Manchester, Manchester, UK

GS Stavrakakis
Technical University of Crete, Chania, Greece

R Steim
Konarka Technologies GmbH, Nürnberg, Germany

BJ Stokes
CNJV LLC, Washington, DC, USA

L Sun
*KTH—Royal Institute of Technology, Stockholm, Sweden;
Dalian University of Technology (DUT), Dalian, China*

L Suo
Science and Technology Committee of the Ministry of Water Resources, Beijing, China

DT Swift-Hook
*Kingston University, London, UK;
World Renewable Energy Network, Brighton, UK*

A Synnefa
National and Kapodistrian University of Athens, Athens, Greece

S Szabo
Joint Research Centre, European Commission, Institute for Energy and Transport, Ispra, Italy

MJY Tayebjee
The University of Sydney, Sydney, NSW, Australia

A Tesfai
University of St Andrews, St Andrews, UK

P Thornley
The University of Manchester, Manchester, UK

Y Tripanagnostopoulos
University of Patras, Patras, Greece

L Tsakalakos
General Electric – Global Research Center, New York, NY, USA

JWG Turner
Lotus Engineering, Norwich, UK

E Tzen
Centre for Renewable Energy Sources and Saving (CRES), Pikermi, Attica, Greece

T Unold
Helmholtz Zentrum für Materialien und Energie GmbH, Berlin, Germany

J van der Heide
imec vzw, Leuven, Belgium

P van der Vleuten
Free Energy Consulting, Eindhoven, The Netherlands

F Van Hulle
XP Wind Consultancy, Leuven, Belgium

GC van Kooten
University of Victoria, Victoria, BC, Canada

WGJHM van Sark
Utrecht University, Utrecht, The Netherlands

I Waller
FiveBarGate Consultants Ltd, Cleveland, UK

I Walsh
Opus International Consultants Ltd., New Zealand

Y Wang
Newcastle University, Newcastle upon Tyne, UK

T Wizelius
Gotland University, Visby, Sweden; Lund University, Lund, Sweden

LL Wright
University of Tennessee, Knoxville, TN, USA

H Xie
Changjiang Institute of Survey, Planning, Design and Research, Wuhan, China

M Yamaguchi
Toyota Technological Institute, Tempaku, Nagoya, Japan

P Yianoulis
University of Patras, Patras, Greece

EH Yu
Newcastle University, Newcastle upon Tyne, UK

H Yu
Newcastle University, Newcastle upon Tyne, UK

DP Zafirakis
Technological Education Institute of Piraeus, Athens, Greece

G Zaragoza
Centro de Investigaciones Energéticas Medioambientales y Tecnológicas (CIEMAT), Plataforma Solar de Almeria, Almeria, Spain

M Zeman
Delft University of Technology, Delft, The Netherlands

PREFACE

Comprehensive Renewable Energy is the only multivolume reference work of its type at a time when renewable energy sources are increasingly in demand and realistically sustainable, clean, and helping to combat climate change and global warming. Renewable energy investment has exceeded US$10 billion per year during the past 5 years. The World Renewable Energy Network (WREN) predicts that this figure is set to increase to US$20 billion per year by 2015.

As Editor-in-Chief, I have assembled an impressive world-class team of 154 volume editors and contributing authors for the eight volumes. They represent policy makers, researchers, industrialists, financiers, and heads of organizations from more than 80 countries to produce this definitive complete work in renewable energy covering the past, explaining the present, and giving the ideas and prospects of development for the future. There are more than 1000 references from books, journals, and the Internet within the eight volumes. *Comprehensive Renewable Energy* is full of color charts, illustrations, and photographs of real projects and research results from around the world. Each chapter has been painstakingly reviewed and checked for consistent high quality. The result is an authoritative overview that ties the literature together and provides the user with reliable background information and a citation resource.

The field of renewable energy research and development is represented by many journals that are directly and indirectly concerned with the field. But no reference work encompasses the entire field and unites the different areas of research through in-depth foundational reviews. *Comprehensive Renewable Energy* fills this vacuum, and is the definitive work for this subject area. It will help users apply context to diverse journal literature, aiding them in identifying areas for further research and development.

Research into renewable energy is spread across a number of different disciplines and subject areas. These areas do not always share a unique identifying factor or subject themselves to clear and concise definitions. This work unites the different areas of research and allows users, regardless of their background, to navigate through the most essential concepts with ease, saving them time and vastly improving their understanding so that they can move forward, whether in their research, development, manufacturing, or purchase of renewable energy.

The first volume is devoted to Photovoltaic Technology and is edited by Mr. Wilfried G. J. H. M. van Sark from the Netherlands. It consists of 38 chapters, written by 41 authors from Europe, the United States, Japan, China, India, Africa, and the Middle East. The topics covered range from the smallest applications to MW projects. A brief introduction and history is followed by chapters on finance and economics, solar resources, up- and downconversion, crystalline photovoltaic (PV) cells, luminescent concentrators, thin-film and multiple-junction plastic solar cells, dye-sensitized solar cells, bio-inspired converters, application of micro- and nanotechnology, building integrated photovoltaics (BIPV) application in architecture, and very large-scale PV systems. Without doubt, this is an impressive tour of an immense field.

Volume 2 is devoted to Wind Energy and is edited by Professor John K. Kaldellis from Greece. It consists of 22 chapters written by 22 authors, again from various parts of the world, covering all aspects of wind energy from small wind mills to very large wind farms. The volume includes chapters on the history of wind power, the potential of wind power, wind turbine development, aerodynamic analysis, mechanical and electrical loads, control systems, noise and testing, onshore and offshore wind systems, policy, industry, and special wind power applications.

Volume 3 is devoted to Solar Thermal Applications and the editor is Professor Soteris A. Kalogirou from Cyprus. It consists of 19 chapters written by 17 authors. All aspects of solar thermal energy and its applications are covered. The volume begins with solar energy as a source of heat and goes on to describe the history of thermal applications, low-temperature and high-temperature storage systems, selective coating, glazing, modeling and simulation, hot water systems, space heating and cooling, water desalination, industrial and agricultural applications, concentration power, heat pumps, and passive solar architecture. The authors have looked at the Sun from the thermal energy aspect and put together a very informative and up-to-date volume from which every interested person, no matter what their level of knowledge, can benefit.

Volume 4 is on Fuel Cells and Hydrogen Technology and is edited by Dr. Andrew Cruden from the United Kingdom. It consists of 14 chapters covering the following topics: introduction and perspectives on hydrogen and fuel cells; theory and application of alkaline fuel cells; application of proton exchange membrane (PEM) fuel cells; molten carbonate fuel cells; solid oxide fuel cells; microbial and biological fuel cells; storage of compressed gas and hydrogen; the economy and policy of hydrogen technology; hydrogen safety engineering and future progress; the use of hydrogen for transport; and hydrogen and fuel cell power electronics. The 14 chapters were written by 16 authors. All aspects of practice, innovative technology, and future guidelines for researchers and industry have been addressed in this definitive volume.

Volume 5 deals with the huge field of Biomass and Biofuels and is edited by Professor Dermot J. Roddy from the United Kingdom. This work consists of 21 chapters written by 23 authors, again covering all aspects of biomass and biofuels, including their past, present, and future. The volume explains the history and prospective future of biofuels; bioethanol development in Brazil; power generation from biomass; biomass co-firing stations; biomass world market; a critical assessment of biomass – combined heat and power (CHP) energy systems; the ethics of biofuel production – issues, constraints, and limitations; greenhouse gases life cycle analysis; six different solutions from gasification and pyrolysis; new processes in biomass-to-liquid technology; new processes in biofuel production; biofuels from waste materials; novel feedstocks and woody biomass; feedstocks with the potential of yield improvement; renewable fuels – an automotive prospective; and novel use of biofuels in a range of engine configurations. Under Expanding the Envelope, there are chapters on biochar, extracting additional value from biomass, and biomass to chemicals. Finally, the chapter on bioenergy policy development concludes the volume.

Volume 6 is concerned with Hydro Power and is edited by Professor André G. H. Lejeune from Belgium. This is the oldest of all the renewable energy applications and has progressed over the ages from pico-hydro of a few hundred watts to large- and mega-scale dams generating more than 3000 MW with innovative civil engineering capability. This volume consists of 18 chapters prepared by 21 authors. It contains introduction – benefits and constraints of hydropower, recent developments and achievements in hydraulic research in China, and the management of hydropower and its impacts through construction and operation. The volume then assesses nine hydropower schemes around the world: the Three Gorges Project in China; large hydropower plants of Brazil; hydropower in Iran – vision and strategy; the recent trend in developing hydropower in India; the evolution of hydropower in Spain; hydropower in Japan; hydropower in Canada; an overview of institutional structure reform of the Cameroon power sector and assessment; and hydropower reliability in Switzerland. Other important issues are covered: pumped storage power plants; simplified generic axial-flow microhydro turbines; the development of a small hydroelectric scheme at Horseshoe Bend, Teviot River, New Zealand; concrete durability in dam design structure; and long-term sediment management for sustainable hydropower.

Volume 7 deals with Geothermal Energy. The editor of this volume is Professor Thorsteinn I. Sigfusson from Iceland. The volume consists of 10 chapters, which are written by 15 different authors. It covers the following areas: introduction and the physics of geothermal resources and management during utilization; geothermal shallow systems – heat pumps; geothermal exploration techniques; corrosion, scaling, and material selection in geothermal power production; direct heat utilization of geothermal energy; geothermal power plants; geochemical aspects of geothermal utilization; geothermal cost and investment factors; and the role of sustainable geothermal development.

Volume 8 is devoted to Generating Electricity from the Oceans, edited by Professor AbuBakr S. Bahaj from the United Kingdom. It consists of six chapters written by five authors. The volume covers the historical aspects of wave energy conversion, resource assessment for wave energy, development of wave devices from initial conception to commercial demonstration, air turbines, and the economics of ocean energy.

One chapter is totally devoted to Renewable Energy Policy and Incentives. It is included in the first volume only. The author of this chapter is Mr. David Porter, Chief Executive of the Association of Electricity Producers in the United Kingdom, an author who has had vast experience of dealing with electricity generation in the United Kingdom over many years. He has advised the British Government on how to meet supply and demand of electricity and coordinate with all electricity producers regarding their sources and supply. The chapter outlines the types of mechanisms used to promote renewable energy and their use, the impact on their deployment, ensuring investor certainty, the potential for harmonizing support schemes, and the conclusion.

In short, my advice to anyone who wants to acquire comprehensive knowledge concerning renewable energy, no matter which subject or application, is that they should acquire this invaluable resource for their home, research center and laboratory, company, or library.

Professor Ali Sayigh BSc, DIC, PhD, FIE, FIEE, CEng
Chairman of WREC (World Renewable Energy Congress)
Director General of WREN (World Renewable Energy Network)
Chairman of IEI (The Institution of Engineers (India))
Editor-in-Chief of *Renewable Energy*
Editor-in-Chief of *Renewable Energy Magazine*

CONTENTS

Editor-in-Chief	v
Volume Editors	vii
Contributors for All Volumes	xi
Preface	xix

Volume 1   Photovoltaic Solar Energy

Renewable Energy

1.01	Renewable Energy Policy and Incentives D Porter	1

Photovoltaic Solar Energy

1.02	Introduction to Photovoltaic Technology WGJHM van Sark	5
1.03	Solar Photovoltaics Technology: No Longer an Outlier LL Kazmerski	13
1.04	History of Photovoltaics LA Lamont	31

Economics and Environment

1.05	Historical and Future Cost Dynamics of Photovoltaic Technology GF Nemet and D Husmann	47
1.06	Feed-In Tariffs and Other Support Mechanisms for Solar PV Promotion D Jacobs and BK Sovacool	73
1.07	Finance Mechanisms and Incentives for Photovoltaic Technologies in Developing Countries M Moner-Girona, S Szabo, and S Rolland	111
1.08	Environmental Impacts of Photovoltaic Life Cycles VM Fthenakis and HC Kim	143
1.09	Overview of the Global PV Industry A Jäger-Waldau	161
1.10	Vision for Photovoltaics in the Future E Despotou	179

| 1.11 | Storage Options for Photovoltaics
VM Fthenakis and T Nikolakakis | 199 |

Resource and Potential

| 1.12 | Solar Radiation Resource Assessment for Renewable Energy Conversion
DR Myers | 213 |
| 1.13 | Prediction of Solar Irradiance and Photovoltaic Power
E Lorenz and D Heinemann | 239 |

Basics

| 1.14 | Principles of Solar Energy Conversion
LC Hirst | 293 |
| 1.15 | Thermodynamics of Photovoltaics
V Badescu | 315 |

Technology

1.16	Crystalline Silicon Solar Cells: State-of-the-Art and Future Developments SW Glunz, R Preu, and D Biro	353
1.17	Thin-Film Silicon PV Technology M Zeman and REI Schropp	389
1.18	Chalcopyrite Thin-Film Materials and Solar Cells T Unold and CA Kaufmann	399
1.19	Cadmium Telluride Photovoltaic Thin Film: CdTe TA Gessert	423
1.20	Plastic Solar Cells L Sims, H-J Egelhaaf, JA Hauch, FR Kogler, and R Steim	439
1.21	Mesoporous Dye-Sensitized Solar Cells A Hagfeldt, UB Cappel, G Boschloo, L Sun, L Kloo, H Pettersson, and EA Gibson	481
1.22	Multiple Junction Solar Cells M Yamaguchi	497
1.23	Application of Micro- and Nanotechnology in Photovoltaics L Tsakalakos	515
1.24	Upconversion TW Schmidt and MJY Tayebjee	533
1.25	Downconversion MJY Tayebjee, TW Schmidt, and G Conibeer	549
1.26	Down-Shifting of the Incident Light for Photovoltaic Applications Y Jestin	563
1.27	Luminescent Solar Concentrators JC Goldschmidt	587
1.28	Thermophotovoltaics J van der Heide	603
1.29	Intermediate Band Solar Cells E Antolín, A Martí, and A Luque	619
1.30	Plasmonics for Photovoltaics S Pillai and MA Green	641
1.31	Artificial Leaves: Towards Bio-Inspired Solar Energy Converters A Pandit and RN Frese	657

Applications

1.32	Design and Components of Photovoltaic Systems *WGJHM van Sark*	679
1.33	BIPV in Architecture and Urban Planning *TH Reijenga and HF Kaan*	697
1.34	Product-Integrated Photovoltaics *AHME Reinders and WGJHM van Sark*	709
1.35	Very Large-Scale Photovoltaic Systems *T Ehara, K Komoto, and P van der Vleuten*	733
1.36	Concentration Photovoltaics *M Martinez, O de la Rubia, F Rubio, and P Banda*	745
1.37	Solar Power Satellites *GA Landis*	767
1.38	Performance Monitoring *N Pearsall and R Gottschalg*	775
1.39	Standards in Photovoltaic Technology *H Ossenbrink, H Müllejans, R Kenny, and E Dunlop*	787

Volume 2 Wind Energy

2.01	Wind Energy – Introduction *JK Kaldellis*	1
2.02	Wind Energy Contribution in the Planet Energy Balance and Future Prospects *JK Kaldellis and M Kapsali*	11
2.03	History of Wind Power *DT Swift-Hook*	41
2.04	Wind Energy Potential *H Nfaoui*	73
2.05	Wind Turbines: Evolution, Basic Principles, and Classifications *S Mathew and GS Philip*	93
2.06	Energy Yield of Contemporary Wind Turbines *DP Zafirakis, AG Paliatsos, and JK Kaldellis*	113
2.07	Wind Parks Design, Including Representative Case Studies *D Al Katsaprakakis and DG Christakis*	169
2.08	Aerodynamic Analysis of Wind Turbines *JN Sørensen*	225
2.09	Mechanical-Dynamic Loads *M Karimirad*	243
2.10	Electrical Parts of Wind Turbines *GS Stavrakakis*	269
2.11	Wind Turbine Control Systems and Power Electronics *A Pouliezos*	329
2.12	Testing, Standardization, Certification in Wind Energy *F Van Hulle*	371
2.13	Design and Implementation of a Wind Power Project *T Wizelius*	391
2.14	Offshore Wind Power Basics *M Kapsali and JK Kaldellis*	431

2.15	Wind Energy Economics *D Milborrow*	469
2.16	Environmental-Social Benefits/Impacts of Wind Power *E Kondili and JK Kaldellis*	503
2.17	Wind Energy Policy *GC van Kooten*	541
2.18	Wind Power Integration *JA Carta*	569
2.19	Stand-Alone, Hybrid Systems *KA Kavadias*	623
2.20	Wind Power Industry and Markets *PE Morthorst*	657
2.21	Trends, Prospects, and R&D Directions in Wind Turbine Technology *JK Kaldellis and DP Zafirakis*	671
2.22	Special Wind Power Applications *E Kondili*	725

Volume 3 Solar Thermal Systems: Components and Applications

Solar Thermal Systems

3.01	Solar Thermal Systems: Components and Applications – Introduction *SA Kalogirou*	1
3.02	Solar Resource *HD Kambezidis*	27
3.03	History of Solar Energy *VG Belessiotis and E Papanicolaou*	85

Components

3.04	Low Temperature Stationary Collectors *YG Caouris*	103
3.05	Low Concentration Ratio Solar Collectors *SA Kalogirou*	149
3.06	High Concentration Solar Collectors *B Hoffschmidt, S Alexopoulos, J Göttsche, M Sauerborn, and O Kaufhold*	165
3.07	Thermal Energy Storage *LF Cabeza*	211
3.08	Photovoltaic/Thermal Solar Collectors *Y Tripanagnostopoulos*	255
3.09	Solar Selective Coatings *P Yianoulis, M Giannouli, and SA Kalogirou*	301
3.10	Glazings and Coatings *G Leftheriotis and P Yianoulis*	313
3.11	Modeling and Simulation of Passive and Active Solar Thermal Systems *A Athienitis, SA Kalogirou, and L Candanedo*	357

Applications

3.12	Solar Hot Water Heating Systems G Faninger	419
3.13	Solar Space Heating and Cooling Systems SA Kalogirou and GA Florides	449
3.14	Solar Cooling and Refrigeration Systems GG Maidment and A Paurine	481
3.15	Solar-Assisted Heat Pumps DA Chwieduk	495
3.16	Solar Desalination E Tzen, G Zaragoza, and D-C Alarcón Padilla	529
3.17	Industrial and Agricultural Applications of Solar Heat B Norton	567
3.18	Concentrating Solar Power B Hoffschmidt, S Alexopoulos, C Rau, J Sattler, A Anthrakidis, C Boura, B O'Connor, and P Hilger	595
3.19	Passive Solar Architecture D Kolokotsa, M Santamouris, A Synnefa, and T Karlessi	637

Volume 4 Fuel Cells and Hydrogen Technology

4.01	Fuel Cells and Hydrogen Technology – Introduction AJ Cruden	1
4.02	Current Perspective on Hydrogen and Fuel Cells K Burke	13
4.03	Hydrogen Economics and Policy N Hughes and P Agnolucci	45
4.04	Hydrogen Safety Engineering: The State-of-the-Art and Future Progress V Molkov	77
4.05	Hydrogen Storage: Compressed Gas D Nash, D Aklil, E Johnson, R Gazey, and V Ortisi	111
4.06	Hydrogen Storage: Liquid and Chemical P Chen	137
4.07	Alkaline Fuel Cells: Theory and Application F Bidault and PH Middleton	159
4.08	PEM Fuel Cells: Applications AL Dicks	183
4.09	Molten Carbonate Fuel Cells: Theory and Application T Leo	227
4.10	Solid Oxide Fuel Cells: Theory and Materials A Tesfai and JTS Irvine	241
4.11	Biological and Microbial Fuel Cells K Scott, EH Yu, MM Ghangrekar, B Erable, and NM Duteanu	257
4.12	Hydrogen and Fuel Cells in Transport K Kendall and BG Pollet	281
4.13	H_2 and Fuel Cells as Controlled Renewables: FC Power Electronics N Schofield	295
4.14	Future Perspective on Hydrogen and Fuel Cells K Hall	331

Volume 5 Biomass and Biofuel Production

Biomass and Biofuels

5.01	Biomass and Biofuels – Introduction DJ Roddy	1
5.02	Historical Perspectives on Biofuels G Knothe	11

Case Studies

5.03	Bioethanol Development in Brazil A Altieri	15
5.04	Biomass Power Generation A Malmgren and G Riley	27
5.05	Biomass Co-Firing A Nuamah, A Malmgren, G Riley, and E Lester	55

Issues, Constraints & Limitations

5.06	A Global Bioenergy Market O Olsson and B Hillring	75
5.07	Biomass CHP Energy Systems: A Critical Assessment M Börjesson and EO Ahlgren	87
5.08	Ethics of Biofuel Production I Waller	99
5.09	Life Cycle Analysis Perspective on Greenhouse Gas Savings N Mortimer	109

Technology Solutions – New Processes

5.10	Biomass Gasification and Pyrolysis DJ Roddy and C Manson-Whitton	133
5.11	Biomass to Liquids Technology G Evans and C Smith	155
5.12	Intensification of Biofuel Production AP Harvey and JGM Lee	205
5.13	Biofuels from Waste Materials AA Refaat	217

Technology Solutions – Novel Feedstocks

5.14	Woody Biomass LL Wright, LM Eaton, RD Perlack, and BJ Stokes	263
5.15	Potential for Yield Improvement J Spink, E Mullins, and P Berry	293

Technology Solutions – Novel End Uses

5.16	Renewable Fuels: An Automotive Perspective RJ Pearson and JWG Turner	305
5.17	Use of Biofuels in a Range of Engine Configurations A Roskilly, Y Wang, R Mikalsen, and H Yu	343

Expanding the Envelope

5.18	Biochar *CE Brewer and RC Brown*	357
5.19	Extracting Additional Value from Biomass *MF Askew*	385
5.20	Biomass to Chemicals *A Kazmi and J Clark*	395
5.21	Bioenergy Policy Development *P Thornley*	411

Volume 6 Hydro Power

Hydro Power

6.01	Hydro Power – Introduction *A Lejeune*	1

Constraints of Hydropower Development

6.02	Hydro Power: A Multi Benefit Solution for Renewable Energy *A Lejeune and SL Hui*	15
6.03	Management of Hydropower Impacts through Construction and Operation *H Horlacher, T Heyer, CM Ramos, and MC da Silva*	49

Hydropower Schemes Around the World

6.04	Large Hydropower Plants of Brazil *BP Machado*	93
6.05	Overview of Institutional Structure Reform of the Cameroon Power Sector and Assessments *J Kenfack and O Hamandjoda*	129
6.06	Recent Hydropower Solutions in Canada *M Fuamba and TF Mahdi*	153
6.07	The Three Gorges Project in China *L Suo, X Niu, and H Xie*	179
6.08	The Recent Trend in Development of Hydro Plants in India *SP Sen*	227
6.09	Hydropower Development in Iran: Vision and Strategy *E Bozorgzadeh*	253
6.10	Hydropower Development in Japan *T Hino*	265
6.11	Evolution of Hydropower in Spain *A Gil and F Bueno*	309
6.12	Hydropower in Switzerland *B Hagin*	343

Design Concepts

6.13	Long-Term Sediment Management for Sustainable Hydropower *F Rulot, BJ Dewals, S Erpicum, P Archambeau, and M Pirotton*	355
6.14	Durability Design of Concrete Hydropower Structures *S Jianxia*	377
6.15	Pumped Storage Hydropower Developments *T Hino and A Lejeune*	405

6.16	Simplified Generic Axial-Flow Microhydro Turbines A Fuller and K Alexander	435
6.17	Development of a Small Hydroelectric Scheme at Horseshoe Bend, Teviot River, Central Otago, New Zealand P Mulvihill and I Walsh	467
6.18	Recent Achievements in Hydraulic Research in China J Guo	485

Volume 7 Geothermal Energy

7.01	Geothermal Energy – Introduction TI Sigfusson	1
7.02	The Physics of Geothermal Energy G Axelsson	3
7.03	Geothermal Energy Exploration Techniques ÓG Flóvenz, GP Hersir, K Sæmundsson, H Ármannsson, and P Friðriksson	51
7.04	Geochemical Aspects of Geothermal Utilization H Ármannsson	95
7.05	Direct Heat Utilization of Geothermal Energy JW Lund	169
7.06	Shallow Systems: Geothermal Heat Pumps L Rybach	187
7.07	Geothermal Power Plants R DiPippo	207
7.08	Corrosion, Scaling, and Material Selection in Geothermal Power Production SN Karlsdóttir	239
7.09	Geothermal Cost and Investment Factors H Kristjánsdóttir and Á Margeirsson	259
7.10	Sustainable Energy Development: The Role of Geothermal Power B Davidsdottir	271

Volume 8 Ocean Energy

8.01	Generating Electrical Power from Ocean Resources AS Bahaj	1
8.02	Historical Aspects of Wave Energy Conversion AFO Falcão	7
8.03	Resource Assessment for Wave Energy EBL Mackay	11
8.04	Development of Wave Devices from Initial Conception to Commercial Demonstration V Heller	79
8.05	Air Turbines AFO Falcão and LMC Gato	111
8.06	Economics of Ocean Energy T Stallard	151
Index		171

6.01 Hydro Power – Introduction

A Lejeune, University of Liège, Liège, Belgium

© 2012 Elsevier Ltd. All rights reserved.

6.01.1	Introduction	1
6.01.2	Hydroelectricity Progress and Development	4
6.01.2.1	Key Features of Hydroelectric Power	6
6.01.2.1.1	Cost	6
6.01.2.1.2	Ancillary services	6
6.01.2.1.3	Pumped-storage plants	7
6.01.2.1.4	GHG emissions	9
6.01.2.1.5	Environmental and social problems	9
6.01.2.2	Hydropower Development	9
6.01.2.2.1	Where the hydropower potential has been exploited	10
6.01.2.2.2	Where large hydropower potential has still to be exploited	11
6.01.2.2.3	Hydropower in integrated water resources management	12
6.01.2.2.4	International cooperation	12
6.01.2.2.5	Guidelines	12
6.01.3	Volume Presentation	12
6.01.3.1	Contributions and Authors, Affiliations of Volume 6	14
References		14

Glossary

Baseload power plant Baseload plant (also baseload power plant or base load power station), is an energy plant devoted to the production of baseload supply. Baseload plants are the production facilities used to meet some or all of a given region's continuous energy demand, and produce energy at a constant rate.

Energy Energy is the power multiplied by the time.

Gigawatt hour (GWh) Unit of electrical energy equal to one billion (10^9) watt hours.

Hydropower Hydropower, $P = hrgk$, where P is Power in kilowatts, h is height in meters, r is flow rate in cubic meters per second, g is acceleration due to gravity of $9.8\ ms^{-2}$, and k is a coefficient of efficiency ranging from 0 to 1.

Hydropower resource Hydropower resource can be measured according to the amount of available power, or energy per unit time.

Megawatt (MW) Unit of Electrical power equal to one million (10^6) watt.

Pumped storage plant Pumped-storage hydroelectricity is a type of hydroelectric power generation used by some power plants for load balancing. The method stores energy in the form of water, pumped from a lower elevation reservoir to a higher elevation. Low-cost off-peak electric power is used to run the pumps. During periods of high electrical demand, the stored water is released through turbines. Although the losses of the pumping process makes the plant a net consumer of energy overall, the system increases revenue by selling more electricity during periods of peak demand, when electricity prices are highest. Pumped storage is the largest-capacity form of grid energy storage now available.

Tetrawatt hour (TWh) Unit of electrical energy equal to one thousand billion (10^{12}) watt hours.

6.01.1 Introduction

In 2006, 17% of the world's electricity that was generated from hydropower represented nearly 90% of renewable electricity generation worldwide. Thus, it is by far the most widespread form of renewable energy.

Since 1965, the world's total energy consumption from oil, natural gas, coal, nuclear power, and hydropower (of which only hydropower is considered as a renewable resource) increased from 46.52 to 127.93 million gigawatt-hours (GWh). A gigawatt-hour is a measure of the total energy used over a period, equal to 1 million kilowatt-hours; 1 GWh is sufficient to power approximately 89 US homes for 1 year or 198 homes in the European Union for 1 year. As of 2007, the world's primary energy consumption was for oil, followed by coal (at 35.6% and 2\8.6%, respectively), and consumption in those areas has been growing. However, their growth has been curbed by the growth in energy consumption from renewable sources, including hydropower (**Figure 1**).

Due to the growing demand, the use of energy is continuously increasing. While in the 1970–2000 period the rate of increase was almost constant, in the past few years this rate increased. Electricity is growing faster than any other end-use source

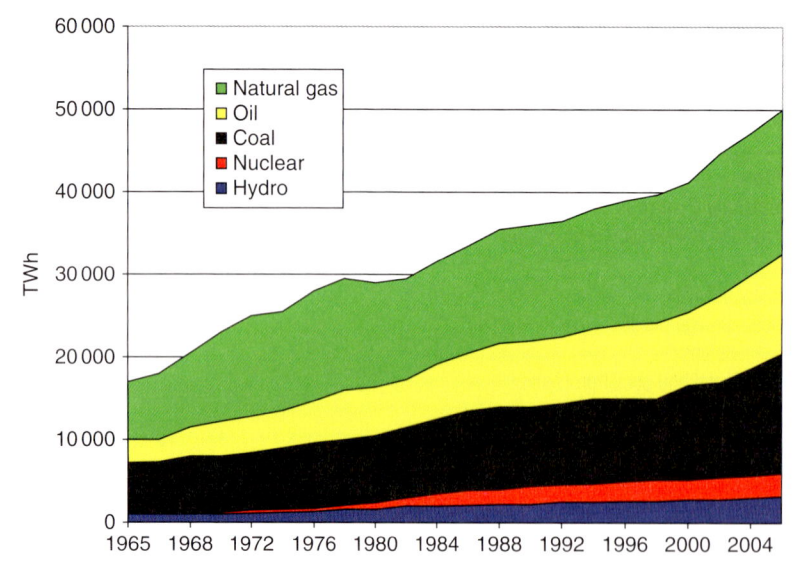

Figure 1 Historical trend in world's primary energy consumption by source, 1965–2006 (1 TWh = 1 terawatt-hour = 1000 GWh). Source: BP (2009) [9].

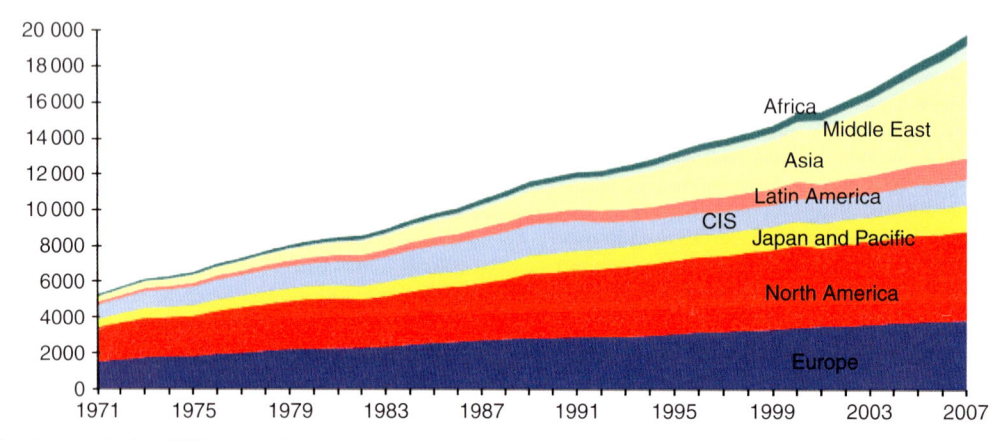

Figure 2 Electricity production (TWh) since 1970.

of energy; the rate of increase is currently in the order of 800 terawatt-hour (TWh) yr^{-1} (more than +4% yr^{-1}), as shown in **Figure 2** [1].

This is related to the high rate of economic growth in emerging economies that mainly contribute to keep up energy needs and soaring prices. While in Organisation for Economic Co-operation and Development (OECD) countries, accounting for half of the total electricity market, the power production continued with the usual historical trend (+2%), Asia and Middle East reported a rapid growth in their energy needs, with a special focus on the Chinese performance. Asian power generation has now exceeded the amount of electricity produced by North America or Europe, and China now accounts for >16% of the world's total electricity output. The current distribution of electricity generation by region is shown in **Figure 3** [2].

Electricity generation from coal and gas has been increasing faster than from any other sources, and counts now for >60% of total generation. The future scenarios for energy have been examined by several agencies.

The latest scenarios about the global energy forecast from 2005–20 are the following:

- World energy consumption will increase by about 30%, with China and India being the two main drivers.
- The power sector will be the biggest contributor to the world's energy demand growth, representing about 40% of the total energy consumption increase by 2020.
- The world's CO_2 emissions will increase up to 30% by 2020, mainly from Asia accounting for >70% of the total increase.

Based on existing and near-commercial technologies, the International Energy Agency [3] examined long-term scenarios and identified two such scenarios:

- 'Reference Scenario' in which renewables will only constitute ~14% of the world's primary energy demand by 2030.

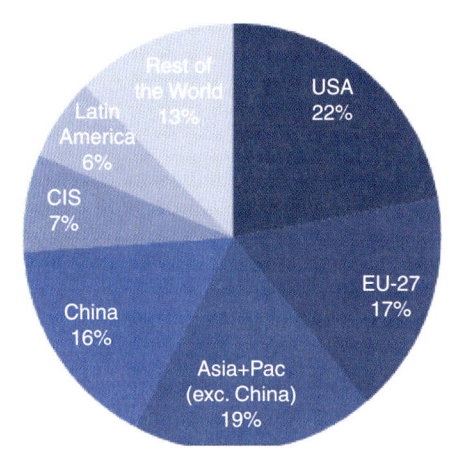

Figure 3 Electricity generation by region.

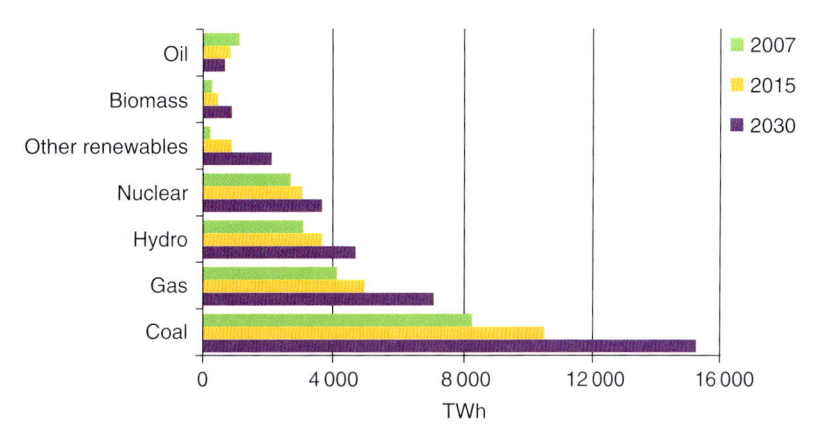

Figure 4 Evolution of global electricity production by fuel, 2007, 2015, and 2030 [3]. Forecasts of International Energy Agency on nuclear power generation will be modified due to the Fukushima accident.

- 'Alternative Policy Scenario' in which renewables share rises to ~16%, assuming the implementation of policies currently being considered by governments to ensure energy security and reduced CO_2 emissions (**Figure 4**).

In addition to the data illustrated in **Figure 4**, it is also necessary to consider the Millennium Development Goals (MDGs) [4] that 189 United Nations (UN) member states and at least 23 international organizations have agreed to achieve by 2015. The following eight development goals have been adopted to improve the social and economic conditions in the world's poorest countries, encompassing universally accepted human values and rights:

1. Eradicate extreme poverty and hunger
2. Achieve universal primary education
3. Promote gender equality and empower women
4. Reduce child mortality
5. Improve maternal health
6. Combat HIV/AIDS, malaria, and other diseases
7. Ensure environmental stability
8. Develop a global partnership for development.

Though energy access is not an MDG in itself, it is evident that adequate provision of energy and energy access to all remain crucial for achieving the MDGs. Furthermore, as stated in the 2010 'Millennium Goals Report', severing the link between energy use and greenhouse gas (GHG) emissions will require more efficient technologies for the supply and use of energy and a transition to cleaner and renewable energy sources. Therefore, it is evident that the world needs energy, clean energy, and cheap energy.

6.01.2 Hydroelectricity Progress and Development

With two-thirds of the world's electricity still coming from fossil fuel, hydropower currently produces the bulk (about 90%) of electricity derived from renewable sources. The evolution of the world's hydroelectricity production (TWh) since 1970 and its distribution by region are shown in **Figures 5** and **6** [1].

In about 60 countries, hydroelectricity is contributing >50% of the national electricity supply. In absolute terms, more than half of the total hydroelectricity production is produced by five countries only: China, Canada, Brazil, United States, and Russia (**Figure 7**) [5].

According to the World Register of Dams, dams were built around the world primarily for irrigation purpose (38%) and secondarily for hydropower purpose (18%). But today, some 8200 large dams are currently in operation having hydropower as the main or sole purpose. Some of them serve very large hydro plants. The largest hydroelectric dams and plants in operation are listed in **Table 1**. About 900 GWh are currently installed and over 150 are under construction, most of them in Asia. The largest and main schemes under construction are listed in **Table 2**. The data greatly emphasize that the major role in hydropower dams construction is played by China and most of the largest hydroelectric dams under construction are constructed by the Chinese.

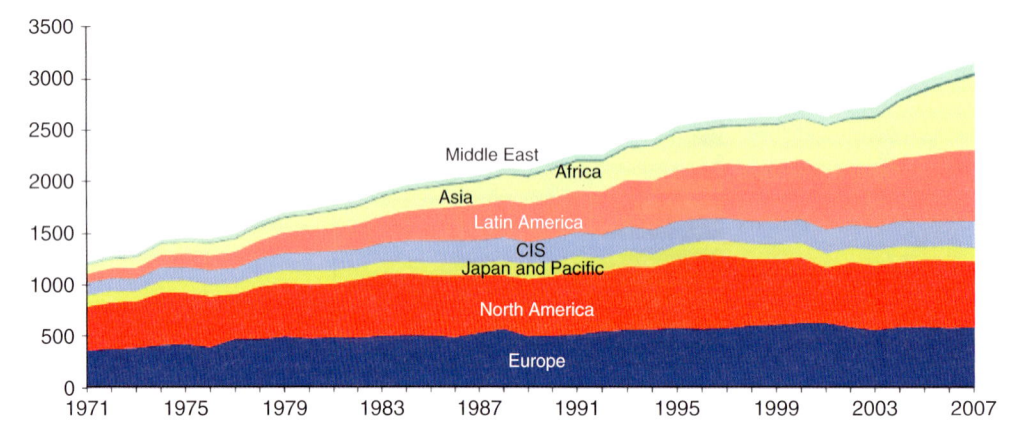

Figure 5 Evolution of hydroelectricity production (TWh) since 1970.

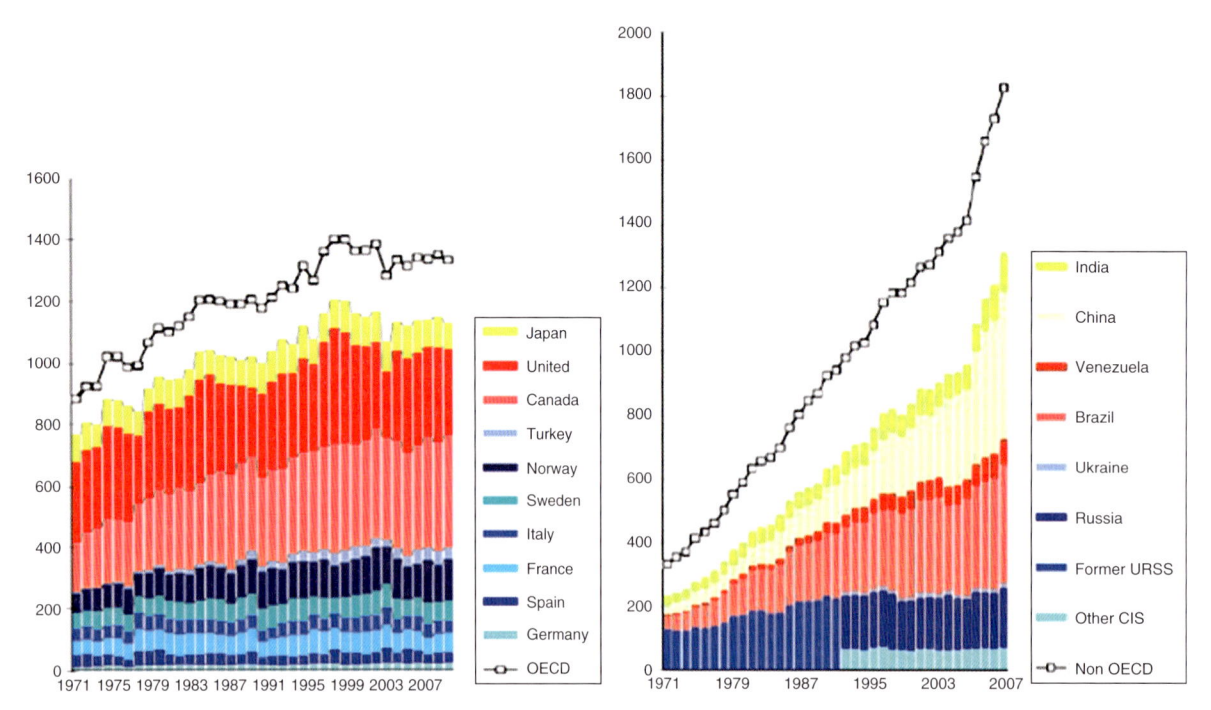

Figure 6 Evolution of hydroelectricity (TWh) in OECD and non-OECD countries.

Producers	TWh	% of world total
Peoples Rep. of China	436	14.0
Canada	356	11.3
Brazil	349	11.2
United States	318	10.2
Russia	175	5.6
Norway	120	3.8
India	114	3.6
Japan	96	3.1
Venezuela	79	2.5
Sweden	62	2.0
Rest of the world	1016	32.7
World	3121	100.0

2006 data

Installed capacity (based on production)	GWh
Peoples Rep. of China	118
United States	99
Brazil	71
Canada	72
Japan	47
Russia	46
India	32
Norway	28
France	25
Italy	21
Rest of the world	308
World	867

2005 data
Sources: United Nations, IEA

Country (based on first 10 producers)	% of hydro in total domestic electricity generation
Norway	98.5
Brazil	83.2
Venezuela	72.0
Canada	58.0
Sweden	43.1
Russia	17.6
India	15.3
Peoples Rep. of China	15.2
Japan	8.7
United States	7.4
Rest of the world*	14.3
World	16.4

2006 data

Figure 7 Hydroelectricity production by region.

Table 1 Largest hydroelectric dams and plants in operation

Dam	Country	Year of completion	Capacity (MW)	Max annual production (TWh)
Three Gorges	China	2009	17 600[a]	>100
Itaipú	Brazil/Paraguay	1984–2003	14 000	90
Guri (Simón Bolivar)	Venezuela	1986	10 200	46
Tucurui	Brazil	1984	8 370	21
Grand Coulee	USA	1942/1980	6 809	22.6
Sayano-Shushenskaya	Russia	1985/1989	6 400	26.8
Krasnoyarskaya	Russia	1972	6 000	20.4
Robert-Bourassa	Canada	1981	5 616	
Churchill Falls	Canada	1971	5 429	35
Bratskaya	Russia	1967	4 500	22.6
Ust-Ilim skaya	Russia	1980	4 320	21.7
Yaciretá	Argentina/Paraguay	1998	4 050	19.2
Longtan	China	2009	3 500[b]	18.7
Tarbela	Pakistan	1976	3 478	13
Ertan	China	1999	3 300	17.0
Ilha Solteira	Brazil	1974	3 200	
Xingó	Brazil	1994/1997	3 162	
Gezhouba	China	1988	3 115	17.0
Nurek	Tajikistan	1979/1988	3 000	11.2

[a] 22 500 when complete.
[b] 6300 when complete.

Table 2 Main schemes under construction

Dam	Country	Maximum capacity (MW)	Construction start	Scheduled completion
Xiluodu	China	12 600	2005	2015
Xiangjiaba	China	6 400	2006	2015
Longtan	China	6 300	2001	2009
Nuozhadu	China	5 800	2006	2017
Jinping-II Hydropower	China	4 800	2007	2014
Laxiwa	China	4 200	2006	2010
Xiaowan	China	4 200	2002	2012
Jinping-I Hydropower	China	3 600	2005	2014
Pubugou	China	3 300	2004	2010
Goupitan	China	3 000	2003	2011
Boguchan	Russia	3 000	1980	2012

6.01.2.1 Key Features of Hydroelectric Power

After more than a century of experience, the strengths and weaknesses of hydropower are equally well understood.

Its weaknesses (possible negative environmental and social impact, high upfront investment, etc.) are often overemphasized by opponents to dams and reservoirs, whereas its numerous and great benefits are not always adequately emphasized.

An analysis of the advantages and disadvantages of hydropower is found in Chapter 3, Constraints of hydropower development (*Hydropower: a multi benefit solution for renewable energy*), from which are derived the comments given hereunder about the key features of hydropower.

6.01.2.1.1 Cost

There are six different sources of renewable electricity. Hydroelectricity is the principle source with an 86.3% share of the total renewable output. Biomass, which includes solid biomass, liquid biomass, biogas, and renewable household waste, is the secondary source with 5.9%, a little ahead of the wind power sector with 5.7%, followed by geothermal power with 1.7%, solar power including electro-solar and photovoltaic plants and ocean energies with 0.01% (**Table 3**).

The cost of producing electricity is one fundamental criterion for decision making. The high realization costs of dams, reservoirs, and hydro plants are sometimes considered to classify hydropower as an 'expensive option'. However, hydropower converts energy from natural moving water directly into electricity and has therefore a very short and efficient energy chain, compared with fossil fuels. It has also a very efficient conversion process: modern plants can convert >95% of moving water's energy into electricity, whereas the best fossil fuel plants are about 60% efficient. Hydropower also has the best performance with respect to energy payback ratio, which is defined as the ratio of energy produced during a plant's life span to the energy required to build, maintain, and fuel the generating equipment. A hydropower plant can produce during its life span >200 times the energy needed to build, maintain, and operate it (**Figure 8**) [6].

Compared with the other renewable energies, hydropower is one of the least expensive sources of renewable electricity (**Figure 8**) [6]. Furthermore, hydro's autonomy from the fuel price variations, in addition to low annual operating costs, contributes significantly to 'energy security' (defined as "uninterrupted physical availability of energy products on the market, at a price which is affordable for all consumers," **Table 4**).

6.01.2.1.2 Ancillary services

Most of the hydropower projects were (and are) built to provide a primary 'base load' power generation. Moreover, this pattern will continue in countries where hydropower occupies a significant share in the power generation mix. As other technologies are introduced, hydro production is mainly used to respond to gaps between supply and demand, allowing the optimization of base load generation from less flexible sources (such as nuclear, thermal, and geothermal plants), which can continue to operate at constant level at their best efficiency. The fast response of hydro plants enables to meet sudden fluctuations due to peak demand or loss of other power supply options.

These benefits are part of a large family of benefits of hydropower in assisting the stability of electricity production (ancillary services):

- Spinning reserve: ability to run at a zero load while synchronized to the electric system; when loads increase, additional power can be loaded rapidly into the system to meet the demand.
- Nonspinning reserve: ability to enter load into the system from a source not on line; other energy sources can also provide nonspinning reserve, but hydropower's quick start capability is unparalleled.

Table 3 Structures of electricity production from renewable sources in 2008

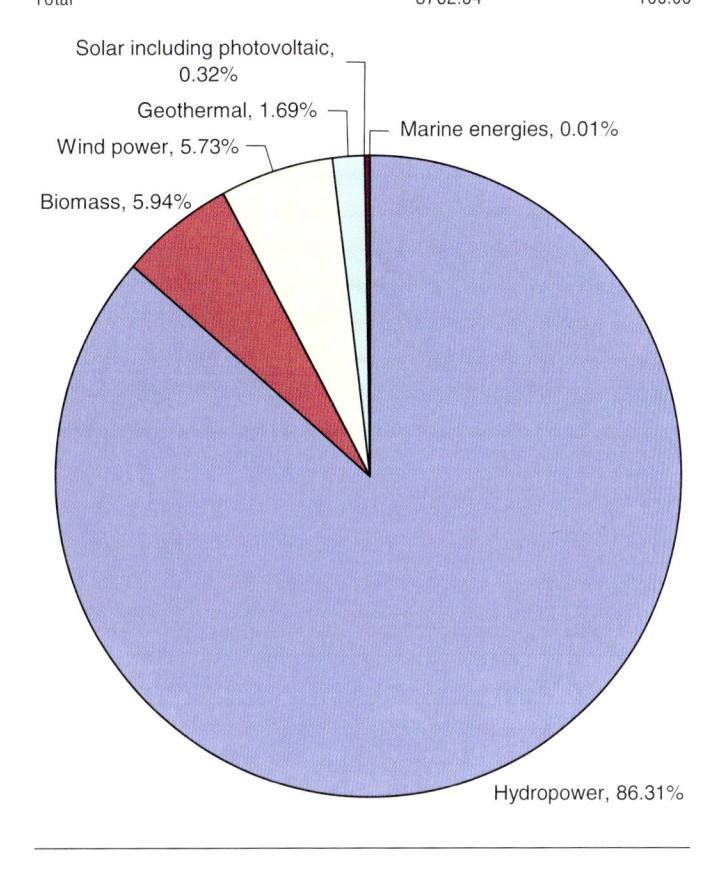

Source	TWh	%
Hydropower	3247.30	86.31
Biomass	223.50	5.94
Wind power	215.70	5.73
Geothermal	63.40	1.69
Solar including photovoltaic	12.10	0.32
Marine energies	0.54	0.01
Total	3762.54	100.00

- Regulation and frequency response: ability to meet moment-to-moment fluctuations in system requirements; when a system is unable to respond properly to load changes, its frequency changes, resulting not just in a loss of power but potential damage to electrical equipment as well.
- Voltage support: ability to control reactive power, thereby ensuring that power will flow from generation to load.
- Black-start capability: ability to start generation without an outside source of power; this service allows to provide auxiliary power to other generation sources that could take a long time to restart.

Of course, the capability of providing these ancillary services depends on the storage capacity. The full set of ancillary benefits described above refers to schemes with reservoirs. Run-of-river schemes, with little or no impoundment, just contribute to the 'base load' generation, producing relatively low-value base power and offering few of the ancillary benefits listed above.

6.01.2.1.3 Pumped-storage plants

Pumped-storage plants are particularly well suited to manage peaks in electricity demand and to assure reserve generation. In this role, they also have a remarkable environmental value: without pumped storage, to cope with unexpected peak demand or sudden loss of generating power, many thermal plants should operate at partial load as reserve generators, with increased fuel consumption and GHG emissions. They also have great capability of load leveling because they can absorb power when the system has an excess. Pumped-storage plants are therefore very effective means of improving ancillary services, thus playing a vital role for the reliability of electricity systems in an increasingly deregulated power market.

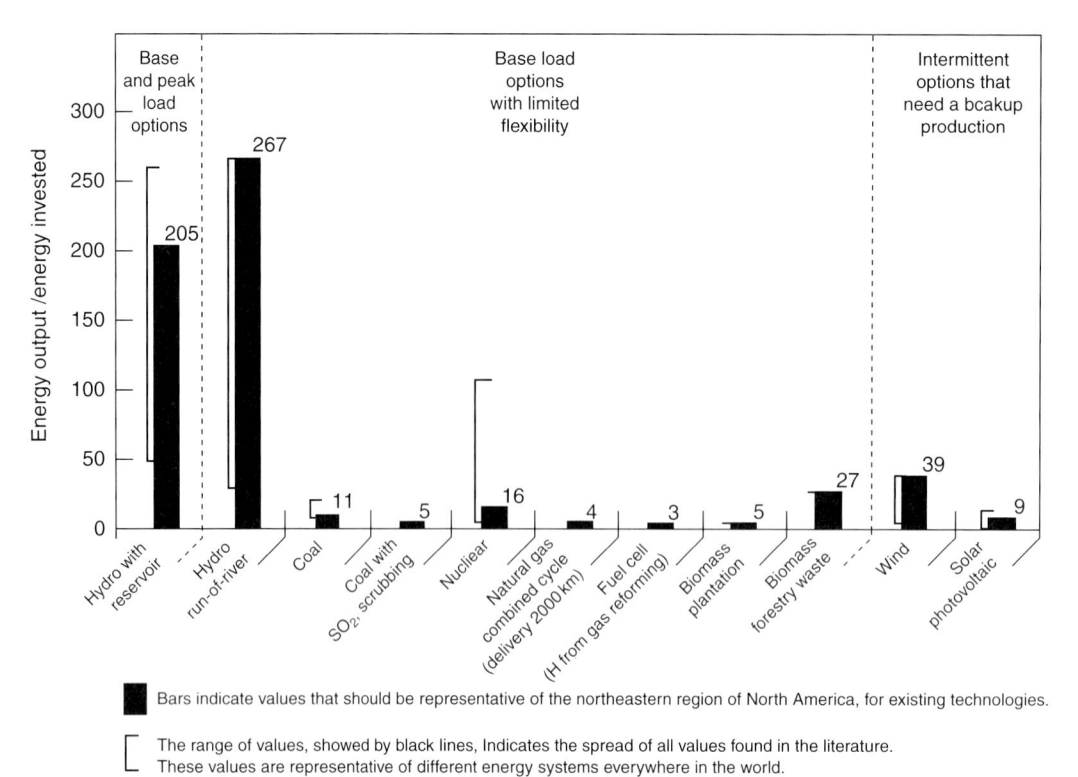

Figure 8 Energy payback ratio: comparison among different options.

Bars indicate values that should be representative of the northeastern region of North America, for existing technologies.

The range of values, showed by black lines, Indicates the spread of all values found in the literature. These values are representative of different energy systems everywhere in the world.

Table 4 Energy technologies and generating costs

Technology	Costs in US$
Biomass energy	
Electricity	5–15 ¢ kWh^{-1}
Heat	1–5 ¢ kWh^{-1}
Ethanol	8–25 $ GJ^{-1}
Wind electricity	5–13 ¢ kWh^{-1}
Solar photovoltaic electricity	25–125 ¢ kWh^{-1}
Solar thermal electricity	12–18 ¢ kWh^{-1}
Low-temperature solar heat	3–20 ¢ kWh^{-1}
Hydroelectricity	
Large	2–8 ¢ kWh^{-1}
Small	4–10 ¢ kWh^{-1}
Geothermal energy	
Electricity	2–10 ¢ kWh^{-1}
Heat	0.5–5 ¢ kWh^{-1}
Marine energy	
Tidal	8–15 ¢ kWh^{-1}
Wave	8–20 ¢ kWh^{-1}
Current	8–15 ¢ kWh^{-1}

Pumped-storage plants have some distinctive features in comparison with conventional hydropower plants:

- Greater output can be obtained with smaller reservoirs.
- They do not need natural inflow to the reservoirs.
- They can be built with considerably fewer hydrological and topographical restrictions.
- Their impact on the surrounding ecosystems is comparatively less.

6.01.2.1.4 GHG emissions

The links between production of energy and climate change are now understood, and GHG emissions, mainly produced by burning fossil fuels, are known to contribute to global warming. Hydropower tends to have a very low GHG footprint. As water carries carbon in the natural cycle, all ecosystems (especially wetlands and seasonally flooded areas) emit GHG. If the watershed contains a man-made reservoir, the preimpoundment emissions of the area would need to be compared with the emissions after the formation of the reservoir.

Studies in North America showed that hydropower reservoirs tend to increase the emissions marginally and a value of 10 000 ton TWh^{-1} of CO_2 equivalent has been allocated to schemes in this region. Because of a lack of data confirming the situation in warmer and tropical climates, a larger value (40 000 ton TWh^{-1}) has been proposed as an international average value for hydropower. Even so, hydropower GHG emissions amount to only a few percent of any kind of conventional fossil-fuel thermal generation (Figure 9) [6].

The evaluation of the net GHG emissions from reservoirs is becoming more and more important for CO_2 credits evaluation, and there is a growing concern to determine the contribution of freshwater reservoirs to the increase of GHG emissions in the atmosphere. Therefore, it is important to continue the efforts for a better understanding and a quantitative definition of the subject.

6.01.2.1.5 Environmental and social problems

Environmental concerns and problems related to dams and reservoirs are one of the main reasons emphasized by the opponents. However, they are now a much-studied process. Great efforts have been taken to understand them and to devise measures to avoid or rectify negative consequences. These efforts resulted in a much greater knowledge and in the development of a broad range of mitigation strategies. The integration of environmental and social considerations in the planning, design, and operation of dams is now a standard practice in many countries. The analyses of possible problems and a comprehensive negotiation processes with all the involved stakeholders greatly improved the development effectiveness of the projects by eliminating unfavorable projects at an early stage.

Even the World Commission on Dams, who concluded that hydropower schemes had often environmental or social unacceptable costs, did not recommend that hydropower should be discouraged, or that only small schemes should be developed. Instead, an inclusive process was recommended in the planning, development, and management of the schemes. It must also be noted that many well-conceived schemes have seen unappreciated service for several generations. Some sites have been chosen as sites of special scientific interest because of the ecosystems that have become established in the reservoir areas.

6.01.2.2 Hydropower Development

Only one-third of the world's potential of hydropower resources have so far been developed. Figure 10 points out that while in Europe and North America almost all the technically and economically feasible hydropower potential has been harnessed, a large unexploited hydropower potential is available in Asia, where the current production is less than one-third of the potential, and in Africa, where the ratio is even smaller.

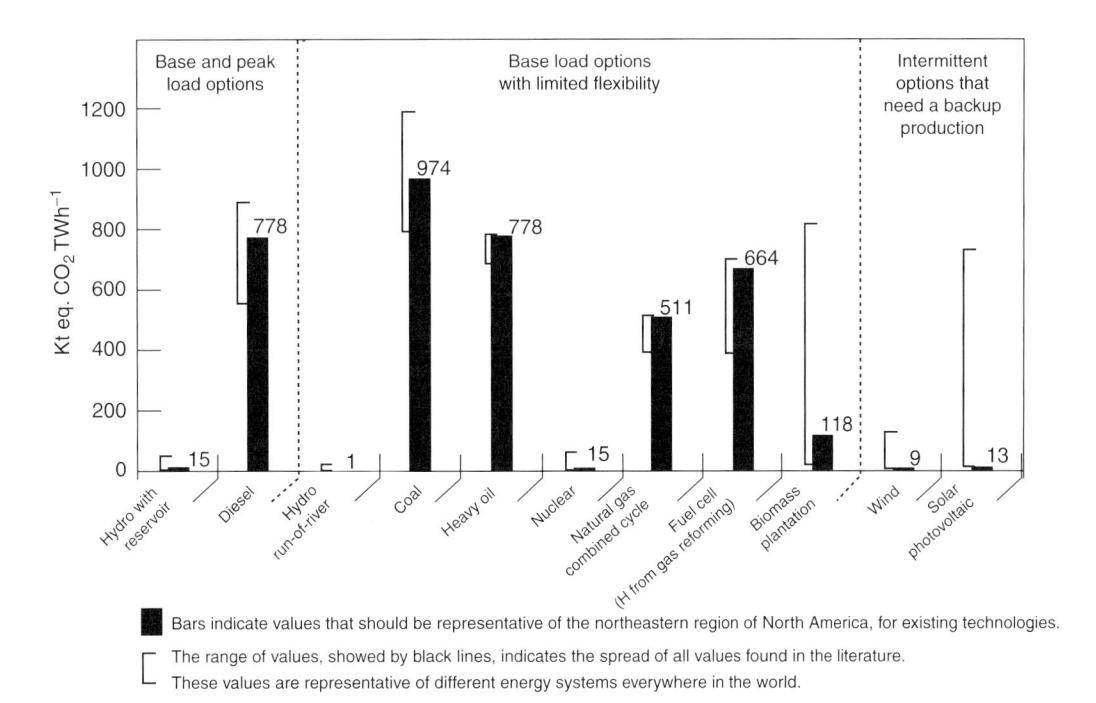

Figure 9 GHG emission: comparison among power generation options.

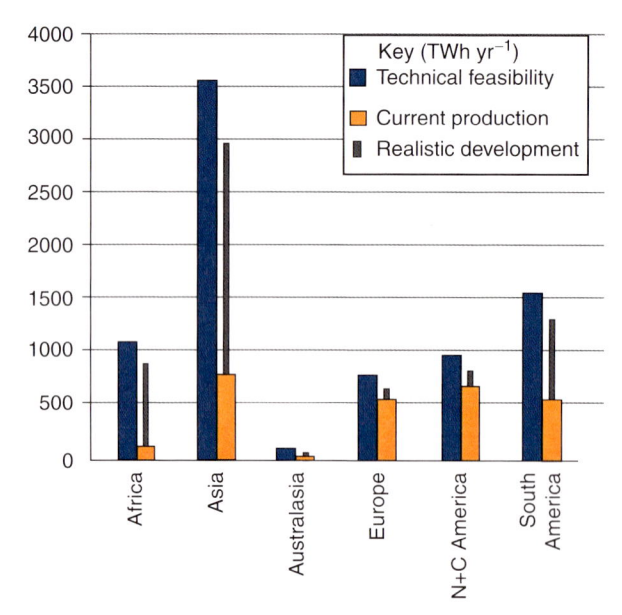

Figure 10 Hydropower potential: feasible vs. exploited.

6.01.2.2.1 Where the hydropower potential has been exploited

In most of the countries where the hydro-potential has been extensively harnessed, the hydropower development started one century ago and many dams and plants are therefore old. In these countries, the focus is therefore on

- maintaining the ageing works in safe and efficient conditions;
- managing new requirements and needs, minimizing the negative impact on the power production; and
- getting the most out of the existing infrastructures.

6.01.2.2.1(i) Safety and efficiency of the existing dams and reservoirs

The modernization of existing power plants is motivated and economically supported by the consequent addition of more efficient production. However, maintaining existing dams and reservoirs in good and safe conditions may require important and expensive remedial works conflicting with the available resources and the duration of the concessions.

The recurring problems are those related to the considerable length of service of many works:

- Obsolete dam typologies, not corresponding to the current state of the art.
- Dams designed using design criteria not fully compatible with current more demanding safety standards.
- Ageing and degradation process, among which expansive phenomena in concrete are having an increasing importance.
- Silting of reservoirs, with problems for the proper working of outlets and intakes, and additional loads applied to the structures. Hydropower reservoirs can generally be filled by sediments to a higher percentage than nonhydropower reservoirs, as they are mainly addressed to maintain the head for the power generation, but silting remains a problem requiring in many cases important works for sediment removal.

Furthermore, many countries have to face the problems of renewing the dam engineering profession, preserving the available experience, and transmitting it to young engineers.

6.01.2.2.1(ii) Additional purposes/requirements

During the operating life of hydroelectric dams and reservoirs, new requirements are often introduced in addition to the initial sole hydroelectric purposes, such as flood protection, irrigation and potable supply, discharge for minimum vital flow, recreational purposes and touristic development, and wetland habitat. The new needs introduce limitations and constraints in the use of the water often conflicting with the optimization of the power production. Some additional requirements apply only to some dams and reservoirs, depending on the capacity of the reservoirs and the local situation and needs. The requirement of a continuous water discharge to assure the minimum vital flow and to improve the downstream ecological condition apply to many dams, potentially to all, and it can reduce the electrical production of a significant amount on a national scale (in Europe, e.g., the reduction could be estimated around 10%). Consequently, the introduction of this requirement is stimulating significant activities for the installation of mini-hydro turbines to generate a continuous discharge, thus mitigating the negative impact on power production. A significant example of additional requirement is the use for flood mitigation of the hydroelectric reservoirs in the Paraná Basin (Brazil) [7]. In this basin, there is a large integrated reservoirs system (46 reservoirs). The installed capacity is >45 000 MW, including the Paraguayan share of Itaipú. Initially, the majority of the reservoirs were dimensioned for hydroelectric purpose only. Flood control operations were not foreseen at that time. Later on, flood control rules were established for all power plants. Maximum outflow constraints were set for each reservoir and a flood-forecast system was developed, thus entailing social and economic benefits through the reduction of flood impacts in the downstream areas. A trade-off between flood control and energy production was consequently defined, since for the electric production it would be desirable to keep the reservoirs at their maximum capacity.

6.01.2.2.1(iii) Getting the most out of existing infrastructures

Where most of the hydro-potential has been harnessed and further development is limited to rather marginal contributions, the current focus is not on building new dams but rather tapping existing ones for their hydroelectric potential and getting the most out of existing infrastructures. This is accomplished through a variety of engineering strategies including:

- Upgrading existing schemes and extending their operational life to take advantage of the long life of the civil structures.
- Optimizing the output of the plant to meet the needs of the power market.
- Adding capacity for extra generation when high flows are available.
- Adding small hydro facilities to generate the discharge for the minimum vital flow.
- Adding hydropower capabilities at nonpower dams.

The addition of hydropower capabilities at nonpower dams is an important option because the large majority of the dams in the world do not have a hydroelectric component. For instance, a resource assessment carried out 10 years ago by the US Department of Energy concluded that in the United States a hydro-capacity of about 20 000 MW could be gained by adding generating units to about 2500 existing dams. More than 70 of such projects are currently in progress, with a collective potential of over 11 000 MW [8].

6.01.2.2.2 Where large hydropower potential has still to be exploited

As far as concerns, the countries with a large hydro-potential are still to be developed; in Asia and in South America, the development is driven by leading countries with important economic growth (China, Brazil, India, etc.).

In Africa, where 65% of the population does not have access to electricity and the needs are consequently very urgent, only a very small amount of the hydroelectric potential has been harnessed. After a period of difficulty, international lenders are now supporting dams and reservoirs and several important declarations have been recently adopted in favor of hydropower. At the World Water Forum in Kyoto 2003, the most substantial effort to address the global warming problem, the Ministerial Declaration of 170 Countries stated "We recognize the role of hydropower as one of the renewable and clean energy sources, and that its potential should be realized in an environmentally sustainable and socially equitable manner." The 2004 Political Declaration adopted at the 'International Conference for Renewable Energies' acknowledged that renewable energies, including hydropower, combined with enhanced energy efficiency, could contribute to sustainable development, providing access to energy and mitigating GHG emission. At the 2004 UN Symposium on 'Hydropower and Sustainable Development', the representatives of national and local governments, utilities, UN agencies, financial institutions, international organizations, nongovernmental organizations, scientific community, and international industry associations have concluded with a strongly worded declaration in support of hydropower. Many important key points are clearly stated in this declaration. Warmly recommending the reading of the full declaration, some points are resumed hereinafter:

- the acknowledgement of the contribution made by hydropower to development, and the agreement that the large remaining potential can be harnessed to bring benefits to developing countries and to countries with economies in transition;
- the need to develop hydropower, along with the rehabilitation of existing facilities and the addition of hydropower to present and future water management systems;
- the importance of an integrated approach, considering that hydropower dams often can perform multiple functions;
- the acknowledgement of the progress made in developing policies, frameworks, and guidelines for evaluation and mitigation of environmental and social impacts, and the call to disseminate them.

Finally, in November 2008, a 'World Declaration – Dams and Hydropower for African Sustainable Development' has been approved by the African Union, the Union of Producers Transporters and Distributors of Electric Power in Africa, the World Energy Council, the International Commission on Large Dams, the International Commission on Irrigation and Drainage, and the International Hydropower Association (IHA). This World Declaration points out that current condition is now ripe for hydropower development in Africa. A new political commitment will exist today, and more projects are under development, as shown in **Figure 11**.

The Grand Inga project is a clear example of the tremendous potential available in Africa: a high power capacity project (up to 100 000 MW) with small impacts on environment, generating >280 TWh yr^{-1} of exceptionally cheap electricity.

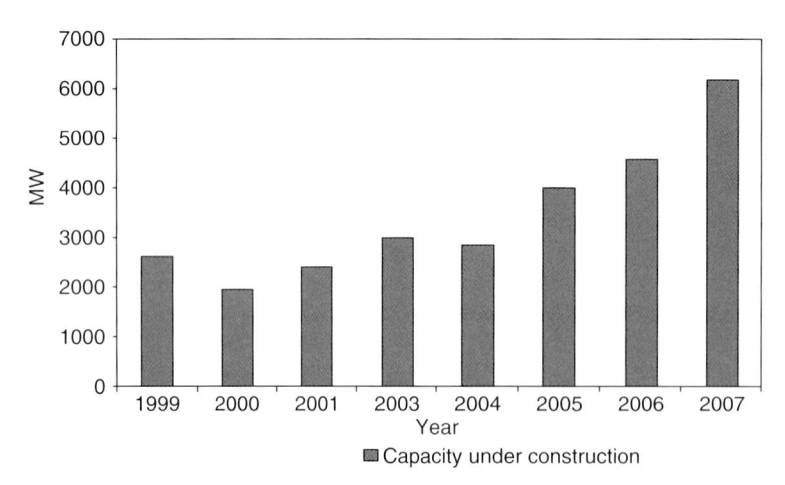

Figure 11 Trend in hydropower capacity under construction in Africa.

6.01.2.2.3 Hydropower in integrated water resources management

Worldwide there is a major focus on integrated water resources management, highlighting the multiple benefits of dams and reservoirs. Nowadays, it is not acceptable simply to maximize the economic profits of a hydroelectric scheme. Closer linkages are required between water and energy resources, and the increasing need for water management is a main driver for hydro development. Integrated Water Resources Management provides both a framework for sustainable reservoir management and a context in which the impacts and true value of a dam may be assessed. It requires that scheme design and operation be considered at the catchment scale. Management must take into account multiple objectives, including both economic and noneconomic benefits.

6.01.2.2.4 International cooperation

There is an increase in international and regional cooperation for hydropower development. For example, companies from some Asian countries, well experienced in hydro development, such as China and Iran, are investing in schemes in Africa. In South and East Asia, a number of binational developments are moving ahead, based on power purchase agreement, enabling some of the less developed countries to gain economic benefits from exporting their hydropower production. In developing markets, interconnection between countries and the formation of power pools will build investor's confidence. The critical importance of international cooperation in the development of the water resources of Africa is evident, considering that Africa has 61 international shared rivers, whose basins cover about 60% of the surface of the continent. As an example, the West Africa Power Pool Project is the vehicle designed to ensure the stable supply of electricity to member countries of the Economic Community of West Africa States, beginning with four member nations, namely Niger, Ghana, Benin, and Togo. The first phase of the project is a 70 km line linking Nigeria to the Republic of Benin.

6.01.2.2.5 Guidelines

In the final declaration adopted at the 2004 UN Symposium on 'Hydropower and Sustainable Development', the dissemination of good practice and guidelines was recommended. With regard to this, it is worthwhile to mention the 'Sustainability Guidelines' developed by the IHA to promote greater consideration of environmental, social, and economic aspects in the sustainability assessment of new projects and in the management of existing power schemes. The guidelines define general principles that need of course to be adapted to the specific context and unique set of circumstances of each particular project. The Sustainability Guidelines were formally adopted in 2003 by the IHA membership, which spans 82 countries. Subsequently, they have been submitted to international funding agencies and UN organizations, with the proposal that they are used in the evaluation of future projects and in the screening of applications for credit relating to existing schemes. Supplementing the guidelines is an 'Assessment Protocol' that sets out a system by which sustainability performance can be measured.

6.01.3 Volume Presentation

Purpose of the volume.

Volume 6 of the Comprehensive Renewable Energy edition is dedicated to Hydropower. The contribution of hydropower in the generation of electricity is important, representing around 18% of the total electricity generation and 80% of the generation of renewable electricity. It needs a volume.

Table 5 List of contributions of authors and affiliations

1	Introduction	André Lejeune	University of Liège	Belgium
2	Constraints of Hydropower Development	Samuel L. Hui	Bechtel	USA
3	Management of Hydropower Impacts through Construction and Operation	Carlos Matias Ramos, Margarida Cardoso da Silva	Laboratório Nacional de Engenharia Civil (LNEC)	Portugal
		Hans B Horlacher Thorsten Heyer	University of Dresden	Germany
4	Large Hydropower Plants in Brazil	Brasil Pinheiro Machado	Intertechne	Brazil
5	Overview of Institutional Structure Reform of the Cameroon Power Sector and Assessments	Joseph Kenfack Oumarou Hamandjoda	University of Yaoundé	Cameroon
6	Recent Hydropower Implementations in Canada	Musandji Fuamba Tew-Fik Mahdi	École Polytechnique de Montréal	Canada
7	The Three Gorges Project in China	Lisheng Suo, Xinqiang Niu Hongbing Xie	Hohai University Changjiang Institute of Survey, Planning, Design and Research	China
8	The Recent Trend in Development of Hydro Plants in India	Siba Prasad Sen	Former Director Technical NHPC Ltd., India Consultant	India
9	Hydropower Development in Iran: Vision and Strategy	Eisa Bozorgzadeh	Iran Water and Power Resources Development Company (IWPCO)	Iran
10	Hydropower Development in Japan	Toru Hino	Electric Power Development J-Power	Japan
11	Evolution of Hydropower in Spain	Arturo Gil Garcia Francisco Bueno Hernandez	Iberdrola University of Burgos	Spain
12	Hydropower in Switzerland	Bernard Hagin		Switzerland
13	Long-Term Sediment Management for Sustainable Hydropower	Benjamin J. Dewals François Rulot, Sébastien Erpicum, Pierre Archambeau, Michel Pirotton	University of Liège	Belgium
14	Durability Design of Concrete Hydropower Structures	Jianxia Shen	Jiangsu Provincial Water Investigation, Design and Research Institute	China
15	Pumped-Storage Hydropower Developments	Toru Hino	Electric Power Development J-Power	Japan
		André Lejeune	University of Liège	Belgium
16	Simplified Generic Axial-Flow Microhydro Turbines	Adam Fuller Keith Alexander	Canterbury University	New Zealand
17	Development of a Small Hydroelectric Scheme	Peter Mulvihill Ian Walsh	Pioneer Generation Opus International Consultants	New Zealand
18	Recent Achievements in Hydraulic Research in China	Jun GUO	China Institute of Water Resources and Hydropower Research (IWHR)	

Instead of rewriting an already existing textbook about hydropower, it was decided to enhance the progress and development of hydropower by remarkable worldwide examples or projects.

The volume is divided into three main sections:

- Constraints of hydropower development
- Hydropower schemes in the world
- Design concept.

6.01.3.1 Contributions and Authors, Affiliations of Volume 6

The list of the contributions of the authors and their affiliations is given in **Table 5**.

All the authors are highly and warmly thanked for their contributions.

References

[1] ENERDATA (2010) *Energy Statistics Yearbook*. France: ENERDATA.
[2] ENERDATA (2010) *The World Energy Demand*. France: ENERDATA.
[3] International Energy Agency (2010) *Energy Technology Perspectives*. France: IEA.
[4] United Nations (2010) *The Millennium Development Goals Report 2010*. New York, NY: United Nations.
[5] World Energy Council (2007) *Survey of Energy Resources 2007*. London: WEC.
[6] International Hydropower Association (2004) *Sustainability Guidelines*. London: IHA.
[7] Carvalho E (2001) *Flood Control for the Brazilian Reservoir System in the Paraná River Basin*. Oxfordshire: IAHS.
[8] Bishop N (2008) Waterways. *International Water Power and Dam Construction*, June.
[9] British Petroleum (2009) *Statistical Review of World Energy 2010*.

6.02 Hydro Power: A Multi Benefit Solution for Renewable Energy

A Lejeune, University of Liège, Liège, Belgium
SL Hui, Bechtel Civil Company, San Francisco, CA, USA

© 2012 Elsevier Ltd.

6.02.1	Introduction	16
6.02.2	How Hydropower Works	16
6.02.2.1	Characteristics of Hydropower Plants	16
6.02.2.1.1	Essential features	16
6.02.2.1.2	Power from flowing water	17
6.02.2.1.3	Energy and work	17
6.02.2.1.4	Essentials of general plant layout	21
6.02.2.1.5	Factors affecting economy of plant	21
6.02.2.1.6	Types of hydropower developments	22
6.02.2.1.7	Typical of arrangements of waterpower plants	22
6.02.2.1.8	Lowest cost power developments	24
6.02.2.1.9	Highest cost power developments	24
6.02.2.2	Types of Turbines	24
6.02.2.2.1	Pelton turbine	25
6.02.2.2.2	Francis and Kaplan turbines	25
6.02.2.2.3	Cross-flow (Banki) turbine	26
6.02.2.2.4	Hydraulienne and Omega Siphon	27
6.02.2.2.5	Comparison of different turbines	27
6.02.2.3	Types of Dams	28
6.02.2.3.1	Embankment dam types	28
6.02.2.3.2	Concrete dam types	29
6.02.3	**History of Hydropower**	31
6.02.3.1	Historical Background	31
6.02.3.1.1	Use of velocity head	31
6.02.3.1.2	Use of potential head	31
6.02.3.1.3	Electricity is coming	32
6.02.3.2	Hydro Energy and Other Primary Energies	33
6.02.3.3	World Examples	33
6.02.3.3.1	China	33
6.02.3.3.2	Brazil	35
6.02.3.3.3	USA	36
6.02.3.3.4	Japan	37
6.02.4	**Hydropower Development in a Multipurpose Setting**	38
6.02.4.1	Benefits of Hydropower	38
6.02.4.1.1	Social	38
6.02.4.1.2	Economic issues	40
6.02.4.1.3	Environmental issues	41
6.02.5	**Negative Attributes of Hydropower Project**	41
6.02.6	**Renewable Electricity Production**	42
6.02.6.1	Recall	42
6.02.6.2	Sources of Renewable Electricity Energy	42
6.02.6.3	Characteristics of Renewable Energy Sources	42
6.02.6.3.1	Solar	43
6.02.6.3.2	Wind power	43
6.02.6.3.3	Hydroelectric energy	43
6.02.6.3.4	Biomass	43
6.02.6.3.5	Hydrogen and fuel cells	43
6.02.6.3.6	Geothermal power	43
6.02.6.3.7	Other forms of energy	43
6.02.6.4	Distribution per Region of the Percentage of Hydroelectricity and Renewable Non-Hydroelectricity Generation in the World	43
6.02.6.5	Findings about Renewable Electricity Production	44
6.02.7	**Conclusion**	45
Further Reading		46

Glossary

Base-load plant Base-load plant (also base-load power plant or base-load power station) is an energy plant devoted to the production of base-load supply. Base-load plants are the production facilities used to meet some or all of a given region's continuous energy demand, and produce energy at a constant rate.

Energy Energy is the power multiplied by the time.

Gigawatt hour (GWh) Unit of electrical energy equal to one billion (10^9) watt hours.

Hydropower Hydropower $P = hrgk$, where P is power in kilowatts, h is height in meters, r is flow rate in cubic meters per second, g is acceleration due to gravity of 9.8 m s^{-2}, and k is a coefficient of efficiency ranging from 0 to 1.

Hydropower resource Hydropower resource can be measured according to the amount of available power or energy per unit time.

Megawatt (MW) Unit of electrical power equal to one million (10^6) watt.

Pumped-storage plant Pumped-storage hydroelectricity is a type of hydroelectric power generation used by some power plants for load balancing. The method stores energy in the form of water, pumped from a lower elevation reservoir to a higher elevation. Low-cost off-peak electric power is used to run the pumps. During periods of high electrical demand, the stored water is released through turbines. Although the losses of the pumping process makes the plant a net consumer of energy overall, the system increases revenue by selling more electricity during periods of peak demand, when electricity prices are highest. Pumped storage is the largest capacity form of grid energy storage now available.

Tetrawatt hour (TWh) Unit of electrical energy equal to one thousand billion (10^{12}) watt hours.

6.02.1 Introduction

Hydropower is currently the most important renewable source of the world's electricity supply and there is still considerable untapped potential in many areas even though this is a relatively old technology. Continued exploitation of this resource is likely as a response to the world's demand for energy. Environmental legislation such as the Kyoto Protocol is putting increasing pressure on all governments to generate 'clean' energy or energy from sustainable sources. Hydropower produces little CO_2, but in other respects may not be truly sustainable.

In many developing countries, electricity usage is widespread in urban areas, but for many rural areas, infrastructure investment is much lower, and many communities rely on batteries or nothing at all. With the current population growth in many developing countries, there is even greater demand for generating more electricity and distributing it to poorer people so that they are not left behind in the race to develop. Electricity provision to rural communities results in a better quality of life for householders, but also has positive impacts on schools, hospitals, businesses, and agriculture/industry.

This chapter will detail how hydropower works, with special attention to its history. Hydropower development in a multi-purpose setting and its position in the renewable sources of electricity will conclude the chapter.

6.02.2 How Hydropower Works

6.02.2.1 Characteristics of Hydropower Plants

6.02.2.1.1 Essential features

A waterpower development is essentially the utilization of the available power in the fall of a river, through a portion of its course, by means of hydraulic turbines, which, as previously explained, are usually reaction wheels except for a very high head site, where impulse wheels may be used. To utilize its power, water must be confined in channels or pipes and brought to the wheels, so as to bring them into action by utilizing the full pressure of the available head or fall, except for such losses of head as are unavoidable in bringing the water to the wheels. The essential features of a waterpower development are as follows (see **Figure 1**):

6.02.2.1.1(i) The dam

A dam is a structure of masonry, compacted earth with impermeable materials, concrete, or other materials built at a suitable location across the river, both to create head and to provide a large area or pond of water from which draft can readily be made. In many cases, the power development is at or close to the dam, and the entire head utilized is that afforded at the dam itself, in which case the development is one of concentrated fall. In other cases, water is conveyed to a downstream location some distance away, via tunnels or penstocks, utilizing the head differential between the dam and the downstream location for power generation.

6.02.2.1.1(ii) The water conveyance structures

More often the development must be by divided fall, utilizing in addition to the head created by the dam an amount obtained by carrying the water in a conveyance structure, which may be a canal, tunnel, penstock (or closed conduit), or a combination of these for some distance downstream.

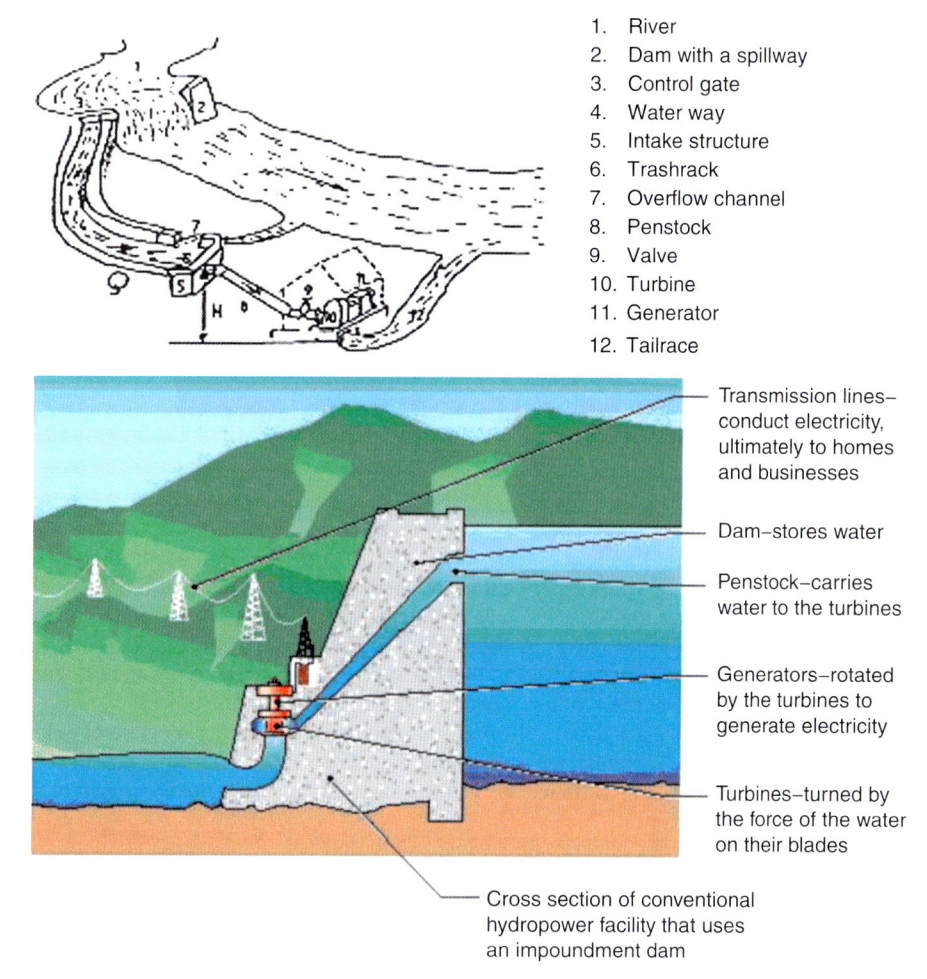

1. River
2. Dam with a spillway
3. Control gate
4. Water way
5. Intake structure
6. Trashrack
7. Overflow channel
8. Penstock
9. Valve
10. Turbine
11. Generator
12. Tailrace

Transmission lines–conduct electricity, ultimately to homes and businesses

Dam–stores water

Penstock–carries water to the turbines

Generators–rotated by the turbines to generate electricity

Turbines–turned by the force of the water on their blades

Cross section of conventional hydropower facility that uses an impoundment dam

Figure 1 Essential features of a hydropower plant.

6.02.2.1.1(iii) The powerhouse and equipment

This includes the hydraulic turbines and generators and their various accessories as well as the building, which is required for their protection and convenient operations. Many existing waterpower developments also utilize the power from the turbines in mechanical drive, that is, operating machinery directly or by belting and gearing.

6.02.2.1.1(iv) The tailrace

This is part of the water conveyance structure that returns the water from the powerhouse back to the river.

6.02.2.1.2 Power from flowing water

We may change the form of energy, but we can neither create nor destroy it. Water will work for us only to the extent that work has been performed on it. We can never realize all the potential energy inherent in the water because there are inevitable losses in converting the potential energy to the form that would be beneficial to us.

In the hydrologic cycle (**Figure 2**), water is evaporated from oceans and carried inland in the form of vapor by air currents. Cooling by adiabatic expansion of these air currents deflected upward by mountain ranges and by other means causes condensation of its vapor and precipitation as rain, snow, or dew onto the land from whence it flows back to the ocean only to repeat the hydrologic cycle. The work done on it by the energies of the sun, winds, and cooling forces places it on the uplands of the world where energies could be extracted from it in its descent to the oceans in a direct correspondence to the energies expended in putting it there.

6.02.2.1.3 Energy and work

Energy is the ability to do work. It is expressed in terms of the product of weight and length. The unit of energy is the product of a unit weight by a unit length, that is, the kilogram-meter. Work is utilized energy and is measured in the same units as energy. The element of time is not involved.

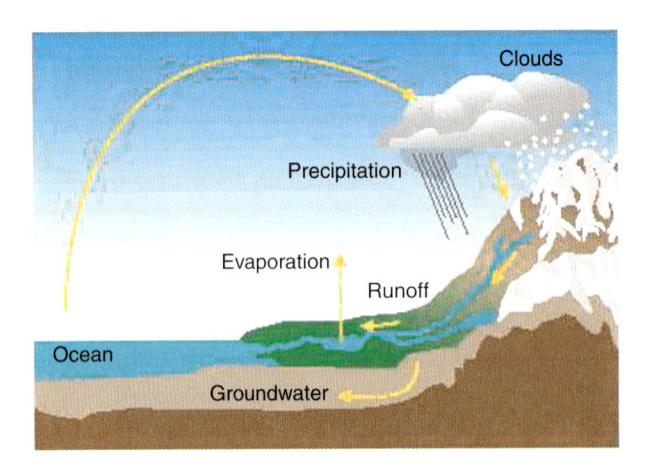

Figure 2 Hydrologic cycle.

Water in its descent to the oceans may be temporarily held in snowpacks, glaciers, lakes, and reservoirs, and in underground storage. It may be moving in sluggish streams, tumbling over falls, or flowing rapidly in rivers. Some of it is lost by evaporation, deep percolation, and transpiration of plants. Only the energy of water that is in motion can be utilized for work.

The energy of water exists in two forms: (1) potential energy, that due to its position or elevation, and (2) kinetic energy, that due to its velocity of motion. These two forms are theoretically convertible from one form to the other.

Energy may be measured with reference to any datum. The maximum potential energy of a kilogram of water is measured by its distance above sea level. The ocean has no potential energy because there is no lower level to which the water could fall. The potential energy of a given volume of stored water with reference to any datum is the product of the weight of that volume and the distance of its center of gravity above that datum.

Power is energy per unit of time, or the rate of performing work, and is expressed in kilowatts.

The potential energy of a stream of water at any cross section must be measured in terms of power, in which time is an indispensable element. It is the product of the weight of water passing per second and the elevation of its water surface (not center of gravity) above the datum considered. The kinetic energy of a unit weight of the stream is measured by its velocity. It must also be measured in terms of power since velocity involves time. It is the product of the weight of water passing per second and the velocity head, that is, the height the water would have to fall to produce that velocity.

The total energy of a stream is the sum of its potential and kinetic energy. In the case of a perfect turbine, all the potential energy would be converted to kinetic energy. Of course, a perfect turbine does not exist. Some of the potential energy is converted into heat by frictions in the conveyance and energy production system so that the useful part is less than the theoretical total.

6.02.2.1.3(i) Energy grade line

The energy head is a convenient measure of the total energy of a stream of constant discharge at any particular section. It is the elevation of the water surface, potential energy, plus the velocity head, kinetic energy, of a unit weight of the stream. Although every unit of the stream has a different velocity, the velocity head corresponding to the mean velocity of the stream is usually considered. If the stream is flowing in a pipe, the energy head is the elevation of the pressure line, or the height to which water would stand in risers, plus the velocity head of the mean velocity in the pipe.

A line joining the energy heads at all points is the energy grade line.

The energy grade lines would be horizontal if the energy converted to heat was included. Energy converted to heat is however considered lost; hence the energy grade line always slopes in the direction of flow and its fall in any length represents losses by friction, eddies, or impact in that length. Where sudden losses occur, the energy line drops more rapidly. Where only channel friction is involved, the slope of the energy grade line is the friction slope.

Figure 3 illustrates the principles of the foregoing example. The potential energy head of the tank full of water without inflow or outflow is that of the center of gravity of the tank of water Z. With inflow and outflow equal, however, the potential energy head is H. As the water passes into the canal, a drop of the water surface equal to the velocity head in the canal $V_1^2/2g$ must occur. At the entrance to the pipeline, an entrance loss h_1 is encountered as well as an additional drop for the higher velocity in the pipe. At any point on the line, the pressure head h_p will be shown in a riser.

The energy head at any point is the pressure head plus the velocity head, and the line joining the energy heads is the energy grade line. The energy lost (converted to heat) is the sum of friction, entrance, bend, and other losses in all the conduits, including the turbine and draft tube. The useful energy is the power developed by the turbine. The sum of the useful energy and the lost energy must equal the original total potential energy.

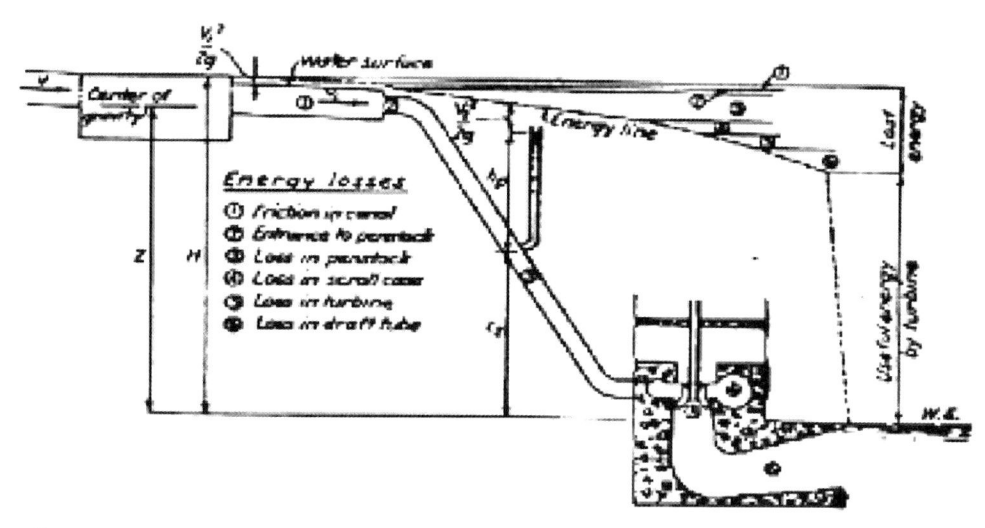

Figure 3 Energy line.

6.02.2.1.3(ii) The Bernoulli theorem

The Bernoulli theorem expresses the law of flow in conduits. For a constant discharge in an open conduit, the theorem states that the energy head at any cross section must equal that at any other downstream section plus the intervening losses. Thus above any datum

$$Z_1 + \frac{V_1^2}{2g} = Z_2 + \frac{V_2^2}{2g} + h_c \qquad [1]$$

In **Figure 4**, Z is the elevation of a free water surface above datum whether it be in a piezometer tube or a quiescent or moving surface of a stream, V the mean velocity, h_c the conduit losses between the two sections considered, and e the energy head above the chosen datum. Obviously, Z may be made up of a number of elements such as elevation of streambed and depth of water in an open channel y.

6.02.2.1.3(iii) Head

There are several heads involved in a hydroelectric plant, which are defined as follows:

- Gross head is the difference in the elevation of the stream surfaces between points of diversion and return.
- Operating head is the difference in elevation between the water surfaces of the forebay and tailrace with allowances for velocity heads.
- Net or effective head has different meanings for different types of development. It can be explained as follows:
  1. For an open-flume turbine, it is the difference in the elevation between (1) the headwater in the flume at a section immediately ahead of the turbine plus the velocity head, and (2) the tailwater velocity head.
  2. For an encased turbine, it is the difference between (1) the elevation corresponding to the pressure head at the entrance to the turbine casing plus the velocity head in the penstock at the point of measurement, and (2) the elevation of the tailwater plus the velocity head at a section beyond the disturbances of the exit from the draft tube.
  3. For an impulse wheel, including its setting, it is the difference between (1) the elevation corresponding to the pressure head at the entrance to the nozzle plus velocity head at that point, and (2) the elevation of the tailwater as near the wheel as possible to be free from local disturbances. When considered as a machine, the effective head is measured from the lowest point of the pitch circle of the runner buckets (to which the jet is tangent) to the water surface corresponding to the pressure head at the entrance to the nozzle plus the velocity head.

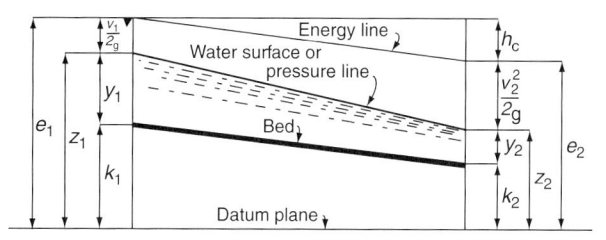

Figure 4 Bernoulli equation in an open conduit.

Strictly speaking, the various heads described above are the differences in the energy heads. For the gross head, the velocities in the stream are generally disregarded, as well as the velocity heads in the tailrace for the operating head. The net head, however, is important in determining efficiency tests of a turbine in its setting; hence it is important to use the difference in the energy heads at the entrance and exit of the plant. The net head includes the losses in the casing of the turbine, and the draft tube, for they are charged to the efficiency of the wheel.

6.02.2.1.3(iv) Efficiency

Efficiencies of the components of a hydroelectric system are measured as the ratio of energy output to input or to total potential energy at the site. No component is perfect, because its functioning involves lost energy (conversion to heat). The efficiency of a plant or system is the product of the efficiencies of its several components; thus,

$$E_s = E_c E_t E_g E_u E_l E_d \qquad [2]$$

where E_s is the over-all system efficiency made up of the product of the several efficiencies of the conduits; E_t is the efficiency of the turbines, including the scroll case and the draft tube; E_g is the efficiency of the generators, including the exciter; E_u is the efficiency of the step-up transformers; E_l is the efficiency of the transmission lines; E_d is the efficiency of the step-down transformers; and E_c is the efficiency of the canal, the tunnel, the penstocks, and the tailrace.

Formula [2] expresses the overall efficiency from the river intake to the distribution switches at the substation. To this could be added the efficiency of the distribution system, even to the customer's meters, his lights, water heaters, ranges, motors, etc.

The overall efficiency of a plant is the product of the instantaneous efficiencies of its several pieces of equipment referred to the gross head on the water wheels. It obviously varies with capacity of units, head, load, and the number of units in service. Plant efficiencies are not always observed and frequently involve many complexities. In general, the plant efficiency is the ratio of the energy output of the generator to the water energy corresponding to the gross head (difference of forebay and tailrace levels) and that discharge and load for which the indicated efficiency of the turbine is maximum. In any case, it should be clearly defined.

6.02.2.1.3(v) Power and energy

From previous paragraphs, the power is defined as follows in kilowatts:

$$9.81 Q H E_s \quad (\text{kW}) \qquad [3]$$

And the energy produced by the plant is defined as

$9.81 Q H E_s t \quad (\text{kWh})$

where Q is the discharge flowing through the unit(s) in m³ s⁻¹; H the net head in meters; and t the time in hours for which the flow and head are constant or for which they are average values. When the flow and head vary continuously, the period considered can be divided into smaller time intervals for which they are sensibly constant.

- Power from any particular plant or system is limited by the capacity of the installed equipment. It may be limited also by the available water supply, head characteristics, and storage.
- Firm power, or primary power, is that load within the plant's capacity and characteristics that may be supplied virtually at all times. It is fixed by the minimum stream flow, having due regard for the amount of regulating storage available and the load factor of the market supplied. In certain cases, it could be the average power/energy, which could be produced, based on stream flow records of a specified time period according to prior agreements among parties for a specific region, such as the northwest of the United States.
- Surplus power, or secondary power, is the available power in excess of the firm power. It is limited by the generating capacity of the plant, by the head, and by the water available in excess of the firm water.
- Dump power is surplus power sold with no guarantee of the continuity of service, that is, it is delivered whenever it is available.

6.02.2.1.3(vi) Load

The average load of a plant or system during a given period of time is a hypothetical constant load over the same period that would produce the same energy output as the actual loading produced.

The peak load is the maximum load consumed or produced by a unit or a group of units in a stated period of time. It may be the maximum instantaneous load or the maximum average load over a designated interval of time.

The load factor is an index of the load characteristics. It is the ratio of the average load over a designated period to the peak load occurring in that period. It may apply to a generating or a consuming station and is usually determined from recording power meters. We may thus have a daily, weekly, monthly, or yearly load factor; it may apply to a single plant or to a system. Some plants of a system may be run continuously at a high load factor, acting as a base-load plant for the system, whereas variations in load on the system are taken by other plants in the system, either hydro or fossil-fuel power plants. Hydro plants designed to take such variations must have sufficient regulating storage to enable them to operate on a low factor. They are often called peak-load plants.

Operating on a 50% load, there must be sufficient storage to enable such a plant, in effect, to utilize twice the inflow for half the time; on a 25% load factor, the plant should be able to utilize 4 times the inflow for a quarter of the time, and so on. The lower the load factor, the greater the storage required.

The utilization factor is a measure of plant use as affected by water supply. It is the ratio of energy output to available energy within the capacity and characteristics of the plant. Where there is always sufficient water to run the plant capacity, the utilization factor is the same as the capacity factor. A shortage of water, however, will curtail the output and may either decrease or increase the utilization factor according to the plant load factor.

6.02.2.1.4 Essentials of general plant layout

The two basic principles to be kept in mind in planning a waterpower development are economy and safety, or in other words a maximum of power output at a minimum of cost, but at the same time a safe and proper construction that can meet the exigencies of operation imposed by structures which control as far as may be, but of necessity interfere somewhat with, natural forces, variable and often large in amount and uncertain in regimen. The hazards due to floods, ice, etc. must be provided for not only from the point of view of safety but also to minimize interruptions in plant operation as far as practicable.

Owing to the uncertainties and irregularities of the forces of Nature to which a hydropower development must of necessity be subjected, fossil-fuel power plants were formerly considered as more dependable prime source of supplying energies. However, because of the interruptions in service at steam plants in the countries during the times of fuel shortage, when for times, hydropower alone was the dependable source of power supply. With continued high fuel costs, it has materially changed our perspective in this respect. The trend of modern hydropower developments toward simple and effective layout and also the greater use of stored water have resulted in a better appreciation of the value and dependability of hydropower, when properly utilized.

6.02.2.1.5 Factors affecting economy of plant

The factors or conditions affecting the relative economy of a hydropower development may be divided into the characteristics of (1) site and (2) use and market.

1. The site characteristics are those particularly affecting the construction and operating cost of the plant and, therefore, the conditions that are most likely to decide first of all whether a site is worthy of development and, if so, the best manner of making this development.

    These include geologic conditions as affecting available foundations for structures, particularly the dam, whose type may be thus determined. The absence of suitable rock foundations for the dam may even prevent the utilization of a power site.

    Topographical conditions are also of great importance in determining the dimensions of the dam and thus largely affecting its cost and the relative proportion of the fall or head to be developed by the dam or by waterway, as well as the manner in which the waterway may be constructed, whether canal or penstock or a combination of these.

    The slope of the river is of importance, as it governs the head, which is available to generate power. This directly affects the length and the cost of the water conveyance structure, as well as the amount of poundage required at the dam to meet the economic objectives of the development.

    The relation of head to discharge also greatly affects the economic objectives of a power development. For a given amount of available power, the greater the head as compared with the discharge, the less costly will be the development, owing to the greater capacity required for all the features except the dam, as discharge increases. In general, therefore, the higher head developments are always less expensive per horsepower of capacity than those of the lower head.

    Storage possibilities at sites upstream are of special importance, where storage cost is reasonable, which will usually require the use of the stored water at several power plants in order to lessen its cost at each plant. This also increases the dependability of the waterpower development, and the proportion of its output, which will be primary of dependable power.

    Operating costs may also be affected by special conditions, which may prevail on a given stream. Thus, a stream subject to frequent floods or high water periods may have the power at a given site frequently curtailed by backwater in the tailrace, and on such a stream, the flashboards on the dam, if present, may also require frequent renewal. The presence of ice, particularly anchor or frazil ice, on streams having numerous falls or stretches of rapids also introduces troublesome problems of operation and often adds to its cost.
2. The characteristics of use and market include the conditions particularly affecting the sale price and value of the developed power; thus, proximity to market is a vital consideration. A hydropower site may be capable of development at low cost, namely, with advantageous natural features. But if it is situated very far from any possible market, it may not be worthy of consideration for development, unless the transmission costs are low, particularly in transmission efficiency. In this respect, the radius of possible transmission of power is constantly growing due to advances in transmission technology, and today lines of more than 2000 km are possible.

On the other hand, to transmit power such distances economically requires relatively large blocks of power, and in any event, the cost of transmission must be included in power cost in competing with fossil-fuel plants at a distance. The transmission of power across state lines is also in some cases hampered or prohibited by state laws.

The cost of other alternative power sources at the available market is of importance as it affects the sale price of hydropower. These other power sources commonly come from fossil fuel, whose cost is largely affected by fuel cost. Hence, much variation in the cost of power may be found in different parts of the country, depending upon the distance that coal (or oil or natural gas, in many cases) must be transported, with freight charges here constituting the important element. Of course, there is nuclear power as well, the licensing of which is greatly affected by government regulations and environmental concerns with its operations and the disposal of the spent-fuel rods.

6.02.2.1.6 Types of hydropower developments
No two hydropower developments that are exactly alike will probably ever be built, and every power site has its special problems of design and construction, which must be met and solved. We may, however, distinguish certain general types of plant layout consistent with the general site characteristics of importance – head, available flow, topography of river, etc., all more or less being interdependent. These characteristics affect the manner of development together with those of market and type of load, which in turn affect the size of plant and number of its units. The general classification could be (1) concentrated fall where the head of the hydropower is mainly due to the height of the dam (**Figure 5**; Three Gorges Power Plant, China); (2) divided fall where the dam acts only as a barrier and the head of the hydropower is due to the local topography and most of the time much more higher than the height of the dam (**Figure 6**; Grande Dixence, Switzerland); in Grande Dixence, the height of the dam is 285 m and the head of the hydropower plant is more than 2000 m.

In the case of a concentrated fall project with penstocks, the ordinary upper limit of head on the turbines is placed at up to 300 m, although a dam of that height would seldom be economical for power development unless it afforded at the same time substantial storage capacity.

Hydropower plants could also be divided as a function of the head, in three ranges: low, medium, and high head.

6.02.2.1.7 Typical of arrangements of waterpower plants
6.02.2.1.7(i) Concentrated fall project
The location of the powerhouse with reference to the dam will depend upon local conditions. Often a low-cost development could be made by placing the powerhouse in the river at one end of the dam (**Figure 7(a)**).

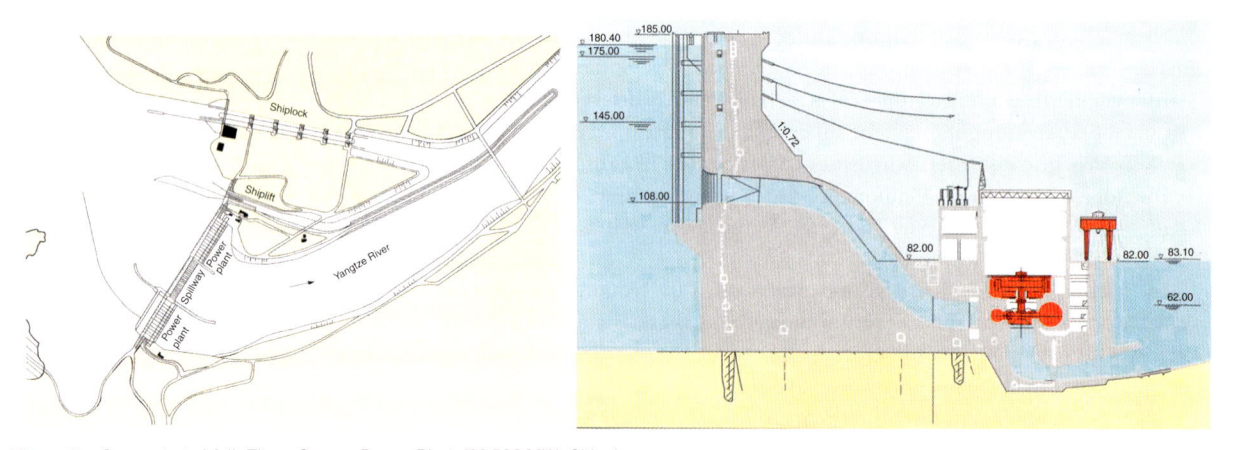

Figure 5 Concentrated fall: Three Gorges Power Plant (22 500 MW, China).

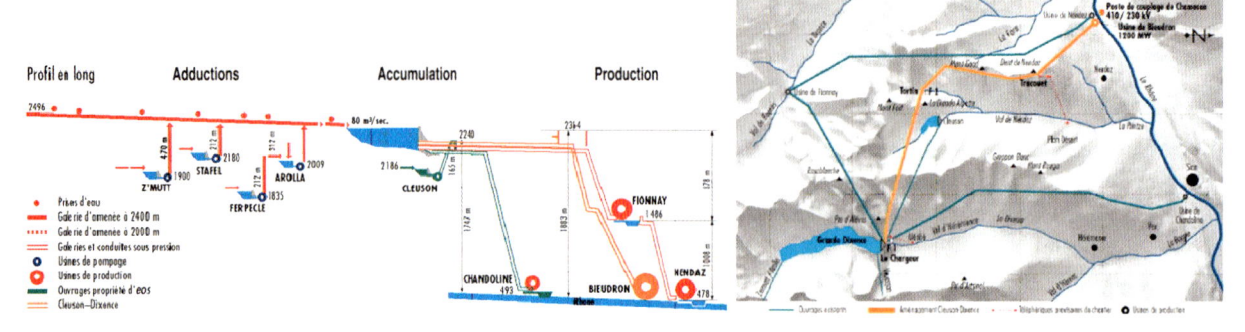

Figure 6 Divided fall: Grande Dixence (2000 MW, Switzerland).

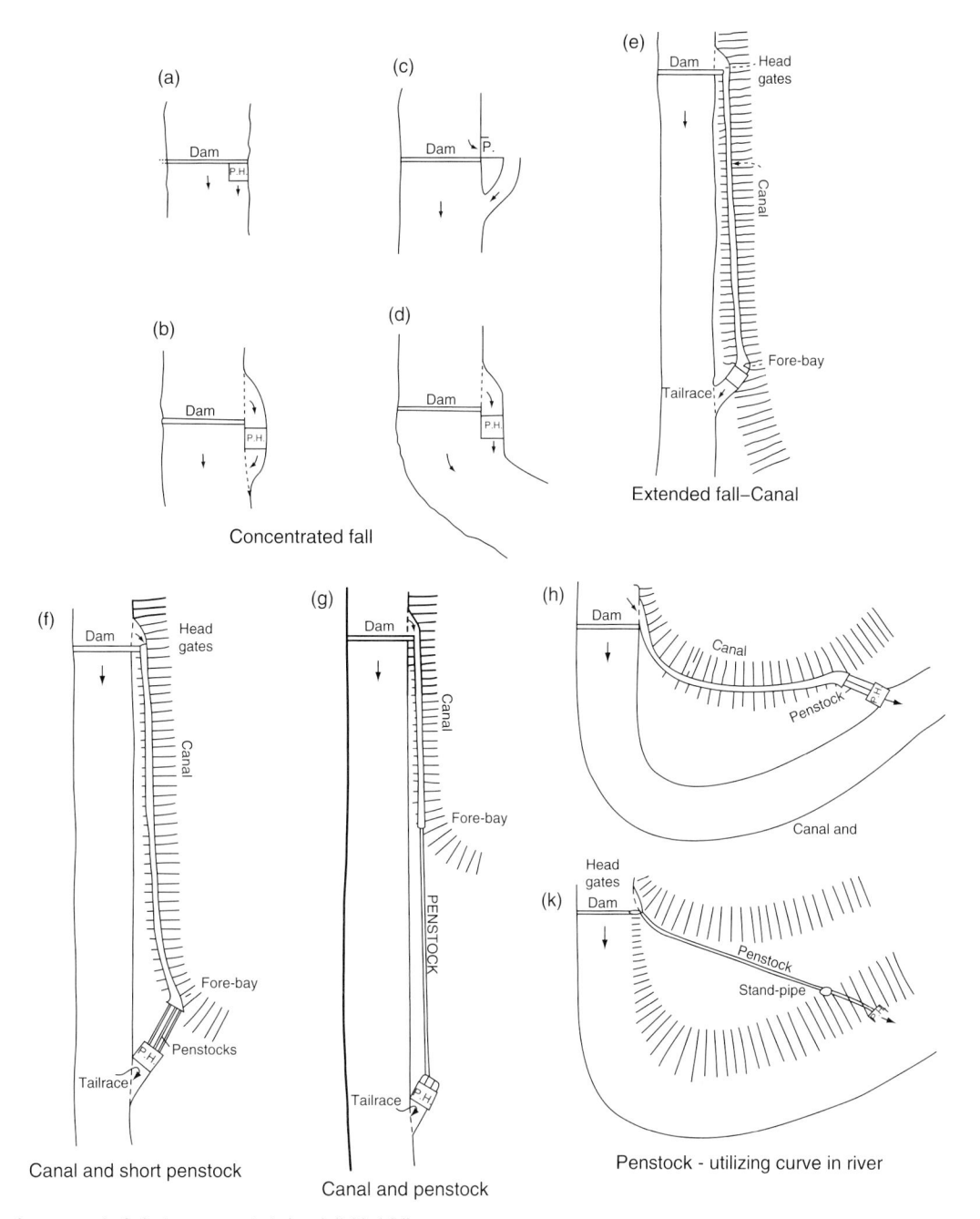

Figure 7 Arrangement of plants – concentrated and divided fall.

This would generally result, however, in an undesirable limitation in the length of spillway and possible subjection of the powerhouse to flood and ice hazards. To obtain the necessary spillway length, the powerhouse must often be located in such a manner as shown in **Figures 7(b)–7(d)**.

6.02.2.1.7(ii) Divided fall projects
Various typical plant arrangements for the divided fall arrangement are shown in **Figures 7(e)–7(k)**. Aside from the capacity to be handled, the dominating feature is the topography of the region adjacent to the river. Thus, in **Figure 7(e)**, the riverbank remains high and affords room for a canal development, which with open wheel pit could utilize a head of only about 7 m, but with concrete flume, settings might make it possible to use a head of 50 m.

The arrangement in **Figure 7(f)** is typical of many developments where flow is relatively large, where the riverbank permits the use of a canal to a forebay near the powerhouse, from whence individual penstock lines run to each turbine unit. The head utilized in such a development will nominally be more than about 100 m and is limited above that amount only by the fall in the river between dam and tailrace level.

In **Figure 7(g)**, the topography is such that a canal can be used for only a part of the distance. If flow is large, it may be necessary to use more than one penstock line, although such a development would result in increased cost, as compared with **Figure 7(f)**, for a given total length of waterway.

In **Figure 7(h)**, the manner of development is similar to that of **Figure 7(g)**, but advantage is taken of a bend in the river to utilize a greater head for a given length of waterway.

In **Figure 7(k)**, the flow is low enough to permit the use of a penstock throughout, which is kept at relatively high level to save cost, until near the powerhouse, where a quick descent is made, usually with individual penstocks to each wheel unit. Here again a curve in the river is utilized to shorten the length of penstock.

A modification of **Figure 7(k)** of service where the riverbank between the dam and powerhouse site is very high, as with a hill, consists in constructing a tunnel penstock with surge tank and individual penstock lines to each unit from the point on the hillside where the tunnel emerges. The material most favorable for tunnel construction is rock, and usually the tunnel would be lined to increase its flow capacity. The tunnel grade would be usually kept relatively flat, the sudden pitch being made with the penstock lines.

6.02.2.1.8 Lowest cost power developments

Keeping in mind the variations in site, use, and market characteristics, it will be seen that the lowest cost development as well as of power produced will be secured with the following conditions:

Conditions favoring low-cost developments (a penstock development) are

1. relatively high head and small flow,
2. discharge assured by storage, the cost of which is carried by several plants,
3. favorable dam site: good foundations, narrow valley, and a minimum of material in dam,
4. good penstock location, fairly straight line with moderate grade for most of the distance, and then a quick drop to the powerhouse site,
5. a few large turbine units,
6. relatively short transmission to market, and
7. high load factor often made possible where the plant is a unit of a large power system.

6.02.2.1.9 Highest cost power developments

Conversely, the highest cost development and of power produced will be for the following conditions:

Conditions resulting in high-cost developments (a canal development) are

1. relatively low head and large flow,
2. variable flow with small minimum or primary power,
3. poor dam site: poor foundations, wide valley, and relatively large material requirements for dam construction,
4. poor canal location deep cut in hard material,
5. a relatively large number of small-capacity turbine units,
6. long transmission to market, and
7. low load factor, as with an isolated plant, and poor load characteristics.

6.02.2.2 Types of Turbines

In water turbines, the kinetic energy of flowing water is converted into mechanical rotary motion. As noted earlier, theoretical power is determined by the available head and the mass flow rate. To calculate the available power, head losses due to friction of flow in conduits and the conversion efficiency of machines employed must also be considered. The formula, thus, is the following:

$$P = H_n Q \rho g E_s \quad (P \text{ in watts}) \qquad [5]$$

where P is the output power in watts; H_n the net head = gross head − losses (m); Q the flow in $m^3 s^{-1}$; g the specific gravity = $9.81\ m\ s^{-2}$; ρ the specific mass of the water; and E_s the overall efficiency.

The oldest form of 'water turbine' is the water wheel. The natural head − difference in water level − of a stream is utilized to drive it. In its conventional form, the water wheel is made of wood and is provided with buckets or vanes round the periphery. The water thrusts against these, causing the wheel to rotate.

A water turbine is characterized by the following parameters:

N rotational speed ($r s^{-1}$)
Q turbine discharge ($m^3 s^{-1}$)
H design head (m)

The so-called kinematic specific speed N_s, a dimensionless number, is deduced from these parameters:

$$N_s = N \frac{Q^{1/2}}{H^{3/4}}$$

In practice, each type of turbine has N_s range for good operation, that is,

Pelton turbine $N_s = 3-14$
Francis turbine $N_s = 20-140$
Kaplan turbine $N_s = 140-300$
Banki turbine $N_s = 20-80$

6.02.2.2.1 Pelton turbine

The principle of the old water wheel is embodied in the modern wheel, which consists of a wheel provided with spoon-shaped buckets round the periphery (**Figure 8**). A high-velocity jet of water emerging from a nozzle impinges on the buckets and sets the wheel in motion. The speed of rotation is determined by the flow rate and the velocity of the water; it is controlled by means of a needle in the nozzle (the turbine operates most efficiently when the wheel rotates at half the velocity of the jet). If the load on the wheel suddenly decreases, the jet deflectors partially divert the jet issuing from the nozzle until the jet needle has appropriately reduced the flow. This arrangement is necessary because in the event of sudden load decrease, or rejection, the jet needle would be closed suddenly, and the flow of water would be reduced too abruptly, causing harmful 'water hammer' phenomena in the water system. In most cases, the control of the deflector is linked to an electric generator. A Pelton wheel is used in cases where large heads of water are available (**Figure 8**).

Pelton turbines belong to the group of impulse (or free-jet) turbines, where the available head is converted to kinetic energy at atmospheric pressure. Power is extracted from the high-velocity jet of water when it strikes the cups of the rotor. This turbine type is normally applied in the high head range (>40 m). From the design point of view, adaptability exists for different flow and head. Pelton turbines can be equipped with one, two, or more nozzles for higher output. In the manufacture, casting is commonly used for the rotor, materials being brass or steel. This necessitates an appropriate industrial infrastructure.

6.02.2.2.2 Francis and Kaplan turbines

In a great majority of cases (large and small water flow rates and heads), the type of turbine employed is the Francis or radial flow turbine. The significant difference in relation to the Pelton wheel is that Francis (and Kaplan) turbines are of the reaction type, where the runner is completely submerged in water, and both the pressure and the velocity of water decrease from inlet to outlet. The water first enters the volute, which is an annular channel surrounding the runner, and then flows between the fixed guide vanes, which give the water the optimum direction of flow. It then enters the runner and flows radially through the latter, that is, toward the center. The runner is provided with curved vanes upon which the water is largely converted into rotary motion and is not consumed by eddies and another undesirable flow phenomenon causing energy losses. The guide vanes are usually adjustable so as to provide a degree of adaptability to variations in the water flow rate in the load of the turbine.

The guide vanes in the Francis turbine are the elements that direct the flow of the water, just as the nozzle of the Pelton wheel does. Water is discharged through an outlet from the center of the turbine. A typical Francis runner is shown in **Figure 9**. The volute, guide vanes, and runner are also shown schematically in **Figure 9**.

In design and manufacture, Francis turbines are much more complex than Pelton turbines, requiring a specific design for each head/flow condition to obtain optimum efficiency. The runner and housing are usually cast, on large units welded housings, or cast in concrete at site, are common. With a large variety of designs, a large head range from about 30 m up to 700 m of head can be achieved.

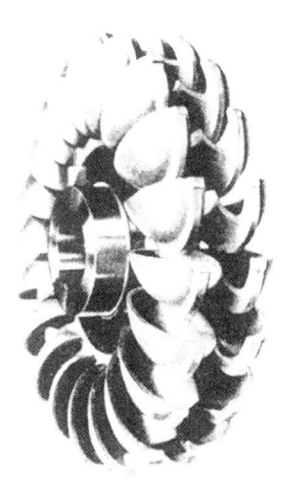

Figure 8 Pelton wheel.

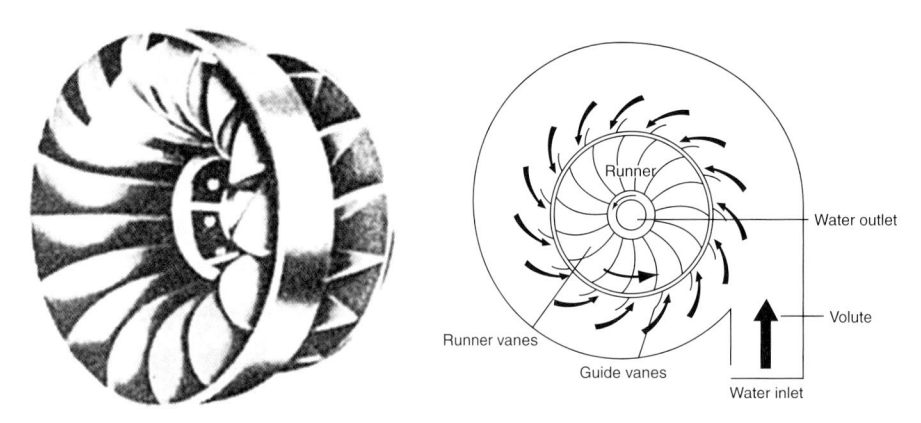

Figure 9 Francis runner and schematic of flow in Francis turbine.

For very low heads and high flow rates – for example, at the run-of-river dams – a different type of turbine, the Kaplan or propeller turbine, is usually employed. In the Kaplan turbine, water flows through the propeller and sets the latter in rotation. Water enters the turbine laterally (**Figure 10**), is deflected by the guide vanes, and flows axially through the propeller. For this reason, these machines are referred to as axial-flow turbines. The flow rate of the water through the turbine can be controlled by varying the distance between the guide vanes; the pitch of the propeller blades must also be appropriately adjusted (**Figure 10**). Each setting of the guide vanes corresponds to one particular setting of the propeller blades in order to obtain high efficiency.

Especially in smaller units, either only vane adjustment or runner blade adjustment is common to reduce sophistication but this affects part load efficiency. Kaplan and propeller turbines also come in a variety of designs. Their application is limited to heads from 1 m to about 30 m. Under such conditions, a relatively larger flow as compared to high-head turbines is required for a given output. These turbines therefore are comparatively larger. The manufacture of small propeller turbines is possible in welded construction without the need for casting facilities.

6.02.2.2.3 Cross-flow (Banki) turbine

The concept of the cross-flow turbine – although much less well known than the three big names Pelton, Francis, and Kaplan – is not new. It was invented by an engineer named Michell, who obtained a patent for it in 1903. Quite independently, a Hungarian professor named Donat Banki reinvented the turbine again at the University of Budapest. By 1920, it was quite well known in Europe, through a series of publications. There is one single company by the name of Ossberger in Bavaria, Germany, which produces this turbine for decades. A very large number of such turbines are installed worldwide; most of them were made by Ossberger.

The main characteristic of the cross-flow turbine is the water jet of rectangular cross section, which passes twice through the rotor blades – arranged at the periphery of the cylindrical rotor – perpendicular to the rotor shaft. The water flows through the blades first from the periphery toward the center (refer to **Figure 11**), and then, after crossing the open space inside the runner, it strikes the blades as it moves from the inside out of the turbine. Energy conversion takes place twice: first upon impingement of water on the

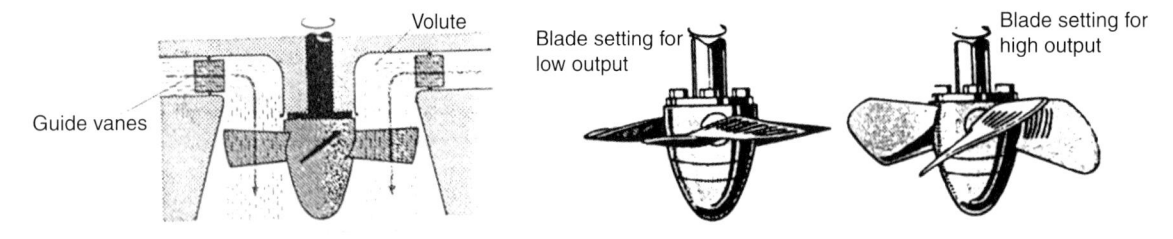

Figure 10 Schematic of Kaplan turbine and propeller.

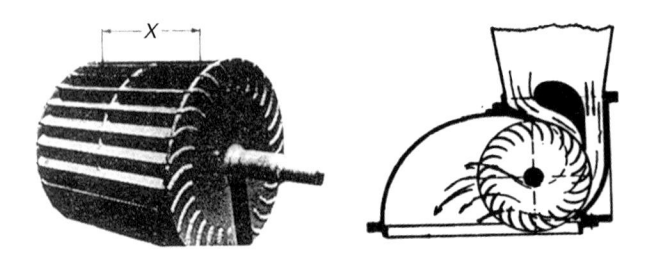

Figure 11 Cross-flow runner.

blades upon entry, and then when water strikes the blades upon exit from the runner. The use of two working stages provides no particular advantage except that it is a very effective and simple means of discharging the water from the runner.

The machine is normally classified as an impulse turbine. This is not strictly correct and is probably based on the fact that the principal design was a true constant-pressure turbine. A sufficiently large gap was left between the nozzle and the runner, so that the jet entered the runner without any static pressure. Modern designs are usually built with a nozzle that covers a bigger arc of the runner periphery. With this measure, unit flow is increased, permitting to keep turbine size smaller. These designs work as impulse turbines only with small gate opening, when the reduced flow does not completely fill the passages between the blades and the pressure inside the runner is therefore atmospheric. With increased flow completely filling the passages between the blades, there is a slight positive pressure; the turbine now works as a reaction machine.

Cross-flow turbines may be applied over a head range from less than 2 m to more than 100 m (Ossberger has supplied turbines for heads up to 250 m). A large variety of flow rates may be accommodated with a constant-diameter runner, by varying the inlet and runner width. This makes it possible to reduce considerably the need for tooling, jigs, and fixtures in manufacture. Ratios of rotor width/diameter, from 0.2 to 4.5, have been made. For wide rotors, supporting discs welded to the shaft at equal intervals prevent the blades from bending.

A valuable feature of the cross-flow turbine is its relatively flat efficiency curve, which Ossberger is further improving by using a divided gate. This means that at reduced flow, efficiency is still quite high, a consideration that may be more important than a higher optimum point efficiency of other turbines.

It is easy to understand why cross-flow turbines are much easier to make than other types, by referring to **Figure 11**.

6.02.2.2.4 Hydraulienne and Omega Siphon

The 'Hydraulienne' provides an electrical power using the velocity of the water in a stream by means of a floating wheel (see **Figure 12**).

Consisting mainly of a float, a rotor, and a stabilizer, the operating mode of the 'Hydraulienne' is of great simplicity. They are floating hydro-generators on a river at a point where the current velocity is up to 2 m s^{-1}. The current turns a wheel, which produces electricity. When the height of water increases or decreases, the float obliges the 'Hydraulienne' to move vertically in concert.

The depth of water must be at least 0.5 m and per each wheel the available power could be up to 15 kW.

The Omega Siphon (see **Figure 13**) is also a floating structure using the head of an existing weir.

6.02.2.2.5 Comparison of different turbines

Figure 14 is a graphical presentation of a general turbine application range of conventional designs. The usual range for commercially available cross-flow turbines is shown in relation (dotted line). In the overall picture, it is clearly a small turbine.

Figure 12 Schematic view of 'Hydraulienne' in operation.

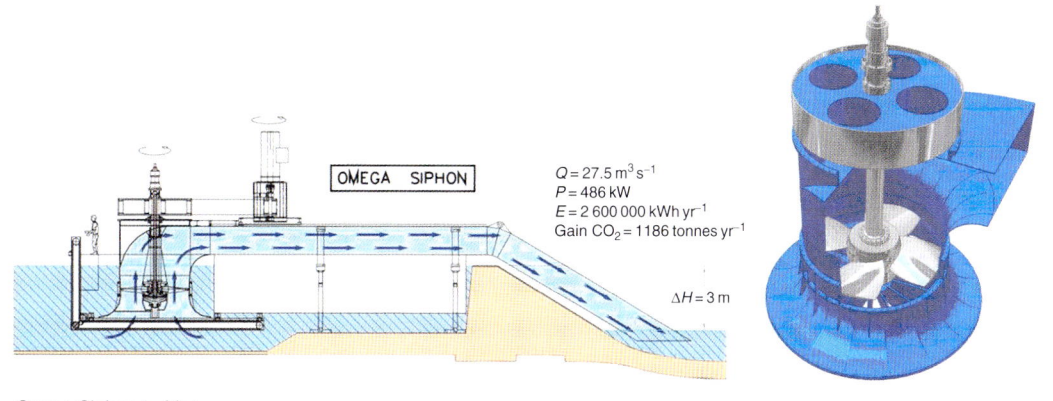

$Q = 27.5 \text{ m}^3\text{s}^{-1}$
$P = 486 \text{ kW}$
$E = 2\,600\,000 \text{ kWh yr}^{-1}$
Gain $CO_2 = 1186$ tonnes yr^{-1}
$\Delta H = 3 \text{ m}$

Figure 13 Omega Siphon turbine.

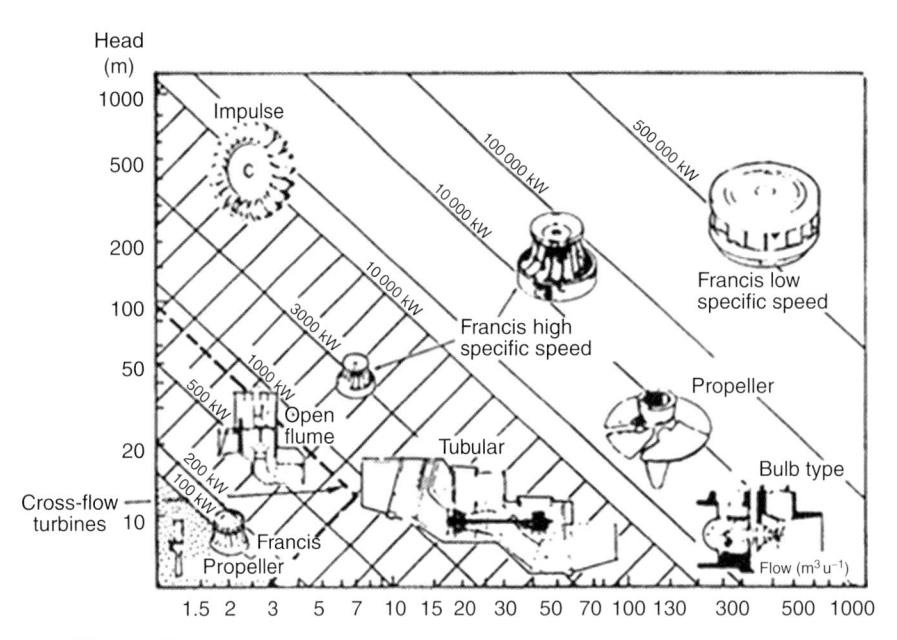

Figure 14 Comparison of different turbines.

6.02.2.3 Types of Dams

The primary purpose of a dam may be defined as to provide for the safe detention and storage of water. There is no nominal structural design life for dams, if the effects of reservoir siltation or similar time-dependent limitations on their operational utility are disregarded. As a corollary to this, every dam must represent a design solution specific to its site circumstances. The design therefore also represents an optimum balance of local technical and economic considerations at the time of construction. Reservoirs are readily classified in accordance with their primary purpose, for example, irrigation, water supply, hydroelectric power generation, flood control, etc. Dams are of numerous types, and the type classification is sometimes less clearly defined. An initial broad classification into two generic groups can be made in terms of the principal construction material employed:

- Embankment dams: constructed of earthfill and/or rockfill. Upstream and downstream face slopes are similar and of moderate angle, giving a wide section and a high construction volume relative to height.
- Concrete dams: constructed of mass concrete. Face slopes are dissimilar, generally steep downstream and near vertical upstream, and dams have relatively slender profiles dependent upon the type.

The latter group can be considered to include also older dams of appropriate structural type constructed in masonry. Embankment dams are numerically dominant for technical and economic reasons. Older and simpler in structural concept than the early masonry dam, the embankment dams utilized locally available and untreated materials. As the embankment dam evolved, it has proved to be increasingly adaptable to a wide range of site circumstances. In contrast, concrete dams and their masonry predecessors are more demanding in relation to foundation conditions. Historically, they have also proven to be dependent upon relatively advanced and expensive construction skills.

6.02.2.3.1 Embankment dam types

The embankment dam can be defined as a dam constructed from natural materials excavated or obtained nearby. The materials available are utilized to the best advantage in relation to their characteristics as bulk fill in zones within the dam section. The natural fill materials are placed and compacted without the addition of any binding agent, using high-capacity mechanical equipment. Embankment construction is consequently an almost continuous and highly mechanized process, equipment-intensive rather than labor-intensive. Embankment dams can be classified in broad terms as being earthfill or rockfill dams. The division between the two embankment variants is not absolute, with many dams utilizing fill materials of both types within appropriately designated internal zones. Small embankment dams and a minority of larger embankments employ a homogeneous section, but in the majority of instances, embankments employ an impervious zone or core combined with supporting shoulders, which may be of relatively pervious material. The purpose of the latter is structural, providing stability to the impervious element and to the section as a whole. Embankment dams can be of many types, depending upon how they utilize the available materials. The initial classification into earthfill or rockfill embankments provides a convenient basis for considering the principal variants employed:

- Earthfill embankments: An embankment may be categorized as an earthfill dam if compacted soils account for over 50% of the placed volume of material. An earthfill dam is constructed primarily of engineering soils compacted uniformly and intensively in relatively thin layers and at a controlled moisture content.
- Rockfill embankments: In the rockfill embankment, the section includes a discrete impervious element of compacted earthfill or a slender concrete or bituminous membrane. The designation 'rockfill embankment' is appropriate where over 50% of the fill material may be classified as rockfill, that is, coarse-grained frictional material. Modern practice is to specify a graded rockfill, heavily compacted in relatively thin layers by heavy plant. The construction method is therefore essentially similar to that of the earthfill embankment.

The terms zoned rockfill dam or earthfill–rockfill dam are used to describe rockfill embankments incorporating relatively wide impervious zones of compacted earthfill. Rockfill embankments employing a thin upstream membrane of asphaltic concrete, reinforced concrete, or other non-natural material are referred to as 'decked rockfill dams'. The saving in fill quantity arising from the use of rockfill for a dam of given height is very considerable. It arises from the frictional nature of rockfill, which gives relatively high shear strength, and from high permeability, resulting in the virtual elimination of pore water pressure problems. The variants of earthfill and rockfill embankments employed in practice are too numerous to identify all individually. The embankment dam possesses many outstanding merits, which combine to ensure its continued dominance as a generic type. The more important can be summarized as follows:

- The suitability of the type to sites in wide valleys and relatively steep-sided gorges alike.
- Adaptability to a broad range of foundation conditions, ranging from competent rock to soft and compressible or relatively pervious soil formations.
- The use of natural materials, minimizing the need to import or transport large quantities of processed materials or cement to the site.
- Subject to satisfying essential design criteria, the embankment design is extremely flexible in its ability to accommodate different fill materials, for example, earthfills and/or rockfills, if suitably zoned internally.
- The construction process is highly mechanized and is effectively continuous.
- The earthfill dams can be designed more economically in areas of high seismic activities.
- Largely in consequence, the unit costs of earthfill and rockfill have risen much more slowly in real terms than those for mass concrete.

The most popular type of rockfill dams used for the moment is the concrete face rockfill dam (CFRD). CFRDs are constructed of permeable rockfill, the impermeable membrane being a concrete slab constructed on the upstream face of the dam wall (**Figure 15**). The CFRD has been greatly advanced in China during the last 10 years. Its value to the hydro resources and electricity supply sectors is shown by the great investment in designing and constructing such dams. So far, more than 50 CFRDs have been built or nearly completed in China. Among these is Tianshengqiao 1, which has a dam height of 178 m, and Shuibuya, which at 233 m is the highest CFRD in the world (**Figure 16**).

The main reason that CFRDs have developed so rapidly in China is that they have advantages such as full use of local embankment materials, simpler construction, a shorter construction period, and a lower construction cost. CFRDs are therefore more suited to both the engineering and the state conditions in China for water resources and hydropower.

6.02.2.3.2 Concrete dam types

The relative disadvantages of the embankment dam are few. The more important disadvantages include an inherently greater susceptibility to damage or destruction by overtopping, with a consequent need to ensure adequate flood relief and a separate spillway, and vulnerability to concealed leakage and internal erosion in dam or foundation. The principal variants of the modern concrete dam are defined below:

- Gravity dams: A concrete gravity dam is entirely dependent upon its own mass for stability. The gravity profile is essentially triangular, to ensure stability and to avoid overstressing of the dam or its foundation. Some gravity dams are slightly curved in

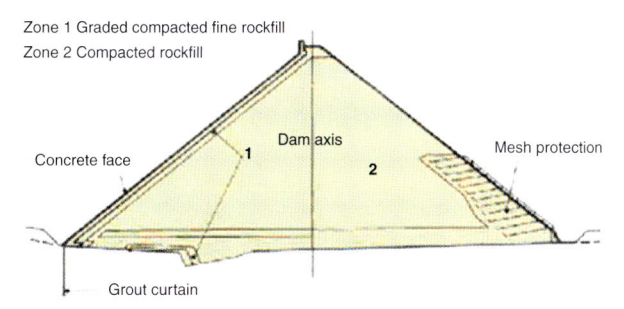

Figure 15 Concrete face rockfill dam.

Figure 16 Shuibuya (CFRD, 1600 MW, China).

plan for aesthetic or other reasons, and without placing any reliance upon arch action for stability. Where a limited degree of arch action is deliberately introduced in design, allowing a rather slimmer profile, the term arch-gravity dam may be employed.
- Buttress dams: In structural concept, the buttress dam consists of a continuous upstream face supported at regular intervals by downstream buttresses. The solid head or massive head buttress dam is the most prominent modern variant of the type, and may be considered for conceptual purposes as a lightened variant of the gravity dam (**Figure 17**).
- RCC (roller compact concrete) dams: The volume instability of mass concrete due to thermal effects imposes severe limitations on the size and rate of concrete pour, causing disruption and delay because of the need to provide contraction joints and similar design features (**Figure 18**). Progressive reductions in cement content and partial replacement of cement with pulverized fuel ash (PFA) have served only to contain the problem. Mass concrete construction remains a semicontinuous and labor-intensive operation of low overall productivity and efficiency. In the construction of RCC dams, the mixture is placed and roller compacted with the same commonly available equipment used for asphalt pavement construction. RCC has low water content, requiring it to be mixed in a continuous flow system. Lifts, which range from 0.2 to 0.4 m in thickness, are then compacted using vibratory steel-wheel and pneumatic tire rollers. Immediately after workers complete compaction, water is applied as a fine mist to cure the concrete. The surface spillway is usually included in the dam itself, with very often a stepped spillway type.
- Arch dams: The arch dam has a considerable upstream curvature. It functions structurally as a horizontal arch, transmitting the major portion of the water load to the abutments or valley sides rather than to the floor of the valley. The profile consists in a relatively simple arch, that is, with horizontal curvature only and a constant upstream radius. It is structurally more efficient than the gravity or buttress dam, greatly reducing the volume of concrete required. A particular derivative of the simple arch dam is the cupola or double-curvature arch dam. The cupola dam introduces complex curvatures in the vertical as well as the horizontal plane. It is the most sophisticated of concrete dams, being essentially a dome or shell structure, and

Figure 17 Itaipu (buttress dam, 12 600 MW, Brazil).

Figure 18 RCC gravity dams.

is extremely economical in concrete. Abutment stability is critical to the structural integrity and safety of both the cupola and the simple arch.

The characteristics of concrete dams are outlined below with respect to the major types, that is, gravity, massive buttress, and arch or cupola dams. Certain characteristics are shared by all or most of these types. However, many are specific to particular variants. Merits shared by most concrete dams include the following:

- With the exception of arch and cupola dams, concrete dams are suitable to the site topography of wide or narrow valleys alike, provided that a competent rock foundation is available at shallow depth.
- Concrete dams are not sensitive to overtopping under extreme flood conditions (cf. the embankment dam).
- As a corollary, all types can accommodate a crest spillway if necessary over their entire length, provided that steps are taken to control downstream erosion and possible undermining of the dam. The cost of a separate spillway and channel is therefore avoided.
- Outlet works, valves, and other ancillary works are readily and safely housed in chambers or galleries within the dam.
- The inherent ability to withstand seismic disturbance without catastrophic collapse is generally high.

Type-specific characteristics are largely determined through the differing structural modus operandi associated with variants of the concrete dam. In the case of gravity and buttress dams, for example, the structural response is in terms of vertical cantilever action. The reduced downstream contact area of the buttress dam imposes significantly higher local foundation stresses than for the equivalent gravity structure. It is therefore a characteristic of the former to be more demanding in terms of the quality required of the underlying rock foundation. The structural behavior of the more sophisticated arch and cupola variants of the concrete dam is dominated by arch action, with vertical cantilever action secondary. Such dams are totally dependent upon the integrity of the rock abutments and their ability to withstand arch thrust without excessive yielding. Consequently, it is characteristic of arch and cupola dams that consideration of their suitability is confined to a minority of sites in narrow steep-sided valleys or gorges.

6.02.3 History of Hydropower

6.02.3.1 Historical Background

6.02.3.1.1 Use of velocity head

Humans have been harnessing water to perform work for thousands of years. The Persians, Greeks, and Romans used water wheels in the old time, starting from 2000 BC. Indeed the use of waterpower by crude devices dates back to ancient times. The primitive wheels, actuated by river current, were used for raising water for irrigation purposes, in mills for grinding corn, and in other simple applications. The Chinese Nora (**Figure 19**), built of bamboo, with woven paddles, is still in use, as well as other forms of current wheel elsewhere. Such devices have a very low efficiency and utilize but a small part of the power available in a stream, that is, the available velocity head (in Bernoulli equation, where V is the velocity of water).

The first watermills recorded about 2000 years ago in Greece, Norway, and Middle Eastland were similar to the scheme shown in **Figure 20**.

6.02.3.1.2 Use of potential head

With the introduction of the overshot or pitchback waterwheel (**Figure 21**), and the use of the potential head (Z in Bernoulli equation, weight of the water or elevation above the reference level), the efficiency of the device increased significantly.

The evolution of the modern hydropower turbine began in the mid-1700s when a French hydraulic and military engineer, Bernard Forest de Bélidor, wrote Architecture Hydraulique. In this four-volume work, he described using a vertical axis versus a horizontal axis machine. During the 1700s and 1800s, water turbine development continued. The impulse turbines (Pelton, cross-flow, etc.) and the reaction turbines (Francis, Kaplan, Bulb, etc.) started to be invented and gradually put to use.

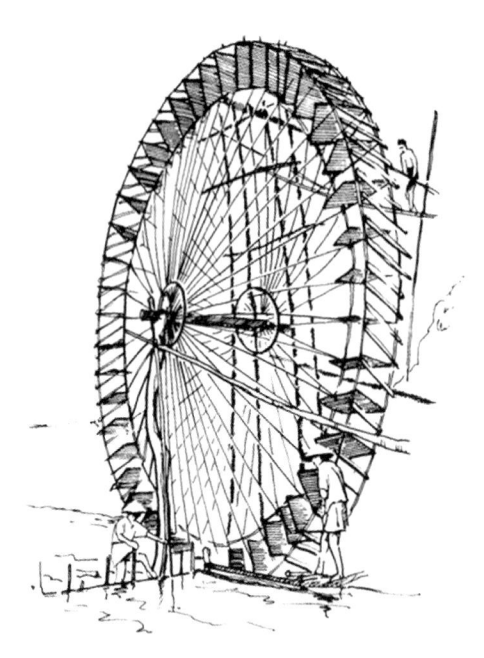

Figure 19 Chinese Nora. www.fao.org/docrep/010/ah810e/AH810E12.htm.

Figure 20 Old watermill scheme (Pippa Miller's drawing of a typical Norfolk watermill). www.norfolkmills.co.uk/Watermills/aldborough.html.

6.02.3.1.3 Electricity is coming

For more than a century, the technology for using falling water to create hydroelectricity has existed. In 1869, Zenobe Gramme, a Belgian electrician, set up the first prototype of a dynamo and an electric engine, and in 1881, a brush dynamo connected to a turbine in a flour mill provided street lighting at Niagara Falls, New York. Starting from that period, the hydropower started to be used mainly for electricity generation. **Table 1** depicts the percentage of waterpower in electric energy production in the world for the last century.

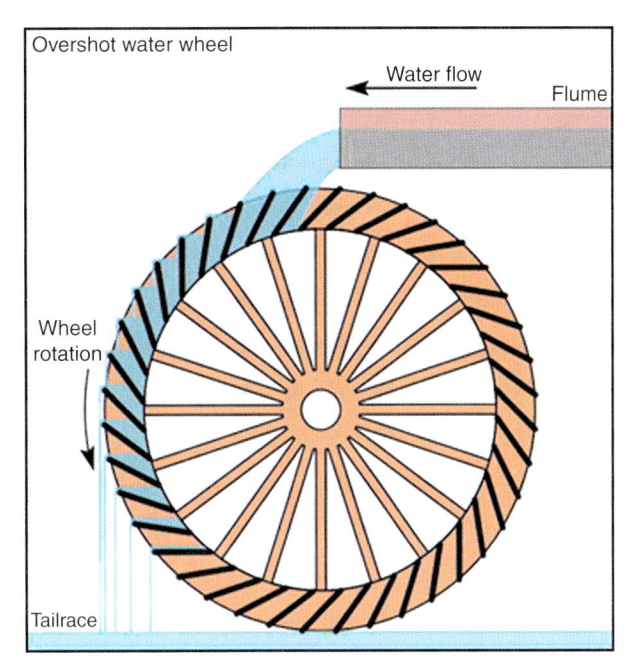

Figure 21 Overshot water wheel. www.nrgfuture.org/Hydro.html.

Table 1 Percentage of waterpower in electric energy production in the world

1925	1950	1963	1974	1985
40%	36%	28%	23%	18.4%

6.02.3.2 Hydro Energy and Other Primary Energies

The historical trend in the world's primary energy consumption is given in **Figure 22**.

Moreover, hydropower plants produce around 16% of world total electricity generation. The current data about main hydroelectricity capacities for the various major producing countries are given in **Table 2**.

6.02.3.3 World Examples

Hydropower production and dams are interconnected. We need dams, large and small, to produce hydropower. They are partners and collaborators in the production of energy. The consequences of probable climate changes could lead to modifications of the electricity generation and short supply in some parts of the world.

Hydropower generation depends on natural conditions, mainly on the availability of water and head. Most of the 'easy' potential sites have already been implemented. Because of the high initial capital costs and the potential 'harm' to the environment associated with hydropower developments and operations, it is necessary to reduce capital costs (by using RCC – roller compacted concrete – dams for instance) and to increase the protection of the environment. Even under these trying conditions, new implementations or studies of hydropower plants are still on the way with a special attention to large- (Inga in Congo, Romaine in Canada) and small-scale projects, but not the medium-scaled ones, which have been postponed or cancelled (Memve'ele in Cameroon), particularly in the new emerging economic powers, such as China, Brazil, and India, and in the 'old' Russian Federation States, and in some African countries, where hydropower resources are plentiful.

6.02.3.3.1 China

The planning of hydropower developments in river basins in China is structured in two levels: the planning of river or river reach for a cascade development and the planning of comprehensive river basin development. The relevant national and provincial departments are responsible for the organization and coordination of the planning activities. The former focuses on the planning of cascade development on the main stem of river or river reach with hydropower generation as the main purpose, while the latter also involves unified development and utilization of water and land resources in the entire river basin (**Table 3**).

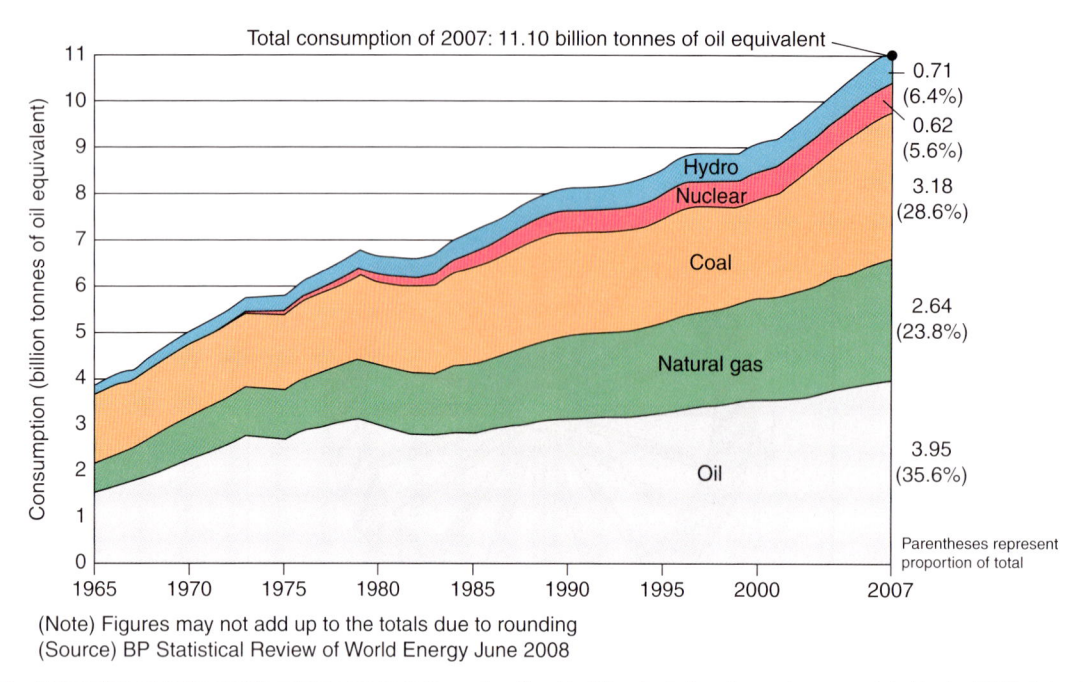

Figure 22 Historical trend in the world's primary energy consumption. Graphical flip-chart of nuclear and energy related topics 2009. Federation of Electric Power Companies of Japan (FEPC).

Table 2 Main hydroelectricity capacities

Country	Annual hydroelectric energy production (TWh)	Installed capacity (GW)	Percent of all electricity
China	585	155	17
Canada	369	88	61
Brazil	363	69	85
USA	250	79	5
Russia	167	45	17
Norway	140	27	98
India	115	33	15
Venezuela	86	-	67
Japan	69	27	7
Sweden	65	16	44
Paraguay	64		
France	63	25	11

Source: BP Statistical Review of World Energy (June 2009). www.usaee.org/usaee2009/submissions/presentations/Finley.pdf

In the last 50 years, in order to comprehensively ascertain hydropower resources and promote their development and utilization, China has carried out general survey, planning, and analysis of hydropower resources four times. Soon after the founding of the People's Republic of China, preparation and organization for the planning studies of development of the Yellow River Basin, the second largest river of the nation, were carried out, and in 1954, the Report on the Technical-Economy for the Multiple Utilization of the Yellow River was submitted. Afterward, the planning of the comprehensive development of hydropower was implemented, in turn, for 112 important main streams and tributaries and 69 major river reaches in the nine major river basins of Yangtze, Pearl, Northeast Rivers, Huaihe, Haihe-Luanhe, Southeast Coastal Rivers, Southwest International Rivers, North Interior Rivers, and

Table 3 Some large hydropower projects in China

Dam	Height (m)	Type	Installed capacity (MW)
Xiaowan	292	Arch dam	4 200
Shuibuya	233	CFRD	1 600
Longtan	216.5	RCC gravity dam	6 300
Xiangjiaba	161	Concrete gravity dam	6 400
Xiluodu	278	Arch dam	12 600
Jinping I	305	Arch dam	3 600
Lianghekou	295	Rockfill dam	3 000
Shuangjiangkou	312	Rockfill dam	2 000

Source: CHINCOLD (2009) Current activities: Dam construction in China 2009. www.chincold.org.cn/newsviewen.asp?s=3483

Xinjiang Rivers. According to the state codes, 263 formal planning reports were prepared. In the reports, thorough initial comparison analysis and screening based on the related technical-economic, social, and environmental conditions, the basic development patterns and the layout schemes of cascade power stations, and the projects to be constructed in the first phase of each river were recommended. This includes 1356 large- and medium-sized hydropower stations each with an installed capacity equal to or more than 25 MW, totaling 404.47 GW, corresponding to an annual energy output of 1911.23 TWh. These reports provided optimized schemes for the large-scale hydropower development and reliable basic data for the study of regional energy composition, formulation of long-term plans, and distribution of construction projects. At the same time, considering the very uneven distribution of energy resources in the country, in order to give priority to the full use of clean and renewable hydropower resources and meet the power needs in energy scarcity areas, based on the planning of rivers and river reaches, it was proposed to establish 12 major hydropower bases in the areas with rich hydropower resources and good conditions for hydropower development. The rich hydropower resources are the Jiansha River, Yalong River, Dadu River, Wujiang River, Upper Yangtze River, Hongshui River, Lancang River, Upper Yellow River, Middle Yellow River, West Hunan Province, Fujian-Zhejiang-Jiangxi, and Northeast. In addition, 41 pumped-storage power stations were also planned and sited in 15 provinces (autonomous regions or municipalities) in Mainland China, mainly in the southeast coastal areas. In Taiwan, a reestimation was carried out during the 1983–94 period on the theoretical hydropower potential of 76 rivers among the provincial 129 rivers of all sizes, and the planned technically exploitable large- and medium-sized hydropower stations each with an installed capacity of more than 20 000 kW had a total installed capacity of 5.05 GW. At the same time, key investigation and planning were carried out on pumped-storage power stations for the nine rivers of Lijiaxi, Zhushuixi, Dajiaxi, etc., with the exploitable pumped-storage power stations having a total installed capacity of 12.80 GW.

China's installed hydro capacity currently stands at about 155 GW, and the aim is to increase this to 300 GW by 2020, and China's total exploitable hydropower potential is estimated to be 542 million kilowatts, ranking first in the world.

6.02.3.3.2 Brazil

In the south and southeast regions of Brazil, the development of dam construction was mainly due to the implementation of hydroelectric projects. The first hydroelectric plant in the country dates back to 1883. It was built on the Inferno River with only two 6 kW under 5 m of head for a diamond mining project. In 1887, a hydroelectric plant was put in operation on the Macacos River, and it provided a gross output of 370 kW under 40 m of head in a gold mining project. The first hydroelectric plant for supplying an industrial plant and a city as a utility was the 252 kW Marmelos power plant on the Paraibuna River, which today is a small museum. The original rockfill dam had an upstream wood face to provide water tightness. All these projects were built in the Minas Gerais state. From 1890 to 1901, the Parnaiba 2 MW power plant was built on the Tietê River to supply power to São Paulo city. Its concrete dam, later named Edgard de Souza, was the first large dam built in Brazil. In those early days, it was almost impossible to imagine that hydropower would develop so much throughout the country. Until the 1950s, all power utilities were private enterprises and small power plants were built mainly in the south and southeast Brazil. Most of the dams were not very high concrete gravity structures. Presently, there is 1206 MW of existing hydro capacity in units more than 50 years old. Several of these units are now being rehabilitated and upgraded.

In 1934, the federal decree n° 24643 known as Code of Waters and the deletion of the clause protecting the utilities from the effects of the national currency devaluations strongly discouraged the power investors. Due to the tariff constraint and weakness of domestic private capital, there was insufficient power supply throughout the country in the following decades. There was no way to provide power other than the federal and some state governments creating power utilities.

Soon after World War II, the private utility Light, in the most developed area of the country, built several dams and large underground power plants in Rio de Janeiro and São Paulo (Table 4). Currently, the projects are very important and **Table 4** shows some large projects under construction.

Table 4 Some large hydro projects in Brazil

Dam	Height (m)	Type	Installed capacity (MW)
Jirau	35	Run of river	3 300
Santo Antonio	55	Run of river	3 150.4
Germano Dam	170	RCC gravity dam	
Belo Monte	114	Concrete gravity dam	11 183

Source: Brazillian Committittee on Dams (CBDB) (2009) In: Piasentin C (ed.) *Main Brazilian Dams III. Design, Construction and Performance*. Paris, France: International Commission on Large Dams.

Table 5 Mean cost of electric power generation in Brazil

Diesel oil	US$214 per MWh
Fuel oil	US$144 per MWh
Wind	US$86 per MWh
Natural gas	US$61 per MWh
Nuclear	US$60 per MWh
Coal	US$59 per MWh
Hydroelectric	US$50 per MWh

With the largest hydropower plants in operation in May 2009, the total installed capacity is 50 TW, 19 TW of which is provided by small-scale hydropower plants. The present costs of the different systems of power generation in Brazil are presented in **Table 5**, which shows that power from hydroelectric plants is by far the most economical, besides being a renewable source of energy.

6.02.3.3.3 USA

The first American hydroelectric power plant for major electricity generation was completed at Niagara Falls in 1881, and is still a source of electric power. In 1882, Nikola Tesla discovered the rotating magnetic field, a fundamental principle in physics and the basis of nearly all devices that use alternating current. He adapted the principle of rotating magnetic field for the construction of alternating current induction motor and the polyphase system for the generation, transmission, distribution, and use of electrical power. The early hydroelectric plants were direct current stations built to power arc and incandescent lighting during the period from about 1880 to 1895. When the electric motor came into being, the demand for new electrical energy started its upward spiral. The years 1895 through 1915 saw rapid changes in hydroelectric design and a wide variety of plant styles being built. The waterfalls in the area make them significant producers of electricity. This includes the 2515 MW Robert Moses Hydroelectric Plant owned by New York Power Authority, which has been in operation since 1957. (Across the Niagara River on the Canadian side there are 1600 MW Sir Adam Beck Hydroelectric Stations owned by the Ontario Power Generation Company.)

In the framework of the Colorado River development, implementations of hydropower plants started around 1910 in Arizona with the Salt River and in Utah with the Strawberry Valley Project. In the early 1920s, hydroelectric power developments in the Colorado River Basin were mostly confined to tributaries of the river. There were 36 power plants with the combined installed capacity of only 37 MW. The largest of these were the one by the United States Bureau of Reclamation (USBR) at Roosevelt Dam on the Salt River in Arizona (10.3–36 MW) and the Shoshone Plant of the Central Colorado Power Company on the main stream of the Colorado River upstream from Glenwood Springs, Colorado (10 MW). Hoover Dam (2080 MW) in 1939 on the lower Colorado River, and Glen Canyon and Flaming Gorge Project on the upper Colorado River in 1964 are the major development in the system. USBR also completed the Shasta Dam on the Sacramento River in northern California in 1944, which later became part of the Central Valley Project in California.

The hydropower development of Columbia with Bonneville dam (total capacity in two stages: 1092.9 MW) in 1938 and Grand Coulee (6809 MW) in 1941 was implemented by the US Bureau of Reclamation (**Figure 23**).

The US Corps of Engineers was one of the main forces of the hydropower development of the Mississippi River and its tributary the Missouri River with the construction of hydropower plants, improvements of navigation conditions, and flood control. Many of the hydropower stations are constructed as part of the lock-and-dam systems on the Mississippi River. From its origin at Lake Itasca to St. Louis, Missouri, the flow of the Mississippi River is moderated by 43 dams. Fourteen of these dams are located above Minneapolis, Minnesota, in the headwaters region and serve multiple purposes including power generation and recreation. One of the starting points was the implementation of a hydropower plant at St. Anthony Falls in Minneapolis in 1882. Now the hydropower plant at St. Anthony Falls generates 12.4 MW for the upper and 8.9 MW for the lower developments.

The Tennessee Valley Authority (TVA) was created in 1933 to provide navigation, flood control, and electricity generation in the Tennessee Valley, a region particularly impacted by the Great Depression. Norris Dam (131.4 MW) on the Clinch River was one of the first dams built and was completed in 1936. The TVA is now the largest US public power company with 29 hydroelectric dams (**Table 6**).

Figure 23 Grand Coulee Dam (USA).

Table 6 Some key events in the history of hydropower in USA

1879	First commercial arc lighting system installed, Cleveland, Ohio
1879	Thomas Edison demonstrates incandescent lamp, Menlo Park, New Jersey
1880	Grand Rapids, Michigan: brush arc light dynamo driven by water turbine used to provide theater and storefront illumination
1881	Niagara Falls, New York: brush dynamo, connected to turbine in Quigley's flour mill lights city street lamps
1882	Appleton, Wisconsin: Vulcan Street Plant, first hydroelectric station to use Edison system
1883	Edison introduces 'three-wire' transmission system
1886	Westinghouse Electric Company organized
1886	Frank Sprague builds first American transformer and demonstrates the use of step-up and step-down transformers for long-distance AC power transmission in Great Barrington, Massachusetts
1886	40–50 water-powered electric plants reported online or under construction in the United States and Canada
1887	San Bernadino, California: High Grove Station, first hydroelectric plant in the west
1888	Rotating field AC alternator invented
1889	American Electrical Directory lists 200 electric companies that use waterpower for some or all of their generation
1889	Oregon City, Oregon: Willamette Falls station, first AC hydroelectric plant. Single-phase power transmitted 13 miles to Portland at 4000 v, stepped down to 50 v for distribution
1891	Ames, Colorado: Westinghouse alternator driven by Pelton waterwheel, 320 foot head. Single-phase, 3000 v, 133-cycle power transmitted 2.6 miles to drive ore stamps at Gold King Mine
1891	Frankfurt am Main, Germany: first three-phase hydroelectric system used for 175 km, 25 000 V demonstration line from plant at Lauffen
1891	60-cycle AC system introduced in the United States
1892	Bodie, California: 12.5-mile, 2500 AC line carried power from hydroelectric plant to ore mill of Standard Consolidated Mining Co.
1892	San Antonio Creek, California: single-phase 120 kW plant, power carried to Pomona over 13 miles on a 5000 V line. Voltage increased to 10 000 and line extended 42 miles to San Bernadino within a year. First use of step-up and step-down transformers in hydroelectric project
1892	General Electric Company formed by the merger of Thomson-Houston and Edison General Electric
1893	Mill Creek, California: first American three-phase hydroelectric plant. Power carried 8 miles to Redlands on 2400 V line
1893	Westinghouse demonstrates 'universal system' of generation and distribution at Chicago exposition
1893	Folsom, California: three-phase, 60-cycle, 11 000 V alternators installed at plant on American River. Power transmitted 20 miles to Sacramento
1889–93	Austin, Texas: first dam designed specifically for hydroelectric power built across Colorado River
1895	Niagara Falls, New York: 5000 HP, 60-cycle, three-phase generators go into operation
1907	Hydropower provided 15% of US electrical generation
1920	Hydropower provided 25% of US electrical generation
1940	Hydropower provided 40% of electrical generation

6.02.3.3.4 Japan

Hydropower production was first developed for in-house use by the spinning and mining industries. The first electric power plant developed to provide commercial electric power was constructed in Kyoto called the Keage Power Plant (1892) and it used water drained from Lake Biwa (conduit type). Its power was used to operate the first electric street cars in Japan. The Lake Biwa Canal project, planned under the leadership of Tanabe Sakuroi, was undertaken to stimulate industry in Kyoto, which had declined since

Table 7 Oldest hydropower plants in each region

Region	Name of power plant	River system	Effective head (m)	Maximum discharge (m^3s^{-1})	Maximum output (kW)	Beginning of operation	Classification	Current state
Tohoku	Sankyozawa	Natori	26.67	5.57	5	July 1888	In-house use	1000 kW operating
Kanto	Shimotsuke Asa Bouseki (Owner)	Tone			17	July 1890	In-house use	Abolition
Chubu	Iwazu	Yahagi	53.94	0.37	50	July 1897	Project use	130 kW operating
Kansai	Keage	Yodo	33.74	16.7	80 × 2	November 1891	Project use	4500 kW operating

Source: From Japan Commission on Large Dams (2009) *Dams in Japan; Past, Present and Future.* Paris, France: International Commission on Large Dams, ISBN 978-0-415-49432-8.

the capital was moved from Kyoto to Tokyo in 1869. The purpose of this project was to construct a shipping canal linking Lake Biwa with the Uji River in Kyoto by cutting a canal to Lake Biwa with its rich water resources and at the same time using water from Lake Biwa to generate hydropower, irrigate farm fields, and fight fires.

The demand for electric power for lighting began in 1887 and electric power demand for factories appeared in 1903, when Japanese industry finally modernized. Early electric power projects were primarily intended to supply electric power for lighting from thermal power plants. During this period, transportation within Japan was inconvenient and transporting coal was costly, so it was difficult to produce thermal power in inland regions of Japan. Therefore, most power produced in such regions was hydropower. In other words, hydropower development began in regional cities close to hydropower zones.

Many water intake systems used at hydropower plants at that time were made by packing boulders obtained on the scene into frames of assembled logs. **Table 7** is a table of the oldest hydropower plants in various regions.

The earliest hydropower plants in Japan were extremely close to their demand regions, and their generator output and transmission voltage were both low. However, in 1899, the transmission of 11 kV for 26 km and the transmission of 11 kV for 22 km were achieved in the Chugoku and Tohoku regions, respectively, permitting longer distances between hydropower plants and consumption regions, thereby contributing greatly to electric power production projects in Japan. Later, electric power companies worked to increase transmission voltages, to lengthen transmission distances, and to develop high-capacity hydropower plants.

During this period, intake facilities used to generate electric power also changed as low fixed water intake weirs that could take in the flow rate in the dry season were replaced by dams with gates, and these were expanded to include dams with regulating ponds. Large-scale hydropower plants were developed in this way.

Of these, the Shimotaki Power Plant in the northern Kanto Region supplied power to Tokyo at that time, supplying almost the entire demand (approx. 40–80 million kWh yr^{-1}) to run trams in Tokyo. The Kurobe Dam (33.9 m), constructed as the water intake dam for the Shimotaki Power Plant, which is Japan's first concrete gravity dam for hydropower, has a total reservoir capacity of 2.366 million m^3 (effective reservoir capacity: 1.160 million m^3).

In addition, the Yatsuzawa Power Plant (Tokyo Electric Power Company, Inc. (TEPCO), 1912) in western Kanto was not only a high-capacity dam, but also a conduit type with a large regulating pond (effective capacity: 467 000 m^3). It was an epoch-making type of dam at that time. The Ono Dam (37.3 m **Figure 3**), which formed this large regulating pond, was the largest earth dam in Japan at that time.

6.02.4 Hydropower Development in a Multipurpose Setting

6.02.4.1 Benefits of Hydropower

After more than a century of experience and services, hydropower's strengths and benefits are equally well understood. The added values due to the implementations of hydropower plants could be presented in social, economic, and environmental terms.

6.02.4.1.1 Social
6.02.4.1.1(i) Multiple use benefits

6.02.4.1.1(i)(a) **Provide irrigation, flood mitigation, water supply, and recreation**   Hydropower projects deliver multiple use benefits over and above electricity generation. They include water supply, flood control, recreation, navigation, as well as reduction of greenhouse gas (GHG) emission compared to other sources of energy production. Of course, these benefits need to be realistically assessed and planned in a holistic fashion. These multiple use benefits differentiate hydro generation from other forms of power generation, and are among the criteria to be considered when evaluating the social, economic, and environmental sustainability of an electricity generation project.

For example, with hydropower, affected communities can benefit from the availability of drinking water supply and sanitation, water for business and industry, water for sustainable food production (both in-reservoir and via irrigation), flood mitigation, water-based transport, and recreation and tourist opportunities. These benefits generate economic activities over and above those of electricity generation, but could also incur some costs. They need to be taken into account in project planning as well as in ongoing management. An example of additional cost might be an operating requirement to maintain water levels in reservoirs for fishing. This may reduce electricity sales.

Optimal delivery of intended multipurpose benefits occurs where a hydropower scheme is developed as part of a regional strategy; where costs and benefits are thoroughly assessed; and where social and environmental assessments are undertaken, implemented, and monitored.

Hydropower schemes also have the capacity to provide additional economic benefits as a result of the synergy between hydropower and other intermittent renewable energy resources such as wind and solar power. Further added benefits are ancillary services such as spinning reserve, voltage support, and black start capability. Perhaps one of the greatest benefits of hydropower projects is the avoidance of greenhouse emissions and particulate pollution associated with fossil-fuel power generation projects. These externalities may be difficult to determine but deserve recognition in the wider economic context of project assessment.

6.02.4.1.1(i)(b) **Leaves water available for other uses** Hydropower is not a consumer of water except in the case of dam with reservoir, which involves water loss due to evaporation at the surface. In function of the reservoir operation management, the downstream discharge is modified compared with the natural discharge of the river, but the total quantity of water flowing from upstream to downstream remains constant.

In the case of run-of-the river hydropower plants, the natural flow and elevation drop of a river are used to generate electricity, and no modification of the downstream discharge is observed.

6.02.4.1.1(i)(c) **Enhance navigation conditions** The run-of-the-river power plants fulfill other functions also such as enhancing navigation conditions. The most common case is the utilization of waterpower in plants built next to navigation locks (**Figure 24**).

There are many examples all over the world of this suitable layout of complex water resources utilization, involving hydropower development and navigation with a better control of the minimum draft of the boats and barges (**Figures 25** and **26**).

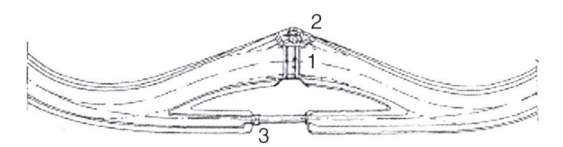

Figure 24 Movable gates (1) with hydropower plant (2) and navigation lock (3).

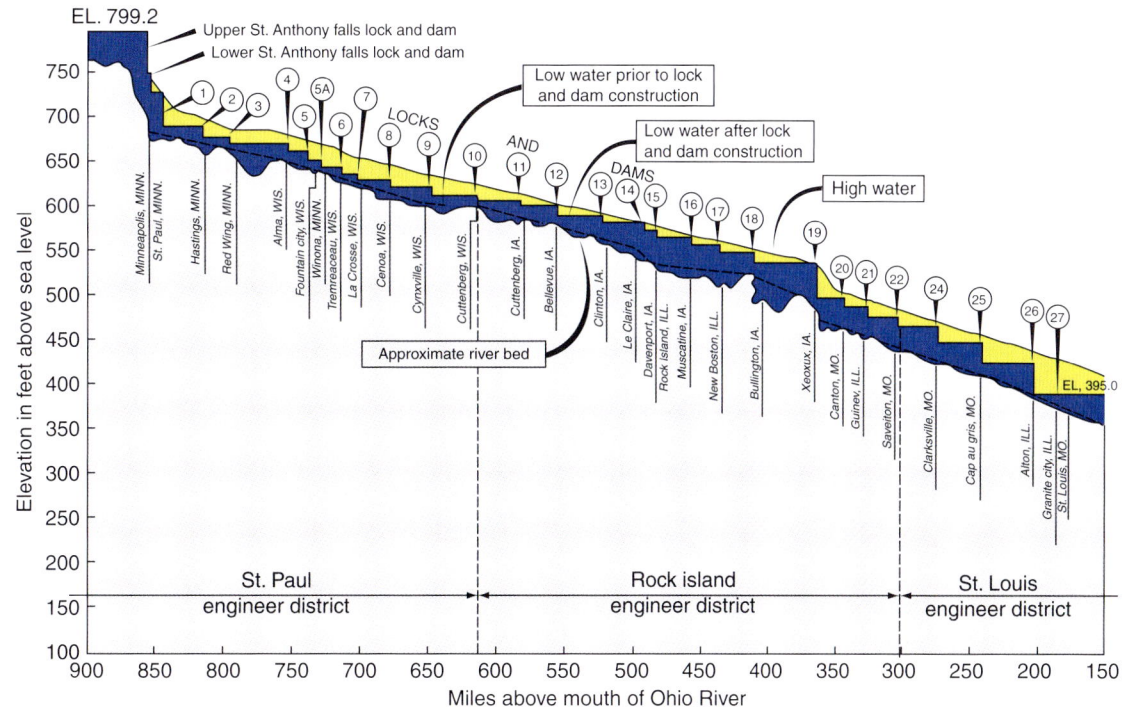

Figure 25 Mississippi River (USA).

Figure 26 Goagang complex (China).

6.02.4.1.2 Economic issues

6.02.4.1.2(i) Sources of hydropower generation

The sources of hydropower generation are widely spread around the world. Potential exists in about 150 countries, and about 70% of the economically feasible potential sites remain to be developed. They are mostly in developing countries.

6.02.4.1.2(ii) Advanced technology

Hydropower is a proven with more than a century of operating experience and construction know-how and well-advanced technology with modern power plants providing the most efficient energy conversion process (>90%). The latter is an important environmental benefit which must be considered in any economic assessment for alternative energy developments.

6.02.4.1.2(iii) Peak load energy

The production of peak-load energy from hydropower is another economic benefit. It allows for the best use of other less flexible electricity-generating sources to produce the base-load power, notably wind and solar power. Its fast response time enables it to meet the sudden fluctuations in demand in the supply electric grids.

6.02.4.1.2(iv) Cost and plant life

Hydropower plant has the lowest operating costs and the longest plant life compared with other large-scale generating options. Once the initial investment has been made in the necessary civil works, the plant life can be extended economically by relatively cheap maintenance and the periodic replacement of electromechanical equipment (replacement of turbine runners, rewinding of generators, etc. – in some cases the addition of new generating units). Typically a hydro plant in service for 40–50 years can have its operating life doubled.

The 'fuel' (water) is renewable, and is not subject to fluctuations in market. Countries with ample reserves of fossil fuels, such as Iran and Venezuela, have opted for a large-scale program of hydro development, by recognizing its environmental benefits. Development of hydropower resources could also represent energy independence for many countries which depend on import of fossil fuels for power generations.

6.02.4.1.2(v) Electrical system benefits

Hydropower, as an energy supply, also provides unique benefits to an electrical system. First, when stored in large quantities in the reservoir behind a dam, it is immediately available for use when required. Second, the energy source can be rapidly adjusted to meet demand. These benefits are part of a large family of benefits, known as ancillary services. They include:

- Spinning reserve – the ability to run at a zero load while synchronized to the electric system. When loads increase, additional power can be loaded rapidly into the system to meet demand. Hydropower can provide this service while not consuming additional fuel, thereby assuring minimal emissions.
- Nonspinning reserve – the ability to enter load into an electrical system from a source not online. While other energy sources can also provide nonspinning reserve, hydropower's quick start capability is unparalleled, taking just a few minutes, compared with as much as 30 min for other turbines and hours for steam generation.
- Regulation and frequency response – the ability to meet moment-to-moment fluctuations in the system power requirements. When a system is unable to respond properly to load changes, its frequency changes, resulting in not only a loss of power, but also

potential damage to electrical equipment connected to the system, especially computer systems. Hydropower's fast response characteristic makes it especially valuable in providing regulation and frequency response.
- Voltage support – the ability to control reactive power, thereby assuring that power will flow from generation to load.
- Black start capability – the ability to start generation without an outside source of power. This service allows system operators to provide auxiliary power to more complex generation sources that could take hours or even days to restart. Systems having available hydroelectric generation are able to restore service more rapidly than those dependent solely on thermal generation.

6.02.4.1.3 Environmental issues
6.02.4.1.3(i) Avoids greenhouse gas emissions
Today, 85% of the primary energy consumption involves fossil fuels (coal, oil, and gas) or traditional sources (wood), with associated large-scale emissions of GHGs to the atmosphere: carbon dioxide from combustion, and methane from processing coal and natural gas. It is well recognized at the international level that this is leading to major climatic changes, and will therefore also have consequences on the hydrologic system (on water supply and agriculture, as well as the sea level rising). Recent research in North America confirms that the GHG emission factor for hydro plants in boreal ecosystems is typically 30–60 times less than those of fossil fuel generation. Studies have also shown that development of even half of the world's economically feasible hydropower potential could reduce GHG emissions by about 13%, and the impact on avoided sulfur dioxide (SO_2) emissions (the main cause of acid rain) and nitrous oxide emissions is even greater. Taking into account the fuel required to build hydropower stations, a coal-fired plant can emit 1000 times more SO_2 than hydropower systems. Each GWh of electricity produced by hydropower would cut CO_2 emissions by 700 tonnes.

6.02.4.1.3(ii) Produces no waste
Contrary to coal, gas, oil, and nuclear power plants, the generation of electricity by hydropower does not produce any atmospheric pollutants or environmentally harmful wastes. The water used for electricity generation purposes is not consumed nor would it pollute. Relative to the other large-scale energy generation options, the emissions of GHG are very limited.

Most of the world electric energy comes from thermal resources and it is reasonable to assume that the replacement energy will come from renewable sources. **Table 8** shows the amount of coal or oil or natural gas that would be required to generate the same amount of electricity in 2008 as all forms of hydropower, including run-of-the-river hydropower plants. Hydropower plants with large dams and reservoirs account for a large share of global hydropower production.

6.02.5 Negative Attributes of Hydropower Project

On the opposite side, implementation of hydropower plants will involve possible resettlement, modification of local land use patterns, management of competing water uses, and waterborne disease vectors. Effects on impacted people's livelihoods as well on cultural heritage will need to be addressed, with particular attention to vulnerable social groups. The negative economic effects could be the dependence on precipitation; in some cases, the decrease in the storage capacity of reservoirs due to sedimentation; and the requirement of multidisciplinary involvement, of long-term planning, and often of foreign contractors and funding. The negative environmental impacts are the inundation of terrestrial habitat and the modification of aquatic habitats and hydrologic regimes. The water quality will need to be monitored/managed as well as the temporary introduction of pollutants into the food chain and the species activities and populations. The hydropower plants will be barriers for fish migration.

We could say that Three Gorges Project has surely some negative impacts (resettlement, fish barriers, etc.), but on the other side, this development will replace a production of 22 000 MW of coal power and will also save the lives of people living downstream and millions of property damages from annual flood hazard.

Table 8 Thermal equivalents to hydropower generation

Hydroelectricity production in 2008	TWh	Tonne of coal equivalent
North America	649.7	7.98E+07
South and Central America	710.0	8.72E+07
Western Europe	535.4	6.57E+07
Central and Eastern Europe	319.5	3.92E+07
Africa	72.0	8.84E+06
Middle East	12.5	1.54E+06
Asia	947.7	1.16E+08
Total	3246.8	3.99E+08

42 Constraints of Hydropower Development

To conclude this section, we also point out the facts that for any major developments, there will always be positive and negative effects on the environment and on the cultural heritage of the area or country. We simply have to balance these opposing effects to arrive at the best solution for the benefits of the people of the region and country in question, rather than doing nothing. Hydropower developments on the Colorado River, Sacramento/San Joaquin River (Central Valley Project in California), Columbia River, Mississippi River, and Tennessee River in the United States have helped, benefited, and propelled the developments of the industrial bases of the United States to their current form.

6.02.6 Renewable Electricity Production

6.02.6.1 Recall

Renewable electricity production (including pumped-storage hydro plants) rose to 3762.6 TWh in 2008, that is, 18.7% of the total electric energy production. This share in electricity output was larger than that of nuclear power (13.5% in 2008), but much less than the fossil fuel electricity (67.7%). The remaining 0.1% was provided by the incineration of nonrenewable waste.

6.02.6.2 Sources of Renewable Electricity Energy

There are six different sources of renewable electricity. Hydroelectricity is the principal source with an 86.3% share of the total renewable output. Biomass, which includes solid biomass, liquid biomass, biogas, and renewable household waste, is the number two source (5.9%) and is a little ahead of the wind power sector (5.7%), followed by geothermal power (1.7%), solar power including electro-solar and photovoltaic plants (0.3%), and ocean energies (0.01%) (**Table 9**).

6.02.6.3 Characteristics of Renewable Energy Sources

Most of these renewable energies depend in one way or another on sunlight. Wind and hydroelectric power are the direct result of differential heating of the Earth's surface, which leads to air moving about (wind) and precipitation forming as the air is lifted.

Table 9 Structures of renewable sources of electricity production in 2008

Source	TWh	%
Hydropower	3247.30	86.31
Biomass	223.50	5.94
Wind power	215.70	5.73
Geothermal	63.40	1.69
Solar including photovoltaic	12.10	0.32
Marine energies	0.54	0.01
Total	3762.54	100.00

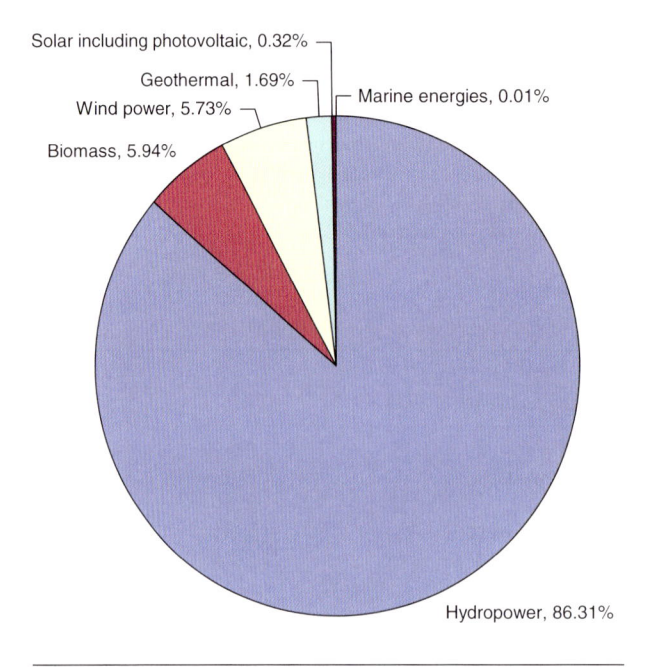

Solar energy is the direct conversion of sunlight using panels or collectors. Biomass energy is stored sunlight contained in plants. Other renewable energies that do not depend on sunlight are geothermal energy, which is the result of radioactive decay in the crust combined with the original heat of the accreting Earth, and tidal energy, which is the result of conversion of gravitational energy.

6.02.6.3.1 Solar
This form of energy relies on the nuclear fusion power from the core of the Sun. This energy can be collected and converted in a few different ways. The range is from solar water heating with solar collectors or attic cooling with solar attic fans for domestic use to the complex technologies of direct conversion of sunlight to electrical energy using mirrors and boilers or photovoltaic cells. Unfortunately, these are currently insufficient to fully power our modern society.

6.02.6.3.2 Wind power
The movement of the atmosphere is driven by differences of temperature at the Earth's surface due to varying temperatures of the Earth's surface when lit by sunlight. Wind energy can be used to pump water or generate electricity, but requires extensive areal coverage to produce significant amounts of energy.

6.02.6.3.3 Hydroelectric energy
As it was stated in the beginning of the chapter, this form of energy uses the gravitational potential of elevated water that was lifted from the oceans by sunlight.

6.02.6.3.4 Biomass
Biomass is the form of energy derived from plants. Energy in this form is very commonly used throughout the world. Unfortunately, the most popular is the burning of trees for cooking and for warmth. This process releases copious amounts of carbon dioxide gases into the atmosphere and is a major contributor to unhealthy air in many areas. Some of the more modern forms of biomass energy are methane generation and production of alcohol for automobile fuel and for fueling electric power plants.

6.02.6.3.5 Hydrogen and fuel cells
These are also not strictly renewable energy resources but are very abundant in availability and are very low in pollution when utilized. Hydrogen can be burned as a fuel, typically in a vehicle, with only water as the combustion by-product. This clean burning fuel can mean a significant reduction of pollution in cities. Or the hydrogen can be used in fuel cells, which are similar to batteries, to power an electric motor. In either case, significant production of hydrogen requires abundant power. Due to the need for energy to produce the initial hydrogen gas, the result is the relocation of pollution from the cities to the power plants. There are several promising methods to produce hydrogen, such as solar power, that may alter the picture drastically.

6.02.6.3.6 Geothermal power
Geothermal power is energy left over from the original accretion of the planet and augmented by heat from radioactive decay, which seeps out slowly everywhere and everyday. In certain areas, the geothermal gradient (increase in temperature with depth) is high enough to be exploited for the generation of electricity. This possibility is limited to a few locations on the Earth and many technical problems still exist that limit its utility. Another form of geothermal energy is Earth energy, a result of the heat storage in the Earth's surface. Soil everywhere tends to stay at a relatively constant temperature year around. It can be used with heat pumps to heat a building in winter and cool it in summer. This form of energy can lessen the need for other power to maintain comfortable temperatures in buildings.

6.02.6.3.7 Other forms of energy
Tides, the oceans, and hot hydrogen fusion are other sources of energy that can be used to generate electricity. Each of these has been considered in some detail. But they all suffer from one or the other significant drawback in that they cannot be relied upon at this time to solve the upcoming energy crunch.

6.02.6.4 Distribution per Region of the Percentage of Hydroelectricity and Renewable Non-Hydroelectricity Generation in the World

As it is confirmed by **Figure 27**, the contribution of hydropower to the generation of electricity is around 14% in every part of the world except in South America where it is 55%. The percentage of renewable sources of electricity derived from hydropower is still, for the moment, about 5% (**Figure 27**).

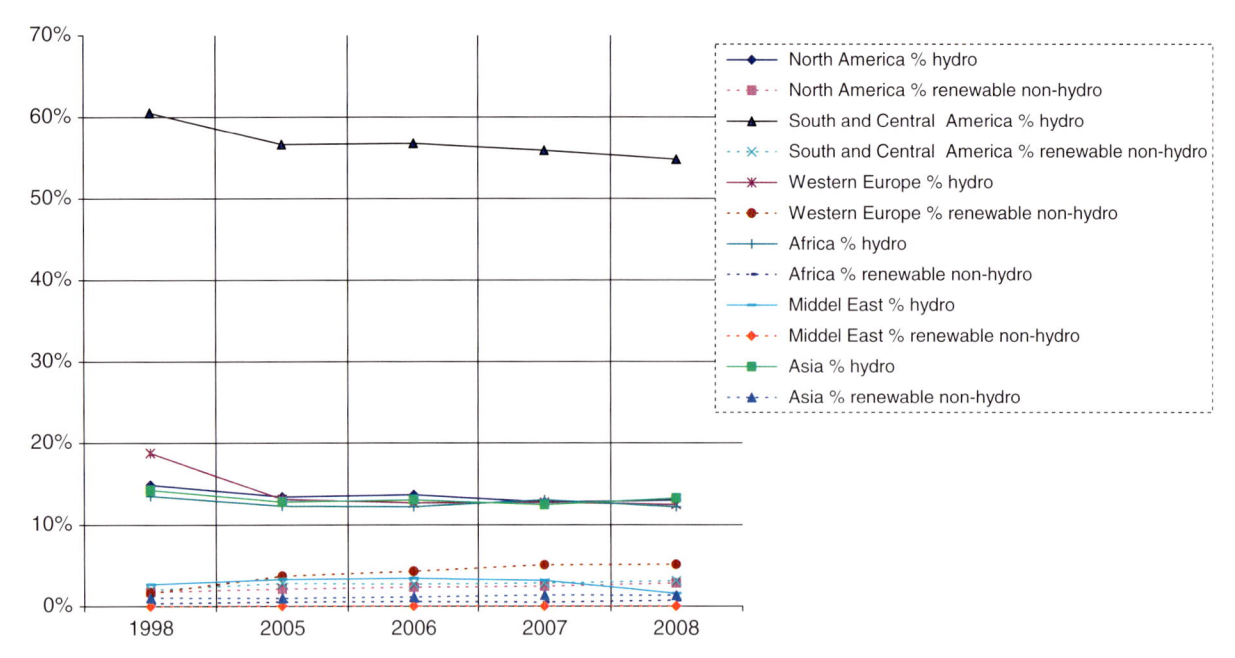

Figure 27 Distribution per region of the percentage of hydroelectricity and renewable non-hydroelectricity generation in the world.

6.02.6.5 Findings about Renewable Electricity Production

Between 1998 and 2008, renewable electricity production in the world rose from 2794.9 to 3762.6 TWh, that is, an additional 967.7 TWh, which equates to almost double the amount of electricity produced in France. Between 2007 and 2008, the renewable sectors gained enough momentum to gain another half percentage point share in the breakdown of total electricity production. China, which leads the field of countries that have supported this growth, is now the leading world producer of renewably sourced electricity with 599.4 TWh in 2008. The commissioning of the last phase of the Three Gorges Dam has largely contributed to the 100 TWh increase in hydropower produced in China in the span of a year. Hydroelectricity, whose limits are far from being reached, is the country's top renewable source of energy. An additional 6900 MW will shortly come online with the country's third largest dam, the Longtan Dam. Hydropower represents 86.3% of all renewable production leaving biomass (5.9%) and wind power (5.7%) trailing a long way behind.

Nonetheless, the wind power sector has continued to put in a remarkable performance with a mean annual growth of 29.4% between 1998 and 2008. The 100 000 MW mark for installed capacity worldwide was passed during the first half of 2008 and the GWEC (Global World Energy Council) forecasts accumulated capacity of 240 300 MW as of 2012. In China, wind power output has risen from 6.5 TWh in 2007 to 14.2 TWh in 2008. It has even been a resounding success in the United States, where production has risen from 34.6 TWh in 2007 to 52.4 TWh in 2008, that is, a 51.5% increase. Furthermore, the United States has become the world's top wind power producer, ahead of Germany, which leads the field in renewable energies in Europe. Its very active policy of supporting these sectors has enabled it to increase renewable electricity share by over 10 points from 5.2% in 1998 to 15.4% in 2008. Renewably sourced electricity production has risen at the same time from 28.8 to 98.1 TWh, about a mean annual increase of 13%.

In Europe, the renewables' share has also increased steadily. It has risen from 14.2% in 1998 to 17% in 2008, once again much of it through wind power, whose mean annual growth in the European Union between 1998 and 2008 was 26.6%. Wind power, especially offshore wind power, growth potential will not peak for a long time. Furthermore, the offshore wind power tests on floating foundations currently under way off the coast of Norway could open up a new high potential development channel for wind power, once its currently prohibitive high installation costs can be brought down.

In contrast, growth of the biomass sector across the world slowed down slightly between 2007 and 2008 as only an additional 8.1 TWh was produced in 2008 over 2007, compared to the increase of 14.1 TWh between 2006 and 2007.

The other renewable sectors (solar, geothermal, and ocean energies) continued climbing up along the growth curve, adding to electricity production at a lower scale. Solar output rose in 2008 to a similar level as that of wind power in 1997, confirming the buildup and organization of the sector. World installed capacity passed the 10 000 MWp (megawatt-peak) mark in 2008, and could exceed 20 000 MWp in 2010. Electricity capacity, which rose to 7910 MWp in 2008, put on an 85% spurt over 2007. The EPIA (European Photovoltaic Industry Association) reckons that even in its 'conservative' scenario, worldwide installed capacity should be in the vicinity of 21 600 MW in 2010 and will embark on a very high growth level after that. The growth of photovoltaic electricity output has actually accelerated as it rose by 49% between 2007 and 2008 compared to the mean annual rate of 39.4% between 1998 and 2008. In 2008, solar power (photovoltaic and electro-solar sectors combined) produced an additional 4.2 TWh over 2007, for 12.1 TWh.

Off-grid photovoltaic also kept up its momentum. The newly installed capacity in the 10 countries surveyed in this inventory (Argentina, Brazil, India, Kenya, Mali, Morocco, Mexico, the Philippines, Senegal, and South Africa) rose to 17.8 MWp in 2008 as against 16.8 MWp in 2007 (up 5.6%). Stand-alone photovoltaic systems were installed in 166 443 homes, bringing the total number of electrified households through photovoltaic in the 10 countries targeted by the survey to over 1.8 million. However, it has to be noted with great regret that the current economic crisis has led to a drop in aid being made available for decentralized rural electrification programs in developing countries. As a consequence, isolated site photovoltaic installation could suffer more from the financial crisis during 2009 than the other renewable electricity sources.

The crisis has also had an impact on global electricity production, whose growth between 2007 and 2008 was only 1.8% whereas the mean annual rate between 1998 and 2008 was 3.5%. Nevertheless, growth in output in a number of countries has been extraordinary. Over the past 10 years, China achieved a mean annual growth rate of 11.5% and in South Korea, it was 7.5%. However, the current rise in worldwide electricity production continuously increases GHG emissions, as electricity production from fossil-fuel plants is still about 4 times higher than that of the renewable electricity sources. The share of nuclear power in global production is shrinking, despite a slight recent increase in production. The sector's mean annual growth over the 1998–2008 period is only 1.1%.

Conditions in the developed and developing world affect the economics of a hydropower project. In the analysis of the cost of the different systems of power generation in the United States, it appears that, besides being a renewable source of energy, hydroelectric plants are by far the most economical. The mean cost of generation by US hydropower plants is only 40% of that by fuel oil.

In Africa, several large hydro projects are in the planning stages or under construction: the proposed Grand Inga complex in the Democratic Republic of Congo – an $80 billion complex that is expected to produce almost twice the electricity of the Three Gorges Project; the construction of the Lom Pangar dam in Cameroon; the rehabilitation and upgrading of the Kariba dam on the Zambezi River between Zambia and Zimbabwe; the construction of the Gibe III hydropower plant in Ethiopia; and the construction of the Gurara Water Transfer Project in Nigeria. These come in spite of the closures of the Tanzanian hydro plants in 2006 and the 14 MW Masinga dam in Kenya in 2009, due to recurrent droughts, and the diminished capacities of the Inga 1 and Inga 2 dams, due to poor maintenance. In 2008, the World Bank invested more than $1 billion in small-scale and micro-hydro projects in the developing world. These projects displace fewer people than the large ones and they also reduce the cost of transmitting electricity to rural areas, across vast distances, and over natural barriers such as the Sahara Desert.

Renewable energies thus still have a lot to offer and many countries have just begun to realize this. China intends to become the leading photovoltaic panel and wind turbine manufacturer. Its highly competitive stance will no doubt force European and American manufacturers to struggle for market share. Renewable energies will become even stronger, for we are no longer witnessing a ripple but a ground swell.

6.02.7 Conclusion

Hydropower is the most important source of renewable energy for the moment and is the subject of much debate. As it was stated, it produces extremely small quantities of carbon dioxide (mostly from power plant construction and from decaying organic matter that readily grows in the stagnant water of reservoirs); the amount is even less than that of the alternative wind, nuclear, and solar energy sources. Hydropower is also clean, and its supply is generally stable since water is abundant in many places. One of the greatest drawbacks of hydropower is the cost. Hydropower's initial investment costs from dam and power plant construction are relatively high (in part this is because project planning is site specific due to the many geographic variables involved). Other costs include the installation of (or hook up to) transmission lines, the operation and maintenance of the facility, and the costs (both financial and social) of resettling people displaced by the dam and its reservoir. The loss of agricultural land and the potential damage to ecosystems are also important factors to be considered.

As a final conclusion, despite hydropower's high initial costs, its long-term overall costs tend to be low because the energy source (flowing water) is renewable and free. The following figures give two overviews of the cost of the electricity generation in 2009 and in 2016. Sources of these figures are compilations from data of EIA (US Energy Information Administration), OECD (Organization for Economic Co-operation and Development), and the Institute for Energy Research. The mentioned cost of electricity production by nuclear power plant does not include the costs of waste treatments and the impacts of future regulations and reviews of licensing. The average cost of electricity production by hydropower is still very attractive for the moment. In the next future, hydropower will keep its position, in front of the other renewable sources of electricity, on the same level with biomass, better than wind power and much cheaper than solar energy (**Figures** 28 and 29).

46 Constraints of Hydropower Development

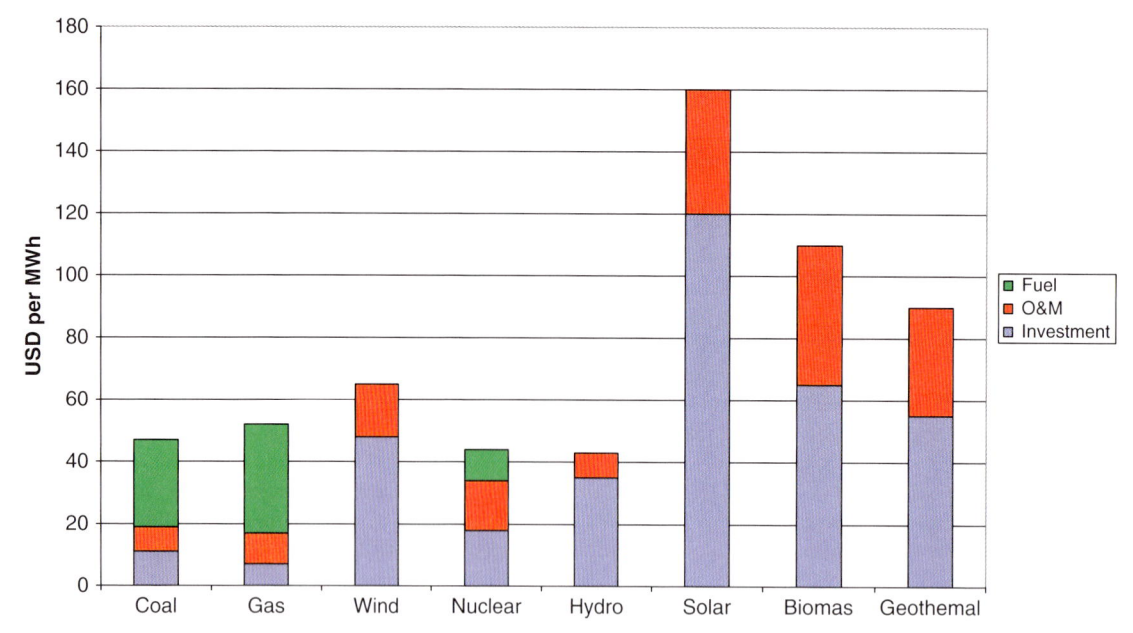

Figure 28 Average cost of electricity production by source in 2008 in USD per MWh.

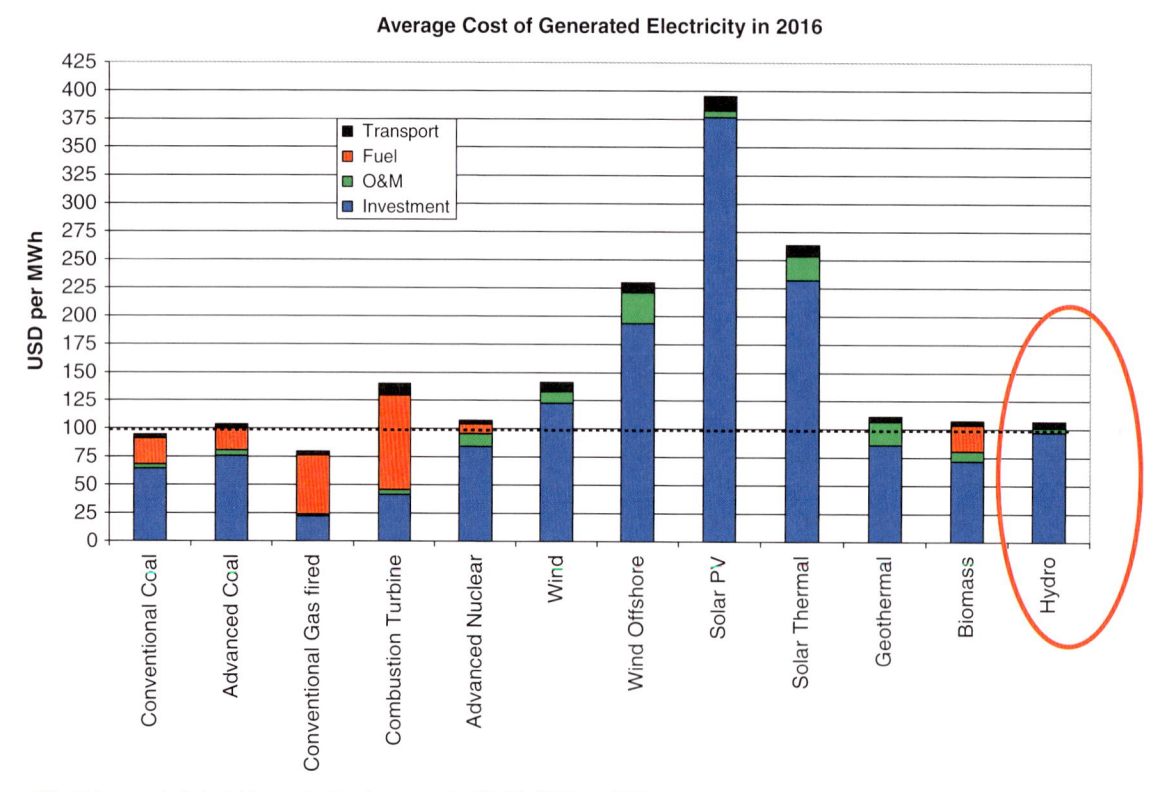

Figure 29 Future cost of electricity production by source in 2016 in USD per MWh.

Further Reading

[1] Wisconsin Valley Improvement Company (n.d.) Facts about hydropower. http://new.wvic.com/index.php?option=com_content&task=view&id=7&Itemid=44a (accessed 27 December 2009).
[2] Wachter S (2007, 19 June) Giant dam project aims to transform African power supplies. *The New York Times*. http://www.nytimes.com/2007/06/19/business/worldbusiness/19iht-rnrghydro.1.6204822.html (accessed 14 January 2010).
[3] Browne P (2009, 30 September) The rise of micro-hydro projects in Africa. *The New York Times*. http://greeninc.blogs.nytimes.com/2009/09/30/the-rise-of-micro-hydro-projects-in-africa/ (accessed 14 January 2010).

[4] Williams A and Porter S (n.d.) Comparison of hydropower options for developing countries with regard to the environmental, social and economic aspects. http://www.udc.edu/cere/Williams_Porter.pdf (accessed 17 December 2009).

[5] Allin SRF (2004) An examination of China's three Gorges dam project based on the framework presented in the report of the World Commission on Dams. http://scholar.lib.vt.edu/theses/available/etd-12142004-125131/unrestricted/SAllin_010304.pdf (accessed 10 January 2009).

[6] Alpiq Group (n.d.) Grande Dixence. http://www.alpiq.com/what-we-offer/our-assets/hydropower/storage-power-plants/grande-dixence.jsp (accessed 12 January 2010).

[7] Asian News International (2008) Rampur labourers, residents demand protection of their rights. *Thaindian News*. http://www.thaindian.com/newsportal/india-news/rampur-labourers-residents-demand-protection-of-their-rights_10035323.html (accessed 14 January 2010).

[8] Asian News International (2009) Farmers protest against hydropower project in Himachal. http://www.thefreelibrary.com/Farmers+protest+against+Hydro+Power+Project+in+Himachal.-a0208182537 (accessed 14 January 2009).

[9] Bertoldi P and Atanasiu B (2007) Electricity consumption and efficiency trends in the enlarged European Union: Status Report 2006. http://re.jrc.ec.europa.eu/energyefficiency/pdf/EnEff%20Report%202006.pdf (accessed 14 January 2010).

[10] BIC (2003) BIC factsheet: The IDB-funded Cana Brava Hydroelectric Power Project. http://www.bicusa.org/Legacy/Cana%20Brava%20PPA.pdf (accessed 13 January 2010).

[11] BP (2009) BP statistical review of world energy: June 2009. http://www.bp.com/liveassets/bp_internet/globalbp/globalbp_uk_english/reports_and_publications/statistical_energy_review_2008/STAGING/local_assets/2009_downloads/statistical_review_of_world_energy_full_report_2009.pdf (accessed 8 October 2009).

[12] Bridle R (2000) China Three Gorges project. http://www.britishdams.org/current_issues/3Gorges2.pdf (accessed 12 January 2010).

[13] Chen L (2009) Developing the small hydropower actively with a focus on people's well-being, protection & improvement: Keynote speech on the 5th Hydropower for Today Forum. http://www.inshp.org/THE%205th%20HYDRO%20POWER%20FOR%20TODAY%20CONFERENCE/Presentations/Speech%20by%20H.E.%20Mr.%20Chen%20Lei.pdf (accessed 21 December 2009).

[14] FAO Corporate Document Repository (1986) FAO Irrigation and Drainage Papers Water lifting devices. Chenese nora www.fao.org/docrep/010/ah810e/AH810E12.htm

[15] United Nations Centre on Transnational Corporations (1992) Climate change and transnational corporations. Analysis and trends. Institute for Energy and Environmental Research, United Nations Centre on Transnational Corporations, Environment Series No. 2, United Nations, New York.

[16] Cotillon J (1982) Place de l'hydroélectricité dans le bilan énergétique [The significance of hydroelectric power in the world energy balance]. *Houille Blanche* 37(5–6): 381–392.

[17] Federation of Electric Power Companies of Japan (2009) Graphical flip-chart of nuclear & energy related topics 2009. http://www.fepc.or.jp/english/library/graphical_flip-chart/__icsFiles/afieldfile/2009/04/02/zumen2009.pdf (accessed 8 October 2009).

[18] Food and Agriculture Organization of the United Nations (2009) Water lifting devices. www.fao.org/docrep/010/ah810e/AH810E12.htm (accessed 8 October 2009).

[19] GESS-CZ, s.r.o (2009) Small water plant (Archimedean screw turbine) from http://www.gess.cz/en/small-water-plant-archimedean-screw-turbine.html (accessed 16 November 2009).

[20] Grande Dixence SA (1950) http://www.grande-dixence.ch/energie/hydraulic/switzerland/grande-dixence-altitude-2365.html (accessed 17 December 2009).

[21] Highlands and Islands Enterprise (2009) Electricity demystified. http://www.hi-energy.org.uk/electricitydemystified.html (accessed 21 December 2009).

[22] International Energy Agency, Small-Scale Hydro Annex (n.d.) What is small hydro? http://www.small-hydro.com/index.cfm?fuseaction=welcome.whatis (accessed 13 January 2010).

[23] International Union for Conservation of Nature (The World Conservation Union) (2006) The future of sustainability: Re-thinking environment and development in the twenty-first century. http://cmsdata.iucn.org/downloads/iucn_future_of_sustanability.pdf (accessed 8 October 2009).

[24] Jia J (2008) Dam construction in China – 2006. http://www.chincold.org.cn/news/li080319-jjs.pdf (accessed 8 October 2009).

[25] Khemani H (2009) Classification of hydroelectric power plant: Part-2: Based on the head of water available. http://www.brighthub.com/engineering/mechanical/articles/7827.aspx (accessed 18 December 2009).

[26] Kozloff N (2009) Blackout in Brazil: Hydropower and our climate conundrum. *The Huffington Post*. http://www.huffingtonpost.com/nikolas-kozloff/blackout-in-brazil-hydrop_b_363651.html (accessed 13 January 2009).

[27] LeJeune A and Topliceanu I (2002) Energies renouvelables et cogeneration pour le development durable en Afrique: Session hydroelectricite [Renewable energy and cogeneration for the development of sustainable development of Africa: Hydroelectricity session]. http://sites.uclouvain.be/term/recherche/YAOUNDE/EREC2002_session_hydro.pdf (accessed 13 January 2010).

[28] Maldonado JK (2009) Putting a price-tag on humanity: Development-forced displaced communities' fight for more than just compensation. *Hydro Nepal: Journal of Water, Energy and Environment* 4: 18–20.

[29] Hydro M (n.d.) McArthur generating station. http://www.hydro.mb.ca/corporate/facilities/gs_mcarthur.pdf (accessed 10 January 2010).

[30] Minister of Natural Resources Canada (2004) Small hydro project analysis. http://74.125.47.132/search?q=cache:u_dTQTzJtyYJ:www.retscreen.net/download.php/ang/107/1/Course_hydro.ppt+canada+minister+of+natural+resources+average+small+hydro+power+construction+cost+2004&cd=4&hl=en&ct=clnk&gl=us (accessed 14 January 2010).

[31] National Geographic (n.d.) Hydropower. http://environment.nationalgeographic.com/environment/global-warming/hydropower-profile.html (accessed 17 December 2009).

[32] Norfolk Mills Old watermill scheme. Pippa Miller's drawing of a typical Norfolk watermill. http://www.norfolkmills.co.uk/watermill-machinery.html

[33] Panda Travel and Tour Consultants (n.d.) Some facts of the Three Gorges dam project. http://www.chinadam.com/dam/facts.htm (accessed 12 January 2009).

[34] PBS (n.d.) Great Wall across the Yangtze. http://www.pbs.org/itvs/greatwall/dam1.html (accessed 12 January 2010).

[35] Pew Center on Global Climate Change (n.d.) Hydropower. http://www.pewclimate.org/technology/factsheet/hydropower (accessed 18 December 2009).

[36] Power-technology.com (2009) Three Gorges dam hydroelectric power plant, China. http://www.ctgpc.com/achievement/achievement_a.php (accessed 18 December 2009).

[37] Scarborough VL (2003) *The Flow of Power: Ancient Water Systems and Landscapes*. Santa Fe, NM: School of American Research.

[38] Hydro Tasmania (2009) Sustainable Hydropower. www.sustainablehydropower.org (accessed 8 October 2009).

[39] Truchon M and Seelos K (2004) Managing the social and environmental aspects of hydropower. http://www.energy-network.net/resource_center/launch_documents/documents/Managing%20the%20social%20&%20environmental%20aspects%20of%20hydropower%2020.pdf (accessed 13 January 2010).

[40] United States Department of Energy (2005) Types of hydropower plants. http://www1.eere.energy.gov/windandhydro/hydro_plant_types.html (accessed 18 December 2009).

[41] United States Energy Information Administration (2008) [Table 6.3]. World total net electricity generation (billion kilowatthours), 1980–2006. http://www.eia.doe.gov/iea/elec.html (accessed 23 December 2009).

[42] United States Energy Information Administration (2008) World's top hydroelectricity producers, 2008 (billion kilowatthours). http://www.eia.doe.gov/emeu/cabs/Canada/images/top_hydro.gif (accessed 23 December 2009).

[43] United States Energy Information Administration (2009) Renewables and alternative fuels: Hydroelectric. http://www.eia.doe.gov/cneaf/solar.renewables/page/hydroelec/hydroelec.html (accessed 22 November 2009).

[44] United States Energy Information Administration (2009) Table 5: U.S. average monthly bill by sector, census division, and state 2007. http://www.eia.doe.gov/cneaf/electricity/esr/table5.html (accessed 14 January 2010).

6.03 Management of Hydropower Impacts through Construction and Operation

H Horlacher and T Heyer, Technical University of Dresden, Dresden, Germany
CM Ramos and MC da Silva

© 2012 Elsevier Ltd. All rights reserved.

6.03.1	Introduction	50
6.03.1.1	Background	50
6.03.1.1.1	The role of hydropower	50
6.03.1.1.2	Hydropower and sustainability (environmental, economic, and social)	50
6.03.1.1.3	Hydropower construction	50
6.03.1.1.4	Hydropower operation	52
6.03.1.2	Upstream Impacts	53
6.03.1.2.1	Water quality	53
6.03.1.2.2	Sedimentation	53
6.03.1.3	Downstream Impacts	56
6.03.1.3.1	Flow regime	56
6.03.1.3.2	Ecological discharge and minimum flow	56
6.03.1.3.3	Surge	56
6.03.1.3.4	Degradation and aggradation in downstream river reaches	56
6.03.1.3.5	Effects of sediments on turbines	57
6.03.1.3.6	Water characteristics	57
6.03.1.3.7	Fish migration	58
6.03.2	**Reservoir Water Quality**	58
6.03.2.1	Introduction	58
6.03.2.2	General Characteristics of Reservoirs	59
6.03.2.2.1	Morphology and hydrodynamics	59
6.03.2.2.2	Thermal stratification	59
6.03.2.2.3	Pollutants and stressors on reservoirs	59
6.03.2.3	Water Quality Processes – Eutrophication and Oxygenation	61
6.03.2.3.1	Introduction	61
6.03.2.3.2	General concepts	61
6.03.2.3.3	Eutrophication symptoms and effects	61
6.03.2.3.4	Growth of aquatic plants	62
6.03.2.3.5	Anoxia	62
6.03.2.3.6	Species changes	62
6.03.2.3.7	Hypereutrophy	62
6.03.2.3.8	Elevated nitrate concentrations	63
6.03.2.3.9	Increased incidence of water-related diseases	63
6.03.2.3.10	Increased fish yields	63
6.03.2.3.11	Nutrient recycling	63
6.03.2.3.12	Assessment of trophic status	63
6.03.2.4	Water Quality Parameters	64
6.03.2.4.1	Behavior in reservoirs	64
6.03.2.4.2	Oxygen	64
6.03.2.5	Nutrient Dynamics	66
6.03.2.5.1	Nitrogen	66
6.03.2.5.2	Phosphorus	66
6.03.2.6	Overview of Water Quality Models of a Reservoir	67
6.03.2.7	Lake Stability	68
6.03.2.7.1	The Wedderburn and lake numbers	68
6.03.2.7.2	Monitoring and control	69
6.03.2.7.3	Real-time data acquisition, modeling, and control	70
6.03.2.8	Water Quality Models	70
6.03.2.8.1	One-dimensional temperature models	70
6.03.2.8.2	One-dimensional water quality models	71
6.03.2.8.3	Multilayer models	71
6.03.2.8.4	Two- and three-dimensional water quality models	72

6.03.2.8.5	Eutrophication models	72
6.03.2.8.6	Special models	72
6.03.2.9	Final Remarks	73
6.03.3	**Management of the Impact of Hydraulic Processes in Hydropower Operation**	73
6.03.3.1	Introduction	73
6.03.3.1.1	Gas supersaturation	73
6.03.3.1.2	Fish passage	73
6.03.3.1.3	Unsteady flow	74
6.03.3.1.4	Sediment transportation	74
6.03.3.1.5	Reservoir operating strategies	75
6.03.3.2	Reduction of Gas-Supersaturated Water	75
6.03.3.2.1	Case histories	75
6.03.3.2.2	Retrofit solutions for spillways with deep stilling basins	77
6.03.3.3	Control of Floating Debris	79
6.03.3.3.1	Type and origin of debris	79
6.03.3.3.2	River transport of debris	81
6.03.3.3.3	Debris transport through flow control structures	81
6.03.3.3.4	Proposed countermeasures	82
6.03.3.4	Hydropower Operating Strategies	83
6.03.3.4.1	Artificial destratification	83
6.03.3.4.2	Management of reservoir filling	84
6.03.3.4.3	Unsteady flow	87
6.03.3.4.4	Population protection measures	87
6.03.3.5	Mitigation Measures	87
6.03.3.5.1	Structural options – Multilevel offtake towers	87
6.03.3.5.2	Floating offtakes with pivot arms or trunnions	88
6.03.3.5.3	Dry multiport intake towers	88
6.03.3.5.4	Shasta Dam Temperature Control Device, California	90
6.03.3.5.5	Glen Canyon Dam, Arizona	90
6.03.3.5.6	Flaming Gorge Dam, Utah	90
References		90

6.03.1 Introduction

6.03.1.1 Background

6.03.1.1.1 The role of hydropower

Humans have been harnessing water to perform work for thousands of years. About 2000 BC, the Persians, Greeks, and Romans all used waterwheels. Indeed, the use of waterpower by crude devices dates back to ancient times. The primitive wheels, actuated by river current, were used for raising water for irrigation purposes, for grinding corn in mills, and in other simple applications.

The historical trends in the world's primary energy consumption are shown in **Figure 1**. Consumption associated with generation from coal, oil, and gas has been increasing faster than that from other sources. Currently, about 88% of energy consumption is derived from fossil sources. Moreover, hydropower plants produce 20% of the world's total electricity. Data about the main hydroelectricity capacities are given in **Table 1**.

6.03.1.1.2 Hydropower and sustainability (environmental, economic, and social)

The positive social aspects of the implementation of hydropower are related to the role of dams in terms of their importance in water resources management. Hydropower dams frequently serve several purposes: water supply, irrigation, flood control, navigation, and recreation.

In addition, as far as environmental impacts are concerned, hydropower plants produce no waste or atmospheric pollutants, avoid the depletion of nonrenewable fuel resources (i.e., coal, gas, and oil), and produce very few greenhouse gas emissions relative to other large-scale energy options. They can also enhance knowledge and improve the management of valued species and increase attention to existing environmental issues in the affected area. Compared with other energy sources, hydropower, being a renewable energy source, contributes significantly to the reduction of atmosphere polluting emissions (**Table 2**). **Table 2** presents levels of major air pollutants from different sources of electricity generation.

6.03.1.1.3 Hydropower construction

Hydropower plants are planned, constructed, and operated to meet human needs: electricity generation, irrigated agricultural production, flood control, public and industrial water supply, drinking water supply, and various other purposes. Hydropower

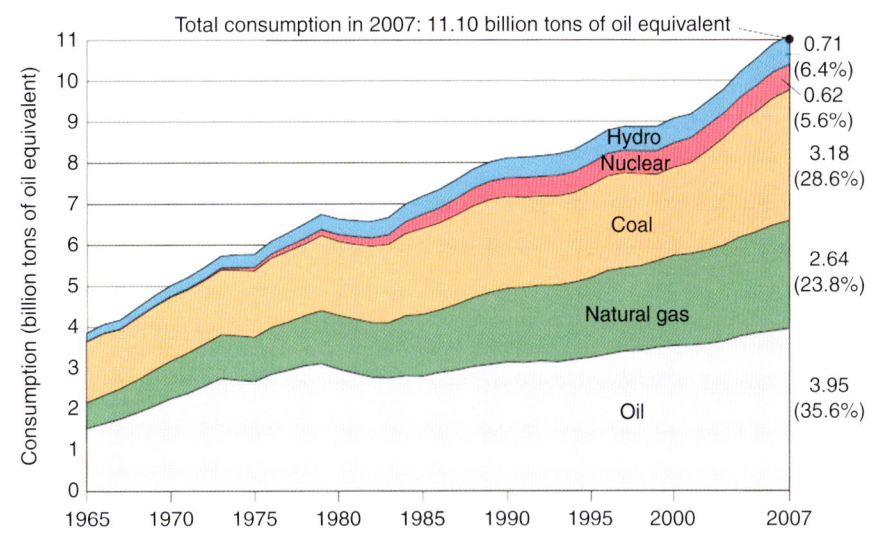

Figure 1 Historical trends in the world's primary energy consumption. From Federation of Electric Power Companies of Japan (FEPC) (2009) Graphical Flip-chart of Nuclear & Energy Related Topics. The percentages within parantheses represent proportion of total. Note: Figures may not add up to the totals due to rounding. Source: BP Statistical Review of world Energy, June 2008.

Table 1 Main hydroelectricity capacities (4) (BP Statistical Review of World Energy (June 2009))

Country	Annual hydroelectric energy production (TWh)	Installed capacity (GW)	Percent of total electricity
China	585	155	17
Canada	369	88	61
Brazil	363	69	85
USA	250	79	5
Russia	167	45	17
Norway	140	27	98
India	115	33	15
Venezuela	86		67
Japan	69	27	7
Sweden	65	16	44
Paraguay	64		
France	63	25	11

dams impound water in reservoirs during times of high flow, which can then be used for human requirements during times of low flow (i.e., when natural flows are inadequate). Positive impacts of dams are improved flood control and improved welfare resulting from new access to irrigation and drinking water. Concerning the role of dams, having in mind their multipurpose functions, it is relevant to refer to Mr. Jamal Saghir, the representative from the World Bank at the Hydro 2004 Conference in Oporto, Portugal, October 2004:

> (…) delivering water for food, water for sanitation, water for drinking, water for power services, is an arm in the fight against hunger and poverty.

Despite this, there remain significant concerns about the environmental impacts of dams. Flood control by dams reduces discharge values during natural flood periods. Altering the pattern of the downstream flow (i.e., intensity, timing, and frequency) may lead to a change in the sediment and nutrient regimes downstream of the dam. Water temperature and chemistry are modified and consequently may lead to a discontinuity in the river system. These environmental impacts are complex and far-reaching, may occur in remote areas far from the dam site, may occur during dam construction or later, and may affect the biodiversity and productivity of natural resources.

Each hydropower plant has its own operating characteristics. Dams are located in a wide array of conditions – from highlands to lowlands, temperate to tropical regions, fast- and slow-flowing rivers, urban and rural areas, with and without water diversion. The

Table 2 Polluting emissions (g kWh^{-1})

Polluting emissions	Biomass	Hydro	Wind electricity	Geothermal	Oak	Oil	Gas
CO_2	15–18	9	7–9	79	955	818	722
SO_2	0.06–0.08	0.03	0.02–0.09	0.02	11.8	14.2	1.6
NO_x	0.35–0.51	0.07	0.02–0.06	0.28	4.3	4.0	12.3

impact of water diversion differs between northern countries, where temperate climates and little irrigation occur, and semiarid countries, which may have extensive out-of-river uses and high evaporation rates. The combination of dam type, operating system, and the context where the dams are located yields a wide array of conditions that are site-specific and highly variable. This complexity makes it difficult to generalize about the impacts of dams on ecosystems, as each specific context is likely to have different types of impacts and to different degrees of intensity. In addition, the height of dams and their reservoir areas are extremely variable.

Dams for flood control serve to moderate peak flow. Usually, hydroelectric dams are designed to provide flow regulation in order to maximize electricity generation, and therefore tend to have a similar effect on the downstream flow pattern. However, if the purpose is to provide power during peak periods, considerable variations in discharge can occur over short periods, creating artificial freshets or floods downstream. Dams for irrigation cause moderate variations in flow regime on a longer timescale, storing water at times of high flow for use at times of low flow. Flows that exceed the storage capacity are usually spilled, allowing some floods to pass downstream, albeit in a routed and hence attenuated form. As dams are often designed to serve multiple functions, their impacts will have a combination of the above forms. It should be noted that hydraulic structures such as barrages and weirs, as well as water diversion structures or interbasin transfer projects, can have similar impacts on dams.

This chapter compiles the advances in knowledge and state-of-the-art technology used to avoid or mitigate the environmental impacts of dams on the natural ecosystem, as well as on the people who depend on them for their livelihood.

6.03.1.1.4 Hydropower operation

Sources of hydropower generation are widely spread around the world. Potential exists in about 150 countries, and about 70% of the economically feasible potential sites remain to be developed. These sites are mostly in developing countries.

Hydropower is a proven technology with more than a century of operating experience and construction know-how and is also a well-advanced technology with modern power plants providing a highly efficient energy conversion process (>90%). The latter is an important environmental benefit, which must be considered in any economic assessment for alternative energy developments.

The production of peak load energy from hydropower is another economic benefit. It allows for the best use of other less flexible electricity generating sources, notably wind and solar power, to produce the base load power. The fast response of hydropower enables it to meet the sudden fluctuations in demand in the electricity supply grids.

Hydropower plants have the lowest operating costs and the longest plant life compared with other large-scale generating options. Once the initial investment has been made in the necessary civil works, plant life can be extended economically by relatively cheap maintenance and the periodic replacement of electromechanical equipment (replacement of turbine runners, rewinding of generators, etc. – in some cases, the addition of new generating units). Typically, a hydro plant in service for 40–50 years can have its operating life doubled.

Hydropower is a renewable energy and is not subject to market fluctuations. Countries with ample reserves of fossil fuels, such as Iran and Venezuela, have opted for large-scale programs of hydro development by recognizing its environmental benefits. Development of hydropower resources could also represent energy independence for many countries which currently depend on imported fossil fuels for power generation.

Hydropower, as an energy supply, also provides unique benefits to an electrical system. First, when stored in large quantities in the reservoir behind a dam, it is immediately available for use when required. Second, the energy source can be rapidly adjusted to meet demand.

The fast response of hydro plants enables them to adjust to sudden fluctuations due to peak demand or loss of power supply. This is particularly important in order to give a correct response to gaps between supply and demand, allowing the optimization of base load generation from less flexible sources (e.g., nuclear, thermal, and geothermal plants) and an adjustment to the energy oscillations associated with random sources (e.g., wind, waves, and sun).

These benefits are part of a large family of benefits, known as ancillary services. They include the following:

- *Spinning reserve* – the ability to run at a zero load while synchronized to the electrical system. When loads increase, additional power can be loaded rapidly into the system to meet demand. Hydropower can provide this service while not consuming additional fuel, thereby assuring minimal emissions.

- *Nonspinning reserve* – the ability to enter load into an electrical system from a source not online. While other energy sources can also provide nonspinning reserve, hydropower's quick start capability is unparalleled, taking just a few minutes, compared with as much as 30 min for other turbines and hours for steam generation.
- *Regulation and frequency response* – the ability to meet moment-to-moment fluctuations in the system power requirements. When a system is unable to respond properly to load changes, its frequency changes, resulting not only in a loss of power but also in potential damage to electrical equipment connected to the system, especially computer systems. Hydropower's fast response characteristic makes it especially valuable in providing regulation and frequency response.
- *Voltage support* – the ability to control reactive power, thereby assuring that power will flow from generation to load.
- *Black start capability* – the ability to start generation without an outside source of power. This service allows system operators to provide auxiliary power to more complex generation sources that could take hours or even days to restart.
- *Quick answer* (dynamic service) of the hydropower is fundamental in
  - the power frequency regulation, adjusting the offer/production to the demand/consumption;
  - the intervention in 'emergency' situations during short periods; and
  - the intervention as 'operational reserve'.

Pumped storage plants are particularly important to assure reserve generation, to manage the increase of other renewable energy sources (wind, waves, etc.) with random production, and to give better balance in the power diagrams.

6.03.1.2 Upstream Impacts

6.03.1.2.1 Water quality

Water stored in deep reservoirs has a tendency to become thermally stratified. Typically, three thermal layers are formed: a well-mixed upper layer (the epilimnion); a cold, dense bottom layer (the hypolimnion); and an intermediate layer of maximum temperature gradient (the thermocline). Water in the hypolimnion may be up to 10 °C lower than in the epilimnion. In the epilimnion, the temperature gradient may be up to 2 °C for each meter.

Thermal stratification depends on a range of factors, including climatic characteristics. Reservoirs nearest to the equator are least likely to become stratified. At higher latitudes, the governing factor is the input of solar energy. Shallow reservoirs respond rapidly to fluctuations in atmospheric conditions and are less likely to become stratified. Strong winds can effect rapid thermocline oscillations. The pattern of inflows, as well as the nature of outflows from the reservoir, also influences the development of thermal stratification.

Current generated from large water level fluctuations in reservoirs caused by operation regimes can also sometimes prevent thermal stratification. Many deep reservoirs, particularly at mid- and high latitudes, become thermally stratified, as do natural lakes under similar conditions. The release of cold water into the receiving downstream river can be a significant consequence of stratification.

Water storage in reservoirs induces physical, chemical, and biological changes in the stored water and in the underlying soils and rocks, all of which affect water quality. The chemical composition of water within the reservoir can be significantly different from that of the inflows. The size of the dam, its location in the river system, its geographical location with respect to altitude and latitude, the storage detention time of the water, and the source of the water all influence the way that storage detention modifies water quality.

Major biologically induced changes occur within thermally stratified reservoirs. In the surface layer, phytoplankton often proliferate and release oxygen, thereby maintaining concentrations at near-saturation levels for most of the year. In contrast, the lack of mixing and sunlight for photosynthesis in conjunction with oxygen being used in the decomposition of submerged biomass often results in anoxic conditions in the bottom layer.

Nutrients, particularly phosphorus, are released biologically and leached from flooded vegetation and fertilized soil. Although oxygen demand and nutrient levels generally decrease over time as the mass of organic matter decreases, some reservoirs require a period of tens of years to develop stable water quality regimes. After maturation, reservoirs, like natural lakes, can act as nutrient sinks, particularly for nutrients associated with sediments. Eutrophication of reservoirs may occur as a consequence of organic loading and/or nutrients. In many cases, these are the consequences of anthropogenic influences in the catchment (application of fertilizers) rather than the presence of the reservoir. However, there are reservoirs, particularly in tropical climates, that have the ability to recycle nutrients from the reservoir sediments through the water column, without any significant addition of new nutrients from the stream flow.

6.03.1.2.2 Sedimentation

Rivers transport particles, from fine ones such as silt in turbid water to coarser ones such as sand, gravel, and boulders associated with bed-load transport. The speed and turbulence of currents enable transportation of these materials. When riverbed gradient or the river flow diminishes, particles tend to drop out. This happens when river flows reach reservoirs.

Large reservoirs store almost the entire sediment load supplied by the drainage basin. The sediment transport into the reservoir depends on the size of the reservoir's catchment, the characteristics of the catchment area that affect the sediment yield (climate,

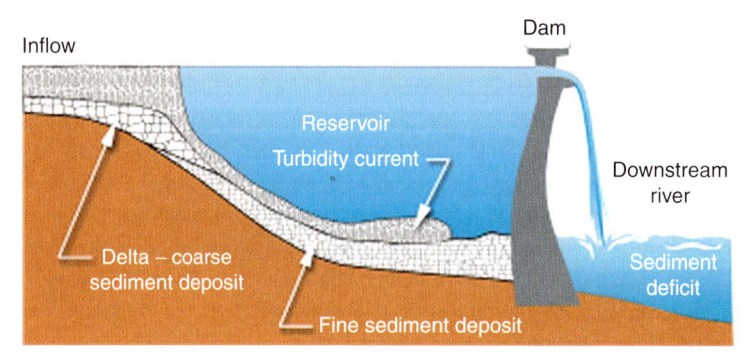

Figure 2 Schematic representation of reservoir sedimentation process.

geology, soils, topography, vegetation, and human disturbance), and the ratio of reservoir size to mean annual inflow into the reservoir. Sediment transport shows considerable temporal variation, seasonally and annually. The amount of sediment transported into the reservoir is greatest during floods.

The main problems associated with reservoir sedimentation are related to volume loss, the risk of obstruction of water intakes, abrasion of conduits and equipment, deterioration of water quality, and bed erosion (bed degradation) downstream of the dam. **Figure 2** presents a schematic representation of the reservoir sedimentation process considering fine sediments, fundamentally transported by turbidity currents, and larger coarse sediments associated with bed-load transport. The turbidity currents result in fine sediment transport in suspension in the reservoir.

Measures to reduce sediment inflow volume (sediment yield) include soil conservation practices based on reasonable land use, which includes agricultural practices and reforestation. Upstream trapping by check dams and vegetation screens can also be adopted to hold back sediments. A sound integrated water resources management in catchment areas should treat water as an integral part of the ecosystem, a natural resource, and a social and economic good.

ICOLD (International Commission on Large Dams) Bulletin 67 (1989) and 115 (1999) present some guidelines related to sedimentation control of reservoirs, including some case studies. Reservoirs can be filled at low or medium flows when sediment concentrations are low. High flows with high sediment concentrations have to be bypassed through channels or tunnels. There are two ways to pass sediments through reservoirs. The sediment-laden flow can be discharged through reservoirs at a reduced water level during flood seasons. This method is called sluicing and is mainly applicable to fine sediments. Under special conditions, density currents may develop and transport suspended sediment underneath a fluid layer of lower density toward the dam. This method is called density current venting.

Mitigation of the accumulation of sediments has been achieved in several ways. Periodic dredging can reduce the accumulation. This method usually requires low water levels for extended periods of time. Dredging is costly and the disposal of large quantities of sediment often creates problems. In other cases, the sediments have been removed through periodic flushing of the reservoir by releasing large volumes of water through the low-level outlet structures (**Figure 3**). This method has the advantage of renewing the sediment load to the downstream channel and also flushing the downstream channel with a high flood event.

The effect of outlet discharges on the mitigation of reservoir sedimentation, particularly the fine sediments transported in suspension, is shown in **Figure 3**. For many dams, sediment accumulation remains a major concern. Due to the configuration and bathymetry of most reservoirs, sediments frequently accumulate at the head of the reservoir, a long way from the dam wall, and the bottom outlet (**Table 3**). Jiroft Dam is a concrete arch dam with height 134 m.

Figure 3 Jiroft Dam (Iran): flood discharge through surface spillways and outlets. From http://www.stucky.ch/en/h_2.php.

An adequate design of outlets with great discharge capacity is particularly relevant in dams located in erodible catchment areas. Figures 4 and 5 present two solutions for dam spillways based on deep orifices in order to minimize the sedimentation process in the reservoirs.

Reservoirs can be filled at low or medium flows when sediment concentrations are low. High flows with high sediment concentrations have to be bypassed through bypass channels or tunnels.

Table 3   Estimates of annual reservoir volume losses in different regions

Region	Number of dams	Estimates of annual reservoir volume losses due to sedimentation (%)
North America	7 205	0.20
South America	1 498	0.10
North Europe	2 277	0.20
South Europe	3 220	0.17
North Africa	280	0.08
Sub-Saharan Africa	966	0.23
China	1 851	2.30
South Asia	4 131	0.52
Central Asia	44	1.00
Southeast Asia	277	0.30
Pacific Border	2 778	0.27
Middle East	895	1.50
World total	25 422	0.5–1.0 (average)

Reproduced from Alves E (2008) Sedimentation in Reservoirs by Turbidity Currents (in Portuguese). PhD Thesis, Laboratório Nacional de Engenharia Civil [1].

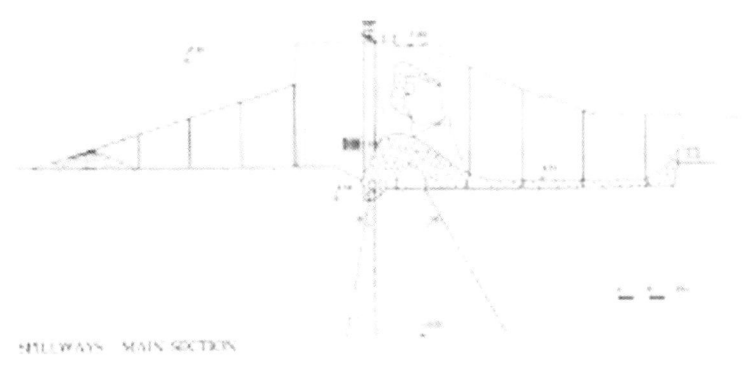

Figure 4   Pequenos Libombos Dam (Mozambique).

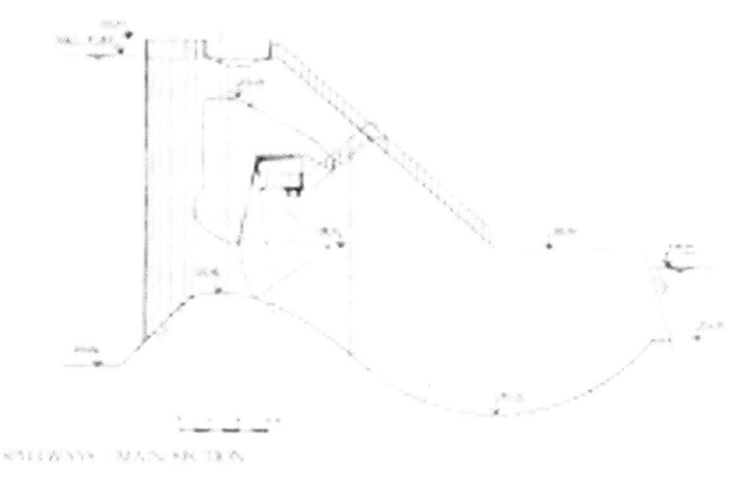

Figure 5   Fagilde Dam (Portugal).

6.03.1.3 Downstream Impacts

6.03.1.3.1 Flow regime

The existence of a reservoir introduces modifications in the hydrological regime downstream of the dam. These modifications are associated with the frequency and magnitude of floods and with the timing to peak (hydrograph). The effect of a reservoir on individual flood flows depends on both the storage capacity of the dam relative to the volume of flow and the management regime. Reservoirs having a large flood storage capacity in relation to total annual runoff can exert almost complete control on the annual hydrograph of the river downstream. Even small-capacity detention basins can achieve a high degree of flow regulation through a combination of flood forecasting and management regime.

The hydrological effects of the dam become less significant at greater distances downstream, that is, as the proportion of the uncontrolled catchment increases. The frequency of the tributary confluence below the dam and the relative magnitude of the tributary streams play an important role in determining the length of the river affected by an impoundment. Catchments with significant storage may never recover their natural hydrological characteristics, even at the river mouth, especially when dams divert water for agriculture or municipal water supply.

Flow regimes are the key driving variable for downstream aquatic ecosystems. Flood timing, duration, and frequency are critical for the survival of plant and animal communities living downstream. Small flood events may act as a biological trigger for fish and invertebrate migration; major events create and maintain habitats. The natural variability of most river systems sustains complex biological communities that may be different from those adapted to the stable flows and conditions of a regulated river.

A sufficient continuous minimum discharge to downstream of a dam is one main prerequisite to reduce the impact on the ecosystem. This may be achieved by adjusting the operation of the reservoir to this objective. This minimum discharge is called ecological discharge. The ecological discharge must be defined in order to guarantee the downstream river ecosystems, that is, to maintain the essentials of their natural biodiversity and productivity. The amount, timing, and conditions under which water should be released have to be carefully determined.

6.03.1.3.2 Ecological discharge and minimum flow

Ecological demands for each month are determined, starting from ecological discharges and taking into account the following issues:

- additional discharges for diminishing the effect of reduced dissolved oxygen (DO) in water in summer time;
- additional discharges for the fish reproduction season;
- flush discharges – artificial floods for washing up of fine sediments laid down, in particular, on water sectors placed downstream of reservoirs; and
- additional discharges to ensure proper dilution when accidental pollution occurs – relying on the methods of ecological discharges lately developed in many countries, laws and standards that set up the methodology to ascertain that the ecological discharges and demands have been established, as well as the priorities to supply water to the users.

Minimum discharges could also be defined by the needs of water downstream, for irrigation, domestic and industrial uses, and so on.

6.03.1.3.3 Surge

The term 'surge' refers to the artificially increased discharge of water during the operation of hydroelectric turbines to satisfy peak demand. Surges are punctuated by low-water phases during periods of low demand, that is, at night and at weekends. This periodic alternation between the two different flow regimes is often referred to as hydro peaking. This operation causes frequent and rapid changes in the water flow. It can create sudden changes in water levels, strong undertows, turbulence, and sudden, powerful surges of water moving downstream in what was once calm-looking surface water. The sudden, unexpected release of water from hydropower generation presents a hazard to anglers, swimmers, and canoeists below the dam. This variation of the power changes the downstream river environment. The flow after the turbining can lead to scouring of riverbeds and loss of riverbanks. This is particularly relevant in dams with daily fluctuations and where turbines are often opened intermittently. The erosion process downstream of the Grand Canyon Dam is associated with the daily cyclic flow variation.

6.03.1.3.4 Degradation and aggradation in downstream river reaches

Changes in the flow and sediment regime initially cause a degradation downstream from the dam, as the entrained sediment is no longer replaced by material arriving from upstream. According to the relative erodibility of the riverbed and riverbanks, the degradation may be accompanied by either narrowing or widening of the channel. A result of degradation is a coarsening in the texture of material left in the riverbed; in many cases, a change from sand to gravel is observed, or even, in an extreme case, the scour may proceed to the bedrock. On most rivers, these effects are constrained to the first few kilometers below the dam.

Further downstream, increased sedimentation (aggradation) may occur because material mobilized below a dam and material entrained from tributaries cannot be moved so quickly through the channel system by regulated flows. Channel widening is a frequent concomitant of aggradation.

The accumulation of sediments in the river channel downstream from the dam due to the altered flow regime may be mitigated through periodic flushing of the river channel with artificial flow events. Flushing requires outlet structures like sluice gates of sufficient capacity to permit generation of managed floods. These outlets should be placed in such a way that the releases can be made when the reservoir storage exceeds 50% of its capacity.

Damming a river can alter the character of the floodplains. In some circumstances, the depletion of fine suspended solids reduces the rate of overbank accretion so that new floodplain takes longer to form and soils remain infertile, or channel bank erosion results in loss of floodplains.

In the Nile Valley, following the closure of the Aswan High Dam in 1969, the lack of sediment in floodwater reduced soil fertility in the Nile Valley downstream of the dam. The reduction in sediment flows has also led to the erosion of the shoreline of the delta and saline penetration of coastal aquifers.

The erosion process is particularly pronounced at alluvial sites with noncohesive sandy bank materials, and has been attributed to the release of silt-free water, the maintenance of unnatural flow levels, sudden flow fluctuations, and out-of-season flooding. However, in some cases, the reduction in the frequency of flood flows and the provision of stable low flows may encourage vegetation encroachment, which will tend to stabilize new deposits, trap further sediments, and reduce floodplain erosion. Hence, depending on the specific conditions, dams can either increase or decrease floodplain deposition/erosion.

Managed flood releases can be a strategy to mitigate the detrimental impact downstream of dams. An objective of these managed flood releases is the conservation or restoration of floodplain ecosystems.

6.03.1.3.5 Effects of sediments on turbines

The erosion of turbines (abrasion) depends on

- *eroding particles* – size, shape, and hardness (associated fundamentally with abrasion);
- *substrates* – chemistry, elastic properties, surface hardness, and surface morphology; and
- *operating conditions* – velocity, impingement angle, and concentration.

Depending on the gradient of the river and the distance traversed by the sand particles, the shape and size of sediment particles vary at different locations of the same river system, whereas the mineral content is dependent on the geological formation of the river course and its catchment area.

To minimize sediment effects on turbines, some excluding devices are adopted, the more frequent being associated with sedimentation chambers. In lateral water intakes located in alluvial bed rivers, solutions based on entry sills, submerged vanes designed to generate transverse bottom velocity components, and sluice channels are adopted.

Run-of-river projects are constructed to utilize the available water throughout the year without having any storage. These projects usually consist of a small diversion weir or dam across a river to divert the river flow into the water conveyance system for power production. Therefore, these projects do not have room to store sediments but should be able to bypass the incoming bed loads to the river downstream. The suspended sediments will follow the diverted water to the conveyance system.

6.03.1.3.6 Water characteristics

Water temperature is an important quality parameter for the assessment of reservoir impacts on downstream aquatic habitats because it influences many important physical, chemical, and biological processes. In particular, temperature drives primary productivity. Thermal changes caused by water storage have the most significant effect on in-stream biota. The level in the reservoir from which the discharge is drawn, for example, cool deep temperatures or warm surface temperatures, may affect temperatures downstream of the dam, which in turn may affect fish spawning, growth rate, and length of the growing season. Cold water releases from high dams of the Colorado River are still measurable 400 km downstream, and this has resulted in a decline in native fish abundance. Even without stratification of the storage, water released from dams may be thermally out of phase with the natural temperature regime of the river.

The quality of water released from stratified reservoirs is determined by the elevation of the outflow structure relative to the different layers within the reservoir. Water released from near the surface of a stratified reservoir will be well-oxygenated, warm, and nutrient-depleted. In contrast, water released from near the bottom of a stratified reservoir will be oxygen-depleted, cold, and nutrient-rich, which may be high in hydrogen sulfide, iron, and/or manganese. Water depleted in DO not only is a pollution problem in itself, affecting many aquatic organisms (e.g., salmonid fish require high levels of oxygen for their survival), but also has a reduced assimilation capacity and so a reduced flushing capacity for domestic and industrial effluents. The problem of low DO levels is sometimes mitigated by the turbulence generated when water passes through turbines.

Water passing over steep spillways may become supersaturated in nitrogen and oxygen, and this may be fatal to the fish immediately below a dam. Fish with a swim bladder are particularly affected.

Measures to mitigate the potential effects of nutrient accumulation in an impoundment have focused on reducing the inflow of nutrients to the reservoir and increasing the removal of nutrients from the water. Reduction of inflow of nutrients has been accomplished through the construction of wastewater treatment facilities at communities along the margins of the impoundment as well as in the watershed upstream. Other methods include seasonal flushing of the reservoir or the training of local farmers in the use of fertilizers. The effectiveness of this process, however, is dependent on the volume of the reservoir relative to the inflow.

6.03.1.3.7 Fish migration

The changes in the aquatic fauna regime can be quite far-ranging. One of the most significant indicators of these changes can be the impact on the migratory patterns and relative abundance of fish species. The effects of changed temperature regimes on fish abundance have been previously referred to.

Fish species have several different migratory patterns. The well-known species of fish that migrate are the anadromous fishes such as salmon or steelhead trout and the catadromous fishes such as eels. Adult salmon migrate up the river to spawn and the young descend to the ocean where they spend much of their adult life. The reverse occurs with the catadromous fishes. Preservation of the fisheries resource is extremely important in planning a dam project on these rivers. The blockage of fish movement can be one of the most significant negative impacts of dams on fish biodiversity.

The river continuum includes the gradual natural change in river flow, water quality, and species that occur along the river length from the source to the coastal zone. A dam breaks this continuum and can stop the movement of species unless appropriate measures are taken.

Effective measures to mitigate the blockage of upstream migration of fish include the installation of fish passage facilities to allow movement of fish from below the dam to the reservoir and further upstream. The types of fish passage facilities include fish ladders, fish elevators, and trap-and-haul techniques.

6.03.2 Reservoir Water Quality

6.03.2.1 Introduction

From the beginning of the twentieth century, technological progress and a greater need for energy, water supply, and flood control have motivated an increase in the number of dams constructed all over the world. Although lakes and reservoirs contribute to only 0.35% of the whole volume of freshwater in our planet [2] as a response to this enormously increased demand, more than half of ICOLD's registered 45 000 large dams have been built in the period of 1962–97 [3]. The storage capacity of the total registered large dams is about 6000 km^3.

The construction of dams, although initially motivated for power generation, creates reservoirs with multipurpose uses and functions, which include the availability of water to urban water supply and agriculture, the mitigation of devastating floods, navigation, and the support of leisure activities. The new habitats these water bodies create and their scenic value attract activities that produce waste.

All dams and reservoirs become a part of the environment, which they influence and transform to a degree and within a range that varies from project to project. Frequently seeming to be in opposition, dams and their environment interrelate with a degree of complexity that makes the task of the dam engineer particularly difficult [4].

Reservoirs can become the receiving body for urban, agricultural, and industrial wastewater. These wastes and the evolution of the water quality in the reservoir, due to the fact that the prevailing processes and characteristics change when water is stored and not flowing, cause changes in the quality of water discharged downstream.

In the 1960s, along with increased recognition of water quality problems, a large number of relevant technical publications started to be produced [5]. Nevertheless, in contrast to flowing waters, lakes and impoundments were not a priority subject in the early years of water quality modeling. This is because, with notable exceptions such as the Great Lakes of North America, they have not historically been a major focus of urban development.

Research activities on the water quality of reservoirs not only followed the great development of dam construction but also aimed at answering the challenges of sustainable use and the preservation of the newly created ecosystems. The often conflicting uses of reservoirs require the introduction of management systems, and these created the need to have management tools that have the ability to model water quality.

The 'guidance' for the implementation of the European Union Water Framework Directive (EU WFD; Directive 2000/60/EC of 23 October 2000 establishing a framework for Community action in the field of water policy) advises the classification of reservoirs as "heavily modified water bodies" on which a "good ecological potential" has to be maintained or achieved. The environmental quality objectives for the characteristics of such water bodies will be as similar as possible to the ones that would prevail in similar 'natural' water bodies (in terms of, e.g., morphology and location) in pristine conditions.

This chapter presents an overview of the pressures and processes that affect reservoir water quality. A general description of the basic characteristics that have a direct connection with the water quality of such water bodies is also presented, as well as a description of the behavior of the chemical entities that characterize water quality. Particular attention is paid to the eutrophication phenomenon. A review of the general issues related to water quality modeling of reservoirs, modeling methodology, and types of models most commonly used is also presented. Finally, a summary of the process of identification of heavily modified water bodies in the context of EU WFD is also presented.

6.03.2.2 General Characteristics of Reservoirs

6.03.2.2.1 Morphology and hydrodynamics

Water quality characteristics as well as ecological features of reservoirs are strongly interconnected and are a function of their morphology and hydrodynamics as well as of the energy fluxes driven by the climatic factors. They are also a function of the morphology and hydrology of the region.

The phenomena that occur in a reservoir are complex, and their interpretation and analysis is a difficult task that must take into consideration the context provided by the morphology and physical processes. An overview of those factors influencing the quality of water in reservoirs is presented below.

As most reservoirs are created by damming a river, they generally tend to be elongated or dendritic. For water quality purposes, the most important morphological features are connected with the ratio of area to volume, that is, with the average depth (H) of the reservoir. This parameter will contribute to the tendency for stable stratification and will determine the relative importance of interface processes such as reaeration and benthonic nutrient recycling. Some authors (e.g., Chapra [6]) propose the classification of shallow reservoirs (or lakes) as those with $H < 7$ m and deep water bodies as those with $H > 7$ m.

The hydrodynamic regime of a reservoir is one of the most important factors to control its behavior and water quality. Average retention time, defined as the ratio of mean annual inflow to the net reservoir volume, is a relevant characteristic that allows water quality characteristics to be anticipated. A 'run-of-the-river' type of reservoir will have a relatively small retention time, in the order of days or weeks, while a 'large' reservoir, with capacity for flow regulation, will have a long residence time with values of the order of years or even decades.

Also relevant in the control of water quality characteristics and behavior are the physical processes that occur, taking into account the characteristics of the various inflows and withdrawals, as well as the circulation induced by the wind, which has particular relevance in shallow water bodies. **Figure 6** presents a diagram of the main physical processes present in a reservoir.

6.03.2.2.2 Thermal stratification

The thermal energy exchange at the water surface is a relevant factor in the control of water quality in a reservoir, especially if the water column is deep. Other important climatic factors are the wind and the precipitation regime in the catchment, which determine the regime of the runoff to the reservoir and its hydrodynamics [7].

Stratification is of major importance for water quality of reservoirs throughout the year. Most reservoirs are well mixed during winter. As spring progresses and the temperature rises, thermal stratification will be established in the near surface of water and continues until mixing is confined to the upper layer. The attainment of persistent stratification leads to the establishment of three circulation regimes: the upper (epilimnion) and the lower (hypolimnion), separated by a narrow region of sharp temperature change (thermocline or metalimnion; **Figure 7**). In late summer and fall, the unstable situation returns and strong vertical convection mixing occurs, with a progressive deepening of the thermocline, creating the event called the autumn/fall turnover. However, in some tropical areas, where there is less temperature variation, these processes may not be as dominant.

6.03.2.2.3 Pollutants and stressors on reservoirs

As with any other type of water body, water quality of reservoirs is greatly affected by the different pressures that are exerted on it. Pressures are derived from its uses, and the most relevant in the present context are polluting loads.

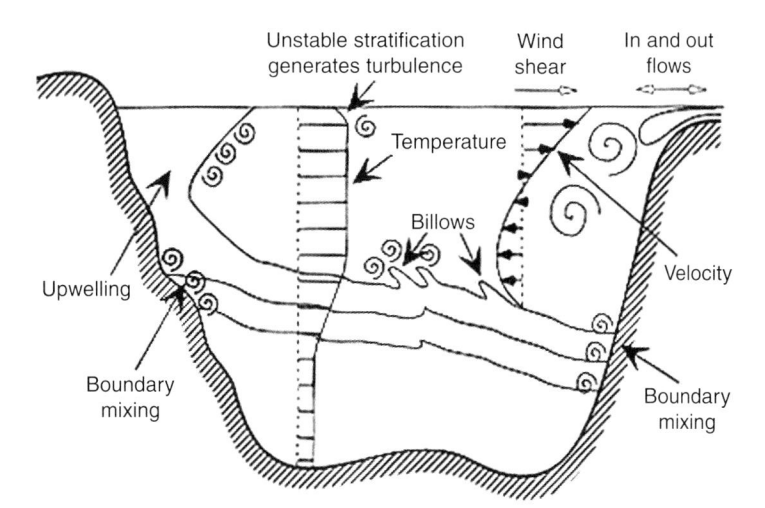

Figure 6 Physical processes in reservoirs. Adapted from UNEP-IETC (n.d.) Planning and management of lakes and reservoirs: An integrated approach to eutrophication. Newsletter and Technical Publications of the IETC. http://www.unep.or.jp/ietc/publications/techpublications.

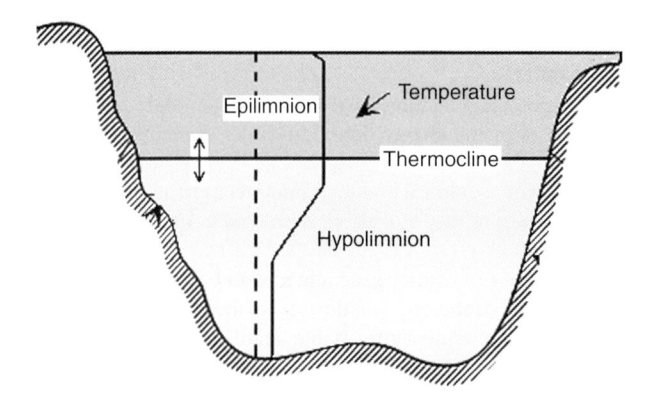

Figure 7 Vertical structure of the water column in a stratified reservoir.

A study presented by the United States Environmental Protection Agency (US EPA; http://www.epa.gov/owow/lakes/quality.html) identified the pollutants and stressors that cause water quality degradation in US lakes and reservoirs. The most common cause of water quality deterioration is associated with excessive nutrient (nitrogen and phosphorus) input, followed by metals. Third in the ranking of pressures is solids input causing siltation. Also important as a cause for water quality degradation is the input of carbonaceous organic matter, in general from sewage, with a high oxygen demand.

The same study identified agriculture as the leading source of pressures; also important are the inputs from urban runoff and storm sewers. General nonpoint sources and municipal point sources have an equivalent contribution in relative terms. The database used in the study referred to not only pertains to reservoirs but also includes natural lakes and other impoundments, which suggests that the relative importance of agricultural sources may still be more relevant when only reservoirs are considered, as fewer urban settlements are established on their direct drainage basin.

The same study also proposes a qualitative classification of reservoirs, using as criteria their capability to support traditional or desired uses, as follows:

- *good* – fully supporting all of their uses or fully supporting all uses but threatened for one or more uses
- *impaired* – partially or not supporting one or more uses
- *not attainable* – not able to support one or more uses

In the context of the previously mentioned EU WFD, five quality classes must be defined, as explained in the paragraph dedicated to the issues associated with this directive.

Figure 8 is a representation of the conditions observed in a reservoir impacted by different polluting sources and those observed in a healthy ecosystem. The figure addresses the issue of nutrient enrichment and the effects of inputs of metals that accumulate in

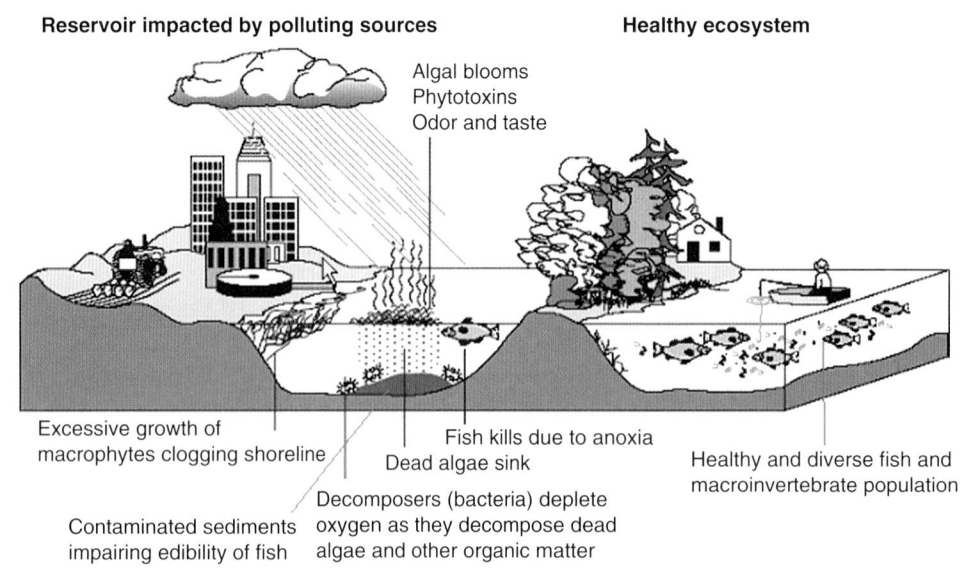

Figure 8 Comparison between a healthy ecosystem and one impacted by polluting loads. Adapted from http://www.epa.gov/owow/lakes/quality.html.

sediments and, later, contaminate biota. In some circumstances, the aquatic fauna will accumulate xenobiotics in such quantities that their life cycles and their edibility are impaired. The eutrophication process, its effects and symptoms, and assessment criteria are addressed in detail in the next section.

When reservoirs are used as potable water sources, contamination by fecal pathogens is a major issue and is becoming more relevant as urban settlements, in many cases associated with the growing interest of reservoirs as tourist centers and places for water sports, become more common around reservoirs. Urban settlement, on the one hand, requires high-quality water and, on the other hand, has the potential to cause significant degradation of the value of the resource, representing a paradigmatic situation for the need to implement clear user rules and codes of practice, as required to harmonize uses and to preserve the health of the ecosystems.

The contamination by xenobiotics, metals, and microorganic pollutants, although not a very widespread problem, may be of local relevance. An example of a situation where that type of pollution may be relevant is the reservoirs that have mining zones in their catchment, either in exploitation or abandoned.

6.03.2.3 Water Quality Processes – Eutrophication and Oxygenation

6.03.2.3.1 Introduction

The physical, chemical, and biological behavior of stored surface waters has been the subject of research in the domain of limnology. Stored water may improve water quality, but in some cases this water may be more susceptible to deterioration. These aspects have to be taken into account during the design phase of dams and later, when management plans for the reservoir and its catchment are in place.

As nutrient inputs are the most frequent and serious pressures on reservoirs, the resulting eutrophication and the related influence on oxygenation status are the most important water quality processes to be taken into consideration. They will be treated in some detail in the following paragraphs.

6.03.2.3.2 General concepts

Eutrophication can be defined as the process of enrichment of water with organic matter, caused by an increase of nutrients for plants (such as nitrogen and phosphorus) that stimulate primary production [8–10].

Lakes and reservoirs can be broadly classified as ultraoligotrophic, oligotrophic, mesotrophic, eutrophic, or hypereutrophic, depending on the concentration of nutrients in the body of water and/or based on ecological symptoms of the nutrient loading, although strict boundaries for these classes are often difficult to define.

There are commonly three main criteria for the degree of eutrophication:

- total phosphorus concentration,
- mean chlorophyll concentration, and
- mean Secchi disk visibility.

In general terms, oligotrophic lakes and reservoirs are characterized by low nutrient inputs and primary productivity, high transparency, and a diverse biota. In contrast, eutrophic waters have high nutrient inputs and primary productivity, low transparency, and a high biomass of fewer species with a greater proportion of cyanobacteria.

Although the fundamental characteristics of eutrophication are similar in all water bodies, differences in basin shapes and flow patterns may lead to longitudinal variations in the degree of eutrophication in reservoirs (**Figure** 9). In addition, water supply and power generation requirements often lead to large variations in water level in reservoirs. These changes in level usually expose or inundate littoral regions, which may enhance nutrient supply.

6.03.2.3.3 Eutrophication symptoms and effects

The process of eutrophication in all water bodies causes a series of effects that are visible by symptoms that often impair some or most of the uses of the water. A brief description of these eutrophication consequences is presented below.

6.03.2.3.3(i) Harmful algal blooms

A common result of eutrophication is the increased growth of algae. Cyanobacteria are an especially harmful group causing the formation of surface scum, severe oxygen depletion, and fish mortalities. The ingestion of freshwater toxins (neurotoxins, hepatotoxins, cytotoxins, and endotoxins), which are produced almost exclusively by cyanobacteria, may lead to death of cattle and other animals. Gastrointestinal disorders in humans can also be associated with the drinking of water that contained blooms of cyanobacteria.

Cyanobacteria and filamentous species of chlorophytes (green algae) can cause odors and clogging of filters in water treatment or industrial facilities. Dinoflagellates, the so-called red tides, are another group of concern that is known to develop, which can include toxic strains. One by-product of dense algal blooms is high concentrations of dissolved organic carbon (DOC). When water with high DOC is disinfected by chlorination, potentially carcinogenic and mutagenic trihalomethanes are formed.

Narrow, channelized basin	Broader, deeper basin	Broad, deep, lakelike basin
Relatively high flow	Reduced flow	Little flow
High suspended solids; low light availability at depth	Reduced suspended solids; more light availability at depth	Relatively clear; more light availability at depth
Nutrient supply by advection; relatively high nutrient levels	Advective nutrient supply reduced	Nutrient supply by interval recycling; relatively low nutrient levels
Light-limited primary productivity	Primary productivity relatively high	Nutrient-limited primary productivity
Cell losses primarily by sedimentation	Cell losses by sedimentation and grazing	Cell losses primarily by grazing
Organic matter supply primarily by allochthonous source	Intermediate	Organic matter supply primarily by autochthonous source
More eutrophic	Intermediate	More oligotrophic

Figure 9 Longitudinal zones of environmental factors controlling trophic status in reservoirs. Adapted from Ryding S-O and Rast W (eds.) (1989) *The Control of Eutrophication of Lakes and Reservoirs.* Paris, France: UNESCO [11].

6.03.2.3.4 Growth of aquatic plants

Dense mats of floating aquatic plants such as water hyacinth (*Eichhornia crassipes*) can cover large areas near the shore and can float into open water. These mats block light from reaching submerged vascular plants and phytoplankton, and often produce large quantities of organic detritus that can lead to anoxia and emission of gases such as methane and hydrogen sulfide. Accumulations of aquatic macrophytes can restrict access for fishing or recreational use of lakes and reservoirs and can block irrigation and navigation channels and intakes of hydroelectric power plants.

6.03.2.3.5 Anoxia

Another symptom of eutrophication is the depletion of oxygen concentration in the water column. Anoxic conditions are not suitable for the survival of fishes and invertebrates. Moreover, under these conditions, ammonia, iron, manganese, and hydrogen sulfide concentrations can rise to levels deleterious to the biota and to hydroelectric power facilities. The anoxic conditions also increase the rate of redissolution of phosphate and ammonium, which increases the nutrient availability in the water column, creating a positive feedback loop in the eutrophication process.

6.03.2.3.6 Species changes

Shifts in the abundance and species composition of aquatic organisms often occur in association with the alterations of ecosystems caused by eutrophication. Reduction in underwater light levels because of dense algal blooms or floating macrophytes can reduce or eliminate submerged macrophytes. Changes in food quality associated with shifts in algal or aquatic macrophyte composition and decreases in oxygen concentration often alter the species composition of fishes. For example, less desirable species, such as carp, may become dominant. However, in some situations, such changes may be deemed beneficial.

6.03.2.3.7 Hypereutrophy

Hypereutrophic water bodies are in the upper end of the eutrophication process. A water body becomes hypertrophic when reductions in nutrient loading are not feasible or will have no effect at reversing the trophic enrichment. Hypereutrophic systems usually receive uncontrollable diffuse and nonpoint sources of nutrients, originating from overfertilized or naturally rich soils.

Nevertheless, these systems may constitute a valuable and integral part of the landscape, providing sanctuaries for birds and an important aquatic habitat, and, if properly managed, can provide valuable and highly productive fisheries.

6.03.2.3.7(i) Enhanced internal recycling of nutrients

When the eutrophication process is well established, internal loading of nutrients from benthonic resolubilization may become the dominant source, in addition to external loading of nutrients from both point and diffuse sources. This process is of particular relevance when the average depth is small and near-bottom anoxic and nutrient-rich layers of water frequently mix with surface layers. Once a eutrophic or hypereutrophic state is reached, the dependence on external sources of nutrients is diminished and the water body will function as a system with positive feedback, the sediments providing an adequate supply of nutrients, even when the external sources are reduced.

6.03.2.3.8 Elevated nitrate concentrations

High concentrations of nitrate resulting from nitrate-rich runoff or nitrification of ammonium within a lake can cause public health problems. Methyl-hemoglobinemia occurrence in infants results from nitrate levels above 10 mg l^{-1} in drinking water. By interfering with the oxygen-carrying capacity of blood, the high nitrate levels can lead to a life-threatening deficiency of oxygen.

6.03.2.3.9 Increased incidence of water-related diseases

In some situations where a portion of the population producing sewage suffers from infections transmitted directly or indirectly via water, the spread of human diseases can be a very significant impact of sewage entering a water body. While such situations are especially prevalent in tropical countries, avoiding the spread of disease via water is a concern for all countries.

6.03.2.3.10 Increased fish yields

In some circumstances, the eutrophication process, up to a certain point, can have a positive impact on fisheries, as yields of fish tend to increase as primary productivity increases. Greater increases in fish yields occur for smaller increments in primary productivity in oligotrophic or mesotrophic waters than in eutrophic systems. However, when the undesirable effects of eutrophication are present, namely, oxygen depletion or significantly altered (as in alkaline or reduced as in acid) pH and elevated ammonia levels, the increases in fish yields as primary production rises will be reduced. In this situation, the edible and marketable condition of the fish catch may also be threatened.

6.03.2.3.11 Nutrient recycling

Aquaculture of fishes can be an effective way to obtain benefits from nutrients that cause eutrophication. The fish in an aquaculture system can take up a large portion of the nutrients and transform them into a harvestable, marketable form.

Phytoplankton and floating aquatic macrophytes can be very effective at nutrient uptake and are capable of reducing dissolved inorganic nutrient concentrations to very low levels. Hence, if the plants are subsequently removed from the water, they may function as tertiary municipal wastewater treatment or as sources of organic matter for other uses (e.g., biogas generation or agro-fertilizers).

6.03.2.3.12 Assessment of trophic status

There is no established methodology to determine what the trophic state of a water body is. As previously referred to, there are commonly three main criteria for the degree of eutrophication:

- total phosphorus concentration,
- mean chlorophyll concentration, and
- mean Secchi disk visibility.

Many simple empirical models have been developed to predict the concentration of total phosphorus in a lake as a function of annual phosphorus loading. Extensions of such models offer predictions of chlorophyll concentration, Secchi disk visibility, or pH or DO levels. The values predicted by these models can have uncertainties from as low as ±30% to as high as ±300%, and usually require modifications for different regions.

The best known and most widely applied model is the Vollenweider method [10, 12, Vollenweider 1976, OECD 1982]. This method relates the trophic condition of the reservoir (or lake) to nutrient loading, on the basis of the relationships presented in **Figure 10**. Originally, the abscissa was H, the average depth of the lake, but later it was recognized that the flushing rate of the lake also played a relevant role in the tendency for eutrophication, and the 'Vollenweider plot' was transformed with the consideration of the flushing time (τ_w) to q_s, the hydraulic overflow rate (m yr^{-1}). Refinements and adaptations of Vollenweider's approach have improved correlation and added or substituted nitrogen loading for some regions. Further research is required to incorporate responses of aquatic macrophytes into these models.

The trophic state is also dependent on knowing which of the macronutrients is the limiting factor of primary productivity, and this is a function of

- the ratio of nitrogen to phosphorus in the inputs and in the vertical fluxes of dissolved nutrients in the water column;
- preferential losses from the euphotic zone by processes such as denitrification, adsorption of phosphorus to particles, and differential settling of particles with different nitrogen:phosphorus ratios;

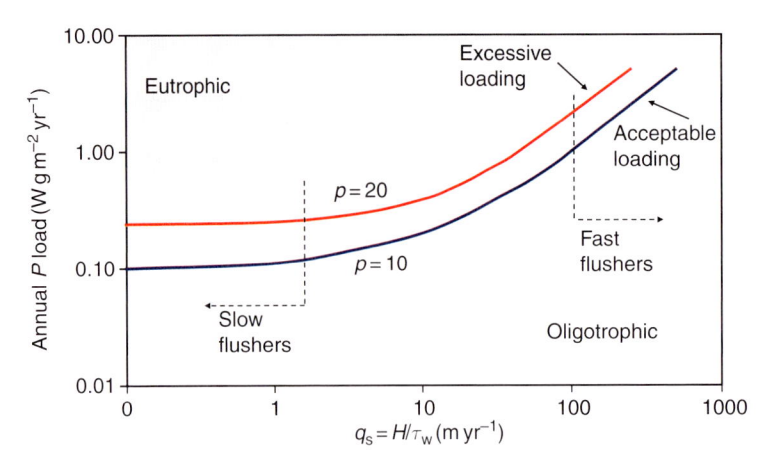

Figure 10 Nutrient loading and trophic condition. Redrawn from Chapra (1997) and Thoman RV and Mueller JA (1987) *Principles of Surface Water Quality Modeling and Control.* New York: Harper & Row [13].

Table 4 Trophic state classification

Variable	Oligotrophic	Mesotrophic	Eutrophic
Total phosphorus ($\mu g\, l^{-1}$)	<10	10–20	>20
Chlorophyll *a* ($\mu g\, l^{-1}$)	<4	4–10	>10
Secchi disk depth (m)	>4	2–4	<2
Hypolimnion oxygen (% sat.)	>80	10–80	>10

- the relative magnitude of external supply to internal recycling and redistribution; and
- the contribution from nitrogen fixation.

Unfortunately, these processes have been measured in a coordinated manner in only very few lakes. Instead, inferences from several indicators of nutrient limitation must be made. The nitrogen:phosphorus ratio in suspended particulate matter is a potentially valuable index of the nutritional status of the phytoplankton, if contamination from terrestrial detritus can be discounted. Healthy algae contain approximately 16 atoms of nitrogen for every atom of phosphorus. Ratios of nitrogen to phosphorus less than 10 often indicate nitrogen deficiency and ratios greater than 20 indicate phosphorus deficiency. When phosphorus is the limiting nutrient, criteria for the classification of reservoirs are as presented in **Table 4** (Chapra 1997).

6.03.2.4 Water Quality Parameters

6.03.2.4.1 Behavior in reservoirs

The ecological and water quality relationships in a reservoir are complex. The succession of trophic states within an aquatic system is characterized by quality parameters that include DO, nutrients, suspended solids, detritus, and sediments. The transformations of mass and energy are associated with the processes of primary production, growth, respiration, mortality, predation, and decomposition, which in turn are governed by environmental parameters such as temperature, light availability, and nutrients. In the following paragraphs, an overview of the processes that govern oxygen and nutrient dynamics in lotic water bodies is presented.

6.03.2.4.2 Oxygen

Among water quality parameters, oxygen is of key importance, not only because its concentration, presence, or absence dictates the type of living organisms present, as in its absence only anaerobic microbial activity is possible, but also because it rules some of the chemical processes such as the oxidation of organic matter. The oxygen cycle in a reservoir is a complex phenomenon with important differences in its distribution, as a function of diurnal and seasonal cycles and of the trophic state of the system.

Horizontal variation in oxygen content can be great in reservoirs where the photosynthetic production of oxygen by littoral vegetation exceeds that of open water algae, that is, when benthic and infralittoral processes associated with algae and riparian vegetation dominate the photosynthetic pelagic production. A schematic division of the reservoir is presented in **Figure 11**. The profile of DO concentration at surface will vary strongly with the horizontal morphology of the reservoir as well as with its bathymetry.

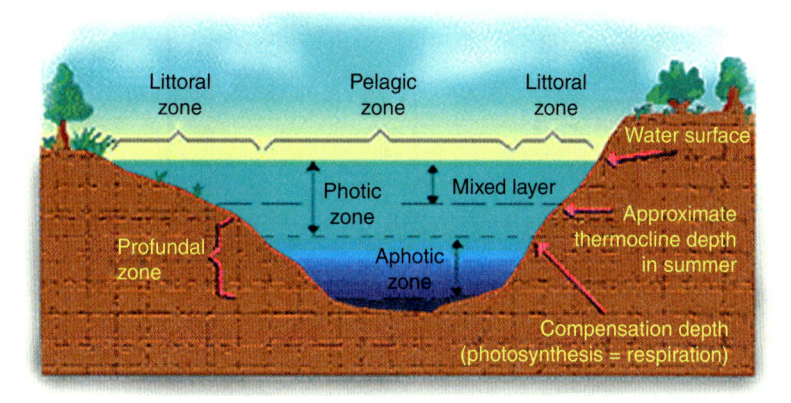

Figure 11 Horizontal variation of DO concentrations.

Extensive and rapid decay of littoral plants or phytoplankton can result in large reductions in the oxygen content, in particular in small, shallow reservoirs, leading to the death of large numbers of aquatic animals. This process is often known as 'summerkill'.

Vertical distribution of DO concentrations in the water column has a series of typical patterns. As diffusion of oxygen from the atmosphere into and within water is a relatively slow process, turbulent mixing of water is required for DO to be distributed in equilibrium with that of the atmosphere. Subsequent distribution of oxygen in the water of thermally stratified water bodies is controlled by a number of solubility conditions, hydrodynamics, photosynthetic activity, and sinks due to chemical and biochemical oxidation reactions.

In summer, in stratified oligotrophic reservoirs, the oxygen content of the epilimnion decreases as the water temperature increases due to the decreased solubility and often due to the more quiet wind conditions that also decrease the rate of reaeration in the water–atmosphere interface. The oxygen content of the hypolimnion is higher than that of the epilimnion because the saturated colder water from spring turnover experiences limited oxygen consumption. This oxygen distribution is known as an 'orthograde oxygen profile' (**Figure 12**).

In eutrophic reservoirs, the loading of organic matter and sediments to the hypolimnion increases the consumption of DO. As a result, the oxygen content of the hypolimnion of thermally stratified lakes is reduced progressively during the summer stratification period – usually most rapidly at the deepest portion of the basin where a lower volume of water is exposed to the intensive oxygen-consuming processes of decomposition at the sediment–water interface. This oxygen distribution is known as a 'clinograde oxygen profile' (**Figure 13**).

Oxygen saturation, at existing water temperatures, returns throughout the water column during fall overturn. The oxygen concentrations at lower depths in productive water bodies are reduced, but not to the extent observed in the summer, because of colder water temperatures throughout the water column, resulting in greater oxygen solubility and reduced respiration by aquatic organisms. In the spring, the water is mixed and oxygen becomes saturated throughout the water column.

The metalimnetic oxygen maximum distribution occurs when the oxygen content in the metalimnion is supersaturated in relation to levels in the epilimnion and the hypolimnion. The resulting positive heterograde oxygen curve is usually caused by extensive photosynthetic activity by algae in the metalimnion.

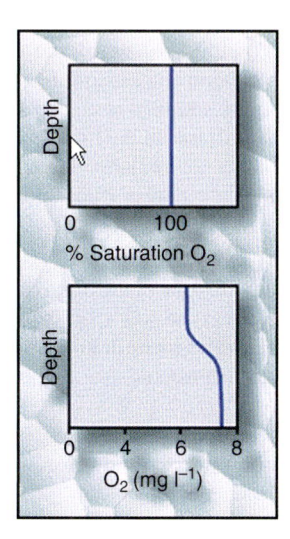

Figure 12 Orthograde oxygen profile.

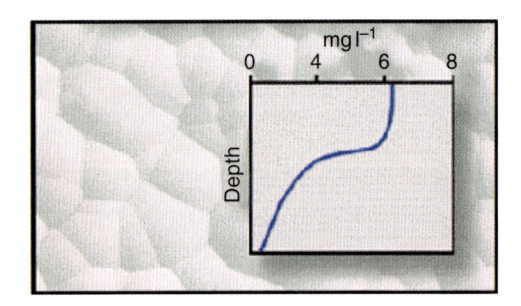

Figure 13 Clinograde oxygen profile.

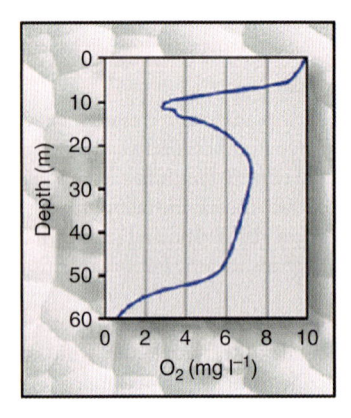

Figure 14 Metalimnetic oxygen maximum.

Epilimnetic oxygen concentrations vary on a daily basis in productive lakes. Rapid fluctuations between supersaturation and undersaturation of oxygen can result when daily photosynthetic contributions and night respiratory oxygen consumption exceed turbulent exchange with the atmosphere (**Figure 14**).

6.03.2.5 Nutrient Dynamics

6.03.2.5.1 Nitrogen

Figure 15 presents the nitrogen cycling that occurs in a reservoir. The dissolved inorganic forms present in the water column are ammonia (NH_4), nitrite (NO_2), and nitrate (NO_3), all derived from organic nitrogen compounds by a series of chemical reactions presented in a simplified form in **Figure 16**.

Nitrification is the process that transforms ammonia, directly input into the water body from sewage or produced by the ammonification of organic nitrogen compounds, into nitrite and nitrate, in the presence and with the consumption of oxygen. If DO concentrations are depleted creating anaerobic conditions, denitrification occurs with the production of molecular nitrogen, which is diffused to the atmosphere. This is a process occurring predominantly in sediments, although it may also occur in the deoxygenated hypolimnia of some reservoirs. In eutrophic stratified reservoirs, concentrations of N_2 may decrease in the epilimnion because of reduced solubility as temperatures rise and increase in the hypolimnion from denitrification of nitrate (NO_3) to nitrite (NO_2) to molecular inorganic nitrogen (N_2). NO_2 rarely accumulates except in the metalimnion and hypolimnion of eutrophic systems. Concentrations of nitrite are usually very low unless organic pollution is high.

6.03.2.5.2 Phosphorus

Although phosphorus is needed in only small amounts, it is one of the more common growth-limiting elements for phytoplankton in freshwater. These shortages arise as there is no biological pathway enabling phosphate fixation similar to the process of nitrogen fixation and due to a geochemical shortage of phosphorus in many drainage basins. The anthropogenic addition of phosphorus to freshwater bodies is one of the causes of the increase of their trophic state, as previously mentioned. **Figure 17** presents the dynamics of phosphorus in the aquatic environment.

In deep stratified systems, surface waters may have limited sources of phosphate and the quantity of 'available' phosphorus in late winter may determine the level of phytoplankton primary production in summer. Intensive algal growth in spring usually depletes phosphate levels in the surface waters. Hence, phytoplankton growth during the summer usually consumes recycled phosphate, excreted by animals feeding on phytoplankton. Direct benthonic fluxes from the sediments may be the most important source of this nutrient in the summer in shallow areas.

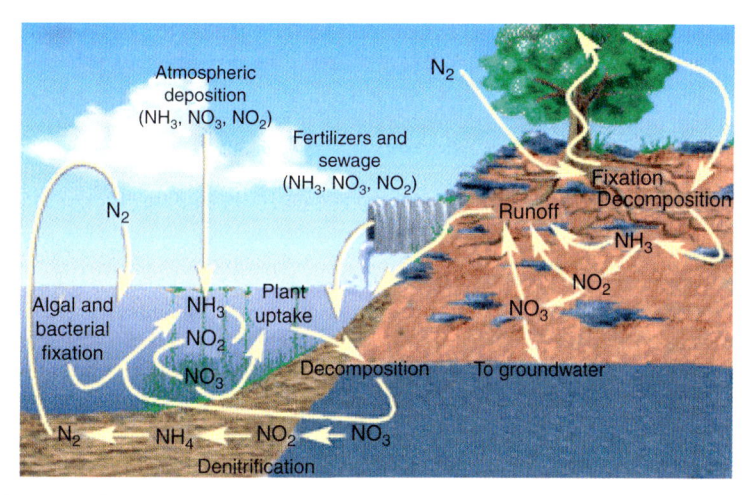

Figure 15 Nitrogen cycling in a reservoir.

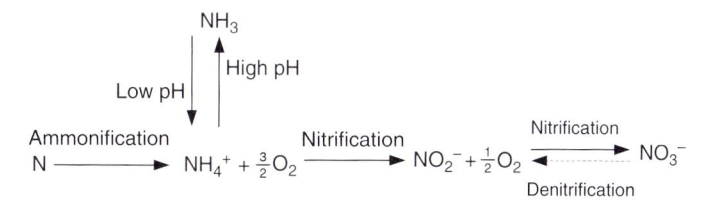

$$\text{Ammonification} \qquad \text{Low pH} \updownarrow \text{High pH } NH_3$$

$$N \xrightarrow{\text{Ammonification}} NH_4^+ + \tfrac{3}{2}O_2 \xrightarrow{\text{Nitrification}} NO_2^- + \tfrac{1}{2}O_2 \underset{\text{Denitrification}}{\overset{\text{Nitrification}}{\rightleftarrows}} NO_3^-$$

Figure 16 Nitrogen chemical transformations.

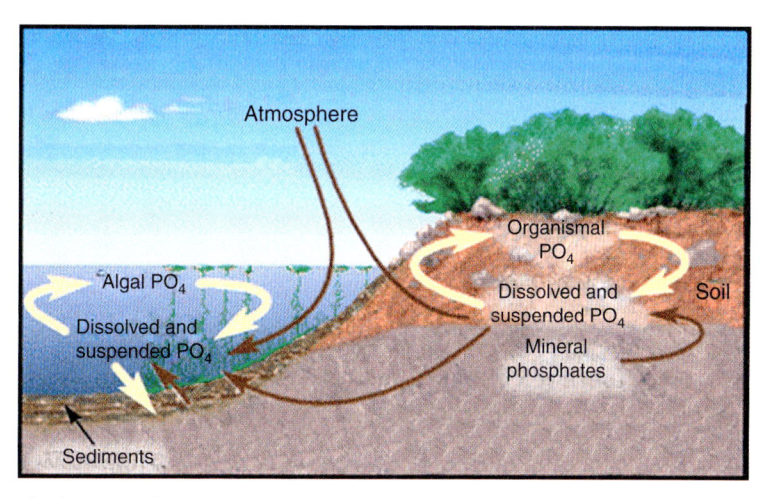

Figure 17 Phosphorus dynamics in a reservoir.

Rooted aquatic plants get phosphorus from sediments and can release large amounts of this element to the water column. Phosphate (in contrast to nitrate) is readily adsorbed to soil particles, and high inputs of total phosphorus are due to erosion of erodible soils and from runoff. Agricultural, domestic, and industrial wastes are the major sources of soluble phosphate and frequently contribute to an increase of the trophic state and to the occurrence of algal blooms.

6.03.2.6 Overview of Water Quality Models of a Reservoir

... in science, a model has as objective to uncover what structure or what set of relationships are a genuine representation although partial of reality.

This definition (McFague 1982), cited by Thoman and Mueller [13], enhances three characteristics of models:
- Models are about 'discovery'.
- Models are about behavior.
- Models are true and not true at the same time.

In fact, a model is no more than a representation of reality that contains some of the characteristics of a system, representing, in a more or less detailed way, our understanding of the system and of the processes that govern its state and of the relations between its components [14].

Water quality models are built for three main reasons (Schooner 1996):

- to get a better understanding of the destiny and transport processes of substances present in the aquatic environment;
- to determine concentrations of substances to which humans and aquatic organisms are exposed; and
- to forecast future environmental state under different scenarios of pressures as a consequence of the adoption of alternative courses of action and management measures.

The growing capability of models to forecast the behavior of aquatic systems was the main reason for presenting these techniques as decision support tools. Prognostic modeling is the use of models to simulate consequences of alternative courses of action, in one of the most attractive roles of modeling. Another use of models is made in the context of diagnostic modeling, where the conceptual and mathematical representation aims to help the understanding of available information in order to better identify cause–effect relationships for the observed phenomena.

Although diagnostic modeling does not possess the appeal of the capability of prognosis, it is not less relevant (Baptista 1994). The credibility of a forecast will be dependent on the degree of calibration and validation of the model that produced them.

Dynamic simulation models incorporate mathematical descriptions of physical, chemical, and biological processes in lakes or reservoirs. If properly designed and calibrated, these models can assist with management decisions that require considering alternative scenarios. Moreover, they often offer sufficient spatial and temporal resolution to model algal blooms and other responses to eutrophication. Conversely, the data requirements and process-level understanding demanded by dynamic models can be formidable. While such models have been used for decades and continue to be developed, it is prudent to be skeptical of their predictive power and realism. If a model is to be used, it should be selected based on the information available about the lake or reservoir and the questions to be answered. The most complex model is seldom necessary. Therefore, and although models never replace observations, they can be very useful to guide in the definition of strategies to design monitoring programs and contribute to increasing efficiency of fieldwork.

Adequate management of water resources and, in particular, aspects related to water quality should not exclusively depend on modeling. In fact, due to the complexity of the problem, and although the models constitute an important tool, management should always result from a global, weighted, and multidisciplinary analysis of several aspects.

A new predictive technique for remediation of aquatic environment, which comes from the field of information technology, was recently described. This technique, known as the 'knowledge-based' (K-B) approach, faces the problem from a different perspective to mathematical modeling. Prediction by mathematical modeling is a common choice in countries that have a rich, reliable database, the scientific capacity for the modeling, and experienced management. These are usually not available in developing countries. On the other hand, the 'knowledge-based' prediction focuses on the use of local and domain knowledge. As the use of mathematical models in developing countries usually requires a foreign expert, the use of the K-B approach builds local expertise in predictive techniques. Ongley and Booty (1999) recently discussed details and advantages of the K-B technique.

An overview of the types of models more commonly used for the study of environmental problems in reservoirs is presented below.

6.03.2.7 Lake Stability

6.03.2.7.1 The Wedderburn and lake numbers

The simplest model of a stratified lake comprises a warm surface layer (epilimnion) overlying a cooler bottom layer (hypolimnion), separated by a sharp thermocline. In this model, wind blowing over the lake moves the surface water, tilting the thermocline. The response of the lake is determined by the relative strength of the restoring baroclinic force, due to the density difference between the two layers, and the overturning force of the wind. This ratio is the Wedderburn number [15]:

$$W = \frac{g'h^2}{u_*^2 L}$$

where $g' = \Delta\rho/\rho g$ (with $\Delta\rho$ being the density difference between the layers), h the depth to the thermocline, L the length of the lake (in the direction of the wind), and u_* is the shear velocity induced by the wind.

According to this model, if $W < 1$, the baroclinic restoring force is insufficient to prevent the thermocline tilting so far that the hypolimnetic water upwells to the surface at the windward end of the lake, accompanied by significant mixing.

In many lakes, however, this two-layer model is too simple and a much thicker gradient layer, the metalimnion, separates the epilimnion and the hypolimnion. In these lakes, some upwelling of metalimnetic occurs even when $W > 1$.

For a continuous stratification, a more useful measure of stability is the lake number [16]:

$$L_N = \frac{gS_t(1-h/D)}{\rho_0 u_*^2 A^{3/2}(1-z_g/D)}$$

where A is the surface area of the lake, h the depth to the center of the thermocline, D the depth of the lake, z_g the height of the center of volume of the lake, and S_t the stability of the lake, given by

$$S_t = \int_0^D (z-z_g)A(z)\rho(z)\mathrm{d}z$$

For large lake numbers ($L_N \gg 1$), the stratification is so strong that the lake is very stable and there is no upwelling and little mixing. When the lake number is very small ($L_N < 1$), cold hypolimnetic water will upwell and will be accompanied by significant mixing. There is an intermediate regime in which $L_N > 1$ but $W < 1$ and the wind will bring the metalimnetic water to the surface, but not the deeper hypolimnetic water.

The lake number generally follows a seasonal trend reflecting the stratification and wind conditions, increasing to a maximum in late summer (in temperate lakes) when the stratification is most stable. The lake number has been used as an indicator of mixing and vertical transport in lakes and reservoirs and as a predictor of water quality parameters such as DO, nutrient, and metal concentrations. The lake number is typically calculated using profiles of temperature and is well suited to automated calculation from thermistor chains or CTD profiles.

6.03.2.7.2 Monitoring and control

6.03.2.7.2(i) Thermistor chains

Since the thermal stratification of a reservoir is central to vertical fluxes, and hence to the biological and chemical processes that determine water quality, it is surprising that the evolution of the temperature profile is often overlooked in regular monitoring programs. Many reservoir operators include temperature profiles in their monitoring program, but this is often restricted to quarterly measurements to coincide with other water quality parameters. The usual technique during such sampling exercises is to drop an instrument through the water column, continuously measuring temperature and depth (and often conductivity) at a spatial resolution of the order of 1 cm. The relatively high cost of collecting and analyzing water samples for chemical composition usually ensures that any monitoring is restricted to the absolute minimum necessary.

An alternative to obtaining temperature profiles using a single thermistor on a probe is to employ an array of thermistors permanently fixed at depths in the reservoir – a thermistor chain. A single thermistor chain might include thermistors at a vertical spacing of 1–2 m near the surface and at a greater spacing at depth. The thermistor chain is fixed to a mooring that allows for the anticipated changes in water level. Where large operating ranges are expected, systems of weights and floats are necessary to ensure that the thermistor chain remains approximately vertical. Each thermistor measures the temperature at periods of typically several minutes, although some applications allow sampling periods of as little as 10 s. The individual thermistors are connected to a data logger that either stores the data locally on the chain for manual retrieval or relays them to a shore station via telemetry.

A permanent thermistor chain allows a reservoir manager to measure a wide range of physical processes, from the seasonal stratification to internal waves. In this way, it is possible to understand important issues that affect water quality such as how the seasonal thermocline evolves, when autumn turnover is likely, and the amplitude of large-scale internal waves. When a thermistor chain is linked to a shore station, by telemetry, the temperature data can be made available in real time. This aids reservoir managers in deciding operating strategies such as the choice of offtake or the use of an artificial destratifier.

In addition to the advantages of greater temporal resolution, thermistor chains can provide a cost-effective monitoring program where the cost of manual profiling is high, for example, in remote locations.

6.03.2.7.2(ii) Weather stations

We have described how the dynamics of a reservoir are determined by the balance between the stabilizing effects of thermal stratification, caused by solar radiation, and the destabilizing effects of wind and cooling. The measurement of thermal stratification, ideally using a thermistor chain, provides only part of the story; it describes the net effect of meteorological forcing on the thermal stratification but provides no record of the forcing itself. The major meteorological data of relevance to water quality in reservoirs are air temperature, wind speed, solar radiation, and humidity. All of these contribute to the thermodynamics of the surface layer and the wind speed also contributes energy and momentum for driving internal waves and mixing.

In many locations, high-quality meteorological data are collected at a nearby station by the relevant government agency. However, in some instances, it is desirable to measure at the reservoir site, preferably on the lake itself. Since wind plays such an important role in mixing, and it is often local, it is becoming more common to include at least a wind anemometer at the site, often on a thermistor chain mooring. The combination of temperature data from a thermistor chain and wind data from an anemometer allows the reservoir operator to calculate the lake number, from which reservoir dynamics, mixing, and even water quality can be inferred. The collection of more complete meteorological data is usually reserved to those sites where numerical models are used.

6.03.2.7.2(iii) Lake number correlation model

This is a simple computer model that uses temperature data measured by the thermistor chain in a lake and wind speed over the lake. This allows the lake number to be computed. From the lake number and correlations with historical records of biological and chemical variables, it is possible to predict oxygen, manganese, and iron levels and, most recently, phytoplankton biomass. The correlation model is based on the premise that if the lake stability is weak, then the geochemical variables vary only due to mixing, and if the stability of the lake is strong, then the variation in the variables is predominantly due to changes in the rate of biogeochemical fluxes. This simple correlation model has been applied to a number of lakes and yields excellent results. This technique could be extended to potentially cover the transport of other chemicals and microorganisms by inflows. The model has to be calibrated for each reservoir.

The main objective of such techniques is to provide continuous, rapid, simple indicator measurements, which are then used to control the operations of the lake such as water level and offtake level and possibly alert the operators of the treatment plant downstream of changes in water quality.

6.03.2.7.2(iv) Inflow characteristics

In some circumstances, it is important to be able to predict the depth at which an inflow will insert in a reservoir; an inflow may be of poor water quality or even contaminated by an event in the catchment. As we have described earlier, the depth at which an inflow inserts depends on the stratification in the reservoir and the temperature of the inflow. The stratification can be measured by a temperature profile, or preferably a thermistor chain, but this information must be combined with the temperature of the inflow. Although it is possible to infer inflow temperatures from air temperatures during rainfall events or to directly measure the stream water temperature at the time of interest, it is now possible to install small self-logging thermistors to continuously record the inflow water temperature. This is particularly important if numerical models are being used to predict the dynamics of the reservoir.

6.03.2.7.2(v) Autosamplers and autoanalyzers

Recent advances in instrument development have resulted in robust automatic sampling and analysis systems that are able to provide a continuous record of chemical and nutrient concentrations at selected locations. Although still relatively expensive, this technology has a place in the management of critical drinking water resources and in the collection of high-quality data for the calibration and validation of numerical models. Recent advances in sensor technology also allow measurement of various additional water quality parameters, including light at depth, fluorescence, DO, and pH.

6.03.2.7.3 Real-time data acquisition, modeling, and control

Recent advances in our understanding of reservoir dynamics, in instrumentation, and in techniques to control reservoir dynamics provide us with all the elements necessary to develop an integrated real-time data acquisition and control system for the management of reservoirs. As the value of some water resources increases and the threat to the quality of those water resources also increases, the need for such a system may not be as far in the future as we might think.

Real-time data acquisition and display systems are already widely available. Useful data would include reservoir stratification and water quality, meteorological forcing, and inflows. The data would be automatically transferred to a database that could be accessed through a computer network. The data acquisition system could also include some simple instrument checks and alarms to notify operators and managers of sensor failure. This would allow timely maintenance, repair of instruments, and minimization of gaps in an otherwise valuable data set.

The next step would be to link the data acquisition to hydrodynamic and water quality models of the reservoir. Access to real-time continuous temperature, inflow, and meteorological data provides the opportunity to continuously check the validity of the hydrodynamic and water quality models and to adjust calibration coefficients if necessary. Such checks could be automated at a regular (weekly) interval to ensure that the models remain well calibrated. The same real-time data can also be used to initialize the models, allowing predictive simulations to commence from real time. The models would use a historical database of inflows and meteorological forcing to step forward from any given initial condition. This would allow a reservoir manager to predict future temperature structure and water quality in a reservoir following a particular event and to investigate a range of operating strategies.

Finally, telemetry would allow the reservoir manager to implement the operating strategy recommended by the modeling, activating control measures required, such as a bubble plume destratifier, or changing the selective withdrawal depth.

6.03.2.8 Water Quality Models

6.03.2.8.1 One-dimensional temperature models

These models simulate the energy balance in a reservoir, forecasting vertical temperature distribution and variation, considering the reservoir as a one-dimensional system.

When stratification is strong and the reservoir is deep with a relatively reduced surface, results are, in general, satisfactory. On the contrary, when stratification is weak or even inexistent, or when the reservoir is long and narrow, the hypothesis of horizontal homogeneity is sometimes far away from reality. For such cases, two- and three-dimensional representations are required.

Simulation of vertical temperature variations has been achieved by the use of the one-dimensional advection–diffusion equation and the energy conservation equation.

There are basically two mathematical models with very similar structures that although developed in the 1960s or early 1970s are still widely used: one was developed by the company Water Resources Engineers, Inc. (WRE) [17], and the other by the Parsons Laboratory, Massachusetts Institute of Technology (MIT) [18]. Both use the heat budget procedures that were developed by the Engineering Laboratory of the Tennessee Valley Authority [19], and both are well documented.

6.03.2.8.2 One-dimensional water quality models

Once the annual thermal cycle in stratified reservoirs could be represented, the next step in modeling was to extend the one-dimensional temperature models toward the characterization of the corresponding water quality cycles. This was first achieved using the one-dimensional advection–diffusion equation and the same conceptual structure as used for temperature models, adding terms for other processes related to water quality.

According to Orlob [7], a possible criterion about the applicability of a one-dimensional representation is based on the calculation of the reservoir densimetric Froude number. This dimensionless parameter compares the inertia forces, represented by an average flow velocity, with the forces that tend to maintain the densimetric stability.

Modeling water quality parameters in reservoirs was a logic sequence of temperature simulation, and some authors carried it out during the 1970s (e.g., WRE and MIT). An ecological water quality model was developed by Chen for the US EPA [20] and later became the LAKECO model. This model includes 22 different biotic and abiotic state variables.

The MIT group made an extension of the temperature model to include the simulation of DO and biochemical oxygen demand (BOD), and demonstrated its application at the Fontana reservoir, in the water resources system of the Tennessee Valley Authority [21].

Baca and Arnett [22] introduced an improvement in the solution technique for the one-dimensional water quality models, incorporating the finite element method. The resulting model avoids problems of numerical diffusion, instability, and adaptation to high gradients.

The one-dimensional dynamic model DYRESM, developed by Imberger (1981), appeared in the 1980s and was successfully applied for temperature and salinity forecast in lakes and reservoirs of small and average size. More recently, the Environmental Laboratory in Vicksburg developed the CE-QUAL-R1 model, which describes the vertical distribution of temperature and chemical and biological substances of a reservoir along the time. More recent examples of the application of these models, using improved versions, were reported by, for example, Hamilton and Schladow (1997), Han *et al.* [23], and Gal *et al.* [24].

Currently, other one-dimensional ecological water quality models for stratified reservoirs are in use all over the world. One model representative of these was originally developed by WRE [20], and it formed the basis of the WQRRS (Water Quality for River-Reservoir Systems) model (HEC (Hydrologic Engineering Center) family of models). This model describes the vertical distribution of thermal energy and of concentrations of substances, and is meant to be used as a planning tool to study water quality before and after the construction of a certain dam, as well as for the evaluation of the effects of reservoir operation. The model also incorporates the water quality issues associated with eutrophication and anaerobic conditions.

6.03.2.8.3 Multilayer models

The final goal of the dynamic modeling of reservoirs or large lakes is to allow an adequate description for the simulation of ecological and water quality balances of a limnological system. Due to the stratification introduced by the effects of temperature, salinity, or suspended solids, one- and two-dimensional models are sufficient in only special cases.

A large number of reservoirs and lakes that are relatively small and with a clear thermal stratification may be well modeled in one dimension. However, when the reservoir is long and narrow, or when stratification is strongly affected by the momentum transferred by large inflows, the one-dimensional approach is not satisfactory anymore.

In shallow lakes and reservoirs, this problem is not so relevant once vertical homogeneity is ensured. In this case, the two-dimensional (in the horizontal) circulation models may be adequate to describe the current fields and the mass transport. That is the case of models by Leendertse [25] and Masch *et al.* [26], originally developed to simulate the circulation in shallow water systems like estuaries.

However, even the shallow systems may require a vertical resolution, for instance, to deal with the biological cycles that are related to the solar radiation at the air–water interface, as well as with the benthic processes at the lower part of the water column. Stratification due to water density requires a greater accuracy on the mathematical representation of the hydrodynamic phenomena of the reservoir.

Models where hydrodynamics of the thermally stratified flow are the main issue are generally included in the multilayer models [27, 28]. In these models, the thickness of the layers may be constant or variable and the number of layers may also differ. The multilayer models developed by Simons [28] are well representative of such models, and were applied with reasonable success to some of the Great Lakes in North America and to Lake Vanern in Sweden [29, 30].

The two-dimensional circulation models in stratified flows used to simulate the behavior of long-shaped and narrow reservoirs are well represented by the finite element model RMA2, developed by King *et al.* [31], and by the finite difference model of Edinger and Buchak [32]. The RMA2 model has been improved along the years and is currently available in a commercial version.

6.03.2.8.4 Two- and three-dimensional water quality models

The modeling of transport and conservative substances in shallow lakes is represented by the model of Lam and Simons [33], which was applied to Lake Erie. The problem of nonconservative substances, including nutrients and phytoplankton, was solved in the Green Bay model by Patterson *et al.* [34] and in the phytoplankton productivity model of Lake Erie developed by DiToro *et al.* [35].

In each of these examples, models were run from a known field of currents obtained by field measurements or a circulation model. The Lam and Simons model treated the lake system as having a vertical mixture (one layer) or having stratification (two layers), while the other models assumed vertical homogeneity.

The eutrophication phenomena have been modeled using the principle of general nutrient balance, which was first presented by Vollenweider [12] and later by Snodgrass and O'Melia [36]. The models of Thomann *et al.* [37] and Chen *et al.* [38], for Lake Ontario, are examples of a more comprehensive two- and three-dimensional representation of the time-varying interactions between nutrients and biota in lake systems.

Three different types of models were identified by Thomann *et al.*, ranging from a simple three-layer model (epilimnion, hypolimnion, and benthos) to a seven-layer model with 67 segments and up to 15 variables.

CLEANER, an ecological model for lakes, which includes up to 34 state variables, reduces the water body to a 1 m^2 water column that may be divided up to 10 cells to allow vertical resolution. This model was used in a variety of situations and in countries like the United States, Scotland, and Scandinavia; it was also applied to Lake Balaton in Hungary and also to other lakes in the Czech Republic and Italy.

The state of the art of ecological or water quality modeling is probably well represented, even today, by the phytoplankton productivity models of DiToro *et al.* [35], Thomann *et al.* [37], and Chen *et al.* [38]. The abovementioned model CLEANER adds to those models the biological characterization in reservoirs.

Edinger and Buchak [32] developed the model LARM2 for the simulation of hydrodynamics and transport of pollutants in reservoirs. This model is laterally averaged, being two-dimensional in the $X–Y$ plane (longitudinal–vertical), and has the possibility of adding or eliminating longitudinal segments during the rising or the falling of the reservoir water level. More recently, Cole and Buchak [39] developed the CE-QUAL-W2 model, an extension of LARM2, which includes the possibility of simulating reservoirs with several branches.

Several versions of RMA models have been used for different water quality studies (e.g., Reference 31). For example, the abovementioned RMA2 model was used as the hydrodynamic basis for a water quality model, and the RMA4 model has been used successfully both in estuarine systems and in reservoirs. RMA7 was developed to simulate water quality variables in a two-dimensional, laterally averaged reservoir.

6.03.2.8.5 Eutrophication models

Nutrient enrichment in lakes and reservoirs has been a growing concern among the pollution control experts, biologists, and environmentalists in general. However, it has been only recently that, in some countries, a combined effort has been made to quantify the effects and to establish alternative control strategies in terms of nutrient balances in reservoirs.

The previously mentioned model of Vollenweider [12] for the phosphorus is among the first eutrophication models of nutrient mass balances to have a wide application. Taking the balance of phosphorus as the sum of external, effluent and sedimentation sources, as well as the reservoir residence time, Vollenweider proposed a relatively simple equation to evaluate the time evolution of the phosphorus concentration.

Jørgensen [40] made an analysis of various approaches to the eutrophication problem and concluded that it is crucial that eutrophication models include at least three trophic levels: phytoplankton, zooplankton, and fish. They should also allow the nutrient exchange between sediments and water. Such models may give a more accurate description of the system response to the seasonal variations of nutrient inputs.

A comprehensive review of the available models of this type is presented by Reckhow and Chapra [41].

6.03.2.8.6 Special models

There are a relatively large number of special models for reservoirs, or models used for particular purposes. Among them are those that simulate water quality in a system of reservoirs, or models that simulate water quality in a reservoir with back-pumping in a pumped storage reservoir system. The latter may be used to define pumping rules that can be used to improve reservoir water quality [42]. One model of this type was used for the recently built Alqueva reservoir in Portugal [43].

Some other special models refer to sedimentation, or even to the forecast of the consequences of landslides into the reservoir. HEC-6, Scour and Deposition in Rivers and Reservoirs, developed by the US Army Corps of Engineers [44], is one of the models that allow the simulation of sediment deposition in reservoirs. This model was developed to forecast the long-term hydromorphological behavior of fluvial systems and is not adequate to evaluate short-term responses due to certain events, like floods.

Turbidity currents are the main mechanism of fine sediment transport in suspension in reservoirs. Computational models have been developed for the simulation of turbidity currents in reservoirs. These models adopt numerical schemes of Godunov type, based on HLL and HLLC Riemann solvers and on the second-order TVD version of the WAF scheme [1].

Tests performed in the laboratory (**Figure 18**) confirm the adequacy of the computational models.

Figure 18 Evolution of turbidity currents in reservoirs. Physical model tests [1].

6.03.2.9 Final Remarks

It seems to be evident that the state of the art of water quality mathematical modeling in lakes and reservoirs is well represented by certain simulation models, which range from the one-dimensional temperature and water quality models in stratified systems to the multidimensional, wind-driven, and ecological models, for large water bodies, as well as those used to simulate turbidity currents.

Water quality models may be extremely useful during the planning phase of a dam, anticipating some measures that will contribute to minimize negative impacts or even to promote some potential benefits both in the reservoir and in the downstream river reach. These models may also play an important role as part of the management tools used for the reservoir operation, including the potential to forecast the water quality response to different pressure scenarios or remediation works.

In general, model calibration and validation is not a precise science. For this reason, it is crucial that the most appropriate data are collected in order to produce a good water quality forecast.

6.03.3 Management of the Impact of Hydraulic Processes in Hydropower Operation

6.03.3.1 Introduction

6.03.3.1.1 Gas supersaturation

Water in streams and rivers can become supersaturated with the gases that make up the atmosphere because of both natural and man-made actions that cause air to be entrained in the flow at great depths. The solubility of gas increases with pressure, and thus when air is introduced at depths of several meters below the water surface, a higher amount of total gases will be dissolved than at atmospheric pressure. This condition, when the amount of dissolved gas exceeds the maximum amount of dissolved gas at atmospheric pressure, is called 'total gas supersaturation'.

While the condition of supersaturation requires that air be introduced into water at an elevated pressure, the condition is not easily reversed when the supersaturated water mass moves to shallower depths or near-atmospheric conditions. Water in a river may remain supersaturated for many kilometers downstream of the location where the supersaturation condition is generated.

Natural conditions that can result in gas supersaturation include deeply plunging waterfalls. Masses of air drawn into the plunge pool by the plunging jet of water contribute to the supersaturated condition. Hydraulic structures can also contribute to gas supersaturation. Deeply plunging spillway discharges and deep stilling basins operating with submerged hydraulic jumps have been known to cause supersaturation that contributed to fish mortality.

Fish that are exposed to supersaturated water accumulate the dissolved gases in their bloodstream in their natural respiration process. Symptoms of gas bubble disease occur when the fish swim at shallower depths where the gases expand in their circulatory system and cause ruptures that are often fatal. The amount of supersaturation that can be tolerated by fish depends on the species and the environment to which they are exposed. Fish can adapt to some level of supersaturation by sounding down to deeper water if a deep channel exists in the river. If, however, the downstream river channel is relatively shallow, little supersaturation can be tolerated. Case studies and a discussion of the issues are given in Section 6.03.3.2.1.

6.03.3.1.2 Fish passage

The conservation of fish life in rivers is a concern which is reflected in the policy of sustainable development adopted by ICOLD. The ICOLD's *Position Paper on Dams and the Environment* specifies that

> ... more and more we also recognize an urgent need to protect and conserve our natural environment as the endangered basis of all life.

The conservation of fish life in rivers and lakes can only be considered from a holistic point of view, taking into account the entire life cycle of the species concerned: reproduction, feeding, movement, and migration; with the mitigation, one has to consider the whole river system from the source to the sea. It is thus preferable to adopt a comprehensive approach to the aquatic environment and develop a piscicultural plan for the entire river in question.

The impact of dams on the environment, and in particular on fish life, has been discussed in some detail at several ICOLD congresses and in various bulletins of ICOLD, particularly Bulletin 116, Dams and Fishes. Bulletin 116 provides detailed information on fish life and draws up an outline of the existing knowledge and experience of dam constructors and operators. However, given the diversity and range of environmental, climatic, and hydrological conditions, it was not intended to be an exhaustive treatise on the subject. The bulletin gives general information to engineers and dam owners, providing an understanding of the studies and investigations to be developed before the implementation of a successful fish management program. Three areas are considered:

- *the reservoir* – the conditions necessary for fish life, including food and reproduction;
- *the dam* – fish passage techniques; and
- *the river downstream of the dam* – flow conditions required to maintain fish life.

Some techniques described in the bulletin are relatively recent and the described approaches should be regarded as a first step, to be further refined as research into particular applications.

The maintenance and development of fish life is one of the important aspects to be considered in dam projects. The bulletin notes that the dam design must be developed with the following objectives:

- to conserve the diversity of living species,
- to enable riverside communities to fish for food, and
- to provide for the development of water-based recreational activities.

The bulletin goes on to note that these objectives must be adapted to the size of the dam, the particular environmental and social situations that exist, and the regulations in force in the country concerned.

6.03.3.1.3 Unsteady flow

Hydropower projects can generate energy that is more valuable by turbining during periods of peak daily power demands and storing water during off-peak periods. This peaking mode of operation is possible at retrofit projects built at dams with at least minimal storage capacity. The daily flow cycles that result can have adverse impacts downstream, such as stranding fish (including spawning nests and juvenile fish), posing hazards to recreational users, and increasing bank erosion. These impacts can be mitigated by (1) not allowing daily flow cycles (a common requirement) or (2) building some kind of reregulation structure (such as another small reservoir or a low-head weir) downstream to even out daily flow cycles. Since such fluctuating flows can have adverse impacts and can conflict with the original uses of a dam, they are often not allowed. Frequent switching of turbines on and off has been determined to be the reason of intensive and long-lasting riverbed scour and also the cause of significant reduction of fish communities. Each switching on and off of the turbines is found to cause a sudden change of water discharge and level in the downstream reach. Water level suddenly drops down after the turbine switches off. Uplift force of groundwater flowing from a riverbed destructs a reinforcing layer of large ground particles formed during the self-lining process. Scour of small particles from the bottom sets in. The riverbed deepens significantly until a new reinforcing layer forms. Suggestions are to slow down turbine switching within technical possibilities. This simple measure allows increasing the length of a reflux wave, to reduce the speed of water level drop and the length of river reach under the scour danger.

6.03.3.1.4 Sediment transportation

Conversion of sedimenting reservoirs into sustainable resources requires fundamental changes in design and operation. It requires that the concept of a reservoir life limited by sedimentation be replaced by a concept of sustainable management. The suitable reservoir sedimentation management is a key factor for the sustainability of water resources. Five basic strategies can be considered in sediment management:

1. *Sediment yield reduction.* Apply erosion control techniques to reduce sediment yield from tributary watersheds. These techniques will typically focus primarily on soil stabilization and revegetation.
2. *Sediment storage.* Provide sediment storage volume adequate for the anticipated sediment yield over a 'long' period of time either in the reservoir itself or in upstream impoundments or debris basins.
3. *Sediment routing.* Pass sediments around or through the storage pool to minimize sediment trapping by employing techniques such as offstream storage, temporary reservoir drawdown for sediment pass-through, and release of turbid density currents.
4. *Sediment removal.* Remove deposited sediment by dredging or hydraulic flushing.
5. *Sediment focusing.* Here the techniques are designed to tactically rearrange sediments within the impoundment to solve localized problems such as impacts from delta deposition. Any washout of sediment from the reservoir that may occur is incidental to the primary objective.

In reviewing options, a full range of management alternatives should be analyzed. Harrison *et al.* (2000) for Lake Solano, California, describe an example of this approach. Optimal management may include two or more strategies applied simultaneously or at different points in the reservoir life. The applicability of different strategies varies at different stages of reservoir life, being a function of the reservoir's hydrologic size (capacity:inflow ratio), beneficial uses, and other factors such as environmental regulations. Techniques such as sediment routing require significant pool drawdown and use part of the natural inflow to transport sediment beyond the storage pool, making it impossible to capture and regulate 100% of the flow. Consequently, some types of routing

Table 5 Current hydropower 80% depletion date

Region	Hydropower dams: Date by which 80% is filled with sediment
Africa	2100
Australasia	2035
Europe and Russia	2070
North America	2060
Asia	2080
Central America	2060
Middle East	2060
South America	2080

techniques will not be feasible at hydrologically large reservoirs. However, sedimentation will eventually convert large reservoirs into small ones, and sediment routing techniques may become feasible at a future date. The raising of a dam could be an alternative for compensating for the loss of the reservoir, especially in arid regions; however, it does not provide a long-term solution to the sedimentation problem.

The impact of sedimentation is by no means the same for hydropower as for other dam functions. For hydropower, corresponding to more than 80% of the total storage, part of the sedimentation is in the dead storage, with little or no impact, and part affects the live storage, where a reduction of 50% means a much lower reduction in power production. A reduction of storage of 0.3% per year means a reduction of power of much less than 0.1% of production, that is, less than 10% in a century.

Hydropower dams can generally be filled to a higher level than non-hydropower dams, as it is mainly necessary to maintain the head for power generation and to provide a storage capacity sufficient to meet all expected demands for power. It is expected that hydropower dams will be severely impacted when they reach a level of sedimentation of 80%. Global data on hydropower dam depletion can be seen in **Table 5**.

6.03.3.1.5 Reservoir operating strategies

This section outlines a range of measures that can be used to manage water quality in reservoirs. These range from various means of implementing artificial destratification and mixing in reservoirs to structural methods of managing water quality such as selective withdrawal through various forms of offtake works. The section proceeds further to discuss the options for monitoring and representation of real-time data with the prospects for further integration of the data gathering and modeling to enable real-time forecasts of reservoir behavior and its response to management strategies.

The reservoir operating strategy of the Three Gorges is as follows: the 39 km^3 Three Gorges reservoir on the Yangtze River has been designed to achieve sediment balance across the impounded reach after approximately 100 years, allowing the project to operate indefinitely while passing 530×10^6 t yr^{-1} of sediment and 451 km^3 of water. This is achieved by designing a hydrologically small reservoir (C:I 0.087) with adequate low-level outlet capacity to operate in seasonal drawdown mode. A conventional impounding reservoir of the same capacity on the Yangtze River would have a half-life of less than 100 years. The reservoir is gradually drawn down during the dry season by making releases for hydropower and downstream navigation. Outlets and turbines are operated during the initial part of the flood season to maintain the reservoir pool at a low level. This empties the flood storage pool generating high flow velocities along the reservoir, which is generally not more than 1 km wide along its 600 km length. These high velocities will transport most suspended sediment and sandy bed material through the reservoir and beyond the dam. Once equilibrium conditions have been reached, gravels will continue to be trapped and must be removed by dredging. About 2×10^6 m^3 of sand and silt is also expected to be dredged annually near the navigational locks at the dam.

6.03.3.2 Reduction of Gas-Supersaturated Water

6.03.3.2.1 Case histories

6.03.3.2.1(i) Western United States

Spillway operation at several dams on the Pacific Northwest contributed to gas bubble disease-related fish mortality in the 1970s. The majority of problems are related to the operation of spillways with hydraulic jump energy dissipators. In these cases, the stilling basin elevations were selected to provide sufficient tail water depth to contain the hydraulic jump during the maximum design flood. Since the conjugate depth curve is typically steeper than the tail water curve, the hydraulic jump is drowned for all spillway discharges lower than the design flood. Higher saturation rates generally occur when the ratio of tail water depth to the conjugate depth increases.

Measurements in the Columbia and Snake rivers, starting in 1968, showed supersaturation values over 130%. The highest concentrations occurred during high flow years when the greatest volumes of water passed over the spillways. Large fluctuations in spillway operations, which caused significant decreases in depth following supersaturation conditions, caused fish to lose the ability to sound down to depths that would protect them, thus increasing fish mortality.

Laboratory experiments were conducted to determine the tolerance of both adult and juvenile salmonids to the level of supersaturation. Studies showed that when both adults and juveniles were confined to shallow water (1 m or less), substantial mortality occurred at 115% saturation of total dissolved gas. When salmonids were allowed to sound to deeper water to obtain higher hydrostatic compensation, significant mortality did not occur until saturation reached 120%.

6.03.3.2.1(ii) Australia – Lower Pieman Power Scheme
The Lower Pieman Power Scheme near the west coast of Tasmania, Australia, was built in the period 1973–86. Two short power tunnels convey water from the storage to two 119 MW turbines in the Reece power station. Below the power station, the Pieman River flows a further 30 km to the sea.

In 1989, a number of dead trout were reported at Corinna, 15 km downstream of the power station. Immediate investigations by the Inland Fisheries Commission determined that the trout died from gas bubble disease, while native species were largely unaffected. The source of the gas bubbles was traced to supersaturated water emerging from the power station tailrace.

Each turbine in the station is fitted with a system to admit or inject air below the runner, to combat rough running of the turbine during start-up and at certain power outputs. It was the use of these systems that caused supersaturation of the water and the demise of the fish. The degree of supersaturation was increased by the aeration of the water from the reservoir. The unusually low level of the reservoir (drawn down for maintenance reasons) and the passage of water through an accumulation of logs and debris at the tunnel intakes caused this aeration.

As a temporary measure to protect the fish, the air admission and injection systems were disconnected. This action, while necessary, had an adverse impact on station operation. The station is very suitable for frequency control of the Tasmanian electricity grid, as the station has a relatively large capacity and its response time to changes in demand is rapid. Operation of the station in frequency control mode involves a fluctuating station output with a high probability of extended periods requiring either air injection or air admission. If its role in frequency control were continued without air, the machines would be subjected to severe vibration and additional wear and tear.

Studies were commissioned to determine both the conditions under which supersaturation of the water occurs and the tolerance of the fish to various levels of supersaturation. These studies are outlined below.

Gas saturation investigations were conducted on-site. The power station was run under a variety of conditions, and gas saturation levels were measured in the tailrace with a tensionometer (to measure total gas saturation) and a DO meter. The readings confirmed that supersaturation occurred only when the air admission/injection facility was used. Passive air admission increased total gas saturation to about 110%, and forced air injection increased total gas saturation to about 120%. The water was generally less saturated with DO than dissolved nitrogen, probably indicating that the DO concentration in the water leaving the storage was low.

Fish exposure tests on-site were primarily designed to establish whether relatively short-term exposures to the range of supersaturated conditions produced by the operation of the power station were harmful to fish. Rainbow trout were held in a flow-through tank fed with water pumped from the tailrace. Fish held in an identical tank fed with water from the intake side of the power station were used as a control. Measurements of total gas saturation, DO, and temperature were made in both tanks.

The autostart procedure takes about 17 min to bring the machine up to efficient load, during which air is admitted passively or forcibly into the draft tube for about 12 min. As the tensionometer is slow to react to changes in saturation, it is estimated that the actual level of saturation may have reached 120%, but no mortalities or signs of stress were noted in the fish.

- Two hours of operation in the range requiring air admission, producing about 110% saturation, caused no mortalities or signs of stress.
- Two hours of operation in the range requiring air injection, producing about 120% saturation, also did not result in any mortalities or signs of stress.
- Six hours of frequency control mode, during which the machine was predominantly in the range requiring air injection, produced no mortalities. However, there were some signs that the fish were becoming stressed, for example, loss of balance and erratic swimming behavior.

Fish exposure tests were conducted at the Salmon Ponds hatchery. Here again, two identical tanks were set up, an experimental one in which the degree of supersaturation of the inflow could be varied and a control tank fed from the same water source.

- When the trout were exposed to a saturation level of 120%, mortalities began after about 6 h and 91% of the fish had died after 24 h.
- When the fish were exposed to a saturation level of 115% for 48 h, mortalities began after 30 h and the final mortalities were 12% and 24% in the two tests.
- Finally, the fish were subjected to a cycle of 6 h at 120% saturation, followed by 6 h at 100% saturation, repeated 4 times over a 48 h period. None of the fish died from gas bubble disease.

It was concluded that the fish would be unaffected if

- the saturation level was generally below 110% and
- any period of 120% saturation was limited to 6 h and was followed by a period of exposure to 100% saturation in which to equilibrate.

These conclusions were also consistent with overseas experience that 110% is a tolerable saturation level in natural streams and lakes, where depth compensation for the effects of supersaturation is normally possible for fish. The results of the tests provided considerable scope for relaxing the restrictions imposed after the original fish kill.

Under normal conditions, passive air admission operates between about 15 and 75 MW output for each turbine. Within this range, air injection is required between ~40 and 70 MW. With two machines in the power station, it was realized that the only times during which the saturation level is likely to exceed 110% are when both machines are operating on frequency control, or when one machine is on frequency control and the other is shut down. Therefore, provided that the station load exceeds the output of one machine at its most efficient load (>75 MW), the other machine may be operated indefinitely on frequency control, furnishing up to 100 MW of output to meet fluctuating demand.

The power station is again able to operate as an efficient frequency control station. No further fish kills have occurred, and the possibility of a recurrence due to gas bubble disease is considered unlikely.

6.03.3.2.1(iii)   Australia – King River Power Development

The King River Power Development was constructed in the period 1983–93 near the west coast of Tasmania, Australia. The scheme comprises a large storage covering 54 km^2, a 7 km long headrace tunnel, and a single-machine power station with an installed capacity of 143 MW. The tunnel intake is at a relatively low level in the reservoir because the tunnel was excavated at a rising grade from the upstream end for economic and environmental reasons.

As soon as the power station was commissioned, the foul smell of hydrogen sulfide gas from the tailrace water was immediately apparent. Although the production of hydrogen sulfide gas was a temporary phenomenon caused by rotting vegetation in the newly filled reservoir, the gas was also a health hazard. Being heavier than air, the gas could concentrate in the confined valley immediately downstream of the station. While the presence of hydrogen sulfide is initially all too apparent from the smell, that sense becomes dulled by exposure and heightened awareness is essential to avoid the risk of a fatality.

Monitoring of the water quality then found that the tailrace water was low in DO. The concern was that slugs of oxygen-depleted water would be discharged into Macquarie Harbor 20 km downstream, where fish farming is an important industry. While the farms are normally located kilometers away from the river mouth, fish pens in transit could pass through the danger zone.

Various methods of increasing the level of oxygen in the water were considered. The adopted solution made use of the existing air injection system installed on the turbine. When required, jet pumps inject air immediately below the turbine runner, to combat rough running conditions during start-up and at particular power outputs. Operation of the jet pumps was not without cost, as the pumps absorb about 3 MW of the station output.

The injection of air also helps to reduce the release of hydrogen sulfide by precipitating the sulfide as iron sulfide and the oxidation of hydrogen sulfide to sulfate, which is essentially nontoxic to the aquatic environment. Air injection is now utilized on a seasonal basis to increase DO concentrations during periods of stratification in the lake, and is one of the formal operating rules of the power station. Continuous monitoring of the water quality discharged by the turbines ensures the timely utilization of the air injection facility.

6.03.3.2.1(iv)   Aeration weir in Nam Theun 2

Due to a lack of DO and excess dissolved methane in the reservoir of the Nam Theun 2 dam (Laos), an aeration structure was implemented for transfer of water of 375 m^3 s^{-1}. The dimensions of the basin were 205 m × 50 m. The effectiveness of aeration through analysis of formation and repartition of air bubbles was tested. A model test at scale 1/20 was undertaken (see **Figures 19** and **20**). **Figure 21** gives the aeration weir in operation for half of the nominal discharge.

6.03.3.2.2   Retrofit solutions for spillways with deep stilling basins

The physical process that causes supersaturation is associated with a submerged hydraulic. The shear force at the air–water interface along the upper nappe of the flow over the spillway crest and chute combined with the reverse roller of the submerged jump causes air to be drawn to the bottom of the stilling basin where the hydrostatic pressure is high. In a free jump, air is not carried out in such large quantities to areas of elevated hydrostatic pressure. Redirecting the flow along the surface so that air is not dragged to the bottom of the stilling basin can avert saturation caused by the submerged jump condition.

The US Army Corps of Engineers designed flow deflectors for seven spillways with stilling basin energy dissipators that contributed to the supersaturation problems, using physical hydraulic model studies to determine the dimensions. The purpose of the deflectors, also called 'flip lips', is to direct flow for lower, more frequent discharges along the water surface. The deflectors are of simple step geometry with a horizontal floor and a vertical downstream face.

The location of the deflector on the spillway surface and the dimensions (length, height) are dependent on the depth of flow on the spillway at the location of the ramp and the variation of tail water level over the range of flows for which it is intended to be effective. If the deflectors are positioned too low with respect to the tail water level, the flow will penetrate too deeply in the basin and supersaturation will not be averted. If the deflectors are set too high, the flow will plunge into the basin with the same effect. If

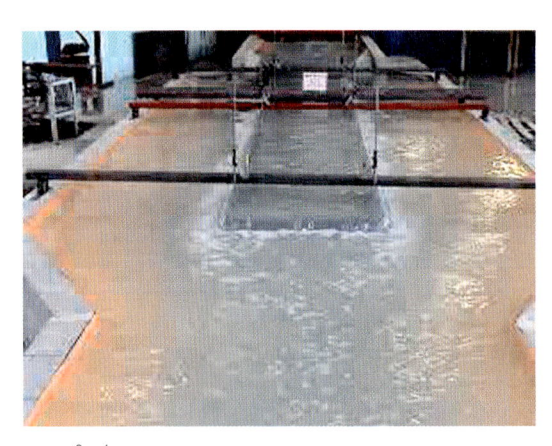

Figure 19 Plan view of the model test for 150 m³ s⁻¹ at the scale 1/20 (Hydraulic Laboratory of Constructions, University of Liège, Belgium).

Figure 20 View of the flow details on the aeration weir model test for 150 m³ s⁻¹ (Hydraulic Laboratory of Constructions, University of Liège, Belgium) and on-site implementation.

Figure 21 On-site view of the aeration in operation for 150 m³ s⁻¹ (Electricitié de France (EDF)).

the length (and height) of the deflectors is too small in comparison with the thickness of the flow on the chute, the deflectors will not effectively turn the flow. If the deflectors are too large, they will compromise the energy dissipation during the spillway design flood. The optimum dimensions are best determined by physical model studies.

Deflectors were designed for flows equivalent to the 10-year flood or less for spillways at the Bonneville, John Day, McNary, Ice Harbor, Lower Monumental, Little Goose, and Lower Granite dams. These devices were installed at all the abovementioned projects except John Day and Ice Harbor. The deflectors are installed below the water surface and proportioned using the physical model to deflect the flows in the design range along the water surface, but allow the hydraulic jump to form normally for the spillway design flood.

Similar deflectors were designed according to physical model studies of the Brazo Principal and Brazo Ana Cua spillways of the Yacyretá Hydroelectric Project in Argentina. Deflectors were installed and they perform effectively in the Brazo Ana Cua spillway. Other design considerations include lower unit discharge, divider walls, and low discharge bays with higher basin elevations. Operational considerations include the avoidance of abrupt change in spillway flow and nonuniform gate operation to provide the balance of best operation and maintenance practice with best environmental practice.

6.03.3.3 Control of Floating Debris

6.03.3.3.1 Type and origin of debris

Rivers carry not only water and sediments but also various kinds of debris, which may constitute both an operational problem and a dam safety problem. On a number of occasions, floating debris has blocked spillway openings and led to significant reductions in effective discharge capacity at the very time that capacity was needed. The possibilities and consequences of spillway blockage with floating debris therefore need to be considered. In some cases, action also needs to be taken to stop, divert, pass, or otherwise remove floating debris.

Precipitation, type of terrain, vegetation, reservoir treatment, and other human activities around reservoirs and rivers are factors governing the potential amounts of floating debris. During major floods, both the debris flux and the size of individual items of debris tend to increase, which may affect cooling water intakes, trash racks, and even large structures like spillways. The debris may be floating or transported in deeper zones. It may comprise diverse bits and pieces of vegetation, such as grass, bushes, sunken logs, or entire trees, and manufactured items, such as boats, piers, and houses. Ice runs may cause similar problems in some rivers.

However, the role of smaller debris in clogging intakes cannot be ignored. Professor Guo reported that they have experienced clogging of hydropower plant intakes with debris comprising tree branches, logs, brush or grasses, stalks, straw, and ice that floats toward the intakes and accumulates on the trash racks. The accumulation of trash can be several meters deep and can cause the collapse of trash screens in extreme cases. Significant head losses occurred, and in some cases, the trash screens were damaged. The head loss in the intakes to power stations can lead to significant losses in energy generation, resulting in substantial economic losses.

Mires are the source of another type of floating debris in some countries. On occasion, large chunks of mires may lift to form floating islands covering several hundred square meters each and with a depth of a few meters. Floating mires tend to be released either when the ice cover is melting in the spring or when the water is getting warmer in the summer, apparently lifted by expanding gas bubbles previously dissolved in the cooler water.

Floating rafts of reeds such as bulrushes and other aquatic plants and materials such as peat bog can also break loose and cause problems with the operation of hydraulic structures. There are numerous examples around the world, but only a few specific examples are referred to in this report.

Figure 22 shows floating rafts of bulrushes (*Typha domingensis*) growing out from the shoreline, which together with floating pondweed in deeper water outside combine to block access to the shoreline on Lake Kununurra in Western Australia.

6.03.3.3.1(i) Case histories

Case histories of clogged or damaged spillways come from northern countries with temperate climates, but it would be reasonable to assume that similar problems occur in other climates.

Figure 22 Floating rafts of bulrushes along shoreline.

1. Norway experienced a large flood in 1789. The flood covered a number of bigger rivers in the southeastern part of the country and is estimated to have been of a size of present-day spillway design floods (possible max flood (PMF)) for major structures. Witness accounts from the time reported that normally clean rivers were "thick as gruel and dead animals and houses, timber and trees floated in the current."

   > Norway's biggest lake, the "Mjosa, was almost entirely covered with bushes and trees and the water was so dirty that the fish died and became uneatable. In May 1790, the water had not yet cleared. Rivers and streams fell over the steep valley sides and brought mill houses and bridges along. – People thought it was Armageddon."

2. In November 1955, the Alouette Dam in British Columbia, Canada, was exposed to a flood which caused the water to rise 1.5 m above the ungated fixed weir of a concrete spillway [1]. A large tree got stuck on the weir and damaged a concrete weir panel, probably by the changed flow conditions. The resulting seepage lifted a number of panels and finally caused 25 m of the weir to fail and the underlying clay foundation was severely scoured. The failure occurred toward the end of the flood, which prevented a catastrophe. The reservoir banks are, however, steep and heavily forested. The same storm also caused the 5.2 m wide spillway openings of the nearby Jordan Dam to become clogged with floating debris from the poorly cleared reservoir. The dam was overtopped by 0.6 m, causing erosion at the base of the dam. The Jordan Dam is 40 m high.

3. In 1978, the Palagnedra Dam in Switzerland suffered a major flood, which caused an embankment dam adjoining the main concrete arch dam to fail due to overtopping after all the 13 spillway openings measuring 5 m × 3 m had clogged up with floating debris, mostly logs. The amount of debris carried during the flood was estimated at 25 000 m^3 (**Figure 23**).

4. In October 1987, part of southeastern Norway was hit again by a flood estimated to have a return period of around 100 years. The rivers carried a lot of debris. Significant blockage of spillways by floating debris occurred at six dams [45], most of which were equipped with several smaller spillway openings. At one of the dams with a number of 2 m wide openings, 20 men equipped with chainsaws, two excavators, two forest harvest machines, and five trucks could not keep the spillways clean. At another dam, floating debris collected on top of the partially open radial gate in the early part of the flood before the gate was fully opened. The debris got wedged in between the gate and the walkway on top of it and could not be cleared away with manpower and chain saws.

5. The example in **Figure 24** shows the effect of a substantial flood in the Derwent River in Australia that carried with it a large number of logs which clogged the river diversion openings during the construction of Catagunya Dam. Catagunya Dam is a concrete gravity dam on the Derwent River in the Australian state of Tasmania. When the dam was under construction in 1960, a 1-in-100 AEP (annual exceedance probability) flood occurred. At that time, the structure consisted of a series of alternate high and low blocks across the valley, with normal river flows diverted by an upstream cofferdam through four 5 m × 3 m openings in the dam. The peak flow at Catagunya greatly exceeded the diversion capacity, and the excess water passed over the low blocks of the dam itself. The upstream cofferdam and the formwork erected for the next concrete pours on the dam were damaged. Upstream of the dam, the Derwent Valley is heavily timbered, and the flood brought with it a vast assortment of trees, logs, and branches of all sizes. Tasmanian eucalypts are quite dense and many logs travel at or below the surface. Much of this material passed over the dam, but when the flood subsided, a great mass of timber had built up across the diversion openings (see **Figure 24**).

   At first sight, it was thought that the diversion capacity had been reduced to about 25%, but a check on the pond level and river flow produced the surprising result that about 70% of the design flow was still finding its way through the maze of logs. Removal of the timber, log by log, was a slow and somewhat dangerous task.

Figure 23 Debris in Palagnedra Dam, Switzerland, following record floods.

Figure 24 Catagunya Dam, Australia, April 1960: accumulation of logs at diversion openings after a major flood.

6. Reports from China [46] indicate that the Gezhouba Power Plant, which is located on the Yangtze River 43 km downstream from the Three Gorges Project, has suffered from loss of energy production due to clogging of the intake screens. The energy loss due to clogging of the intake screens in the period 1982–84 was 79.1 GWh per annum. The clogging was sufficient to stop some units from running. The clogging of intakes with debris causing head losses of up to 6.2 m was reported during the initial operations at the Yantan Hydropower Station, located on the Hongshui River in southwest China.

7. In the Australian state of New South Wales, the structural failure of the Wingecarribee Swamp peat bog in a storm event in early August 1998 resulted in almost 6000 megaliter (Ml) of peat and sedimentary material being deposited in Wingecarribee Reservoir, which previously had a storage capacity of 34 500 Ml. The peat flowed into the reservoir as floating blocks several meters thick and ranging in size from individual tussocks to clumps of several hectares. Increases in turbidity in the water body forced the cessation of raw water supply to the treatment plant. However, the floating peat posed a significant threat to the security of the dam, having the potential to block the narrow single-gated spillway. In order to contain the peat, a 1.2 km long steel mesh barrier was built across the reservoir.

6.03.3.3.2 River transport of debris

While it might be tempting to try and describe debris transport with formulae developed for sediment transport, the mechanism of initiation of motion is quite different as logs are often delivered into the stream by slides in the banks rather than direct erosion. Moreover, debris tends to be transported midstream at the water surface rather than along the bed or throughout the whole body of water. Although the surface velocity is usually slightly higher than the mean stream velocity, the debris transport velocity measured over substantial distances may be only a fraction of the mean stream velocity.

Floating uprooted trees tend to align themselves with the stream with the larger of the root wad and the canopy at the prow. However, not all trees float like that. Trees with heavy root wads have also been known to be transported standing up.

Generally, the transport of individual debris pieces is subject to a significant random element. Some debris may get temporarily stranded at a bend or other obstruction, while other debris may pass the same point. However, with some debris stuck, there is an increased probability for more debris to get stuck at the same point. When a significant amount of debris has gathered, it will obstruct the flow and eventually may become unstable so that a slug of debris is released. Where floating debris is transported over longer distances, there is accordingly a tendency for it to be transported entangled in slugs or rafts, especially when there is much debris in the river.

6.03.3.3.3 Debris transport through flow control structures

The behavior of floating material approaching flow control structures such as spillways has been the subject for hydraulic model testing on many occasions over the years. Most of the earlier studies dealt with logs floating through specially designed log outlets designed to extract only surface water and to line up the logs to avoid blockage. More recent hydraulic model studies have also been made to investigate the ability of common types of spillways to discharge floating debris.

In Scandinavia, hydraulic model tests [47, 48] have been made involving single trees, pairs of trees traveling together, and larger slugs of trees. The tests dealt with the passage of debris over both gated and ungated spillways and the possibilities of stopping slugs of trees with the help of floating boom arrangements.

The models used young plants of spruce, *Picea abies*, of around 0.3 m length, with root systems less than 0.05 m in diameter, to simulate grown-up trees of 25–30 m length, typical for Scandinavia. It was noted that the model trees were proportionally stiffer and

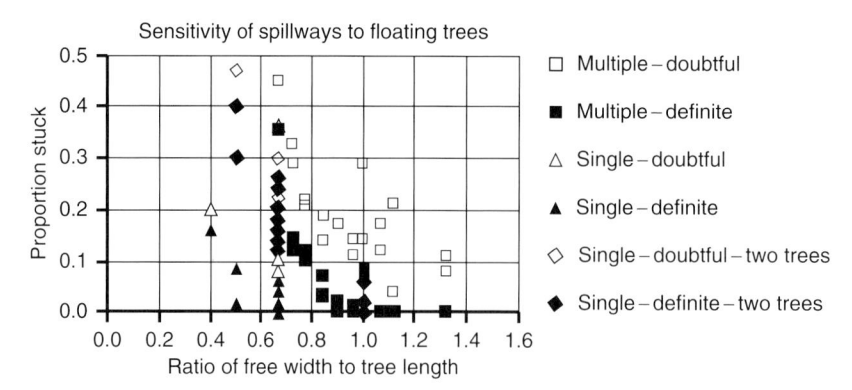

Figure 25 Model test results for floating trees at spillways.

stronger than the prototype trees, especially at the top ends, and the model trees may therefore have stuck easier at spillways. It was suggested therefore that not only the root but also the top portion of prototype trees having a trunk diameter less than 0.05 m should be disregarded to establish an effective tree length, L. **Figure 25** gives some test results for single and multiple spillway openings with a single tree or, where indicated, two trees together approaching a spillway. Trees stuck across a spillway opening from one pier to the next are denoted as 'definite'. Some trees were caught by a combination of actions involving also roots and branches caught by bridges and spillway sills; they are marked with open symbols and are denoted as 'doubtful'.

The approach flow to the spillway was found to be important in two respects. High flow velocities in the approach zone tend to increase the momentum of the trees, which reduced the risk of jamming. On the other hand, a certain acceleration of the flow velocities tends to line up the trees parallel to the flow, which also increased passage rates. As can be seen **Figure 25**, a somewhat larger free width may be required where there are multiple spillway openings next to each other so that the flow acceleration upstream is less pronounced.

The following dimensions were required to allow single trees a 95–100% probability of passing through fixed sill spillway structures:

- a free distance between piers not less than $0.75L$ for single spillway openings and $1.0L$ for multiple spillway openings separated by piers and
- a head of the upstream pool over the fixed sill and a free height between sill and overlying bridge not less than $0.15L$.

Passage of 80% of the tested slugs of trees required a minimum head of the upstream pool over the fixed sill of $0.15L$–$0.20L$ and a free distance between piers of $1.1L$.

The capability of bottom outlets to pass trees sucked down from the water surface to outlets placed in a vertical front of a dam was tested. The outlet had a rectangular shape with the free height equal to half the free width and a slightly bell-shaped approach with no sharp corners where trees could get stuck. Single trees were safely passed as long as the free width of the outlet exceeded $0.5L$. The higher flow velocities and the marked flow acceleration in front of the outlet may be the reason for the improved performance compared to that of the surface spillways.

The results are relevant only to passage of trees of the species used in the model tests. Other species of trees with different sizes, shapes, and strengths require separate investigations.

6.03.3.3.4 Proposed countermeasures

The first step would be to try and assess if a potential for debris problems exists at a particular dam. If the upstream river runs through forested terrain and there are no lakes or reservoirs upstream where the debris is collected and removed, such a potential usually exists unless spillway dimensions are extremely large. If the terrain around the reservoir is steep and prone to erosion, the problem may be severe.

A debris management plan may be developed to limit the amounts of floating debris. There are a number of different methods [49] that may be employed to counteract clogging of spillways:

Control of debris inflow by

1. cooperation with forestry companies to promote suitable practices such as
   - leaving standing timber barriers,
   - providing adequate drainage of slopes,
   - minimizing strip clearing, and
   - rapid replanting;
2. identification and protection of reservoir slopes prone to slides, especially those influenced by human activities such as road construction, logging, and mining operations;

3. creation of debris traps on streams entering the reservoir;
4. cooperation and joint approach to debris management with other dam projects in the same river; and
5. management of the inflow of debris to reservoirs from areas around the rim and from tributaries, which has not been very successful.

Collection and removal of debris on and around the reservoir by

1. construction of bag shear and containment booms,
2. construction of containment dykes in shallow water,
3. clearing of snags and stumps in shallow parts of the reservoir, and
4. controlled raising of reservoir to float off debris around reservoir rim.

Protection of spillways by

1. booms to restrain, deflect, and stop debris;
2. diverting debris to other weirs; and
3. construction of visor structures at spillways.

The design of booms is critical, as gathered debris may be released in slugs after boom failure or after reaching a depth sufficient to pass under booms. Boom arrangements are therefore presently not favored as a single line of defense [50]. The concept of visors is based on the idea of allowing the spillways to function, perhaps at some reduced capacity, although large amounts of debris have been collected against some visor structure just upstream.

Existing spillways' ability to pass debris can be checked by

1. model testing spillways to assess their sensitivity to the expected debris;
2. increasing free width or height of spillway, for instance, by removal of piers, lowering of crest, or raising/removal of bridge or gate lip in top position;
3. modifying spillway approach zone to improve debris passage;
4. revising the operating procedure to reduce the likelihood of debris jams, for instance, by early complete raising of gates; and
5. introducing a new spillway with better capability to pass debris, perhaps a bottom outlet.

A number of possible spillway approach improvements have been model tested in Germany and Switzerland. These include improved pier shapes and patterns of piers constructed upstream [51] to better align floating debris with the flow.

6.03.3.4 Hydropower Operating Strategies

6.03.3.4.1 Artificial destratification

One approach to mitigating the adverse water quality caused by prolonged stratification of a reservoir is to artificially mix or destratify the water column. By removing the stratification, DO concentrations are maintained throughout the water column and the depth of the photic zone is increased, reducing algal growth. By preventing anoxia, iron and manganese levels can be reduced with a consequent reduction in phosphorus release. The two main destratification techniques are bubble plume mixers and mechanical stirrers.

6.03.3.4.1(i) Bubble plume mixers

Bubble plumes are the most common method of destratification and involve the release of compressed air from a series of diffusers at the bottom of the reservoir. The resultant buoyant bubble plume entrains water as it rises, transporting colder water to the surface where it is released into the surface layer. A well-designed bubble plume destratifier will introduce sufficient buoyancy to lift the coldest water just to the surface, resulting in an efficiency of the order of 5–10%. A bubble plume destratifier does not increase the DO concentration by dissolution of gas from the bubbles, but by allowing atmospheric oxygen transfers to be mixed through the full depth of the reservoir.

6.03.3.4.1(ii) Mechanical stirrers

A less commonly used technique is the use of mechanical stirrers, which are usually large low-speed impellers designed to pump the surface water downward. These systems use either an open impeller or an impeller in a draft tube. An open impeller system creates a jet that impinges on the thermocline, gradually eroding it. A draft tube enables lower velocity impellers to transport surface water to depth, where it forms a positively buoyant plume.

Until recently, mechanical stirrers were considered to be less efficient than bubble plumes, but the use of low-speed impellers makes this technique potentially more efficient. It has been suggested that downward impellers may have the advantage of allowing oxidized metals to settle from the water column at a lower depth than would be the case using a bubble plume since the latter transports the anoxic hypolimnetic water to the surface. There remain some important unanswered questions as to the relative effectiveness of each of these techniques.

6.03.3.4.1(iii)   Curtains

Flexible curtains can be used to control mixing and to separate inflows or withdrawals. For example, surface-suspended curtains can separate cold inflows from the epilimnion of the main reservoir, preventing the entrainment of the warmer water as the inflow plunges. This technique has been used to reduce the hypolimnetic temperature in a reservoir in which cold environmental releases were required for sustaining downstream fish populations.

Typically, temperature control curtains are positioned around intake structures where they control withdrawal elevation. Curtains may also be positioned at other locations within a reservoir or downstream of outlets, particularly in the tailraces of hydropower stations, to control hydrodynamics that might otherwise affect reservoir water quality. Curtains potentially offer substantial cost savings over traditional selective withdrawal structures. However, the considerable uncertainties about their performance were examined in one recent study of three reservoirs [52]. This study concluded that the performance of curtains was complex and not easily characterized.

6.03.3.4.1(iv)   Hypolimnetic aeration and oxygenation

In some instances, it is desirable to maintain the thermal stratification and yet increase the DO concentration in the hypolimnion. For example, some fish require cold water temperatures but high DO concentrations. Although destratification would increase the DO at depth, it would also increase the temperature. Another important example is when the low DO concentration leads to increased nutrient release from the sediments. In such a case, destratification would mix the high nutrient concentration water with the surface layer, increasing the possibility of an algal bloom.

The DO concentration in the hypolimnion can be increased by the introduction of air or pure oxygen. The use of pure oxygen is significantly more efficient, although a supply of compressed oxygen is required. In shallow systems, low-DO water is pumped from the reservoir and oxygen is injected using a venturi and then returned to the reservoir. In deeper reservoirs, the oxygen is introduced directly into the hypolimnion, although usually through a venturi to ensure dissolution.

6.03.3.4.2   Management of reservoir filling

Study case: Turbidity generated during the filling of a reservoir – The Péribonka case (Québec, Canada)

The Péribonka hydroelectric installation comprises a dam of 80 m height by 700 m crest length, two closing dykes, one underground power plant equipped with three Francis turbines of 385 MW total installed power generating 2.2 TWh annual energy, a dual-pass spillway with a maximum capacity of 5300 m^3 s^{-1}, and a reservoir of 35 km length with a surface area of 32 km^2.

Filling the reservoir required 37 days, from 27 September to 3 November 2007. Due to the proximity of a large reservoir upstream (**Figure 26**), the water naturally contains very little suspended load at the Péribonka power plant site. (Péribonka is derived from the Montagnais word *pelipaukau*, which means "a river digging in or removing the sand.")

During the filling process of the reservoir, landslides occurred on the river (**Figures 27 and 28**), resulting in a temporary increase in water turbidity. The plume of brown water could be followed from day to day by satellite images (**Figure 29**) and by punctual measurements. Without immediate and appropriate actions undertaken by Hydro-Québec, this turbidity would have an important impact on the drinking water supply of the surrounding inhabitants living downstream. Their water treatment systems were not designed to take account of high turbidity.

The Péribonka river is the main source of freshwater for two municipalities located next to the river mouth. The municipality of Sainte-Monique draws its drinking water from the Chute-à-la-Savane power plant reservoir (**Figure 26**), whereas the municipality of Péribonka draws water directly downstream of this last power plant. Both water treatment systems (using chlorination) became inefficient when the suspended load increased.

Special measures using tanker trucks were used during the period of time required to reach the normal concentration of suspended load. Setup of these corrective actions was facilitated by the delay of the reservoir filling and the time taken by the turbidity plume to progress downstream. Normal concentrations were reached 2 months after the turbidity front had reached the water supply installations of the two municipalities (**Figures 30(a) and 30(b)**).

The following lessons were drawn from this experience:

- Despite the geological surveying, the geomorphologic study, and the deforestation of the reservoir banks, it had been impossible to predict such a level of turbidity. Indeed, sources of suspended load were limited to a few zones with silt and clay content, and were very hard to detect.
- Emphasis was laid on the importance of the communication system between the developer and the concerned population, which allowed for excellent cooperation in order to limit adverse effects.
- Concerning erosion and landslides, this experience shows the importance of having an efficient environmental follow-up program during the filling phase of the reservoir so as to ward off all eventualities. It was surely through such an efficient environmental follow-up that the impact on drinking water supply was controlled.

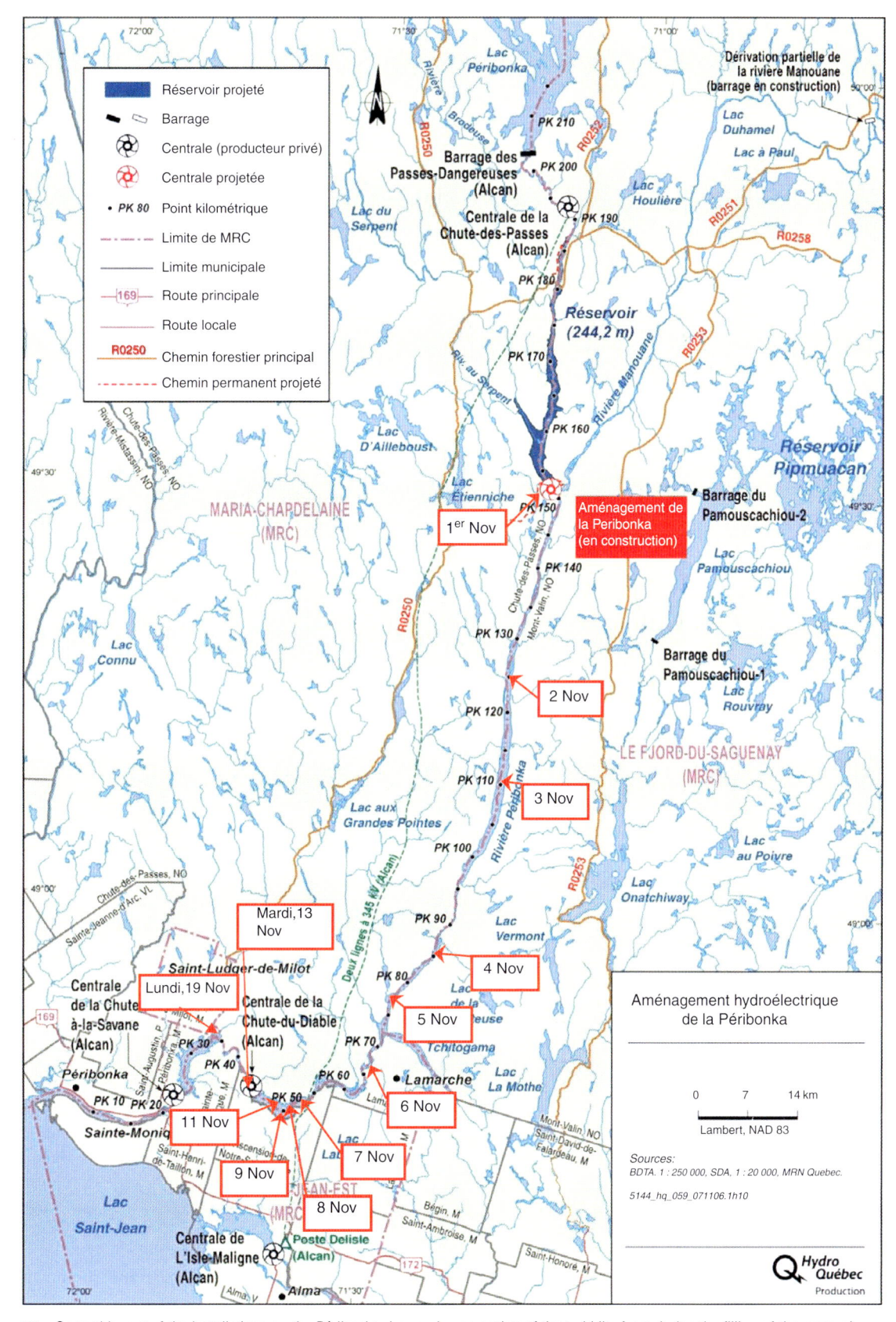

Management of Hydropower Impacts through Construction and Operation 85

Figure 26 General layout of the installations on the Péribonka river and progression of the turbidity front during the filling of the reservoir.

Figure 27 Landslides in a sandy bank.

Figure 28 Landslides in a sandy bank with silt content.

Figure 29 Satellite image taken on 5 November 2007.

- The importance of observation of the great capacity of the marine fauna to temporarily tolerate and sustain unusual environmental conditions was recognized.

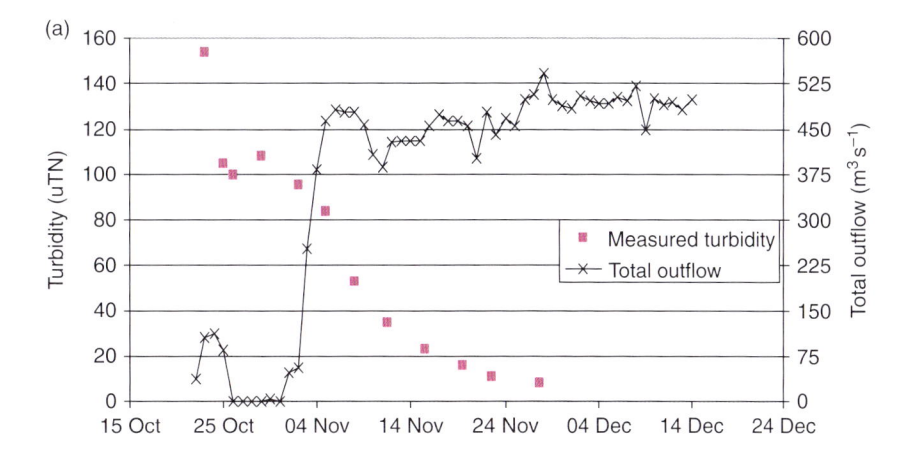

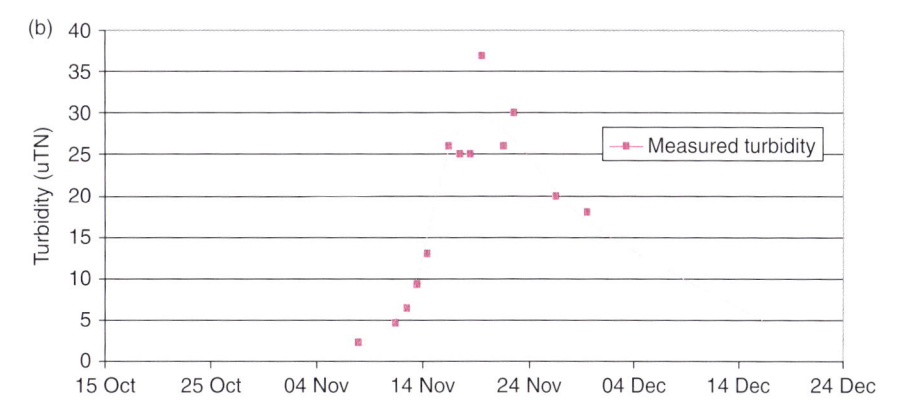

Figure 30 (a) Turbidity progression in the Péribonka power plant reservoir. (b) Turbidity in the Chute-du-Diable reservoir (105 km downstream).

6.03.3.4.3 Unsteady flow

To reduce the downstream unsteady flows of the operating turbines (see Section 6.03.3.1.3), a regulating pond could be implemented to minimize fluctuations in daily discharge. Water discharged from the turbines is conveyed through a concrete transition stilling structure into an excavated tailrace channel. The tailrace channel will convey the water to a regulating pond downstream from the power station.

In the Nam Theun 2 hydropower plant, the maximum discharge from the power station into the regulating pond is $330 \, m^3 \, s^{-1}$. The regulating pond enables the project to be operated as an intermediate peaking facility by regulating the downstream flows for environmental reasons. The purpose of the regulating pond is to limit water level fluctuations in the Xe Bang Fai, in particular during start-up, shutdown, and load-changing operations. It consists of the construction of an additional dam with two contiguous concrete structures, one spilling into the Nam Kathang and the other into a downstream channel. An earth and rockfill embankment was constructed to complete the downstream closure of the regulating pond. The regulating pond will have an active storage volume of 8 million m^3.

6.03.3.4.4 Population protection measures

A basin-wide flood forecasting and warning system would be useful to ensure that all downstream power projects and local towns and villages receive adequate warning in the event of a flood or upstream dam break. Developing such a system is beyond the capacity of any individual developer and should be coordinated by state and central agencies. This effort would require upgrading remote data gathering sites, strengthening telemetry communications, and developing a central database and data processing capacity.

6.03.3.5 Mitigation Measures

6.03.3.5.1 Structural options – Multilevel offtake towers

There are several structural options available for the selective withdrawal of specific layers of water from the reservoir, ranging from floating offtakes to multilevel fixed offtakes and continuous screen systems.

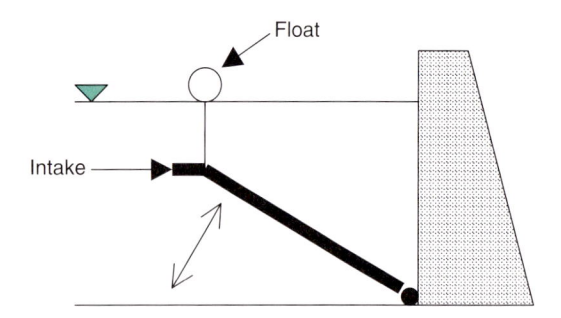

Figure 31 Pivoting arm option.

6.03.3.5.2 Floating offtakes with pivot arms or trunnions

The basic concept consists of a pipe offtake that is attached to a float. The concept is shown diagrammatically in **Figure 31**. Generally, for this type of option, the diameter of the intake is limited to about 1000 mm or flow rates of up to about $2\,m^3\,s^{-1}$, which does not provide sufficient capacity for the bulk water discharges required for major water storages, hydropower stations, or irrigation dams. In addition, there is a practical difficulty that limits the length of the pipe to about 25 m, and hence it is only suitable for withdrawals at shallower depths. This option would be suitable for smaller-volume town water supply.

6.03.3.5.3 Dry multiport intake towers

In Australia, a recent survey indicated that the preferred method for achieving selective withdrawal is via a 'dry' tower consisting of multileveled bell-mouth inlet ports connected to an internal conduit that passes vertically down inside the tower. A butterfly valve or a penstock gate, which is either fully open or closed, controls inflow into each inlet port. Operation of the valves and maintenance of the system are easily carried out from within the tower structure, with access being either from the top platform or via a tunnel under the embankment. Many of these structures can also be operated remotely from the tower platform or from control rooms by supervisory control and data acquisition. In Australia, the majority of these types of dry intake structures are used for drinking water supply; however, some authorities do operate this type of inlet for irrigation water (**Figure 32**).

Dry intake structures have limited flexibility in being able to selectively withdraw from the specific levels at which the ports are set. Typically, dry intakes may have no more than six draw-off levels. The acceptability of the arrangement will depend on the specific conditions within the storage and the objectives for the withdrawal conditions. The main drawback, however, for the dry-type intakes is the limited draw-off capacity. This capacity is limited by the cost to provide sufficiently large intakes, valves, and conduits. For this reason, dry intake structures are typically not suited to flow rates in excess of $10-12\,m^3\,s^{-1}$.

6.03.3.5.3(i) Continuous balk and screen options

These structures incorporate a method of selective withdrawal by using a trash rack and balk system. The trash racks and balks are positioned vertically within a slot located on the upstream side of the intake tower and line up with the corresponding inlet ports, depending on the withdrawal level or depth that has been selected. This type of intake tower is considered to be the best design and practice for the required discharge volumes, having been used to control discharges up to $50\,m^3\,s^{-1}$. In practice, however, design

Figure 32 Two intakes exposed on a dry multiport intake tower.

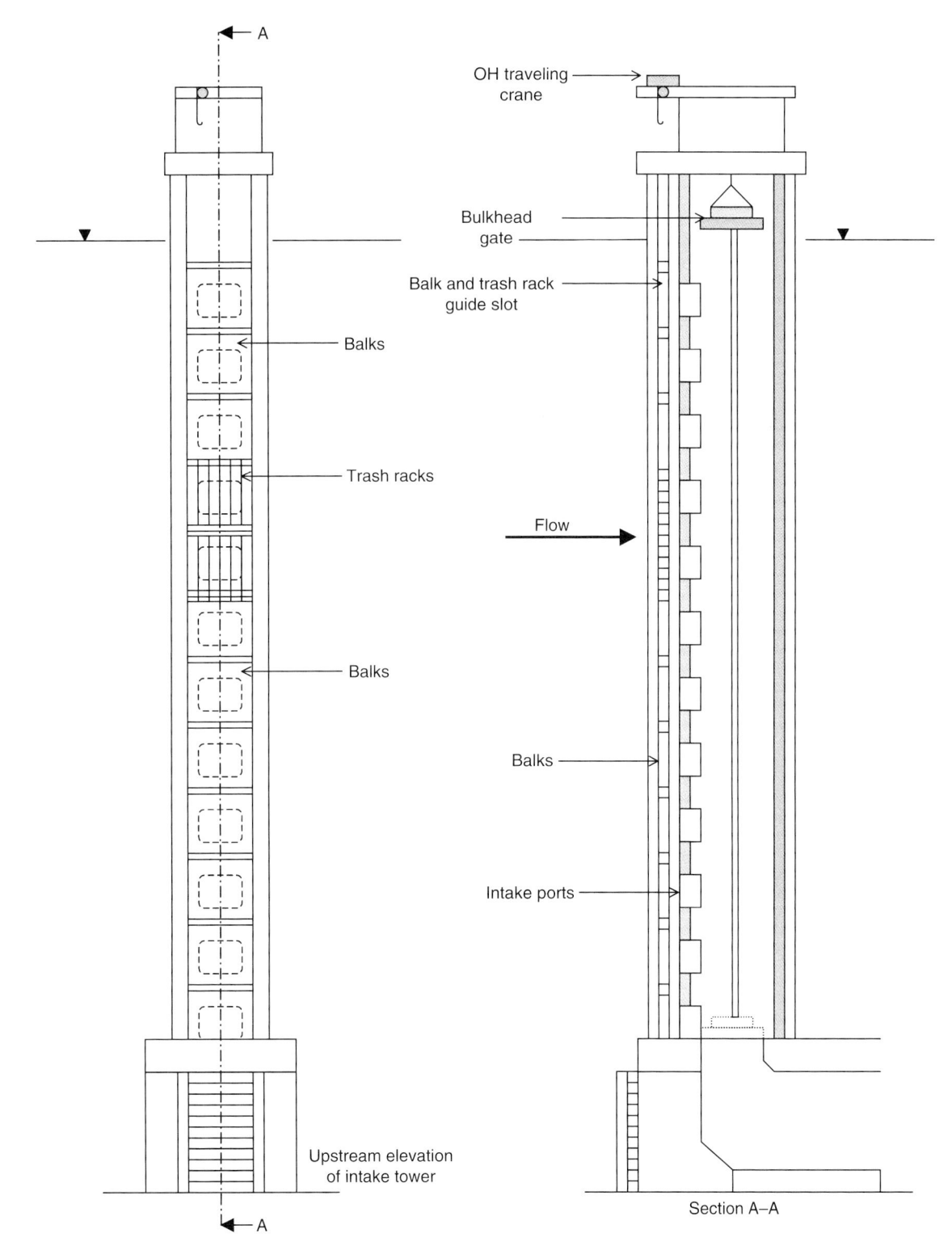

Figure 33 Basic operation of selective withdrawal.

limitations pose potential significant constraints to operating these structures for effective downstream thermal and water quality management. Changing the withdrawal level is a slow and manually intensive task involving some significant occupational safety issues.

All the structures consist of either one or two vertical columns of intake ports on the upstream side of the tower. Positioned in front of each column of intake ports is a single slot that permits the trash racks and balks to be vertically stacked one on top of the

Figure 34 Shasta Dam multilevel intake structure.

other in line with the port openings. The balks prevent water entering the intake structure at the corresponding depth and are positioned above and below the desired release depth. The trash racks screen coarse material and reservoir debris and are set at a height corresponding to the desired intake level. The trash racks on some dams have been retrofitted with finer screens suitable for use with mini hydro schemes. The intake structures are described as being 'wet' since water fills the entire internal cavity and gravitates down to the base of the tower, through the bulkhead, and into the outlet tunnel. Flow through the intake structure is controlled with a penstock valve. Lowering of the main bulkhead gate enables the penstock to be dewatered. **Figure 33** is a diagrammatic representation of the system.

In the United States, a survey of selective withdrawal systems undertaken by the US Bureau of Reclamation (USBR) (2003) gathered basic design and operational data for large selective withdrawal dams in the United States. Many of the dam operators canvassed in the USBR survey indicated that it would not be practical to automate the operation of the selective withdrawal gates at their dams. The most common reason given for not automating the operation of systems was that the infrequency of operation made it difficult to justify the cost. The majority of respondents indicated that intake level change was undertaken on average once every month.

A number of intake structures in the United States have undergone major retrofitting to add selective withdrawal capability to improve released water quality. A selection is briefly described below.

6.03.3.5.4 Shasta Dam Temperature Control Device, California

Completed in 1998, this is a retrofitted multilevel water intake structure. Water withdrawal is controlled by a 91 m tall and nearly 80 m wide shutter structure that was added to the upstream face of this concrete dam. The shutter extends about 15 m upstream from the face of the dam, and is open between units to permit crossflow in front of the existing trash rack structures. It was manufactured off-site, lowered into the water, assembled by divers, and attached to the upstream face of the dam. The total cost of the project was US$80 million (**Figure 34**).

6.03.3.5.5 Glen Canyon Dam, Arizona

This is the fourth highest dam in the United States. The proposal is for an uncontrolled overdraw design, where flow enters the top of the intake tower (built on the upstream face of the dam) 50 m above the existing intake. The operational flexibility of this design is limited due to reservoir elevation fluctuation.

6.03.3.5.6 Flaming Gorge Dam, Utah

Completed in 1978, the retrofit consists of electrically controlled gates that allow the release of water from different depths in the reservoir.

References

[1] Alves E (2008) Sedimentation in Reservoirs by Turbidity Currents (in Portuguese). PhD Thesis, Laboratório Nacional de Engenharia Civil.
[2] Baumgartner A and Reichel E (1975) *The World Water Balance*. Munich, Germany: R. Oldenbourg Verlag.

[3] International Commission on Large Dams (ICOLD) (1998) *World Register of Dams*. Paris, France: ICOLD.
[4] International Commission on Large Dams (ICOLD) (1997) *Position Paper on Dams and Environment*. Paris, France: ICOLD.
[5] Petts GE (1984) *Impounded Rivers: Perspectives for Ecological Management*. New York: Wiley.
[6] Chapra SC (1996) *Surface Water-Quality Modeling*. New York: McGraw-Hill.
[7] Orlob GT (ed.) (1983) *Water Quality Modeling: Streams, Lakes and Reservoirs*. IIASA State of the Art Series. London: Wiley Interscience.
[8] Dodds W, Jones JR, and Welsh EB (1998) Suggested classification of stream trophic state: Distributions of temperate stream types by chlorophyll, total nitrogen and phosphorus. *Water Research* 32(5): 1455–1462.
[9] Nixon SW (1995) Coastal marine eutrophication: A definition, social causes and future concerns. *Ophelia* 41: 199–219.
[10] Vollenweider RA, Rinaldi A, Viviani R, and Todini E (1996) *Assessment of the State of Eutrophication in the Mediterranean Sea*. Athens, Greece: MEDPOL/FAO/UNEP.
[11] Ryding S-O and Rast W (eds.) (1989) *The Control of Eutrophication of Lakes and Reservoirs*. Paris, France: UNESCO.
[12] Vollenweider RA (1975) Input–output models with special reference to the phosphorus loading concept in limnology. *Schweizerische Zeitschrift fur Hydrologie – Swiss Journal of Hydrology* 37: 53–83.
[13] Thoman RV and Mueller JA (1987) *Principles of Surface Water Quality Modeling and Control*. New York: Harper & Row.
[14] Cardoso da Silva M (2003) Tools for the Management of Estuaries: Environmental Indicators (in Portuguese). PhD Thesis, New University of Lisbon.
[15] Imberger J and Hamblin PF (1982) Dynamics of lakes, reservoirs and cooling ponds. *Annual Review of Fluid Mechanics* 14: 153–187.
[16] Imberger J and Patterson JC (1990) Physical limnology. In: Wu T (ed.) *Advances in Applied Mechanics*, vol. 27, pp. 303–475. Boston, MA: Academic Press.
[17] Water Resources Engineers, Inc. (WRE) (1968) Prediction of thermal energy distribution in streams and reservoirs. Report to California Department of Fish and Game. Walnut Creek, CA: WRE.
[18] Huber WC, Harleman DRF, and Ryan PJ (1972) Temperature prediction in stratified reservoirs. *Journal of the Hydraulics Division, ASCE* 98(HY4): 645–666.
[19] Tennessee Valley Authority (TVA) (1972) Heat and mass transfer between a water surface and the atmosphere. Engineering Laboratory, Report No. 14. USA: TVA.
[20] Chen CW and Orlob GT (1975) Ecologic simulation for aquatic environments. In: *Systems Analysis and Simulation in Ecology*, vol. 3, ch. 12. New York: Academic Press.
[21] Markofsky M and Harleman RF (1973) Prediction of water quality in stratified reservoirs. *Journal of the Hydraulics Division, ASCE* 99(HY5): 729–745.
[22] Baca RG and Arnett RC (1976) A finite element water quality model for eutrophic lakes. *Proceedings of the International Conference on Finite Elements in Water Resources*. Princeton, NJ, USA.
[23] Han B-P, Armengol J, Garcia JC, et al. (2000) The thermal structure of Sau Reservoir (NE: Spain): A simulation approach. *Ecological Modeling* 125: 109–122.
[24] Gal G, Imberger J, Zohary T, et al. (2003) Simulating the thermal dynamics of Lake Kinneret. *Ecological Modeling* 162: 69–86.
[25] Leendertse J (1967) Aspects of a computational model for well-mixed estuaries and coastal seas. R. M. 5294-PR. Santa Monica, CA: The Rand Corporation.
[26] Masch FD, et al. (1969) A numerical model for the simulation of tidal hydrodynamics in shallow irregular estuaries. Technical Report HYD 12-6901. Austin, TX: Hydraulic Engineering Laboratory, University of Texas.
[27] Cheng RT, et al. (1976) Numerical models of wind-driven circulation in lakes. *Applied Mathematical Modeling* 1: 141–159.
[28] Simons TJ (1973) *Development of Three-Dimensional Numerical Models of the Great Lakes*. Scientific Series No. 12. Burlington, ON: Inland Waters Directorate, Canada Centre for Inland Waters.
[29] Orlob GT (1977) *Mathematical Modeling of Surface Water Impoundments*, vol. 1. Lafayette, CA: Resource Management Associates.
[30] Simons TJ, et al. (1977) Application of a numerical model to Lake Vanern. NrRH09. Suécia: Swedish Meteorological and Oceanographic Institute.
[31] King IP, Norton WR, et al. (1975) A finite element solution for two-dimensional stratified problems. In: *Finite Elements in Fluids*, ch. 7, pp. 133–156. London: Wiley.
[32] Edinger JE and Buchak EM (1975) A hydrodynamic, two-dimensional reservoir model: The computational basis. Report to US Army Corps of Engineers, Ohio River Division, Cincinnati, OH, USA.
[33] Lam DCL and Simons TJ (1976) Numerical computations of advective and diffusive transports of chloride in Lake Erie, 1970. *Journal of the Fisheries Research Board of Canada* 33: 537–549.
[34] Patterson DJ, et al. (1975) Water pollution investigations: Lower Green Bay and Lower Fox River. Report to EPA, Contribution No. 68-01-1572, USA.
[35] DiToro DM, et al. (1975) Phytoplankton–zooplankton–nutrient interaction model for Western Lake Erie. In: Patten BC (ed.) *Systems Analysis and Simulation in Ecology*, ch. 11, vol. 3. New York: Academic Press.
[36] Snodgrass WJ and O'Melia CR (1975) *A Predictive Phosphorus Model for Lakes: Sensitivity Analysis and Applications*. USA: Environmental Science and Technology.
[37] Thomann RV, et al. (1975) Mathematical modeling of phytoplankton in Lake Ontario. National Environment Research Center, Office of Research and Development, EPA, Corvallis, OR, USA.
[38] Chen CW, Lorenzen M, and Smith DJ (1975) A comprehensive water quality: Ecologic model for Lake Ontario. Report to Great Lakes Environment Research Laboratory. USA: Tetra Tech, Inc.
[39] Cole TM and Buchak EM (1995) CE-QUAL-W2: A two-dimensional, laterally averaged, hydrodynamic and water quality model, version 2.0, user manual – Draft version. U.S. Army Engineer Waterways Experiment Station, Vicksburg, MI, USA.
[40] Jørgensen SE (1976) A eutrophication model for a lake. *Ecological Modeling* 2(2): 147–165.
[41] Reckhow KH and Chapra SC (1999) Modeling excessive nutrient loading in the environment. *Environmental Pollution* 100: 197–207.
[42] Chen CW and Orlob GT (1972) Ecologic simulation for aquatic environments. Final Report. Walnut Creek, CA: Water Resources Engineers, Inc. (WRE).
[43] Diogo PA and Rodrigues AC (1997) Two-dimensional reservoir water quality modeling using CE-QUAL-W2. *IAWQ Conference on Reservoir Management and Water Supply – An Integrated System*. Prague, Czech Republic, 19–23 May.
[44] Hydrologic Engineering Center (HEC) (1991) *HEC-6, Scour and Deposition in Rivers and Reservoirs: Users Manual – Generalized Computer Program*. Davis, CA: HEC, US Army Corps of Engineers.
[45] Svendsen (1987) Flood discharge at dams (in Norwegian) [Flomavledning ved dammer, Erfaringer fra oktoberflommen]. NVE Report No. V18.
[46] Guo J and Liu ZP (2003) Field observations on the RCC stepped spillways with the Flaring Pier Gate on the Dachaoshan Project. *Proceedings of the IAHR XXX International Congress*, August.
[47] Godtland K and Tesaker E (1994) Clogging of spillways by trash. *ICOLD 18th Congress*, R 36. Durban, South Africa.
[48] Johansson N and Cederstrom M (1995) Floating debris and spillways. *Water Power '95 Conference*. San Francisco, CA, USA.
[49] Canadian Dam Safety Association (1995) *Dam Safety Guidelines*. Edmonton, AB, Canada, January.
[50] Rundqvist J (2006) *Debris in Reservoirs and Rivers – Dam Safety Aspects*. Canada: CEATI.
[51] Strobl T (2005) *Wehranlage Baierbrunn*. Germany: Versuche Technische Universität Munchen.
[52] Vermeyen T (1997) The use of temperature control curtains to control reservoir release temperatures. Report No. R-97-09. Denver, CO: Water Resources Research Laboratory, Technical Services Centre, Bureau of Reclamation.

6.04 Large Hydropower Plants of Brazil

BP Machado, Intertechne, Curitiba, PR, Brazil

© 2012 Elsevier Ltd. All rights reserved.

6.04.1	Introduction and Background	93
6.04.1.1	Historical Evolution of the Electric Sector in Brazil	94
6.04.1.2	Main Hydroelectric Projects	96
6.04.2	The 14 000 MW Itaipu Hydroelectric Project	96
6.04.2.1	General Description of the Project	96
6.04.2.2	The Dam	98
6.04.2.3	The Spillway	99
6.04.2.4	The Power Plant	99
6.04.3	The 8125 MW Tucurui Hydroelectric Project	101
6.04.4	The 6450 MW Madeira Hydroelectric Complex	104
6.04.4.1	The Santo Antonio Project	105
6.04.4.2	The Jirau Project	106
6.04.5	The Iguaçu River Projects	108
6.04.5.1	The Foz do Areia Project	108
6.04.5.2	The Segredo Project	110
6.04.5.3	The Salto Santiago Project	111
6.04.5.4	The Salto Osorio Project	114
6.04.5.5	The Salto Caxias Project	116
6.04.6	The Uruguay River Projects	119
6.04.6.1	The Machadinho Project	119
6.04.6.2	The Itá Project	121
6.04.6.3	The Campos Novos Project	123
6.04.6.4	The Barra Grande Project	124
6.04.7	The Belo Monte Project	125
References		127

6.04.1 Introduction and Background

Brazil is located in South America where it occupies 47.7% of the territory of this continent. Brazil has the fourth largest territorial area in the world. Its 8.5 million km^2 spans from latitude 4° north to 33° south, and from longitude 75° to 40° west. Its population is of about 190 million people. Economically, it is the eighth largest economy of the world with a gross national product equivalent to about US$1.6 trillion [1]. Politically, it is a federation of 27 states with a diversified legislation on the use of natural resources giving, in general, to the central government primary (but not exclusive) prerogatives on the licensing to build and operate infrastructure undertakings, including hydraulic and hydroelectric projects.

The right to explore the use of water resources is granted to public and/or private agents through concessions. In the case of hydroelectric projects, concessions for selected projects are offered by the federal agency for electric power (Agencia Nacional De Energia Elétrica, ANEEL) for interested parties under a competitive tendering process. The concessionaires are supposed to sell the electric power to the retailing companies with a preestablished tariff, which is the basis for the competitive tendering process. The National System Operator, which manages the National Interconnect Transmission System, daily defines the generation level of each plant, so as to optimize the overall availability of hydrological resources and the use of regulating reservoirs. The compensation for the concessionaire is not dependent on the power produced by his plant but he receives a fixed amount coresponding to a virtual 'firm energy' associated with his plant which was established by ANEEL prior to the concession tendering process.

Brazil is a country extremely rich in water resources. Although certain areas of the country can be classified as having a semiarid environment, for its seasonal intermittent rainfall pattern, the Brazilian territory is well endowed with tropical and subtropical humid climates, with a predominance of perennial drainages projected by tablelands and lower plateaus. This of course favors the rather extensive use of water resources for the development and well-being of its population, with hydroelectric power generation, urban water supply, and river flow regulation being the main objectives of projects carried on.

Figure 1 depicts schematically the main river basins on the Brazilian territory.

As a result, the construction of hydroelectric projects, dams, and reservoirs was the object of an important effort by the Brazilian people, through both government and private initiatives. The most important dams in Brazil were built in relation to hydroelectric projects. Presently (August 2009), 74.3% of the electric power installed capacity in the country originates from hydroelectric developments. This, of course, reflects not only the abundance of hydroelectric potential but also the scarcity of fossil fuels, which are responsible elsewhere by the bulk of the electric power generation needs.

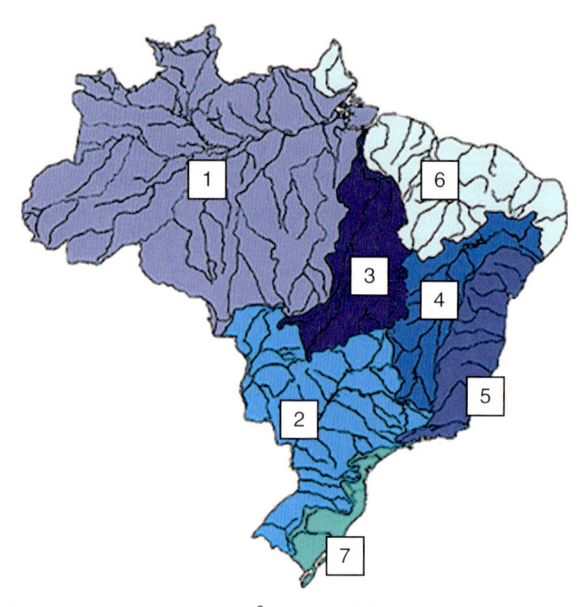

Figure 1 Main Brazilian river basins. 1, Amazon Basin – 3.8 million km^2; 2, Paraná-Uruguay Basin – 1.4 million km^2; 3, Tocantins-Araguaia Basin – 0.97 million km^2; 4, San Francisco Basin – 0.63 million km^2; 5, Eastern Atlantic Basin – 0.57 million km^2; 6, Northeastern Atlantic Basin – 1.0 million km^2; 7, Southeastern Atlantic Basin – 0.23 million km^2.

Actually the hydroelectric prevalence and the existence of different hydrologic regimes in different areas of the country and the existence of an interconnected, countrywide, HV transmission system make the Brazilian generation system different from that of any other country in the world. In July 2009, the total installed capacity of hydroelectric generating plants was 78 126.3 MW, not considering 7000 MW, which is the Paraguayan share of the Itaipu project, which however is mostly sold to the Brazilian system [2]. The total production of electric power generated from hydroelectric projects offered to the Brazilian market in 2008 was 363.8 TWh, corresponding to 73.1% of the total from all sources. These figures are expected to grow at an annual rate of 3–4%, in spite of the fact that efforts toward building more fossil-fueled plants, mainly using natural gas, are underway. The main reason for the accelerated growth of thermal plants is the increasing opposition of environmentalists to the realization of hydraulic regulating reservoirs.

6.04.1.1 Historical Evolution of the Electric Sector in Brazil

The first hydroelectric plant in Brazil was built in the industrial state of Minas Gerais, in southwest Brazil in 1889. It was a 252 kW power plant, which provided electric power for public lighting for the town of Juiz de Fora [3]. In the early 1900s, electricity utility companies controlled by private foreign capital started developing the hydroelectric potential to supply electric power to São Paulo and Rio de Janeiro, the main urban and industrial centers of the country. In 1901, the Canadian company São Paulo Light and Power Company inaugurated the first major plant in the São Paulo area, in the Tietê River, a dam that today is practically located within the boundaries of the city (Edgard de Souza dam). In 1907, the Rio de Janeiro Light and Power Company completed the construction of the Fontes hydroelectric project, with dam and power plant generating 24 000 kW, then one of the largest hydroelectric projects in the world.

From the beginning of the century to the mid-1930s, the construction of dams and plants for electric power generation remained with private companies, practically without government interference. In 1934, however, the federal government issued new legislation considering the country water resources as public property and started to issue concessions for the use of these resources by private agents, for any purpose including power generation, urban supply, and irrigation.

With the end of World War II, in Brazil as elsewhere, the idea of a direct participation of governments in the economic activities related to infrastructure works, creating conditions for an accelerated industrial development, started to flourish within government planning offices and leading industry entrepreneurs. The first major centrally formulated economical development plan for the country was created by the federal government administration that took office in 1946. This prepared the setting for the following administration, which started in 1950 with a deep nationalistic view of government and public participation in promoting development, to act and implement a number of public-owned companies that were given the responsibility of building infrastructure works in such diverse areas as electric power, oil, roads and highways, irrigation, and land development. The first major state-owned company established to develop the electric power potential of Brazilian hydroelectric resources, Chesf (Companhia Hidrelétrica do São Francisco), was created in 1948 with the specific responsibility of building a major dam and power plant at the Paulo Afonso Falls, in the São Francisco River, in the southern limits of the Brazilian northeast. The Paulo Afonso plant began operation in 1954 and presently four powerhouses have been built in the site, with a total installed capacity of 3400 MW.

The initiative of the federal government with Chesf triggered similar moves in the some major states. Minas Gerais created CEMIG (Companhia Energética de Minas Gerais) in 1953 and Paraná, COPEL, in 1954, two of the most successful state-owned companies that played major roles in the development of dam and hydroelectric engineering in Brazil. These organizations evolved competing with foreign private electric power operators that dominated, by that time, the hydroelectric production and distribution of power in the major industrial southcentral and southern states of the country. By the end of the 1950s, a new and important central government-owned company – Furnas – was created to build a new and large plant in the Grande River, between Minas Gerais and São Paulo. This company eventually grew to build a large number of dams and generating plants and was a key player in developing and securing dam and hydroelectric technologies required to implement larger and more powerful plants in rivers of the southcentral region.

In São Paulo, the most industrialized state of the Brazilian Union, by the mid-1950s, dam construction and electric power generation was primarily done by São Paulo Light and Power Company, which, as mentioned above, was established in the area since the beginning of the century. To promote further development in the more distant areas of the interior of the state, the state government set up companies with responsibilities of developing the hydro potential of the main state river basins, following to a certain extent the successful example of the American TVA – Tennessee Valley Authority – that during previous decades was the major example of the government interference in a private-dominated sector of the economy. The development of the three major rivers of the state, the Tietê, the Paranapanema, and the Paraná, was assigned to new companies, and these rivers, in less than 20 years, were completely transformed with dams, reservoirs, and hydro plants, some of which were benchmarks in the development of dam engineering in Brazil.

Until the end of the 1950s, the expansion of generation and transmission facilities in Brazil, as in many parts of the world, was carried out by separate utilities operating on a local basis. As the better hydro sites close to the local loads were developed, and annual growth began to approach 400 MW or more, regional planning of the expansion of generating became important, and some of the state utilities began to realize that a broader survey of hydro resources in their area became mandatory [4, 5]. In 1961 a survey of the hydro potential of the south central area of the country, followed by a similar survey of the southern region, was carried out. This resulted in one of the major systematic surveys of hydro resources ever carried out anywhere in the world with specific technological and methodological procedures developed for the study. The study covered an area of about 1.3 million km^2 and identified and appraised hundreds of potential hydro sites. Most of them were implemented during the following 40 years and became the backbone of the Brazilian electric system.

The growth of public utilities in Brazil had, by the early 1970s, in practical terms completely eliminated the private competitors, which were either absorbed or extinguished along the process. Each state created its own public company responsible for supplying electric power, either by purchasing from other state-owned companies or building their own hydraulic projects. On the federal level, the central government assigned Eletrobras as their holding company, with four subsidiaries – Chesf, Furnas, Eletronorte, and Eletrosul – covering the whole of the Brazilian territory and responding for the construction and operation of large dams and power projects and for the interconnected transmission system.

Major projects built between the early 1950s and 1980s, under the sponsorship of state-owned companies, are among the most important ever built in the country. Among these were the 14 000 MW Itaipu project, then the largest hydroelectric project in the world; the 2500 MW Foz do Areia project, with a 160 m high dam, at the time the highest concrete-face rockfill dam in the world; the 3200 MW Ilha Solteira project; the pioneering 1216 MW Furnas project; and the 2680 MW São Simão project are significant examples of the diversified engineering and construction achievements of the Brazilian hydro engineering of the period.

By the beginning of the 1980s, the Brazilian economy entered into a period of stagnation resulting from various factors, including increases in the international price of oil, of which the country was extremely dependent, and instabilities in the world financial markets, in general. This period caused the halting of about a dozen dam projects that suffered a lack of funds for proceeding with construction already started while some others, in spite of concessions to some utilities, could not even have their works started.

In spite of this unfavorable situation some major dam projects, such as the 8000 MW Tucurui project, the first large dam project in the Amazonian area, had their first phase (4000 MW) completed. Other important projects, such as the 5000 MW Xingó project and 1200 MW Segredo project, proceeded and were completed during the decade. However, the poor financial and economical situation of some of the state-owned utilities prevented the increased raising of capital resources required to keep up with the very large needs of the country in hydropower and dam construction. The consequence was the return to the private market to finance the expansion of the sector.

Privatization of the electric power sector and dam construction in Brazil, during the 1990s, actually meant the complete reformulation of rules of operation and access to concession of hydropower sites. One major federal electric generating utility and some large state-owned power companies were sold to private parties, some of them belonging to international corporations. This has brought in a reasonable inflow of badly needed capital and, as a consequence, the resumption of dam and power plant projects previously halted.

Presently, the rules for owning and operating electric power plants in Brazil allow the participation of private and public parties, either independently or in association. The federal government produces an inventory of possible sites, evaluates their technical feasibility and defines the technical and environmental requirements, and organizes the priorities for development. Concessions for building and operating plants during 30 years are granted to interested parties under competitive dispute on public auctions in which the winner is the party that offers the lowest price for selling the energy (kWh) to the integrated system that is responsible for transmitting and distributing, through local companies, the electric power.

6.04.1.2 Main Hydroelectric Projects

Presently (2009), there are 517 hydroelectric projects in operation in Brazil with an aggregated installed capacity of 78 218.4 MW. There are also 91 hydroelectric projects under construction that will have a combined capacity of 11 537.8 MW [6]. These figures do not include projects that are being studied and for which concessions have not been granted. It does not include, for example, the 11 000 MW Belo Monte project that will be offered for concession early in 2010.

In continuation, some representative hydroelectric Brazilian projects are presented as samples of the type of projects built in the country.

6.04.2 The 14 000 MW Itaipu Hydroelectric Project

The Itaipu hydroelectric project is presently the second largest hydroelectric generating installation in the world. It is a joint undertaking between Brazil and Paraguay, located in the Paraná River, in a reach in which this river constitutes the international border between the two countries (**Figure 2**). Construction of the project started in May 1975 and the first 700 MW unit entered into commercial operation in May 1984. The installation of 18 units was carried out from 1984 to 1991 and the last 2 units were only added in 2006 completing the full capacity of the project.

The realization of this major binational hydroelectric project, which until recently was the largest in the world, was made possible by extensive diplomatic negotiations between Brazil and Paraguay. These negotiations culminated with the signing by the two countries, in 1966, of a document setting up the intention of jointly studying and evaluating the hydroelectric potential of the international reach of the Paraná River. Furthermore, the agreement established that the hydroelectric power produced in this stretch would be equally divided between the two countries and that each country would have the preferential right to acquire the power owned by the other country that it would not use for its own domestic consumption.

Based on the Brazil–Paraguay agreement, a Joint Technical Commission was created in 1967 and feasibility studies were carried out that concluded by the recommendation of a single project to be implemented to develop the full power potential of the international reach of the river. As a result a treaty was signed in 1973, and a binational entity – Itaipu Binacional – was formed by both countries to conduct the construction of the project and, subsequently, operate it [7].

The project was essentially financed by international loans guaranteed by the Brazilian government. It has been in continued and very successful operation since the first unit entered on line in 1984. About 95% of the energy produced is fed into the Brazilian electric system and the 5% balance represents the domestic Paraguayan consumption.

6.04.2.1 General Description of the Project

The information and description that follows is essentially based on the book. The Paraná River basin covers an area of about 3 million km^2, of which 899 000 km^2 are in the Brazilian territory with the remaining in Paraguay, Argentina, and Uruguay. The drainage area at the Itaipu site is 820 000 km^2. The upper stretches of the Paraná basin are located in the central mainland of Brazil, with elevations between 600 and 700 m. When the river reaches the international border, at Guaira, the elevation is about 215 m and from this point onto the Itaipu site it drop 140 m. Practically, all this drop was concentrated at the Sete Quedas Falls, now flooded by the Itaipu reservoir.

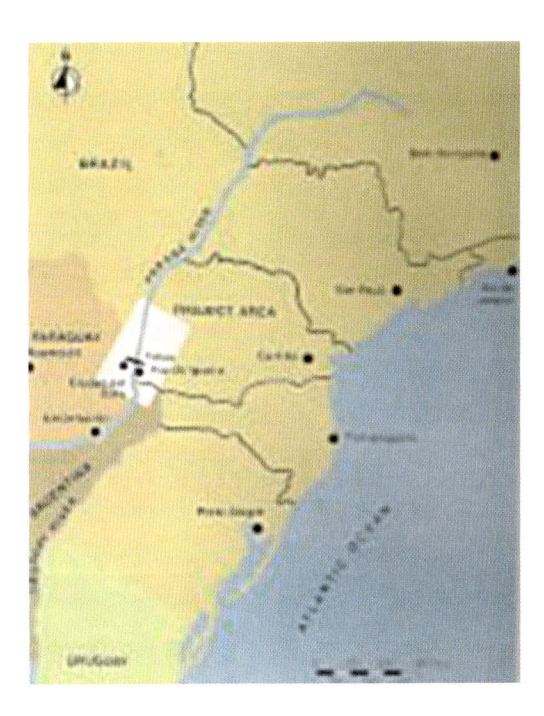

Figure 2 Project location.

The average natural annual flow at the Itaipu site, computed without consideration of the existing upstream reservoirs, is 9700 m^3 s^{-1}. Most of the upstream reaches of the Paraná River and of its main tributaries are already developed for hydroelectric generating projects and this has created a significant regulating capacity that naturally influenced the power studies. Another important factor in these studies is the fact that Iguaçu River discharges into the Paraná immediately downstream of the Itaipu site, creating large variations on the elevation of the water level depending on the flow regime of both rivers, and therefore affecting the head available for power generation.

The final basic installation was defined with 18 units with a nominal capacity of 700 MW each, corresponding to 12 600 MW. Two additional units were considered to allow more flexibility in the plant operation and maintenance. These units were installed in 2006, and presently the plant has a total capacity of 14 000 MW.

The Itaipu site has rather good geological characteristics for the construction of a hydro project. It is located in an area underlain by the basalt flows that cover the upper Paraná basin. The basalt flows at the site are essentially horizontal with thickness varying from 20 to 60 m with breccia layers between the flows, with thickness from 1 to 30 m. The massive basalt rock has excellent mechanical properties and is suitable both as foundation for the structures and as construction material. The breccia, however, is relatively weak and heterogeneous. The formulation of the project involved very extensive studies and investigations, including more than 30 000 m of core drilling, almost 400 m of shafts, 1600 m of tunnels, and 660 m of trenches.

The extensive hydrologic studies performed were based on data collected from 59 stream gauging stations and 65 meteorological stations located in Brazil, Paraguay, and Argentina and on data from the 136 planned and existing reservoirs that affect flow at the site. Flood studies were computed during feasibility studies based on the Probable Maximum Flood (PMF) concept and were eventually recomputed and confirmed with data from the unprecedented high floods that occurred in 1982–83 over the Paraná basin. The maximum peak inflow at the site was established as 72 020 m^3 s^{-1} and the hydrograph was routed through the reservoir to define the spillway design flood. For the design of the river diversion structures, the 100-year flood corresponding to 35 000 m^3 s^{-1} was selected.

The layout locates the powerhouse at the middle of the river that is flanked on the right bank by a curved (in plan) concrete buttress dam up the spillway site, continuing after the spillway by an earthfill dam closing the valley (**Figure 3**). On the left bank the powerhouse is crossed by the diversion channel, where three units are located, continues with a concrete buttress dam followed by a rockfill dam and an earthfill section closing the valley.

River diversion was done through a channel excavated on the left bank. A concrete gravity structure aligned with the main dam, and ultimately becoming part of it, housed the 12 sluiceways, 6.7 m wide by 22 m high, controlled by diversion gates. Concrete arch cofferdams were built for the construction of the diversion structure. These cofferdams were later blasted and removed to allow the flow of the river through the sluiceways. The sequence and key dates of the river diversion scheme are depicted in **Figure 4**.

The whole sequence and features of the diversion operation were tested in hydraulic models carried out by the Federal University of Paraná hydraulic laboratory, in Curitiba, Brazil. Final closure of the diversion gates was carried out on 13 October 1982 and was completed successfully in 8 min. The flow of the Paraná River was 12 000 m^3 s^{-1} and the reservoir was filled in 15 days. As provided by the design, the diversion gates were recovered and used for the power intake. Storage in reservoirs on the Iguaçu River provided the riparian flow for downstream reaches of the Paraná River, during the filling of the Itaipu reservoir.

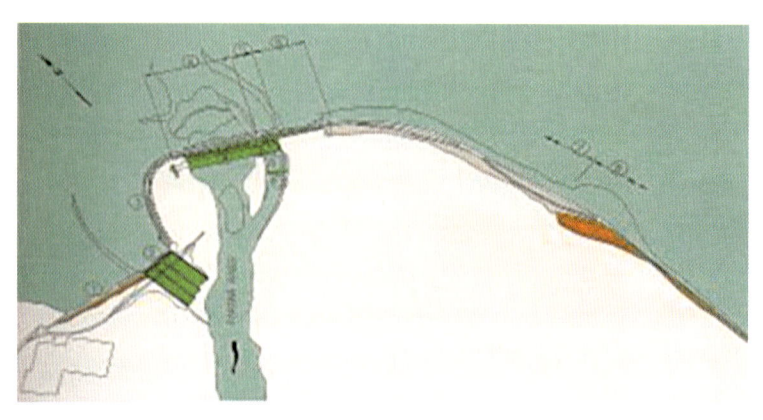

Figure 3 Project general layout. 1, Right bank earthfill dam; 2, Spillway; 3, Right bank hollow gravity dam; 4, Main dam and power intakes; 5, Diversion channel dam and intakes; 6, Left bank concrete dam; 7, Left bank rockfill dam; 8, Left bank earthfill dam; 9, Powerhouse at river channel.

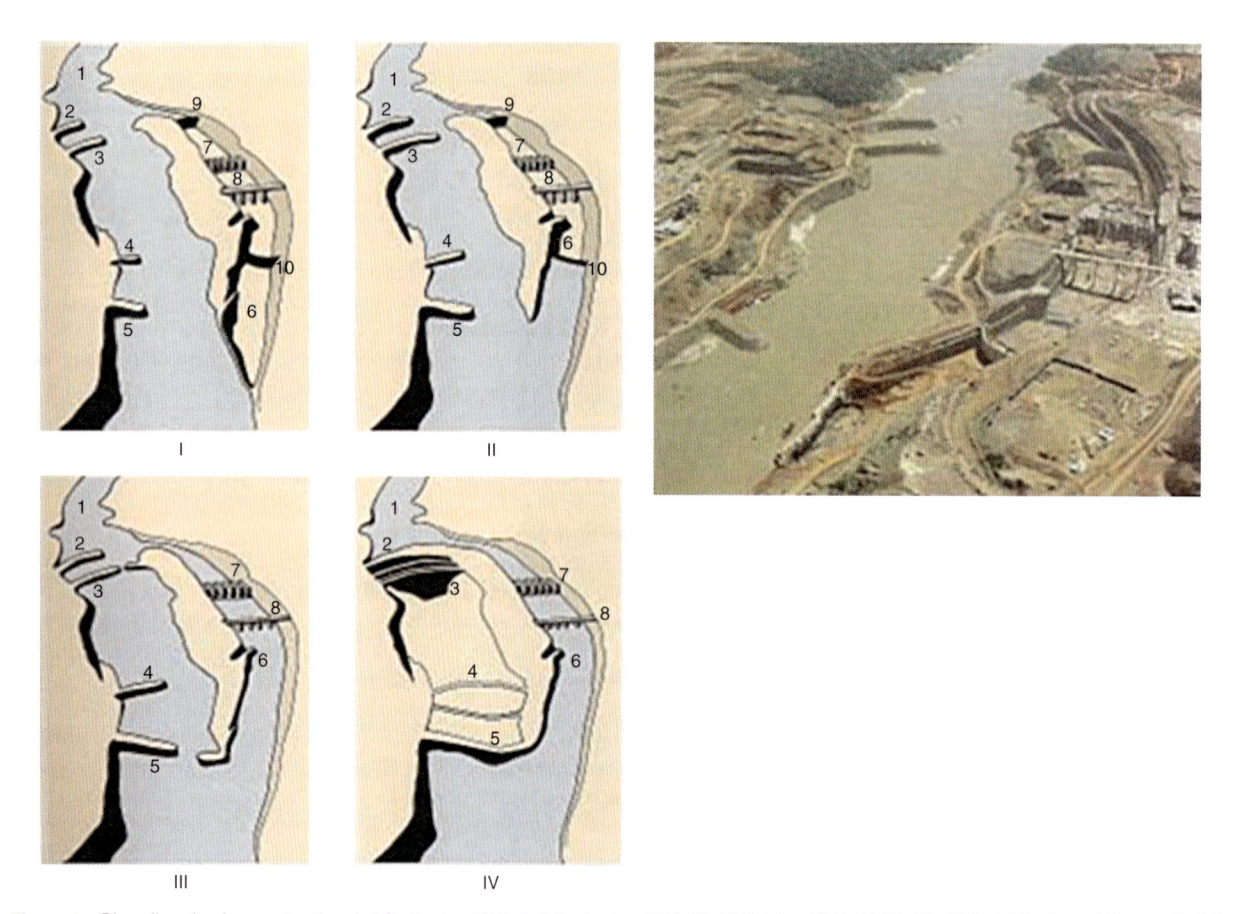

Figure 4 River diversion for construction. I, 2 September 1978; II, 6 September 1978; III, 20 October 1978; IV, 30 July 1979; 1, The Paraná River; 2, 3, 4, 5, dikes for main cofferdams; 6, diversion channel; 7, diversion structure; 8, service bridge; 9, upstream arch cofferdam; 10, downstream arch cofferdam.

6.04.2.2 The Dam

The dam at Itaipu is made up by a central stretch of a hollow gravity concrete dam, a mass concrete gravity section housing the river diversion facilities, and two wings to the right and left of the central stretch formed by concrete buttress structures, adding up to a length of 3472 m and, on both sides, earthfill and rockfill dams closing the valley. The length of the earthfill and rockfill reaches is 4728 m, adding up to a total length of 7750 m. The maximum height of the dam is 196 m, measured from the foundation at the central part of the river.

The central stretch concrete dam is formed by 18 hollow gravity blocks. The 16 blocks located immediately upstream of the powerhouse support the power intake. The blocks are monolithic cells, each consisting of an upstream head supported by two buttress stems and enclosed by a downstream face slab. Adjoining blocks abut against each other at the upstream head, at the downstream face slab, and in the upper portion but are separated by transverse contraction joints.

The buttress dam portion is made up, on the right bank, by 64 blocks and on the left side, by 19 blocks. All blocks are identical in structural configuration and profile and are 17 m wide at the axis. Height of the blocks range from 35 to 85 m. Except for the galleries near the crest and the foundation, there are no major openings crossing the buttress dam portion.

The closing portion of the dam at the right side of the valley and a part of the left wing dam (numbered 7 in **Figure 3**) are earth-core rockfill dams. The earthfill stretch of the dam on the left bank was selected because of the availability of adequate soil material in the area. The maximum height of this stretch is 30 m and its length is 630 m.

6.04.2.3 The Spillway

The Itaipu spillway is a gated surface chute spillway with capacity of passing 62 200 $m^3 s^{-1}$ with the reservoir at the full supply level at El. 223. It is located on the right bank of the Paraná River and is divided into three independent chutes to allow operational flexibility and capability to safely handle emergencies, and for that, each chute can discharge about twice the average natural flow of the river.

The spillway has an ogee-shaped control structure with 14 segment-type (tainter) gates, 20 m wide by 21.34 m high supported by 5 m wide piers. Its total width is 380 m and maximum length is 483 m. Each chute has a different length and longitudinal profile, fitted to the location and foundation surface. The chutes end in a flip-bucket configuration to provide energy dissipation without damage to permanent works and without significant surcharge of the powerhouse tailwater. The specific discharge of the spillway is 183 $m^3 s^{-1} m^{-1}$ and the exit velocity is 40 $m s^{-1}$. The design of the spillway including the geometry of the buckets was extensively tested with hydraulic models in Curitiba, Brazil.

The hydraulic performance of the spillway was satisfactory and essentially free of major problems. Some lateral erosion in the rock was observed in the downstream plunge pool. This was, to a certain degree, forecasted in the model studies, although in one case it did affect the left side of the chute immediately downstream of the bucket [8]. This has been associated with the unusually intense operation of the spillway in the first years after its commissioning. In fact, due to the schedule of the installation of generating equipment, during the first 3 years all three chutes of the spillway operated almost continuously after reservoir impounding. Thereafter, for the next 3 years one or two of the chutes operated continuously. This is rather unusual in hydroelectric projects, but it represented a unique opportunity to check spillway design and the result confirmed the excellence of it. It is estimated that during the five initial years of operation about 500 TWh of energy passed over the Itaipu spillway.

6.04.2.4 The Power Plant

The Itaipu power plant is formed by 20 generating units, each with a capacity of 700 MW. The powerhouse is located immediately downstream of the dam, in the central part of the river. The power intake is located on top of the hollow gravity dam and allows short penstocks to reach the generating units.

A special characteristic of the Itaipu power plant is that half of the units generate power in 60 Hz and half in 50 Hz, respectively, according to frequencies of the Brazilian and Paraguayan electrical systems. The power generated at 18 kV is transformed at the GIS step-up substation, located immediately upstream of the powerhouse, to 500 kV, and from there connected to the respective systems in each country. As mentioned earlier, each country has the right to purchase and use the excess power not used for domestic supply. For that reason the Brazilian side is also connected to the 50 Hz generating system and in Brazil is converted into direct current, transmitted to the São Paulo area, reconverted to AC 60 Hz and fed into the country integrated transmission system.

Figure 5 depicts a typical transversal section of the powerhouse with indication of its main installation features.

All power intakes are identical in configuration, design, and equipment. The Itaipu plant was planned to operate as a run-of-river plant, with a normal maximum drawdown of 1 m with possibility, in an emergency situation at the spillway, to deplete the reservoir level to the elevation of the spillway sill. **Figure 6** shows a typical cross section of the power intake.

The penstocks are made of welded steel, with an internal diameter of 10.5 m, and feed directly to the turbines as indicated in **Figure 5**. They are anchored to the dam and embedded in second-stage concrete placed in a large blockout in the face of the dam.

The powerhouse is an independent 968 m long structure located at the toe of the main dam. It contains the 20 bays of the units, along with two equipment erection and maintenance areas, and miscellaneous areas for technicians and operators. The central control room is located downstream from the powerhouse in an independent building. **Figure 7** shows a sketchy representation of the powerhouse arrangement and an external view of the powerhouse and administration building.

Each unit bay is 34 m wide and is 94 m high, from El. 50 to El. 144. It houses a turbine-generator unit, three main unit single-phase step-up transformers, switchgear, and mechanical and electrical auxiliary equipment.

The right-bank erection area has an unloading, unpacking, and preassembly area at El. 144 and is served by two 2.5 kN cranes accessing the main assembly area at El. 108. This main assembly area is 141.3 m long and 29 m wide. The central erection area has also an unpacking and preassembly area at El. 144 with another two 2.5 kN cranes that can also access the main assembly area.

The central control room is located downstream at El. 135, between units 9A and 10 with a viewing area above El. 139.

The turbines are of Francis type, and were specified to develop 715 MW at the rated head of 112.9 m. The head for overall best efficiency was 118.4 m. Performance of the turbines so far has been excellent. They have been commissioned without any problem

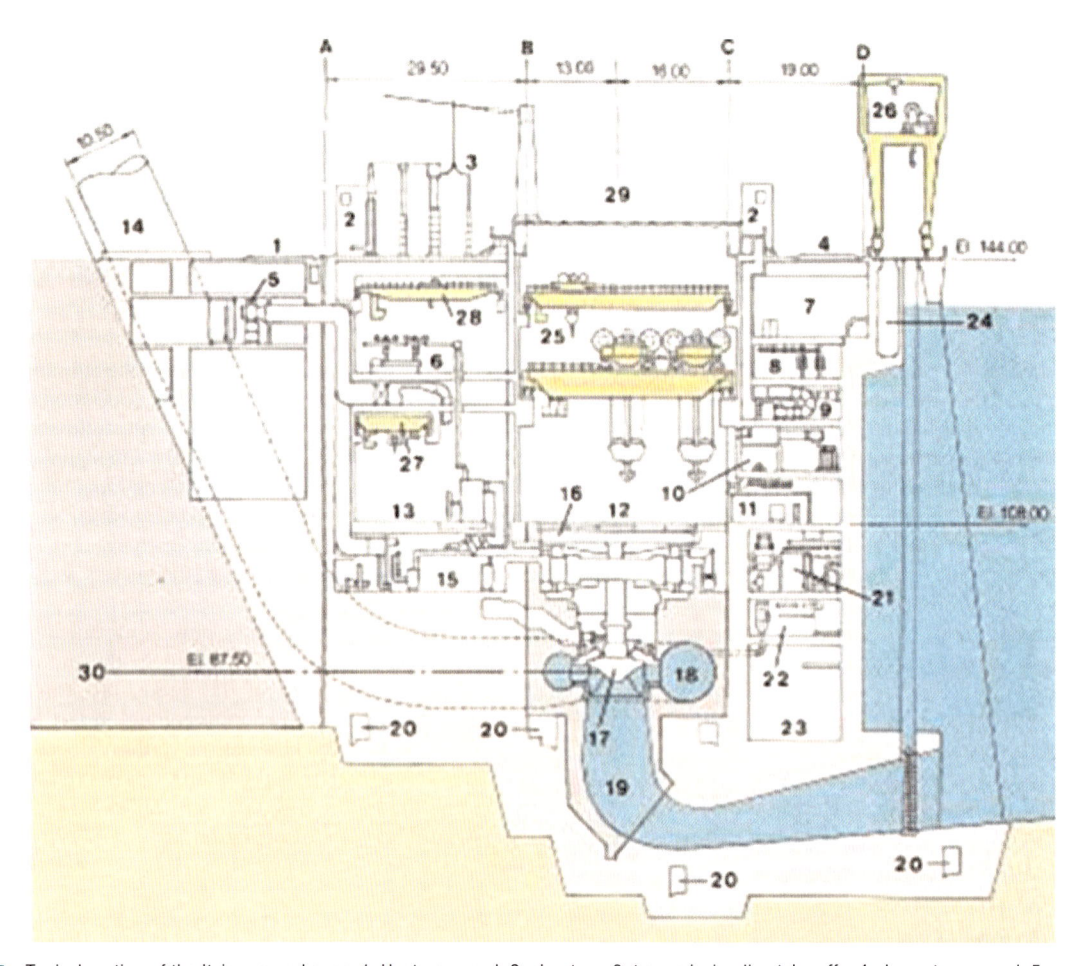

Figure 5 Typical section of the Itaipu powerhouse. 1, Upstream road; 2, elevators; 3, transmission line take-offs; 4, downstream road; 5, powerhouse upstream ventilation rooms; 6, GIS; 7, electrical equipment gallery; 8, electrical cable gallery; 9, ventilation equipment gallery; 10, battery room; 11, local unit control room; 12, generator hall; 13, main transformers gallery; 14, penstock; 15, electrical auxiliary and excitation equipment gallery; 16, generator; 17, turbine; 18, spiral case; 19, draft tube; 20, drainage gallery; 21, mechanical equipment gallery; 22, pumps, strainers, and piping gallery; 23, anti-flooding gallery; 24, draft tube stop-log storage; 25, main powerhouse crane (10 MN); 26, gantry crane 1.4 MN; 27, main transformers crane 2.5 MN; 28, GIS equipment crane.

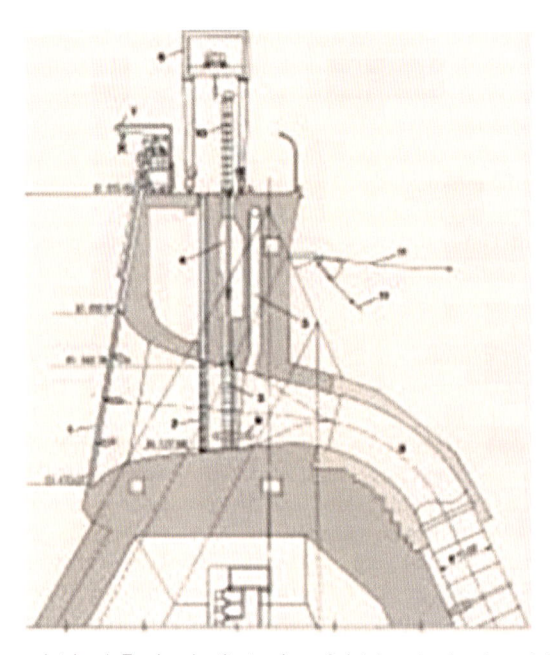

Figure 6 Typical cross section of the power intake. 1, Trashracks; 2, stop logs; 3, intake gate; 4, gate maintenance chamber; 5, air vent; 6, 1100 kN gantry crane; 7, trashrack cleaning machine; 8, penstock; 9, bypass valve; 10, intake-gate servomotor; 11, transmission line.

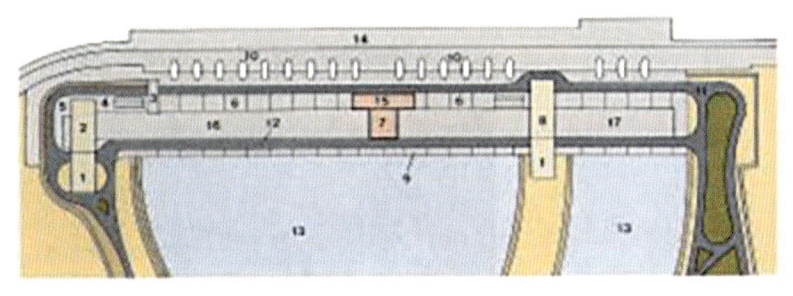

Figure 7 Powerhouse layout and external view. 1, Equipment unloading building; 2, right-bank erection area; 3, transformer unloading area; 4, auxiliary service transformers; 5, vertical circulation access; 6, transmission line take-offs; 7, central control room; 8, central erection area; 9, draft-tube stop-log hatches; 10, penstocks; 11, upstream road; 12, downstream road; 13, tailrace; 14, dam and power intakes; 15, operation and administration building; 16, river-bed powerhouse; 17, diversion-channel powerhouse.

and have operated in a satisfactory way for many years. The only repair work carried out was related to minor cavitation damage in the runners, probably associated with low load operation.

Because of the need to produce electric current at different frequencies, half of the generators generate at 50 Hz and the other half at 60 Hz, all of them driven by identical turbines of 715 MW rated capacity. The 60 Hz generators have a rated power factor of 0.95, which corresponds to a rated output of 737 MVA considering a generator efficiency of 0.98. The power factor of the 50 Hz generators is 0.85, corresponding to a rated output of 823.6 MVA.

6.04.3 The 8125 MW Tucurui Hydroelectric Project

The Tucurui project is the second largest hydroelectric project in the Brazilian territory and the largest installation that is 100% Brazilian [9–11]. It is located in the Tocantins River in the state of Pará, in northern Brazil. It was built, is owned, and is operated by Eletronorte – a federal government public utility for electric power responsible for the bulk supply of electric power in the northern region of Brazil. The project was designed by the Brazilian consulting firms Engevix and Themag and built by contractor Camargo Correa. The construction supervision was done directly by Eletronorte.

The Tocantins River and its main tributary, the Araguaia River, is one of the major river systems in the Brazilian territory (see **Figure 1**). Its total drainage area is 967 059 km^2, of which 758 000 km^2 are upstream of the Tucurui site. The Tocantins River headwaters are located in the central part of Brazil, at an elevation of about 800 m above sea level (m asl), where the country's capital, Brasilia, is located. Its course runs essentially in a south–north direction for a length of about 2500 km discharging in the estuary of the Amazon River near the city of Belém, capital of the state of Pará.

The Tucurui project is the furthest downstream hydroelectric project contemplated in the cascade of projects of the Tocantins River, which include five other projects presently in operation (Serra da Mesa, 1275 MW; Canabrava, 465 MW; São Salvador, 243 MW; Peixe Angical, 452 MW; and Lageado, 903 MW), one under construction (Estreito, 1087 MW), and four being studied (Ipueiras, 480 MW; Tuparitins, 620 MW; Serra Quebrada, 1328 MW; and Marabá, 2160 MW).

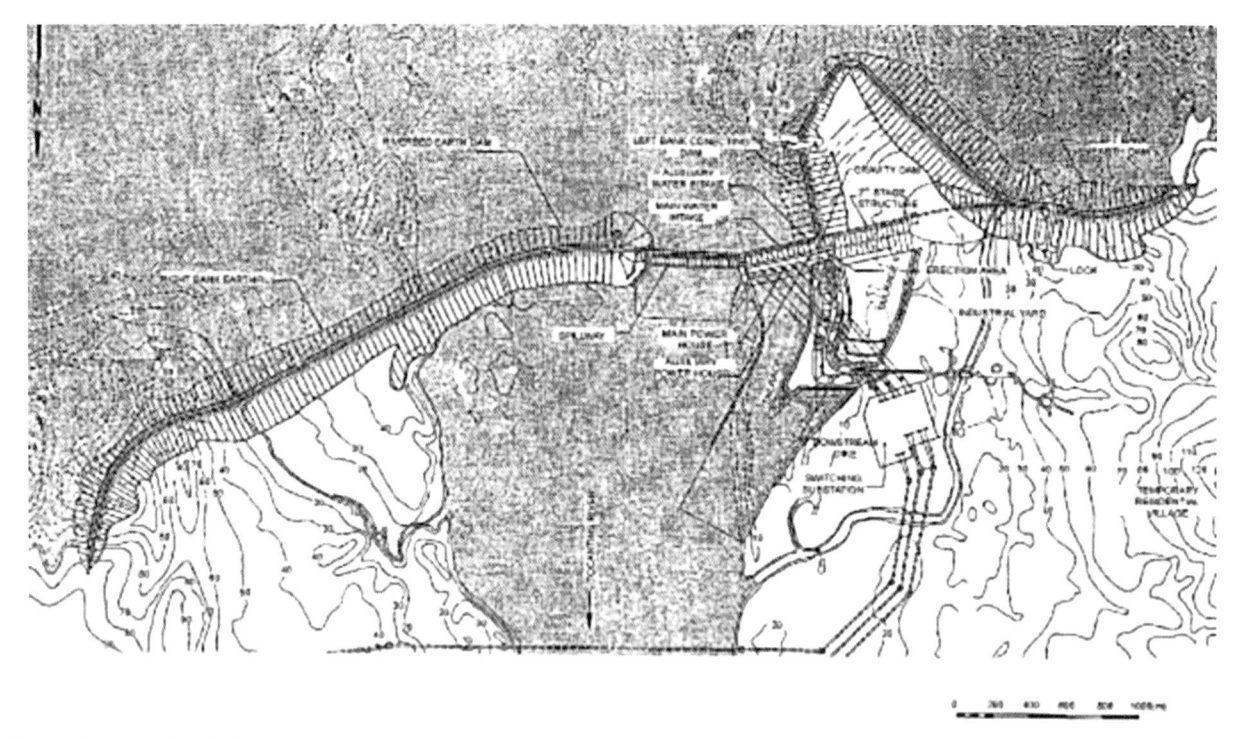

Figure 8 General layout of the Tucurui project.

The Tocantins River at Tucurui has a wide valley with low topography. The project layout displayed the structures in sequence, with the spillway followed by the power plant on the riverbed area near the left bank and the remainder of the valley closed to the right and to the left by rockfill dams. This arrangement has allowed the isolation of the area for the second-stage power plant and provided initial structures for the future incorporation of navigation locks. Figure 8 shows the general layout of the project.

The full installed generating capacity of the Tucurui project is 8125 MW, which was achieved in two stages. The works corresponding to the initial stage included the rockfill dams in both margins, the spillway, and half of the power plant with an installation of 12 units, each with rated output of 330 MW and two 20 MW auxiliary ones, totaling 4000 MW. This was done between 1976 and 1984. To the left of the first-stage power plant, an area was isolated by cofferdams to allow the later second-stage power plant, which was completed in 2007, increasing the project capacity to 8125 MW.

For the construction of the project, river diversion and control during the works posed major challenges, not only because of the magnitude of the flows to be managed but also because the river bottom was found to be extremely irregular with rock channels and sand deposits that complicated considerably the construction of impervious cofferdams. The initial construction sequence considered a two-phase diversion, which consisted of earth-rockfill cofferdams isolating areas in both margins, the construction of the spillway structure with sluiceways underneath to handle the river during the second phase and final closing of sluiceways to start reservoir impounding. The initial studies, including the project basic design, considered that cofferdams were to be designed for the 50-year recurrence flood of $51\,000\,\text{m}^3\,\text{s}^{-1}$. However, in 1980, with cofferdams built in the left margin and construction in progress, a major flood of $68\,400\,\text{m}^3\,\text{s}^{-1}$ occurred, exceeding by 33% the diversion design flood. This size of flood had never occurred in the historical record of 100 years [12]. Exceptional circumstances as the widening of the constricted river channel due to previous flood erosions on the opposite margin and a conservative freeboard in the cofferdams luckily prevented the construction site to be flooded. The event forced a modification of the river diversion scheme, which was changed from a two- to a three-phase sequence. The flood event changed the hydrologic series and the 50-year design flood was recalculated to be equal to $58\,600\,\text{m}^3\,\text{s}^{-1}$. Except for this exceptional event, which fortunately did not affect the construction area and did not change the construction schedule, the realization of the project was accomplished successfully.

The Tucurui spillway is one of the largest in the world, with a design capacity of $110\,000\,\text{m}^3\,\text{s}^{-1}$. It is a gated spillway structure incorporated into the mass concrete of the dam. It is equipped with 23 radial gates, each 20.0 m wide by 20.75 m high. The discharge of the spilled flow into the river is done through a cylindrical shaped bucket that issues a jet hitting the water surface between 80 and 130 m away from the toe of the structure and over an excavated plunge pool. The maximum specific flow over the bucket, $207.0\,\text{m}^3\,\text{s}^{-1}\,\text{m}^{-1}$, is also a very high figure in comparison with other projects elsewhere.

Underneath the spillway, there were 40 diversion sluiceways, 6.5 m wide by 13.0 m high, which were used to close the river and start reservoir impounding. Figure 9 shows a view of the spillway structure with the diversion sluiceways. Closure operation used 20 steel recoverable gates to close the upstream entrance of each sluiceway and precast concrete stop logs to close the downstream end. These were lowered from the downstream bridge after the flow in each passage was interrupted by the upstream recoverable steel gate.

Figure 9 View of the Tucurui spillway during construction.

The Tucurui power plant has two powerhouses. **Figure 10** shows the cross section of the first power plant.

The first power plant built with the initial project works includes 12 main units and 2 auxiliary ones, as indicated before. The second one that was built while the first one was operating contains 11 units with an individual capacity of 375 MW. Except for the difference in unit capacity, the arrangements of the two powerhouses are similar.

The power intake is a gravity-type structure divided into blocks corresponding to the generating units. The penstocks are imbedded in the structure as shown in **Figure 10**. The powerhouses, located at the toe of the power intakes, are essentially similar to each other and are of the sheltered type with auxiliary service galleries placed downstream whose structure supports the power transformers.

The Tucurui project incorporates locks that allow the navigability of the river linking the agriculture productive areas of the central plateau of Brazil to the port of Belém.

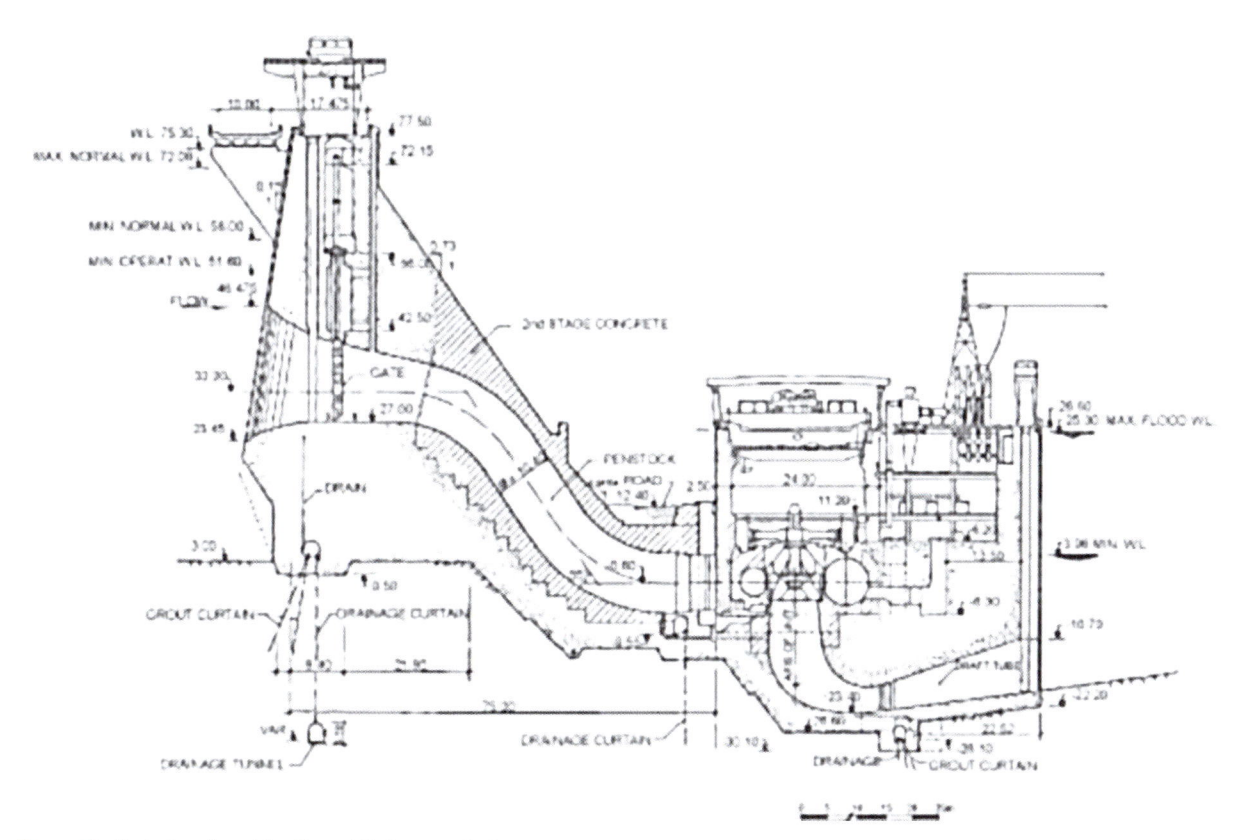

Figure 10 Typical section of the Tucurui first power plant.

6.04.4 The 6450 MW Madeira Hydroelectric Complex

The Madeira hydroelectric complex is formed by two projects located in sequence on the Madeira River, in the state of Rondonia in northwest Brazil. These two projects are presently (2009) under construction and are expected to be on line by 2012. The upstream project is the Jirau project, with an installed capacity of 3450 MW and the other is the Santo Antonio project with an installed capacity of 3150 MW.

The Madeira River is the main tributary of the southern bank of the Amazon River. Its course is 1450 km long. Its headwaters are located on the Andes Mountains in Bolivia, with the name of Beni River. Its course follows initially the south–north direction changing to the SE–NE direction after receiving the waters of the Guaporé River and entering the Brazilian territory first as a border river and then traveling inland.

The hydroelectric potential of the Madeira River was evaluated by a comprehensive study carried out under the sponsorship of Eletrobras, the Brazilian federal agency for electric power with a view to developing power projects and extending inland navigation along the river. The intention was to define these projects in the context of the Initiative for the Integration of the Regional Infrastructure of South America (IIRSA), a combined effort by several South American countries [13]. As a result of this study, two projects were defined to develop the head between El. 90, where the river starts to run inside Brazil, and El. 50 downstream of the Santo Antonio rapids. This stretch of the river is 250 km long and after its lower end it will still run for about 1000 km before reaching the main course of the Amazon River.

Figure 11 shows a map with the location of the Jirau and Santo Antonio projects, which constitute the Madeira hydroelectric complex. The Santo Antonio project is located 10 km upstream of the city of Porto Velho, capital of the state of Rondonia, and Jirau 110 km upstream. Both undertakings are very low head projects and were designed to have a minimum increase in the natural flood level of the river. They incorporate navigation locks and separate passageways for migrant fishes that abound in the river.

The concession for construction and operation of the projects was the object of a competitive tendering process in two independent auctions. As a result the Jirau project was awarded to a group of private- and state-owned companies led by GDF-Suez Energy and including Camargo Correa – a Brazilian contractor – and Chesf and Eletrosul – state-owned utilities. Similarly, the Santo Antonio project was awarded to a consortium led by Odebrecht – a Brazilian contractor – and Furnas – a state-owned utility.

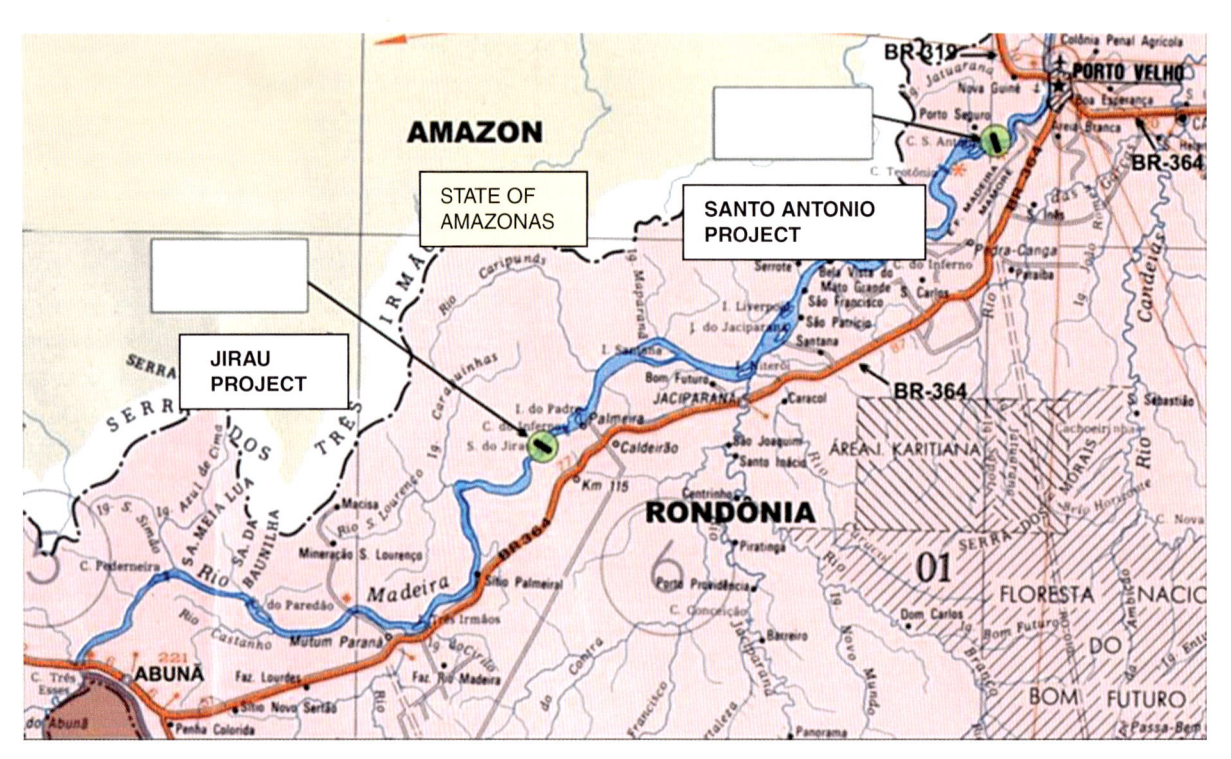

Figure 11 Location of the Madeira hydroelectric projects.

6.04.4.1 The Santo Antonio Project

The Santo Antonio project was the first awarded of the two projects. It is being built by an EPC contractor, led by Odebrecht and including equipment suppliers Alstom, Voith, and Andritz. It is being designed by two Brazilian engineering consultants, PCE and Intertechne. Its construction started in September 2008 and the program calls for the first generating unit be on line in May 2012 and the 44th unit, in June 2015, encompassing 81 months to complete the whole project.

The Madeira River at Santo Antonio has an average flow of $18\,000\,\text{m}^3\,\text{s}^{-1}$. The spillway design flood is $84\,000\,\text{m}^3\,\text{s}^{-1}$, which makes it the second largest in Brazil, after the Tucurui spillway.

The project utilizes the head of 13.9 m created by the dam. It contains 44 generating units of the bulb type each driven by horizontal-shaft turbines. The project layout was defined taking into consideration, besides local topographical and hydrological consideration, the construction sequence and the maximum possible anticipation of the production of power. The layout placed the various structures in a linear sequence but considered three separate powerhouses, one near the right bank, one near the left bank, and one in the middle of the river (**Figure 12**). Since the project head is low, and its dams are of course of modest height, it will be possible to generate power (and income) before the third powerhouse is completed while being protected by cofferdams.

The two powerhouses close to the left bank will have 24 units (units 9–32) and three assembling areas for the equipment installation. Following these powerhouses is the main spillway with 15 passages controlled by radial gates, each 20.0 m wide by 24.18 m high, supported by 5.0 m wide pillars. In sequence the third powerhouse located at the middle of the river contains 12 units (units 33–44) and 2 assembling areas. This third powerhouse is connected to the main spillway on its right side. To the left side, concrete gravity dams make the connection to the complementary spillway. This last spillway will have three passages also controlled by radial gates of the same size as the main spillway. After this spillway, in the direction of the right bank, the fourth powerhouse will contain eight units (units 1–8) and one assembling area. On both margins the concrete structures are complemented by earth dams to close the section.

The total installed capacity of the project is 3150 MW, formed by 24 units with nominal capacity of 73.28 MW and 20 units with 69.59 MW. The powerhouses are formed by typical modules, each including four units sharing the same step-up transformer and formed by two structurally independent blocks. Each four-unit module is 85 m wide by 72.9 m long by 58.2 m high. **Figure 13** depicts a typical section of the unit block.

The river diversion and control during construction took into consideration the marked seasonality of the Madeira River, which has wet and flood period from July to November concentrating, on the average, 45–60% of the larger annual floods. The river diversion sequences were programmed with isolating areas on both margins, and, after the construction of the main spillway, divert the river through it. The construction sequence and the natural river conditions allowed the selection of different design floods for the protection of different parts of the works, from 100-year recurrence flood for channel and major excavation works to 300-year floods for relevant concrete structures. The 100-year flood is of the order of $40\,000\,\text{m}^3\,\text{s}^{-1}$ (June–November) and the 300-year flood, for the same period, is approximately $45\,000\,\text{m}^3\,\text{s}^{-1}$.

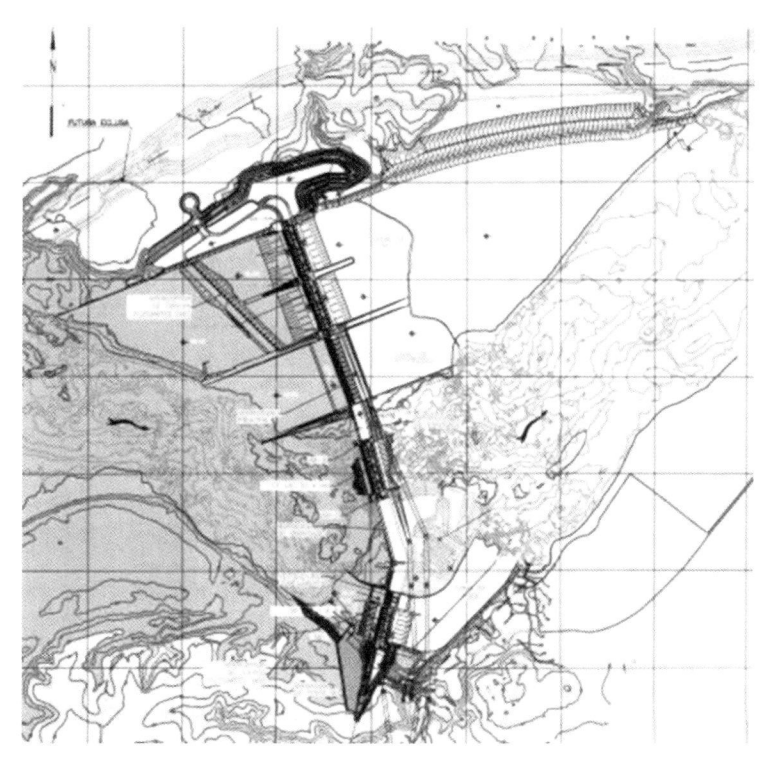

Figure 12 General layout of the Santo Antonio project.

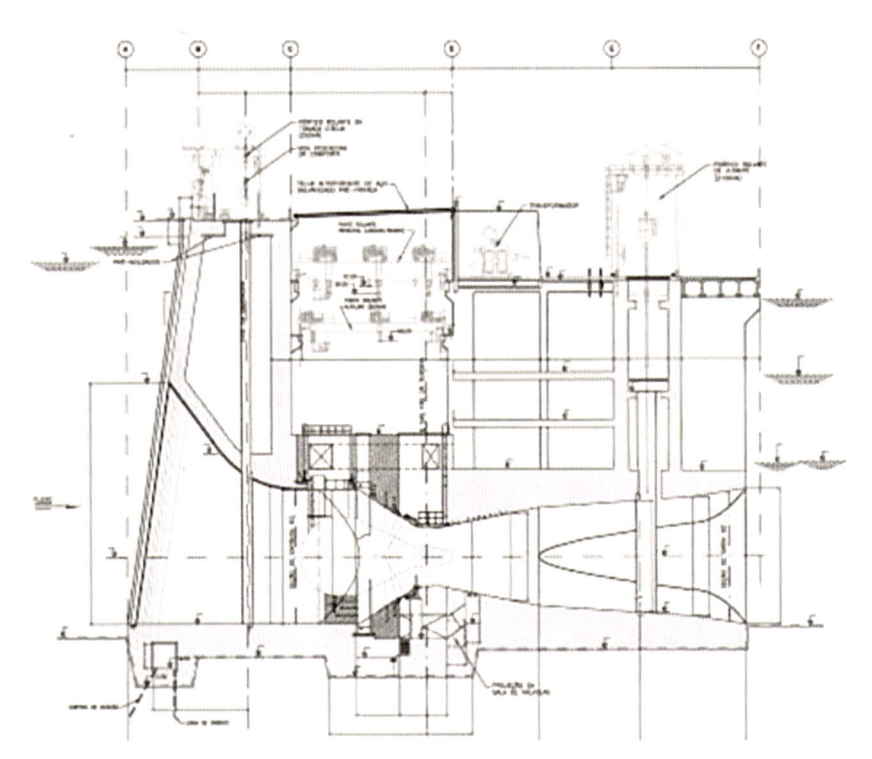

Figure 13 Typical section of the Santo Antonio powerhouse.

Figure 14 View of the construction site of the Santo Antonio project in September 2009.

The main construction quantities involved in the construction of the Santo Antonio project (**Figure 14**) are the following:

Soil excavation 38.4 million m^3
Rock excavation 21.2 million m^3
Concrete (conventional) 2.3 million m^3
RCC (rolled compacted concrete) 0.8 million m^3

6.04.4.2 The Jirau Project

The Jirau project, located upstream of the Santo Antonio project, is being built, under direct owner coordination, by Brazilian contractor CamargoCorrea. Main electromechanical equipment is being supplied by an association of major suppliers, including Alstom, Voith, Andritz, and DEC. Engineering design of the works is being carried out by Brazilian consulting firm Themag. Leme Engenharia is acting as Owner's Engineer. Its construction started in April 2009 and the first unit is programmed to be on line in March 2012.

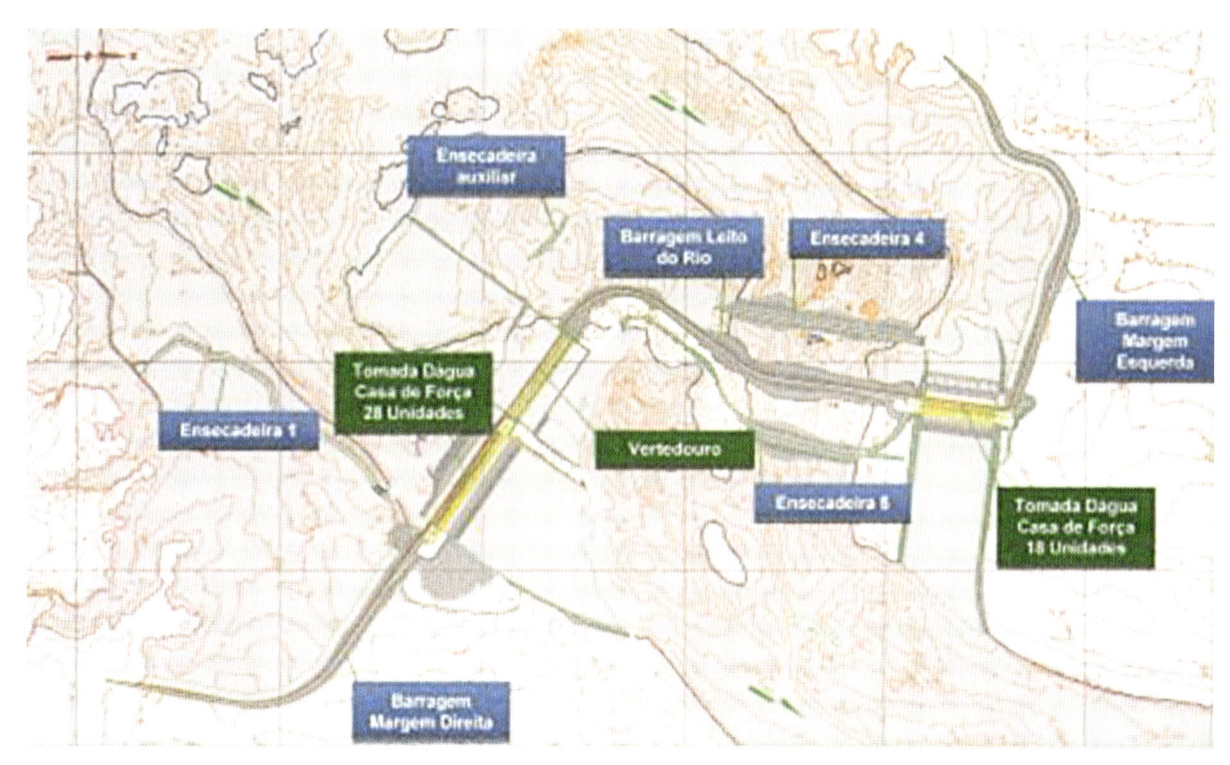

Figure 15 General layout of the Jirau project.

The project site is located in a wider stretch of the river where an island in the middle of the river will be used to facilitate the river diversion during construction. **Figure 15** shows the project layout.

The project layout displays the various structures aligned, forming a 'V' with its vertex on the river island. The project will utilize the maximum head of 19.9 m created by the dam and install 46 generating units with unit capacity of 75 MW driven by bulb turbines with horizontal axis. The total installed capacity will be of 3450 MW.

The project will have two powerhouses: one (PH 1) in the main course of the river, on the right-hand side of the island, with 28 units; and the other (PH 2) on the channel excavated on the left bank, with 18 units. There will be only one spillway to discharge the maximum design flood of 82 600 m^3 s^{-1}, with 18 passages controlled by radial gates 20.0 m wide by 21.82 m high. Between the spillway and PH 2, there will be an earth-rock dam 575 m long and with a maximum height of 53 m.

River diversion will be done isolating the area between the right bank and the middle-river island by two roughly parallel cofferdams. In this area the spillway and powerhouse PH 1 will be built while the river is flowing on the channel close to the opposite margin. The second-phase diversion will be through the spillway, while the left bank is being protected by cofferdams. **Figure 16** shows a view of the construction stage in November 2009.

Figure 16 View of the construction site of the Jirau project in November 2009.

6.04.5 The Iguaçu River Projects

The information for this section was based on material compiled in Reference [14]. The Iguaçu River is one of the main tributaries of the middle course of the Paraná River. It drains a basin of about 69 000 km^2 and runs essentially in an east–west direction through a length of about 1000 km, from its headwaters, near the city of Curitiba, in the state of Paraná, to its mouth in the Paraná River, immediately downstream of the Itaipu project. Along its course it drops more than 800 m, frequently through concentrated steps, among which the internationally famous, Iguaçu Falls. For about 90% of its length, it runs inland in the Brazilian territory, in the state of Paraná, but for the last 120 km downstream it constitutes the border between Argentina and Brazil.

The natural conditions of the basin favor the hydroelectric development. The topography of the area with concentrated drops of the rivers, a geologic configuration formed by basalt flows, and a subtropical climate with no definite dry seasons allowing an average of more than 1300 mm of annual rainfall throughout the basin provide a substantial power potential that was implemented in six major projects cascading along the Brazilian reach of the river. There are no projects in the international reach that is relatively flat upstream of the Iguaçu Falls. At the site of these falls, so far no project has been seriously considered as well, and surely there will never be one.

The hydroelectric projects built on the Iguaçu River are the following:

Project name	Ultimate capacity (MW)	Present installed capacity (MW)	Date commissioned
Foz do Areia	2511	1674	1980
Segredo	1260	1260	1992
Salto Santiago	2000	1332	1980
Salto Osorio	1050	1050	1972
Salto Caxias	1240	1240	1998

There is one other project located downstream of Salto Caxias, which is presently under initial construction. It is called Baixo Iguaçu, and is located immediately upstream of the border between Brazil and Argentina and is planned in such a way as to clear the area of the Iguaçu Falls Natural Park. The installed capacity in this project will be of 350 MW.

All the Iguaçu River projects are connected to the Brazilian National Interconnect System mainly through 500 kV links.

6.04.5.1 The Foz do Areia Project

The Foz do Areia project is the most upstream of the Iguaçu River projects. It was built and is owned by COPEL the electric power utility of the state of Paraná. It was designed to generate power at the site and to provide a regulating reservoir to benefit downstream projects. Presently, its full regulating capacity is no longer used because the integrated system provides electrically this regulation and the avoidance of extreme depletion of the reservoir allows a larger generation of energy.

The project was built between 1975 and 1980. It was designed by the Brazilian firm Milder-Kaiser Engenharia and built with two sequential construction contracts, one for diversion by Andrade Gutierrez S.A. and rest of the project by CBPO – Companhia Brasileira de Projetos e Obras – of the Odebrecht Group.

The project is located in the upper part of the middle course of the Iguaçu River at about 240 km from the city of Curitiba. The local topography of the site does not present a concentrated step, as in other sites of the river. The river, at the site, is relatively narrow and the abutments rise in step-type features, reflecting the superposition of basaltic flows. The local sequence of rocks has a marked predominance of dense basalts with basaltic breccia making up the balance. The soil and weathered rock mantle is unusually thick in the area. This would have favored a soil-rockfill dam, but the excessive humidity of the site would have made it expensive and would have affected the project schedule.

Figure 17 shows a view of the completed project and **Figure 18** depicts the general project layout. The Foz do Areia project includes a concrete-face rockfill dam, a spillway and a power plant formed by a power intake, and six power tunnels feeding the external powerhouse containing six bays for 418.5 MW units, of which only four have been installed, and a GIS step-up substation. River diversion for constructing the dam was done through two unlined tunnels, 12 m in diameter, excavated in the right abutment.

A significant feature of the Foz do Areia project is its dam. When it was completed in 1980, it was a world record for concrete-face rockfill dams, with its height of 160 m, and remained as such until the mid-1990s, when Aguamilpa, in Mexico, reached the height of 190 m. The dam included many engineering advancement details in the concrete-face design and construction that made it a reference milestone for this type of structure. A view of the dam and appurtenant facilities immediately before reservoir impounding is shown in **Figure 19**.

The spillway of the Foz do Areia project is a gated chute spillway, located on the left abutment, designed for a discharge of 11 000 m^3 s^{-1} corresponding to the 1/10 000 year maximum project flood. The chute is 70.6 m wide and 400 m long, ends in a flip bucket to dissipate energy in the plunge pool excavated in the channel downstream. The chute was provided with three aeration devices to prevent cavitation, which showed excellent results both in model and prototype tests and along the life of the structure. As a result of revised hydrological studies carried out after very large floods observed during the 1980s, the reservoir has been systematically operated below the normal operating level to provide additional volume to allow the safety discharge of the spillway.

Large Hydropower Plants of Brazil 109

Figure 17 The Foz do Areia project.

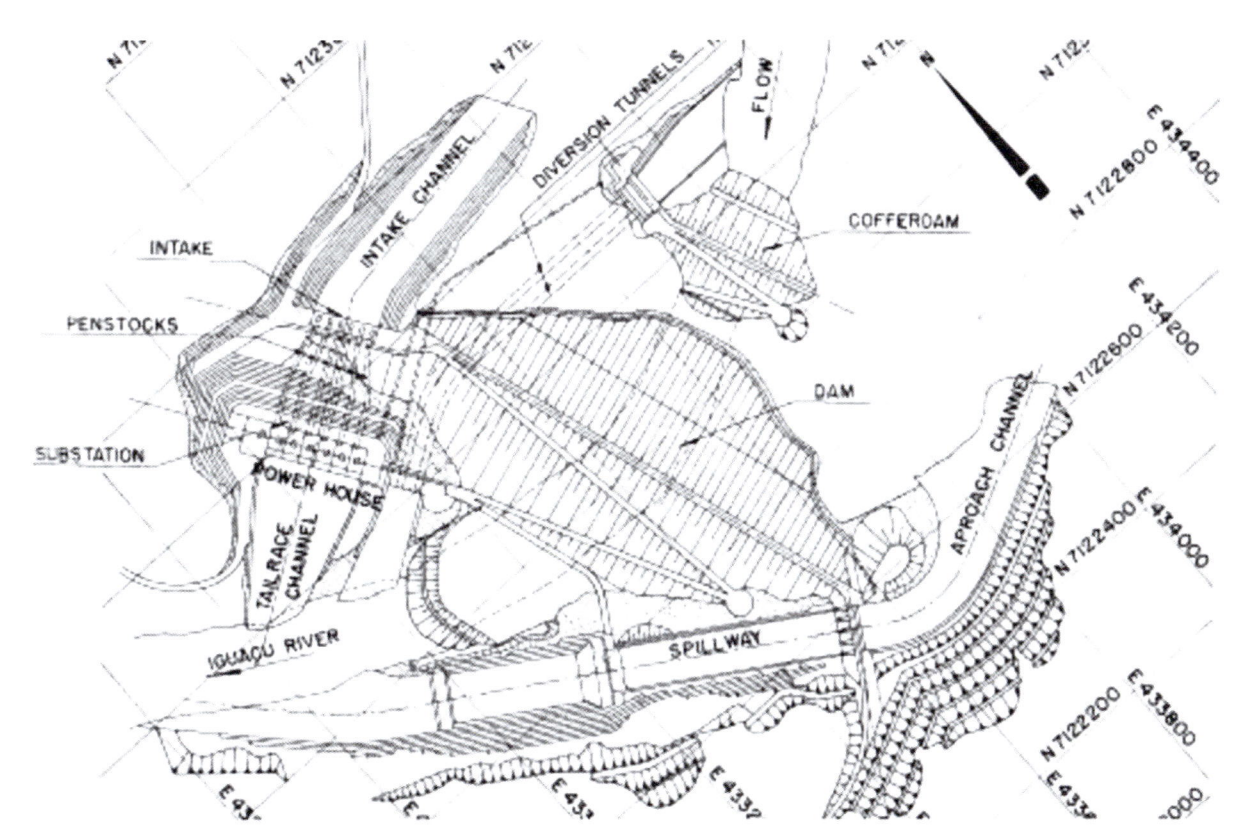

Figure 18 General layout of the Foz do Areia project.

The power plant is depicted in **Figure 20**. The power intake is 72 m high, deeply excavated in the rock and sustained by a rock ledge left between the intake and the deep powerhouse excavations. The power tunnels were excavated below this rock ledge reaching the external powerhouse downstream. An extensive system of drainage was provided to assure the stability of the rock ledge.

The powerhouse is an external structure, of the semi-outdoor type with four installed Francis-driven units, each with 418.5 MW capacity adding up to 1674 MW. The powerhouse contains also two additional bays for future installation, which so far have not been equipped. The GIS substation, as shown in **Figure 20**, is installed immediately upstream of the powerhouse.

The performance of the project during almost 30 years of continuous operations has been excellent without any major problem.

Figure 19 View of the Foz do Areia dam and spillway before reservoir impounding.

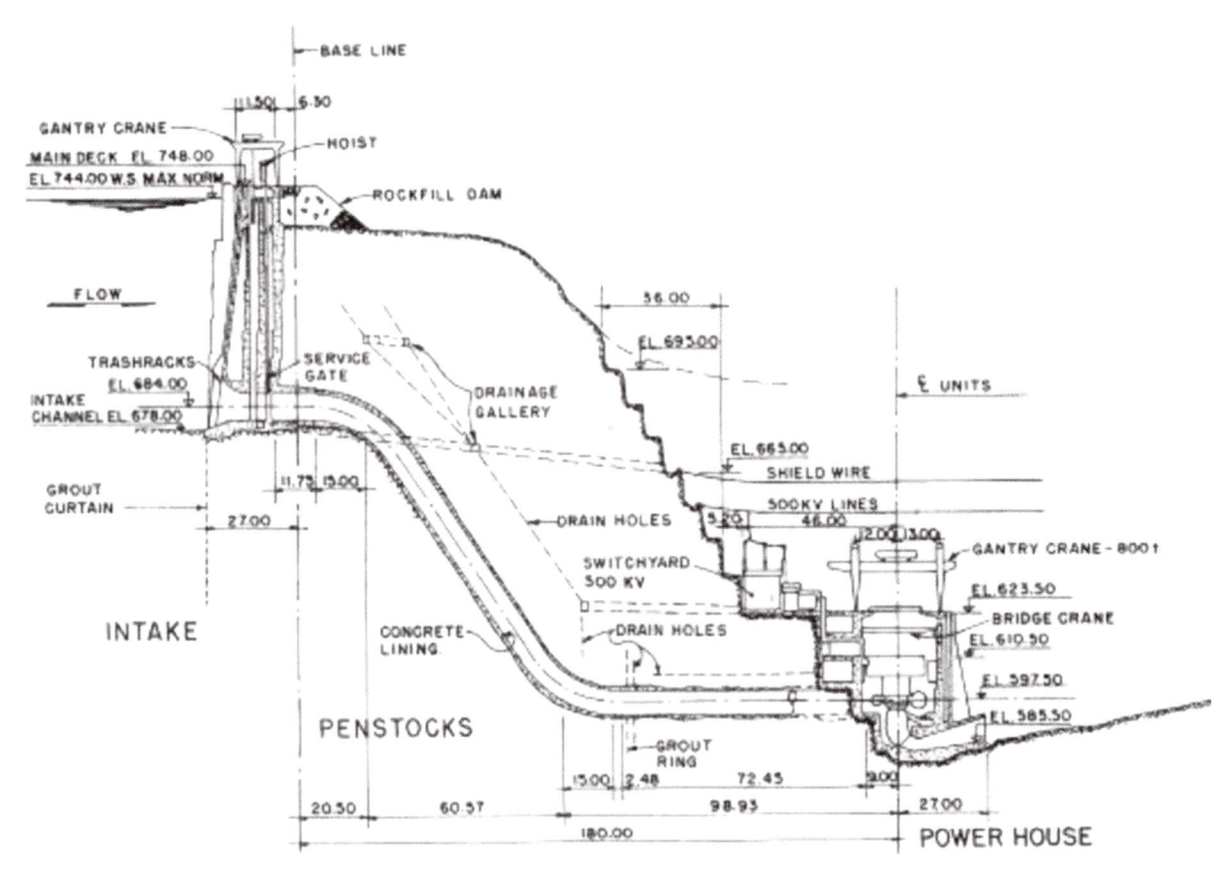

Figure 20 Typical cross section of the Foz do Areia power plant.

6.04.5.2 The Segredo Project

The Segredo hydroelectric project is located on the Iguaçu River, immediately downstream of the Foz do Areia project. It is also owned and was built by COPEL. It is a run-of-river plant with an installed capacity of 1260 MW and no provision for later additional installation. It contains a concrete-face rockfill dam, 145 m high and a power plant formed by four units (**Figure 21**).

Figure 21 The Segredo project.

The project was built in two stages. The first one included river diversion tunnels and construction of part of the cofferdam and of the dam, between 1986 and 1988, and was carried out by a Brazilian contractor, C.R. Almeida S.A. The second one for the completion of the project was carried out by the consortium of three Brazilian contractors, DM Engenharia, CESBE, and SINODA. It carried out the works from August 1988 up to July 1993. The first unit entered operation in September 1992. The project was designed by the Brazilian consulting firms, MDK Engenharia and CENCO.

The project site is exceptionally fit for a hydroelectric project. It is a very narrow valley with steep abutments with the dam axis located immediately upstream from a river curve. The local geology is formed by successive basalt flows, with thickness varying from 10 to 80 m, showing no striking geological discontinuities. The soil cover, although quite variable, is not thick.

The drainage area at the site is 34 100 km^2 and the long-term average flow is 700 m^3 s^{-1}. The hydrologic studies were very much influenced by the 1983 floods observed in south Brazil, which caused floods in the Iguaçu River considered of the order of 1/1000 years of recurrence. With data from these events (which did not exist when Foz do Areia was designed) the maximum design flood for the project was computed based on the PMF concept and checked by normal probabilistic analysis. The value adopted was 16 000 m^3 s^{-1}.

Figure 22 shows the general arrangement of the project. River diversion was accomplished through three unlined tunnels excavated on the left abutment with 13.5 m in diameter and 600 m in length. The upstream cofferdam was 60 m high and the downstream one, 40 m high, because the downstream project, Salto Santiago, was already built and has a reservoir level that can be higher than the tailrace level of Segredo, to allow combined operation.

One peculiarity of the diversion tunnels is that one of them had its entrance placed in a higher elevation than the others and differently from the others, had no entrance-controlled structure. This was done for economy reasons and followed the assumption that this upper tunnel would be (and in fact it was) closed by a soil cofferdam some time before final reservoir closing, allowing construction of the tunnel concrete plug before the date of final closure. The concept was that during the short period when only two tunnels were in service, the dam had been completed and the risk of a major flood reaching the site would be less. It is interesting to note that exactly during this two-tunnel period an unexpected major flood peaking about 7000 m^3 s^{-1} occurred, and the level at the empty reservoir rose substantially and was retained by the dam. Except for that, no other problem occurred.

The concrete-face rockfill dam at Segredo follows the design of the previous successful Foz do Areia dam. The rockfill material is sound basalt mainly from required excavation, with a small portion of basaltic breccia and amygdaloidal basalt. The spillway at Segredo is a gated chute spillway. There are four passages controlled by radial gates 14 m wide by 21 m high. Its design capacity at reservoir full level is 15 800 m^3 s^{-1}. The unusual feature is the concrete lining of the chute, which ends 280 m before reaching the river, allowing a cascading flow over the bare rock. The quality of the local rock was the reason for this solution. Performance along the project life has been good and only localized repairs on the rock chute have been necessary.

Figure 23 shows a typical section of the Segredo power plant. The power intake is 38 m high and feeds four external steel penstocks, 7.5 m internal diameter and 168 m long. The powerhouse is of the semi-outdoor type, and houses four Francis-driven units, each one rated with 315 MW capacity. The switchyard is a conventional-type substation, located on the opposite margin of the powerhouse.

The performance of the project along the 15 years of operation was excellent, without any relevant incident.

6.04.5.3 The Salto Santiago Project

The Salto Santiago project is located downstream of the Segredo project as the next project in the Iguaçu River cascade. It is presently owned by Tractebel Energia, a private utility of the GDF Suez International Group, who bought it from the original owner, a state-owned utility Eletrosul, who also built and operated the plant from 1981 to 1997.

The project was designed and built for an installed capacity of 2000 MW with six generating units, but only four have been installed so far, corresponding to an installation of 1333 MW.

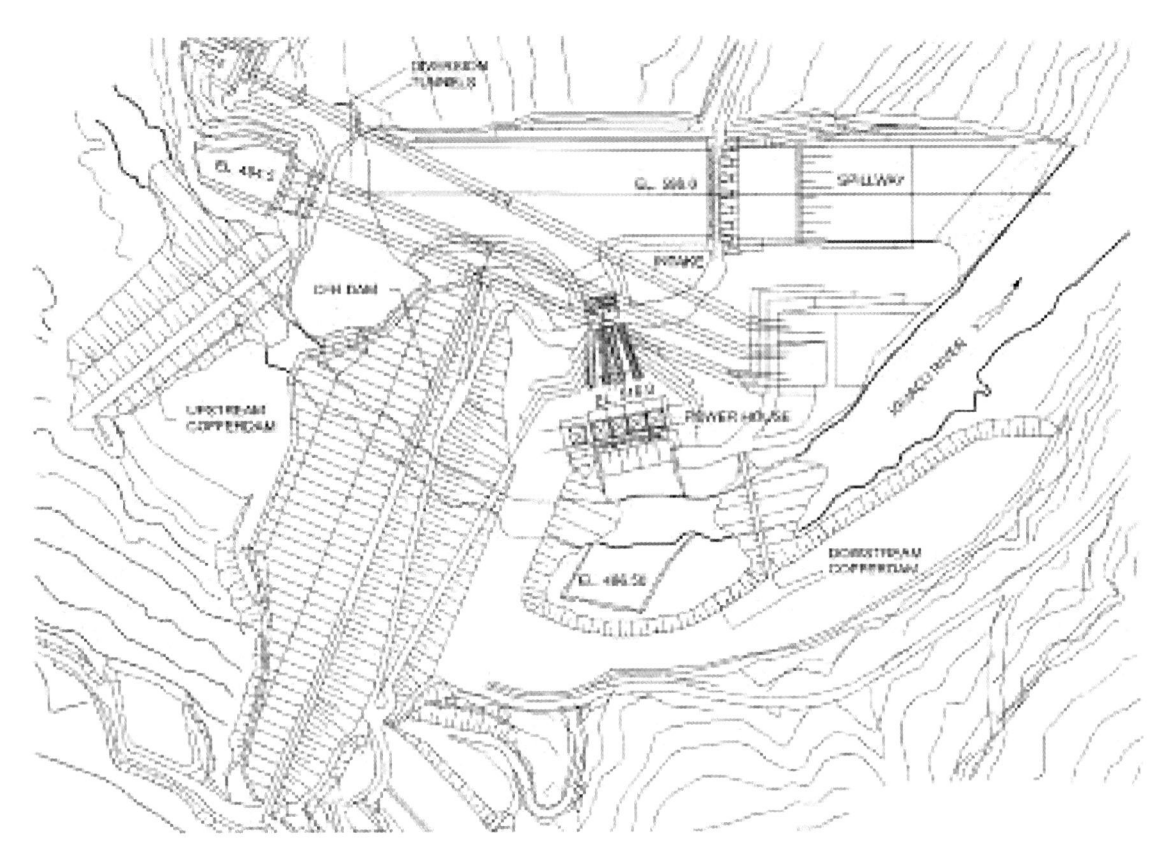

Figure 22 General layout of the Segredo project.

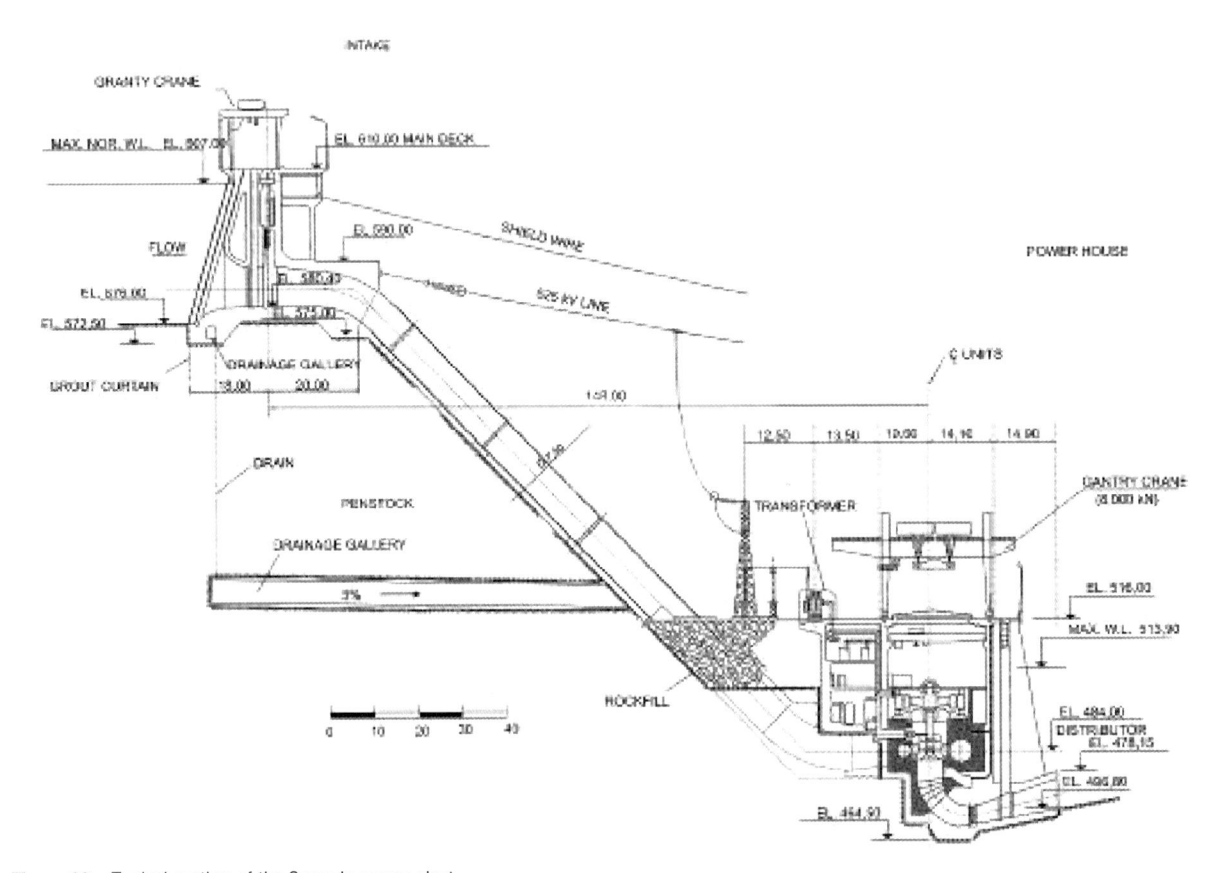

Figure 23 Typical section of the Segredo power plant.

Figure 24 The Salto Santiago project.

The project's construction began in January 1976 and started commercial operation on 31 December 1980. It was designed by Milder-Kaiser Engenharia S.A. and built by Construções e Comércio Camargo Correa S.A. under general coordination of the original owner, Eletrosul.

The project site is very favorable to a hydroelectric development. It is located on a rather closed curve of the river, in which there is an abrupt level difference of about 40 m, the Santiago Falls. The project layout placed an earth-core rockfill dam upstream of the falls and had the power plant bypassing them to discharge downstream and thus creating a useful head of about 110 m used for power generation. The dam created a flow regulation reservoir benefiting the local generation and downstream projects. **Figure 24** shows a view of the completed project and **Figure 25** the project layout.

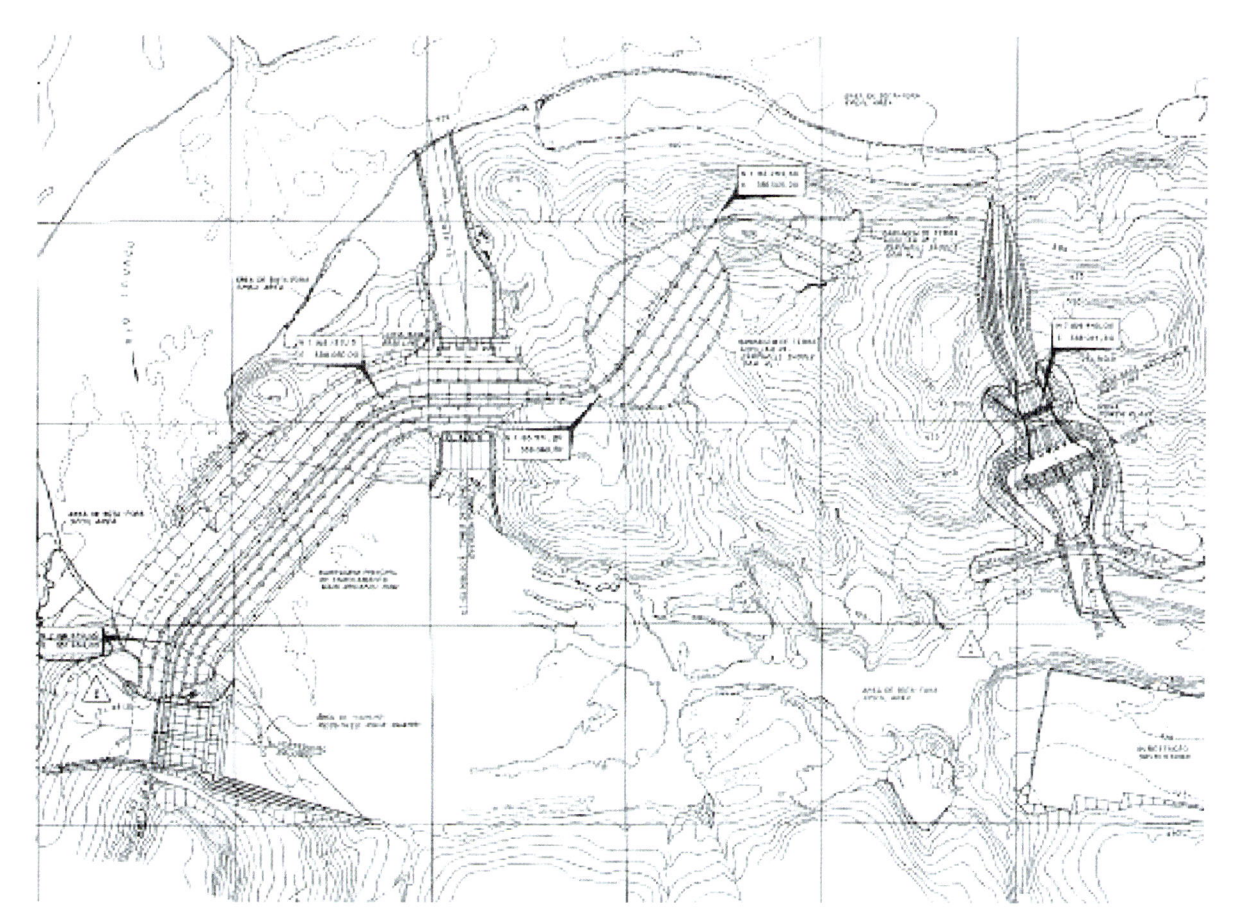

Figure 25 General layout of the Salto Santiago project.

The earth-core rockfill dam at Salto Santiago is 80 m high, with a crest length of 1400 m. Its total volume is about 10 million m³, composed basically of compacted sound basalt rockfill, filters, and an impervious core. **Figure 25** depicts the typical section of the dam. The project has also three saddle dams closing lower points in the reservoir rim. Saddle dam no. 1 is much larger than the other two, and is an earthfill structure, with a residual clayey basalt-soil core, saprolitic material shells, and vertical drains. It has a maximum height of 65 m. The other two saddle dams are less important earth structures with 28 and 9 m of height.

River diversion was accomplished through four 13.5 m diameter horseshoe-shaped tunnels excavated across the left abutment, as shown in **Figure 25**. The diversion scheme comprising the tunnels and the cofferdam upstream of the dam axis was designed for a maximum flow of 10 700 m³ s⁻¹ corresponding to 100-year recurrence period. During the period in which the diversion tunnels operated, the actual maximum flow observed was 6300 m³ s⁻¹. The diversion tunnels were unlined and protected, with considerable success, with shotcrete and rock bolts along the crown and walls.

The diversion tunnels had individual tunnel intake structures for the final closure of the river and reservoir impounding. This closure was designed to be achieved by lowering three 150 ton reinforced concrete gates in each tunnel mouth. The concrete gates were built at the site and handled by a 250 ton gantry crane previously used in the Salto Osorio project. The concrete gates measured 1.2 m thick by 5.3 m wide by 7.4 m long. Immediately upstream from the concrete gates, there were guides for lowering an emergency auxiliary wheel gate that could be used in any opening if required. In addition to these facilities, the intake block for tunnel no. 1 was provided with a passage for allowing compensation discharge downstream during reservoir impounding. The actual closure of the tunnels was successfully accomplished as planned.

The spillway was placed in the right bank, next to the dam, as also shown in **Figure 25**. It is a gated concrete structure with chute and flip bucket, designed to pass the 10 000-year flood corresponding to 24 530 m³ s⁻¹. The spillway is equipped with eight 15.3 m wide by 21.57 m high radial gates.

The power plant is located away from the dam, across the ridge separating the up- and downstream portions of the river curve. It is formed by an intake channel leading to the power intake, six penstocks, a six-unit powerhouse, and the tail water channel. The step-up substation is located on the right margin of the river.

The power intake is formed by three gravity-type blocks, 58 m high and 81 m wide. Each block has two gate-controlled intake openings feeding two individual penstocks. The penstocks are made of steel and have a diameter of 7.6 m.

The powerhouse is an external indoor-type structure, designed for six generating units, a service area, and a control building. It measures 215 m long, 67 m wide, and 64 m high. It is presently equipped with four 333 MW generating units, driven by Francis turbines.

The project has been in operation since 1980. In 1983, extreme floods happened in the Iguaçu basin and indicated that the backwater curve from the downstream Salto Osorio project for extreme floods could lead to a higher downstream flood level at Salto Santiago as the design anticipated. To provide protection of the powerhouse area, a concrete wall was built along the external area of this structure. This was the only major operational problem with the plant.

In 1997, as mentioned the Salto Santiago project was acquired by Tractebel Energia, in a privatization process. This company has also other generating plants in Brazil and decided to install its operation center at Salto Santiago.

6.04.5.4 The Salto Osorio Project

The 1050 MW Salto Osorio project was the first hydroelectric project built on the Iguaçu River cascade. The construction started in 1970 and its first unit was commissioned in 1975. As the other projects in this river, its site is very favorable to receive a hydroelectric project both topographically and geologically. Besides, it drains a basin of 45 200 km² with an average flow of 940 m³ s⁻¹ without definite dry season.

The project started to be implemented by COPEL, the state of Paraná electric power utility who during mid-construction transferred the ownership to Eletrosul, an agent of the federal government in charge, at that time, to supply power to the southern region of Brazil. But, as part of the transfer agreement, COPEL remained the manager of the construction and installation of the project. Eletrosul operated the project from 1975 to 1997, when, as a result of the privatization process, Tractebel Energia, a company of the GDF Suez Group acquired the plant, as it has done with the upstream Salto Santiago project.

The Salto Osorio project was designed by Kaiser Engineers Inc. of the United States, operating in association with Serete Engenharia, a Brazilian consulting firm. It was built in a two-construction contract scheme, the first one for building the cofferdam, with CBPO, and the second one, for the rest of the job, with Andrade Gutierrez, both Brazilian contractors. During this construction period, only four units of the six considered in the project were installed. The last two units were added in 1977, and in March 1978 the full capacity of the project was put online.

The project includes an earth-core rockfill dam spanning the river immediately upstream of the original Salto Osorio Falls, a power plant encroached on the right abutment and two spillways, one between the power plant and the main dam and the other between the dam and the right abutment (**Figure 26**). The reason to divide the spillway capacity into two structures was that the one placed near the power plant incorporated the diversion structures and the other one was purely for discharge of flood flows. **Figure 27** shows the project general layout.

The Salto Osorio project is a run-of-river project with a 1050 MW power plant formed by six generating units each with 175 MW of capacity. The earth-core rockfill dam has a maximum height of 56 m, a crest length of 750 m, and a total volume of 4.2 million m³. The two gated spillways have a combined capacity of discharging 27 000 m³ s⁻¹, and contain a total of nine radial gates, 15.3 m wide by 20.77 m high. The power intake has six controlled passages leading to six steel penstocks with an internal diameter

Figure 26 The Salto Osorio project.

of 7.4 m. The powerhouse is of the semi-outdoor type and houses the blocks for the six units. The tailrace is a rather long excavated channel, running parallel to the river bank and absorbing the powerhouse and the left spillway flow. The step-up substation is located close to the powerhouse, on the left bank of the tailrace, as shown in **Figure 27**.

The river diversion scheme was a major factor for defining the project layout. In fact, the Iguaçu River has no definite dry season and major floods can occur during any month of the year. Based on records existing at the time (1931–70 series) the project was

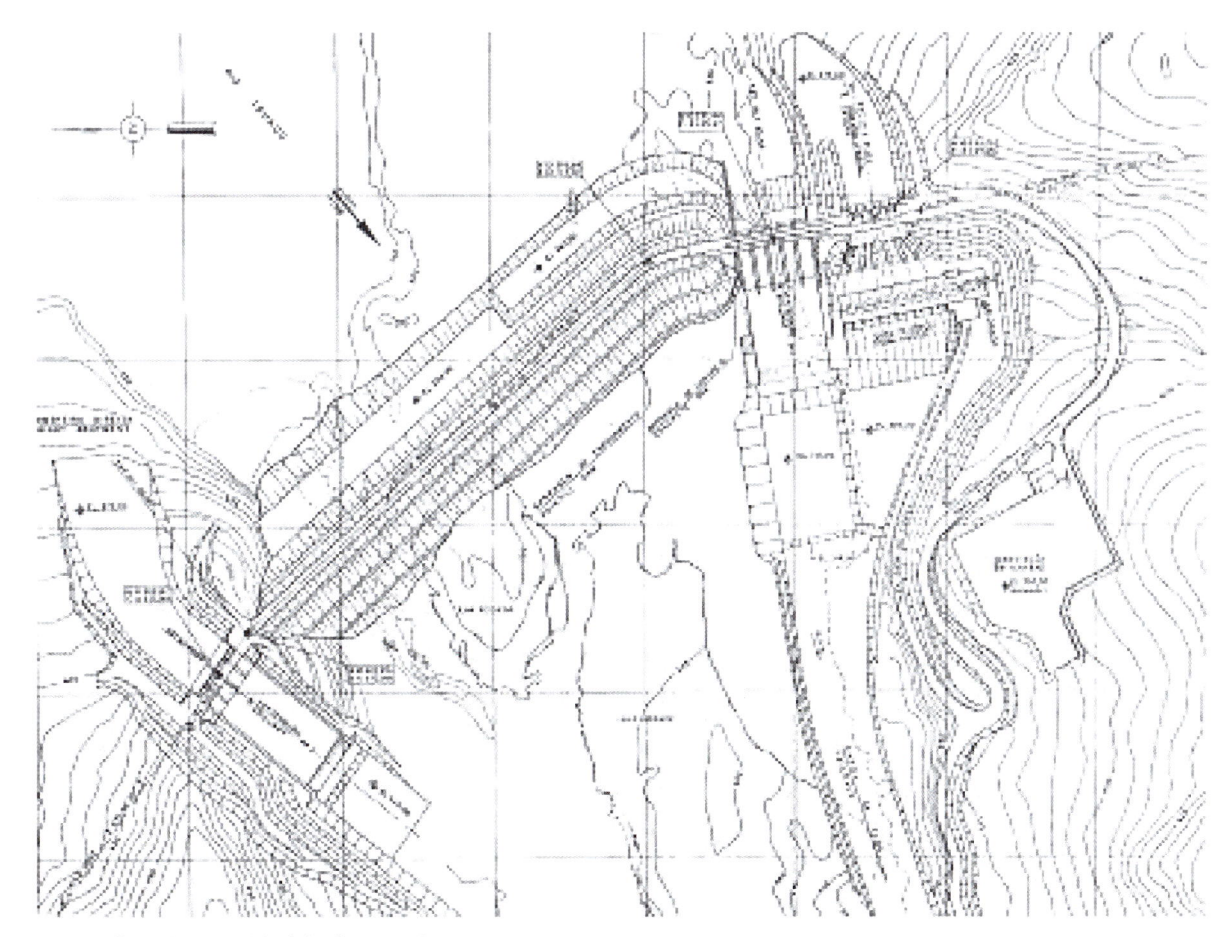

Figure 27 General layout of the Salto Osorio project.

designed, the diversion design flood was established in 13 000 m³ s⁻¹, corresponding to the 1/100 year flood. Although geology is favorable, for this size of flood the topography of the site prevented the economic use of diversion tunnels and diversion sluiceways have been provided in the concrete structure of spillway no. 1. Therefore, river diversion for construction was done by constructing cofferdams in the left-hand part of the river width and creating an area where part of the dam, the left spillway with diversion sluiceways, and the power intake were built. When these structures were completed or reached an elevation compatible with the diverting flow through the sluiceways, the right-hand natural channel of the river was closed with a cofferdam tying into the part already built of the dam, and right spillway and the remaining part of the dam built.

The 10 sluiceways, 6.5 m wide by 14.0 m high, under the left spillway were designed to handle the maximum flow of 10 700 m³ s⁻¹, corresponding to the maximum observed flow on record. They were designed to be closed with reinforced concrete gates. Each gate weighted 250 tons and was lowered into place with the intake gantry crane operating across the spillway bridge. The diversion sluiceways operated as planned along the construction time and after completion of the dam were successfully closed. Four sluiceways were closed some days before total closure to test the procedure and acquaint construction personnel. The final closure of six sluiceways was done in 10 h without incidents.

The project has been in operation since commissioning without any relevant problem.

6.04.5.5 The Salto Caxias Project

The Salto Caxias project is the last built project in the Iguaçu River cascade. It is a 1240 MW project and is owned and operated by COPEL, the state of Paraná electrical utility. It was built between January 1995 and 1999 with its first unit coming on line on 1 February 1999. The project was designed by an association of four Brazilian consulting engineering firms, led by Intertechne Consultores S.A. Construction was carried out by the Brazilian contractor DM Engenharia de Obras Ltda.

The Salto Caxias site is located about 90 km downstream of the Salto Osorio project and about 80 km upstream from the point where the Iguaçu River becomes binational and marks the border between Brazil and Argentina and about 190 km from the internationally famous Iguaçu Falls.

At Salto Caxias the river drains a basin of 57 000 km² and has an average flow of 1240 m³ s⁻¹. The site has a peculiar morphology with the river turning a sharp 180° bend, with two narrow rock noses protruding from each bank. The width of the river upstream and downstream from this section is about 600 m. A low 5 m high waterfall (the 'Salto' Caxias) crosses the river width upstream of the right bank nose. The dam axis was located immediately upstream of this feature. The geology of the site is made up of basaltic rocks occurring in nearly horizontal flows. Individual flows range in thickness from less than 5 m to more than 50 m.

A view of the completed project is shown in **Figure 28** and the project layout is depicted in **Figure 29**. The reservoir's normal maximum operating level is at El. 325 with no drawdown for flow regulation, except daily pondage. At the dam axis, the average elevation of the rock foundation is El. 258 resulting in a maximum height for the dam equal to 67 m. Main features of the project layout are the following:

- An RCC dam, 1100 m long, incorporating in its right end a surface spillway and underneath it, sluiceways for river diversion.
- A surface spillway built on top of the RCC dam, formed by 14 radial gates, each 16.5 by 20.0 m, capable of discharging 48 307 m³ s⁻¹ with a reservoir level at El. 326, corresponding to the PMF inflow hydrograph routed through the reservoir.

Figure 28 The Salto Caxias project.

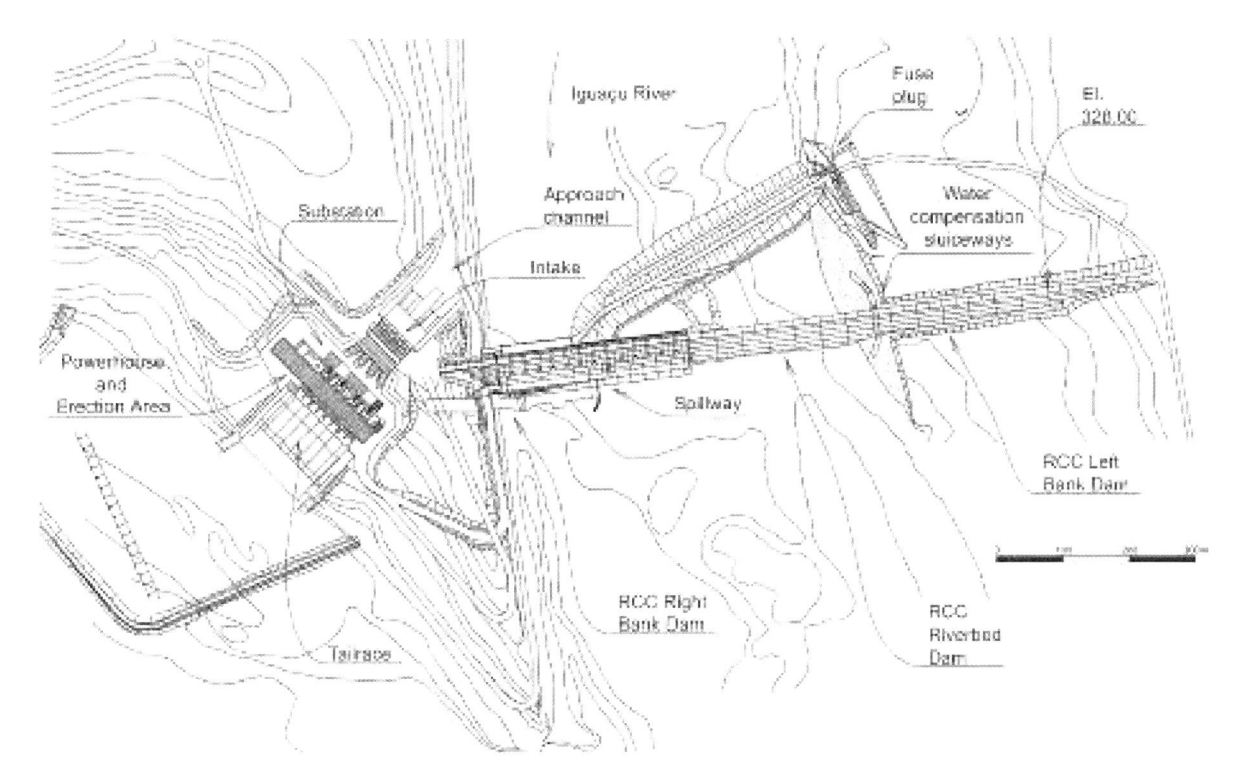

Figure 29 General layout of the Salto Caxias project.

- Fifteen sluiceways placed under the rightmost five gated passages of the surface spillway, each 4.35 m wide by 10.0 m high, used for river diversion during second phase of dam construction.
- A power plant placed across the rock ridge that forms the right abutment of the site comprising an intake channel excavated in rock, the intake structures with four independent water passages, four 11 m diameter exposed steel penstocks feeding the powerhouse, and an outdoor structure sheltering four 310 MW generating units.

The dam comprises the RCC dam proper and the spillway structure. This is a conventional concrete structure built on top of an RCC body. The total concrete volume of the dam is 1 000 000 m^3, of which 950 000 m^3 correspond to RCC. This made Salto Caxias the largest RCC-volume dam in Brazil, besides being unique in incorporating the largest gated surface spillway placed on top of an RCC body.

The river diversion was carried out in two phases. In the first phase the natural river channel was restricted by a U-shaped cofferdam built from the right bank. This cofferdam allowed the construction of the spillway and diversion sluiceways, part of the dam, and the RCC right-hand blocks that connect the spillway to the right abutment.

After the completion of this part of the structure, the river was diverted through the sluiceways. A second-phase cofferdam connected the left bank to the already built spillway blocks, so that the left part of the dam could be built.

During this second-phase construction, exceptional events happened that are described in continuation. The construction period, from January 1995 to September 1998, presented a pluviosity index significantly above the historical record. The average flow of the Iguaçu River at the site during this period was about 2500 m^3 s^{-1}, which is roughly twice the computed historical long-term average flow.

The second stage of the river diversion for construction, corresponding to diverting the river flow through the sluiceways provided under the spillway, was revised and replanned considering the possibility of overtopping part of the RCC blocks under construction. This was done because it became desirable to start the second-phase diversion five months earlier than originally planned and this would cause an increase in the upstream water level before the date anticipated for the relocation of the population affected by the reservoir flooding.

To harmonize the schedule of the population relocation program and the desirability of anticipating the second-stage construction, it was necessary to maintain the maximum flood level upstream of the construction site, for the same design floods as originally planned. To achieve this, the heightening of the RCC dam, for a stretch of 280 m long in the river area, was stopped at a lower elevation, about 20 m above natural river bottom. The second-stage cofferdam was built with the crest at this same elevation, and a side channel with a fusible soil dike provided a means of controlled filling of the space between the dam and the cofferdam, before the overtopping of the cofferdam structure.

Figure 30 River diverted through sluiceways at Salto Caxias.

Work proceeded normally in the dam in the left abutment reach. During this phase the construction site was flooded and the dam overtopped five times, with the largest of such events corresponding to a flood of about 14 000 m^3 s^{-1} with 5000 m^3 s^{-1} spilling over the partially built dam. In all cases, the flooding of the site was kept under control, no significant damages were observed, and normal construction work resumed 3–4 days after the flood receded (**Figure 30**).

The power plant at Salto Caxias includes a conventional concrete gravity structure 41 m high and 64.4 m long, with four water passages controlled by wheel gates. Four steel penstocks 11 m in diameter and 107 m long convey the water to the generating units. The powerhouse, of the indoor type, houses the four generating units each rated 310 MW and driven by Francis turbines. There is a GIS substation immediately upstream of the powerhouse and outgoing 550 kV lines connecting the plant to National Integrated Transmission System. **Figure 31** shows a typical section of the power plant.

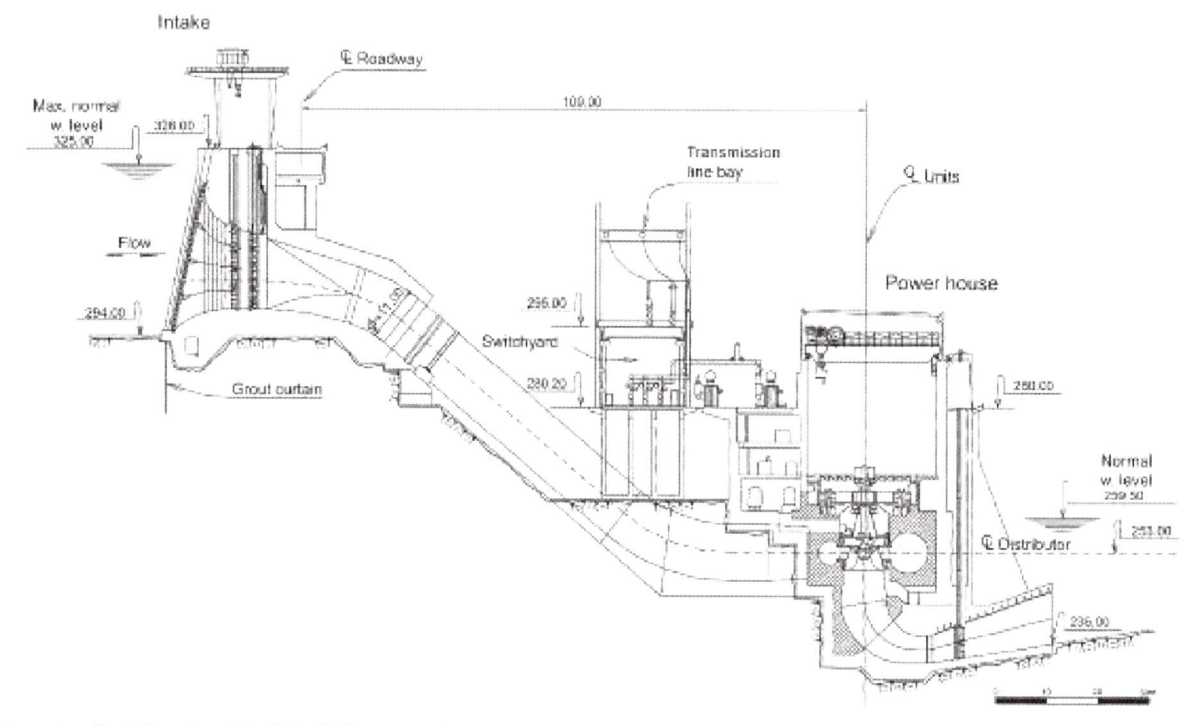

Figure 31 Typical section of the Salto Caxias power plant.

Figure 32 The Brazilian basin of the Uruguay river.

6.04.6 The Uruguay River Projects

The information for this section was based on material compiled in Reference [15]. The Uruguay River is part of the Paraná River basin in the sense that it discharges in the estuary of the Paraná, which is also called the La Plata River, separating the countries of Argentina and Uruguay. In the Brazilian territory, the Uruguay River has its headwaters in the southern part of the country, in the state of Santa Catarina, formed by two important rivers, the Canoas and the Pelotas, which joining their waters form the Uruguay main course. This stretch of the river separates the states of Santa Catarina and Rio Grande do Sul, and then turns south and constitutes the border between Brazil and Argentina and in continuation, Uruguay and Argentina, reaching finally the La Plata estuary. **Figure 32** shows schematically the configuration of the course of the river in the Brazilian territory.

Before reaching the Brazil–Argentina border, the Uruguay River and the two main rivers that form it have a significant hydroelectric potential that up to now has been developed in four major hydroelectric projects. Additional projects in the national stretch of the Uruguay River and of the Canoas and Pelotas tributaries have been defined but have not yet been the object of concessions. On its international stretch, along the border between Brazil and Argentina, a major binational project has been conceived and is being studied. Along the border between Uruguay and Argentina, another important binational project, Salto Grande (1890 MW), has been built and is in operation.

The two rivers that form the Uruguay River, the Pelotas and the Canoas rivers, have their head waters at the mountains of the Brazilian range, the Serra Geral, at an elevation of about 1800 m asl. These rivers and the Uruguay River, after their junction, run essentially in an east–west direction until they reach the international border, and then turn NE–SW until they reach the Uruguayan territory. Afterward, between Uruguay and Argentina, their course is essentially north–south. The total length of the Uruguay River, including the Pelotas, is 2150 km, of which 940 km corresponds to the full Brazilian stretch before reaching the international border.

A very significant characteristic of the Uruguay River and its tributaries is the magnitude and frequency of the floods. The headwater part of the basin (hatched in **Figure 32**) has steep exposed rock slopes relatively devoid of forests and subject to a high rate of rainfall throughout the year. This generates very high runoff, with a very rapid increase in river flows, and major floods that are a determinant fact for the design of the hydroelectric projects.

6.04.6.1 The Machadinho Project

The Machadinho project is located on the main course of the Uruguay River in a stretch that is still known as the Pelotas River, which means that its location is downstream from the confluence of the Canoas River. It is a major hydroelectric project with an installed capacity of 1140 MW using the head developed by a concrete-face rockfill dam, 125 m high and 673 m long. The project is 94.5% owned by an association of private parties and 4.5% owned by CEEE the public utility company of the state of Rio Grande do Sul. The private parties include Tractebel Energia, belonging to the GDF-Suez Group; DME of the town of Poços de Caldas, Minas Gerais;

Figure 33 Aerial view of the Machadinho project.

three aluminum producers, Alcoa, CBA, and Valesul; and two cement producers, Votorantim and Camargo Correa Cimentos. The project was designed by Brazilian consulting engineering company, CNEC and was built by Brazilian contractor Camargo Correa. Construction started in 1998 and power operation was accomplished in 2002. **Figure 33** shows an aerial view of the project, and **Figure 34** shows a plan of the general layout of the project.

The drainage basin at Machadinho has an area of 32 000 km^2. Hydrologic studies based on data from 1914 to 1992 indicated the importance of construction floods and the possibility of their occurrence in any month of the year, although summer months

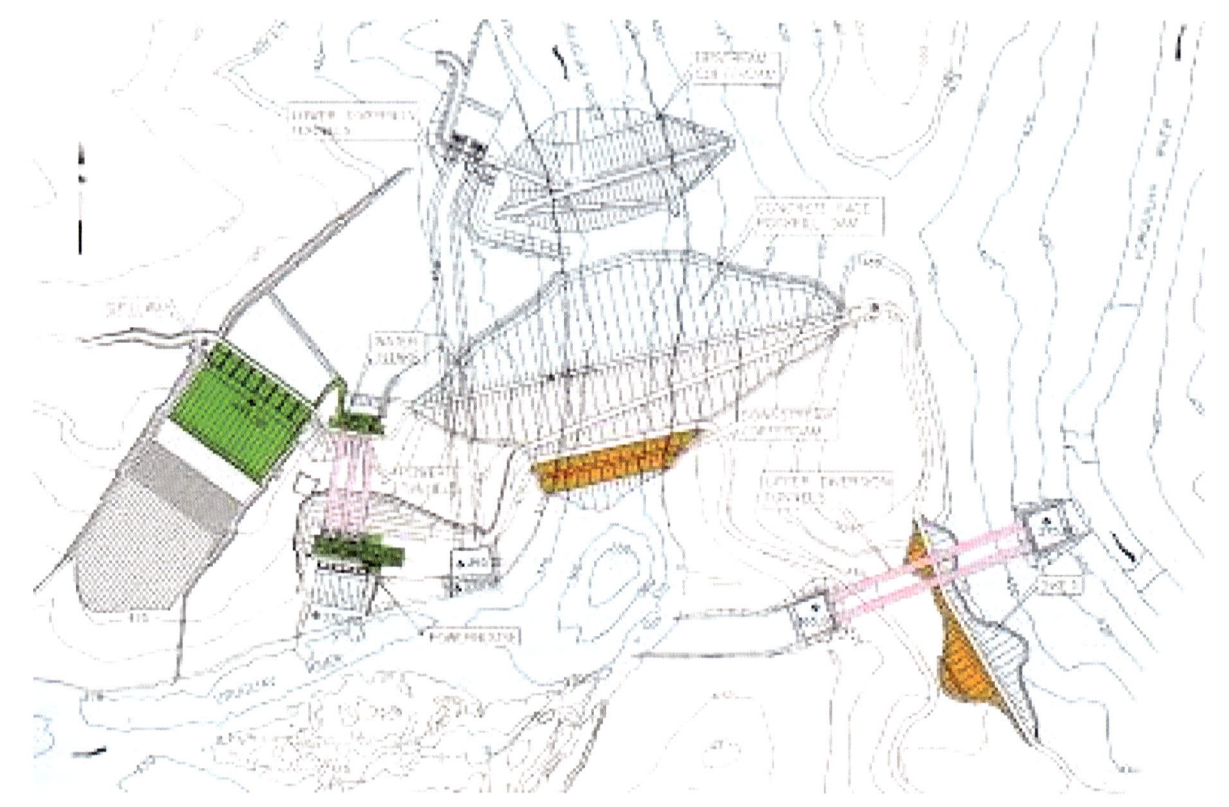

Figure 34 General layout of the Machadinho project.

(November–April) were found less susceptible to present large floods. The project layout was then very much influenced by the scheme for diversion and control of the river during construction. The adopted solution was based in constructing cofferdams on the river and diverting it through four diversion tunnels, two at lower elevation and two placed at a higher elevation using the unusual configuration of a natural channel of a tributary on the left bank of the river, as shown in **Figure 34**. The tunnels have rectangular sections, 14 m wide by 16 m high, and only the lower tunnels had structural intakes with mechanical closing facilities. The upper elevation tunnels were closed with cofferdams at their upstream end during low flow season when only the lower tunnels could take care of the river flow. The tunnels and cofferdams were designed to initially protect the site for the 10-year flood of 14 110 $m^3 s^{-1}$ and finally, considering the progress of the rockfill dam construction, to the 500-year flood equal to 24 700 $m^3 s^{-1}$.

The spillway was designed for discharging the 10 000-year flood, equal to 37 350 $m^3 s^{-1}$ with a flood level 4.38 m above full supply level of the reservoir. It is a gated structure, with eight passages controlled by radial gates, 18.0 m wide by 20.0 m high. It is a chute spillway with the unusual characteristic that the chute is concrete lined only in the first one-third of its length, and the discharge flow is left running on top of the excavated very sound basalt rock. It is interesting to mention that during the final stage of the reservoir impounding a major flood occurred on the river and forced the operation of the spillway with a discharge of 16 000 $m^3 s^{-1}$, almost 43% of the design flood. Some local erosion was observed in the unlined chute, but otherwise performance was good.

The power plant is formed by a power intake with three passages, three independent underground penstocks, and an external powerhouse accommodating three generating groups and compact GIS 500 kV. The power intake is a hollow gravity concrete structure founded on rock, 47.0 m high and 61.2 m long. The power tunnels have an average length of 147 m, are concrete lined along the initial 94.5 m with an internal diameter of 9.4 m, and are steel lined along the final stretch with 8.0 m diameter. The powerhouse is of the indoor type, formed by five independent blocks of reinforced concrete, housing the generating units and the erection and maintenance areas.

The project has been in full operation since July 2002 without any major problem.

6.04.6.2 The Itá Project

The Itá project is located on the Uruguay River, immediately downstream of the Machadinho project. The project is owned by Itá Energética S.A., a company formed by private utilities and large industrial electric power consumers, which include Tractebel Energia (of the GDF-Suez Group), Votorantim Cement, Alcoa, and others. The project was designed by Brazilian engineering consultant Engevix, built by CBPO (of the Odebrecht Group) and furnished with equipment supplied by ABB, Alstom, Voith, and COEMSA. The Itá project has an installed capacity of 1450 MW, a 125 m high concrete-face rockfill dam and a spillway with a capacity of 49 940 $m^3 s^{-1}$.

The Uruguay River at Itá has a peculiar configuration in the sense that it follows very abrupt and U-shaped curves (which would look like meanders in a flood plain) running in a basaltic rock topography. **Figure 35** shows an aerial view of the site illustrating this characteristic feature.

Figure 36 shows the project general layout. The 125 m high dam creates a gross head of 105 m across the branches of the U-shaped curve and combined with a long-term average flow of 1100 $m^3 s^{-1}$ allows the installation of five generating units each with a capacity of 290 MW. The project is provided with two spillways to discharge the design flow of 49 940 $m^3 s^{-1}$. One spillway is built next to the dam and the other across the ridge separating the river stretches.

Figure 35 Aerial view of the Itá project site.

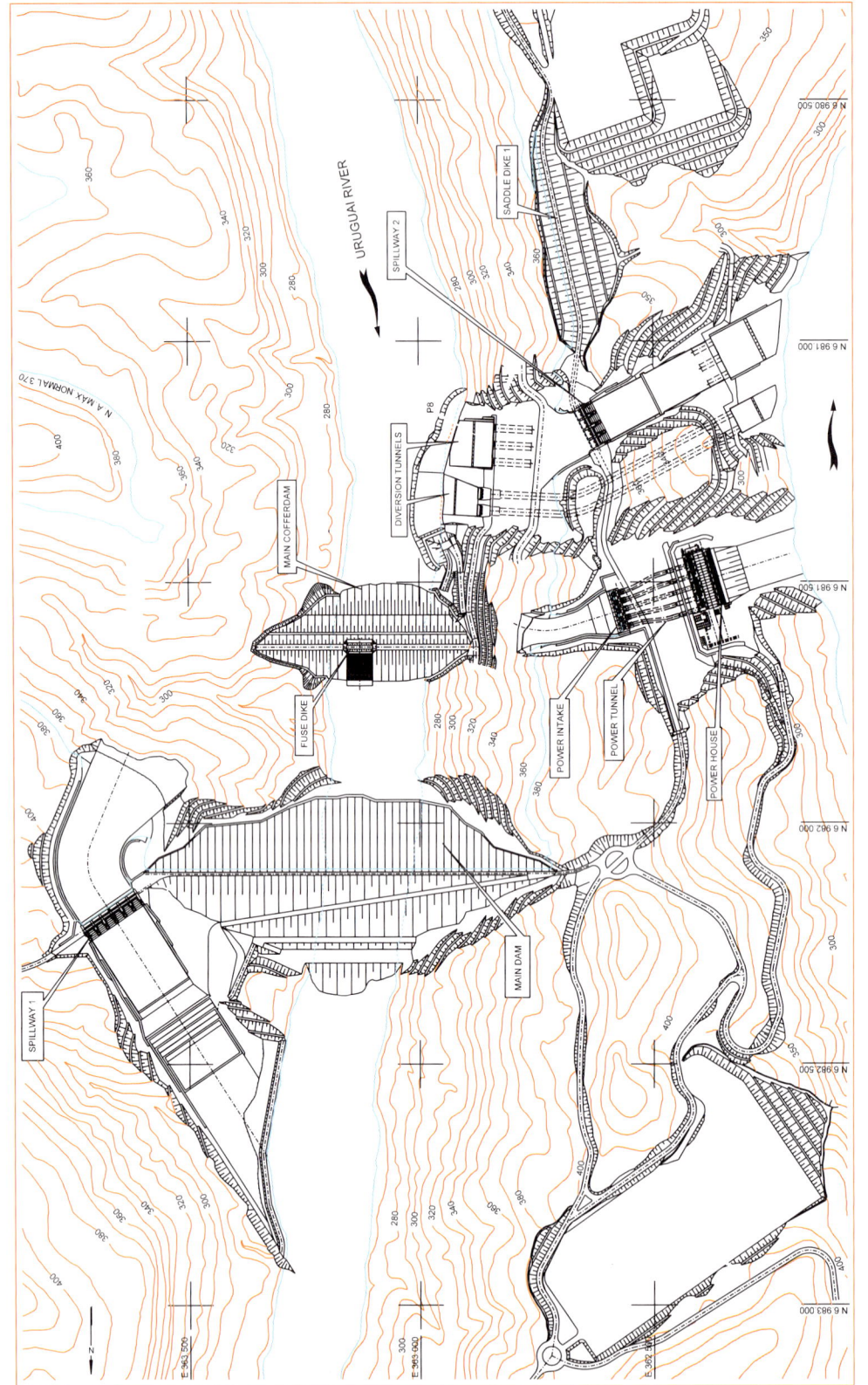

Figure 36 General layout of the Itá project.

Figure 37 The Itá project: Diversion tunnel entrances.

River diversion for construction was accomplished with five tunnels, two of them with a cross section 14 by 14 m, and the other three, 15 by 17 m. A 51 m high cofferdam was built to initially protect the site against the 10-year flood of 19 000 m^3 s^{-1} and later combined with the construction of the upstream part of the rockfill dam to reach a height compatible with protection against the 500-year flood. To cope with the risk of overtopping, the cofferdam was provided with a fuse-plug controlled spillway to allow a smooth overtopping of it and a smooth filling of the construction site upstream of the dam, if this was necessary. However this situation did not occur and although major floods happened during construction, the cofferdam was never overtopped.

The diversion tunnel entrances were set at different elevations and only two of them had concrete intake structures with control gates. This can be seen in **Figure 37**. Immediately before closing the river for reservoir impounding, the river was flowing through the two lower tunnels and an RCC cofferdam was built upstream of the entrance of the three unprotected tunnels. This allowed the construction of the definitive concrete plugs inside the tunnels. Then, for final closure, the gates of the lower tunnels were closed.

The power plant at Itá is similar to the one described for the Machadinho project. The project has been in operation since 1999.

6.04.6.3 The Campos Novos Project

The Campos Novos project is located on the Canoas River, about 20 km upstream from where it joins the Pelotas River to form the Uruguay River (**Figure 38**).

The project is owned by a joint venture of a private public utility, CPFL, two private industries large consumers of electric power of the Votorantim Group, and two regional state-owned public utilities. The project was designed by Brazilian consultants Engevix and CNEC, built by contractor Camargo Correa, and furnished by equipment supplier GE-Inepar.

Figure 38 The Campos Novos project.

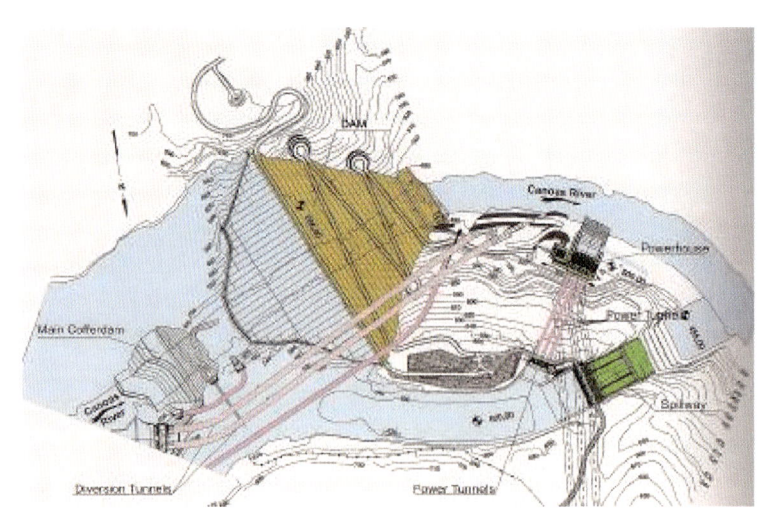

Figure 39 General layout of the Campos Novos project.

The project includes a 205 m high concrete-face rockfill dam, a 18 300 m³ s⁻¹ spillway, and a power plant with a capacity of 880 MW. Diversion was done through two tunnels, with a cross section 14.5 by 16.0 m and about 900 m long. **Figure 39** depicts the project layout.

The significant fact associated with this project was a series of cracks in the concrete face of the dam and a major leakage in the diversion tunnels after closure and impounding of the reservoir. The two facts are independent of each other but ended up by being related because the reservoir had to be lowered to fix the tunnel problem and exposed the concrete face cracks. The occurrence of the cracks in the face in a regularly built dam was the object of many technical considerations and speculations and was associated with the deformation of the rockfill in a high dam built in a narrow valley.

Both problems had been fixed and the project is in satisfactory operation since 2007.

6.04.6.4 The Barra Grande Project

The Barra Grande project is located on the Pelotas River, upstream of the mouth of the confluence with the Canoas River.

The project is owned by a joint venture of private industries that are large consumers of electric power and one public municipal utility. The project was designed by engineering consultant Engevix, built by contractor Camargo Correa, and furnished by Alstom who supplied the main equipment. **Figure 40** shows a view of the completed project and **Figure 41** the project layout.

The project has a power plant with a capacity of 708 MW that uses the head created by a 185 m high concrete-face rockfill dam. The project spillway has a capacity of 21 800 m³ s⁻¹.

Figure 40 The Barra Grande project.

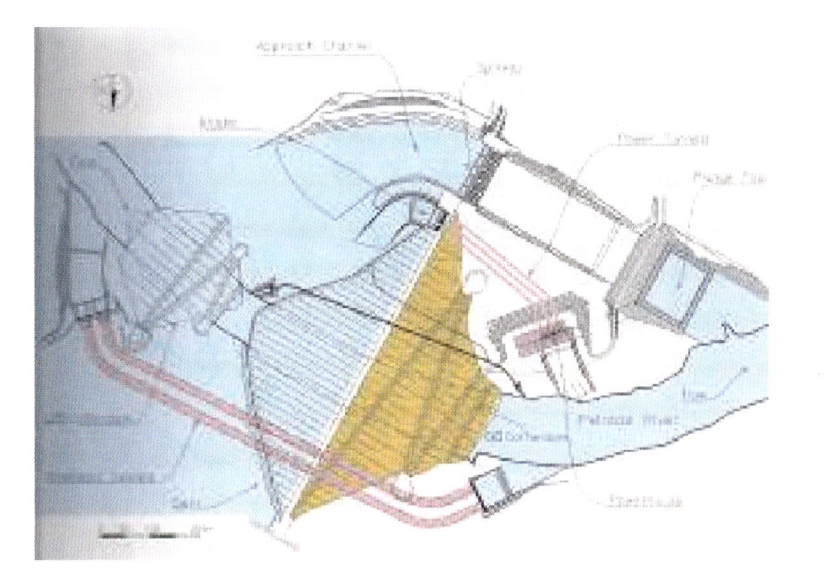

Figure 41 Project layout.

The reservoir was impounded between July and September 2005. After reaching its full operational level, excessive leakage was recorded, and after investigation a problem in the upper part of the central concrete face joint was observed. This problem was considered similar to the one observed at the higher Campos Novos dam, associated with compression forces resulting from rockfill deformation in very narrow valleys. The joint problem was repaired and the project is in satisfactory operation since that date.

6.04.7 The Belo Monte Project

The description that follows is based on the project feasibility study carried out by Eletrobras [16] and published to assist investors in the prospective concession auction scheduled for the early 2010. The Belo Monte project is an 11 181 MW hydroelectric project that will be built in the Xingu River, in the northern region of Brazil. Its concession to private and/or public investors is scheduled to be awarded during the first months of 2010 and construction started by the end of this year to have the first unit on line by 2015.

The project has been studied since the beginning of the 1980s and has had very strong opposition by various groups, both Brazilian and international, based on the alleged negative impact on the Amazonian environment and on Indian tribes that live in the area. The project conception has been adjusted to reduce to a minimum the area flooded by the reservoir and to avoid interference with areas occupied by the Indian tribes. Very thorough and long environmental impact surveys have been carried out and are presently (November 2009) being discussed in public hearings according to Brazilian legislation. Environmental license to proceed with the concession process allowed the award to the successful bidder in April 2010. A very large number of mitigation measures, essentially on the social and anthropological areas, are required from investors.

In any case, it is a very large and important undertaking. The Xingu River is a tributary of the lower stretch of the Amazon River. Its course is essentially south–north, running about 1900 km from its headwaters in the state of Mato Grosso, at elevation 600 m asl, to about sea level in its mouth. Although various sites with possibilities for hydroelectric development have been identified along the river, only the Belo Monte site is presently considered for construction, essentially as a result of the negotiations between the government and the habitants of the river basin area.

The Belo Monte project is located in an area where the Xingu River forms a large curve and drops about 95 m through a series of rapids along about 100 km of river. **Figure 42** shows the location of the Belo Monte project. It can be seen from this figure that the town of Altamira is immediately upstream of the 'large curve' of the Xingu River. The project will develop the hydroelectric potential of the site by damming the river in the upstream branch of the curve and deriving the flow through excavated channels directly to the downstream branch, bypassing the rapids, and installing a 11 000 MW power plant in this downstream location.

The general arrangement of the project is shown schematically in **Figure 43**. At the Pimental site, some 40 km downstream of the town of Altamira a low dam is built to allow the river flow diversion through the two diversion channels. At this dam the main spillway is located and an auxiliary power plant (181 MW) is built to use the minimum flow of 300 $m^3 s^{-1}$ that is left running along the river natural course. This minimum flow is a requirement set forth by the environmental authorities to grant the environmental license. The reservoir created by the Pimental Dam has an area of 440 km^2, which is only twice the area flooded by the natural river during the wet season.

Figure 42 Location of the Belo Monte project.

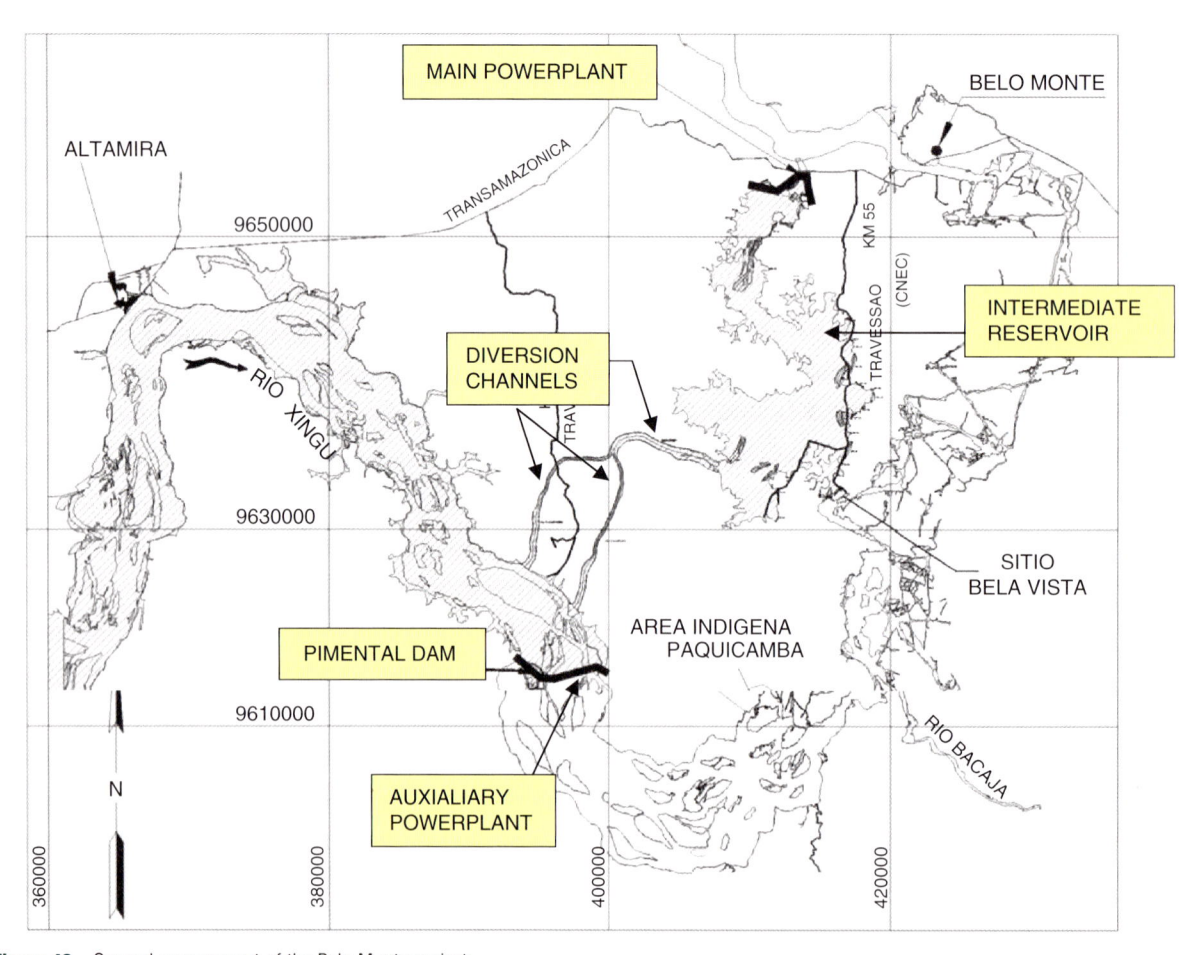

Figure 43 General arrangement of the Belo Monte project.

The two diversion channels join in a single channel midway between their entrance and the intermediate reservoir. They have been laid-out in using lower spots on the ground to diminish required excavation. Nevertheless, because of channel lengths and volume of water to be conveyed (14 000 m^3 s^{-1}), the expected excavation volume is of the order of 200 million m^3.

The channels discharge into an intermediate reservoir, located in a general favorable area but ensured by a series of saddle dikes. In one of the dikes, an intermediate spillway will discharge the excess flow generated at the area during flood seasons.

At the downstream end of this intermediate reservoir, the main power plant will be located. It will house 20 generating units each 550 MW, adding up to 11 000 MW. When completed the Belo Monte project will be the largest hydroelectric project in the Brazilian territory. It will be connected to the Integrated Brazilian Transmission System through direct current lines reaching directly the São Paulo area, 2500 km away.

References

[1] *The World Bank: World Development Indicators – Database*. 1 July 2009.
[2] Agencia Nacional de Energia Elétrica (ANEEL) and Brazilian Agency for Electric Power (2009) *Banco de Informações de Geração [Electricity Generation Data Bank]*. 8 August.
[3] de Melo FM (1978) A century of dam construction in Brazil. In: *Topmost Dams of Brazil*. São Paulo: Novo Grupo Editora Técnica Ltda.
[4] Canambra Engineering Consultants Limited (1967) *Survey of Hydroelectric Resources in South Central Brazil*. Technical Paper No. 1. Prepared for the Steering Committee of the Power Study of the South Central Region, January.
[5] Canambra Engineering Consultants Limited (1967) *The Power Study of South Central Brazil*. Technical Paper No. 2. Prepared for the Steering Committee of the Power Study of the South Central Region, January.
[6] ANEEL, Brazilian Agency for Electric Power. Website: www.aneel.gov.br. November 2009.
[7] Itaipu Binacional (1994) *Itaipu Hydoelectric Project – Engineering Features*. Foz de Iguaçu.
[8] Sucharov M and Fiorini AS (2002) The Itaipu Spillway. *Large Brazilian Spillways*. Rio de Janeiro: Brazilian Committee on Dams (CBDB).
[9] Eletronorte (1989) *Usina Hidrelétrica Tucuruí – Memória Técnica*. Brasilia DF.
[10] Brazilian Committee on Dams (CBDB) (2000) *Main Brazilian Dams – Design, Construction and Performance*, vol. II. Rio de Janeiro: CBDB.
[11] Brazilian Committee on Dams (CBDB) (2002) *Large Brazilian Spillways*. Rio de Janeiro: CBDB.
[12] Vieira de Carvalho R, Magela G, Mello H, and Araújo A (1988) Historic flood during the 2nd phase of the Tocantins River diversion for the construction of the Tucuruí power plant, Question 63. In: *Proceedings of the Sixteenth Congress on Large Dams*. San Francisco, CA, USA.
[13] Arantes Porto CMA, *et al.* (2006) The Madeira hydro complex: Regional integration and environmental sustainability. *The International Journal on Hydropower & Dams* Issue 2.
[14] Brazilian Committee on Dams (CBDB) *Main Brazilian Dams*, vols I, II, III. Rio de Janeiro, 1982, 2003, 2009.
[15] Brazilian Committee on Dams (CBDB) *Main Brazilian Dams*, vol. III, Rio de Janeiro, 2009.
[16] Eletrobrás (2009) *Feasibility Study of Belo Monte Project*. Brazilia.

6.05 Overview of Institutional Structure Reform of the Cameroon Power Sector and Assessments

J Kenfack and O Hamandjoda, University of Yaounde, Yaounde, Republic of Cameroon

© 2012 Elsevier Ltd. All rights reserved.

6.05.1	Introduction	129
6.05.2	Hydro Potential	130
6.05.2.1	The River System	130
6.05.2.2	Existing Hydro Plants	131
6.05.2.2.1	Production and transportation of electricity	132
6.05.3	Dams	134
6.05.3.1	Storage Dams Under Operation	134
6.05.3.2	Hydrology	135
6.05.4	Mid-Term Development Plan for Hydro Plants in Cameroon	135
6.05.4.1	Objectives	135
6.05.4.2	Context of the Development Plan	137
6.05.4.3	Future Outlook	137
6.05.4.3.1	Lom Pangar project	137
6.05.4.3.2	Dam characteristics	138
6.05.4.3.3	Outcome of the Lom Pangar project	138
6.05.4.3.4	Project justification and other alternatives	139
6.05.4.3.5	Optimizing the reservoir	139
6.05.4.4	Memve'Elé	139
6.05.4.4.1	Project area and location	139
6.05.4.4.2	Initial cost of the project	139
6.05.4.4.3	Project layout and structures	140
6.05.4.4.4	Dam-reservoir on the Ntem	141
6.05.4.4.5	Environmental impact	141
6.05.4.4.6	Funding	141
6.05.4.5	Mekin Hydropower Project	141
6.05.4.5.1	Introduction	141
6.05.4.5.2	Investment estimate	144
6.05.4.5.3	Economic assessment	145
6.05.4.6	Bini Warak Project	145
6.05.4.7	Colomines Project	145
6.05.4.8	Ngassona Falls 210 Project	146
6.05.4.9	Overview of Institutional Structure Reform	147
6.05.4.9.1	Previous assessments of the power sector reforms	147
6.05.4.9.2	Historical overview of the sector	147
6.05.4.9.3	Current status	148
6.05.4.10	Weaknesses of Institutions	149
6.05.4.11	Investing in the Electric Power Sector	149
6.05.5	Conclusion	150
References		150
Relevant Websites		151

6.05.1 Introduction

In the African continent, the Republic of Cameroon is situated between 2° and 12° latitude north and meridian 8° and 16° east from the Atlantic Ocean to Lake Chad. The country has an area of 475 000 km^2 and a population of about 17 million.

Cameroon offers a wealth of hydropower opportunity and owns the fourth largest hydro potential in Africa. Although 722 MW of this has already been developed, about 19 GW of hydropower still remain untapped. To overcome the important energy deficit, the country has initiated several studies and projects. Some of the projects require the improvement of the current ongoing reform in the sector which is really changing.

Before 1974, electricity was supplied by many different companies in the country. Then all those companies were nationalized and merged into a single vertically integrated company that had the responsibility for production, transmission, distribution, and retail sales of electricity. This monolithic organization had, however, shown its limits (among which are low productivity, planning, etc.). In order to increase the productivity of the company and satisfy the needs in the future, new measures have been taken since 1998 to overcome these limits. The national company was privatized, the sector was opened to competition, and new institutions were set up to manage this new competitive environment. Different regimes now apply to actors of the sector, depending on the type of their activity and on the power produced for the electricity producer. We distinguish the concession regime, the license regime, the authorization regime, the declaration regime, and the liberty regime.

The weaknesses of the current institutional structure of the power sector have already been proved, and nevertheless, we will examine the ongoing projects and conditions for introducing the private sector in transmission, distribution, system operations, and retail sales. The country has adopted a new national energy plan to reduce poverty and studies on a mid-term development plan in the sector have been done. Cameroon is therefore looking forward to having the means from international financing institutions to implement the plan. This issue will also be examined. Electricity production in the country and in the subregion does not meet the demand; therefore, there are real opportunities.

6.05.2 Hydro Potential

Cameroon has a gross theoretical potential of 294 TWh. However, only 115 TWh is considered to be technically feasible. The country hence has the fourth largest potential in Africa behind the Democratic Republic of Congo (1397 TWh), Madagascar, and Ethiopia. Of the country's total installed capacity of 1018 MW in 2009, 722 MW was from hydropower plants. Compared to the potential and to the needs, the hydro sector is hence underexploited up to the point where the country experiences energy shortage during low water periods and is obligated to install and run several important thermal plants (**Figure 1**).

6.05.2.1 The River System

The river system of Cameroon is made of main catchments, namely the Atlantic basin, the Congo basin, the Niger basin, and the tributaries of Lake Chad. The catchments are made of many rivers from the south to the north. The main river Sanaga has a pluriannual flow that can reach 2000 m^3 s^{-1}. Other rivers have a pluriannual flow less than 500 m^3 s^{-1}. Water flows to the Atlantic Ocean and Lake Chad.

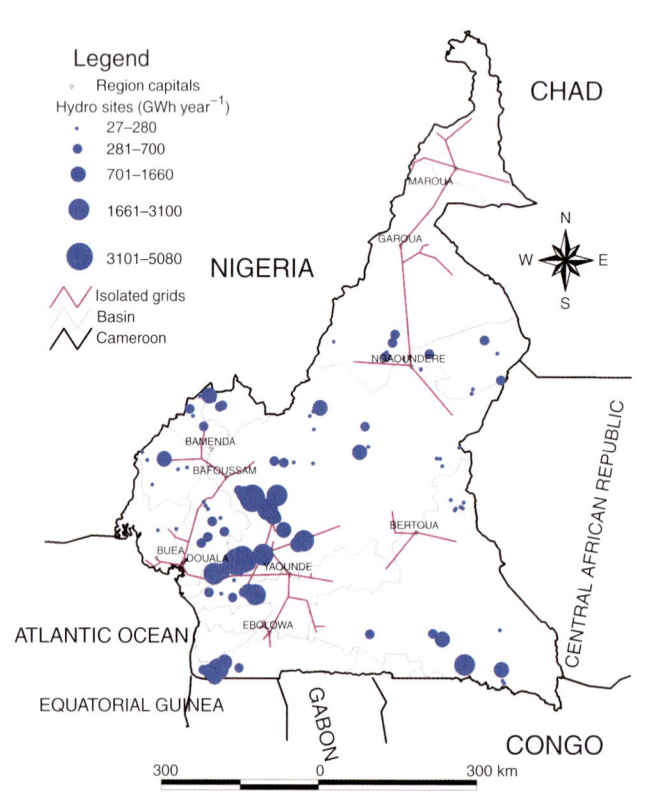

Figure 1 Cameroon hydro potential.

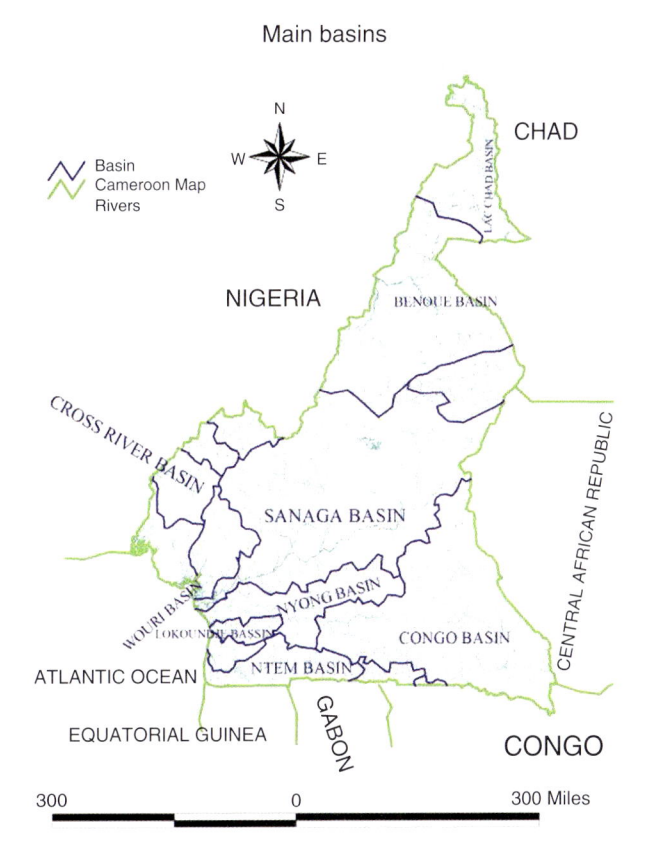

Figure 2 Main catchments in Cameroon.

The river system can then be broken down into four clearly distinct subsystems of different sizes as shown in **Figure 2**.

1. The Atlantic catchment is the largest of the four subsystems, with the Sanaga river draining alone a catchment area of 135 000 km^2 and a pluriannual flow that can reach 2000 m^3 s^{-1} at Edea. This vast river is formed by the union of the Lom, the Pangar, and Djerem rivers south of Adamaoua Region. Downstream, the Mbam and its tributary the Noun bring in waters from western chains on the right bank. To the south of the Sanaga, the Nyong with a pluriannual flow of around 420 m^3 s^{-1} also flows toward the Atlantic and has no major tributary. The Ntem with a pluriannual flow of around 440 m^3 s^{-1} is the last large river. It springs up in Gabon. The small rivers such as Dibamba, Lokoundje, Lobe, Mungo, and Wouri drain all western chains.
2. For the Sangha catchment, we have three tributaries of the Sangha river, for example, Dja, Boumba, and Kadei, which in turn is a tributary of the Congo river. The Dja and Boumba have at their confluence flows of 500 and 280 m^3 s^{-1}, respectively.
3. For the Benoue catchment area, the Benoue river is the largest of the Niger river's tributaries with a pluriannual flow of 250 m^3 s^{-1}. West of this chain, the Donga, the Katsina Ala, and the Cross rivers also run into the Benoue, but in Nigeria.
4. The tributaries of Lake Chad consist of the Vina in the north and the Mbere. Both rivers form the western branch of the Logone that runs into the Chari that feeds Lake Chad.

Altitudes are from 0 m to more than 2600 m. Annual rainfall varies from 400 mm to more than 10 000 mm. This situation enables Cameroon to have an important hydrographical network.

6.05.2.2 Existing Hydro Plants

Three hydro plants are currently under operation.

- Edea
  Edea hydro plant was developed on the Sanaga river in three stages: Edea I in 1953 with three units of 11.5 MW each, Edea II in 1958 with 6 units of 121.8 MW each, and Edea III in 1975 with 5 units of 107.5 MW each. Some old equipment is currently under replacement with more efficient products (**Figure 3**).

Figure 3 Edea hydro plant at final stage. Reproduced from Atlas du Potentiel Hydroélectrique du Cameroun.

Figure 4 Song Loulou hydro plant. Reproduced from Atlas du Potentiel Hydroélectrique du Cameroun.

- Song Loulou
  Song Loulou hydro plant was built on the Sanaga river in two stages from 1977 to 1988. It consists of eight units of 48 MW each. Edea and Song Loulou are currently the only hydro plants supplying electricity for the southern grid (**Figure 4**).
- Lagdo
  Lagdo hydro plant was developed in 1983 and consists of four units of 18 MW each. The Lagdo plant is the only hydro plant supplying electricity for the northern grid.

The hydro plants so far developed in Cameroon are represented in **Table 1**. It shows that the last hydro plant was developed in 1988 and no other hydro plant has been developed to date. **Figure 5** shows the location of the hydro plants under operation.

Since 1988, Cameroon has not developed any other hydro plant.

6.05.2.2.1 *Production and transportation of electricity*

The total thermal and hydro installed capacity in Cameroon is presented in **Table 2**. It shows the important growth of the thermal plants.

The country has low- to high-voltage power lines. Three high-voltage levels are used for transportation, 225 and 90 kV for the south interconnected grid and 110 and 90 kV for the northern interconnected grid. Energy distribution is done through several

Table 1 Evolution of hydro plants in Cameroon

Hydro plant	Year of completion	Number of units	Total installed power (MW)	Cumulative capacity (MW)
Edea I	1953	3	35.3	35.3
Edea II	1958	6	121.8	157.1
Edea III	1975	5	107.5	264.6
Song Loulou I	1981	4	193	457.6
Lagdo	1983	4	72	529.6
Song Loulou II	1988	8	193	722.6

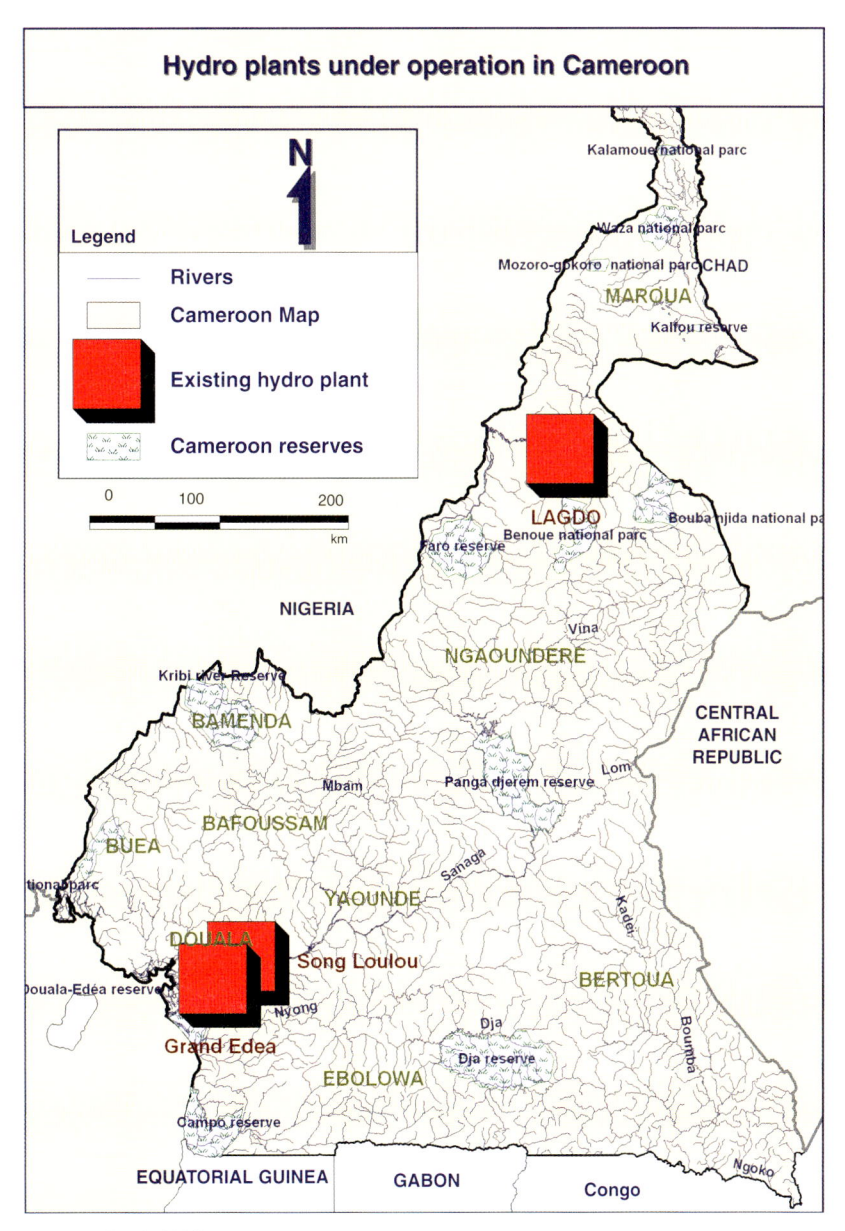

Figure 5 Hydro plant under operation in 2008.

medium-voltage levels, namely 30, 17.3 kV for single-wire earth return; 15, 10, and 5.5 kV. In 2008, the transmission and distribution lines were as presented in **Table 3**.

The overall production, taking into account the production of all hydro plants and all thermal plants including standalone systems managed by the private utility AES-SONEL, is presented in **Table 4**.

Table 2 Installed capacity in Cameroon in 2009

Grid	Locality	Installed power (MW) Hydro	Thermal	Fuel
South grid	Edea	264		Water
	Song Loulou	384		Water
	Limbe		85	HFO
	Yassa		85	HFO
	Bassa		19	LFO
	Bafoussam		14.3	LFO
	Logbaba		17.6	HFO
	Oyomabang I		16	HFO
	Oyomabang II		19.5	HFO
	Ebolowa		1	LFO
	Meyomessala		1	LFO
	Mefou		2.6	LFO
North grid	Lagdo	72		Water
	Djamboutou		17.2	LFO
East grid	Bertoua		6.4	LFO
Standalone systems	30 Thermal plants		14	LFO
Total		720	298.6	
	Total		1018.6	

LFO, light fuel oil; HFO, heavy fuel oil.
Reproduced from ARSEL and AES-SONEL data.

Table 3 Transmission and distribution lines

Lines	Length (km)
High voltage, 225 kV	480
110 kV	337
90 kV	1 210
Medium voltage, 30/15/10/5.5 kV	12 089
Low voltage	13 605

Reproduced from AES-SONEL Annual Report (2008).

Table 4 Overall production of energy

Plants	Production (kWh)	Availability ration (%)
Edea	1 584 871	76.19
Song Loulou	2 425 543	96.99
Lagdo	222 063	81.66
Limbe	179 335	98.11
Other interconnected systems	20 352	57.28
Standalone thermal	70 595	50.55

6.05.3 Dams

6.05.3.1 Storage Dams Under Operation

The production of Edea and Song Loulou hydro plants is sustained by three storage dams: Bamendjin dam (**Figure 6**), Mape dam, and Mbakaou dam (**Figure 7**). All the three dams contribute to regulate the flow rate of the Sanaga river to lower the impact of the dry season, that is, the low water level. Lagdo power plant has a dedicated dam located immediately upstream. **Figure 8** shows the location of the four storage dams under operation.

Figure 6 Bamendjin dam. Reproduced from Atlas du Potentiel Hydroélectrique du Cameroun.

Figure 7 Mbakaou dam. Reproduced from Atlas du Potentiel Hydroélectrique du Cameroun.

6.05.3.2 Hydrology

Table 5 shows that on 18 December 2009, the filled percentage ratio was 97% for Bamendjin dam, 78.33% for Mape dam, and 100% for Mbakaou dam, giving a total capacity of 7.007 35 billion cubic meter out of 7.779 billion cubic meters expected.

Mape dam is not often full and studies were made to find solutions. For the year 2006, 2007, and 2008, the total volume inside the storage dams at the beginning of the regularization were 7605 million cubic meters, 6383 million cubic meters, and 7204 million cubic meters, respectively, as show in Table 6. The table shows that Cameroon still experience important deficit in terms of water storage for the optimal use of the hydro plants under operation. To overcome this situation, the country is among other initiatives planning to construct new dams and new hydro plants.

6.05.4 Mid-Term Development Plan for Hydro Plants in Cameroon

6.05.4.1 Objectives

Cameroon government has made several studies aiming at providing Cameroonian authorities (represented by the Minister of Water and Energy) as well as Cameroon's development partners, in particular, the World Bank, the African Development Bank, and others with an adequate analysis of existing options and their financial implications for the development of the next generation of hydropower plants in the country. The studies suggested the selection and timing of hydro generation investment projects in the electricity sector at a medium and long term (2025–2035). Elements from the development of several thermal plants were taken into account, namely the Limbe, Kribi, and Yassa thermal plants. Among the studies are the energy sector development program (PDSE 2030) and strategies on the sector.

The political objectives of the government is to enhance the fight against poverty by increasing the gross national product *per capita* from around US$1000 in 2005 to more than US$5000 in 2030. This ambitious program requires an implementation of a long-term development plan in the energy sector (PDSE 2030).

In order to attain the goals, Cameroon authorities have decided to rely on important least-cost available resources, mainly hydropower and gas.

Most of the potential hydro generation facilities have been identified on different basins.

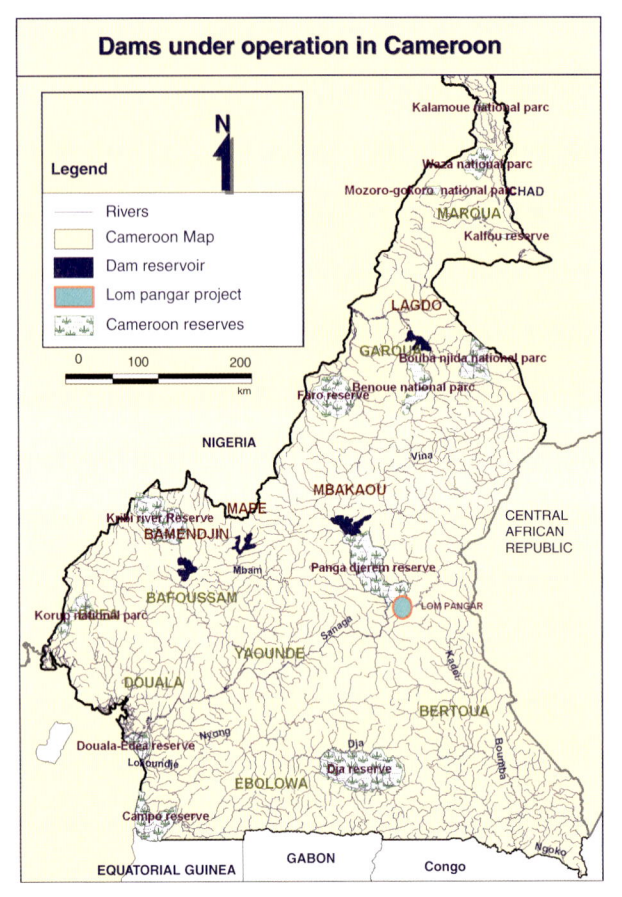

Figure 8 Dams under operation in 2010.

Table 5 Fill level of Mape, Mbakaou, and Bamendjin dams on 17 and 18 December

		Filled level (billions of cubic meters)	
Dam designation	Nominal capacity	17 December 2008	18 December 2009
Bamendjin	1.879	1.574	1.822
Mape	3.300	2.958	2.585
Mbakaou	2.600	2.600	2.600
Total	7.779	7.132	7.007

Table 6 Evolution of the low water level during the years 2006–2008

	Parameter	Unit	Year 2006	Year 2007	Year 2008
1	Regularization start date	Day	10 December	10 December	30 December
2	Regularization end date	Day	14 June	24 June	27 June
3	Regularization period	Days	186	176	150
4	Maximum volume in reservoirs	$10^6 \, m^3$	7 605	6 383	7 204
5	End low water level volume	$10^6 \, m^3$	682	515	1 627
7	Total volume released	$10^6 \, m^3$	7 919	6 598	6 262
9	Observed volume	$10^6 \, m^3$	12 884	12 700	11 953
10	Targeted regulated volume	$10^6 \, m^3$	11 866	11 634	10 854
11	Potential deficit	$10^6 \, m^3$	6 471	5 543	4 712
12	Efficiency	%	81.71	84.01	75.2

For the Sanaga basin, we have:

More equipment at Song Loulou (100–150 MW)
Nachtigal upstream (230–300 MW)
Kikot (around 500 MW)
Song Mbengué (around 900 MW)
Song Ndong (200–300 MW)
Lom Pangar dam (5.5–7 km^3)
Pont Rail dam (3.5–4 km^3).

For the south west basin, we have:

Memvé Elé on the Ntem river (201 MW)
Njock on the Nyong river (120 MW).

For the north basin, we have:

Bini Warak on the Vina du Nord river (75 MW).

For the east basin, we have:

Colomines on the Kadei river (12 MW).

For Nachtigal, Njock, Memve'Elé, Song Ndong, Song Loulou extension, and Noun (1&2), cost estimates already exist and are taken into account for future generation options though the studies are not limited to these projects. Projects have been compared to other options in terms of size and development cost in order to find other realistic alternatives.

The studies focused on analyzing hydro generation options that could be developed in Cameroon by the year 2025. Projects that are clearly inferior were eliminated by using a screening analysis. Those that are not feasible for any other reasons were also eliminated.

Studies were made to satisfy the generation supply options required to meet the demand in the southern interconnected network up to and including the year 2025. Export of electricity to Equatorial Guinea, Gabon, Congo, Nigeria, and Chad are still to be seriously discussed, even though Cameroon, Chad, and Nigeria are already under discussion and have gone a bit further.

6.05.4.2 Context of the Development Plan

It is currently a crucial time in the medium-term development of the electricity sector in Cameroon, as decisions with significant and long-lasting consequences will need to be taken within a relatively short period. Cameroon wants to ensure these decisions are made on the basis of solid and realistic technical, economic, and financial analyses.

Concerning the future demand of electricity, a key issue to be taken into account in the country is the supply options to the aluminum smelter company (ALUCAM). ALUCAM currently accounts for approximately 40% of the south interconnected grid demand. ALUCAM's co-shareholder, Rio Tinto, has indicated that the cost and security of electricity supply to ALUCAM is a key factor in their decisions on the future of ALUCAM's activities. ALUCAM has carried out in-depth studies on an increase of the capacity of ALUCAM's smelter capacity, and is currently proposing to increase the current annual production capacity of around 90 000–120 000 tons to an annual production capacity of 250 000 tons or 1 000 000 tons depending on the availability and cost of energy. Based on this hypothesis, ALUCAM would have an annual electricity demand of at least 450 MW. The other possible options being considered for ALUCAM are either a complete halt in activities or maintaining the current capacity.

ALUCAM's co-shareholder has indicated that failure to a long-term electricity supply contract might lead to the closure of the smelter. It has indicated conditions to fulfill in order to be sufficiently productive to continue with ALUCAM's activities with the existing smelter throughout the year.

6.05.4.3 Future Outlook

6.05.4.3.1 Lom Pangar project
Lom Pangar storage dam project is an ongoing project with a capacity between 5 and 7 billion cubic meters. The aim of Lom Pangar project is to mitigate the severe energy crisis the country has been undergoing since the early nineties. The current hydro production capacity of the country is below the peak demand. The growth of the demand together with severe low water levels during the last decade have convinced the country to envisage right at the early nineties the study and construction of Lom Pangar storage dam. This project has two main objectives:

Enhance the regulation capacity of the Sanaga river
Obtain full production of Song Loulou hydro plant and increase the production of Edea hydro plant.

The increased regulation capacity of the Sanaga river will benefit many other plants expected to be developed in the future, among which are Natchtigal and Song Dong.

This project is a follow up of many other projects on the Sanaga river catchment, after Edea hydro plant in the year 1950, Song Loulou hydro plant between 1981 and 1988, Mbakaou storage dam in 1969, Bamendjin storage dam in 1974, and Mapé in 1988.

A 50 MW hydro plant will be installed at the toe of the dam to cover the needs of the eastern region grid and replace the actual light fuel oil (LFO) thermal plant.

The first study on Lom Pangar project started in 1990, funded by the public utility, SONEL, before privatization, followed by a feasibility report by Coyne and Bellier in 1995 and updated in 1999. The first environmental study was done by INGEROP in 1998.

The updated studies done in 1999 served as a guide to other studies aiming at:

Analyzed other alternatives
Detailed description of the project and the description of the initial state of the project zone
Identify the stakes of the project zone and assess the impacts of the project
Define measures to manage impacts.

6.05.4.3.2 Dam characteristics

Lom Pangar site is on the river Lom at about 4 km downstream of the junction with the river Pangar, about 120 km north of Bertoua town in the east region (**Figure 9**). The site is accessed via the left bank, trough Deng-Deng locality and after 30 km of unpaved road. The location of the site is shown in **Figure 9**.

Latitude north 5° 24′
Longitude east 13° 30′.

At the selected location of the dam, the valley is narrow, 120 m wide.

The dam is 45.55 m high and is mixed type, comprising concrete on one section and earth on another section.

The work is scheduled for 44 months, starting with building of the road on the left bank. The filling of the dam is scheduled for the middle of the final year. The reservoir will cover a maximum area of 590 km^2 under the water level of 674.50 m and the total storage capacity is 7.5 billion cubic meters for a useful capacity of 7 billion cubic meters. The water level will be above the mean level around 6 months yr^{-1}. The marling will be around 10 m under in a normal year and 20 m under in a dry year.

6.05.4.3.3 Outcome of the Lom Pangar project

The project will allow the current regulated flow of the Sanaga river during low water level, which is currently 600 m^3 s^{-1}, to be 925 m^3 s^{-1}. Given the 3.5 hm cube available at Song Loulou, this flow will allow the Song Loulou hydro plant to run under full

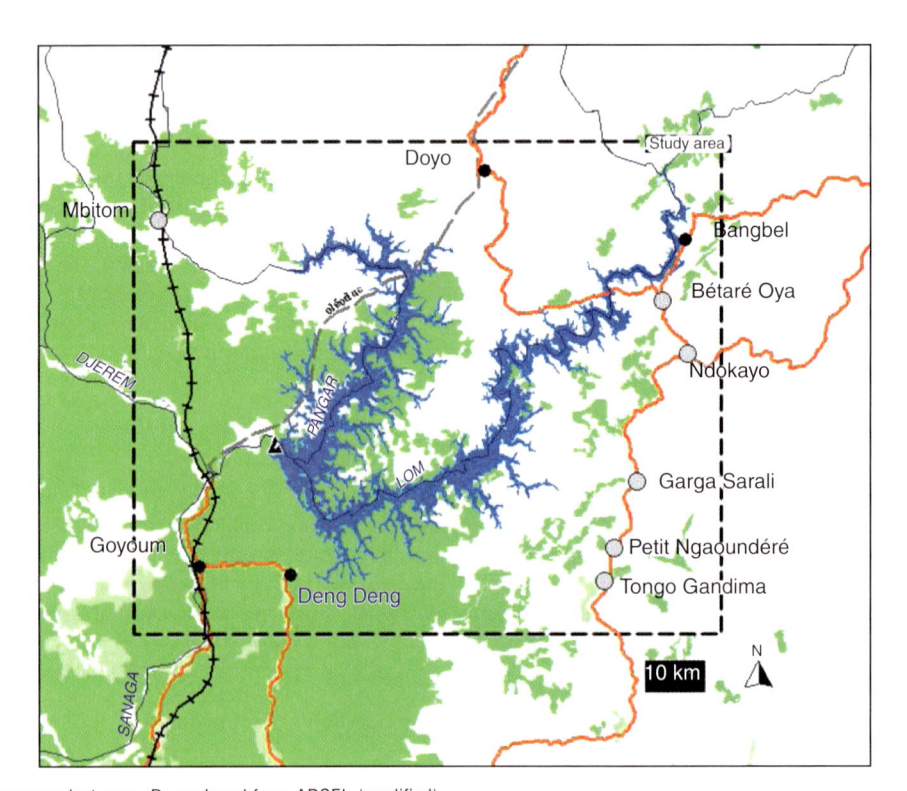

Figure 9 Lom Pangar project zone. Reproduced from ARSEL (modified).

capacity during the 5 h peak of electric consumption. The Lom Pangar storage dam will bring 120 MW more guaranteed power the existing Edea and Song Loulou hydro plant on the Sanaga river and will yield an average of 250 GWh yr^{-1}.

This mean production will be raised to 675 GWh with the development of Natchtigal hydro plant (230–300 MW) and 775 GWh with the development of Song Dong hydro plant (200–300 MW). The development in the future of Kikot (500 MW) and Song Mbengue (900 MW) will also benefit from the Lom Pangar project.

This shows that the Lom Pangar storage dam project is a long-term project that will sustain electricity production of current and all future hydro plants installed along the Sanaga river in Cameroon, making more energy available for the upcoming Inga-Calabar high-voltage power line to be built in central Africa.

6.05.4.3.4 Project justification and other alternatives

The extension of the Kribi thermal plant under construction is presently the project that might economically compete with the Lom Pangar project.

It has been established that the Lom Pangar project will produce within a century 21 million tons of carbon dioxide compared to 17 million tons for the thermal plant. But in the long run, the situation will reverse as soon as another hydro plant is developed on the Sanaga river, giving advantage to the Lom Pangar project. In fact, Lom Pangar and Nachtigal projects will produce seven times less gas emission than the thermal plant within a century. Furthermore, the cost of 1 hydro kWh produced is estimated at €1.98, compared to around €4.57 for the thermal.

Another point is that the cost per stored cubic meter is €1.22 for Lom Pangar, which is more than two times less expensive compared to other concurrent solutions, namely Litala on the Lom river, and Bankim and Nyanzom on the Mbam river.

Among other alternative hydro plant that might meet the short- and mid-term demands, the unit cost of energy (kWh), the energy yielded, and the impacts of Bankim/Nyanzom are closer to Lom Pangar. But the Lom Pangar/Nachtigal complex has two disadvantages compared to Bankim/Nyanzom. The first is the gas emission, which is higher, and the second is the Cameroon–Chad pipeline, which is on the dam site and should be moved. But Lom Pangar has a great advantage because it is in a region where very few people live in and will avoid important displacement of population compared to Bankim/Nyanzom.

Studies have demonstrated that the optimal size of the dam might be 5.5 billion cubic meters. This issue is still to be refined during detailed studies.

Based on the current studies and others, the development of the important hydro potential of the Sanaga river and the Lom Pangar storage dam is the best option for the country. It will cover all the needs of the country and minimize the gas emission.

6.05.4.3.5 Optimizing the reservoir

Given the importance of the flow regulation impact of the Lom Pangar project on the existing and forthcoming hydro plants on the Sanaga river, Cameroon is really concerned by the optimization of the reservoir. Several options have been envisaged for a capacity storage varying from 5 to 7 billion cubic meters. The evolution of the climate change context is an uncertain issue for the optimization of the size of the dam, though it has been considered that the actual tendency will stop. After taking into account the contribution of other dams in the Sanaga basin for a guaranteed flow of 750 m^3 s^{-1} at Nachtigal and 1040 m^3 s^{-1} at Song Loulou during low water level, the studies made by ISL – Oréade-Brèche – Sogreah, in 2007 found an optimal reservoir capacity around 6 km^3.

6.05.4.4 Memve'Elé

6.05.4.4.1 Project area and location

The site of the Memve'Elé hydro plant project shown on **Figure 10** is on the Ntem river, south west of Cameroon, not far from the Equatorial Guinea border, as shown in the figure. The river is one of the largest in the country with a mean annual discharge at the dam site of 398 m^3 s^{-1} and a catchment of around 30 000 km^2 (**Figure 10**).

This project was successively studied in the framework of:

1. The 'Inventory of Hydropower Resources' of Cameroon published in 1983 by SONEL with Electricité de France (EDF)
2. The 'Feasibility Study on Memve'Elé hydroelectric power development project' carried out by Nippon Koeï in October 1993 and funded by Japan International Cooperation Agency
3. Feasibility studies updated by Coyne and Bellier in February 2006
4. Detailed studies by Electricé de France, Globeleq and Sud Energie in Reference [1].

The project is a run-of-river type with a low head dam and its related structures, a headrace channel, a power station, and a high-voltage transmission line from the site to Yaounde or Edea.

6.05.4.4.2 Initial cost of the project

The cost estimate of the project is €217.7 millions divided as shown in Table 7 excluding the power line.

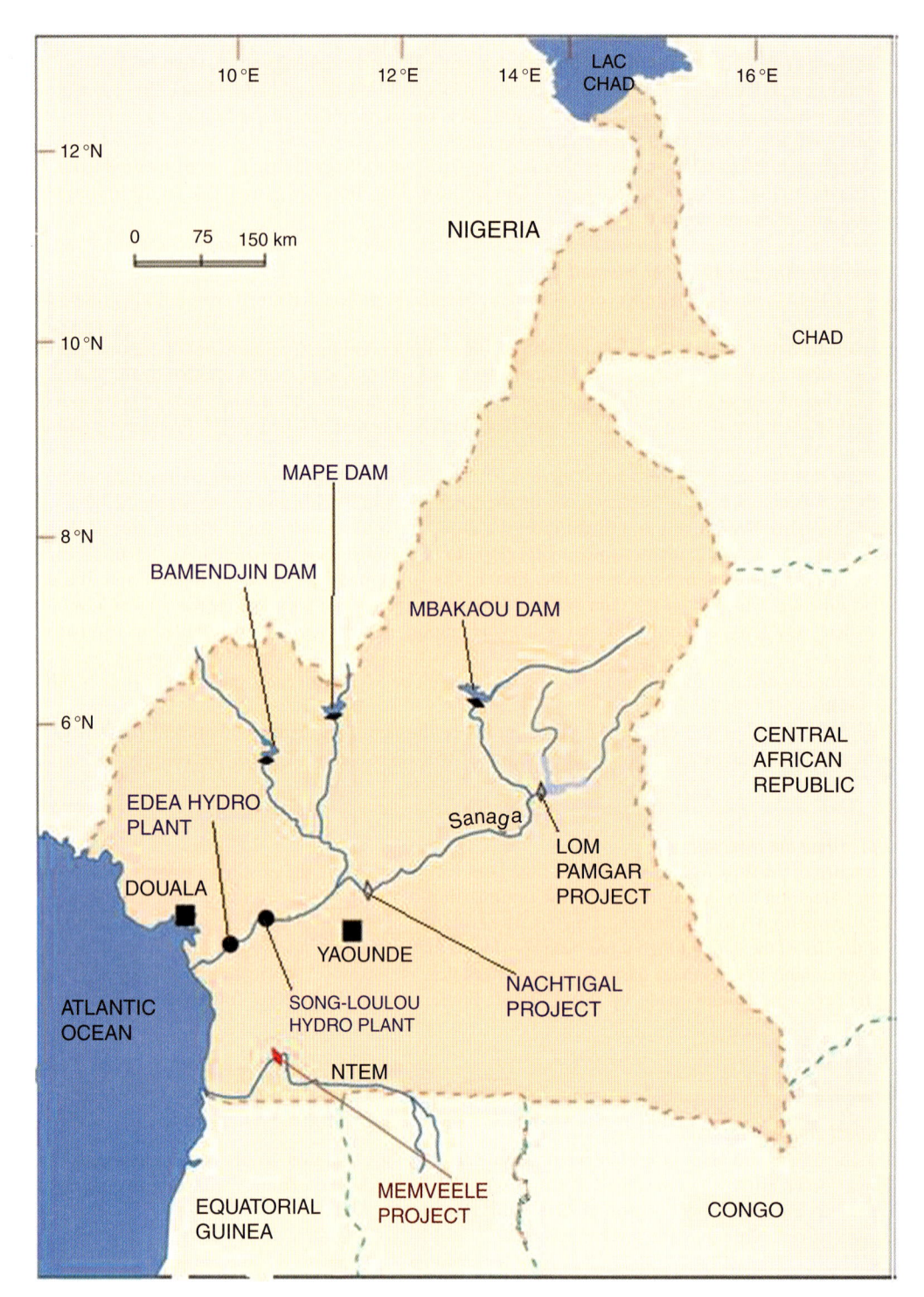

Figure 10 Memve'Elé project location. Reproduced from Ministry of Energy and Water (modified).

6.05.4.4.3 *Project layout and structures*
- Dam: The dyke will be a low head one and made of homogeneous material. The normal pool level is 392 m.
- Spillway: The spillway is arranged on the left side of the main riverbed with six sluice gates (height of 10.50 m and width of 11 m), allowing a peak of flow up to 3300 m^3 s^{-1}.
- Intake: It is situated on the left side and is 3.4 km long.
- Hydro plant capacity: The plant is expected to have an installed capacity of 201 MW.

Table 7 Cost per item as scheduled in 2006

Item	Cost (million €)
Civil engineering	61.6
Hydromechanical equipment	16.6
Electromechanical equipment	47.3
Engineering and administrative cost	12.5
Others	1.3
Risks and unexpected	27.8
Additional works	50.6
Total	217.7

- Power line: Energy will be injected to the southern grid in the capital city Yaoundé or Edea through a 225 kV power line. The base variant is a 285 km 225 kV power line between Memve'Elé and Yaounde. The second variant is to connect Memve'Elé hydro plant to Edea through Kribi. This variant is shorter and might supply energy to important medium-voltage customers like HEVECAM and the aluminum smelting industry ALUCAM. One opportunity is to interconnect Memve'Elé hydro plant to the Equatorial Guinea grid through a 40 km power line to be constructed.

After an environmental impact assessment, countermeasures have been set and all stakeholders have been taken into account. Special attention will be paid to the Campo national park. Less than 30 persons will be displaced and a budget of €2.13 million for a special socioeconomic program will be set up for the population living in the area (around 13 000 persons).

The total construction period is scheduled from 7 to 10 years, including more than 110 km of an access road.

6.05.4.4.4 Dam-reservoir on the Ntem
In order to increase the guaranteed output during the dry season and with respect to the needs in peak and off-peak hours, several dam sites have been assessed. Among all the sites, Nyabibak appears to be the best. The storage capacity estimated with SRTM is 1.8 km^3 at elevation 560 with a dam height of 32 m.

6.05.4.4.5 Environmental impact
This aspect has also been studied and many impacts on the environment were identified. For all possible impacts identified, a number of countermeasures were suggested. For the direct compensatory measures, recommendations were made for the conservation of the Memve'Elé falls, the forest resources to be submerged, the protection of animals in the reservoir area before impounding, the preventive measures in socio-sanitation, the compensation, the restoration of public infrastructures, and the resettlement. For the indirect compensatory measures, other recommendations were suggested, namely the conservation of fauna and flora, the improvement of health services and sanitation including water supply, the agricultural program, the organization of fishing development, the general organization, and follow up of compensatory measures and other recommendations.

6.05.4.4.6 Funding
The government of Cameroon and the Chinese company SINOHYDRO signed a contract for the funding and the development of the Memve'Elé project on 25 September 2009. The project will be realized for €555 894 879 after updating some studies and taking into account additional work.

6.05.4.5 Mekin Hydropower Project

6.05.4.5.1 Introduction
6.05.4.5.1(i) Project significance and assessment purpose
Mekin hydropower project is located on the Dja River in the south of Cameroon. The power station will be installed at the toe of the dam site with a catchment of 10 800 km^2 and a normal impounded water level of 613 m. The total installed capacity will be 12 MW in three sets of 4 MW. The power generation capacity is expected to be 70 GWh.

The energy produced will be injected to the south grid to reinforce the energy access of the grid in the whole southern area and help the utility to remove the installed LFO thermal plants in the localities around, namely Bengbis (229 kW), Djoum (357 kW), and Meyomessala (1 MW). This project will also help to avoid frequent energy shortages in the area and attract investors.

The project is near the Dja Faunal Natural Reserve as shown in **Figure 11** and might seriously impact the reserve.

The following four important goals are targeted:

1. Enhance the rural electrification in the project areas through grid-connected solution around the cities of Djoum, Mintom, Oveng, Zoetele, Bengbis, Sangmelima, Meyomessala, Endom Akonolinga, Somalomo, and Meyomessi on both sides of Dja River

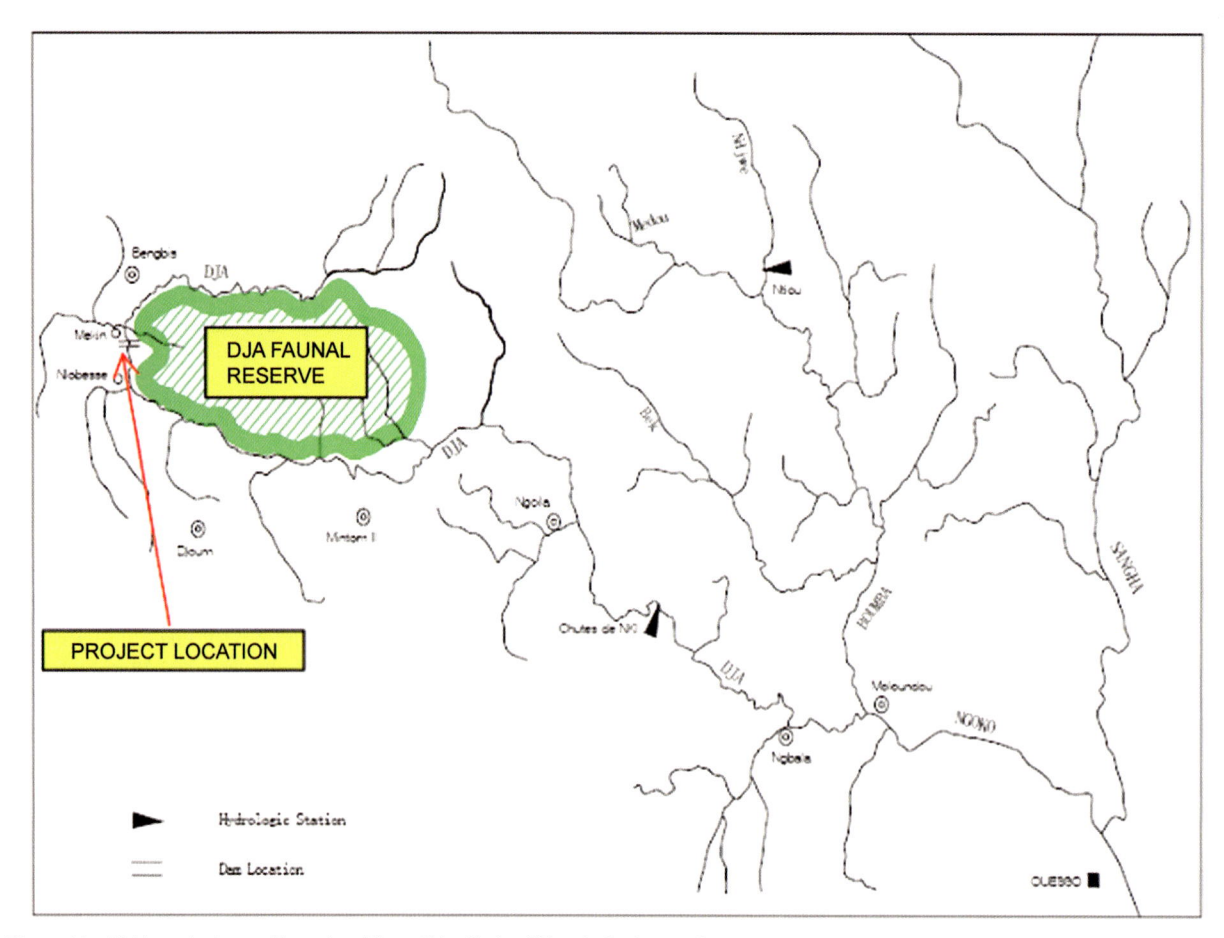

Figure 11 Mekin project zone. Reproduced from China National Electric Equipment Corporation (modified).

2. Sustain the grid and make electricity more available to allow normal operation of power grid connected with the southern part including Ekombitie relay station through 90/30 kV power lines
3. Regulate the fishing and hunting activities in the Dja Faunal Natural Reserve by enhancing the fishery cultivation and irrigation in the reservoir area and promoting tourism in the area
4. Supply electricity for Mubanlan iron mine project 350 km away from Mekin.

6.05.4.5.1(ii) Project layout and structures

6.05.4.5.1(ii)(a) Project layout The Mekin hydropower project mainly includes:

- Dam
- Spillway
- Powerhouse
- Switchgear
- Access roads
- Living and plant areas, and so on.

- **Dam**

The dam consists of two sections, the left dam section with a total length of 1000 m and the right dam section with a total length of 1500 m. For the left dam section, a 2 m wide berm will be provided at the elevation of 608 m upstream of the dam. The upstream dam slope shall be protected with concrete slabs, while its downstream slope is protected with wood-latticed turfing. The relevant data are as follows:

- Crest elevation: 615 m
- Crest width: 3 m
- Maximum height: 11 m
- Waterside dam slope gradient: 1:2.75 and 1:3
- Landside dam slope gradient: 1:2

For the right dam section, it will be constructed in combination with the construction of the roads within the project site.

- **Spillway**

The spillway is 42 m wide in total with the net flow section of 40 m wide. The No. 1 Spillway is on the left side of the main riverbed and right side of the power station. The spillway is 42 m wide in total with the net flow section of 40 m wide. The spillway has a trapezoidal broad crest weir and is 21.25 m long with a height of 9.25 m and a crest elevation of 608.25 m. Four removable hydraulic lifting dams (10 m wide × 5 m high) will be provided on top of the crest weir. These four hydraulic lifting dams are removable, reinforced concrete gates. So, the total elevation will reach 613.25 m.

The No. 2 Spillway is arranged on the left dam section where the construction diversion open channel is located. The crest overflowing width is 100.00 mm. The upstream slope gradient is 1:3, while the bottom elevation of the overflowing protective face is 608.5 m. The downstream slope gradient is 1:3, while the bottom elevation of the overflowing protective face is 599 m. The crest is 8 m wide. The overflowing protecting face will be made of reinforced concrete with a thickness of 0.25 m. Composite geomembrane will be used for seepage-proof bottom linings.

- **Powerhouse**

The main powerhouse is 64.74 m wide and is located on the right side of the main riverbed. An auxiliary powerhouse is also provided on the downstream side of the main powerhouse. The Mekin hydropower plant will be built in the riverbed and the main powerhouse will be on the ground level. The units are vertical. The powerhouse has a turbine floor and a generator floor. The powerhouse has four functional areas, including the main powerhouse, erection bay, auxiliary powerhouse (the high-voltage switch cabinet room and the auxiliary switchboard room), and central control room. The main powerhouse is 64.74 m wide and 15.50 m long in total, where three generating units will be installed. The powerhouse includes two sections and an erection bay. The generator room is on the right side and the spacing distance between generators is 13 m and 14 m, respectively. The erection bay is on the right side and is 18.08 m wide. The net height of the main powerhouse is 31.73 m.

The main powerhouse and the auxiliary powerhouse will be arranged on the upstream and downstream sides. The auxiliary powerhouse with the plane dimensions of 64.74×8.05 m^2 is on the downstream side of the main powerhouse and has two floors in total, which will be connected to the turbine floor and generator floor, respectively.

6.05.4.5.1(iii)   Electromechanical equipment and hydromechanical works

Given the water head and capacity parameters, three movable propeller turbines (Model: ZZ536-LH-330) are chosen and the generator type is SF4000-44/4250, model: SF4000-44/4250. In order to match the turbine type, computer-controlled WST-100 governors are chosen. The model of the supporting oil pressure devices is HYZ-4.00 and the operating oil pressure is 2.5 MPa.

The lifting equipment has a capacity of 50/10 tons and span of 13.5 m. According to the heaviest part and the width of the powerhouse, an overhead travelling crane was chosen.

6.05.4.5.1(iv)   Electrical work

Three hydraulic turbine generators with a single capacity of 4 MW will be installed for the station. The generator output voltage is 10.5 kV, which will be boosted to 110 kV by two main transformers (Model: SF9-8000/110; transformation ratio: $110 \pm 2 \times 2.50\%$/10.5 kV, and capacity: 8000 kVA) for power transmission. Single bus bars will be used to connect medium voltage (10.5 kV), and another single bus bar will also be used to connect the high voltage (110 kV).

The auxiliary electrical equipment includes:

Oil pumps for the oil pressures device of the governors

Various water pumps

Lifting equipment

Ventilators

Auxiliary power supply (APS) for the excitation units

Charging devices for continuous-current plant

Uninterruptible power supply

Lighting for the powerhouse and its surroundings

Intake and outlet gate hoists

Spillway gate hoists.

The above items will be powered through low voltage of 0.4 kV from a 315 kVA, 10.5 kV/0.4 kV. Additionally, an LFO generator will be provided as the standby power supply in case the station fails.

The areas around the power station where low-voltage power will be supplied include:

Reservoir management zone

Office building

Multiple living buildings.

For the neighboring villages, electricity will be supply through a 33 kV power line since this voltage level is widely used in the country. This should be done through a 1000 kVA transformer to be installed inside the booster station. The low-voltage side of the transformer will be connected to the 10.5 kV generator terminal bus bar.

6.05.4.5.1(v) Hydromechanical works

The generating units for the hydropower station are of the axial flow type. The axial flow units consist of three units with the following characteristics: opening 4.33 m wide × 8 m high, design head 2 m, and sill elevation of 602.5 m. There are also six openings with six trash racks.

6.05.4.5.1(vi) Ventilation and air conditioning

The powerhouse will be naturally ventilated. Natural ventilation and mechanical exhaust will be used to ventilate the oil depot, oil treatment room, air compressor room, transformer room, and maintenance and drainage gallery. The ventilation system consists of axial fans and ventilating pipes. The central control room, high-voltage switch cabinet room, and the auxiliary switchboard room will be naturally ventilated. The auxiliary transformer will be installed inside the cable gallery, where high-light windows are provided for natural ventilation. The turbine oil treatment room will be arranged inside the cable interlayers. For safe ventilation and fire vent, mechanical exhaust will be adopted.

6.05.4.5.1(vii) Fire fighting

The fire control design for the power station is based on such a policy of "Fire Prevention First, Prevention and Control Combined and Self-control and Self-rescue".

The general fire extinguishing scheme is as follows:

The water fire extinguishing devices prevails
Chemical fire extinguishing is complementary
The other fire extinguishing method is combined.

A number of chemical fire extinguishers will be provided for oil products. The hoist chambers located on the dam will be equipped with a number of chemical fire extinguishers.

6.05.4.5.1(viii) Inland fish farming and tourism

Fish farming in the Mekin reservoir is expected to generate income for the villagers and sustain the living standards in the area. It has a special meaning of utilizing water resources for comprehensive benefits such as power generation, aquiculture, irrigation, and prosperousness of rural economy in the forest zone. To achieve these goals, a restrictive water level for the inland fish farming has been set at 608.5 m, meaning when the water level is lower than 608.5 m, the generators should be shut down. It is expected that the Mekin reservoir can yield 50×10^4 kg of fish per annum.

The Mekin area enjoys exceptional advantages for tourism such as primeval natural scenery, primeval wild animals, and aboriginal culture. Mekin is a piece of pure land for development. This aspect will be developed to attract tourists.

6.05.4.5.1(ix) Assessment of environmental impact

Among other positive impacts, Mekin Hydropower Project may help develop a multipurpose use of water resources in the Dja river valley, provide clean energy to replace part of the biomass, and may also help to mitigate the power shortage that the neighboring areas are currently experiencing. This project will also contribute to sustain local economic development. Plans to mitigate negative impacts have been developed for water and soil conservation including 150 000 Chinese RMB yuan for the environmental compensation.

6.05.4.5.2 Investment estimate

It the estimated that the total project investment is 293 833 900 Chinese RMB yuan only, including:

- 184 758 400 Chinese RMB yuan for civil works
- 52 862 100 Chinese RMB yuan for eletromechanical equipment and installation
- 5 143 500 Chinese RMB yuan for hydromechanical works and installation
- 24 276 400 Chinese RMB yuan for temporary works
- 18 207 300 Chinese RMB yuan for other expenses
- 7 131 200 Chinese RMB yuan for preparatory cost

- 1 155 000 Chinese RMB yuan for land occupation compensation
- 300 000 Chinese RMB yuan for environmental protection and water and soil conservation.

Additionally, the loan interest amounts to 7 776 000 Chinese RMB yuan.

6.05.4.5.3 Economic assessment

According to the calculations provided, the financial internal rate of return before deducting tax is 13.95%, which is higher than 7%, the basic gross profit for the power industry. The financial net present value is 253 490 000 Chinese RMB yuan, which is bigger than 0. The return on investment period will be 8.7 years and the financial internal rate of return after deducting tax will be 11.84%. Therefore, it is realistic and feasible to implement the project.

6.05.4.6 Bini Warak Project

Bini Warak project (75 MW/300 GWh) is in the northern part of the country on the Vina du Nord river where interannual rainfall varies a lot. Detailed studies are currently being finalized after feasibility studies done by the French company Electricité de France (EDF). The project is the only one to be developed in a short-term period in the northern part of the country. It will be interconnected to the northern grid to sustain the network and permit interconnection with Chad through a power line to be constructed. The two countries have already agreed on this issue and made several studies.

The site is 70 km from Ngaoundere city and has the characteristics shown in Table 8.

The Bini Warak project requires interconnection with the northern grid and the upgrade of the 300 km Ngaoundere–Garoua power line from 110 to 225 kV.

6.05.4.7 Colomines Project

Colomines hydro plant is on the Kadei river in the eastern part of the country, 60 km from Batouri town. The capacity envisaged is units of 6 MW each with a 100 km of 30 kV power line to supply energy to the eastern isolated grid. After updating in 2003 the feasibility studies made by EEI and Decon in 1986, the French company MECAMIDI proposed to the Cameroon government the development of the plant on a build–operate–transfer basis. This proposal was based on a study which was done in September 2003 by MECAMIDI. The site characteristics are as shown in Table 9.

Table 8 Characteristics' of Bini Warak hydro plant project

Designation	Value
Turbines	3 Francis
Nominal power per unit	25°MW
Net height	210°m
Installed capacity	75°MW
Mean energy production	300°GWh
Total flow	40 m^3 s^{-1}
Rainfall	1560 mm (50 l s^{-1} km^{-2})
Catchment	1385 km^2
Specific flow	22 l s^{-1} km^{-2}
Dam area	80 km^2
Total dam volume	560 hm^3
Useful volume	530 hm^3
Design flow level	1046/1049

Table 9 Characteristics of Colomines hydro plant project

Designation	Value
Installed capacity	13.6 MW
Turbine	4 Horizontal axis Francis
Design flow	34 m^3 s^{-1}
Height	48.3 m

The mean flow of the river is 50 m^3 s^{-1}, alllowing the production of up to 90 GWh yr^{-1}. The project was estimated around €61 million in 2006.

6.05.4.8 Ngassona Falls 210 Project

The site is located on the Uve river, a tributary of the Meme river in the southwest of Cameroon, around 30 km from Kumba town. The project is ongoing, co-funded by the 2007 Energy facility from the European Union and the government of Cameroon. It is part of Electricity for Rural Development in Rumpi area (ERD RUMPI) project aiming at electrifying around 100 localities. The installed capacity is 2367 kW under a height of 44 m and a design flow of between 7.4 and 9 m^3 s^{-1}. The power plant is divided into two units, each equipped with a Francis turbine. The design will yield 13 584 MWh yr^{-1}. Energy will be transported through a 30 kV power line to the localities and also be injected to the south grid to sustain the Kumba–Ekondo Titi medium-voltage power line under construction. The project is owned by the Rural Electrification Agency supported by Innovation Energie Développement in France.

The hydro projects and dam projects under development or scheduled for mid-term period are located on the Cameroon map shown in **Figure 12**. **Figure 13** presents all the storage dams, hydro plants, and mid-term projects in the country. Some mid-term projects are ongoing and the construction of others are yet to start.

Figure 12 Current generation assets and future projects.

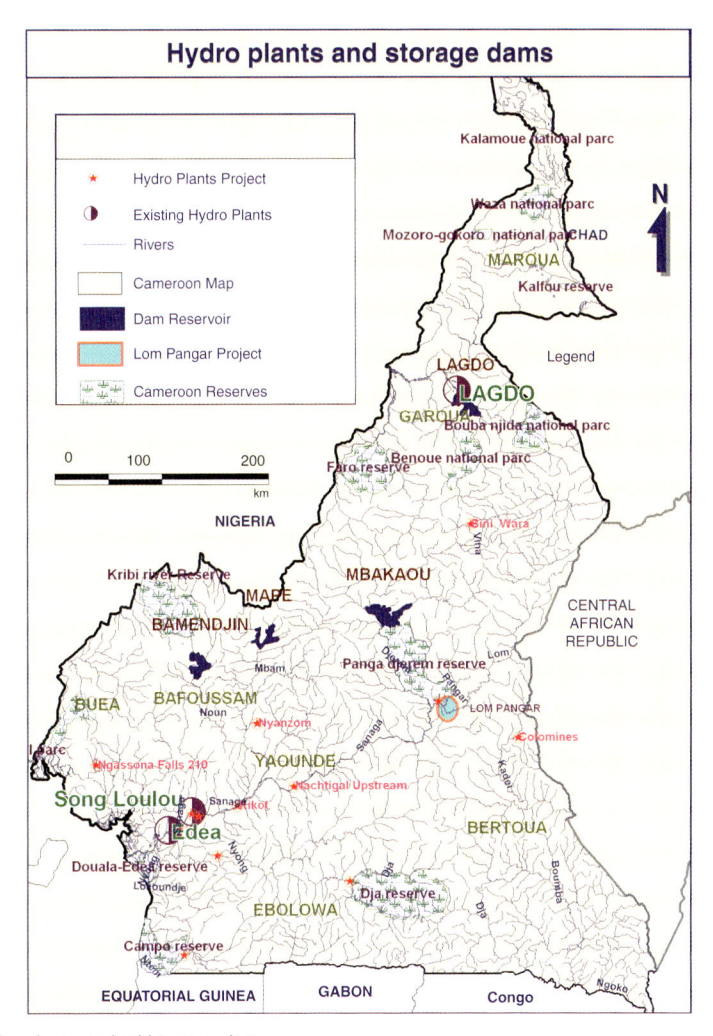

Figure 13 Storage dams, hydro plants, and mid-term projects.

6.05.4.9 Overview of Institutional Structure Reform

6.05.4.9.1 Previous assessments of the power sector reforms

To the best of the author's knowledge, Pineau [2] made a general assessment of the power sector, reviewing the previous assessments. Assessments of reforms are more difficult because the exact power sector situation is not documented and benchmark indicators are not easily available. Efforts are made by the electricity regulatory agency (Agence de Régulation du Secteur de l'Électricité (ARSEL)) for public monitoring and reporting.

6.05.4.9.2 Historical overview of the sector

Before 1974, electricity was supplied by many different companies. Then all those companies were nationalized and merged into a single vertically integrated company that has the responsibility for production, transmission, distribution, and retail sales of electricity.

6.05.4.9.2(i) 26 November 1983 law

This law nationalized electricity. The vertically integrated company, SONEL, was responsible for generation, transmission, distribution, system operations, and sales. Nevertheless, the private sector had the opportunity to generate electricity for their own needs for power under 1000 kW. The state still had the possibility to nationalize generation of above 100 kW if necessary.

6.05.4.9.2(ii) Limits of the 26 November 1983 law

The state-owned company SONEL was expected to make a profit (economic aspect), on the one hand, and relieve the living conditions of populations (social aspect), on the other hand. Since these two main objectives were not compatible, the company started facing problems up to the point where subvention was no longer possible.

In order to tackle the problem, the Government decided to

ameliorate the performance of SONEL in order to save jobs,
ameliorate the productivity and the competitiveness of SONEL,
relieve the heavy subvention weight,
avoid state interference in day-to-day management,
stop diversion of funds.

Based on the World Bank recommendation, the power sector reform has now a new look. Through the program of restructuration of the electricity sector, the Cameroon government wanted to increase the budget resources and concentrate on regulation.

The legislative overhaul made in 1998 to introduce competition was aimed at solving two main problems:

Socioeconomic. In fact, liberalization or even partial liberalization of a strategic sector like electricity cannot be made without a state hold regulatory agency (ARSEL). After privatization in 2001, AES-SONEL is responsible for generation, transmission, distribution, system operations, and sales in the Cameroonian power sector. The company is structured as a regulated private monopoly.

Social. In order to solve the rural electrification problem (remote areas) and fight against poverty, the government created the rural electrification agency (AER standing for Agence de l'Electrification Rurale).

In order to manage important projects in the future, Electricity Development of Cameroon, a new public company, was created in 2006.

Given these objectives, the structure of the power sector significantly changed to fit the new environment.

6.05.4.9.3 Current status

The current status is made of 15 texts among which the most important are law no.°98/022 of 24 December 1998, decree no. 99/125 of 15 June 1999, and decree no. 2000/464/PM of 30 June 2000.

6.05.4.9.3(i) Law no.°98/022 of 2 December 1998

This law governs the electricity sector by clearly setting its structure. It aims at inciting investments from the private sector by

ameliorating generation efficiency, transmission, distribution, and retail sales,
ameliorating service quality and the growth of distribution, and
providing enough electricity at best price to local industries.

The production, transmission, and distribution are authorized under different regimes depending mainly on power.

- *Concession regime.* For hydraulic electricity generation, transmission, and distribution activities. Since electricity service to the population is defined as a public service, production and distribution concessions have some public service obligations whereas transmission concessions have some transparency and third-party access requirements to allow other companies to use the power line. Production concession defines conditions of management of specific installations for electricity generation from any primary source for sales or to a third party. Transport concession defines conditions of network management and distribution concession defines conditions of exclusivity in a given area.
- *License regime.* For independent power producers, medium- and high-voltage energy sales, and international power brokers.
- *Authorization regime.* For self-generation above 1 MW, distribution networks for power less than 100 kW or transport and distribution in areas where there is lack of production means.
- *Declaration regime.* For self-generation consumers between 100 and 1000 kW.
- *Free regime.* For any power generation less than 100 kW, no administrative procedure is required.
- *Special regime.* For rural electrification (namely microhydroplant) where authorization can be given for transport, distribution, and retail sales for power less than 1000 kW.

6.05.4.9.3(ii) Decree no. 99/125 of 15 June 1999

This decree sets up the organization and functioning of the electricity sector regulatory agency. The regulatory agency (ARSEL) controls all electricity sector operators. It has many responsibilities among which are the regulation, the monitoring of the entire sector, the promotion of competition, and the private sector participation. Fifty-six percent of its board of directors is appointed by the government, 22% private sector, 11% consumer, and 11% electricity employee. Revenue of the agency is from the 1% levy on revenue of all electricity companies equally shared with the rural electrification agency according to decree no. 2001/21/PM of 29 January 2001 concerning the taxes on activities of the electricity sector.

Decree no. 2000/464/PM of 30 June 2000 governing the activities of the electricity sector.

This decree is the third major in the sector before the main concession with AES-SONEL was signed. It highlights the monopolistic nature of transmission, distribution, and retail sales. This monopoly was scheduled to end in 2006 in the case of high-voltage consumers (above 1 MW). **Table 10** lists the official texts governing the Cameroon electricity sector.

Table 10 Texts governing the Cameroon electricity sector

Reference of text	Date	Object
Law no. 98/013	14 July 1998	Relating to competition
Law no. 98/015	14 July 1998	Relating to establishments classified as dangerous, unhealthy, or obnoxious
Law no. 99/210	22 September 1999	To admit some enterprise of the public and para-public sectors to the procedure of privatization
Law no. 098/022	24 December 1998	Governing the electricity sector
Law no. 98/019	24 December 1998	Fiscal regime of public concession
Decree no. 99/125	15 June 1999	Organization and functioning of the electricity sector regulatory agency
Decree no. 99/193	8 September 1999	Organization and functioning of the rural electrification agency
Law no. 99/016	22 December 1999	General status of public companies
Decree no. 2000/462	26 June 2000	Renewing concessions, licences, authorizations, and declarations in validity before law no. 098/022 of 24 December 1998 governing the electricity sector
Decree no. 2000/464	30 June 2000	Governing activities of the electricity sector
Decree no. 2001/021	29 January 2001	Fixing the rate, modalities of calculation, recovery, and sharing of dues on the activities of the electricity sector
Order no. F061S/CAB/MINMEE	30 January 2001	Fixing the composition of documents and fees for the study of application of concessions, authorization, and declaration in order to carry out activities leading to production, transport, distribution, importation, exportation, and sales of electricity energy
Decision no. 0017-DG/ARSEL	25 January 2002	To fix prices exclusive of tax of electricity sold by AES-SONEL
Decision no. 0023 ARSEL/DG	27 May 2002	To set up an electricity consumer's advisory committee
Decree no. 2004/320	29 November 2006	Creation of Electricity Development Corporation

After adopting the Strategic Document for Poverty Reduction DSRP (French acronym for Document de Stratégie de Réduction de la Pauvreté) in April 2003, the Cameroon government adopted in December 2005 the National Energy Plan for poverty reduction PANERP (French acronym for Plan National Énergie pour la Réduction de la Pauvreté) in order to meet the millennium development goals. This document clearly stated the will to invest in the energy sector and has been submitted to some financial institutions to raise the funds. The African Development Bank and the World Bank, for instance, are ready to finance some aspects of the project.

Investment opportunities exist in Cameroon and energy can be sold to neighboring countries. The first concession has been awarded to KPDC for the production of up to 300 MW through a thermal plant under construction. Many other concessions are under discussion in the RUMPI area for production and distribution. Discussions are also on the way with Nigeria and Chad for grid interconnection allowing Cameroon to supply electricity to the two countries. Other transborder projects are under study with Equatorial Guinea, Gabon, Congo, and Central African Republic.

6.05.4.10 Weaknesses of Institutions

ARSEL is responsible for the application of environmental regulation. However, it has acquired little experience in energy regulation since its creation in 1999 and has even less experience in environmental issues in the energy sector [2]. Potential investors are still looking for requirements concerning interconnection, cost of transport, and cost of energy for producers. Transport capacity of power lines is not available. Furthermore, its board of directors is made of nine members and is politically appointed (56% government, 22% private sector, 11% consumer, 11% electricity employee). The private sector at this stage does not really exist. Consumers are not organized and conditions to select the electricity employee are not known.

This can only result in a weak institution, which cannot use its independence and expertise to lead the sector toward a more integrated and sustainable stage. At the current stage, ARSEL, in particular, does not have the capacity to play their full role as written in the policy. Studies are on the way to reinforce the monitoring capacity of ARSEL. Much has to be done as far as data transparency is concerned.

Discussions are on the way to set up a single office where the private investor can have anything he might want. For interconnection, the power lines capacity should be examined and injection point identified.

6.05.4.11 Investing in the Electric Power Sector

The private investor should provide a file containing a certain number of papers to be obtained from various institutions. The complexity of the file will depend on the type of activity or the capacity of the power plant. For transport and distribution, discussion should be carried out with the Cameroon authorities, ARSEL and AES-SONEL. The AES-SONEL monopoly ended in the year 2006 for generation, transport, and distribution. For generation, a site should be chosen among those identified or a new one. Power generated can be sold to private inside/outside the country or distributed. An agreement should be found for pricing with ARSEL, because all tariffs are not yet available, including transport and injection. Important power plants can be developed in collaboration with the national company Electricity Development Corporation (EDC).

French company MECAMIDI is still under negotiation to finalize administrative and technical procedure for a small hydro plant of 12 MW at Colomines on the Kadei river in eastern Cameroon. This project will be held out of the AES-SONEL concession area. An agreement on tariff is not yet found. A 2 MW hydropower plant (Ngassona Falls 210) co-funded by the European Union and the government of Cameroon is under construction and will be run through a new concession owned by a company still to be selected.

A subsidiary of AES-SONEL, KPDC has completed a 86 MW thermal plant at Yassa near Douala and has found an agreement with AES-SONEL. The same company is developing another gas thermal plant around Kribi, 200 km from Douala, with a capacity up to 300 150 MW.

The 200 MW Memve'Elé hydro plant, under build–operate–transfer basis is expected to be developed at any time by the Chinese SINOHYDRO company. China will also develop the Mekin hydro plant. Plans for the Aluminum Company Rio Tinto to develop the Nachtigal 300 MW hydro plant after the completion of the Lom Pangar dam are under serious consideration. Terms of the contract are not yet known.

With privatization of SONEL, the question of electrifying the remote areas becomes much more complicated. Given the low grid coverage, many areas would hardly have access to electricity even if need be. The country lacks electricity up to the point where some private investors in the industrial sector are still waiting for more electricity generation to implement their projects.

6.05.5 Conclusion

We have found out that the country has a great potential but most of it is still unexploited, up to the point where Cameroonians lack electricity during the dry season. Many projects and development plans are ongoing to solve the problem, through the construction of new plants and dams. The Cameroon electricity sector is really changing. Institutions in charge of regulating the sector are available, but still much has to be done. They are still experiencing some difficulties in many institutional aspects and this might be an advantage for private investors who have already started operation and might influence future decisions governing the sector. With the end of the AES-SONEL monopoly since 2006, one private producer has already obtained the second license and some are now submitting their files. The great potential [3] allows development of a number of all sizes of hydro plants for Cameroon and neighboring countries.

References

[1] Globeleq, Sud Energie (2008) Electricité de France. Memve'Elé Hydropower Project: Generation Planning Study. Yaoundé, Republic of Cameroon.
[2] Pineau P-O (2005) Making the African Power Sector Sustainable: Cameroon, United Nations Economic Commission for Africa (UNECA).
[3] Kenfack J (2004) Hydro potential and development in Cameroon. *Proceedings of International Conference and Exhibition on Hydropower and Dams*. Porto, Portugal: Hydropower and Dams.
[4] AES Corporation (2005) Form 10-K Annual Report Pursuant to Section 13 or 15(D) of the Securities Exchange Act of 1934 for the fiscal year ended 31 December 2004, Arlington, TX.
[5] AES-SONEL (2008) Compte Rendu de Gestion, Année 2008, Douala, Republic of Cameroon.
[6] Bagui Kari A (2001) Regards sur les Privatisations au Cameroun Suivi d'un Recueil de Textes. Yaoundé, Republic of Cameroon: IPAN.
[7] Bamenjo Jaff N (2003) Energy sector privatisation in Africa: Perspectives for rural electrification. *ESI Africa* 3: 52.
[8] Cadwalader (2005) *Africa Yearbook 2005: Project Finance*. New York: Cadwalader, Wickersham & Taft LLP.
[9] Demenou Tapamo H, (2004) "La Situation de l'Électrification Rurale au Cameroun", Premier Atelier des Agences d'Électrification Rurale Ouagadougou – 13–15 May.
[10] DFAIT-MAECI (Canadian Department of Foreign Affairs and International Trade) (2005) Ministère des Affaires Étrangères et du Commerce International, Canada–Cameroon Relations. http://www.dfait-maeci.gc.ca/africa/cameroon-canada-en.asp (accessed 23 June 2005).
[11] Haman Adji G (1998) Pré-mémoires d'un Homme Public–Entretien avec Laurent Mbassi. Yaoundé, Republic of Cameroon.
[12] Herbert B (2000) Six companies left in tender for Cameroon's SONEL power utility. *BridgeNews* 6 July.
[13] Independent Expert Panel (2004) Mission du Panel des Experts Indépendants Chargés du Contrôle des Études d'Impact Environnemental du Projet de Barrage de Lom Pangar: Rapport de Mission, 29 March to 17 April.
[14] IRIN (2003) Cameroon: Privatization provides no instant solution for electricity company. *IRIN News*, 4 September.
[15] IUCN-BRAC (2005) (World Conservation Union – Bureau Régional pour l'Afrique Centrale), Panel Des Experts Lom-Pangar (accessed 19 June 2005).
[16] Kenfack J, Tamo Tatsietse T, Fogue M, and Lejeune AGH (2006) Overview of institutional structure reform of the Cameroon power sector and assessments. *Proceedings of International Conference and Exhibition on Hydropower and Dams*. Porto, Portugal: Hydropower and Dams. 2006.
[17] Republic of Cameroon (1996) Loi no. 96/12 du 5 Août 1996 Portant Loi-Cadre Relative à la Gestion de l'Environnement.
[18] Republic of Cameroon (2004) la Réforme Institutionnelle du Secteur de l'Électricité et la Privatisation de la SONEL, Ministère de l'Economie et des Finances, Ministère des Mines, de l'Eau et de l'Énergie, Commission Technique de Privatisation et des Liquidations, Government of Cameroon. http://www.gcnet.cm/CITE/privatisations/privatisation%20sonel.htm (accessed 11 March 2004).
[19] Republic of Cameroon (2002) Rapport Principal de l'ECAM II: Conditions de Vie des Populations et Profil de Pauvreté au Cameroun en 2001, Direction de la Statistique et de la Comptabilité Nationale, Yaoundé, Republic of Cameroon: Ministère de l'Économie et des Finances.
[20] Republic of Cameroon (2003) Électricité: Soutien Chinois avec 30 Milliards FCFA, Actualiés: Energie. Yaounde, Republic of Cameroon: Services du Premier Ministère.

[21] Republic of Cameroon (2003) Poverty Reduction Strategy Paper. Yaoundé: Republic of Cameroon.
[22] Republic of Cameroon (2004) Progress Report on the Implementation of the PRSP, April 2003–March 2004, Volume II: Implementation and Follow-Up Mechanisms. Yaoundé, Republic of Cameroon: Ministry of Economic Affairs, Programming and Regional Development.
[23] Kenfack J, Ngundam J, Fogue M, et al. (2002) Inventaire des Sites Hydroélectriques du Cameroun. Séminaire International EREC 2002. Yaoundé, Republic of Cameroon.
[24] Ministère de l'Énergie et de l'Eau (2006) Aménagement Hydroélectrique de Memve'Ele sur le Ntem: Actualisation des Etudes de Faisabilité. Yaoundé, Republic of Cameroon.
[25] Japan International Cooperation (1993) The Republic of Cameroon-Société Nationale d'Electricité du Cameroun-Nippon Koei Co. Feasibility Study on Memve Ele Hydro Power Development Project: Final Report. October.
[26] China National Electric Equipment Corporation (2008) Mekin Hydropower Project, Feasibility Study Report. November.
[27] ARSEL (2005) Etude Environnementale du Barrage de Lom Pangar, Rapport de Synthèse. Yaounde, Republic of Cameroon. October.
[28] ARSEL (2009) Newsletter No. 004. December.

Relevant Websites

http://www.creditsel.com – Annuaire financier: Le répertoire des sites de finance de CreditSel.
http://www.camnews24.com – Camnews24: Accès direct au site.
http://www.chine-informations.com – Chine Informations.
http://www.iucn.org – IUCN, International Union for Conservation of Nature, helps the world find pragmatic solutions to our most pressing environment and development challenges.
http://beaugasorain.blogspot.com – Le Blog de Beaugas-Orian DJOYUM
http://www.thefreelibrary.com – The free library by Farlex.
http://www.bba.org.uk – The voice of banking and financial services.
http://www.izf.net/izf/EE/pro/cameroun/5020_elec.asp – IZF.net: investir en zone franc.

6.06 Recent Hydropower Solutions in Canada

M Fuamba and T-F Mahdi, École Polytechnique de Montréal, Montreal, QC, Canada

© 2012 Elsevier Ltd. All rights reserved.

6.06.1	Introduction	153
6.06.2	Hydroelectric Power in Canada	153
6.06.2.1	Hydroelectric History in Canada	153
6.06.2.2	Hydroelectric Opportunities in Canada	154
6.06.3	Recent Hydropower Solutions in Manitoba	156
6.06.3.1	General	156
6.06.3.2	Wuskwatim Generating Station Project	159
6.06.3.2.1	Technical aspects	159
6.06.3.2.2	Environmental aspects	160
6.06.4	Recent Hydropower Solutions in Quebec	161
6.06.4.1	Existing Hydropower Solutions	161
6.06.4.2	More Recent Hydropower Solutions	161
6.06.4.3	Hydropower Plants Under Construction	163
6.06.4.3.1	La Romaine Complex	163
6.06.4.3.2	Eastmain–Sarcelle–Rupert Diversion Project	167
6.06.5	Recent Hydropower Implementations in British Columbia	173
6.06.5.1	General	173
6.06.5.2	Revelstoke Complex	175
6.06.5.2.1	Technical aspects	176
6.06.5.2.2	Sustainable aspects	177
6.06.6	Conclusion	177
References		177

6.06.1 Introduction

North America is a world leader in the production of hydroelectricity, with an annual average production of 350 TWh (368 TWh evaluated in 2008). Canada has the largest percentage of its electricity generation mix from hydro, at almost 62%, followed by Mexico and the United States at 19% and 8%, respectively [1]. It is one of the world's largest producers of hydroelectricity, producing about 13% of the world's total. In fact, 77% of electricity comes from sources that do not emit greenhouse gases (GHGs). The federal government aims to increase this to 90% by 2020, committing to making its energy production and use cleaner by increasing energy efficiency, expanding renewable energy production, and reducing environmental impacts from conventional sources [2]. The majority of hydroelectricity is produced in the five provinces of Quebec, British Columbia, Newfoundland and Labrador, Ontario, and Manitoba. The La Grande complex in Quebec is the largest hydroelectric development in the world, with a capacity of over 15 000 MW.

Impoundment of hydroelectric reservoirs induces decomposition of a small fraction of the flooded biomass (forests, peatlands, and other soil types) and an increase in the aquatic wildlife and vegetation in the reservoir. The result is higher GHG emissions after impoundment, mainly CO_2 (carbon dioxide) and a small amount of CH_4 (methane). However, these emissions are temporary and peak 2–4 years after the reservoir is filled. During the ensuing decade, CO_2 emissions gradually diminish and return to the levels given off by neighboring lakes and rivers. Hydropower generation, on average, emits 135th of the GHGs that a natural gas generating station does and about 170th the GHGs that a coal-fired generating station does [19–21].

Canada is expanding its hydroelectricity capacity with several projects under consideration such as in Quebec and in Manitoba. Environmental concerns, however, place a significant constraint on the potential for further hydro development. The present chapter ensures how Canadian electricity provider companies are securing a reliable and sustainable electricity supply by promoting the Green Energy, electricity conservation, and planning the electricity demand for the long term. In Canada, while developing new energy sources, the governments, the electricity producers, distributors, transmitters, industry, businesses, and academics communities work together to promote renewable energy sources and environmental sustainability.

6.06.2 Hydroelectric Power in Canada

This section portrays the overall electricity production in Canada, especially hydroelectric power sources.

6.06.2.1 Hydroelectric History in Canada

Canada is one of the largest hydroelectricity producers in the world. A major part of its economic history throughout the twentieth century consisted of the development of sites with large hydroelectric potential. This sequence started with sites found close to

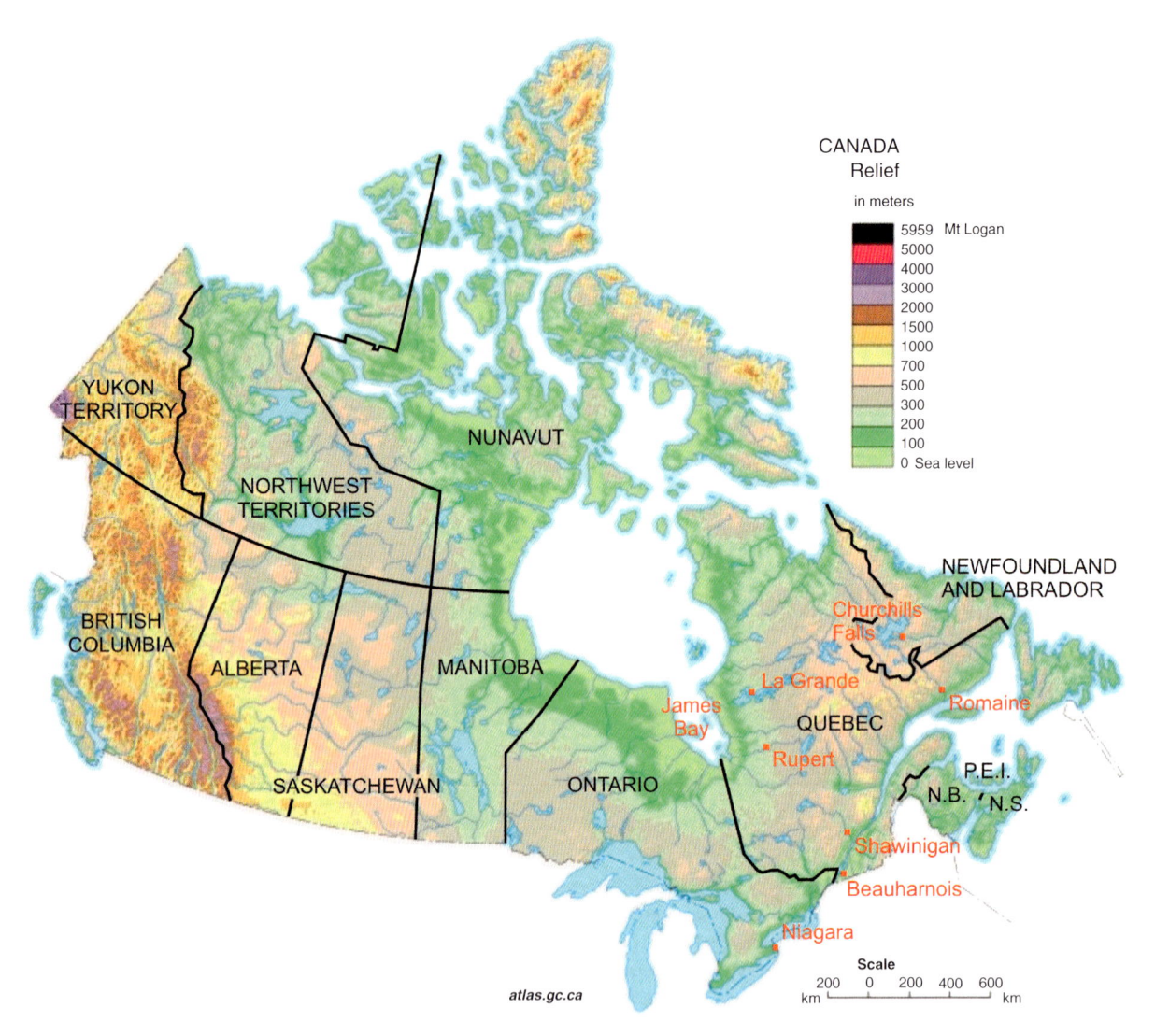

Figure 1 Administrative Canada map and large hydro sites locations [3].

populated areas (Niagara, Beauharnois, and Shawinigan) around the year 1900, and largely ended with the development of huge sites in the northern parts of several provinces in the 1960s and 1970s (James Bay, Churchill Falls, etc.). There have been relatively few large hydro sites developed since that time (La Grande, Rupert, Romaine, etc.) as the environmental and human impacts to be avoided or mitigated in such large projects make them increasingly difficult and costly to plan and build. **Figure 1** shows the administrative Canada map and a few large hydro site locations.

Faced with the challenges of climate and territory, the Canadian energy industry has developed expertise in the generation and transmission of renewable hydroelectric power. Over the years, Canada has also become world-renowned for its hydropower project design and construction. There are approximately 450 operating hydroelectric power. More than 200 are small hydro plants (<10 MW). Canada also has more than 800 dams that are used for hydroelectric power generation, irrigation, and flood control [4, 5]. The largest hydroelectric power development is the James Bay project in Quebec, which started producing electricity in 1979. Its eight dams and 198 dikes contain five reservoirs covering 11 900 km^2, which is about 20 times the size of Lake Geneva in Switzerland.

In 2008, Canada generated 598 TWh, 61.6% of which originates from hydroelectricity. The mean Canadian hydro production per capita was estimated at 11 322 kWh. The most important part of this energy per capita was produced by four provinces: Newfoundland (81 682 kWh), Manitoba (29 254 kWh), Quebec (23 913 kWh), and British Columbia (13 884 kWh). Quebec appeared to be the most important hydroelectricity producer in Canada with 182 TWh [1]. **Table 1** shows all details on the Canadian provinces' hydroelectric production.

6.06.2.2 Hydroelectric Opportunities in Canada

Further hydroelectric opportunities remain for new hydropower project development across Canada. It is estimated that around 118 000 MW of hydropower, twice the amount that is currently in operation, could technically be developed. Every province, except Prince Edward Island, has some remaining potential. The provinces of Quebec, Manitoba, and British Columbia, in particular, hold

Table 1 Provincial energy generation in Canada 2008

Province	Inhabitants	Hydro (TWh)	Steam (TWh)	Nuclear (TWh)	Internal combustion (TWh)	Combustion turbine (TWh)	Wind (TWh)	Tidal (TWh)	Total (TWh)	Hydro/capita (KWh per inhabitant)
Alberta	3 408 975	2.3	48.1		0.1	9.7	0.1		60.3	675
British Columbia	4 235 151	58.8	4.4		0.1	2.5			65.8	13 884
Manitoba	1 182 731	34.6	0.5						35.1	29 254
New Brunswick	746 056	3.5	8.1	1.1		1.4			14.1	4 691
Newfoundland	510 515	41.7	1.1		0.1	0.3			43.2	81 682
Nova Scotia	938 020	1.1	10.7			0.3	0.1		12.2	1 173
Ontario	12 641 497	39.9	28.0	83.5		7.7	0.5	0.0	159.6	3 156
PE Island	137 754						0.1		0.1	0
Quebec	7 623 482	182.3	1.1	4.0	0.3	0.2	0.5		188.4	23 913
Saskatchewan	991 490	4.0	13.0			1.3	0.5		18.8	4 034
Territories	105 999									0
Total	32 521 670	368	115	89	1	23	2	0	598	11 322

significant potential for development as new hydropower projects have been identified. They are, however, faced with a lengthy regulatory process, including comprehensive environmental assessment and public consultations. Despite significantly slowing down development of new hydropower projects, this regulatory process is necessary since environmental concerns must be somehow taken into account. In fact, growing concern over GHG emissions should lead to the development of new hydro capacity to address climate change while helping to meet growing electricity demand. About 6000 MW of additional hydropower capacity are planned for the upcoming years. This includes projects such as Eastmain–Rupert (1200 MW), Wuskwatim (21.5 m earthfill dam, 200 MW), Notigi (14 m earthfill dam, 100 MW), Gull Rapids (22 m earthfill dam, 620 MW) in Manitoba, and Gull Island (99 m earthfill dam) in Labrador [5].

The three following sections (6.06.3–6.06.5) describe new hydropower solutions in Manitoba, Quebec, and British Columbia, respectively. The choice of these provinces was made according to the generated hydro energy per capita and new hydropower projects recently developed or under development. Ontario's hydro energy per capita is smaller compared to the selected three provinces (11% to Manitoba, 13% to Quebec, and 23% to British Columbia).

6.06.3 Recent Hydropower Solutions in Manitoba

This section summarizes the hydroelectric history of Manitoba. It shows how hydroelectricity was gradually developed as early as 1900. It then describes one of the most important hydroelectric achievements currently under development.

6.06.3.1 General

From the early 1900s, Manitoba's power planners were aware of the hydroelectric potential of the province's northern rivers. The major stumbling block to such development was how to transmit the power from the north to the more populated areas of the south. By the early sixties, technological advances in the field of power transmission were such that northern power projects could be considered in earnest. In 1963, the Government of Canada and the Government of Manitoba had entered into an agreement to investigate hydroelectric development on the Nelson River, and to equally share the cost of research. After exhaustive investigations, a new agreement was signed between both governments in 1966 to cooperate and proceed with the development of the Nelson River. As shown in **Figure 2**, the latter is a river of north-central North America, in the Canadian province of Manitoba. Its full length

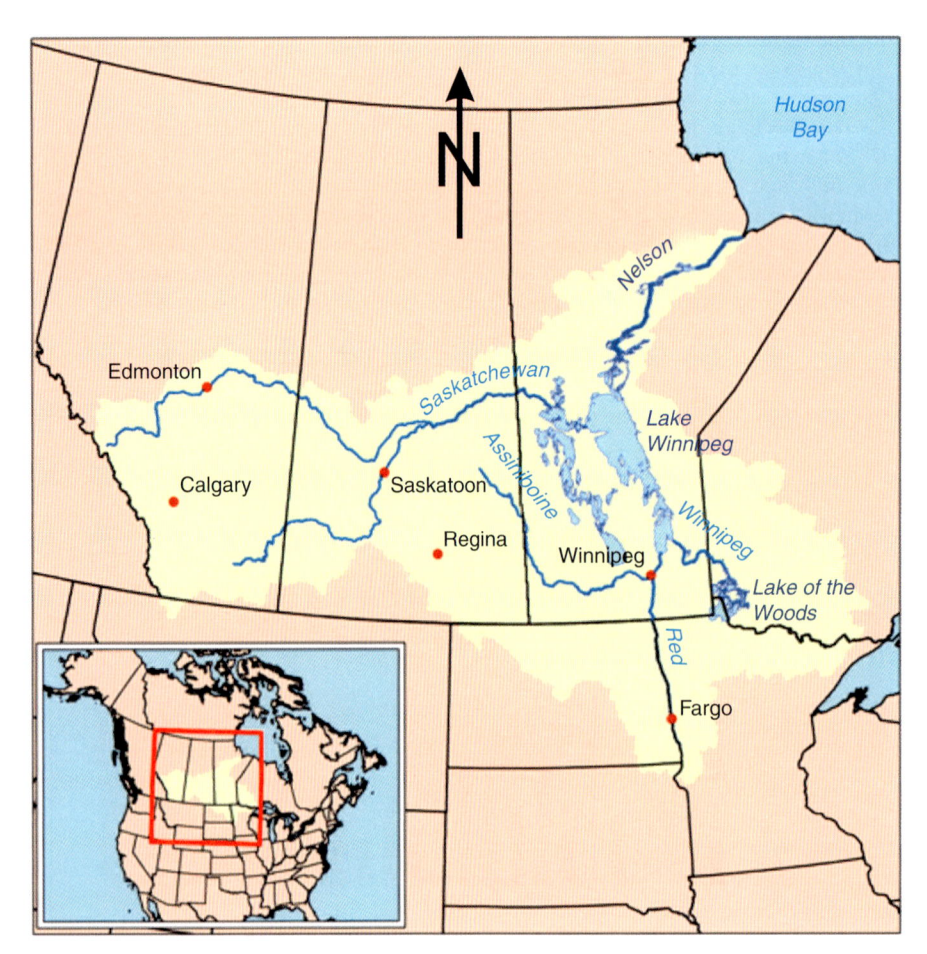

Figure 2 Map of the Nelson River drainage basin [6].

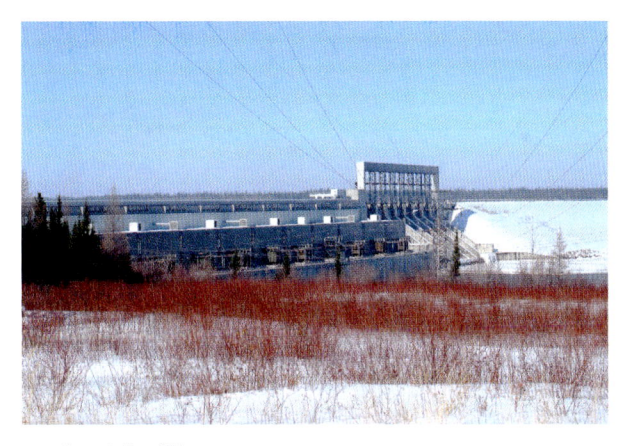

Figure 3 General view of the Kettle generating station [7].

is 2575 km, it has a mean discharge of 2370 m^3 s^{-1}, and has a drainage basin of 982 900 km^2, of which 180 000 km^2 is located in the United States. The river drains Lake Winnipeg and runs 644 km before it ends in Hudson Bay [3].

The main objective of the Nelson River development was to convert the rich natural resource of the Nelson River into a power base for industrial and economic development in Manitoba, and to create a potential power sale outside of Manitoba. The Nelson River development included four main components: (1) the construction of Kettle Generating Station; (2) a high-voltage direct current (HVDC) transmission system from Kettle to Winnipeg; (3) the Churchill River Diversion Project; and (4) the Lake Winnipeg Regulation [7].

- Construction of Kettle began in 1966 and was completed in November 1974. The 1272 MW generating station consolidated a series of rapids into a 30 m operating head (the waterfall created by Kettle's structure). At that time, Kettle was the largest generating station in Manitoba (**Figure 3**).
- One of the keystones to northern hydroelectric development was the transmission system. Manitoba Hydro undertook extensive studies into possible transmission systems as part of the investigations conducted under the 1963 agreement. Alternating current (AC) and direct current (DC) transmission systems were evaluated, and two routes for transmission lines were assessed – one on the east side and the other on the west side of Lake Winnipeg. Converter stations are needed to convert the AC power produced at the northern generating stations into DC power for transmission nearly 900 km south. It is then converted back to AC power for distribution to customers. All these works were completed by October 1978.
- The portion of the Churchill River in Manitoba, which is downstream of Southern Indian Lake, has a hydroelectric potential of more than 3000 MW. However, instead of developing generating stations on the Churchill River itself, it was believed that diverting the Churchill River into the Nelson River would be more economical. It was felt that while the Lake Winnipeg Regulation would assure a more dependable water flow in the Nelson River, the Churchill River Diversion Project would increase the power-producing potential of the Nelson by as much as 40%.

  Three major components accomplished the diversion. The first one is a control structure set up at Missi Falls, the natural outlet of Southern Indian Lake which controls the outflow and raises the lake level by 3 m. The second one is an excavated channel from Southern Indian Lake to Isset Lake to allow Churchill River water to flow into the Rat–Burntwood–Nelson River system. The third one is a control dam at Notigi on the Rat River to regulate the amount of water being diverted.

  Under terms of the license granted in 1972, Manitoba Hydro is permitted to divert up to 850 m^3 s^{-1} of water from the Churchill into the Nelson. The outflow from the Missi Falls control dam was fixed at 14 m^3 s^{-1} during open water season and 43 m^3 s^{-1} during the ice-cover period. The project went into operation in 1977 (**Figure 4**).

- The regulation of Lake Winnipeg was deemed necessary because in its natural state, the water outflow into the Nelson River occurs mainly during spring and early summer months and less in the fall and winter months. The problem of hydroelectric generation in Manitoba is that a greater volume of outflow is needed in the fall and winter. With regulation, the outflow from the lake would be reduced in the spring and early summer, and increased during the fall and winter to meet the province's demand for electricity. Work on the Lake Winnipeg Regulation project began in 1970 and consisted of three main elements. Manitoba Hydro was licensed to regulate Lake Winnipeg between the upper storage limit of 715 feet above sea level (asl) and the lower storage limit of 711 feet for power production purposes. Without HVDC technology and the Lake Winnipeg Regulation, development of the hydroelectric potential of the Nelson River could not have proceeded.

Manitoba Hydro is the electric power and natural gas utility in the province of Manitoba, Canada. Today, the company operates a few interconnected generating stations. Since most of the electrical energy is provided by hydroelectric power, this utility has low and stable rates. Stations in Northern Manitoba are connected to customers in the south by the Nelson River Bipole HVDC system. Main hydroelectric generating stations produce an annual average energy of 28.76 TWh. **Table 2** gives details on the total power and annual average energy.

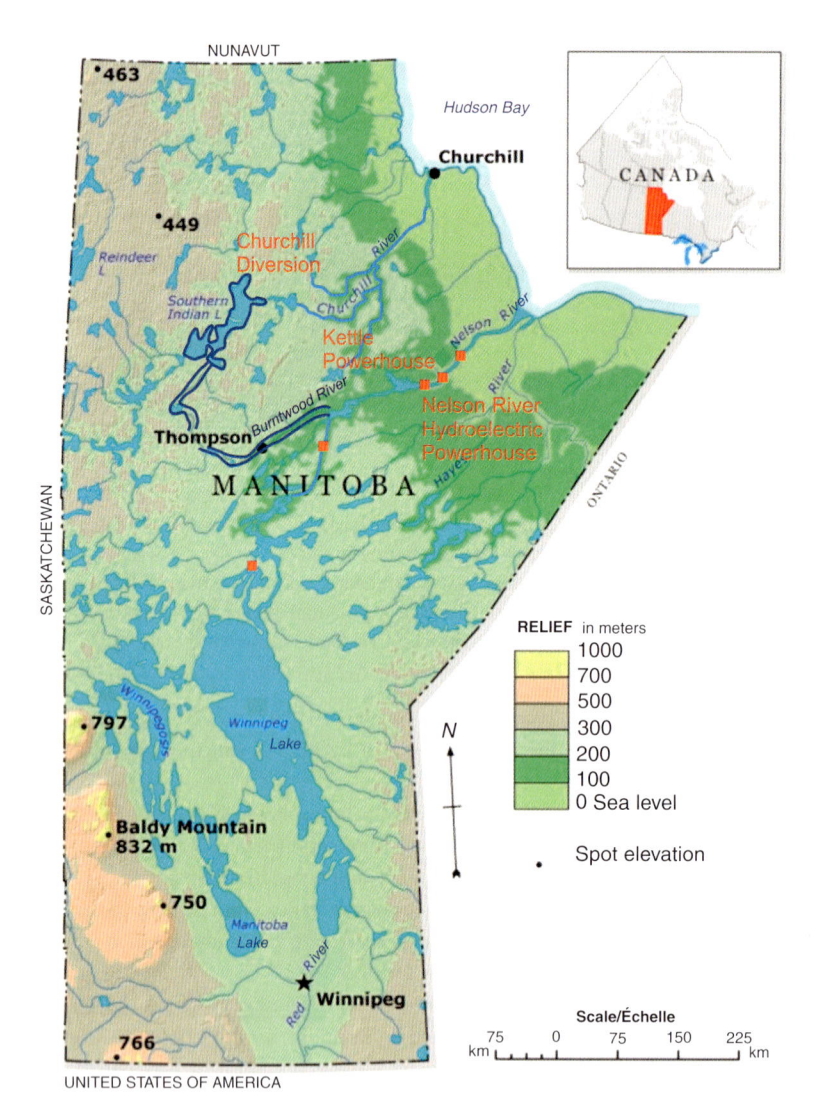

Figure 4 Derivation of the Churchill River [3].

Table 2 Annual average energy and total power produced by Manitoba Hydro

Station	Started	Location	Units	Power per unit (MW)	Total power (MW)	Average annual generation (TWh)
Pointe du Bois	1911	Winnipeg River	16	Various	78	0.6
Great Falls	1922	Winnipeg River	6	Various	131	0.75
Seven Sisters	1931	Winnipeg River	6	25	165	0.99
Slave Falls	1931	Winnipeg River	8	8	67	0.52
Laurie River 1	1952	Laurie River	1	5	5	0.03
Pine Falls	1952	Winnipeg River	6	14	82	0.62
McArthur Falls	1954	Winnipeg River	8	7	56	0.38
Kelsey	1957	Nelson River	7	30	211	1.8
Laurie River 2	1958	Laurie River	2	2.5	5	0.03
Grand Rapids	1965	Saskatchewan River	4	120	472	1.54
Kettle	1970	Nelson River	12	103	1228	7.1
Long Spruce	1977	Nelson River	10	100	1010	5.8
Jenpeg	1979	Nelson River	6	16	97	0.9
Limestone	1990	Nelson River	10	135	1340	7.7

More than 5000 MW of hydroelectric potential could be further developed in Manitoba. This includes 1380 MW at the Conawapa site, 630 MW at the Gull (Keeyask) site, and 1000 MW at the Gillam Island site, all on the lower Nelson River. Other sites have been assessed but are not currently under study for development. All of these developments would require a large increase in electric power exports, since Manitoba's load growth will not require this capacity for a generation or more. Nevertheless, some additional hydroelectric power is needed to prevent any short- and medium-term demand. This is why Manitoba Hydro has started construction of the Wuskwatim generating station (WGS).

6.06.3.2 Wuskwatim Generating Station Project

The WGS project is located on the Burntwood River near Thompson City, as shown in **Figure 5**. The general civil works contract was awarded in 2008, and first power from the project is planned in 2012 [7]. Thompson is a city in northern Manitoba. Considered the 'Hub of the North', it serves as the regional trade and service center of northern Manitoba. Thompson is located 830 km north of the Canada–United States border and 739 km north of the provincial capital of Winnipeg. It is the third largest city in the province with a population of over 13 446 residents. It also serves as a trade center for an additional 36 000–65 000 Manitobans. As such, it has all of the services and amenities that would be expected in a much larger, urban center [7].

6.06.3.2.1 Technical aspects

The WGS involves the development of 200 MW at Taskinigup Falls on the Burntwood River (**Figure 6**). This plant will have three hydraulic turbine generator units, will cause less than one-half square kilometer of additional flooding, and will have only a small reservoir. The project had the most extensive environmental review of any generating project in Manitoba. Participation of the Nisichawayasihk Cree Nation (NCN), First Nation, was passed by a June 2006 referendum by NCN members. This partnership between NCN and Manitoba Hydro will allow advancement of the in-service date to 2012 and opportunities for additional export revenue since the domestic load growth would not require this new capacity until several years from now. **Figure 6** shows the WGS's specific location in the province of Manitoba. **Figure 7** shows the specific locations of the station, camp area, and work area at the WGS site.

As of November 2009, most excavation and primary concrete is completed, and superstructure steel for the powerhouse and service bay is being erected. The access road and construction camp site and construction power have been completed along with services such as water supply and sewage lagoons for the camp. A new 230 kV gas-insulated substation has been constructed adjacent to the dam site. It will distribute power from the Wuskwatim generators to the transmission network.

The Wuskwatim's main dam will be an earth and rockfill embankment with a maximum height above foundation level of approximately 14 m and a 9 m wide crest at an elevation 236.6 m asl. Its overall length will be 330 m. A channel will also be excavated immediately north and adjacent to Wuskwatim Falls through the peninsula. The Wuskwatim's powerhouse will be approximately 120 m long, operating at a head of about 22 m. It will contain three water passages through each intake structure, each with a set of vertical trash racks. Each water passage will lead to a generating unit containing a blade propeller turbine.

Formed by a concrete structure of 43 m in length with three bays, the Wuskwatim's spillway lies north of the powerhouse. It will have a vertical lift gate in each bay measuring 9 m wide by 16 m high. It will be separated from the powerhouse by a 42 m long reinforced concrete gravity nonoverflow dam. Serving as a diversion channel during construction of the main dam, it will provide

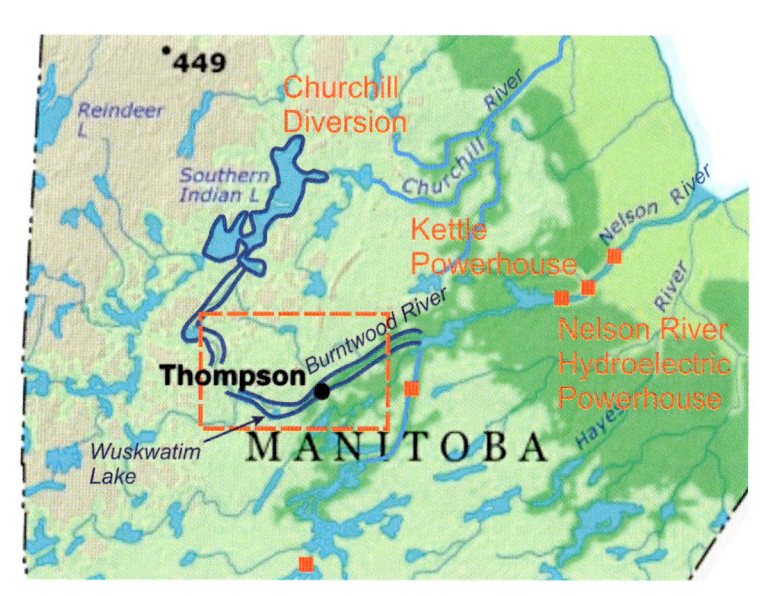

Figure 5 Location of the Burntwood River near Thompson City where the Wuskwatim generating station takes place [7].

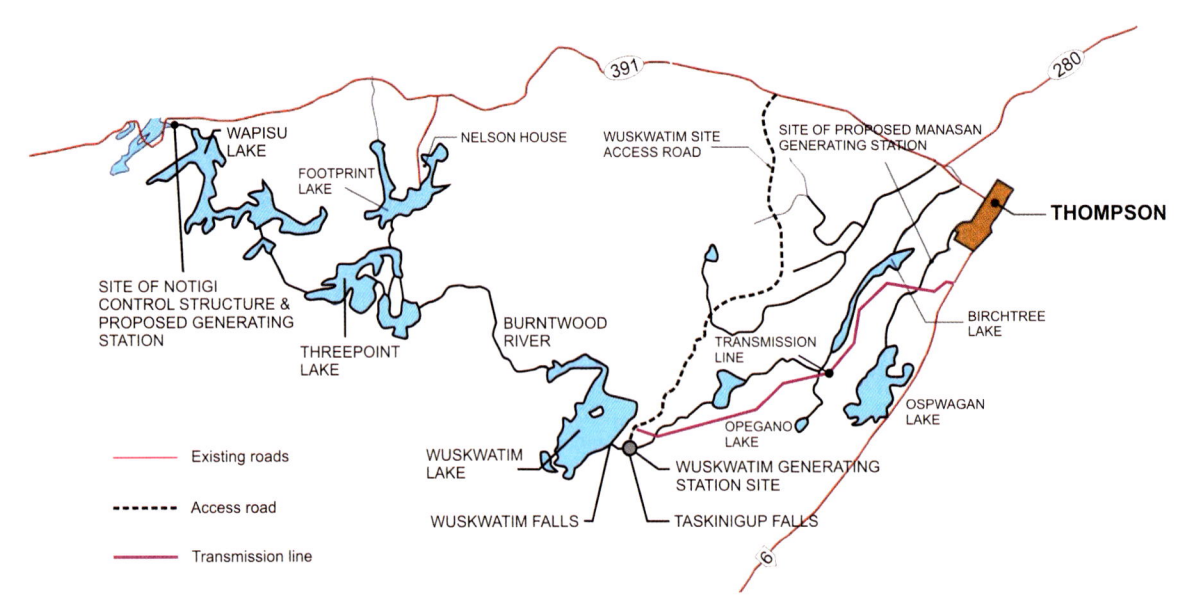

Figure 6 Regional location of the Wuskwatim generating station [7].

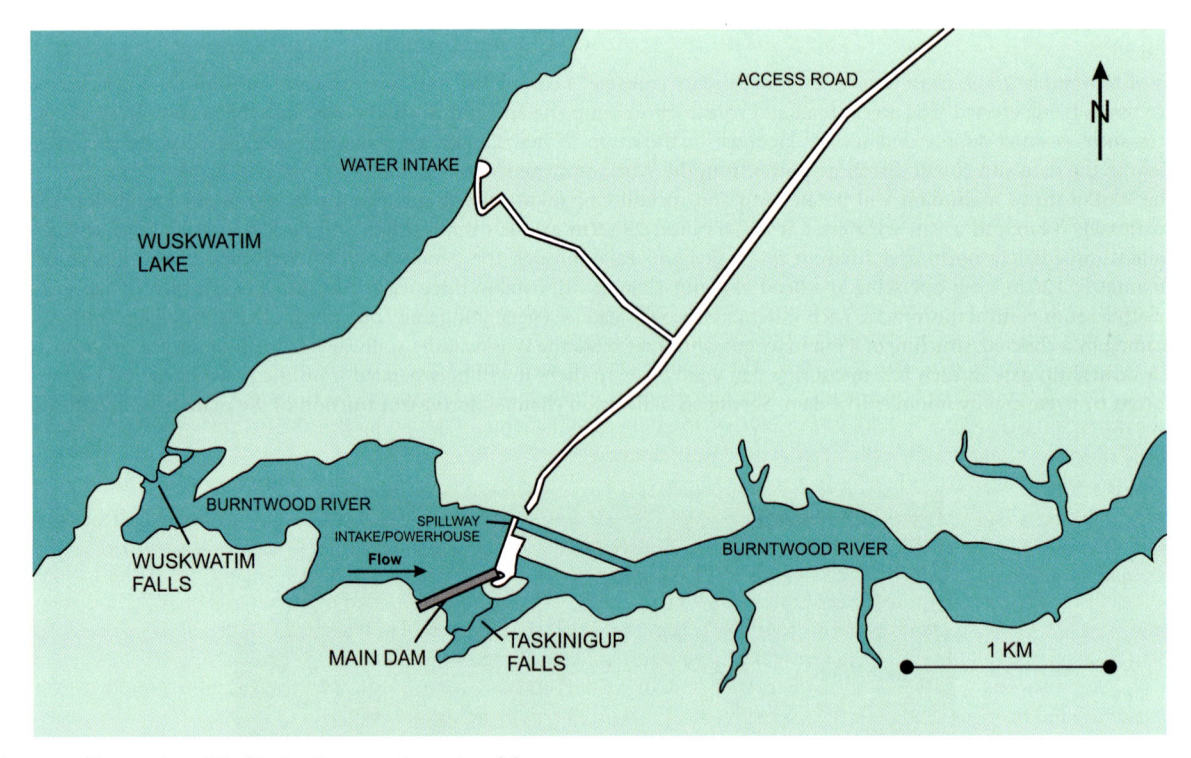

Figure 7 Site overview of the Wuskwatim generating system [7].

flood routing for the reservoir during operation. Additionally, this spillway will be capable of safely passing a peak discharge of about 2700 m³ s⁻¹ (the probable maximum flood rate).

6.06.3.2.2 Environmental aspects
The WGS project was subjected to a public hearing under the Environment Act of Manitoba and an environmental assessment under the Canada–Manitoba Agreement on Environmental Assessment Cooperation. Following intensive discussions between project partners, a low head design was chosen over a high head design for the WGS. The low head design produces 200 MW but minimizes environmental impacts by reducing the amount of flooding to less than a half square kilometer (about 37 ha) compared to about 140 sq km from a high head design generating an estimated high head of 350 MW. Following the decision for a low head design,

rigorous environmental studies were conducted and monitoring plans were set up. A study done by the Pembina Institute for Appropriate Development determined that the total life cycle of GHG emissions from Wuskwatim to be 290 times less than coal, 130 times less than the most efficient natural gas technology, and similar to wind generation.

NCN and Manitoba Hydro are working together to understand the effects that a proposed hydroelectric development may have on the environment. A number of environmental studies regarding the WGS project are currently underway to address the concerns and issues raised by NCN and Manitoba Hydro, one of which is the amount of mercury in fish.

No endangered bird species were observed in the study area in 2000 or 2001. The study results from 2000 to 2001 suggest that the Wuskwatim and Opegano lake areas do not provide regionally important breeding or staging areas for waterfowl as compared to other water bodies outside the Rat–Burntwood River system.

6.06.4 Recent Hydropower Solutions in Quebec

This section gives an overview of the hydroelectric history in the province of Quebec. It shows how hydroelectricity was gradually developed in the James Bay region. In addition, two important hydroelectric projects under development are described.

6.06.4.1 Existing Hydropower Solutions

Quebec has a great asset in the form of major reserves of water power, which provide a renewable source of energy with low environmental and climatic impacts compared to other conventional forms of electricity generation (**Figure 8**). Since the nationalization of electricity, the development of hydroelectric potential has taken on symbolic importance for the Quebec population as a whole. Currently, the available total capacity generated by water power is around 40 250 MW [8].

The 893 km long La Grande River rises in the rugged forest highlands of central Quebec and drains west into the James Bay. Its 97 600 km^2 drainage basin is the third largest in Quebec. During the 1970s, the river was transformed by the James Bay Project, a scheme to divert major rivers flowing into eastern James Bay for hydroelectric development. The project resulted into the erection of huge dams on the La Grande basin and flooding low-lying areas to create reservoirs, and eight plants were therefore constructed by 1996 for an installed power of 16 020 MW: La Grande-1, La Grande-2A, Robert-Bourassa, La Grande-3, La Grande-4, Laforge-1, Laforge-2, and Brisay. **Table 3** gives details on installed electrical power according to each plant. The Robert-Bourassa Complex (RBC) has the largest installed power capacity in North America with 5616 MW. It has 22 turbines Francis type, 16 of which are active. With its 484 m length, the RBC is the longest underground powerhouse in the world. It is located 137 m under ground level. Its main dam is 162 m high; the equivalent of a 53-storey building. The RBC reservoir meanwhile covers an area of 2835 sq km, almost five times the size of Switzerland's Lake Geneva. Regarding the spillway structure, each of its 10 steps is twice the area of a football field (**Figure 9**).

Manicouagan and Outardes river basins also represent significant hydroelectric potential in northeastern Quebec. Eleven plants whose construction has been spread over 50 years from the end of the 1950s make up the entire complex Manic–Outardes for an installed power of 8536 MW: Manic-1, Manic-2, Manic-3, and Manic-5 (Daniel-Johnson dam), and Outardes-2, Outardes-3, Outardes-4, Manic-5PA, Toulnustouc, McCormick, and Hart-Jaune. **Table 4** gives details on installed electrical power according to each plant. Manic-2 is the largest hollowed joint dam in the world. Manic-5 is the largest buttresses and arches dam in the world (**Figure 10**). Thermal improvements were installed in 1991 in 9 of the 13 arches of the dam in order to prevent cracking of the concrete and protect the Daniel-Johnson dam against the effects of large temperature gradients that are common at such latitudes.

6.06.4.2 More Recent Hydropower Solutions

Quebec's demand in electrical power will reach an estimated 39 282 MW during winter 2016–17. That represents an increase of 2730 MW in comparison with the 36 552 recorded MW during the winter peak of 2008–09, 860 MW of which were attributable to colder temperatures than average. Hydro-Quebec anticipates a 1615 MW reduction due to planned actions in energy efficiency. Meanwhile, Hydro-Quebec maintains a power reserve to face hazards such as of unforeseen increases in demand. This reserve is established according to the reliability criterion in power of the Northeast Power Coordinating Council, which demands that the diversion probability in an adjustability zone does not exceed a day by 10 years. More details on power demand are listed in **Table 5** [10].

Quebec still has major undeveloped hydroelectric capacity, estimated at 45 000 MW (35 126 MW from Quebec high-potential power reservoirs and 5128 MW from Churchill Falls), of which almost 20 000 MW has economic potential in the current context of sustainable development [10]. Five hydroelectric plants, with a total capacity of 1108 MW, have been recently developed in accordance with environmental protection regulations and with respect to the local and Aboriginal communities affected: Mercier, Eastmain-1, Peribonka, Rapides-des-Coeurs, and Chute-Allard (**Figure 11**). In each case, the project has proved to be economically viable.

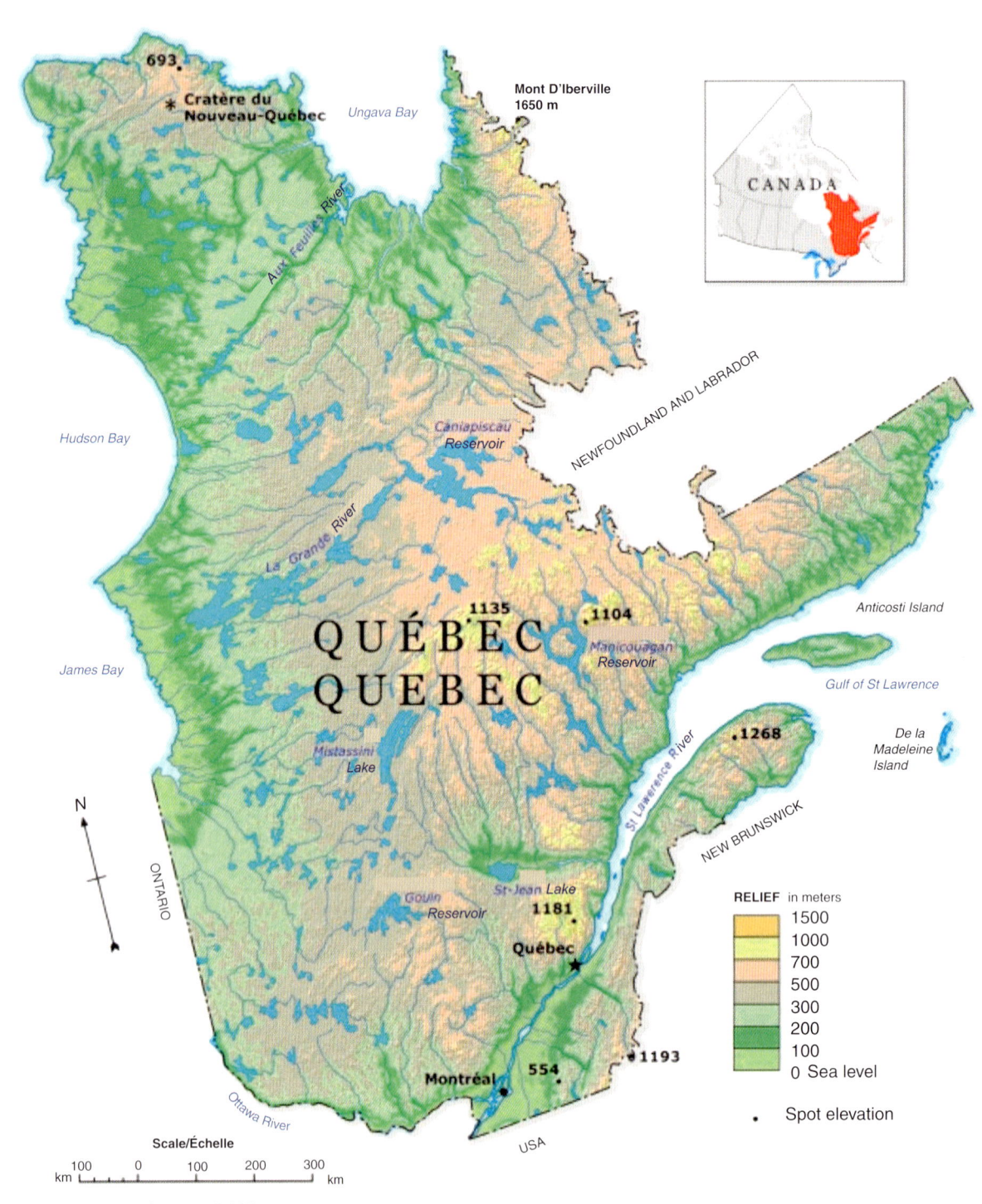

Figure 8 Map of the Quebec relief [3].

'Mercier' is a hydroelectric plant built on the Gatineau River in Quebec (**Figure 12**). Operations started in 2008. Built in 1927 by the Government of Quebec to regulate water in the Outaouais region, Mercier dam is located at the outlet of Baskatong reservoir. 'Eastmain-1' powerhouse is located on the left bank of the Eastmain River (**Figure 13**). The generating station has an average annual output of 2.7 TWh. The design flow is about 840 m^3 s^{-1}, whereas a minimum flow of 140 m^3 s^{-1} will be maintained at all times at the powerhouse outlet. 'Peribonka' is a hydroelectric facility of approximately 450 MW on the Peribonka River, north of Lake Saint-Jean (**Figure 14**). The facility is located at 152 km of the Peribonka River, immediately upstream of its confluence with the Manouane River. Its reservoir is approximately 33 km long. The mean annual flow at the dam site is 438 m^3 s^{-1} for a watershed of approximately 19 450 km^2. The gross head is estimated at 70 m. The design flow of the plant is 630 m^3 s^{-1}. 'Chute-Allard'

Table 3 Hydroelectric power of plants of La Grande basin in Quebec, Canada

Plant	Production type	Total power (MW)	Number of units	Head (m)
Robert-Bourassa	Reservoir	5 616	16	137.2
La Grande-1	Over water	1 436	12	27.5
La Grande-2A	Reservoir	2 106	6	138.5
La Grande-3	Reservoir	2 417	12	79
La Grande-4	Reservoir	2 779	9	116.7
Laforge-1	Reservoir	878	6	57.3
Laforge-2	Over water	319	2	27.4
Brisay	Reservoir	469	2	37.5
Total		16 020		

Figure 9 The Robert-Bourassa spillway and dam, Quebec, Canada [9].

Table 4 Hydroelectric power of plants of Manicouagan Basin in Quebec, Canada

Plant	Production type	Total power (MW)	Number of units	Head (m)
Manic-1	Over water	184	3	36.6
Manic-2	Over water	1145	8	70.1
Manic-3	Over water	1244	6	94.2
Manic-5	Reservoir	1596	8	141.8
Outardes-2	Reservoir	523	3	82.3
Outardes-3	Reservoir	1026	4	143.6
Outardes-4	Over water	785	4	120.6
Manic-5PA	Reservoir	1064	4	144.5
Toulnustouc	Reservoir	526	2	152
McCormick	Reservoir	392	7	37.8
Hart-Jaune	Reservoir	51	3	39.6
Total		8536		

(**Figure 15**) and 'Rapides-des-Coeurs' (**Figure 16**) are producing a total capacity of 138 MW in Haute-Mauricie region. They are located on the Saint-Maurice River, respectively, at a 102 and 99 km drive, north of La Tuque city. These plants have a total installed capacity of 141 MW and started operating in 2008. **Table 6** gives details on the five recent plants. The next section presents the two most important hydropower plants under construction.

6.06.4.3 Hydropower Plants Under Construction

6.06.4.3.1 La Romaine Complex

Started in 2009 and planned to be completed in 2020, La Romaine Complex (LRC) is currently the biggest construction project in Canada. This new complex has been designed to serve as a powerful lever for Quebec's economic development by contributing to

Figure 10 The Daniel-Johnson Dam on Quebec's Manicouagan River, Canada [9].

Table 5 Electric power demands from 2010 to 2017 in Quebec

Winter	2008–09[a]	2009–10	2010–11	2011–12	2012–13	2013–14	2016–17
Energy efficiency (MW)	458	582	704	883	1 069	1 251	1 615
Peak needs (MW after energy efficiency)	36 552	35 353	36 367	36 914	37 355	37 706	39 282
Required stock needed to respect reliability criterion (MW)	NA[b]	3 279	3 630	3 887	4 088	4 127	4 351
Power needs (MW)	36 552	38 632	39 997	40 801	41 443	41 833	43 633

[a]Real data.
[b]A stock of 3485 MW, established from the anticipated normalized needs of 36 040 MW, has been planned for the advancement state of the supply plan of October 2008.

the fight against climate change. A joint federal–provincial review panel approved construction of the project after a rigorous and transparent environmental assessment process.

The project meets three objectives of the Quebec Energy Strategy – enhancing energy security, making greater use of energy as a lever for economic development, and giving a greater role to local and regional communities and the First Nations. The project shows that the interests of economy, environment, and social acceptability can be harmoniously reconciled [11, 12].

6.06.4.3.1(i) Technical aspects
Hydro-Quebec obtained the authorization to build a hydroelectric complex of 1550 MW on the Romaine River, north of the city of Havre Saint-Pierre on the north shore of the Saint-Lawrence River. The LRC will be composed of four power plants (RO-1, RO-2, RO-3, and RO-4) as listed on **Figure 17** with an average annual production of 8 TWh. A permanent road of 150 km will link Highway to the Romaine site area. The LRC project, whose construction has started from 2009 to 2020, will allow Hydro-Quebec to ensure the future of Quebec's energy while taking advantage of business opportunities in foreign markets.

Due to the river's very steep banks, the total area of all four reservoirs will only reach 279 km^2. **Figure 18** shows the reservoir and all the components of RO-1. Each plant of the LRC project includes a rockfill dam, a powerhouse with two generating units, and a spillway. **Table 7** lists the plant components main features including the number and types of generating units, the installed power capacity, the average annual output, the net head, the reservoir size, the number of dams, and the dam elevation. **Figure 19** shows the Romaine River profile from the river estuary to the RO-4 reservoir.

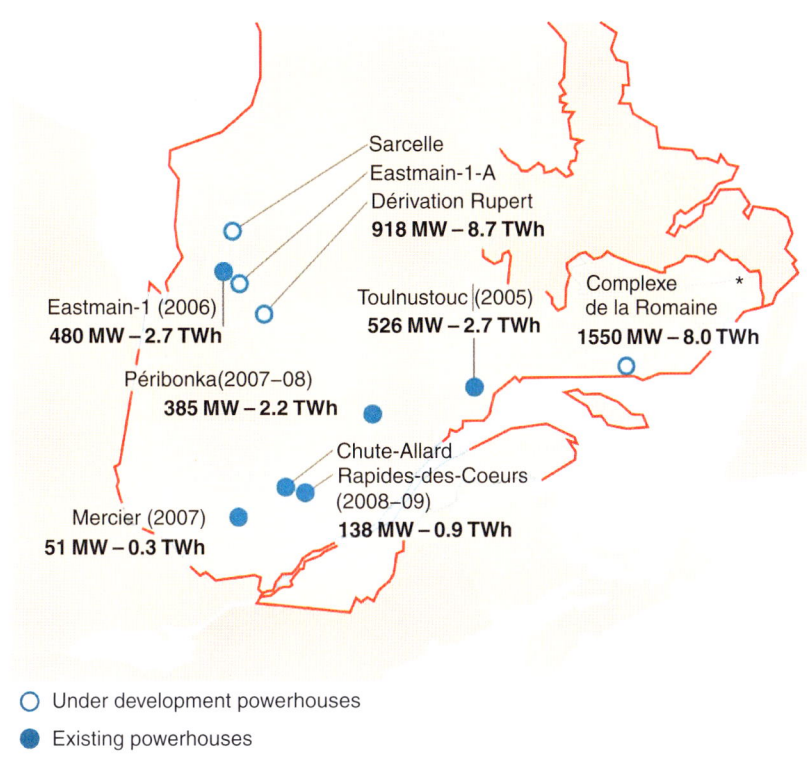

Figure 11 Location of more recent hydropower plants in Quebec [10].

Figure 12 The Mercier plant in Quebec, Canada [9].

6.06.4.3.1(ii) Sustainable aspects
All significant impacts on the environment induced by the LRC project have been analyzed in order to make adjustments and to find sustainable solutions. The components of the physical, biological, and human environment likely to be affected by the project have been analyzed. Hydro-Quebec has developed measures to preserve the ecological environment.

The LRC project was evaluated by a committee composed of cooperative environmental representatives according to the environmental impact assessment process under the 'Quebec *Act on Environmental Quality*' and the 'Canadian Environmental Assessment Act'. According to the Canada-Quebec agreement on environmental assessment, the governments of Canada and Quebec agreed to undertake a cooperative environmental assessment of the project to facilitate the coordination of the two assessment processes.

Figure 13 The Eastmain-1 spillway (left) and dam (right) in Quebec, Canada [9].

Figure 14 The Peribonka spillway (left), dam (center), and outlet tunnel (right) in Quebec, Canada [9].

Figure 15 The Chute-Allard plant in Quebec, Canada [9].

6.06.4.3.1(ii)(a) **Protecting the environment** Several measures have been deployed to protect the human environment and heritage, so that land users (aborigines' peoples: Minganois, Innus, etc.) may safely continue their activities during and after construction. Examples range from fitting three extra widths along roads, installing launch ramps, collection of woody debris (logs, branches, etc.), and construction of booms at the foot of dams and floating barrier to restrict access for safety reasons.

6.06.4.3.1(ii)(b) **Protecting species diversity and the natural environment** A monitoring of fish populations and their habitat will run until 2039 in order to evaluate the maintenance of diversity and abundance of species in reservoirs and in sections of the

Figure 16 The Rapides-des-Coeurs plant in Quebec, Canada [9].

Table 6 Hydroelectric power of more recent implemented plants in Quebec, Canada

Plant	Production type	Total power (MW)	Average annual production (TWh)	Number of units	Head (m)
Mercier	Reservoir	55	0.3	5	18
Eastmain-1	Reservoir	507	2.7	3	63
Peribonka	Over water	405	2.2	3	67.6
Rapides-des-Coeurs	Over water	79	0.5	6	22.7
Chute-Allard	Over water	62	0.4	6	17.8
Total		1108			

Romaine River and to fix the fish habitat in the reservoir. Collected data can then help to evaluate the success of seeding and facilities to improve fish abundance.

- The reserved ecological flow rate system downstream of the installation of the RO-1 will be modulated according to the specific needs of the Atlantic salmon. During the reproduction period, Hydro-Quebec will maintain a constant flow rate to optimize the conditions for spawning of brood stock. In addition, the operator will limit the daily flow fluctuations to protect juvenile salmon. To compensate for the loss of salmon habitat, Hydro-Quebec settled an ambitious program including depositing in incubators, seeding of the river the following spring, and monitoring of population trends in the river.

6.06.4.3.1(ii)(c) **Social acceptance** The project is being carried out in partnership with local and regional communities. Partnership agreements have been concluded with the Regional County Municipality of Minganie and with the four Innu communities: Ekuanitshit, Natashquan, Unamen Shipu, and Pakua Shipi. The agreement with the Minganie community is to support projects of an economic, recreational, social, or cultural nature within the municipality's limits. The agreements with the Innu communities are to finance projects of an economic, community, or cultural nature as well as to foster traditional practices and encourage vocational training. These communities also took part in the local impact studies and will be closely involved in project construction and environmental monitoring.

6.06.4.3.2 Eastmain–Sarcelle–Rupert Diversion Project

Resulting from extensive consultations with the involved public, the Eastmain–Sarcelle–Rupert Diversion (ESRD) project is characterized by sustainable solutions intended to preserve the environment and to address concerns expressed by this public [13, 14].

Located in the James Bay region, the ESRD project consists of three main components (**Figure 20**):

- Construction of 'Eastmain-1A', a 768 MW powerhouse near existing Eastmain-1
- Construction of 'Sarcelle', a 150 MW powerhouse at the outlet of Opinaca reservoir
- 'Partial diversion' of the 'Rupert' River to these two generating stations and from there to Robert-Bourassa, La Grande-2A, and La Grande-1 generating stations.

The 'Rupert diversion' includes the following structures and facilities (described in **Table 8**):

- Four dams
- A spillway

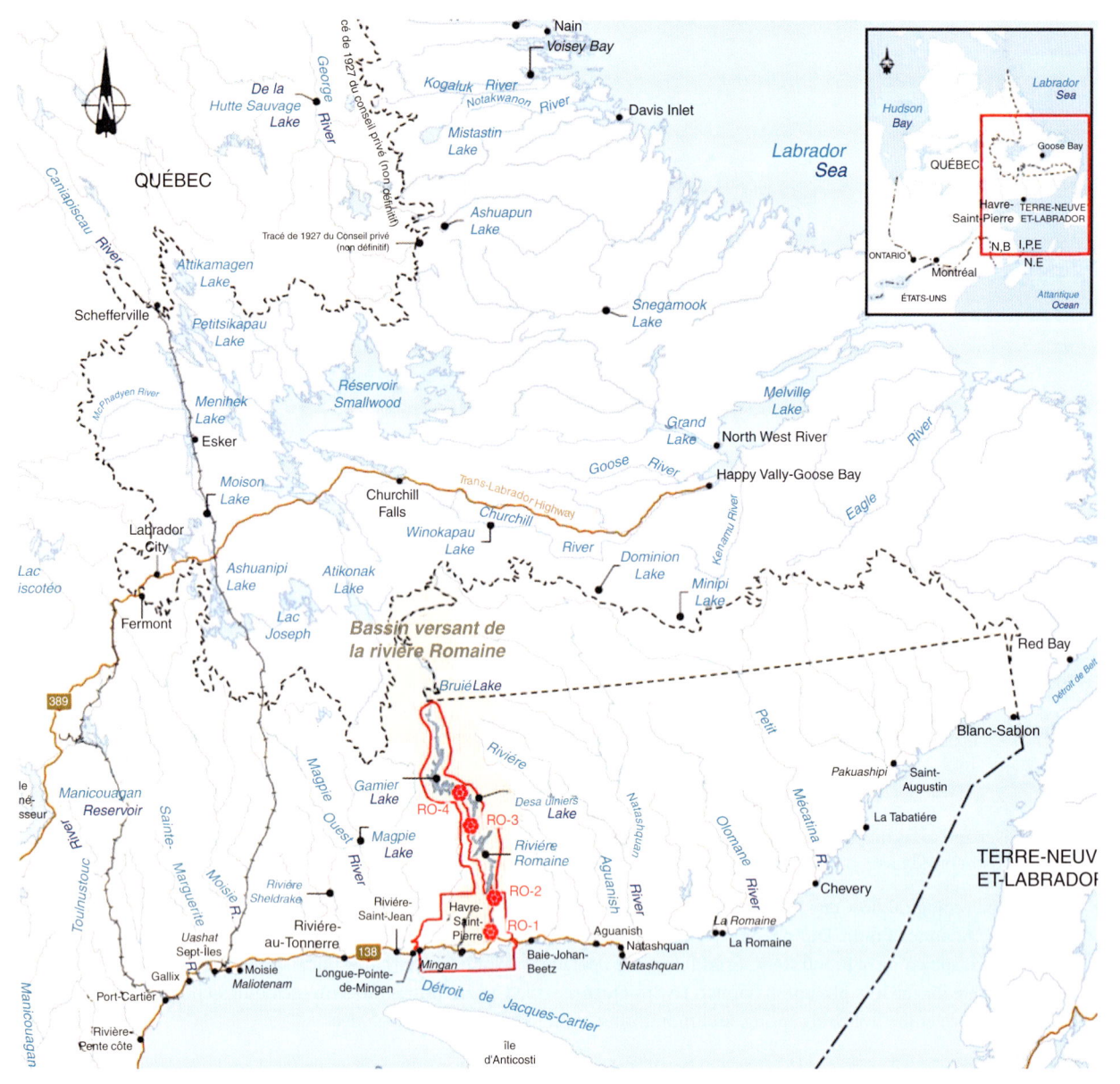

Figure 17 Layout of the Romaine power plant complex in the Quebec North Shore [13].

- 75 dikes
- Two diversion bays (forebay and tailbay) with a total area of about 346 km², at maximum operating level
- A 2.9 km long transfer tunnel between the Rupert forebay and tailbay
- A network of canals with a total length of about 12 km to facilitate flow in the various portions of the diversion bays
- Hydraulic structures on the Rupert River to maintain postdiversion water levels along approximately 48% of the river's entire length.

Scheduled for commissioning in 2011–12, the project will give Hydro-Quebec's generating fleet an additional capacity of 918 MW and an additional output of 8.7 TWh yr^{-1}, distributed as follows:

- Marginal gain at Eastmain-1A and Eastmain-1 powerhouses: 2.3 TWh
- Sarcelle powerhouse: 1.1 TWh.

6.06.4.3.2(i) Technical aspects

Mainly, the project area may be divided into two parts: the increased-flow reach, where the flow rate derivated from the Rupert River will increase, and the Rupert reduced-flow stretch.

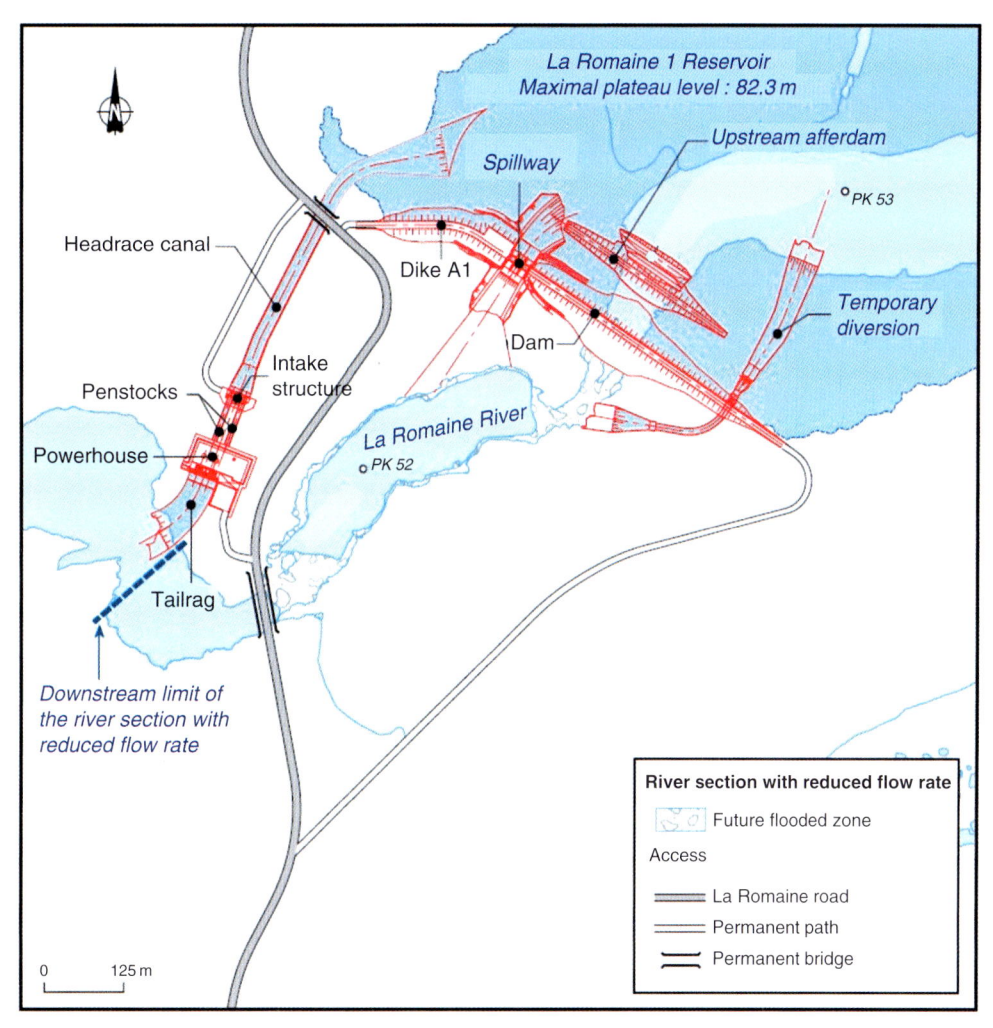

Figure 18 The reservoir and all the components of RO-1 [11].

Table 7 The Romaine plant components main features

	RO-1	RO-2	RO-3	RO-4
Dam site (milepost)	52.5	90.3	158.4	191.9
Dam elevation (m)	37.6	121	92	87.3
Number of dams	1	5	1	-
Size of reservoir (km^2)	12.6	85.8	38.6	142.2
Net head (m)	62	158	119	88
Number and type of generating units	2 Francis	2 Francis	2 Francis	2 Francis
Installed capacity (MW)	270	640	395	245
Average annual output (TWh)	1.4	3.3	2	1.3

6.06.4.3.2(i)(a) Description of the increased-flow section

- **Eastmain-1A powerhouse**

The Eastmain-1A powerhouse will be located east of the Eastmain-1 powerhouse. It is an above-ground facility used to generate power from the additional inflow to the Eastmain-1 reservoir resulting from the partial Rupert diversion. Its superstructure will be made of concrete components prefabricated at the jobsite and then transported to the powerhouse site for assembly (**Figure 21**). One of the advantages of this method is that it speeds up work delays. The powerhouse comprises the following structures: headrace, intake, penstocks, powerhouse with three generating units, and tailrace.

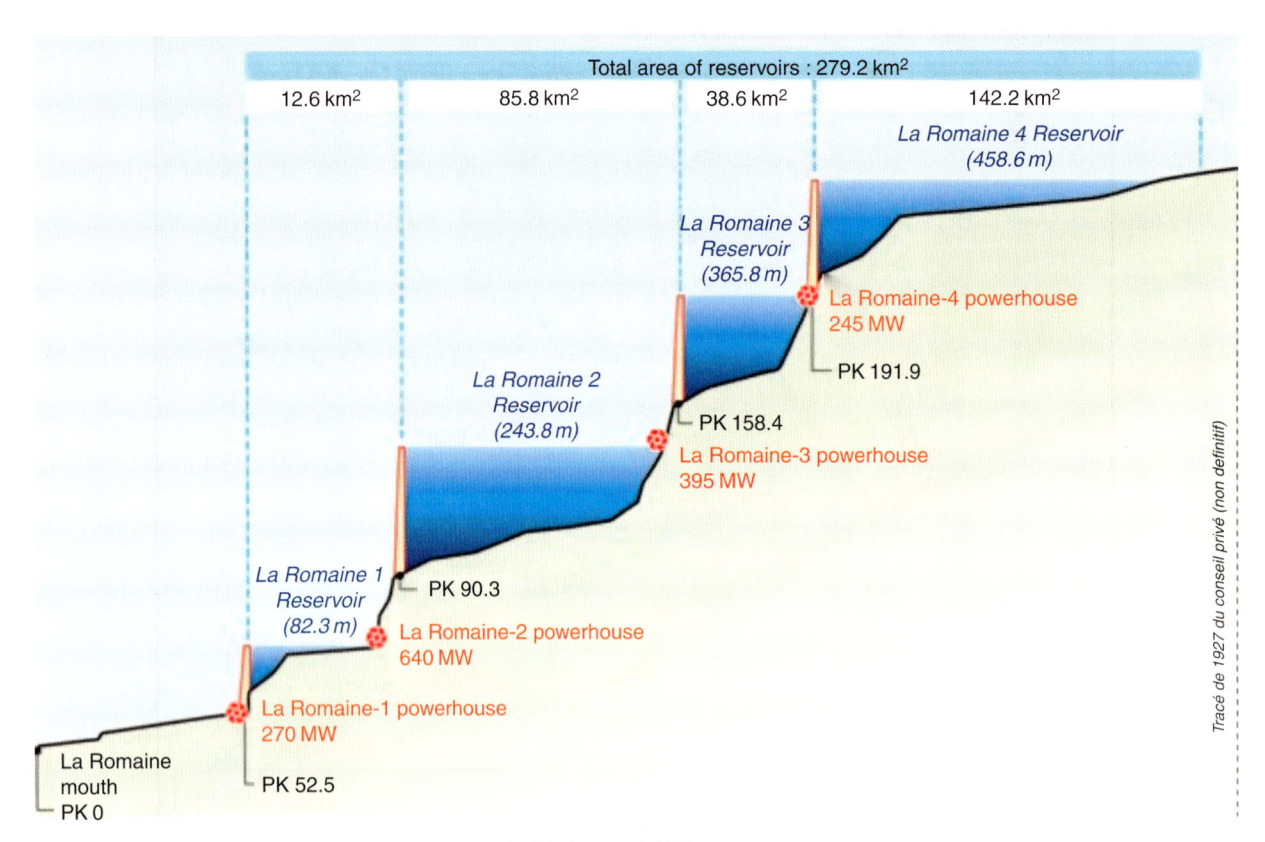

Figure 19 The Romaine River profile from the estuary to the RO-4 reservoir [11].

Eastmain-1A is equipped with three vertical Francis units, each with a rated capacity of 256 MW, for a total installed capacity of 768 MW. Each unit has a rated flow of 448 m³ s⁻¹, giving the powerhouse a design flow of 1344 m³ s⁻¹. The net head is 63.0 m (see **Table 9**).

The combined mean annual output of both Eastmain-1A and Eastmain-1 plants will average 5.1 TWh, assuming Rupert's instream flow of about 28%. The combined capacity factor will be 0.47. Adding Eastmain-1A powerhouse will increase annual output by an estimated 2.3 TWh.

- **Sarcelle powerhouse**

Sarcelle powerhouse is to be built east of the existing control structure. It is a surface powerhouse in an open-cut excavation and configured as a typical run-of-river plant. A canal will be excavated to extend the tailrace and avoid losing head between the powerhouse and Lake Boyd. The powerhouse is supplied by Opinaca reservoir, whose operating level fluctuates between 215.8 and 211.8 m. Net head is 10.3 m. The powerhouse has a mean 1050 m³ s⁻¹ turbine flow and the control structure a mean 235 m³ s⁻¹ spill flow.

Sarcelle powerhouse is equipped with three 40 MW bulb turbine units, for a total installed capacity of 120 MW. The design flow is 435 m³ s⁻¹ per unit or 1305 m³ s⁻¹ for the powerhouse. Average annual output will be about 0.9 TWh. The powerhouse will be remotely controlled from the Regional Control Center. A 101 km long 315 kV line will connect the Sarcelle substation to Eastmain-1 substation.

Bulb-type turbine generating units will be installed for the first time by Hydro-Québec. Bulb-type units, where both the turbine and generator are housed within a watertight metal casing, work well in situations where there is little hydraulic head (about a dozen meters at the Sarcelle site) and high flow.

- **Nemiscau-1 dam**

The core of the Nemiscau-1 dam is to be constructed of asphalt concrete. Well known in Europe, this technique is used in the absence of till to ensure that a structure is watertight. Although there is no lack of till in the Nemiscau area, Hydro-Québec wishes to adopt this technique for use in future projects in northern regions, such as those being carried out on the Romaine River, where till is harder to find (**Figure 22**).

6.06.4.3.2(i)(b) Description of the Rupert reduced-flow stretch
- **Instream flow regime**

The instream flow regimes for the Rupert, Lemare, and Nemiscau rivers have been established in accordance with the Quebec government policy on instream flow for the protection of fish and fish habitat.

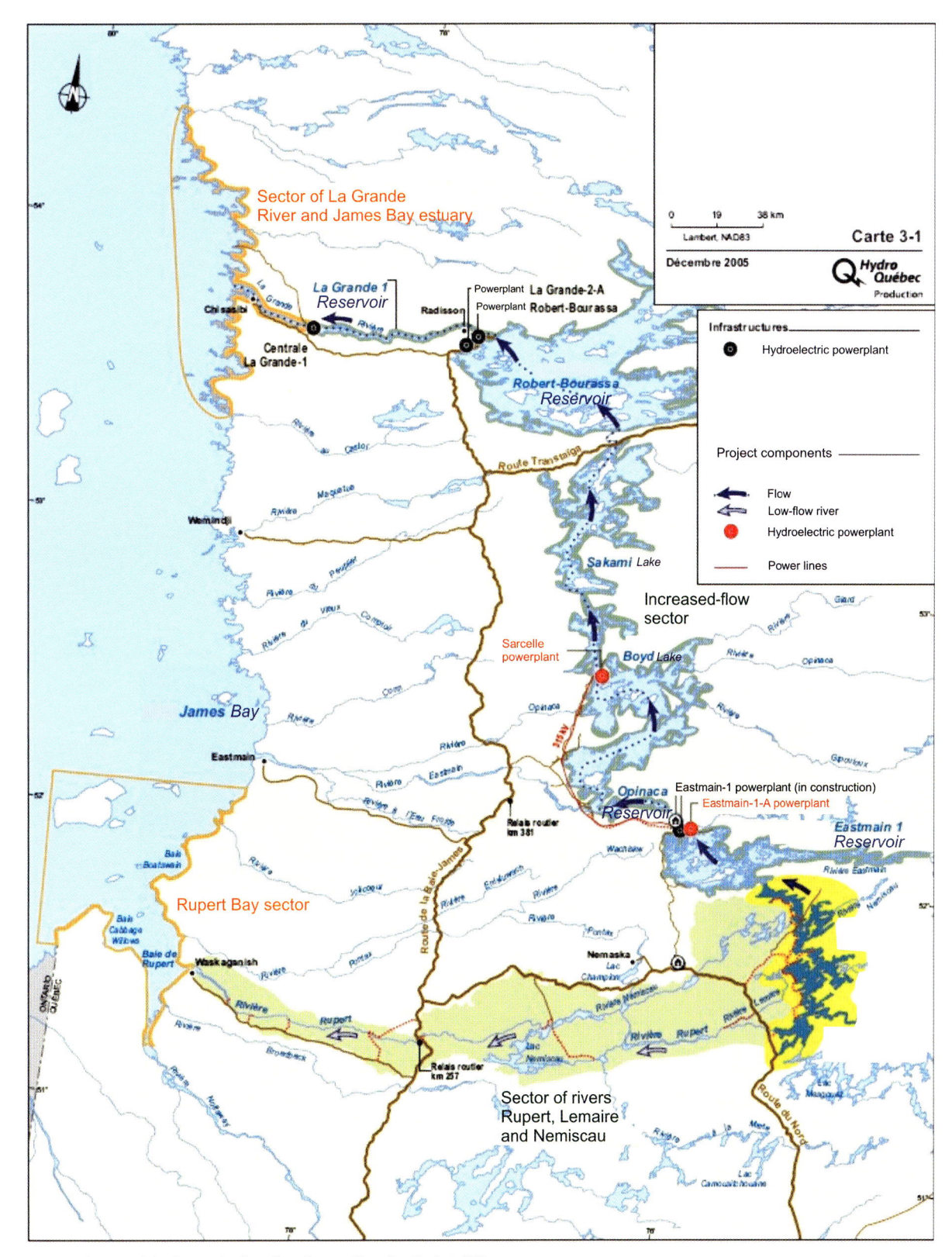

Figure 20 Layout of the Eastmain–Sarcelles–Rupert Diversion Project [13].

On an annual basis, the proposed Rupert instream flow at KP 314 is about 28% of the mean natural flow; this is higher than the percentage stipulated in the 'Boumhounan Agreement'. The instream flow will be adjusted for each of the four major biological periods: 416 m^3 s^{-1} during spring spawning (from about mid-May to early July), 127 m^3 s^{-1} during summer feeding (early June to

Table 8 The Eastmain–Sarcelle–Rupert project components and permanent structures

Rupert diversion bays	
Dams	4 (total fill: 1 100 000 m³)
Dikes	75 (total fill: 4 200 000 m³)
Spillways	1 (on the Rupert)
Instream flow release structures	4 (1 on the Lemarre, 3 on the Nemiscau)
Transfer tunnel	2.9 km
Canals	8 (total excavated material: 3 900 000 m³)
Relocation of transmission lines	8.3 km (19 towers)
Diversion bay access roads	132 km (3 bridges and 1 prefabricated culvert)
Circuits 7069 and 7070 roads	5 km (1 bridge)
25 kV lines	60 km
Increased-flow section	
Eastmain-1A powerhouse	768 MW
Sarcelle powerhouse	120 MW
315 kV Eastmain-1A–Eastmain-1 line	1 km
315 kV Sarcelle–Eastmain-1 line	101 km
Sakami lake outlet	Canal and weir
Muskeg–Eastmain-1 road	40 km
Rupert reduced-flow section	
Rock blanket	KP 20.4
Weirs	KP 33, 39, 85, 110.3, 170, and 223
Rockfill spur dikes	KP 290
Waskaganish drinking water supply	New plant with increased intake pump capacity
Bank stabilization	Near the Waskaganish water intake

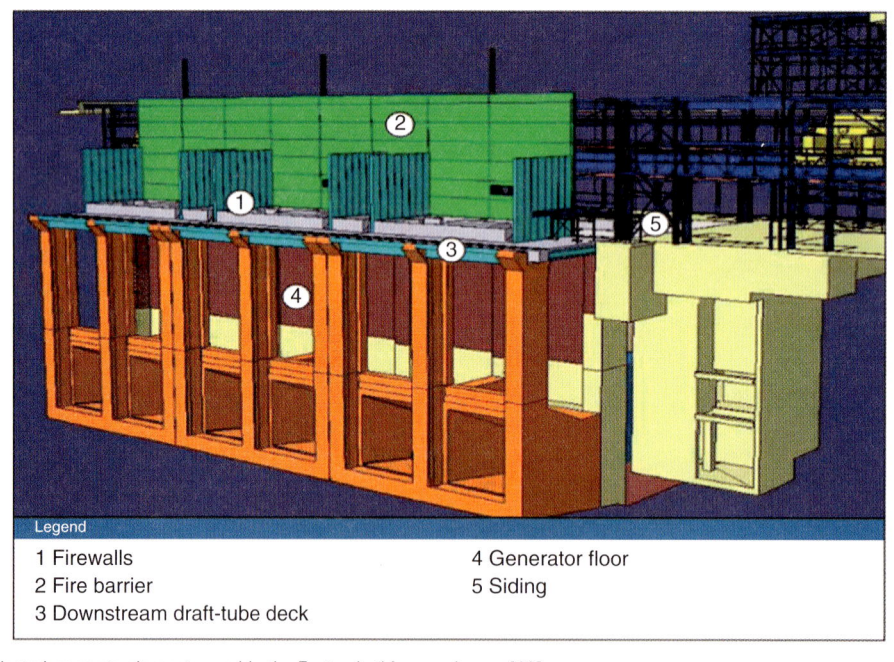

Legend
1 Firewalls
2 Fire barrier
3 Downstream draft-tube deck
4 Generator floor
5 Siding

Figure 21 Prefabricated concrete elements used in the Eastmain-1A powerhouse [13].

late September), $267 \, m^3 \, s^{-1}$ during fall spawning (early October to early November), and $127 \, m^3 \, s^{-1}$ during winter incubation (early November to late May).

- **Rupert River hydraulic structures**

The Rupert River has a step-pool profile with a difference in elevation of about 300 m over a distance of some 314 km. Weirs will keep water levels from falling at six of the eight locations, a rock blanket at KP 20.4 will preserve fish migration, and a spur dike will close part of the river at KP 290. The structures are sized to make the postdiversion water levels resemble mean natural levels previously recorded in

Table 9 Key powerhouse parameters

Parameter	Eastmain-1A	Sarcelle
Maximum operating level (m)	283.11	215.8
Minimum operating level (m)	274.11	211.8
Generating units	3 vertical Francis	3 bulb units
Design flow per unit ($m^3 s^{-1}$)	448	435
Rated net head (m)	63	10.3
Installed capacity (MW)	768	120
Output (TWh yr^{-1})	23 [a]	0.9
Capacity factor	0.47 [b]	0.82

[a]Additional output at Eastmain-1 site.
[b]Capacity factor for both Eastmain-1 and Eastmain-1A sites.

Figure 22 Nemiscau-1 dam with asphalt concrete core under construction [13].

August and September. The structures at KP 20.4 and 170 are also designed so that a 100-year flood will not inundate Gravel Pit (KP 21.35) or Old Nemaska (KP 187). The other structures are designed so that levels corresponding to a spring flood exceeded 10% of the time and are the same as under present conditions. Maintaining these levels serves several environmental purposes.

6.06.4.3.2(ii) Sustainable aspects

Right from the design stage, the ESRD project has incorporated many environmental protection measures that take into account the concerns of host communities. Owing to a combination of dikes and canals that will improve water flow, the creation of the Rupert diversion bays will flood a minimal land area. Moreover, a substantial ecological instream flow and a series of weirs in the Rupert River will protect fish habitat, preserve the landscape, and maintain navigation and other activities in the area.

To ensure a sustainable development, the following aspects were considered during the design phase:

- Minimal flooding of land
- Introduction of an ecological instream flow regime at the Rupert River closure point
- Maintenance of flow equivalent to the natural flow in the Lemare and Nemiscau rivers
- Building of hydraulic structures on the Rupert River to protect fish communities and habitat, preserve the natural appearance of the river, and support navigation and land use in some of its stretches
- Preservation of the natural levels in Mesgouez and Champion lakes and maintenance of the level in Lake Nemiscau
- Secure drinking water supply for Waskaganish
- Maintenance of the water level in Sakami Lake, as stipulated in the 'Sakami Lake Agreement'.

6.06.5 Recent Hydropower Implementations in British Columbia

This section briefly recounts the hydroelectric history in the province of British Columbia. It shows how hydroelectricity was gradually developed on the Peace and Columbia Rivers. The construction of the Revelstoke Unit 5 Project is also described.

6.06.5.1 General

Hydroelectric facilities provide 90% of the total electricity produced in British Columbia by BC-Hydro, a Canadian electric utility in the province and one of the largest electric utilities in Canada. Between 43 000 and 54 000 GWh of electricity are produced per year to supply more than 1.6 million residential, commercial, and industrial customers. Located throughout the Peace, Columbia, and coastal regions of British Columbia, this power is delivered using an interconnected system of over 73 000 km of transmission and distribution lines (BC-Hydro, 2009).

Between 1960 and 1980, BC-Hydro completed six large hydroelectric generating projects. The first large dam was built on the Peace River near Hudson's Hope (Figure 23). The Bennett Dam was built to create an energy reservoir at the Williston Lake for the Gordon M. Shrum Generating Station, which has a capacity of 2730 MW of electric power and generates 13 000 GWh yr^{-1} in energy. The Williston Lake reservoir is the largest reservoir in British Columbia with a surface area of 1773 km^2. The dam is 186 m high and 2068 m long along its crest [15]. A second smaller concrete dam was later built downstream, closer to Hudson's Hope for the Peace Canyon Generating Station which was completed in 1980.

Under the terms of the Columbia River Treaty with the United States, BC-Hydro built a number of dams and hydroelectric generating stations including two large projects at Mica and Revelstoke on the Columbia River. The Keenleyside Dam on the Columbia River north of Castlegar and the Duncan Dam north of Kootenay Lake were also built under the same treaty and are used mainly for flow control, although two generators were installed at Keenleyside in 2002.

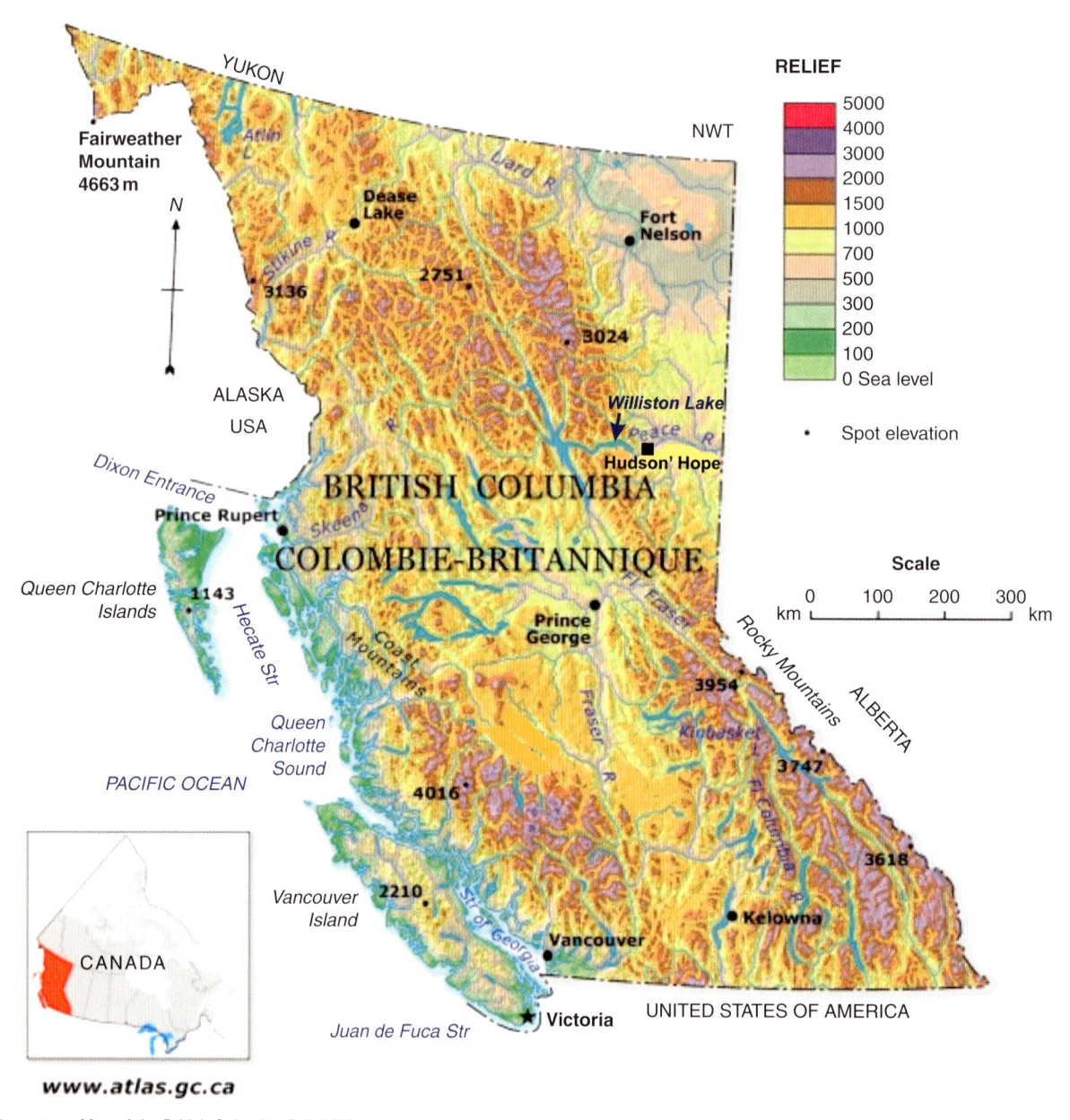

Figure 23 Map of the British Columbia Relief [3].

6.06.5.2 Revelstoke Complex

Revelstoke (population 7500 municipal estimated in 2005) is a city in southeastern British Columbia, Canada. It is located 641 km east of Vancouver, and 415 km west of Calgary, Alberta. The city is situated on the banks of the Columbia River just south of the Revelstoke Dam and near its confluence with the Illecillewaet River.

The Columbia River is the largest river in the Pacific Northwest region of North America (**Figure 24**). The river rises in the Rocky Mountains in British Columbia, Canada, and flows northwest and then south into the US state of Washington. It then heads west to form most of the border between Washington and the state of Oregon before reaching the Pacific Ocean. The river is 2000 km long, and its largest tributary is the Snake River. Its drainage basin is roughly the size of France and extends into seven US states and a Canadian province.

By volume, the Columbia is the fourth largest river in the United States, and it has the greatest flow of any North American river draining into the Pacific. The river's heavy flow and its relatively steep gradient give it tremendous potential for the generation of electricity. The 14 hydroelectric dams on the Columbia's mainstream and many more on its tributaries produce more hydroelectric power than those of any other North American river [17].

Figure 24 Map of the Columbia River watershed with the Columbia River highlighted [16].

6.06.5.2.1 Technical aspects

The Revelstoke Complex, also known as Revelstoke Canyon Dam, is a hydroelectric dam spanning the Columbia River, 5 km north of Revelstoke, British Columbia, Canada. The powerhouse was completed in 1984 and has a generating capacity of 1980 MW. Four generating units were installed initially, with room for two more, which would bring capacity to 2880 MW as shown by **Figure 25** [24]. The reservoir behind the dam is named Lake Revelstoke. All technical characteristics are listed in **Table 10**.

BC-Hydro is currently upgrading the dam by adding a fifth unit. The unit is set to be brought online in October 2010 and will add a 500 MW generating capacity, bringing the total capacity of the dam to around 2480 MW [22]. The additional capacity will allow BC-Hydro to service an equivalent of 40 000 additional homes during peak demand periods, such as cold winter nights when lights, stoves, baseboard heaters, televisions, and computers are all in use. As the Revelstoke Complex was originally designed to house six generating units, the project will not involve significant changes to the facility. The construction of the Revelstoke Unit 5 Project includes:

- Installation of a fifth penstock
- Placement of concrete to form the draft tube and house the generating unit
- Design, fabrication, and installation of the generating unit
- Electrical work to tie the generating unit into switchgear
- Upgrades to existing generating station equipment.

Areas inundated by the dam include the 'Dalles des Morts' or 'Death Rapids', which was the stretch of canyon just above the dam's location, and various small localities located along the preinundation route of the Big Bend Highway, which was the original route of the Trans-Canada Highway until the building of its Rogers Pass section. Just below the dam was the location of La Porte, one of the boomtowns of the Big Bend Gold Rush and the head of river navigation via the Arrow Lakes and Columbia River from Marcus, Washington. Just below the site of La Porte is the former town of Downie Creek, also a Big Bend Gold Rush boomtown but which survived into the days of the Big Bend Highway as a stopping point.

Figure 25 Aerial view of the Revelstoke Dam [18]. Reproduced with permission from BC Hydro.

Table 10 Technical characteristics of the Revelstoke complex

Impounds	Columbia River
Local	Revelstoke, British Columbia
Height	175 m
Opening date	1984

Reservoir information

Creates	Lake Revelstoke
Surface area	11 534 ha

Power generation information

Turbines	4 (till October 2010)
Installed capacity	1980 MW
Maximum capacity	2480 MW

6.06.5.2.2 Sustainable aspects

In 1996, the British Columbia Government established the British Columbia Utilities Commission (BCUC) to regulate public energy utilities and to act as an independent, quasi-judicial regulatory agency regarding energy rates. In 2003, the British Columbia Government passed several pieces of legislation to redefine and regulate power utilities in British Columbia. The Transmission Corporation Act created the British Columbia Transmission Corporation that plans, operates, and maintains the transmission system owned by BC-Hydro. The BC-Hydro Public Power Legacy and Heritage Contract Act, while recognizing the value of low-cost electricity produced by BC-Hydro's existing 'heritage assets,' requires BC-Hydro to meet the province's future needs for additional power through private developers and operators. These acts in conjunction with the government's 2002 BC Energy Plan have allowed Independent Power Producers to sell power to BC-Hydro from facilities which to date have typically been small-scale run- of the river projects [23].

The BCUC granted the Unit 5 Project a Certificate of Public Convenience and Necessity (CPCN). The CPCN established that Revelstoke Unit 5 is the preferred option for meeting customer electricity demand [24]. Before the environmental review, BC-Hydro convened a 26-member Revelstoke Unit 5 Core Committee consisting of regional stakeholders, First Nations, and government representatives. This Committee reviewed socioeconomic and environmental information, and made recommendations on how potential project effects could be addressed.

As recommended by the Revelstoke Unit 5 Core Committee, BC-Hydro made the following commitments to mitigate/compensate for Revelstoke Unit 5 Project effects:

- Adopt fish entrainment strategy to include the effects of the Unit 5 at the Revelstoke power plant
- Advance several Columbia River Water Use Plan studies and physical works projects, River bank protection/enhancement at specific downstream locations
- Support local hire and equity hire for First Nations through Columbia Hydro Council Agreement
- Support local training opportunities/employment through Okanagan College grant, Staff a community liaison position
- Restore Westside Road to preproject condition at the end of the Revelstoke Unit 5 Project construction.

The Committee also recommended an addendum to the Columbia Water Use Plan in order to identify and address any incremental operational impacts of Unit 5. The addendum was approved by the provincial Controller of Water Rights [25].

BC-Hydro's Power Smart program encourages energy conservation among its residential, commercial, and industrial customers. It also promotes energy saving retail products and building construction, and includes a Sustainable Communities Program. Its in-house Resource Smart program is used to identify and implement efficiency gains at existing BC-Hydro facilities. BC-Hydro also practices energy conservation at its generating facilities through the continuous monitoring and efficient use of the water resources used to power its generators.

BC-Hydro is committed by the BC government's Energy Plan to achieve electricity self-sufficiency by 2016, with all new generation plants having zero net GHG emissions by the same year. Over half of the power needed to satisfy the demand gap is to come from conservation and retrofitting of existing facilities; the remaining 50% will come from both private and public sources.

6.06.6 Conclusion

A few Canadian hydroelectric plants (existing and under development) are described in this chapter to demonstrate that hydroelectricity is one of the most economical and environmentally friendly ways of generating electricity. It produces virtually no smog and is a renewable energy source.

The recent development of hydropower facilities in Canada has been done by reducing the impacts on the environment. The key to sustainable hydropower lies in developing well-planned and well-managed projects that work to optimize economic, social, and environmental benefits while minimizing adverse effects. In Canada, environmental legislation covers all stages of hydropower development, from planning through construction to operation. The legislation requires the participation of stakeholders, including the population that might be directly affected by the project. A facility is only developed in Canada when it has gone through a complete environmental assessment process and is deemed economically viable, environmentally sound, and socially acceptable. The Canadian hydropower industry, more and more, makes efforts to work closely with host communities in the planning, construction, and development of projects to alleviate some of the associated negative impacts and ensure that local communities benefit from the project, through improved quality of life, employment, business opportunities, and long-term revenues.

Finally, research on hydropower plants performance is very important not only for the lifetime evaluation of present structures, but also for the durability design of new structures. In the future, research should shed more light on the material of hydropower structures.

References

[1] Canadian Hydropower Association (CHA) (2008) *Hydropower in Canada: Past, Present and Future*, p. 24. Montreal, Canada: CHA.
[2] Canada (2010) Web site scanned in July 2011. http://www.canadainternational.gc.ca/can-am/bilat_can/energy-energie.aspx?lang=eng.
[3] NRC (2011) The Atlas of Canada. Web site scanned in July 2011. http://atlas.nrcan.gc.ca/site/english/aboutus/index.html.
[4] Canadian Energy Research Institute (CERI) (2008) *World Energy: The Past and Possible Futures*, p. 142. Calgary, Canada: CERI.

[5] IWPDC (2010) International Water Power and Dam Construction. Web site scanned in July 2011. http://www.waterpowermagazine.com/story.asp?sectionCode=165&storyCode=2019652.
[6] Wikipedia (2010) Map of the Nelson River drainage basin. Web site scanned in July 2011. http://en.wikipedia.org/wiki/Nelson_River.
[7] Halim (2011) Kettle Rapids Generating Station, Northern Manitoba. Web site scanned in July 2011. http://www.hydroquebec.com/romaine/index.html.
[8] MRNF (2010) Ministère des Ressources Naturelles et de la Faune (Quebec). Web site scanned in July 2011. http://www.mrnf.gouv.qc.ca/.
[9] Hydro-Quebec (HQ) (2010) Hydro-Quebec. Hydroelectric powerhouses in Quebec. Web site scanned in July 2011. http://www.hydroquebec.com/production/centrale-hydroelectrique.html.
[10] HQ (2009) *Strategic Plan 2009–2013. Hydro Quebec. Renewable Energies, Energy Efficiency, Technological Innovation*, p. 94. Montreal, Canada: Hydro-Quebec Affaires.
[11] HQ (2010) Hydro-Quebec. Web site scanned in July 2011. http://www.hydroquebec.com/romaine/index.html.
[12] HQ (2007) *Complex La Romaine: Environmental Impact Statement. Hydro-Quebec Production*, Vols. 1–10. Montreal, Canada. (Report in French): Hydro-Quebec Production.
[13] HQ (2010) Hydro-Quebec. Web site scanned in July 2011. http://www.hydroquebec.com/rupert/en/index.html.
[14] HQ (2008) *Eastmain-1-A and Sarcelle Powerhouses and Rupert Diversion: A Hydroelectric Project for Present and Future Generations, Hydro Quebec Production*. Montreal, Canada: Hydro-Quebec Production.
[15] ILEC (2010) International Lake Environmental Committee. Web site scanned in July 2011. http://www.ilec.or.jp/database/nam/nam-29.html.
[16] Wikipedia (2010) Map of the Columbia River drainage basin. Web site scanned in July 2011. http://en.wikipedia.org/wiki/Columbia_River.
[17] USGS (1990) *Largest Rivers in the United States*, p. 2. Reston, VA: US Geological Survey, Department of the Interior.
[18] BC-Hydro (2010) Web site scanned in July 2011. http://www.bchydro.com/.
[19] Plourde Y (2003) *Review of Knowledge about Greenhouse Gas Emissions from Aquatic Environments*, pp. 93 and apps. Report submitted to Hydro-Québec by GENIVAR. (Report in French).
[20] Therrien J (2003) *Review of Knowledge about Greenhouse Gas Emissions from Aquatic Environments (Carbon Dioxide, Nitrous Oxide and Methane)*, p. 144 and apps. Report submitted to Hydro-Québec by GENIVAR. (Report in French).
[21] Tremblay A (2007) *Reservoirs and Greenhouse Gas*, pp. 48–52. Revue Pêche à la Mouche Destinations, April–May 2007. (Paper in French).
[22] BC-Hydro (2005) *Overview of the Proposed Revelstoke Unit 5 Project*, p. 31. Vancouver, Canada: BC Hydro.
[23] BC-Hydro (2009) *British Columbia's Annual Report*, p. 128. Vancouver, Canada.
[24] BCUC (2007) *In the Matter of British Columbia Hydro and Power Authority Application for a Certificate of Public Convenience and Necessity for Revelstoke Unit 5*, p. 82. Vancouver, Canada: British Columbia Utilities Commission.
[25] BC-Hydro (2007) *Columbia River: Project Water Use Plan. Revised for Acceptance by the Comptroller of Water Rights*, p. 65. Vancouver, Canada: BC Hydro.

6.07 The Three Gorges Project in China

L Suo, Science and Technology Committee of the Ministry of Water Resources, Beijing, China
X Niu and H Xie, Changjiang Institute of Survey, Planning, Design and Research, Wuhan, China

© 2012 Elsevier Ltd. All rights reserved.

6.07.1	Introduction	180
6.07.1.1	Location and Natural Condition	180
6.07.1.1.1	Location	180
6.07.1.1.2	Natural condition	180
6.07.1.2	Project Scale and Main Objectives	182
6.07.1.2.1	Key features of the Three Gorges Project	182
6.07.1.2.2	Main objectives	183
6.07.2	**Hydraulic Complex Structures**	183
6.07.2.1	Dam	184
6.07.2.1.1	Concrete gravity dam	184
6.07.2.1.2	Structures for water discharge and energy dissipation	186
6.07.2.1.3	Structure for sediment flushing and floating debris sluicing	188
6.07.2.2	Powerhouse	188
6.07.2.2.1	Dam toe power plant	188
6.07.2.2.2	Underground power plant	191
6.07.2.3	Navigation Structures	193
6.07.2.3.1	Dual-way five-step ship lock	193
6.07.2.3.2	Vertical ship lift	198
6.07.2.4	Maopingxi Dam	200
6.07.2.4.1	Objective and scale	200
6.07.2.4.2	Rockfill dam	200
6.07.3	**Project Construction**	200
6.07.3.1	Demonstration and Construction	200
6.07.3.1.1	Full demonstration and cautious decision	200
6.07.3.1.2	Milestones of construction	201
6.07.3.2	Construction by Stages	201
6.07.3.2.1	Staged construction	201
6.07.3.2.2	Water diversion during construction	205
6.07.3.2.3	Major temporary structures	206
6.07.3.3	Construction Management	208
6.07.3.3.1	System, mechanism, and relevant regulations of management	208
6.07.3.3.2	Supervision and control on quality	208
6.07.3.3.3	Financing and investment control	209
6.07.4	**Challenges and Achievements**	210
6.07.4.1	Resettlements	210
6.07.4.1.1	General situation in the Three Gorges reservoir area and index of main inundated practicality	210
6.07.4.1.2	Resettlement planning	210
6.07.4.1.3	Implementation of relocation and resettlement	211
6.07.4.1.4	Social and economic development in the reservoir area and further work	211
6.07.4.2	Sediment	212
6.07.4.2.1	Sedimentation of reservoir	212
6.07.4.2.2	Clean water discharging and river channel degradation	212
6.07.4.2.3	Direction of sediment study	213
6.07.4.3	Protection of Ecosystem and Environment	213
6.07.4.3.1	General condition and the effect of TGP	213
6.07.4.3.2	Study on the reservoir operation scheme favorable for environmental protection	214
6.07.4.3.3	Prospect	216
6.07.4.4	Prevention and Control of Geological Hazards	216
6.07.4.4.1	Introduction of geological hazards control and prevention project	216
6.07.4.4.2	Reservoir-induced seism	217
6.07.4.5	Protection of Cultural Relics	217
6.07.4.5.1	General introduction	217
6.07.4.5.2	Case study	217

6.07.4.6	Analysis of Dam Break	219
6.07.4.6.1	Study outcomes	219
6.07.4.6.2	Safety guarantee	220
6.07.4.7	Benefits	220
6.07.4.7.1	Benefit of flood control	220
6.07.4.7.2	Benefit of power generation	221
6.07.4.7.3	Benefit of navigation	221
6.07.4.7.4	Other benefits	221
6.07.4.8	Technical Advancement	221
6.07.4.8.1	Hydraulic structure design	221
6.07.4.8.2	Construction technology	223
6.07.4.8.3	Equipment manufacture	225
6.07.4.8.4	Hydraulic steel structures	226
References		226
Further Reading		226
Relevant Websites		226

6.07.1 Introduction

6.07.1.1 Location and Natural Condition

6.07.1.1.1 Location

Yangtze River, at more than 6300 km, is the longest river in China. The upstream of the river, generally referring to the reach above Yichang City in Hubei Province, is over 4500 km long, controlling a drainage area of 1 million km². Flowing out from the Three Gorges, the river stretches to a wide plain area, including 955 km long middle reach from Yichang to Hukou County in Jiangxi Province (controlling a drainage area of 680 000 km²) and the 938 km long downstream from Hukou to the estuary (controlling a drainage area of 120 000 km²). Embankments are constructed along the middle and downstream reaches. The world-known Three Gorges Project (TGP) is located at the boundary between the middle and downstream of the Yangtze River with its dam situated in Sandouping of Yichang City, Hubei Province, about 40 km upstream of the Gezhouba Project completed in the 1980s. The maximum transmission distance from the TGP to electrical load centers is within 1000 km, as shown in **Figure 1**. (All the figures are provided by the Changjiang Institute of Survey, Planning, Design and Research (CISPDR) except indicated otherwise).

6.07.1.1.2 Natural condition

6.07.1.1.2(i) The Yangtze River basin

With the main stream of the Yangtze River flowing through 11 provinces, autonomous regions, or municipalities (Qinghai, Tibet, Sichuan, Yunnan, Chongqing, Hubei, Hunan, Jiangxi, Anhui, Jiangsu, and Shanghai; see **Figure 2**) and tributaries extending to eight provinces or autonomous regions (Gansu, Shanxi, Guizhou, Henan, Zhejiang, Guangxi, Fujian, and Guangdong), the total drainage area controlled by the Yangtze River reaches 1.8 million km², accounting for 18.75% of the total area of the country. In the basin, the long-term mean annual precipitation is 1100 mm, and the average annual inflow into the sea is 960 billion m³. The basin is characterized with higher altitude in the west than in the east. The water level difference between the river source and the estuary amounts to over 5400 m. Abundant hydraulic energy resources can contribute to the generation of 268 GW power in total, of which 197 GW power is developable, mainly distributed in the upstream of the Yangtze River, accounting for 53.4% of the total developable hydropower resources in China.

6.07.1.1.2(ii) Hydrometeorology

The TGP controls a drainage area of 1 million km², accounting for 55% of the total drainage area of the Yangtze River. Within the reservoir area, although there is plenty of rainfall with the long-term mean annual precipitation of 1100 mm, the maximum daily rainfall is fairly low, only around 150 mm. According to the records from Yichang Gauging Station, the long-term average discharge is 14 300 m³ s⁻¹ and the annual runoff is 451 billion m³. There is no significant variation in the annual runoff, with the coefficient of variation (Cv) being 0.11. Most of the runoff comes from its main tributaries – the Jinsha River, Min River, Jialing River, and Wu River, especially in the flood season. During the flood season (from June to October), the runoff at the Yichang Station can be equal to 72.3% of the total runoff, while the runoff during the dry season accounts for only 27.7% of the total runoff. According to the flood records of more than 100 years at the Yichang Gauging Station, the maximum flood in the history of the Yichang Station occurred in 1870 when the flood discharge was 105 000 m³ s⁻¹. The long-term mean annual sediment yield (sediment load) is 530 million tons and the mean annual sediment concentration is 1.2 kg m⁻³, and these are showing a dramatically decreasing trend in recent years.

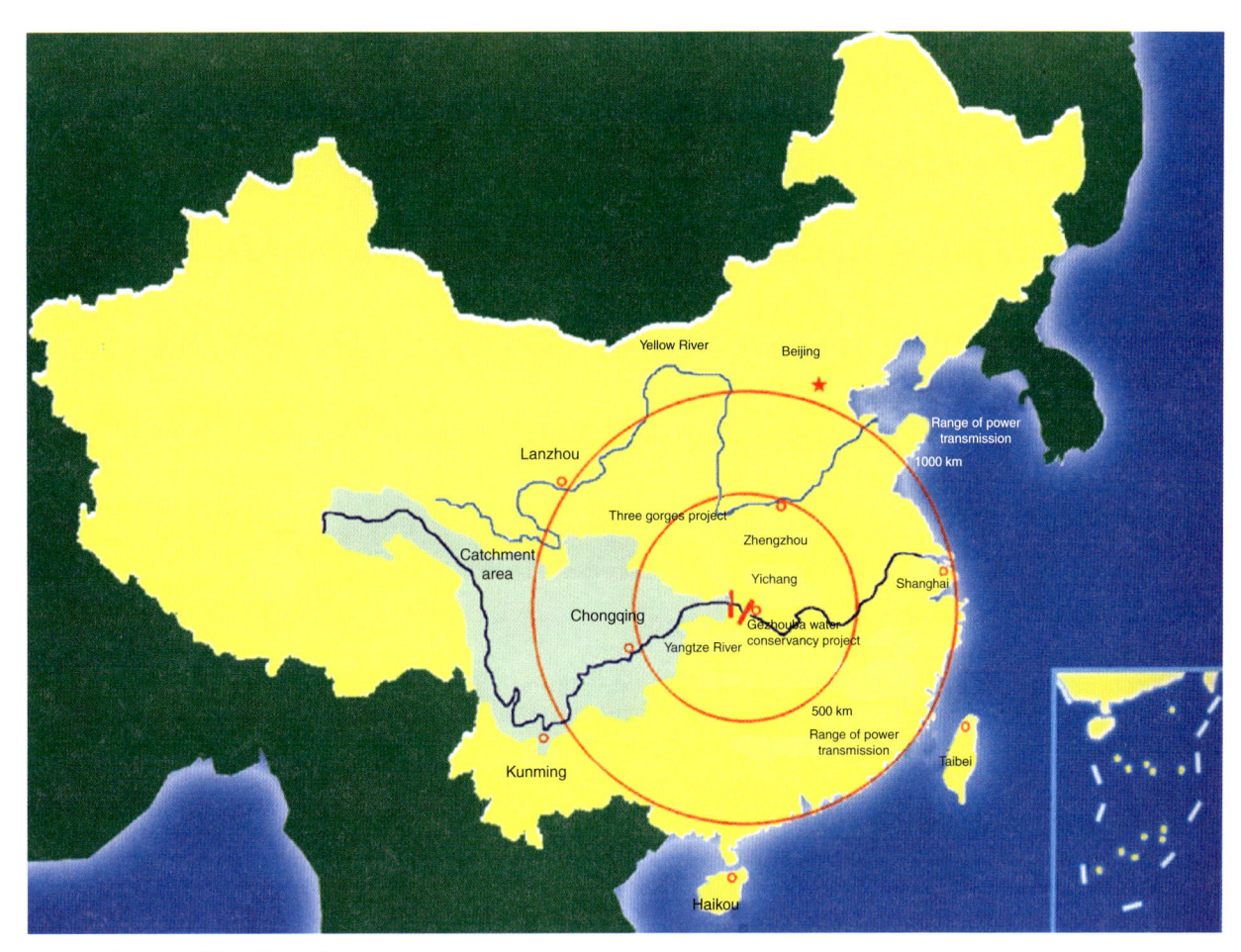

Figure 1 Location of Three Gorges Project.

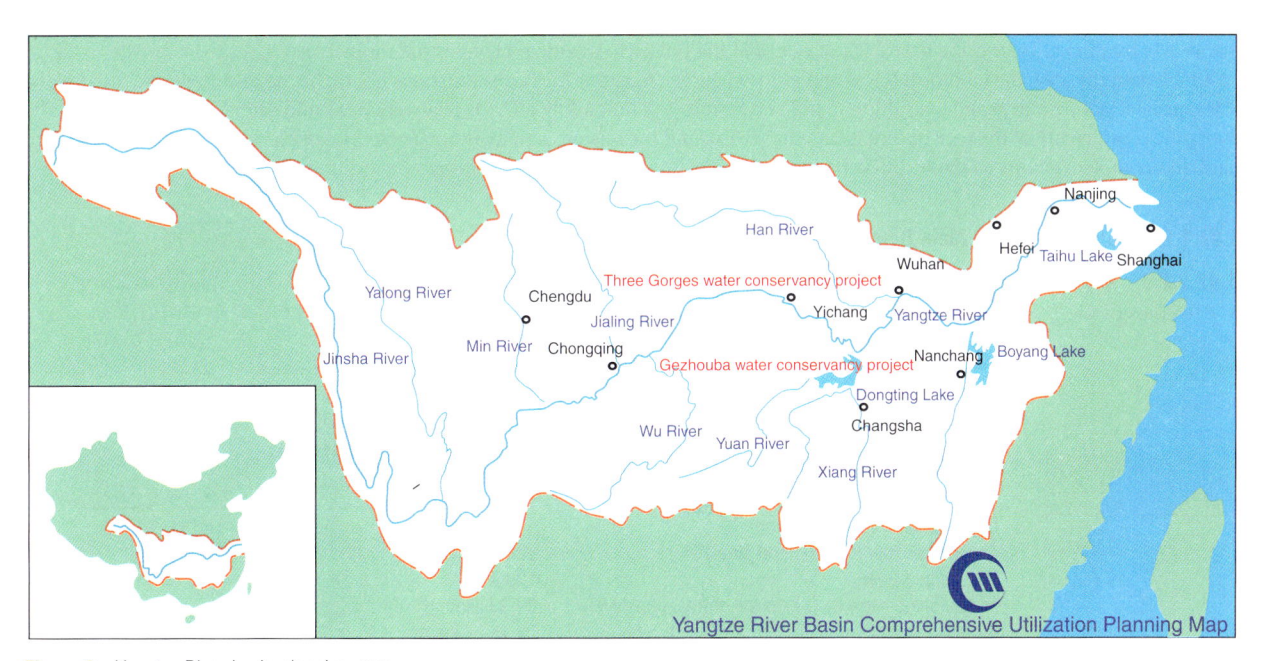

Figure 2 Yangtze River basin planning map.

6.07.1.1.2(iii) Geology

The TGP dam site, as shown in **Figure 3**, is located near the Sandouping Town of Yichang City, Hubei Province. The Yangtze River here is wide with a small island, Zhongbao Island, which divides the river into two channels – the main channel and the back river.

Figure 3 Original landscape of Three Gorges Project site.

This is favorable for staged river diversion (see text below). The river channel within the project area is 9 km long, including the navigation structures. The dam axis is located between the Tanziling Mountain on the left bank and the Baiyanjian Mountain on the right bank. The dam crest is at an elevation of 185 m, where the valley is around 2300 m wide. The rock mass at dam shoulder on the left bank is 250 m wide while that on the right bank is 400 m wide.

The bedrock at the dam site is mainly pre-Sinian period crystalline rock, most of which is porphyritic granite. Gneiss xenoliths and fine-grained diorite inclusion can be found in part of the area. The dam site is located on the Huangling block, where the fault structure was quite developed but the scale of the fault is not big. Fissures in the dam site area are also developed very well, where the direction and properties of fissures are consistent with those of the fault.

Within the dam site area, the weathered crust, including the full weathered, intensely weathered, and weakly weathered layers, is varied in thickness, being generally thicker in the ridges on both the banks and attenuating gradually toward the riverbed. The thickest weathered crust is located on the approach channel within the shiplock area and the next is on the dam section No. 1 of the left bank powerhouse, while the thinnest weathered crust is on the riverbed. In the weathered crust, the full weathered layer is the thickest one, the weakly weathered layer is the next thickest one, and the intensely weathered layer is the thinnest.

In the dam site area, the landform is low and even, where the magnitude of crustal stress is not high. The low-angle structural plane was hardly developed, in general, resulting in little extension. The unloading effect is relatively weak and unloading zone is thin. The hydraulic conductivity of the rock bodies is extremely weak and the seismic activity is not active. The open valley at the dam site, with hard and complete granite as the bedrock, has provided ideal topographical and geological conditions for dam construction.

6.07.1.2 Project Scale and Main Objectives

6.07.1.2.1 Key features of the Three Gorges Project

Key features of the Three Gorges Project:

Normal pool level	175 m (156 m at the initial stage)
Limit level for flood control	145 m (135 m at the initial stage)
Low level in dry season	155 m (140 m at the initial stage)
Pool level with 1% flood	166.9 m
Design peak flow (0.1% flood)	98 800 m^3 s^{-1}
Design flood level (0.1% flood)	175 m
Check peak flow (0.01% flood plus 10%)	124 000 m^3 s^{-1}
Check flood level	180.4 m
Total storage capacity (below the normal pool level)	39.3 billion m^3
Storage capacity for flood control (145–175 m)	22.15 billion m^3
Regulating storage (155–175 m)	16.5 billion m^3
Crest elevation	185 m
Installed gross capacity/guaranteed output	22.4 GW/4.99 GW
Average annual output	90.0 billion kWh
Single unit capacity/number of units	700 MW/32 units

6.07.1.2.2 Main objectives

The TGP is a multipurpose development project with great comprehensive benefits mainly in flood control, power generation, navigation, and supplying water to the downstream during the dry season [1, 2].

6.07.1.2.2(i) Flood control

The primary purpose of building the TGP is to protect the middle and downstream of the Yangtze River from the floods as well as to improve the sustainable development of the middle and downstream area along the Yangtze River. The unique location and topography of TGP, due to it being located at the boundary of middle and downstream of the Yangtze River, in conjunction with enormous storage capacity for flood control, make the TGP capable of effectively controlling the floods that result from the storms in the upstream. It is critically important for protecting the Jinjiang Plain area, of 1.5 million hectare of farmland and towns, and 15 million people, from the floods and plays an important role in controlling the whole-basin flood as well as floods occurring in the middle and downstream. The reservoir of the TGP can control a drainage area of 1 million km^2. When raising the water level in the reservoir to 175 m, the storage capacity for flood control can reach 22.15 billion m^3, which will bring significant benefits from flood control and environmental protection, such as

- the flood control standard at the Jingjiang River section (about 400 km long river section downstream of Yichang) can be upgraded from the level of preventing 10-year floods to that of preventing 100-year floods;
- if a 1000-year flood or an extraordinary flood similar to that of the 1870s occurs, the TGP can relieve both the banks of the Jingjiang River section, the Dongting Lake area, and the Jianghan Plain from fatal disaster by regulating water storage in conjunction with operation of other flood detention area.

6.07.1.2.2(ii) Power generation

The installed gross capacity of the (left and right) powerhouses at dam toe is 18.2 GW, with the expected annual average power generation accounting for up to 84.7 billion kWh. After the underground power station being put into operation, the total installed capacity will amount to 22.4 GW, with the corresponding power generation of 90.0 billion kWh, which is equivalent to building a super coal mine with an annual production of 50 million tons of coal or a super oil field with an annual production of 25 million tons of oil. It is also equal to building 10 large thermal power plants with an installed capacity of 2000 MW including a relevant railway for coal or oil transportation. In other words, its power generation capacity ranks the largest in the world.

When combined with the Gezhouba Power Station, a reregulating power station of the TGP located 40 km downstream of the TGP, the Three Gorges Hydropower Plant (TGHP) can also work as a peaking plant as well as a frequency regulator for the power system. The huge power energy generated by the TGHP has contributed to the formation of a trans-region power system consisting of the power grids of Central China, East China, and South China, through which benefits can be obtained from regulating peaks between regions, compensating regulation between hydropower stations, and capacity exchange between the hydropower and thermal power plants.

6.07.1.2.2(iii) Navigation

The river channel of the Yangtze River has been always called the golden channel because it is a traffic artery connecting the coastal area in the southeast with the hinterland in the southwest, forming a complete inland navigation system. Before the construction of the TGP, however, the natural condition of the river channel from Yichang to Chongqing was quite complex, characterized by densely distributed torrential flow and dangerous shoal, which limited the navigation capability allowing only 1500-tonnage fleet to pass through. After the project is completed, the backwater of the TGP reservoir goes as far as Chongqing resulting in a 660 m long deep-draft channel from Yichang to Chongqing, which enables 10 000-tonnage fleet to pass through for more than 6 months in a year. One-way carrying capacity through this waterway will be upgraded from 10 million tons to 50 million tons.

Meanwhile, the minimum discharge in the dry season in the river downstream of Yichang can be raised to over 5000 $m^3 s^{-1}$, which will also improve the navigation condition in the dry season for the middle and downstream of the Yangtze River.

Besides the abovementioned functions, the TGP also functions to promote tourism and facilitates transfer of water to the north.

6.07.2 Hydraulic Complex Structures

The hydraulic complex structures of the TGP, as shown in **Figure** 4, consist of dam, powerhouses, navigation structures, and Maopingxi Guard Dam. The dam crossing the river is a concrete gravity dam with the spillway section located in the middle of the riverbed and non-overflow sections on both sides, behind which are the left and right powerhouses. The ship lift and ship lock are laid on the left bank while the underground powerhouse and Maopingxi Guard Dam are disposed on the right bank.

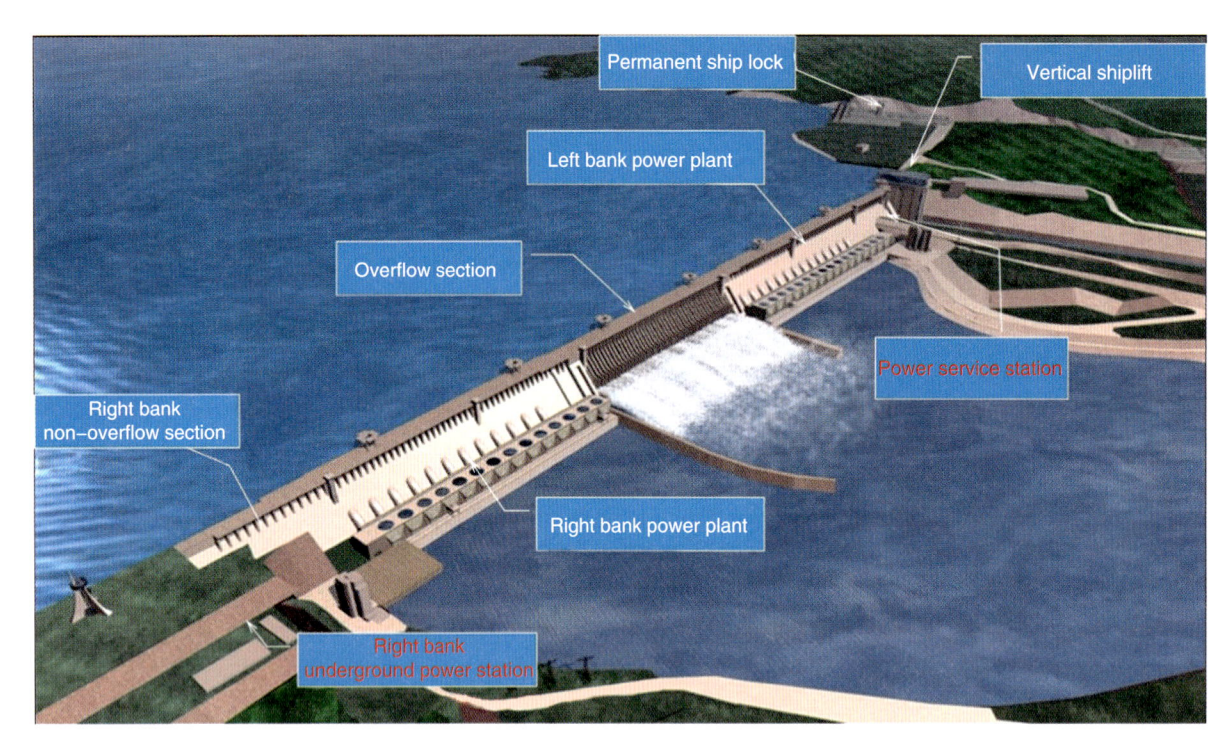

Figure 4 Structures layout of Three Gorges Project.

6.07.2.1 Dam

6.07.2.1.1 Concrete gravity dam

6.07.2.1.1(i) Scale of dam

The dam of the TGP is a concrete gravity dam with crest length 2309.5 m, crest elevation 185 m, and maximum height 181 m. In total 16 million m^3 of concrete is used for the construction of the dam.

The dam consists of four parts: the spillway section in the middle of the riverbed, the powerhouse sections on the left and right sides, the sections of ship lift and temporary ship lock on the left bank, and the non-overflow sections on both left and right banks.

The design standard for earthquake resistance is that the basic seismic intensity is considered as degree VI and the intensity of degree VII is adopted for the dam design.

6.07.2.1.1(ii) Size of dam cross section

The normal pool level of the TGP is finally determined at 175 m, while the dam crest elevation is 185 m, which is 10 m higher than the former one. This is for taking future operation and potential development into consideration. Therefore, the downstream slope of the dam starts at the crest elevation of 185 m. The stabilized stress on the foundation base has been left an appropriate margin for possible future demand of raising the water level.

According to the analysis of fissure extension with fracture mechanics, once fissure occurred at the dam heel, in order to keep the fissure steady and prevent it from extension, the downstream slope of the powerhouse section should be 1:0.72 while the downstream slope of the spillway section should be 1:0.7. The upstream slope of both the powerhouse section and spillway section is vertical (see **Figures 5** and **6**).

6.07.2.1.1(iii) Foundation treatment

The foundation rock mass at the dam site is pyrogenetic-amphibolitic granite (porphyritic granite). Although most of the slightly weathered and fresh rock is of extremely weak hydraulic conductivity, the developed fracture structure zone from upstream to downstream in conjunction with deep grooves of weak to moderate water conductivity exists in the dam foundation and forms the seepage channel in the foundation rock mass. The dam abutment is partly formed by weakly weathered and intensely weathered rock, which results in bypass seepage. Meanwhile, since there is significant uplift pressure on the dam foundation due to the high water level in the downstream of the dam, seepage control measures such as impervious curtain and drainage are adopted based on the field test unwatering the rock mass and the numerical analysis on the seepage field in order to relieve the uplift pressure and mitigate seepage at the dam foundation.

Because the foundation face, on which the spillway section in the middle of riverbed and the powerhouse sections on the left and right sides are based, is at a lower elevation, it is required to adopt closed pumping drainage. In addition, to

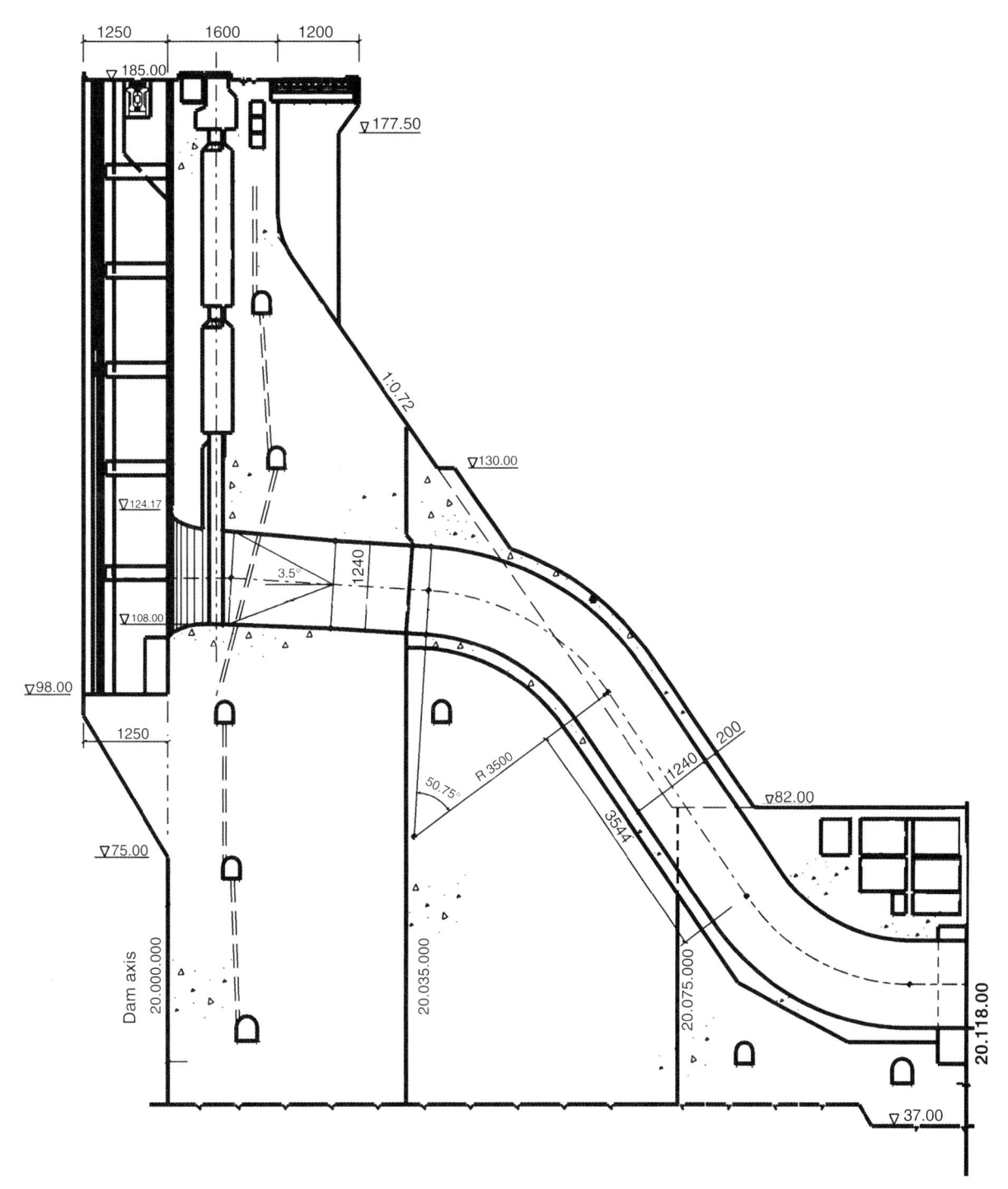

Figure 5 Typical cross section of the powerhouse dam section.

satisfy depth stabilityagainst sliding for those parts close to both banks in the left and right powerhouse sections, the alternative for powerhouse and dam co-bearing load is designed and a closed pumping drainage area is disposed at each side covering the upstream dam and downstream powerhouse. Finally, a huge closed pumping drainage area is formed from the dam section No. 1 of the left powerhouse section to the dam section No. 26 of the right powerhouse section (including units Nos. 1–6 of the left power plant and units Nos. 21–26 of the right power plant). In other words, a main curtain is built in the upstream of the dam foundation for seepage control and drainage while a closed curtain is built in the downstream for the same purpose.

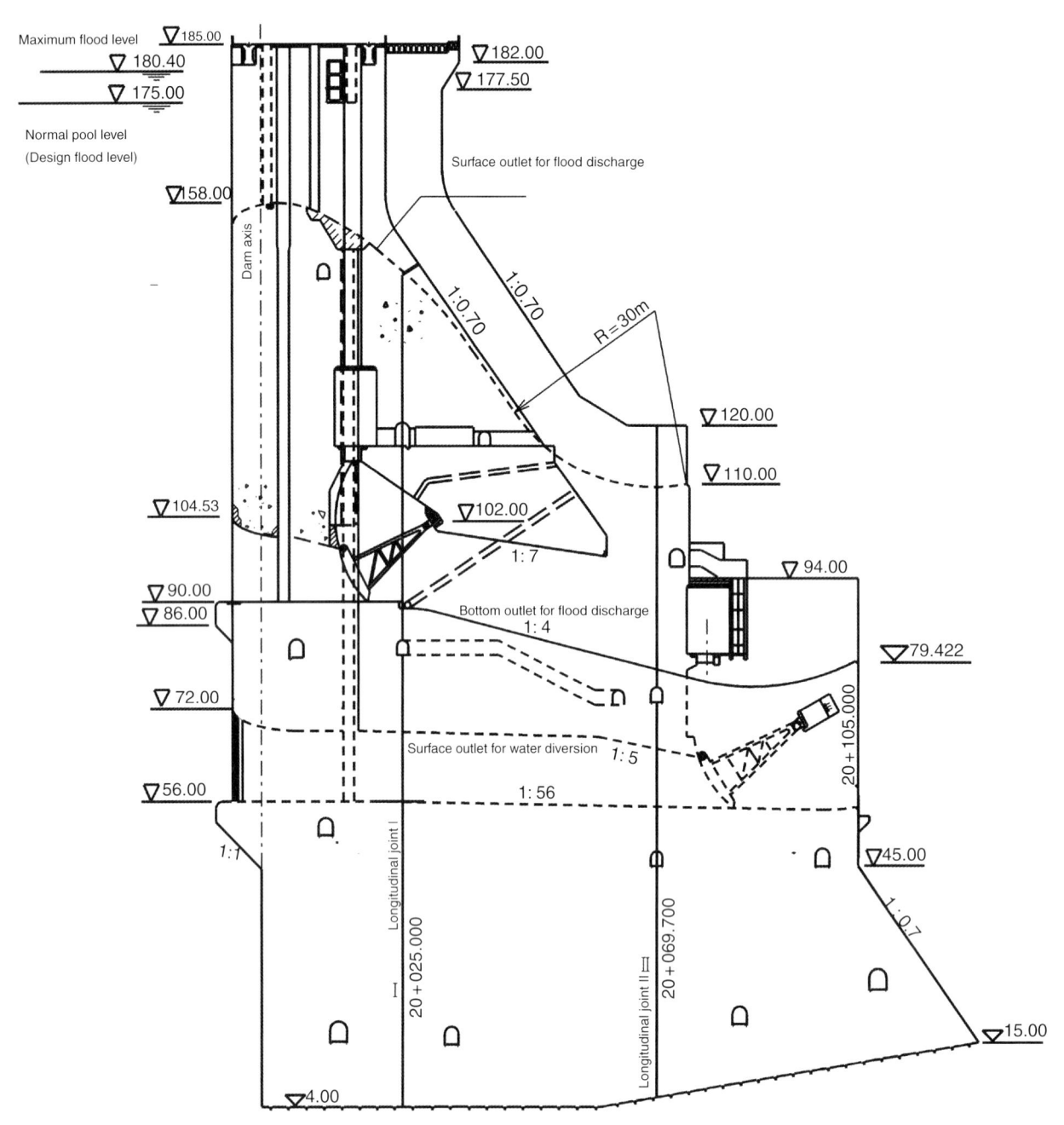

Figure 6 Typical cross section of the overflow dam section.

6.07.2.1.2 Structures for water discharge and energy dissipation

6.07.2.1.2(i) Scale of structures

The primary purpose of the TGP is to mitigate floods in the middle and downstream of the Yangtze River, especially the Jingjiang River section, and the storage capacity for flood control of the TGP is 22.15 billion m³. In order to fully utilize the storage for flood control, water discharging and water storage should be combined in the 'discharging flood operation'. When the inflow from the upstream is less than 56 700 m³ s⁻¹, water discharging should be controlled to ensure that the flow at Zhicheng Station downstream of Yichang is not beyond 56 700 m³ s⁻¹; when the pool level is higher than the design flood level of 175 m and upstream inflow is bigger than that of 1000-year flood, it should discharge water as much as it is capable of, but without exceeding the upstream inflow.

Based on a comprehensive analysis of factors such as flood control, sediment flushing, hydraulic structure protection, and removal of floating debris at the front of power plants, especially due to the characteristics of high water head and enormous amount of flood discharging and sediment flushing, the structures for flood discharging of the TGP are designed as deep outlets (middle-level outlets) alternating with surface outlets (gated spillway openings). In total 23 deep outlets and 22 surface outlets are

Figure 7 Downstream view of overflow dam section.

Figure 8 Flood discharging.

distributed along the spillway section. There are also 22 temporary bottom outlets (low-level sluices) for water diversion and river close-off during the third-stage construction, which will be backfilled later with concrete. **Figure 7** shows the dam section under construction which includes structures for flood discharging. **Figure 8** demonstrates the fine spectacle when a part of the surface outlets is open to discharge flood.

6.07.2.1.2(ii) Deep outlets for flood discharging
The 23 deep outlets are the TGP's main flood discharging structures, through which floods with discharge lower than that of 1000-year floods will be mostly released. The outlets are also used for discharging water during the period of the third-stage river diversion as well as power generation with cofferdam retaining water. These deep outlets are characterized by large numbers, long periods of water discharge, frequent operation, and high water head of great variation. With the design water head being 85 m, the flow velocity at the outlet is about 35 m s^{-1}. In the design, the issues related to aeration and radial gate sealing are taken into specific consideration. Three schemes are studied including a deep outlet without aeration, aeration with sudden drop device, and aeration with sudden enlargement device. The scheme of aeration with the sudden drop device was finally selected to ensure safe operation.

6.07.2.1.2(iii) Hydraulic steel structures
Three layers of outlets at different elevations are arranged within the flood discharge dam section. Included in the top layer are 22 surface spillway openings at the elevation of 158 m, each opening being 8 m wide and 17 m high with two plain gates installed: one for maintenance and one for normal working. At the elevation of 90 m, there are 23 flood discharging deep outlets. The opening of each

outlet is 7 m wide and 9 m high with three gates installed: a backhook stop-log plain gate for maintenance, a fixed-wheel plain gate for emergency use, and a radial gate for normal working. The third layer consists of 22 low-level sluices (bottom outlets), among which 16 sluices located in the central part are at elevation 56 m and 3 sluices at each side, left and right respectively, are at 57 m. Each sluice is 6 m wide and 8.5 m high with four gates installed: a backhook stop-log plain gate for maintenance and intake blocking, a plain gate for emergency use, a radial gate for working, and a backhook stop-log plain gate for maintenance and outlet blocking.

In addition, a trash way outlet is designed at the dam section of the left side training wall and the dam section No. 1 on the right side longitudinal cofferdam, respectively. A radial work gate (tainter gate) is installed in the downstream and a fixed-wheel plain emergency gate is installed in the upstream.

Each of the radial work gates installed in the above-mentioned deep outlets, bottom outlets, and trash way outlet is operated via a hydraulic hoist, while the other gates are operated by two sets of gantry crane located on the top of the dam.

6.07.2.1.3 Structure for sediment flushing and floating debris sluicing

6.07.2.1.3(i) Sand discharge outlet

There are eight sand discharge outlets in the TGHP. In the left bank power plant, outlets No. 1 and No. 2 are arranged on the section of assembly bay II, while outlet No. 3 is arranged on the section of assembly bay III. In the right bank power plant, outlet No. 4 is positioned on the 'outlet section' at the left end of the plant, outlets No. 5 and No. 6 on the section of assembly bay III, which is in the middle of the plant, and outlet No. 7 on the section of assembly bay II at the right end of the plant. Outlet No. 8 serves the underground power plant. Its three inlets are situated at the bottom of the intake tower and three branch tunnels then merge into one toward the outlet located on the section of assembly bay II of the right bank power plant.

6.07.2.1.3(ii) Sediment sluice gate

Sediment may deposit in the upstream and downstream navigation channels after a long term operation of the navigation structures. In order to resolve this problem, a sediment sluice gate is converted from the temporary ship lock used during the construction period.

The temporary ship lock is located between dam sections No. 8 and No. 9 of the left bank non-overflow dam. It is used for navigation purpose during the second-stage construction, in conjunction with the open diversion channel. After the temporary ship lock is abandoned, the navigation channel is blocked and the ship lock is converted to the sediment sluice. It is expected that scouring the channels and flushing away the sediment by operating the sluice can be one of the effective ways to resolve the problem of sediment deposit in the navigation channels.

Two sediment sluice outlets are arranged at an elevation of 102 m in section No. 2 of dam sections converted from the temporary ship lock. Each outlet is 5.5 m wide and 9.6 m high with a plain emergency gate and a radial work gate installed.

6.07.2.1.3(iii) Trash way outlet

Three trash way outlets are designed according to the law of movement of floating debris in front of the dam. Outlets No. 1 and No. 2 are located in the left side training wall and longitudinal cofferdam, respectively, of the spillway dam section. Both outlets are at an elevation of 143 m and are used when the water level in the reservoir is 145 m at the final operation stage. Outlet No. 3 is on the section of assembly bay II at the right end of the right bank power plant. This outlet is at an elevation of 130 m and is used when the water level in the reservoir is 135 m at the initial operation stage.

6.07.2.2 Powerhouse

6.07.2.2.1 Dam toe power plant

6.07.2.2.1(i) Scale of structure

In the two dam toe power plants, 26 units are installed, each with an installation capacity of 700 MW, and the total installed capacity amounts to 18 200 MW. Specifically, there are 14 units installed in the left bank power plant (**Figure 9**) and 12 units in the right power plant (**Figure 10**). Three assembly bays are designed for each power plant.

Each dam toe power plant consists of intakes, penstocks, a main powerhouse, an upstream auxiliary station, a dam-plant deck, a downstream auxiliary station, a draft-tube deck, a tailrace, and front area, as illustrated in **Figure 11**.

6.07.2.2.1(ii) Layout of power plant

As mentioned above, there are 14 turbine generator units in the left bank power plant (see **Figure 12**) and 12 units in the right power plant. In addition, in each power plant, two sets of overhead traveling cranes (bridge cranes) are installed, one small and one big. The main transformer and high-voltage enclosed switchgear are situated in the upstream auxiliary station. Other subsidiary equipment are located in the turbine floor and generator floor in the main powerhouse, as well as in each floor of upstream and downstream auxiliary station and assembly bay. On the sections of assembly bays II and III, sediment flushing outlets are arranged below the downstream water level.

In the upstream–downstream direction, the part of the power plant under the water is 68 m wide and the part above the water is 39 m wide. The unit bay is 38.3 m long. The unit is installed at an elevation of 57 m, while the turbine floor is at an elevation of 67 m and the generator floor at an elevation of 75.3 m. The power plant is 92 m high in total.

Figure 9 Downstream view of the left bank power plant.

Figure 10 Dam and power plant on the right bank.

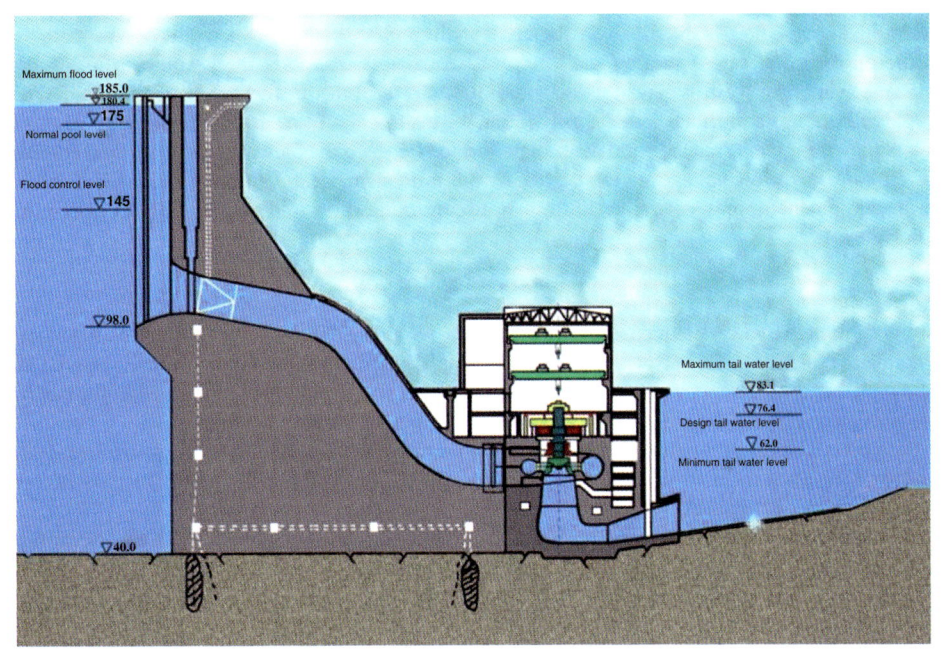

Figure 11 Layout of power plant and conduit system.

Figure 12 Left bank power plant.

Figure 13 GIS in power plant.

The downstream auxiliary station is arranged on the roof of the tailrace, and between the downstream wall of the main powerhouse and the downstream water-retaining wall. It consists of five floors, with an elevation from 50 to 75.3 m, and the draft-tube deck of an elevation of 82 m is served as its roof. The main facilities in the downstream auxiliary station include a technical water supply device, water purifier, water supply room, air treatment unit room, and air exhaust unit room.

The upstream auxiliary station is 17 m wide and 40 m high, located between the dam and the main powerhouse at an elevation from 67 to 107 m. It consists of low-voltage station service distribution switchgear (auxiliary switchgear), main transformers, a central control room, equipment for protective relay, illumination, communication, and maintenance, and gas-insulated switchgear (GIS), as shown in **Figure 13**.

6.07.2.2.1(iii) Layout of conduit system

The type of water diversion of the power plant is 'single tube to single unit'. The penstock is 12.4 m in diameter arranged mainly on the downstream dam surface. The water diversion conduit can be divided into six parts – intake, tapered section, penstock embedded in the dam, penstock on the downstream dam surface, lower flat section (including bedding pipeline section and/or expansion joint section), and exposed pipe in the powerhouse – as shown in **Figure 11**.

The intake of the power plant is at an elevation of 108 m with an opening dimension of 9.2 m wide and 13.24 m high. The tapered section is 15 m long at an inclination of 3.5° between its axis and horizontal line, where the cross section changed from rectangle to a round section with a diameter of 12.4 m.

The 26 penstocks in the dam toe power plant are positioned on the penstock–dam sections, with each section of length 25 m. A trough is reserved in advance on the downstream dam surface for each penstock, which is installed with one-third part embedded in the trough and the other two-third part exposed on the downstream dam surface, as illustrated in **Figures 10** and **11**. The penstock is a steel-lined concrete pipe, and structurally, both parts bear the load together as a whole.

6.07.2.2.1(iv) Structure of the spiral case and the surrounding concrete

The installed capacity of a single turbine generator unit at the TGHP is 700 MW. The unit bay of the power plant is 38.3 m long in the direction of the dam axis. The steel spiral case is of nose angle 345° and its maximum width in the horizontal plane is 34.3 m. The diameter of the inlet of the spiral case is 12.4 m. The central line of the spiral case is at an elevation of 57 m. During the operation, the maximum static water head is 118 m, while the design water head is 143 m.

Three types of techniques are adopted in placing surrounding concrete of the spiral case: placing with internal pressure held, placing with a cushion layer installed, and placing directly.

- *Placing with internal pressure held* is to keep a certain internal water pressure inside the spiral case when placing concrete around it in order to adjust the capacity of bearing internal water pressure between the spiral case and surrounding concrete. This technique is applied for 21 units among the 26 turbine generator units installed in the dam toe power plants, with the internal water pressure held at 70 m and water temperature controlled between 16 and 22 °C.
- *Placing with a cushion layer installed* is to pave a layer of 3 cm thick elastic material of elastic modulus 2.5 MPa on the surface of the spiral case first and then place surrounding concrete on this layer. The elastic layer will determine the proportion of internal water pressure borne by the spiral case and surrounding concrete. This technique is utilized for units Nos.17, 18, 25, and 26 in the right bank power plant.
- *Placing directly* is simply to place concrete directly on the spiral case without either raising internal pressure by filling water or laying a plastic cushion layer. The spiral case is a kind of structure of steel-lined reinforced concrete combined to bear the load. Such method is applied for turbine unit No. 15.

6.07.2.2.1(v) Hydroelectric turbine generation unit

The 14 Francis turbine generator units (as shown in **Figure 14**), each of 700 MW, in the right bank plant are purchased through international bidding, among which eight turbines are supplied by ALSTOM HYDRO, with the hydraulic design and modeling test fulfilled by KE of Norway, and the corresponding generators are supplied by ALSTOM ABB. The other six units are supplied by a joint venture consisting of VOITH, GE Canada, and SIEMENS, whose Chinese partners are Harbin Electric Machinery Company Limited (hereinafter as HEC) and Dongfang Electric Company Limited (hereinafter as DEC). Due to severe competition and the extreme importance of the turbine generation units of the TGP, each supplier pays great attention to the units and, in view of the characteristics of the TGHP, selects special hydraulic and structural optimization design for turbines and generators.

The main parameters of turbines and generators in the left bank power plant are shown in **Tables 1** and **2**, respectively.

Through international bidding, each of HEC, DEC, and ALSTOM supplies four sets of turbine generator units for the right bank power plant, which is a remarkable course of Chinese manufactory starting to design and manufacture gigantic 700 MW turbine generator units.

The main parameters of turbines and generators in the right bank power plant are as shown in **Table 3**.

6.07.2.2.2 Underground power plant

6.07.2.2.2(i) Objective and scale

The underground power station is added, taking advantage of the TGP, to fully utilize the water energy resources of the Yangtze River. Its main task is to provide additional installed capacity and generate more electricity, in conjunction with the 26 generator units in the dam toe power plants, to cope with the demand for energy supply due to the rapid development of the national economy and to relieve the increasing conflict on hump modulation of the power grid. It makes the TGP more beneficial as well.

The underground power station is located in the Baiyanjian Mountain on the right bank of the Yangtze River, where six turbine generator units, each of 700 MW, are installed. See the general layout of an underground power station in **Figure 15**.

Figure 14 Hoisting and installing a runner in a power plant.

Table 1 Main features of turbines in the left bank power plant

Item		Unit	VGS	ALSTOM
Type			Francis, vertical, and single runner	Francis, vertical, and single runner
Number		Set	6	8
Nominal diameter (outlet diameter) of runner		mm	9551.0	9800.0
Head	Maximum head	m	113.0	113.0
	Rated head	m	80.6	80.6
	Minimum head	m	61.0 in the initial stage 71.0 in the final stage	61.0 in the initial stage 71.0 in the final stage
Rated output		MW	710	710
Rated discharge		m^3 s^{-1}	995.6	991.8
Maximum output under continuous operation		MW	767.0	767.0
Maximum output corresponding generator coefficient cos Φ = 1		MW	852	852
Rated rotating speed		rpm	75	75
Specific speed		m kW	261.7	261.7
Specific speed coefficient			2349	2349
Draft head		m	−5	−5
Installation elevation		m	57.0	57.0
Direction of rotation			Clockwise, overlook	Clockwise, overlook

Table 2 Main features of generators in the left bank power plant

Item	Unit	ABB	VGS
Type		Vertical, semi-umbrella	Vertical, semi-umbrella
Cooling		Semi-water cooling	Semi-water cooling
Rated capacity/rated power	MVA/MW	777.8/700	777.8/700
Maximum capacity/maximum power	MVA/MW	840/756	840/756
Rated voltage	kV	20	20
Rated power factor		0.9	0.9
Power factor under maximum capacity		0.9	0.9
Rated frequency	Hz	50	50
Rated speed	rpm	75	75
Runaway speed	rpm	150	150
GD^2	t m^2	450 000	450 000
Rated efficiency	%	98.77	98.75

6.07.2.2.2(ii) Conduit system

The conduit system of the underground power plant includes a headrace, intakes, headrace tunnels, spiral cases and draft tubes, tailrace tunnels of varied roof height, damping shafts, draft-tube outlets, and a tailrace. See the longitudinal section of the conduit system in **Figure 16**.

The headrace is located at the right side of the curve bend near the exit of Maopingxi with the bottom elevation at 100 m. The intake tower contains six floors and is 77 m high, as shown in **Figure 17**. In the flow direction, the intake can be divided into four sections: a trash rack section, a bell mouth section, a gate section, and a tapered section. In the intake, a maintenance gate and a work gate are arranged.

The six headrace tunnels are arranged in parallel, with each tunnel leading to one turbine. And no surge chamber is placed along the tunnel. The distance between tunnel axes is 38.3 m. The axis length of a single tunnel is 244.64 m and the diameter of each tunnel is 13.5 m. The tunnel section downstream of the inclined straight section is steel lined.

A new type of tailrace tunnel with varied roof height is adopted for the underground power plant. The opening of each tailrace tunnel outlet is 13 m wide and 25 m high with a maintenance gate installed. The tailrace deck at the outlets of tailrace tunnels is perpendicular to the tunnel axes and at an elevation of 82 m with its foundation at an elevation of 44 m. The 510 m long tailrace is designed as an arc connected with a straight line. The minimum bottom width of the tailrace is 216 m and the base plate is at an elevation of 52 m.

6.07.2.2.2(iii) Layout of underground power plant

The underground power plant includes a main powerhouse, bus bar tunnels, access tunnels, aeration and vent-pipe tunnels, pipelines and access galleries, a 500 kV booster station on the ground, and the drainage system outside the plant.

The main powerhouse is 311.3 m long with a maximum height of 87.3 m, where the part for unit bays is 231.3 m long and an assembly bay, on the right side of the unit bays, is 80 m long. The turbines are installed at an elevation of 57 m. The turbine floor, generator floor, and the top of the bridge crane track are at elevations of 67, 75.5, and 90.5 m, respectively. The ordinary unit bay is 38.3 m long, of which the part on the left side of unit's central line is 20.9 m long, while the right part is 17.4 m long. The span of the main powerhouse is 32.6 m above the elevation of 88.3 and 31 m below that elevation.

The outgoing line system consists of a bus bar tunnel, bus bar gallery, and bus bar shaft. Each bus bar shaft, in a rectangular cross section, serves two generators. In each shaft, there are two circuits of three-phase large current bus bar, an access stair, a cable shaft, and a positive pressure air shaft. In each of bus bar shafts No. 2 and No. 3, an elevator is installed as well. The bottom of all the shafts is connected with a bus bar gallery at an elevation of 67 m, while the top leads to the booster station at an elevation of 150 m.

The 500 kV booster station is placed on a platform at the top of a mountain 150 m downstream of the underground powerhouse, and is parallel to the longitudinal axis of the main powerhouse. It consists mainly of a bus bar shaft outlet, an auxiliary management building, a GIS room, a main transformer room, a reactor, a ring road, and an unload field. Three takeoff towers are arranged in the downstream of the platform.

6.07.2.3 Navigation Structures

6.07.2.3.1 Dual-way five-step ship lock

6.07.2.3.1(i) Objective and scale

After the completion of the TGP, the navigation condition of the 660 km long waterway from Yichang to Chongqing in the upstream will be greatly upgraded. It will allow the passage of 10 000 tons fleet to Chongqing directly for more than 6 months in a year. The annual single-way traffic capacity of the upstream navigation channel can be raised to over 50 million tons. In the downstream river

Table 3 Main Features of Turbine and Generator of the right bank power plant

	Item	Unit	Nos. 15–18 supplied by DEC	Nos. 19–22 supplied by ALSTOM	Nos. 23–26 supplied by HEC
Turbine	Type		Francis	Francis	Francis
	Number	Set	4	4	4
	Maximum and rated head	m	113.0/85.0	113.0/85.0	113.0/85.0
	Minimum head	m	61.0 in the initial stage 71.0 in the final stage	61.0 in the initial stage 71.0 in the final stage	61.0 in the initial stage 71.0 in the final stage
	Rater discharge	$m^3 s^{-1}$	966	966	966
	Rated output	MW	710	710	710
	Maximum output under continuous operation	MW	767.0	767.0	767.0
	Maximum output corresponding generator coefficient $cos\Phi = 1$	MW	852	852	852
	Rated rotating speed	rpm	75	71.4	75
	Specific speed	M kW	244.9	233.1	244.9
	Specific speed coefficient		2258	2149	2258
	Draft head	m	−5	−5	−5
	Installation elevation	m	57.0	57.0	57.0
	Direction of rotation		Clockwise, overlook	Clockwise, overlook	Clockwise, overlook
Generator	Type		Vertical, semi-umbrella	Vertical, semi-umbrella	Vertical, semi-umbrella
	Cooling		Semi-water cooling	Semi-water cooling	Air cooling
	Rated capacity/Rated power	MVA	777.8/700	777.8/700	777.8/700
	Maximum capacity/Maximum power	MW	840/756	840/756	840/756
	Rated voltage	kV	20	20	20
	Rated power factor		0.9	0.9	0.9
	Rated frequency	Hz	50	50	50
	Rated speed	rpm	75	71.4	75
	Runaway speed	rpm	150	142.8	150
	GD^2	$t m^2$	450 000	450 000	450 000
	Rated efficiency	%	98.75	98.83	98.73

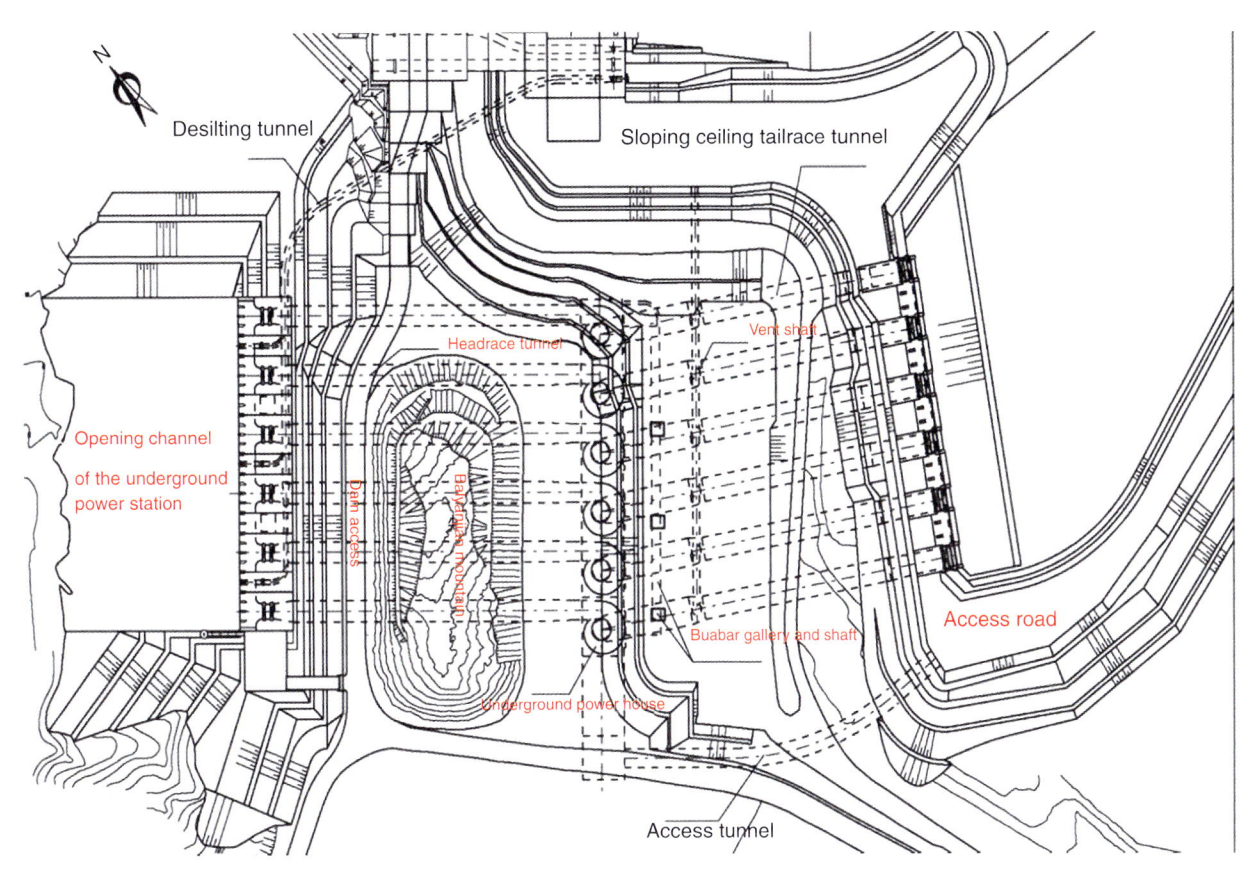

Figure 15 General layout of an underground power plant.

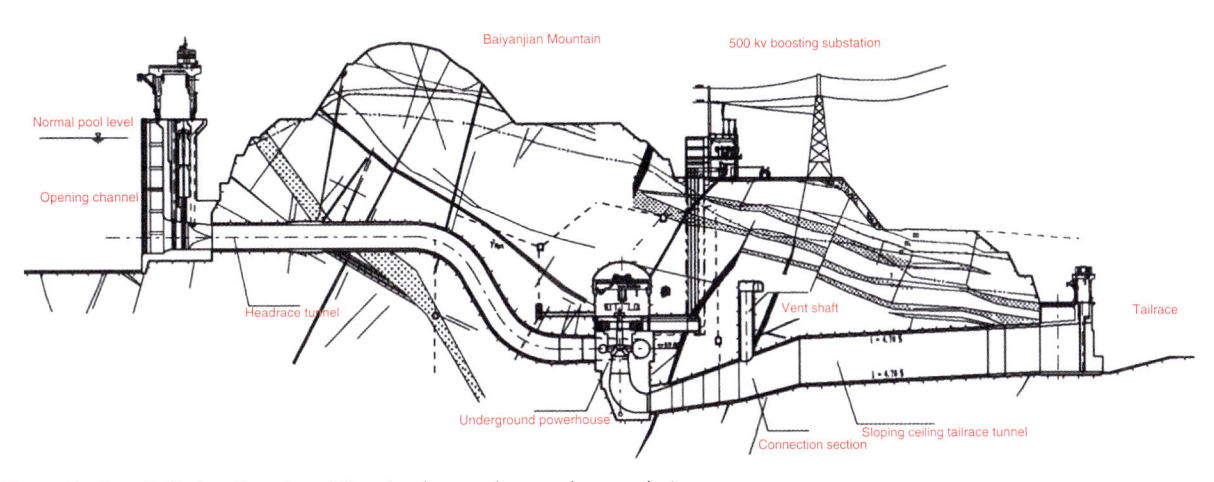

Figure 16 Longitudinal section of conduit system in an underground power plant.

section, the discharge in the dry season can be increased by flow regulation from the current 3000 to over 5000 m^3 s^{-1}, improving the navigation condition considerably.

As required by the development of navigation on the Yangtze River, a dual-way and five-step ship lock is constructed mainly for cargo passing through [3, 4]. A vertical ship lift is constructed as a quick pass mainly for passenger ships and various special engineering ships. By such a way of distributing various ships, the capacity of the ship lock in terms of cargo passing through can be fully utilized. In addition, during the scheduled maintenance period of the ship lock, the ship lift can function as an effective supplement to the traffic capacity of the TGP. The maximum water head of both ship lock and ship lift is 113 m, and the maximum water head between steps of the ship lock is 45.2 m.

Figure 17 Intake of underground power plant.

6.07.2.3.1(ii) Route selection of ship lock

The continuous five-step ship lock is placed on the convex left bank to meet the requirements of the length of the ship lock and the straight section of the approach channel, the bend radius of the navigation channel, and the flow condition at the area of the entrance/exit of the approach channel. The straight length of the main body section is determined by the number of steps of the ship lock and the effective length of each chamber. The length of straight section of the upstream and downstream navigation channels connecting to the ship lock is designed as 3.5 times the maximum fleet length. The route direction is then adjusted through the bend section arranged in the upstream and downstream of straight section, to ensure that the flow condition at the area near the conjunction of the approach channel with the main river course can satisfy the navigation requirement.

The separation levee, constructed at the same time as the navigation work, is arranged in such a way as to contain all the dual-way five-step ship lock, the ship lift, and the temporary ship lock.

6.07.2.3.1(iii) Layout of ship lock and navigation channel

Figure 18 provides the illustration of the dual-way five-step ship lock, while **Figure 19** shows the scene with ships passing the lock. The main body section of the ship lock is 1621 m long. A 2670 m long upstream separation levee and a 3700 m long downstream separation levee are arranged on the right side of the approach channel. The upstream approach channel is 2113 m long while the downstream approach channel is 2708 m long, and both are shared by the ship lift. The route length of the ship lock amounts to 6442 m in total.

The ship lock is placed in a deep excavated channel through the ridge on the left bank. A 57 m wide rock mass is left as central pier between the dual routes of the ship lock. Each route of ship lock contains six chamber heads of Nos.1–6 and five chambers of Nos.1–5. The effective dimension of each chamber is 280 m long and 34 m wide with the minimum water depth on the sill being 5 m.

The approach channel of the ship lock is designed with the following standards: the length of the straight section connecting to the lock is 930 m; the bend radius is not less than 1000 m; the bottom width is not less than 180 m; and the minimum water depth in the upstream channel is 6 m while that in the downstream is 5.5 m.

Figure 18 Sketch of dual-way five-step ship lock.

Figure 19 Navigation through the ship lock.

The minimum width of the approach channel is determined as the width of three designing fleets, in which two fleets are marching and one is berthed, plus three intervals, each being as wide as the designing fleet. A redundant space is taken into consideration in order to eliminate the effect of dredging henceforth in the approach channel on navigation.

6.07.2.3.1(iv) Conduit system

The ship lock of the TGP is a multistep ship lock with the total water head reaching 113 m. The determination of water head for each step should be based on the principles of minimum technical difficulty, cost effectiveness, saving the amount of engineering work, and convenient operation and management of ship lock. The total water head is therefore divided into five steps in such a way that during the water filling in or discharging from the ship lock only supplement water for the second chamber is needed for a short period of time and no overflow will occur at any time. The ship lock may operate in different combinations, based on various water levels, such as five steps, four steps, or three steps, as long as it is consistent basically with the characteristics of neither supplement water nor overflow.

Through the conduit system (**Figure 20**) of the ship lock, water is dispersedly drawn from the upstream and then passing through the main conduit gallery flows into the chamber of each step by gravity. The water in the chamber is mainly discharged to the Yangtze River through two downstream draw-out culverts across the separation levee.

The main conduit gallery is arranged symmetrically on both sides of each route in the shape of 'a rectangular plus a semi-cycle' with the cross section being 4.2 m wide and 4.5 m high. The main conduit gallery is then, through three-dimensional (3D) diversion outlets, connected to the discharging subgallery of the eight subtubes distributed in four sections in the chamber.

6.07.2.3.1(v) Chamber structure and slope

In accordance with the stability of the excavated rock mass as well as the geological condition of the foundation, the chamber is concrete lined with a minimum thickness of 1.5 m. Drain pipes arranged in grids of dimension 4 m wide and 6 m high and high-strength steel anchors of length from 6 to 12 m are placed in the lined chamber wall and the rock mass. See the cross section of the shiplock chamber in **Figure 21**.

After the excavation of the main body section of the ship lock, the man-made high and steep rock slopes are formed on both sides of the ship lock with height varying from 100 to 160 m, with the maximum height being 170 m.

The high slope is characterized with its enormous scale, complex geometrical conformation, deep cut and excavation, and deformation due to the release of crustal stress. Since the slope, as an important part of ship lock, is required to work in combination with the lock head and the chamber, its operational requirement is quite different from that of general slope. Not only the stability and safety of the rock mass during the construction and operation, but also the deformation control of slope rock mass and the

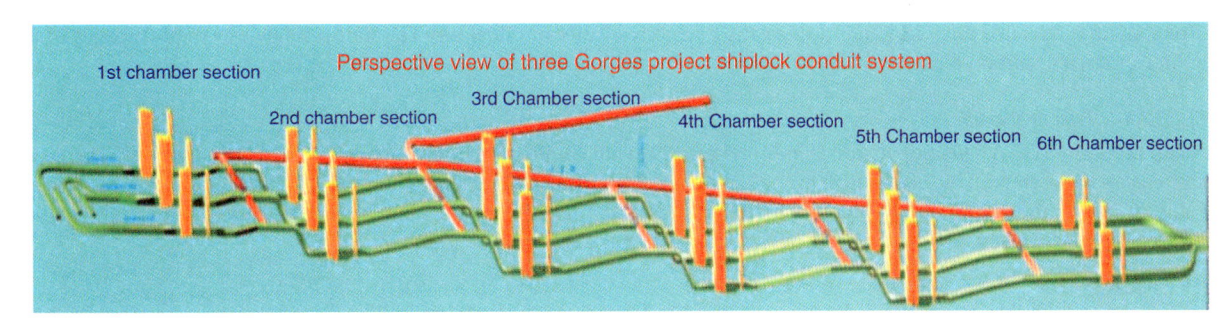

Figure 20 Sketch of the conduit system of the ship lock.

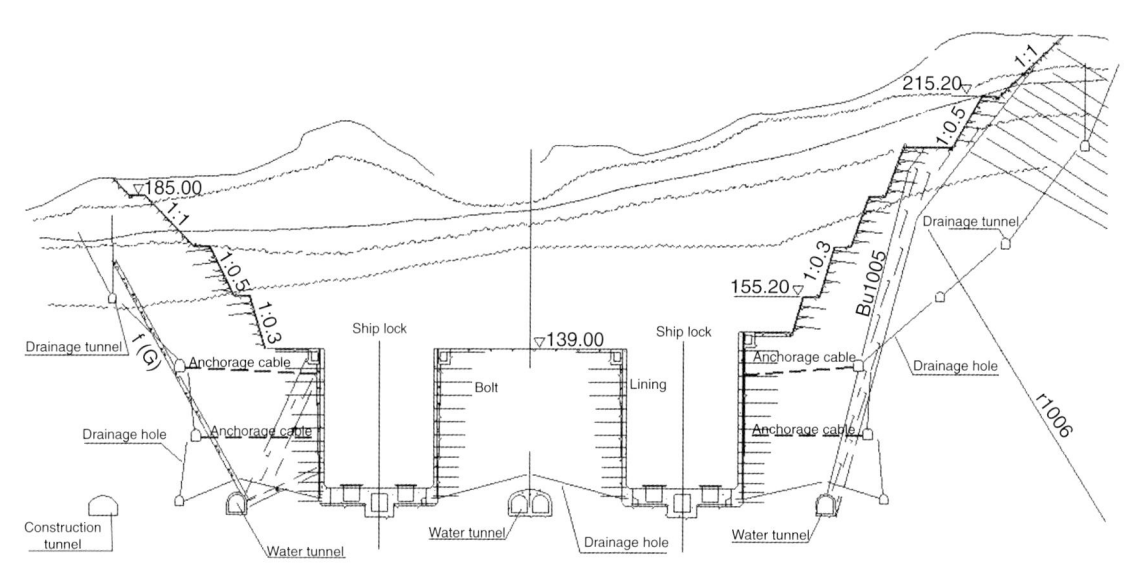

Figure 21 Drainage and anchor support of high-slope excavation.

coordinate work of rock mass with lined structure have to be taken into consideration to ensure the normal operation of the chamber and its gate.

6.07.2.3.1(vi) Layout of hydraulic steel structure

Each lock head is equipped with a miter gate and the corresponding hydraulic hoist. In addition, a maintenance gate and its bridge crane are also installed on lock head No. 1, while a floating maintenance gate is arranged on lock head No. 6. The miter gate is of height varying from 37.5 to 38.5 m and is on the average 20.2 m wide. The maximum inundated depth of the miter gate on lock head No. 1 can reach 35 m. The miter gate of the ship lock is opened and closed by a horizontal hydraulic hoist.

In the filling and discharging valve shaft near the lock head of each step, a reverse radial gate and its hydraulic hoist are arranged in the conduit system between the chamber and the upstream or downstream approach channel, as well as in the conduit system between every two chambers. In the upstream and downstream of the radial gate, a plain maintenance gate is installed. The opening of the reverse radial gate is varied from 4.5 m wide and 5.5 m high to 4.2 m wide and 4.5 m high. The radial gate is operated via vertical hydraulic hoist.

The two sets of hydraulic hoists for the miter gate and the reverse radial gate share one hydraulic system.

6.07.2.3.2 Vertical ship lift

6.07.2.3.2(i) Objective and scale

As a part of navigation structure, the ship lift is a quick pass mainly for passenger ships and other special ships, which can facilitate the full utilization of the dual-way five-step ship lock for cargo passing through.

The ship lift is designed mainly for the purpose of allowing 3000-tonnage large passenger ships sailing between Shanghai and Chongqing or single 3000-tonnage cargo barges passing through the dam. The total weight of the ship container with the water included is around 16 800 tons. The maximum lifting height of the ship lift is 113 m, and the normal lifting speed is 0.2 m s^{-1}. The variation range of water level in the upstream navigation channel is 30 m, while that in the downstream channel is 11.8 m.

The ship lift is located on the left bank of the riverbed and about 1 km away from the dual-way five-step ship lock to the left. The ship lift consists of an upstream lock head, a downstream lock head, a chamber for ship container, an upstream approach channel, a downstream approach channel, an upstream guiding wall, a downstream guiding wall, and a mooring structure. The upstream and downstream approach channel of the ship lift are 6400 m long in total, most of which is shared with the ship lock.

The ship lift is operated every 25 min in a single trip and every 47 min in a dual trip. The annual single-way traffic capacity of the ship lift can reach 4 million tons.

6.07.2.3.2(ii) Type of ship lift

The ship lift of the TGP is a kind of vertical ship lift (**Figure 22**) with counterweight driven by pinion and toothed rack. The major characteristics of the ship lift include its high lifting height, the great uplift volume, the pinion-toothed rack climbing system driving the ship container, the safety mechanism of rotary locking rod and nut post, and the main bearing structure of thick tube-shaped reinforced concrete. Herein, the ship lift is the largest one in terms of scale and technical difficulty.

6.07.2.3.2(iii) Civil works

The civil works of the ship lift mainly include the upstream lock head, the downstream lock head, and the ship container chamber section.

Figure 22 Sketch of the ship lift of the Three Gorges Project.

The upstream lock head is a massive whole structure, which is 130 m long and 62 m wide with its top elevation at 185 m. The central navigation channel in the lock head is 18 m wide with its bottom sill at an elevation of 141 m. In the upstream lock head are arranged a groove for the upstream bulkhead gate, a roadway bridge to the dam crest, a maintenance gate, a service gate at the downstream, and the hoist corresponding to each gate.

The downstream lock head is also a massive whole structure with a length of 37.15 m and a width of 62.9 m. The navigation channel is 18 m wide with its bottom sill at an elevation of 58 m. The left side abutment is 25 m wide while the right one is 19.9 m wide, and their crest is at an elevation of 84 m. The upstream end of the lock head is equipped with a lifting plain service gate and the downstream with a maintenance gate.

The chamber section mainly consists of a bearing structure, a ship container, counterweights, a driving system, a safety mechanism, and electric equipment.

The bearing structure of the ship lift includes four reinforced concrete tube-shaped towers, each measuring 40.3 m high and 16 m wide. Its top elevation is 196 m and the foundation elevation is 48 m. The tower wall is generally 1 m thick.

The pinion for climbing system and the nut post of safety mechanism are installed at the inside of each tower, while stairs and elevators for access and evacuation are placed on the outside of the tower. Meanwhile, a counterweight shaft, excavation stairs, and an elevator are arranged at the inside of each tower as well. A machine room is placed at the top of each tower, which is equipped with a roller for counterweights and a bridge crane for maintenance. The central control room of the ship lift is arranged on the top of the downstream side of the tower.

6.07.2.3.2(iv) Equipment

The upper part of the service gate for the upstream lock head is a 17 m high combined plain gate formed by a large working gate, with a U-shaped opening at its top for navigation, and a small tumble gate, which controls the opening for navigation. The lower part consists of seven 3.75 m high stop-logs. The maximum working head of the service gate is 15 m. The tumble gate is a structure of varied cross section with the supporting span being 18.6 m. The two articulated gate shoes (hinges) between its bottom and the working gate are a kind of spherical sliding bearing. The service gate is operated by a 2×250 kN single-way bridge crane.

The 41.5 m high maintenance gate of the upstream lock head consists of a plain gate as the upper part and eight stop-logs in the bottom. The maintenance gate is operated by a 2×250 kN single-way bridge crane with a maximum lifting height of around 70 m.

The service gate of the downstream head lock, structured with double skin plates and multigirders, is a sunken plain gate with a small tumble gate. The service gate is supported by carriages with a span of 27 m. The opening of the tumble gate is 6.7 m high with a net span of 18 m. The service gate is operated by 2×700 kN hydraulic hoists.

The ship container is a trough type steel construction with an effective dimension of $120 \times 18 \times 3.5$ m (length × width × height), an external dimension of $132 \times 23.4 \times 10$ m, and a free board of 0.8 m. The weight of the ship container with water filled amounts to 16 800 tons. On the top of the main longitudinal girders on both sides, there are 256 steel ropes, which are connected to 16 groups of counterweights through the rope pulley installed on the top of the bearing structure. A balancing chain is suspended at the bottom of each group of counterweights.

The ship container is structured by welding the main longitudinal beams and multitransverse girders with lateral and floor plates. The main longitudinal beams are located on both sides of the container, and at the same height as that of the central structure of the container. The height of the transverse girder equals the thickness of the container floor. Both sides of the main longitudinal girders and the four extending steel cantilevers are equipped with a locking system, ship container driving equipment, and safety mechanism. At the bottom of the ship container, the synchronous shaft of driving electrical motor is installed, whereas the hydraulic hoist and reversible pump are installed at the top of the ship container.

6.07.2.4 Maopingxi Dam

6.07.2.4.1 Objective and scale

Maopingxi River is a small tributary of the Yangtze River with its estuary at the right bank of the Yangtze River, 1 km upstream of the dam axis of the TGP. Within the inundated region, there are three relatively large flat lands which are rare in the Three Gorges reservoir area. Taking into account factors such as less land shared by large population, lack of cultivated land compared with many slopes, and many difficulties in resettlement, it is decided to construct Maopingxi protective works, which is included into the Three Gorges Complex Project.

The same design standard as that for the TGP is adopted for Maopingxi protective dam with the normal pool level at 175 m and the check flood level at 180.4 m. The dam is designed to withstand earthquake intensity of degree VII.

A rockfill dam (**Figure 23**) is built for this protective work, using the earth and rock excavated in the construction of the opening diversion channel nearby in the right bank. In selecting the type of seepage control for the dam body, concrete cutoff wall, clay core, and asphalt concrete core are studied carefully. Due to the poor stress condition and the difficulty to coordinate the construction of the rigid core wall with the rock filling of the dam body, and because of the difficulty to excavate clay near the dam body, the asphalt concrete core is finally adopted for the rockfill dam. The maximum dam height is determined as 105 m.

6.07.2.4.2 Rockfill dam

The dam site is located in an open valley with the river being about 60 m wide. The axis of the dam, the rockfill dam with an asphalt concrete core, is 1070 m long. The core is 0.5 m thick at the top, and increases to 1.2 m thick at an elevation of 94 m, with the slope ratio on both sides being 1:0.004, as shown in **Figure 24**.

There is a transition layer between the asphalt concrete core and the dam shell. The transition layer on the upstream side of the core wall is 2 m thick while on the downstream side it is 3 m thick. The transition layer is formed by sand and gravel aggregate of good grading and with the maximum grain size not bigger than 6–8 times the maximum grain size of the asphalt concrete aggregate. The deformation capability of the transition layer is between that of the asphalt concrete core and that of the dam shell, resulting in compatible deformation and even load transferring between the core and the transition layer. The coefficient of principal stress ratio λ tends to be consistent, and within the permitted range of shear strength of the dam material. The transition layer in the upstream side may also provide grouting area of good grading for the potential seepage treatment in the future.

6.07.3 Project Construction

6.07.3.1 Demonstration and Construction

6.07.3.1.1 Full demonstration and cautious decision

The ambition to build the TGP can be traced back to Mr. Sun Yat-Sen, the great pioneer of Chinese Democratic Revolution, who proposed in his book *Strategy for State, Part II: Industrial Plans* in 1918. Then during the 1930s–1940s, preliminary planning and reconnaissance was fulfilled under the guidance of Dr. John Lucian Savage, a famous American expert in dam construction. After the foundation of new China, with the aim of resolving the flood problem in the middle and downstream of the Yangtze River thoroughly, the Changjiang Water Resources Commission (i.e., the Yangtze River Commission) developed lots of study and demonstration, and proposed various design schemes. In order to ensure the decision on the construction of the TGP being more scientific, more

Figure 23 Maopingxi protective dam.

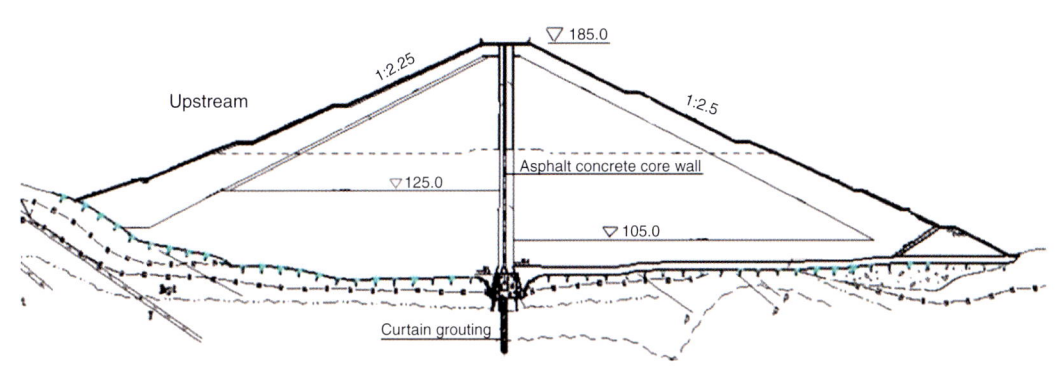

Figure 24 Typical cross section of Maopingxi protective dam.

democratic, more particular, and more accurate, the State Council required, in June 1986, the former Ministry of Water Conservancy and Electricity to organize experts of various disciplines from the whole country to put forward the feasibility study report of the TGP on the basis of extensive comments and deep argumentation. In 1989, the Changjiang Water Resources Commission completed a feasibility study report and 14 subject reports of the TGP with the normal storage level of 175 m. The report was approved by the Three Gorges Project Examination Committee of the State Council in August 1991 and was then approved on the executive meeting of the State Council in September 1991. In March 1992, the fifth meeting of the Seventh National People's Congress adopted the Resolution to Construction of the Three Gorges Project with an affirmative ballot, which indicated that the whole feasibility study and argumentation concluded as well as the project application and approval. So far, the TGP has been the only construction project in China approved by the National People's Congress.

6.07.3.1.2 Milestones of construction

On 29 July 1993, the Report of the Preliminary Design of the TGP forwarded by Changjiang Water Resource Commission was approved by the State Council Three Gorges Project Construction Committee (hereinafter as TGPCC), indicating that the project entered into the overall preparation stage for construction.

On 20 January 1994, the construction of the major structure started, which marked the first-stage construction of the major structure; the TGP was officially started on 14 December 1994.

On 8 November 1997, the river close-off succeeded, which indicated the commencement of the second-stage construction of the major structure.

On 6 November 2002, the close-off of the open diversion channel succeeded while the third-stage construction of the major structure commenced.

In June 2003, with the roller-compacted concrete (RCC) cofferdam retaining water, the reservoir began its storage, raising the water level to 135 m; the dual-way five-step ship lock was put into trial operation; the first turbine generator unit was connected to the power grid and commissioned successfully, which was put into commercial operation in July.

On 20 May 2006, the dam reached the design level at 185 m; then the reservoir started to retain water to level 156 m for the initial operation in September.

On 28 September 2008, the reservoir started to retain water at the normal storage level of 175 m as experimental storage; then the last turbine generator unit (No. 15) in the right bank power plant was officially connected to the power grid and generated power in October.

On 15 September 2009, the reservoir started to operate according to the scheme of normal storage level of 175 m.

6.07.3.2 Construction by Stages

6.07.3.2.1 Staged construction

According to the preliminary design of the TGP, it is planned to construct the project in three stages with the total construction period of 17 years, of which preparation and the first stage are 5 years and the second and the third stages are each 6 years. The project is to commence with its preparation works in January 1993, and to be completed at the end of December 2009.

6.07.3.2.1(i) First-stage works

The first-stage project, as illustrated in **Figure 25**, includes the construction of the first-stage cofferdam formed of rock and earth (**Figure 26**); the concrete placement for the longitudinal concrete cofferdam, as well as for the part below an elevation of 50 m of the upstream RCC cofferdam used in the third stage; the excavation of an open diversion channel and its slope protection and bank protection; the construction of Maopingxi protective dam and water discharging structure; the construction of the temporary ship lock in the left bank; the excavation of the ship lift (postponed afterward), the ship lock, the non-overflow dam section in the left bank, a part of the left bank power plant and the left bank power plant dam section; removing the first-stage cofferdam; and the river close-off. During the construction, works fulfilled are earth and rock excavation of 96.83 million m^3, earth and rock filling of 23.28 million m^3, concrete placing of 3.65 million m^3, and hydraulic steel structures installation of 800 tons without including turbine generator units.

Figure 25 Layout of first-stage flow diversion.

Figure 26 First-stage cofferdam and open diversion channel.

6.07.3.2.1(ii) Second-stage works

The works during the second stage, as illustrated in **Figure 27**, include the construction of the upstream and downstream cofferdams used in the second stage, shown in **Figures 28** and **29**; the concrete placement for the dam section on the longitudinal cofferdam, the spillway dam section, the left bank power plant and non-overflow dam section, the prefabricated part of the intake of the underground power plant in the right bank, as well as relevant metal structure installation; the construction of the ship lift (postponed afterward), the ship lock and its approach channels; the installation of the turbine generator units in the left bank power plant; removing the second-stage upstream and downstream cofferdams; the close-off of the open diversion channel; the construction of cofferdams in the upstream and downstream of the open diversion channel, and the completion of the third-stage RCC cofferdam; and blocking of the temporary shiplock dam section. During the construction, works to be done are earth and rock excavation of 46.95 million m^3, earth and rock filling of 28.76 million m^3, metal structures installation of 157.8 thousand tons, and installation of six turbine generator units.

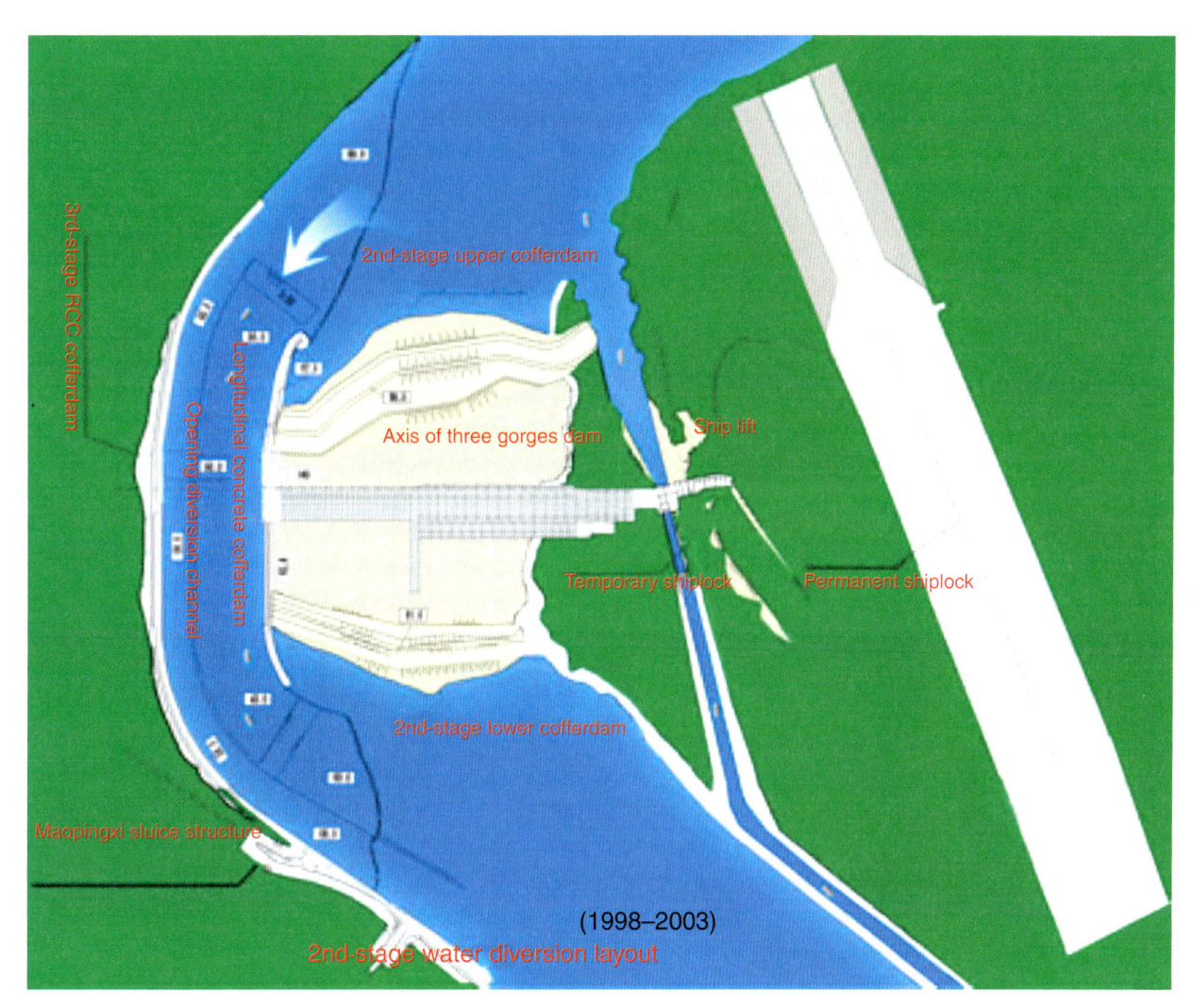

Figure 27 Layout of second-stage flow diversion.

Figure 28 Second-stage river closure.

6.07.3.2.1(iii) Third-stage works
The works in this stage, as illustrated in **Figure 30**, include the concrete placement and metal structure installation on the right bank power plant dam section and non-overflow dam section; the concrete placement of the right bank power plant, as shown in **Figures 31** and **32**; the installation of turbine generator units on both left bank and right bank power plants; the construction

Figure 29 Second-stage construction pit.

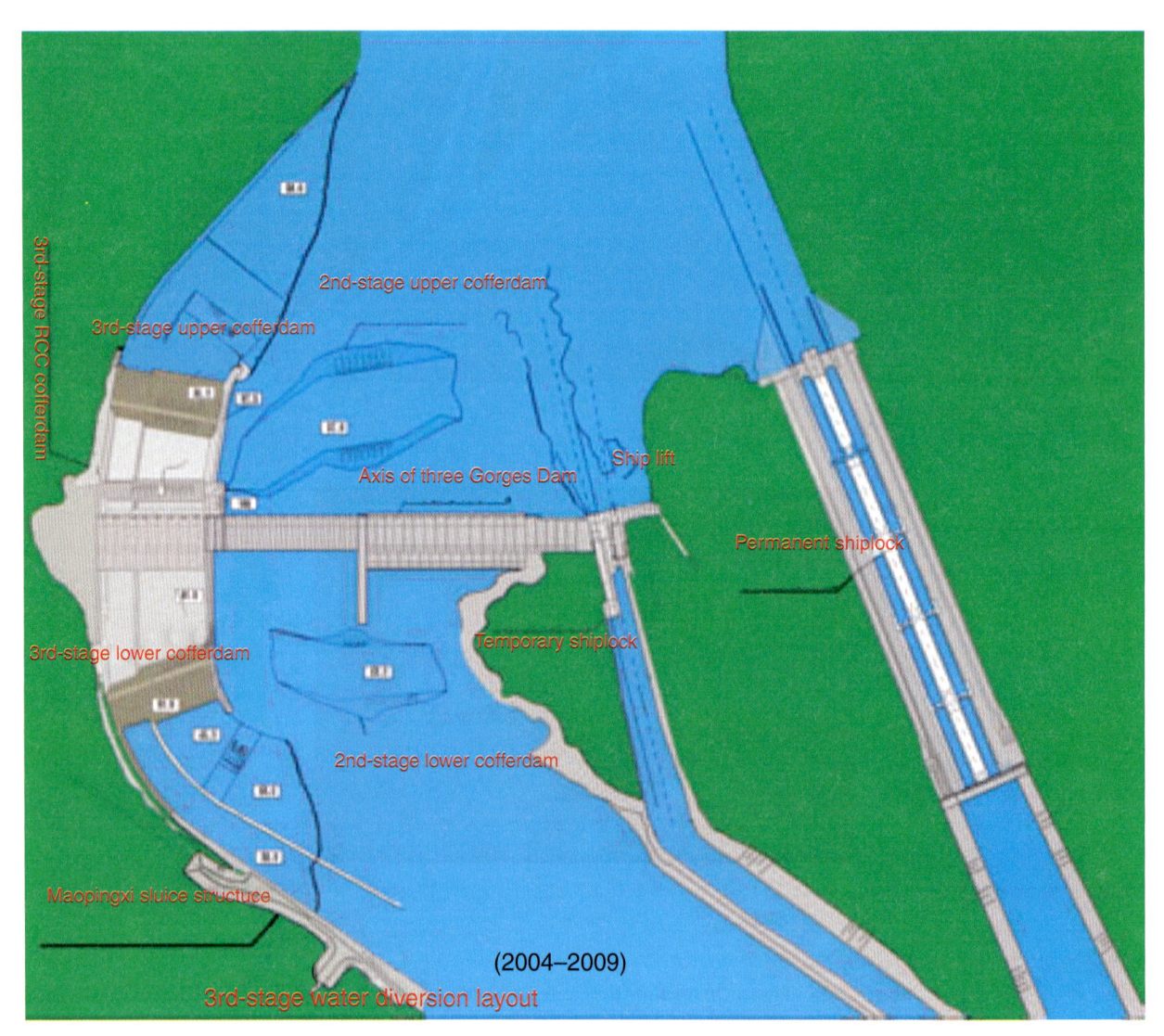

Figure 30 Layout of third-stage flow diversion.

Figure 31 Third-stage construction pit.

Figure 32 Overall perspective of third-stage Three Gorges Project (photo by Zhu Bincheng).

of the service power station; the conversion of the temporary ship lock to the sand sluice gate; the completion of the ship lock; removing the upstream RCC cofferdam and the downstream rock-earth cofferdam used in the third stage; and blocking the deep bottom outlets on the spillway dam section. During the construction, works to be fulfilled are earth and rock excavation of 4.38 million m^3, earth and rock filling of 733.9 thousand m^3, concrete placement of 5.50 million m^3, metal structure installation of 100.1 thousand tons without including turbine generator units, and installation of 20 turbine generation units.

6.07.3.2.2 *Water diversion during construction*
6.07.3.2.2(i) Diversion scheme and navigation during the construction

The river diversion scheme is carefully studied in accordance with the navigation along the Yangtze River, as an important artery of water transportation in China, which must not be closed during construction of the TGP. The construction adopts finally a method of 'diverting water in three stages and navigation through an open diversion channel'. In the first stage, a 350 m wide open diversion channel is built along the right bank while the flood discharge and navigation are ensured still through the main river channel. In the second stage, after the main river channel is obstructed, the left bank dam and the dual-way five-step ship lock are built with flood being discharged through the open diversion channel and ships passing through both open diversion channel and the temporary ship lock on the left bank. In the third stage, the open diversion channel is blocked while the third-stage upstream RCC cofferdam and the power plant dam section in the right bank are built. With the utilization of the RCC cofferdam to retain water, the project's objective of water storage, power generation, and navigation is realized preliminarily. The flow from the upstream of the Yangtze River is released through the spillway dam section completed in the second stage, and the ships pass through the dual-way five-step ship lock. In the later period of construction, after the third-stage RCC cofferdam is removed to an elevation of 110 m and the bottom outlets for water diversion are blocked, the whole dam is used to retain water.

6.07.3.2.2(ii) River closure

A program is adopted to block the main river channel by using a single levee of the upstream cofferdam. As a part of the upstream cofferdam, the levee is set in the downstream of the cofferdam, as seen in **Figure 28**. During river close-off, the design discharge is 14 000–19 400 m^3 s^{-1} (the observed discharge was actually 8480 m^3 s^{-1}), the fall between upstream level and downstream level is 0.08–1.24 m (the observed fall was actually 0.32 m), and the average velocity at the closure gap is 3.3–4.2 m s^{-1} (the observed velocity was 2.6 m s^{-1}).

6.07.3.2.2(iii) Power generation with the third-stage cofferdam

After the RCC cofferdam is built in the third stage, the bottom outlets for water diversion and the deep outlets for flood discharge in the left bank dam are locked. The cofferdam combined with the left bank dam retains water, as shown in **Figure 33**, the first turbine generator units start to generate power, and the dual-way five-step ship lock is put into operation.

6.07.3.2.3 Major temporary structures

6.07.3.2.3(i) Open diversion channel

The open diversion channel is located along the right bank at the dam site. The selected axis of the channel is 3407 m long with the minimum width at the bottom being 350 m. A compound cross section is adopted with the elevations of the bottom and berme at 45, 50, 53, and 58 m, respectively.

6.07.3.2.3(ii) First-stage earth and rock cofferdam

The first-stage earth and rock cofferdam is located in the right side of the main river channel and has an axis length of 2502 m. The area of the construction pit protected by the cofferdam reaches 750 000 m^2. The cofferdam is mainly filled with weathered sand, stone ballast, and block stone. The plastic cutoff wall connected with geomembrane above and curtain grouting on the foundation rock of strong permeable layer is adopted for seepage control.

6.07.3.2.3(iii) Longitudinal concrete cofferdam

The longitudinal concrete cofferdam is applied for the second- and third-stage water diversion, which is located on the right side of the Zhongbao Island, a small island originally in the Yangtze River channel at the dam axis, as shown in **Figure 3**. The cofferdam is divided into three sections: an upper section, a body section, and a down section. The axis of the cofferdam is 1146.4 m long in total, of which the upper section is 462.2 m long, the body section is 115 m long, and the down section is 569.2 m long. The cofferdam is a kind of RCC gravity dam with a crest elevation of 87.5 m and a top width of 8 m.

6.07.3.2.3(iv) Second-stage earth and rock cofferdam

The second-stage upstream and downstream earth and rock cofferdams are of crest elevation 88.5 and 81.5 m, respectively, of top width 15 and 10 m, respectively, and of axis length 1439.6 and 1075.9 m, respectively. The cofferdams are mainly made of weathered sand, stone ballast, ballast mixture, block stone, and transition material. The plastic concrete cutoff wall connected with geomembrane above and curtain grouting on the foundation rock of strong permeable layer is adopted for seepage control.

Figure 33 Power generation after third-stage cofferdam being constructed.

6.07.3.2.3(v) Third-stage RCC cofferdam

The third-stage RCC cofferdam is arranged in parallel with the dam of the TGP. The axis of the cofferdam is located 114 m upstream of the dam axis, and its total length is 580 m. The cofferdam is a gravity dam with crest elevation 140 m, maximum dam height 115 m, top width 8 m, and maximum bottom width 107 m. The cofferdam is to be removed in a scheme of "overturning explosion for the 380 m long middle section of the coffer-dam by use of pre-embedding explosive chambers combined with deep blast holes explosion at both ends." During removal, the water level in front of the cofferdam is 135 m, while the water level behind is 139 m.

6.07.3.2.3(vi) Third-stage earth and rock cofferdam

The third-stage upstream and downstream earth and rock cofferdams are of crest elevation 83 and 81.5 m, respectively, of top width 15 and 10 m, respectively, and of axis length 441.3 and 447.4 m, respectively. The cofferdam is mainly made of weathered sand, stone ballast, ballast mixture, block stone, and transition material. The self-solidifying mortar and high-pressure rotary shotcreting grouting cutoff wall connected with geomembrane above and curtain grouting on the foundation rock of strong permeable layer are adopted for seepage control.

6.07.3.2.3(vii) Temporary ship lock

The temporary ship lock, as shown in **Figure 34**, is set on the left bank and designed as a one-step ship lock for navigation purpose during the construction, to be put into operation in May 1998. The effective size of the lock chamber is $240 \times 24 \times 4.0$ m (length × width × minimum water depth on the sill). The design navigation discharge is $45\,000\,\text{m}^3\,\text{s}^{-1}$, and the maximum design water head is 6 m. According to the calculation based on the scale of typical ship fleet, the annual navigation capacity of the temporary ship lock is 11.6 million tons with 156 ships passing through daily in a single trip.

6.07.3.2.3(viii) Bottom water diversion outlets

There are 22 bottom water diversion outlets in total. Two outlets were sealed for productive experiment in order to accumulate construction experiences for closure of all the other bottom water diversion outlets when 13 turbine generator units in the left bank were put into operation for power generation before the flood season of 2005. The actual construction period to close all bottom water diversion outlets lasted 4.5 months. The six out of the 20 bottom outlets were closed before the flood season of 2006 and the remaining 14 outlets were closed before the flood season of 2007.

Figure 34 Navigation through temporary ship lock.

6.07.3.3 Construction Management

6.07.3.3.1 System, mechanism, and relevant regulations of management

6.07.3.3.1(i) Management system

In January 1993, the TGPCC was set up by the State Council, as the top level decision-making institution for the TGP with responsibilities of organizing and coordinating various stakeholders and making decision on any principle, policy, and key issues related to the construction of the TGP. Since then the Premier has also been the General Director of the TGPCC. Other committee members are principal officers/leaders from relevant governmental departments directly under the State Council, the China Three Gorges Project Corporation, the State Grid Corporation, the Changjiang Water Resources Commission, Hubei Provincial Government, and Chongqing Municipal Government. Under the TGPCC, an executive office is set up, mainly being responsible for carrying out key decisions made by the TGPCC and putting them into effect, as well as doing its duty of organization and coordination, guaranteeing governmental functions such as service, supervision, and inspection (including check, quality examination, and project assessment and acceptance), and implementing the daily administrative work of the TGPCC.

The entity of the TGP is the China Three Gorges Project Corporation, founded in September 1993 and renamed as China Three Gorges Corporation (with abbreviation CTGPC unchanged) on 27 September 2009. The CTGPC takes charge of responsibilities such as organization and implementation of construction of the Three Gorges Complex Project; financing, expenditure, and debt refunding; and running and operating the complex project after its completion. In addition, the CTGPC is also responsible for resettlement financing and credit repayment, although the implementation of resettlement is under the charge of the Migrant Relocation Development Bureau, affiliated to the TGPCC.

The design integration of the TGP is undertaken by the Changjiang Water Resources Commission, including the planning and design of the hydropower complex project and the resettlement.

All construction companies and manufacturers, more than 20 in total, are optimally selected through open bidding, such as the TGP Headquarters of the Gezhouba Group Corporation, the Hubei Yichang No. 378 Associated Corporation for the TGP Construction, the Qingyun Hydraulic and Hydropower Associated Company, the TGP Headquarters of Hydropower Corps of the China Armed Police Force, and the Yichang Three Gorges Three Jointly-Operated Corporation.

Qualified professional institutions are engaged to supervise and manage the processes of construction, equipment manufacture, and equipment installation in terms of construction quality, construction schedule, construction cost, and safety. The main institutions involved in construction supervision and equipment manufacture include but not limit to the Changjiang Water Resources Commission, the HYDROChina Xibei Engineering Corporation, the HYDROChina Zhongnan Engineering Corporation, the HYDROChina Huadong Engineering Corporation, the HYDROChina Dongbei Engineering Corporation, and the Yangtze Three Gorges Technology & Economy Development Co., Ltd.

6.07.3.3.1(ii) Mechanism and regulations of management

In the construction of the TGP, the following principles are implemented such as 'the project entity taking responsibilities' – the core system, 'bidding and tendering system', 'construction supervision system', 'contract management system', and 'capital system'.

Bidding and tendering system. By introducing a market competition mechanism, the project entity may select appropriate contractors for construction, equipment manufacture, substance supply, construction supervision, and other consulting service in order to control the investment.

Contract management system. The project entity achieves project management by signing contracts with various participants in terms of design, construction, supervision, equipment manufacture, and substance supply to define responsibilities and obligations of each contractor in the project construction.

6.07.3.3.1(iii) Assessment and acceptance by stages

Toward the end of each construction stage, the assessment and acceptance committee set up by the TGPCC would inspect, accept, and guarantee every aspect of project construction for the TGP. When the first-stage construction nearly completed, the leading group for assessment and acceptance of the TGP before the river close-off was set up by the TGPCC. The group was divided into two teams, the complex project team and the resettlement team, to implement assessment and acceptance of project and resettlement, respectively. In October 1997, the first-stage construction passed the assessment and was accepted successfully. Toward the end of the second- and third-stage construction, the TGPCC organized the second-stage and third-stage project assessment and acceptance committee, respectively. Each committee consists of three teams (panels) – the complex project team, the resettlement team, and the power transmission and transformation team. The complex project team was led by the Ministry of Water Resources. The assessment and acceptance of the second-stage construction was passed in April–May 2003, and that of the third-stage construction passed in August 2009.

6.07.3.3.2 Supervision and control on quality

6.07.3.3.2(i) Quality examination system

The quality management system contains five-level examinations. The self-examination by the contractor is carried out in three different levels, that is, the level of group, of working team, and of project management department (or company). Meanwhile, the whole process is supervised and controlled by experienced technicians employed by the supervision agency. Then the recheck and sampling inspection are performed by the measurement center, the safety monitoring center, the experiment center, and the metal structure detecting center, which are organized specifically for the purpose, as well as by the project management department of the CTGPC. In this procedure, domestic and international experienced and authorized professionals are retained by the CTGPC and form a general supervision office to guarantee project quality in terms of civil construction, metal structure, and mechanical and electrical equipment. In addition, the construction quality is also evaluated by the expert panel for quality inspection that is set up by the TGPCC.

There are in total 116 quality standards formulated and implemented for the TGP, including 35 quality standards for civil works, 70 quality standards for metal structure, and 11 quality standards for the installation of mechanical and electrical equipment. Also adopted are some management measures and incentive system.

6.07.3.3.2(ii) Major quality problems and countermeasures

Although quality problems were found in some civil work construction and equipment manufacture during the construction of the TGP, no adverse effect was caused on the safety operation of the complex project due to timely and appropriate treatment. One of these quality problems was the unsmooth flow surface of some bottom water diversion outlets. After mending and treatment, the flow surface became level and smooth with no cavitations and scaling occurred after long time flood discharge. Another quality problem was the insufficient density of some concrete under the bottom slab of the bottom water diversion outlets. After reinforced by cement grouting and tested by pressurizing water in drilled holes, it was shown that the quality could meet the design requirement. The third problem was the vertical cleavage cracks on the upstream and downstream slopes of the spillway dam section, where in total 78 cracks were found. In the treatment of these cracks, as many as five safeguard measures were adopted mainly to prevent seepage from reservoir water. The treatment was proved effective by the operation practice. The fourth quality problem was that of concrete run-out from the moldboard when placing the north side wall of valve well No. 9 in the underground conduit tunnel of the permanent ship lock. The run-out concrete was removed and the concrete wall was leveled and contact grouting was performed, which ensured the installation and normal operation of the reverse radial gate. The fifth quality problem was that some continuous horizontal cracks were found in 2004 on the upstream slope of the third-stage concrete cofferdam at an elevation from 111 to 78 m. The treatment to these cracks included caulking the cracks with SR flexible material and sticking SR cover plate across the cracks for underwater crack sealing. An additional deep-hole drainage curtain in the gallery at an elevation of 107.5 m was increased and extended to the gallery at an elevation of 40 m for reducing the seepage quantity passing through the RCC cofferdam and relieving uplift pressure on the surface with cracks, and preventing cracks from expansion. After these treatments, the seepage quantity turned within the design range.

6.07.3.3.2(iii) General quality evaluation

The expert panel for quality inspection organized by the State Council TGPCC evaluates the quality of the second- and third-stage construction in what follows:

- The quality of the second-stage project is excellent in general. In detail, the quality of the concrete works of the second-stage dam and powerhouse construction is excellent in general, so is that of the main body of the ship lock, the conduit system, and the drainage system for the mountain completed in the second stage; that of Maopingxi protective dam; that of the preconstruct works for the underground power station.
- The quality of the third-stage project is excellent. In detail, the quality of the concrete works of the third-stage dam and powerhouse construction is excellent, so is that of the works for completing the ship lock; that of the blocking works of the bottom water diversion outlets; that of the works for completing the surface outlets; that of the works for converting the temporary ship lock into the sand sluice gate; that of the service power station; and that of the completed part of the underground power station.
- The metal structure installation is of excellent quality, so is the installation of the 26 turbine generator units.

6.07.3.3.3 Financing and investment control
6.07.3.3.3(i) Financing

The 'capital system'. The capital system was adopted, that is, the TGP Construction Fund was set up by the State as a self-possessed fund for the TGP. The State Council made a decision to add additional charge on electricity for various usages in the whole country since 1992, except for Tibet and power usage for agricultural irrigation in a depressed area. The added charge on electricity was utilized as the TGP Construction Fund and allocated in a given proportion to the CTGPC and the State Grid Corporation for the construction of the TGP.

The power generation revenue of Gezhouba Hydropower Station. In June 1996, the Gezhouba Hydropower Station with an annual power generation capacity of 15.7 billion kWh was put under the management of the CTGPC, which meant that the benefits from power generation of the Gezhouba Hydropower Station were fully used for the construction of the TGP.

Policy-related loan. The China Development Bank provided a policy-related loan of RMB 3 billion yuan every year from 1994 to 2004 for construction of the TGP with the maturity of the loan being 15 years.

Commercial bank loan. Middle-term and short-term commercial loans were used in order to supplement the fund required at the peak period of construction.

Other financing sources. Special bonds, power generation revenue of the TGP, income due to sale of generator units, returning of income tax, and foreign investment were included in this.

6.07.3.3.3(ii) Investment control

The officially approved static budget on the TGP is approximately RMB 90.09 billion yuan based on the price level at the end of May 1993, of which RMB 50.09 billion yuan is for the project construction and the other RMB 40 billion yuan for the reservoir and migrants resettlement. Taking into consideration the factors of price rises and loan interest that occurred in the long construction period, the dynamic investment may reach RMB 203.9 billion yuan.

Till the end of June 2009, the dynamic investment of RMB 149.44 billion yuan in total was completed on the TGP, of which the equivalent static investment of 47.47 billion yuan is for the complex project, accounting for 94.76% of the budget of 50.05 billion yuan. The difference caused by price rise is 32.65 billion yuan, and the loan interest is 15.14 billion yuan. The predicted total investment can be controlled within 180 billion yuan. Compared with preliminary design, both the static and dynamic investment on the TGP can be controlled within the approved budget for preliminary design and the predicted final investment.

6.07.4 Challenges and Achievements

6.07.4.1 Resettlements

6.07.4.1.1 General situation in the Three Gorges reservoir area and index of main inundated practicality

The Three Gorges reservoir area is located in the west of Hubei Province and the middle part of Chongqing Municipality with southeast and northeast marching upon with the west of Hubei Province, southwest bordering on Sichuan and Guizhou Provinces, and northwest neighboring to Sichuan and Shanxi Provinces. Once a 20-year flood occurs, the reservoir will be 667 km long from the dam site at Sandouping to the end of backwater at Jiangjin District of Chongqing Municipality.

Taking the backwater area for a 20-year flood into account, the inundated area of the Three Gorges reservoir is 1084 km^2, of which the land area is 632 km^2. Twenty districts (counties) of Hubei Province and Chongqing Municipality will be involved and 129 cities and towns will be inundated, including 2 medium-sized cities, 11 counties, 27 towns, and 89 townships. A total of 1624 enterprises will be flooded. The population under the resettlement line is 847 500 (including rural population of 348 700) with the housing area of 34.73 million m^3 and cultivated land and garden area of 27 887 hectare.

6.07.4.1.2 Resettlement planning

The resettlement planning has to adhere to the following general guiding ideology and principles. The first is to insist the policy of development-oriented resettlement and to implement the principles of 'national support, preferential policies, aids from various aspects, and self-dependence' so that the relationship between the State, the community, and the private can be handled appropriately. The second is to determine the scale of each resettlement project and compensation criteria by following the standards defined in the 'Resettlement Statute for Changjiang Three Gorges Project Construction' and 'Outline of Inundation Management and Resettlement Planning of the Yangtze Three Gorges Project Reservoir' with scientific attitude and in a practical and realistic spirit. The third is to implement the principles of 'payment within investment and planning for the extra-expenditure'. The fourth is to combine the resettlement planning with social economic development planning and environmental protection planning of the reservoir area.

- A pupulation of 405 000 in rural areas is planned to be relocated with a production arrangement, of which 322 000 habitants will be relocated still in the districts and counties affected by the impoundment of the reservoir area, accounting for 79.6% of the total relocatee, and the ramaining 83 000 habitants (accounting for 20.4% of the total relocatee) will be relocated outside the districts and counties. It is also planned to build houses in rural areas for a population of 440 000, including building houses in the original districts and counties for 357 000 habitants, accounting for 81.2% of the total, and building houses outside those districts and counties for 83 000 habitants, accounting for 18.8% of the total.
- Experiencing the stages of site selection, general planning, geological survey, and detailed planning, it is planned to move and build 12 cities and counties and to construct infrastructures for a population of 640 100, requiring a land area of 46.35 km^2. It is also planned to move and build 116 towns and townships for a population of 179 500, calling for land area of 11.39 km^2. In total, infrastructures to be constructed are for a population of 819 600 and land needed is 57.74 km^2.
- In total, a population of 1.1056 million (154 300 from Hubei Province and 951 300 from Chongqing Municipality) will be relocated and resettled, of which 440 000 habitants are from rural area and 665 600 habitants are from cities and towns.
- Based on the price of May 1993, a budget of RMB 40 billion yuan is to be invested as a compensation for the resettlement due to the TGP. In detail, there is 6.019 billion yuan for resettlement in rural areas; 13.655 billion yuan for moving and building cities and towns (of which 4.272 billion yuan is for cities, 6.868 yuan for counties, and 2.515 yuan for towns and townships); 6.827 billion yuan for rebuilding industry and mineral enterprises; 4.261 billion yuan for rebuilding special projects; 292 million yuan for environmental protection; 885 million yuan for protective works; 600 million yuan for treating and monitoring landslide; 3.111 billion yuan for other purposes; 3.5 billion yuan for basic preparation; and 852 million yuan for the cultivated land occupation tax.

- The resettlement planning and its budget were adjusted in an all-round way in 2006. Upon the modified planning, a population of 139 900 and 6.78 million m^2 houses were increased for the resettlement of the TGP. Therefore, the total population of resettlement and that of house building in the Three Gorges reservoir area increased from the original 1.1056 million to 1.2455 million without including the dam site area, and the total area of houses to be moved and built increased from the original 368.8 million m^2 to 436.6 million m^2. The increased budget due to modification of planning amounted to 7.951 billion (based on the price of May 1993).

6.07.4.1.3 Implementation of relocation and resettlement

In accordance with the construction program of 'storing water by phases and resettling continuously' for the TGP, the habitants in the Three Gorges reservoir area were relocated and resettled by four phases starting from 1993. In the first phase, the habitants living under elevation 90 m were relocated and resettled in order to meet the requirement of river close-off, which was fulfilled in September 1997 and passed the assessment and acceptance. In the second phase, the habitants living between elevation 90 m and elevation 135 m were relocated and resettled, which was completed in April 2003 and passed the assessment and acceptance, satisfying the need of storing water to 135 m, navigation, and power generation of the first turbine generator units. In the third phase, the habitants living between elevation 135 m and elevation 156 m were relocated and resettled. The task was accomplished and passed the assessment and acceptance in August 2006, fulfilling the demand of storing water to 156 m. In the fourth phase, the habitants living between elevation 156 m and elevation 175 m were relocated and resettled to meet the requirement of experimental storage of water in 2008 by the TGP. The mission passed the assessment and acceptance in August 2008.

In the procedure of resettlement, in accordance with the environmental capacity and the demand for sustainable development of the reservoir area, the State adjusted resettlement policy in time and increased resettlement population going outside of the reservoir area. A population of 196 200 were relocated and resettled in 12 provinces and municipalities such as Shandong Province, Anhui Province, Shanghai Municipality, Zhejiang Province, Fujian Province, Guangdong Province, Jiangxi Province, Hunan Province, Jiangsu Province, Sichuan Province, Hubei Province, and Chongqing Municipality. Among the 196 200 relocatees, 160 000 were organized by the government and 36 200 chose the destination to relocate by themselves.

6.07.4.1.4 Social and economic development in the reservoir area and further work

6.07.4.1.4(i) Social and economic development in the reservoir area

The construction of the TGP brought to the reservoir area not only resettlement of one million people, but also billions of investments and preferential policy. Through scientific planning and adjusting the overall arrangement of cities–towns and fundamental infrastructures, the structure of production was optimized and the urbanization level was greatly raised, which provided an excellent opportunity and strong motivity for social and economic development in the reservoir area, and boosted the construction of socialist new country and the social and economic development of local area. **Figure 35** demonstrates one of the new counties in the Three Gorges reservoir area.

According to the statistics from the annual report of main districts and counties in the Three Gorges reservoir area in 2007, the society and economy in the reservoir area has been developed fully and fast since the resettlement. First, the comprehensive economic strength has been improved rapidly. Since the resettlement of the TGP, the per capita GDP of Hubei Province and Chongqing Municipality has increased at a mean annual speed of 14.36% and 12.22%, respectively, and the per capita fiscal revenue has increased at a mean annual speed of 17.80% and 14.03%, respectively. Both the indices are higher than that of the whole country at the same period. Second, the geographical distribution of different sectors of the economy and the structure of production have been gradually optimized. In the reservoir area of Hubei Province, the production value of primary industry, secondary industry, and tertiary industry in proportion changed from 38:31:31 before the resettlement to 24:36:40 after the resettlement, while in the reservoir area of Chongqing Municipality, the proportion changed from 45:30:25 to 25:33:42. This

Figure 35 Resettled Zigui County.

indicates that the economy in the reservoir area has transformed from agriculture oriented to secondary and tertiary industries oriented due to the improved proportion of production value of manufacturing industry and service industry. Third, the infrastructures and commonweal facilities in the urban and rural areas have been obviously updated, and the town and country have taken on an entirely new look. Fourth, the living standard of habitants in urban and rural areas has been improved significantly. The mean annual increase in the speed of per capita saving deposit at the end of a year is 21.65% and 18.88% in Hubei Province and Chongqing Municipality, respectively; that of per capita disposable income of urban inhabitants is 12.38% and 10.85%, respectively; that of per capita net income of farmers is 12.52% and 8.35%, respectively; that of per capita dwelling space of urban inhabitants is 46.26% and 62.82%, respectively; while that of farmers is 79.26% and 26.17%, respectively.

6.07.4.1.4(ii) Subsequent planning

The resettlement from the Three Gorges reservoir area has achieved a milestone in outcome at the stage of being, and has realized the established objective of removing and stability. However, the mission is still arduous and will take a long time to achieve a new type of reservoir area with prosperous economy, harmonious society, beautiful environment, and people living and working in peace and contentment. First, the Three Gorges reservoir area is located in the poverty region of the State with social and economic development level lagging behind. The conflict between large population and less land will exist for a long time. Second, some rural residents are relocated backward near the reservoir, but the land resource is not adequate and resettlement quality is not high. Third, the fundamental condition for industry development in the reservoir area is poor and the agricultural productivity lags behind. Fourth, the infrastructure in terms of public service, transportation, and water supply in the reservoir area cannot meet the requirement of social development. Fifth, the tasks of controlling geological hazard and protecting ecosystem and environment are hard and difficult. All these problems will be resolved by taking them into account in the preparation of subsequent planning, which is in progress right now.

6.07.4.2 Sediment

6.07.4.2.1 Sedimentation of reservoir

The long-term mean annual sediment coming into the reservoir is 530 million tons, which, if not properly dealt with, may harm the function of the reservoir, shorten its life, and influence the navigation of the Yangtze River. Based on the sediment research continued for more than 30 years, it is proposed that the reservoir should be dispatched in the manner of 'storing clean water and emitting more-sediment water', which is deemed capable of maintaining the reservoir for long periods of operation. It is also estimated that the sediment into and out of the reservoir, after 100 years, may reach a balance, and the capacity of the reservoir for flood control will still be kept at about 86%, and that for regulation will be kept at 92%.

Compared with the data applied in the preliminary design of the TGP, the water quantity in the upstream of Yangtze River during the period from 1991 to 2006 varied little, but the sediment yield decreased significantly, particularly in the Jialing River (a main tributary merged into Yangtze River at Congqing). From 1991 to 2006, the mean annual sediment yield recorded at Beibei Station on Jialing River was 33.8 million tons, that is, it had decreased by 75%.

The mean annual runoff at Cuntan Station on Yangtze River at Chongqing and Wulong Station on Wu River from 1991 to 2006 was 331.8 billion m^3 and 50.1 billion m^3, respectively, while suspended sediment load was 301 million tons and 17 million tons, respectively. Compared with the data before 1990, although the runoff did not decrease obviously, the sediment yield reduced by 35% and 43%, respectively. In 2007, the water quantity at Cuntan Station decreased by 11%, while that at Wulong Station increased by 6%, but the sediment yield decreased by 54% at Cuntan Station and by 66% at Wulong Station.

After the Three Gorges reservoir commenced to store water, the water quantity and sediment yield in the upstream of Yangtze River decreased to a certain degree from 2003 to 2007, with the decrease in sediment yield being more evident. The mean annual runoff and suspended sediment yield at Cuntan Station from 2003 to 2007 were 323.3 billion m^3 and 194 million tons, respectively, where the runoff decreased by 8% and sediment yield decreased by 58%, compared with the data used in the preliminary design of the TGP. The mean annual runoff and suspended sediment yield at Wulong Station from 2003 to 2007 were 43.1 billion m^3 and 8.7 million tons, respectively, where the runoff increased by 13% and sediment yield decreased by 71%, compared with the value in the preliminary design of the TGP.

Many years of observation and investigation on hydrology and sediment conditions have led to an understanding of the factors that cause a decrease in sediment yield in the upstream of Yangtze River:

- climate factor, mainly the variation in the temporal and spatial distribution of rainfall;
- interception of sediment by the reservoir;
- water and soil conservancy and returning the cultivated sloped land to forests; and
- excavation of sand in the river channel.

6.07.4.2.2 Clean water discharging and river channel degradation

6.07.4.2.2(i) Scouring in the downstream

The sediment observation data show that the sediment yield from the upstream decreased significantly since the 1990s. After the Three Gorges reservoir was put into operation in June 2003, the sediment concentration in the flow downstream of the TGP's dam decreased further. When the amount of sediment into the reservoir from the upstream of the Yangtze River decreased by 50%, the annual sediment yield at Yichang Station decreased by 80%, compared with the mean annual data. The sediment transport capability of river flow is in an unsaturated state and obvious degradation has occurred in the downstream river channel of the TGP's dam.

The observation data [5] show that both the beach and riverbed in the river section from Yichang to Hukou are scoured with a total scouring capacity of 614 million $m^3 s^{-1}$ and an average scouring intensity of $643\,000\,m^3\,km^{-1}$. The scouring in the river channel has mainly occurred on the riverbed with a scouring capacity of 499 million m^3, accounting for 81% of the total.

Comparison of the observation data of 4 years from 2003 to 2006 with the forecasted value calculated by the numerical model during the technical design stage of the TGP shows that the observation value of total scouring capacity in the river section from Yichang to Wuhan is bigger than the forecasted value by 4.6%, implying that the difference is little. It is also shown that the local scouring in each shorter section is qualitatively consistent, although the quantitation is somewhat different, which indicates that the forecasting result with the numerical model is fairly reliable. Since the Three Gorges reservoir was put into operation, the degradation of the middle and downstream river channels can be controlled in the range of original forecasting.

6.07.4.2.2(ii) Reason and countermeasures

The degradation of the middle and downstream river channels was intensified due to three reasons. The first one is the dramatic decrease of the sediment yield from the upstream of the Yangtze River. The second reason is the sediment dammed by the Three Gorges reservoir since the water storage in June 2003. The third reason is that the excavation of sand in the river channel contributed to the local scouring on the riverbed. The scouring on the riverbed and variation of river regime in a part of the river sections may affect the flood control and navigation to a certain extent. Such influence on the downstream of the dam will be further magnified, from the consideration of the decrease in the amount of inflowing sediment in the reservoir because of the construction and operation of reservoirs on the tributaries and main river channel upstream of the Three Gorges reservoir.

Great attention has been given to this issue in the procedure of designing the TGP. On the one hand, a special study was organized for the degradation in the middle and downstream river during the technical design stage. The 1D, 2D, and 3D water–sand numerical models for calculating scouring in river and lakes as well as the physical movable-bed model for simulating the Yangtze River flood control and sediment movement were built to develop study on scouring. Systematic observation and detailed analysis of the variation of water, sediment, and scouring in the river channel were in progress. Since the TGP was put into operation, the reservoir was operated in such a way to maintain the water level in the dry season at Yichang, and the riverbed downstream of Gezhouba Hydropower Project was protected and consolidated. On the other hand, the study on the river regime control in the Jingjiang River section was enhanced, and relevant projects were implemented. For example, the collapsed bank was monitored and protected, and the existing bank protection works were consolidated.

6.07.4.2.3 Direction of sediment study

For a long time, a large amount of scientific research has been done on the sediment issue of the TGP by sediment experts and designers. Since the reservoir was put into operation, a series of field observations, study, and analysis of the sediment issue have been developed, with rich outcomes. The sediment experts figured out that, due to the technical progresses in the field of sediment study, there has not been effect of sediment on the normal functions of the TGP in terms of flood control, power generation, navigation, and water supply to the downstream in the dry season since the TGP was put into operation.

However, because the knowledge on the motion law of sediment requires support from long-term observation and gradual accumulation, the judgment on the sediment issue will also need to be verified through various situations of inflow and sediment yield. The countermeasures against sediment problems should be adjusted and modified in accordance with the results of prototype observation and operational experiments. Based on the suggestions by sediment experts, further study on sediment issue of the TGP should be developed toward the following direction:

- Strengthen field observation on sediment and relavent analysis in terms of the inflow and sediment yield from the upstream, the sedimentation in the dam area, the harbor area at the end of the reservoir and the fluctuated backwater area, the degradation of the downstream river channel and the variation of water level in the dry season, and the sediments in the estuary.
- Focus on the study of evolution tendency of the middle and downstream river channel and estuary with the TGP and upstream reservoir group put into operation, as well as the study on countermeasures.
- Optimize the operation scheme of the Three Gorges reservoir, from the view of sediment and aiming at various water and sediment conditions in the upstream and downstream.

6.07.4.3 Protection of Ecosystem and Environment

6.07.4.3.1 General condition and the effect of TGP

6.07.4.3.1(i) General condition of biodiversity and water quality in the reservoir area

There are 47 kinds of precious botanical species that are near extinction and among the national-level list of protecting plants in the Three Gorges reservoir area, but most of them grow in the area of elevation 300–1200 m and will not be influenced by the project. Also there is little virgin vegetation to be submerged. Meanwhile, there are 26 rare wild animals among the first and second classes of national list of protecting animals, but most of them live in high mountain areas and will not be affected by the project. Even so, several natural zones have been established by the State to protect wild animals and botanical species of this area.

Among the above rare animals, most attention has been given to *Acipenser sinensis* (Chinese sturgeon, or Zhonghuaxun in Chinese), a kind of large and rare fish. As early as the Gezhouba Dam was completed in the 1980s, its traveling route in Yangtze River was cut by the dam. From then on, research programs have taken 4 years and have developed a number of sophisticated technologies for artificial breeding, sex gland induction, and artificial spawning induction. Now, the artificial propagation of Chinese sturgeons has succeeded and some 100 thousands of baby fishes are poured into the river every year. And, fortunately, some new spawning areas are also found in the downstream of the Gezhouba Dam. All these mean that the preservation of Chinese sturgeon goes well.

A large amount of investment (close to 40 billion yuan) and great efforts have been made to protect the environment in the Three Gorges reservoir area through constructing wastewater treatment plants, removing solid wastes and garbage, establishing the ecological and environmental monitoring system, and so on. Monitoring data show that, since the TGP was put into operation in 2003, the environment quality in the Three Gorges reservoir and relevant area has been good in general and has continued to remain stable. The water quality in the main stream of the Yangtze River in the TGP area is still kept almost the same as before. The water quality in the main tributaries remains similar to that before impoundment but has turned eutrophic, which induces algal bloom sometimes. In addition, the local climate behaves normal except a little rise in temperature probably due to the global climate change. The precipitation in the reservoir area is equivalent to the normal value.

So far, the effect of the TGP on the ecosystem and environment has not exceeded the range forecasted in the argumentation and preliminary design.

6.07.4.3.1(ii) The favorable and adverse effect on the ecosystem and environment

The TGP can not only contribute to enormous comprehensive benefits in flood control, power generation, and navigation, but also reduce emission, which is favorable to improve the ecological and environmental conditions of the reservoir area as well as the middle and downstream of the Yangtze River. For the middle and downstream area, the ecological benefit is manifested in the fact that the dikes along both banks of the Jingjiang section are protected from collapsing caused by extraordinary flood, protecting the life and property of 15 million people there and effectively avoiding breakout and widespread incidence of pestilence and schistosomiasis. Meanwhile, the reduction of the operation probability of flood into the detention and retention areas in the middle and downstream of the Yangtze River eliminates or relieves the adverse impact on the eco-environment due to flood diversion and retention. The interception of sediment by the Three Gorges reservoir mitigates the sedimentation in the Dongting Lake, extending the life of Dongting Lake and improving the lake's ecological environmental condition. During the dry season, the increase of water discharge from the reservoir may raise the pollutant-carrying capability of downstream river channel, improving and stabilizing the water quality in the downstream of the dam. It may also reduce the salinity at estuary during the high-salinity period and improve evidently the water quality by diluting the salt tide at the estuary.

Once the TGP and upstream reservoir group are put into operation, the runoff and inflow hydrograph will change significantly due to water storage by reservoirs, which will impact the ecological process and integrity of ecosystem due to the variation in the habitat environment of some life forms and the community structure of hydrobioses. The change in the physical characteristics of discharged water, such as low temperature and gas oversaturation, will have a major impact on the condition of fish culture and the habitat of fauna and flora. The variation in runoff due to reservoir operation (clipping flood peak, water storage, and water discharge) will lead to a slow gentle rise in water level in the middle and downstream during the flood season. Therefore, it will take longer for the Boyang Lake and Dongting Lake to exchange water quantity with the Yangtze River during the non-flood season, impacting the habitat environment of fauna and flora and a number of biological resources.

In order to take full advantage of the function of the TGP in maintaining the ecosystem and forming a new ecosystem in the reservoir area, aiming at the impact by the TGP, a series of studies are being developed to improve the environment. A number of countermeasures are also implemented gradually. In the design aspect, in conjunction with the research results from environment experts, the reservoir operation scheme that is favorable for ecosystem protection is studied to ensure the ecological functions of the reservoir ecosystem, to satisfy the hydrological requirement of habitat and reproduction of fauna and flora, and to protect the biodiversity in the middle and downstream of the Yangtze River.

6.07.4.3.2 Study on the reservoir operation scheme favorable for environmental protection

Besides functions such as flood control, power generation, navigation, and water supply, the TGP may also be operated in an appropriate way to protect and improve environment and ecosystem. That is, by controlling water level and discharge flexibly, the adverse impact on the environment and ecosystem due to the construction of the TGP may be relieved or mitigated. Based on the actual situation of the TGP, an operation scheme aiming at controlling the invasion of salt tide at the Yangtze estuary, and a water discharge mode simulating real hydrological condition and therefore favorable for spawning of fish are studied.

6.07.4.3.2(i) Improve aqua-ecosystem and control invasion of salt tide at the Yangtze estuary with the Three Gorges Project

The Yangtze estuary, as illustrated in **Figure 36**, is a delta estuary as wide as 90 km with characteristics of abundant water, much sediment, moderate tide intensity, and multibranches. The water area is vast with very complicated hydrological characteristics and riverbed evolvement. Downstream of the Xuliujing Station, the main stream is divided by Chongming Island into south branch and

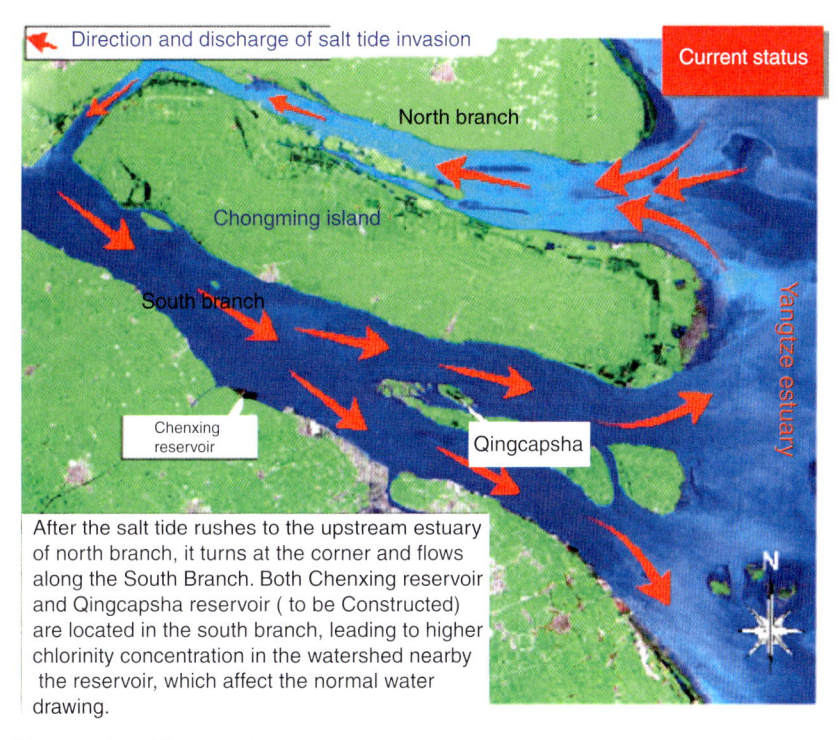

Figure 36 Sketch of salt tide direction at Yangtze estuary.

north branch. The south branch is then divided at Wusongkou into south stream and north stream, and the south stream is further divided into south channel and north channel by Jiuduansha Bar, resulting in the topographical features of three-level branches and four openings into the sea. The four encloses to the sea, that is, north branch, north stream, north channel, and south channel, compose four invasion passes of salt tide, among which the north branch has evolved to a channel with flood current taking absolute advance since 1958. Because of the strong drive of flood current, the salinity in the north branch ranks the first among the four passes. The south branch (including north stream, north channel, and south channel) is the main enclose for discharging Yangtze River's runoff with less invasion of salt tide, but has been affected by reverse flow in north branch in recent years. The invasion of salt tide at the Yangtze estuary has significant influence on industrial and agricultural production and people's life on both banks as well as on the water resouces site along the south branch.

The invasion of salt tide at the Yangtze estuary is a natural phenomenon caused by tide activity and has been in existence for a long time; it is formed due to complicated reasons and is affected by multifactors such as shape of estuary, tidal range, and upstream runoff. The analysis of salinity, an index of saltwater invasion, shows that the salinity varies with time and space, of which the variation with time is consistent with the seasonal variation of runoff. The flood season is the period of low salinity, while the dry season from December to April is the period of high salinity. Therefore the invasion of salt tide generally occurs during the dry season from November to April of the next year. Depending on the characteristics of salt tide in terms of occurrence time, a proper increase in water discharge from the Three Gorges reservoir may play a role against the invasion of salt tide to some extent.

Since the Three Gorges reservoir has a large regulating storage, according to the designed regulation scheme, the reservoir should be operated to meet the requirement of flood control during the flood season and retain water to 175 m after the flood season. In the dry season, the water level in the reservoir will gradually descend to 155 m as required by power generation and navigation. The pool level will then drop down to the flood control limiting level of 145 m before the flood season. During the dry season from December to April of the next year, the Three Gorges reservoir will discharge 16.5 billion m^3 water to the downstream, which may function against the invasion of salt tide besides contributing to power generation, navigation, and water supply to the downstream. In order to increase the role of the TGP in protecting from the invasion of salt tide during the driest period, it is critial to study how the reservoir can retain water to 175 m under various inflow and sediment yield conditions from the upstream so that the reservoir is capable enough to discharge compensative water to the downstream. Meanwhile, an optimized regulating scheme should be studied to increase water discharge during the driest period from January to February by modifying the discharge process in the whole dry season. Through a detailed analysis of the flood characteristics and sedimentation of the Yangtze River, it is proposed to adjust the starting time of water storage from October after the flood season to the middle or last 10 days of September when the inflow is still relatively abundant. The optimized water storage and regulation scheme not only diminishes the adverse effect of decreased water discharge during the storage period of the TGP in October on invasion of salt tide, but also guarantees the capability of the TGP in supplementing water to the downstream during the dry season by improving the probability of full storage in the reservoir.

Taking advantage of the favorable condition of the large regulating storage of the Three Gorges reservoir, changing the regulation and operation mode appropriately on the premise of satisfying the basic requirement of flood control, power generation, and navigation is an effective countermeasure and method for the TGP to maintain the ecosystem and environment.

6.07.4.3.2(ii)   Regulation mode of the Three Gorges Project coping with the propagation of 'Major Four Carps'
Mylopharyngodon piceus, Ctenopharyngodon idellus, Hypophthalmictuthys molitrix, and *Aristichthys nobilis* (or black carp, grass carp, bighead carp, and silver carp), the so-called 'Major Four Carps', are four kinds of carp and are traditional excellent and economic fish to breed in the Yangtze River basin. Their natural spawn and reproduction requires adequate water temperature and flow condition. In general, the lowest water temperature for reproduction is 18 °C and optimal water temperature is from 21 to 24 °C. Under the natural condition, such temperature can be reached generally from the late 10 days of April to the first or middle 10 days of July. Meanwhile, the spawn of 'Major Four Carps' requires stimulation by water rising in the river channel, which generally occurs in river sections with curve bend, rapid streams, narrow river surfaces, and shoals in the center of the river. The construction of the TGP changed the condition of flow and water temperature of the original channel in the reservoir area, with the associated impact on the reproduction of 'Major Four Carps'. According to experts' analysis, a certain measure of reservoir regulation may be taken to discharge water from the reservoir in a way similar to the natural hydrological condition, that is, discharge a man-made flood to create appropriate flow condition for the spawn of 'Major Four Carps'. The field investigation shows that the favorable flow condition for spawn may be achieved if the water level in the river channel can rise 2–3 m within 4–5 days.

According to the regulation mode of the Three Gorges reservoir, the pool level shall drop to flood control limiting level on 10 June; therefore a large quantity of retained water in the reservoir will be released from May to the first 10 days of June. Since the natural inflow during this period is fairly abundant, it is possible for the reservoir to create a 'man-made flood peak'. Implementation of such a scheme should be in coordination with power generation by reducing the output of power generation one day in advance and then gradually increasing, for example, the discharge from 15 000 $m^3 s^{-1}$ (corresponding to an output of 11 million kW) to 24 000 $m^3 s^{-1}$ (corresponding to an output of 18 million kW) within 4–5 days; the water level can correspondingly rise about 3 m, which can support a large scale of seeding. Such 'man-made flood peak', if released 2–3 times during the period from May to June, will facilitate the reproduction of 'Major Four Carps'.

Except for the above-mentioned regulation measures, we can resolve issues such as prevention and relief from crucial environmental accidents, eutrophication in the tributary or branch within the reservoir area, and water quality and protection of Chinese sturgeon through developing a study on a certain regulation scheme aiming at ecosystem protection.

6.07.4.3.3   Prospect

As a grand trans-century project, it is the TGP's responsibility and obligation to maintain the ecosystem and environment, and to promote harmonious coexistence between human beings and nature. Under the guidance of scientific concept of development, studies are developed for the following purposes: maintaining the integrity of ecosystem structure and ecological process within the Three Gorges reservoir; ensuring the ecological functions of ecosystem in the reservoir area; optimizing the regulation of reservoir in order to satisfy the hydrological requirements by habitat and reproduction of fauna and flora; protecting the biodiversity in the middle and downstream of the Yangtze River, and so on. The environment departments have undertaken studies to find out countermeasures for issues such as pollutant-carrying capability of the reservoir area, phenomenon of water bloom and variation in water quality, treatment of solid wastes and falling zone, and controlling and prevention of water pollution.

It is believed that the ecological and environmental issues related to the TGP may be controlled by taking various countermeasures capable of minimizing the adverse effect. The TGP will become an environmentally friendly project with efforts from all aspects.

6.07.4.4   Prevention and Control of Geological Hazards

6.07.4.4.1   Introduction of geological hazards control and prevention project

In 2001, the State set up a special fund of RMB 4 billion yuan (in the second-stage geological hazard prevention planning) to control the related geological hazards before the reservoir retaining water to 135 m in 2003. Later, a fund of 7.3 billion yuan was allocated (in the third-stage geological hazard prevention planning) to control the geological hazards on the reservoir banks which might occur before the reservoir retaining water to its initial level of 156 m after the flood season in 2006, or to 175 m as experimental storage after the flood season in 2008 (the pool level was actually raised to 172.8 m). Meanwhile, the 'Third Stage Geological Hazard Prevention Planning (Protection of High Cutting Slope) in the Three Gorges Reservoir Area (2004)' was proposed and treatment was implemented to resolve the high cutting slope issue existing in the resettlement works of the TGP.

The second and third planning proposed 646 removing and giving-way projects due to geological hazards, with 69 900 people involved. In total 441 places of slope sliding and 2874 places of high-cutting slope were harnessed and 175.05 km long reservoir banks were protected. There are 3113 distributed monitoring points to monitor and protect the area of 600 000 people. Among these monitoring points, 254 are for special purpose, including three points to monitor the ultra-deep layer of reservoir banks and 251 points for collapse observation.

After the implementation of projects proposed in the second- and third-stage planning, most of the endangerment due to collapse, slope sliding and bank collapse on the rebuilt towns, important rebuilt location, and navigation was relieved, which improved the geological environment generally.

6.07.4.4.2 Reservoir-induced seism

The possible earthquake issue due to the Three Gorges reservoir has been emphasized by the government for a long time, and extensive researches have been conducted on the issue in relation to the rock, geological structure, osmosis, and so on. The deep-hole crustal stress observations with hole depth reaching 300–800 m have been carried out at the dam and reservoir site and intensive observations of earthquakes have been made on some fracture zones around the dam. The research results reveal that the geological structure in this area is stable, and there is no geological background for a future heavy earthquake. Researches predict that, after the water rises, the possible maximum earthquake intensity will not exceed degree VI, and therefore will not threaten the project structures designed on the basis of an earthquake of degree VII.

The first seismic network for engineering purpose in China was set up in the Three Gorges area in 1958. The network was updated to wireless telemetry seismic network in 1996, which operated continuously up to now, and accumulated a lot of valuable information. The Reservoir-induced Seism Monitoring and Forecasting System for the TGP was then set up in October 2001. Meanwhile, monographic studies related to reservoir-induced seism and construction-induced seism by the TGP were both listed in the State Key Programs for Science and Technology Development of the 'Seventh Five-year Plan' and in that of the 'Eighth Five-year Plan'.

The monitoring and analysis of the reservoir-induced seism shows that the frequency of micro-earthquake in the dam and reservoir area is increased obviously after water storage, with a certain correlation to the pool level, which indicates the characteristics of reservoir-induced seism. However, most of the areas where reservoir-induced seism occurred so far are included in the expected range. In addition, most micro-earthquakes are shallow earthquakes occurring in karst and mine areas and with a magnitude less than Class 3. The strongest earthquake recorded so far has a magnitude of Class 4.1, which is far lower than that anticipated.

6.07.4.5 Protection of Cultural Relics

6.07.4.5.1 General introduction

The protection of cultural relics is an important task in the planning and design of the Three Gorges reservoir area. With the coordinated effort of the cultural administration department and other departments, the 'Cultural Relics Protection Planning in the Three Gorges Reservoir Area' was completed in 2000 and was then approved by the TGPCC of the State Council. It is planned to implement 1097 protection projects, of which 733 are projects protecting underground cultural relics, 360 are projects protecting cultural relics on the ground, and four are crucial and special itemized projects. Over the past more than 10 years, great efforts have been made at each level by the cultural administration department, the cultural relic protection department, and the archaeological administration department. With regard to the underground cultural relics, the exploration covering an area of 12.14 million m^2 has been fulfilled, of which 1.718 million m^2 has been excavated. All the cultural relics on the ground are mapped for recording information, and projects accomplished so far include 58 protection projects at the original sites and 109 moving and rebuilt projects. There are three projects under construction, two protection projects at the original sites are still waiting for commencement, and 22 rebuilt projects are not started yet. As to the four special itemized protection projects, the rebuilt Zhanghenghou Temple (i.e., Zhangfei Temple, see text below) was completed in July 2003; Shibaozhai Camp (**Figure 37**) was protected at the original site in April 2009; the underwater protection of Baiheliang inscriptions and carvings was completed in May 2009; only the protection project for Quyuan Temple is at the end of completion. In the past over 10 years, in total RMB 749 million yuan was invested in the Three Gorges reservoir area for the protection of cultural relics.

6.07.4.5.2 Case study

6.07.4.5.2(i) Baiheliang inscriptions and carvings

Located on the south bank of the Yangtze River in Fuling District, Chongqing Municipality, Baiheliang is a natural stone girder of 1600 m long and 15 m wide, which is approved by the State Council as a national-level cultural relic protection site. There are more

Figure 37 Shibaozhai Camp at Zhong County.

Figure 38 Cultural relic – Baiheliang inscriptions and carvings.

than 100 inscriptions on the stone girder, which recorded 72 water levels of the Yangtze River in the dry years for more than 1200 consecutive years. These inscriptions and carvings were submerged when the river rose up and emerged again when the water dropped to a certain level in the dry season. For example, the carved rockfish shown in **Figure 38** symbolized a drought in the history when it emerged out of water. Therefore it is regarded as an ancient hydrological station. A number of ancient litterateurs (Huangtingjian, Zhuxi, Wangshizhen, etc.) left inscriptions and carvings on it, leaving calligraphy of more than 30 000 characters which was called 'underwater forest of steles'.

When the pool level rises to 175 m in the reservoir area, the Baiheliang inscriptions and carvings will be fully submerged at a depth of 30 m under the water and cannot emerge on the water surface any more. Therefore in 2001, the academician Ge Xiurun of China Academy of Science proposed an alternative to protect the relic under the water at the original site with a 'non-pressure container', as illustrated in **Figure 39**. This alternative was adopted and an underwater museum was constructed. A non-pressure shield with water both inside and outside was constructed so that the tourist can enter the museum through the underwater pass from the bank to view those inscriptions and carvings. The underwater museum was opened on 18 May 2009, indicating that the hydrological station with a history of a thousand years can see the daylight again.

6.07.4.5.2(ii)   Zhangfei Temple

The Zhangfei Temple was built at the end of the kingdom of Shuhan (AD 221–263) and was repaired and expanded in successive dynasties. It has a history of more than 1700 years. The original site of the Zhangfei Temple was located at Feifeng Mountain, where a rich collection of calligraphy, paintings, inscriptions, and carvings were kept, including over 200 rare cultural relics. Therefore, the Temple is praised as "a fairy spot in the west China territory and a scenic spot most fully depicted in literature". The Zhangfei Temple was evaluated as the national-level cultural relic protection site and national famous scenic spot of China, being one of the important spots along the golden tourism route of Three Gorges.

Due to the construction of the TGP, the Zhangfei Temple turns into the only cultural relic site in the reservoir area which will be moved totally for a long distance. The temple was closed on 8 October 2002 for relocation to a site 30 km upstream – it was

Figure 39 Protection project of Baiheliang inscriptions and carvings.

Figure 40 Zhangfei Temple.

removed from Feifeng Mountain opposite to the Old Town of Yunyang County to Long'an village in Panshi Town, Yunyang County. The new relocated Zhangfei Temple (Figure 40) as original as the old one opened on 19 July 2003. The relocation of Zhangfei Temple ranked the first 'relocatee' in the Three Gorges reservoir area in terms of expenditure and scale.

6.07.4.5.2(iii) Baidi City

The Baidi City was located on the Baidi Mountain east of Fengjie County at the entrance of the Qutang Gorge on the north bank of the Yangtze River. It was 451 km away from Chongqing Municipality and used to be called Ziyang City. A huge mud sculpture of 'Liu Bei's entrustment' was set in the Baidi Temple. In addition, a whole set of cultural relic collection from the coffins suspended at the cliff of Qutang Gorges and 73 steles carved with calligraphy and paintings since the Sui and Tang dynasty were kept in the Temple. The other cultural relics included 1000 cultural relics from past dynasties and more than 100 calligraphy and paintings of ancient and current celebrities. Among them, the 'Bamboo Leaves and Calligraphy Stele' were carved with poem and drawing compromised together in a unique style and the 'Three-King Stele' was carved with phoenix, peony, and phoenix tree, being fancy and magnificent.

The Baidi City will be ringed with water on all sides and become an isolated island in the river center after the TGP retains water (Figure 41). With the banks dipped in water and scoured by the river for a long time, the fluctuation of water rising and falling will produce huge pressure, which will affect the environment of the Baidi City seriously. In order to ensure the safety of the Baidi City, treatment of the collapsed banks of the Baidi City was conducted since December 2003. The whole project is like a golden waist band of Baidi Mountain. It is known that the famous poem written by the poet Li Bai will be carved on the project to embody the Poem City's history and characteristics.

6.07.4.6   Analysis of Dam Break

In order to cope with exceptional failures of the dam caused by emergency events such as wars, during the argumentation and design stage, modeling experiments and analysis are developed on dam break.

6.07.4.6.1   *Study outcomes*

Once the dam break appears, not only the functions of the dam will fail, but also the gigantic water quantity released from the dam instantly will destroy the dike, leading to a flood disaster in the downstream. In order to study this issue and find out

Figure 41 Baidi City after reservoir storage.

countermeasures, a 1/500 undistorted model and a distorted model with 1/500 horizontal scale and 1/125 vertical scale were built, which would facilitate the study of dam break during the argumentation and design stage. The experiments include situations of instant total break and instant half break, meanwhile taking into account various dike break situations in the downstream. The focus includes the dam-break flood peak and flood routing process under various dam-break conditions, and flood level along the collapsed dikes with various dike break schemes.

The dikes along the Yangtze River in the downstream of the Three Gorges Dam protect the cities along the banks and Jianghan plain. Once the dam breaks, the discharged water may exceed the discharge capability of the river channel and the water level in the downstream will exceed the design level of dikes, leading to dike breach and flood disaster. The flood routing process and the affected range with the reservoir operating at various levels have been obtained through experiments. The analysis shows the following results:

- The break width can only affect the maximum instant discharge at the river reach near the dam and will not affect the maximum flood level at the flood control points far away from the dam. The capacity of discharging water from the broken dam will be gradually weakened due to the retardation in the flow rate of the narrow valley sections upstream and downstream of the dam.
- The flood disaster in the downstream due to dam break is subject to the pool level when break occurs and the water level in the downstream river channel. With the same pool level, the higher the water level in the downstream, the bigger the influence on the downstream. This implies that the downstream water level is a crucial factor. The influence of dam break is mainly evident in the flood season when flood may occur and the downstream water level is high. The analysis also shows that reducing the pool level in advance will be an effective measure to mitigate downstream losses.
- The experiments also demonstrate that by reducing pool level in advance and diverting flood, the inundated area by dam-break flood can be controlled up to 3000 km^2 so that the area downstream of Shashi City will not be affected.
- The water discharge capability of the TGP is enormous as the bottom outlets with large discharge capability are arranged on the spillway section. If water is released in amounts as large as the downstream river channel conditions allowed, the pool level can be reduced in a fairly short period. Usually it takes only up to 7 days to reduce the pool level from the normal pool level to the flood control limiting level, which means that the water can be released from the reservoir in advance within the prewarning period.

6.07.4.6.2 Safety guarantee

The Three Gorges reservoir is a valley-shaped reservoir, 600 km long and 1100 m wide. The downstream reach of the reservoir near the dam site is a narrow valley section 160 m long and 500 m wide at water surface. Even in this reach, there are still several valleys with river width less than 300 m. The 20 km long valley section from the dam downstream of Nanjingguan is only 200–300 m wide with cliffs on both banks and three right-angle bends. The peak of released flood due to dam break will be rapidly declined as the flood is obstructed by valleys. When the front peak of released flood flows out of Nanjingguan, the peak has been reduced dramatically. Passing through the following 60 km long river valley with hills, the flood will be further mitigated when arriving at the downstream flood control point of Zhicheng Town.

In accordance with the operation characteristics of the Three Gorges reservoir, a regulation scheme of prereleasing water can be set up to cope with an emergency situation, that is, the pool level can be reduced gradually in advance once there is a sign of war occurrence. Based on this scheme, if the reservoir is operating in the condition of pool level 175 m, water storage should be stopped even if the pool level does not reach 175 m. If the pool level has reached 175 m, water discharge should be increased as much as possible to empty the storage and reduce the pool level; if the pool level is already at 145 m, it should be further dropped below 135 m as long as the discharge is acceptable by the downstream river. The strong water discharge capability of the Three Gorges dam ensures that the pool level is able to be reduced rapidly in a short period.

If the downstream water level is high while the pool level of the Three Gorges reservoir is high too, the detention and retention areas should be used to divert dam-break flood. With the above-mentioned countermeasures, the consequence of dam break will only be a local hazard in part of the downstream area.

6.07.4.7 Benefits

6.07.4.7.1 Benefit of flood control

The TGP is a key project for the flood control system in the Yangtze River. Combined with dike and detention areas, the TGP can change the flood control situation in the middle and downstream of the Yangtze River fundamentally. Since the TGP was put into operation in 2003, it has taken the responsibility of flood control gradually.

During the construction period from 2003 to 2005, the reservoir retained water to 135 m and played a role in flood control by obstructing water with cofferdam to stagger flood peaks. During the flood season of 2006, the reservoir used its flood storage between 135 and 150 m to retain water for the middle and downstream of the Yangtze River, playing a role in flood control in advance. After the flood season of 2006, the TGP started to officially play a role in flood control by retaining water to 156 m in the reservoir. When the reservoir retained water to 175 m in 2009, the effective flood control storage would reach 22.15 billion m^3, which realized the design objective and may contribute to greater benefit from flood control. For example, in the summer of 2009, the reservoir encountered the maximum flood for the past 5 years with peak flood of 55 000 m^3 s^{-1}. After the regulation of the TGP, the discharge was cut down to 40 000 m^3 s^{-1} and the pressure of flood control for the downstream area was greatly relieved.

6.07.4.7.2 Benefit of power generation

The first set of turbine generator units started to generate electricity in 2003 and the 26 turbine generator units installed in the dam toe power plant were put into operation at the end of 2008. Till June 2009, the TGHP generated 320 billion kWh of electricity, delivering clean energy to Central China, East China, and Guangdong Province. If each kWh electricity produced is priced at 8 yuan, the accumulated electricity generated from the TGHP is equivalent to 2560 billion yuan. When the TGHP was put into operation, it was the time when the economic construction was developed in full scale, leading to large demands on electricity. The shortage of electricity in the above-mentioned areas was relieved through speeding up the construction of the TGP, completing the installation of generator units before the schedule, and increasing power generation by regulating the reservoir. The power energy with good quality and low price provides strong drive for the rapid development of the national economy.

Apart from supplying electricity to the electrical system, the TGHP also takes the responsibilities of regulating peak load for the electrical system and being a standby for accidents, which played an important role in maintaining the safe operation of power grid.

6.07.4.7.3 Benefit of navigation

The navigation function of the TGP is gradually realized with the water storage in the reservoir, which directly drives the rapid development of freight traffic in the reservoir area as well as the whole basin. It promotes navigation in the Yangtze River, especially in the upstream to allow large, specialized and intensive ship fleets passing through. The freight volume of the main stream of the Yangtze River in 2006 was 990 million tons, which was 2.5 times the volume in 2000, while the freight volume passing through the Gezhouban Dam was 42.32 million tons, which was 3.5 times the volume in 2000. The displacement tonnage in the Chongqing section of the Yangtze River is raised from 420 tons in 2002 to 1300 tons in 2006. During the period of 6 years of normal operation, the freight volume through the Three Gorges ship lock amounts to 310 million tons, which exceeds the freight volume through the Gezhouba ship lock in the period of 22 years before the water storage of the TGP.

6.07.4.7.4 Other benefits

Benefit of water supply. After the Three Gorges reservoir is put into operation, the regulating storage increases with the rising of pool level after the flood season, which means that the capability of supplementing water to the middle and downstream in the dry season is improved gradually. Through the regulation of the Three Gorges reservoir, the average discharge in the river channel downstream of Yichang in the dry season can be increased by 200–300 $m^3 s^{-1}$ in 2003–05 and 800–1500 $m^3 s^{-1}$ in 2006–08. When the pool level rises to 175 m, the discharge in the river channel downstream of Yichang can be increased to 1000–2000 $m^3 s^{-1}$. The construction of Three Gorges reservoir ensured that emergency measures can be adopted to supplement water to the downstream when serious drought occurs.

Benefit of reducing greenhouse gases emission. The accumulated electricity 320 billion kWh generated from the TGHP by the end of 2008 is corresponding to reducing coal equivalent of 102.4 million tons, therefore reducing sulfur dioxide emission of 1.35 million tons, nitrogen oxides emission of 500 thousand tons, smoke gas emission of 700 thousand tons, and CO_2 emission of 200 million tons, bringing some benign influence in improvement of environment, especially preventing acid rain and greenhouse effect in East and Central China.

6.07.4.8 Technical Advancement

6.07.4.8.1 Hydraulic structure design

6.07.4.8.1(i) The type of intake at the power plant

The intake at the TGHP is designed with characteristics of large dimension opening, wide fluctuation range of water level, and narrow dam section. In order to select proper type and dimension for the intake, researches and experiments on a hydraulic structure model have been conducted since 1985. Intakes of large opening and small opening, of single opening and double openings, and of horizontal opening and inclined opening have been studied. The hydraulic experiments on intake with small opening have drawn a similar conclusion as the Bureau of Reclamation, US Department of the Interior, saying that although the speed of inflow through the small opening is larger than that through the large opening, the shape design of the small opening is reasonable because the hydraulic head loss of small opening is even less than that of the large opening. The experimental results indicate that the mall opening is adoptable in the design of intakes for large-capacity units. Therefore, the intake at the dam toe power station is designed as a single inlet with a small opening with the bottom of the intake at an elevation of 108.0 m.

6.07.4.8.1(ii) Method to embed the spiral case

There are three methods to embed the spiral case of a turbine unit into concrete, which are embedding while maintaining pressure, embedding on a cushion layer, and embedding directly [6, 7]. **Figure 42** shows the worksite of embedding spiral cases. Based on a summary of domestic and international engineering practices, the three methods are further studied and applied in the TGHP.

There are several successful examples in both domestic and international countries for embedding the spiral case of 700 MW units with constant internal pressure. However, the spiral case embedded with constant internal pressure in the TGP possesses the following three characteristics. First, the fluctuation range of water level in the Three Gorges reservoir is wide, varying from the lowest operation level at 135 m during the initial stage to the highest operation level at 175 m in the later period. In order to ensure that the spiral case can stick to the concrete closely when the reservoir operates at 135 m so that the generator units can operate

Figure 42 Construction of power plant.

stably and safely, the water head for maintaining constant pressure can only be 0.5 times the designed internal water head of the spiral case. Second, the water temperature in winter and summer at the TGP is 9 and 28 °C, respectively. In order to reduce the impact of temperature variation to ensure that the spiral case can stick to the concrete closely when the generation units are installed in summer but operate at 135 m in winter, the water temperature should be controlled between 16 and 22 °C for maintaining constant pressure. Thus, placing concrete with constant pressure and temperature was actually adopted. Since the water for maintaining constant pressure had to be heated during the winter, a special heating device was designed to control the temperature of 6000 m^3 water inside the spiral case and to ensure even temperature meanwhile. Third, the method of maintaining constant pressure is only to ensure placing concrete with constant internal pressure, without going through the process of checking the quality of welding seam by raising internal pressure by filling water as well as that of eliminating welding stress. The method of maintaining constant pressure and temperature for embedding the spiral case of turbine units and the relevant water heating device are creative design and practice at home and abroad.

Embedding spiral case with a cushion layer is a traditional method used in hydroelectric projects in China. But this has not been practiced in both domestic and international countries for 700 MW gigantic turbine generator units. The TGP is considered to be the first example. The advantages of this method include little internal pressure on the concrete structure, easy construction, short construction period, and low cost. The main concerns may be that the cushion layer cannot grasp the spiral case as tight as the concrete, which will impact the safe and stable operation of turbine generator units. After static and dynamic analysis and argumentation on the embedding scope and rigidity of the cushion layer, it is deemed that the method of placing a cushion layer outside the spiral case can meet the requirement of rigidity and deformation of structures for safe and stable operation of units, as well as that the cushion layer can play a role in reducing oscillation of the power plant structure. Practices show that the units with a cushion layer on a spiral case in the right bank powerhouse operate safely and stably with little oscillation of the power plant structure, which is consistent with the research conclusion.

The method of placing concrete directly on the spiral case forms a kind of steel-lined reinforced concrete co-bearing load structure, which is a technology to resolve over thick steel lining of conduit structure with high H×D value and to ensure the integrated safety of conduit structure. A systematic study has been developed on this method through linear and nonlinear 3D static structural calculation, 3D dynamic structural calculation, simulation model experiment of material, and model calculation. Not only the scale of units and spiral case studied, but also the systematics and depth of studies conducted may be regarded as the first case in both domestic and international countries.

Although all spiral cases of the units are designed as exposed conduits, for spiral cases embedded with the method of placing concrete directly, the surrounding concrete and spiral case form a co-bearing structure. Since the bearing capacity of such spiral cases cannot be fully utilized, the concrete bears most of internal pressure, leading to serious cracks on the concrete and considerable uneven deformation of the lower frame foundation of the generator units. In order to ensure safe and stable operation of the units, the following measures are added for spiral cases embedded with placing concrete directly: placing an elastic cushion layer in a certain range at the end of the spiral case of bigger diameter (from the inlet of the spiral case to the cross section at 45°), enhancing the reinforcement of concrete appropriately, and raising the grade of concrete in part sections.

6.07.4.8.1(iii) Tailrace tunnel at the underground power station

The operational principle of the tailrace tunnel with varied roof height is as follows [8]. The tailrace tunnel is divided by the intersection of the downstream water level with the tunnel roof into two sections: the upstream pressure flow section and the downstream non-pressure free-flow section. When the downstream water level is low, the submerged depth of the turbine is small, the pressure flow section is short, and the non-pressure free-flow section is long; thus the negative waterhammer pressure is small during the hydraulic transients. Therefore the minimum absolute pressure at the inlet section of the draft tube will not exceed the requirement of standards. When the downstream water level rises, although the length of the pressure flow section is gradually extended and that of the non-pressure flow section is gradually shortened, the negative waterhammer pressure becomes bigger and bigger until the tailrace tunnel is full of pressure flow. Fortunately, the submerged depth of the turbine is also increased and the average flow velocity in the pressure flow section is decreased gradually. With the positive function and negative function counteracting with each other, the minimum absolute pressure at the inlet section of the draft tube can be controlled within the range specified by the design standard. The tailrace tunnel with varied roof height can play the role of a surge chamber, which makes the structure more safe and reliable and reasonable in economy. It can also ensure the safe operation of generator units.

The tailrace tunnel with varied roof height adopted at the underground power station of the TGP not only effectively resolved the problem of flow pattern characteristic by transferring the pressure flow and free flow to each other, but also improved the stability of the wall rock for the underground cavern. It was the first trial in the world to develop a large-scale experiment on the transient process in a combined hydromechanical and electrical system with a model turbine generator unit. The experiment quantitatively defined the reasonable regulation parameters for the units and revealed the impact of the main governing parameters on the hydraulic characteristics of the tailrace tunnel with varied roof height.

6.07.4.8.1(iv) Double-line and five-step ship lock

The scale and design water head of the TGP's ship lock are far in excess of those of other existing ship locks in the world. In addition, the fluctuation range of the upstream water level is quite wide. Taking into account the large sediment concentration, complicated river regime at the dam site, and geological condition, the difficulties of the technology in terms of general design of ship lock, long-term operation, conduit system, structure of ship lock, lock miter gate, hoist, and operation monitoring have substantially gone beyond the level of other ship locks constructed in the world.

6.07.4.8.1(v) Chamber structure of ship lock

The ship lock of the TGP is designed as two parallel lines neighboring to each other with multisteps and tall structure. The main body section of the ship lock is basically placed in a deep cut and excavated rock channels. The lock chamber is 40–70 m high and the slopes on both sides of the ship lock are up to 170 m high. The chamber structure, the tall and thin concrete-lined structure working together with the rock mass, is directly connected with the high rock slope. Strictly following the requirement of the lined structure's profile, the lower part of the rock slopes on both sides as well as the slopes on both sides of the central pier should be excavated to a vertical slope, which is not good for its stability. Besides requiring the slope to maintain stability by itself, the load passed by the shiplock structure to the slope has to be taken into consideration as well. In addition, the deformation of the slopes has to be controlled within a range in order to ensure the normal operation of a shiplock structure and corresponding equipment. All these issues have been resolved very well in the construction of the ship lock, providing experience gained for building large-scale ship lock with a light structure on the rock foundation.

6.07.4.8.2 Construction technology

6.07.4.8.2(i) River close-off and construction of the cofferdam in deep water

The dam site of the TGP is located in the Gezhouba reservoir area. The maximum water depth during river close-off is 60 m, ranking the first in the world. The designed discharge during river close-off is 19 400–14 000 m^3 s^{-1} and the difference in water level between the upstream and downstream sides of the cofferdam is 1.24–0.80 m. The riverbed where the river is closed off has complicated topographical and geological conditions: covered on the granite riverbed is a fully intensive weathered layer, above which are sand and pebbles, sphere of residual deposit, and sedimentation layer. The silt newly settled in the deep channel of the Gezhouba reservoir is 5–10 m thick and the left side of the deep channel is a cliff. These conditions are poor and unsafe for building a levee there. According to the construction scheme, a levee in the upstream is built to block the flow, leaving a 130 m wide closure gap. The technology of pre-leveling up the riverbed at the closure gap, by throwing stone ballast, aggregated rock, and sand-gravel aggregate on the riverbed, to reduce the water depth at the closure gap was shown to be fairly beneficial in preventing the levee from collapse during the river close-off, reducing the amount of throwing work and reducing the intensity of throwing work in closing off the gap. The recorded actual discharge during the river close-off reached 11 600–8480 m^3 s^{-1}, ranking the first in the world.

6.07.4.8.2(ii) Construction of the high slopes of the ship lock

The maximum excavation depth in the construction of the permanent ship lock is 176.5 m, which formed the biggest slope of 150 m high, in which 40–68 m high vertical slopes are built as claimed for by the chamber wall structure (**Figure 43**). In order to fulfill the requirement of channel excavation for the chamber and the demand of the slope stabilization, the excavation was conducted in two stages in accordance with the features of the project. The construction procedures are such that, in the first stage, uncovering excavation is applied for the part above the vertical wall, while, in the second stage, slotted excavation is employed for the part below the top of the vertical wall.

In the first stage, the working field was quite open with the minimum width at the bottom being 230 m. The deep-hole stepped blasting was adopted for the middle part, while the lateral protection layer of 5–8 m thickness was reserved on the slopes of both sides. Then, the slope blasting technology was applied to explore and remove the lateral protection layer, forming the designed slope.

The construction procedure in the second stage is as follows. Layer by layer excavation was synchronously conducted on both sides of the central pier. The lateral protection layer corresponding to the first excavation layer was reserved as 5 m thick layer, below which the lateral protection layers were reserved as 3 m thick layer. The presplit blasting was first conducted on the lateral protection layers, followed by the stepped millisecond blasting on the deep groove, and ended finally with smooth blasting on the reserved lateral protection layer and on the slopes.

6.07.4.8.2(iii) Highly intensive construction of dam concrete and its temperature control

The quantity of concrete used for the construction of the TGP is enormous with a total of 27.95 million m^3, of which 16 million m^3 concrete is used for the construction of the dam, especially concentrated on the dam construction in the second stage and featured with tight construction period of concrete placing and high construction intensity. In 2000, the maximum annual concrete construction intensity reached 5.48 million m^3, creating a new world record. Within that period, six sets of tower belt crane placed concrete of 2 million m^3 and played a good role as leading machines.

Due to the importance of the main structures of the TGP and the enormous quantity of concrete, many technologies were adopted to control the construction quality strictly such as the application of secondary air-cooling aggregate and slight

Figure 43 Construction of the permanent ship lock.

Figure 44 Concrete placement on a dam.

expansion concrete, the utilization of water-reducing agent and air-entraining agent, the adulteration of fly ash, the optimization of the mixture proportion of concrete, the reasonable division of the joints and blocks of concrete, the selection of appropriate thickness for each placing layer and suitable period of interval, the improved curing and heat preservation on concrete surface, and the establishment of quality control index for the concrete. **Figure 44** shows the worksite of concrete placement. The series of measures adopted in the whole process from the design to construction ensured the quality of concrete.

6.07.4.8.3 Equipment manufacture

Through international open bidding, the turbine generator units in the left bank power station were supplied by VGS and ALSTOM, and the domestic manufactories participated in the subcontract of equipment manufacture. In 1998, the turbine models supplied by VGS and ALSTOM for the left bank power plant were checked, and it was revealed that the energy and cavitation index met the requirements specified in the contract but the value of pressure fluctuation in part area could not achieve the guaranteed value as set in the contract. In November 1999, based on the summary of parameter selection, design, manufacture, and experiment for the units in the left bank power station, a special study was organized by the owner, with the participation of a scientific research institute, a design institute, and manufactory, on the following aspects:

- the operation stability of the units;
- the optimized hydraulic design of the turbine and its model test;
- the cooling technology for turbine generator of large capacity; and
- the material of the core part of large-scale units.

Through this special study and by learning the imported technology from the units in the left bank power plant and reinnovating, the domestic manufactories mastered the integrated design technique of extra-large units, manufacturing technology, and crucial process, and developed core technologies with self-owned intellectual property rights such as hydraulic design of turbines, fully air-cooling generators, and stator winding insulation. The domestic manufactories became capable of designing and manufacturing 700 MW turbine generator units independently. During the international bidding for the 12 sets of turbine generator units in the right bank power plant in 2004, the HEMC and the DEC each won the contract of four sets of turbine generator units after competing with international enterprises such as ALSTOM, VOITH, and SIMENS.

The domestic manufactories have made the following innovation in the design and manufacture of the units in the right bank power plant.

With regard to the turbine, the results of the comparative model tests conducted on the same experimental equipment for turbines designed by the domestic and international manufactories have shown that the turbine designed by the HEMC and DEC is as good as that by the international manufactories, with the efficiency and stability reaching the international advanced level. The hydraulic design of the turbine has realized self-design with some innovation.

With regard to the generator, the application of fully air-cooling technology in large-capacity turbine generator achieved an important breakthrough. The HEMC developed and manufactured 840 MWA fully air-cooling turbine generator, the largest capacity of the same type in the world. By use of a method that combined the design calculation of advanced ventilation cooling system with ventilation modeling, the design of a ventilation cooling system for the generator was optimized with appropriate total air quantity, reasonable air distribution, and even-distributed temperature, resulting in a fairly good cooling effect. After adopting advanced technologies such as conducting wire corner field intensity treatment and anti-corona technique, the bar insulation dielectric constant, electrical endurance, and anti-corona capability were excellent. All these outcomes break through the original limit of the unit capacity on the adoption of fully air-cooling technology, which opened a bright future for the application of the fully air-cooling technology in large-scale turbine generator units.

In June 2007, the first domestically produced turbine generator unit was put into operation in the right bank power plant of the TGP, symbolizing that the manufacture technology of hydroelectric equipment in China reached the international advanced level and that China made a great leap in development of designing and manufacturing large-scale turbine generator units.

6.07.4.8.4 Hydraulic steel structures

The miter gate, consisting of two single gates, of the permanent ship lock of the TGP is up to 38.5 m high with each single gate in a width of 20.2 m and weight of 850 tons. The maximum operating water head is 36.25 m and the total hydraulic force amounts to 13.64×10^4 kN. The maximum inundated depth is 35 m. Featured with large dimension, high water head, and deep depth of inundation, the lock miter gate ranks the largest one in the world.

The following technologies were mainly adopted in the design, manufacture, and installation of the lock miter gate:

- The introduction of the concept of low-frequency high-stress fatigue convincingly explained the reasons for crack formation on the miter gate of the existing ship lock, providing a theoretical basis for resolving or preventing the problem of crack formation on the structure.
- With regard to the problem frequently occurring in the ever built miter gate, that is, extrusion between the support pad and pillow pad seriously affecting the stress on the pull rod of the top trunnion and the mushroom head of the bottom pintle, countermeasures were adopted, which improved the safety of the miter gate operation.
- Self-lubricating material was utilized for the first time on the bush of the bottom pintle, which resolved the problem of unreliable lubrication of the bottom pintle of the large miter gate due to a passive lubricating system.
- The long-range horizontal-cylinder directly connected hydraulic hoist of stepless speed change was successfully applied for the miter gate installed at the ship lock of the TGP.
- The technology of installing a support wheel at the end of the oil cylinder and adjusting the location of the point where the piston rod pulled the gate was effectively utilized to control the deflection of the piston rod.

References

[1] Changjiang Institute of Survey, Planning, Design and Research (1997) *Yangtze Three Gorges Project Technical Series (in Chinese)*. Wuhan, China: Hubei Science and Technology Press.
[2] Niu X and Wang X (2004) Design and study on the layout of the Three Gorges Project of the Yangtze River. *International Conference on Dam Engineering*. Nanjing, China.
[3] Song W, Niu X, and Dong S (1997) *Study on the Permanent Navigation Structures of Three Gorges Project*. Wuhan, China: Hubei Science and Technology Press.
[4] Niu X and Song W (2004) Navigation structures design of the Yangtze Three Gorges Project. *International Conference on Dam Engineering*. Nanjing, China.
[5] Bureau of Hydrology, Changjiang Water Resources Commission (2008) Hydrology and sediment observation results of TGP in 2007 (No.1–12) (in Chinese), April 2008.
[6] Changjiang Water Resources Commission (1997) *General Introduction of Technical Research on the TGP (in Chinese)*. Wuhan, China: Hubei Science and Technology Press.
[7] Niu X, Xie H, and Liu Z (2008) Study on the direct embedment of spiral case in the right bank power station of TGP, *Yangtze River* (in Chinese), No.1.
[8] Niu X, Yang J, Xie H, and Wang H (2009) Technical research on and application of sloping ceiling tailrace tunnel of Three Gorges Project underground power station, *Yangtze River* (in Chinese), No.23.

Further Reading

[1] Changjiang Water Resources Commission *Technical Research Summary of Three Gorges Project (in Chinese)*. (1997) Wuhan, China: Hubei Science and Technology Press.
[2] Zhang C (2000) Construction of Three Gorges Dam. In: Pan J and He J (eds.) *Large Dams in China – A Fifty-Year Review*. Beijing, China: China Water Power Press.
[3] Niu X, Qiu Z, Wan X, and Tan C (eds) (2003) *Three Gorges Project and Sustainable Development (in Chinese)*. Beijing, China: China WaterPower Press.
[4] Fu X, Zhou S, Yin Z, and Guo Z (eds) (2004) *Reservoir Resettlement (in Chinese)*. Beijing, China: China WaterPower Press.
[5] Niu X and Song W (2006) *Design on Ship Lock and Ship Lift (in Chinese)*. Beijing, China: China Water Power Press.
[6] Yang G, Wong L, and Li L (2007) *Yangtze Conservation and Development Report 2007 (in Chinese)*. Wuhan, China: Changjiang Press.

Relevant Websites

http://www.cjwsjy.gov.cn/ – The Changjiang Institute of Survey, Planning, Design and Research.
http://www.cjw.gov.cn/ – The Changjiang Water Resources Commission.
http://www.ctgpc.com.cn/ – The China Three Gorges Corporation.
http://www.3g.gov.cn/ – The Executive Office of the State Council Three Gorges Project Construction Committee.
http://www.mwr.gov.cn/ – The Ministry of Water Resources of China.

6.08 The Recent Trend in Development of Hydro Plants in India

SP Sen, NHPC Ltd., New Delhi, India

© 2012 Elsevier Ltd. All rights reserved.

6.08.1	Present Status and Future Planning	227
6.08.1.1	World Bank Comments	229
6.08.2	Hydrology and Climate Change	230
6.08.3	Environment Study	234
6.08.4	Reservoir and Downstream Flow	237
6.08.5	Rehabilitation and Resettlement	239
6.08.6	Project Planning and Implementation	240
6.08.7	Storage and ROR Hydroelectric Projects	242
6.08.8	Sediment Transport and Related Issues	244
6.08.9	Socioeconomic Development and Hydropower in the Himalaya Northeast Region	251
6.08.10	Conclusion	252
References		252

6.08.1 Present Status and Future Planning

The installed generating capacity in India as on 31 March 2009 is 147 965 MW. This included thermal (coal, gas, and liquid), hydro, nuclear, and renewable-based generation. Nearly 84.5% of the installed capacity is with State Governments and Central Government-owned companies.

As on 31 March 2009, hydropower constituted 36 878 MW, which is about 25% of the total capacity. The State Organization and Central Government companies have a more prominent role with about 97% of hydropower generation capacity, out of which nearly 73% is in the state sector. India has a federal constitution with 28 states and 7 union territories. Each state has its power utilities producing power and connected through state, region, and country transmission grid.

The energy resources of the country are unevenly distributed with bulk of the hydro resources in the northern, southern, and northeastern parts, and fossil fuel resources in the eastern, central, and western parts.

The Asian Development Bank, in its assessment of hydropower development in India, summarizes as follows [1]. With regard to the generation, particularly the fuel mix, coal is likely to be the mainstay in the near future with focus on clean coal technologies. However, India's coal reserves are limited. There are also problems of high ash content, processing and washing of coal, regulatory issues regarding transportation of coal, environmental issues, and so on. With regard to the option of natural gas, the supplies are very limited and there is a concern of price viability. In case of liquefied natural gas (LNG), it has to be totally imported and is linked to the global price of crude oil; there will be a huge price risk in importing LNG. Presently, there is a renewed focus on nuclear power. However, a very large capacity addition is not likely in the near future. Also there are concerns of the availability of uranium and cost related to its mining. In recent years, the government has been giving special emphasis for promotion of renewable sources of energy, but the contribution for this could be limited, especially if hydropower is not included, considering the large power requirement of the country; hence, keeping in view the country's energy security, accelerated development of hydropower has to be a top priority.

In the present scenario in India, hydro stations are the best choices for meeting the peak demands, which also plays a subsequent role in supplementing and stabilizing a system largely dependent upon thermal sources of energy. Another important role that the hydropower stations are playing and likely to play very effectively in the coming years is as a source for the development of remote and backward areas, especially around the Himalayan belt of northwestern/northeastern India.

The first scientific study to assess the hydroelectric resources in the country was undertaken during the period 1953–59. This study concluded the economical utilizable hydropower potential at 42 100 MW (corresponding to an annual energy generation of 221 billion units).

The reassessment study completed in 1987 by the Central Electricity Authority (CEA) raised this figure to an order of about 84 000 MW (with installed capacity of about 150 000 MW) to be generated from a total of 845 power stations. In addition 56 project sites for development of pumped-storage capacity schemes with aggregate installed capacity of about 94 000 MW were identified.

In the Hydro Development Plan for the 12th 5-year plan, CEA [3] has done a detailed strudy of projects available for implementation. The projects/power stations that have been identified as a potential source of hydropower have been prioritized from the point of view of project implementation and execution by CEA. Based on the present status of preparedness, the potential projects have been classified into category 'A', 'B', and 'C'. Ten major aspects that play a vital role in implementation of all hydro projects were adopted and considered as the criteria for a ranking study. These were rehabilitation and resettlement aspects, international aspects, interstate aspects, potential of the scheme, type of scheme, height of dam, length of conductor system, accessibility to site, status of the project, and status of upstream or downstream hydropower development, but not in the same order of importance as listed here.

Four hundred schemes with a probable installed capacity of about 107 000 MW were prioritized in these categories. Accordingly, Category A has 98 schemes 15 641 MW, Category B has 247 schemes 69 853 MW, and Category C has 54 schemes 21 416 MW that was identified by CEA.

Subsequently in 2003, CEA initiated a process of preparation of a pre-feasibility report of 162 schemes at a cost of US$5 million and awarded to seven Government-owned agencies/State agencies as consultants. The pre-feasibility report was more a desk study based upon data/information already available for such project sites and use of satellite imageries, remote sensing information, and a reconnaissance survey/visit by a multidisciplinary team.

Out of 162 projects and 47 930 MW generating capacity proposed, projects that can be or will be pursued with approach to expeditious development will be about 140 and with installed capacity around 40 000 MW. These projects will have about 700 km of tunnels to be constructed mainly in the difficult terrains of the Himalayas and will have gross storage of about 15–20 billion m^3.

The country is now in the middle of the 11th Plan spanning 2007–12. In the 11th Plan, the total capacity addition of 78 000 MW, out of which 15 627 MW is from hydro projects, is proposed. Up to 31 March 2009, 3431 MW of hydro projects have been commissioned. Balance projects are under active execution. Annual accounting and planning is from 1 April of a year to 31 March of the next year. For ensuring the 12th Plan success spanning years 2012–17 CEA has adopted a strategy of advanced planning [2]. Since early 2008 it has started identifying the shelf-life of projects, which are likely to be a potential candidate for the 12th Plan. Eighty-seven projects with likely benefits of 20 000 MW in the 12th Plan have been identified.

Presently, in India it takes about more than 10 years for developing medium to large size hydro projects, concept to commissioning. The construction period of a reasonable size of hydro project after obtaining all the clearance and financial arrangements varies from 5 to 7 years.

The advance planning and monitoring of these projects have started in right earnest, and required statutory clearances, necessary action to fix the infrastructure bottlenecks, and so on, are being taken up actively. Project owners, both government and private agencies, are being regularly assisted for the purpose of advanced project implementation planning.

This process shall go a long way to achieving the ambitious programme of 20 000 MW development in the 12th Plan.

The total potential in hydroelectricity as assessed by CEA is 140 701 MW, out of which the capacity developed is 36 878 MW and under development is another 13 675 MW. The region-wise hydropower potential in terms of installed capacity is given in **Table 1**. Contributions from the private sector for hydropower development has been small to date with the major developers being the state and central Agencies (**Figure 1**).

Breaking up this potential as per geographical region and basin, the hydropower potential is concentrated mainly in Himalayan river basins, which are the Brahmaputra, Indus, and Ganga. The rest of the potentials is in the peninsular rivers or non-Himalayan rivers (**Table 2**).

About 120 000 MW is presently from the Himalayan rivers, out of which only about 18 500 MW have been developed. In the peninsular rivers, the potential is only 28 000 MW, of which about 19 400 MW has already been developed. In the past few years, more development of hydropower has taken place in peninsular rivers other than Himalayan rivers in relation to the

Table 1 Break up of hydropower potential by region

Region	Potential assessed (MW)	Potential developed (MW)	Potential under development (MW)	Balance potential (MW)
North Eastern	58 971	1 116	3052	54 803
Northern	53 395	13 425	7529	32 441
Eastern	10 949	3934	2307	4708
Western	8928	7449		1479
Southern	16 458	10 954	787	4717
Total	148 701	36 878	13 675	98 118

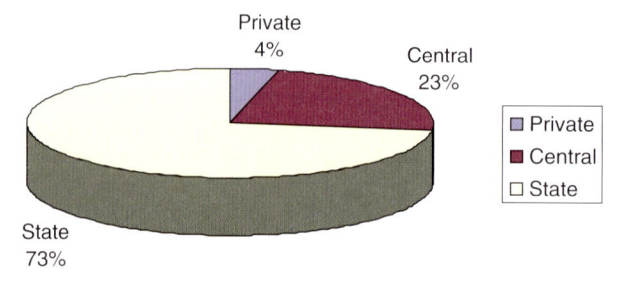

Figure 1 Generation of hydropower by sector.

Table 2 Break up of hydropower potential by river basin

Geographic region	Basin	Hydro potential (MW)	Remark
Himalayan rivers	Brahmaputra	66 065	120 608
	Indus	33 832	About 30 300 already developed and under development
	Ganga	20 711	
Peninsular river	East flowing river	14 511	28 093
	West flowing river	9430	About 20 200 already developed and under development
	Central Indian river	4152	
Total		148 701	

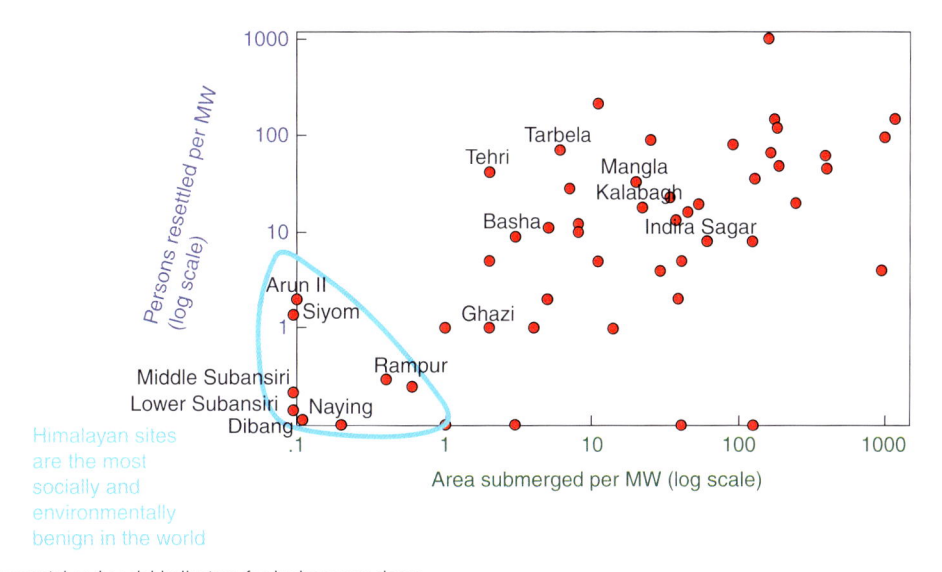

Figure 2 Environmental and social indicators for hydropower dams.

potential available. It is recognized today that majority of the Himalayan sites are the most socially and environmentally benign in the world (**Figure 2**) [8].

Looking at hydropower development from a global point of view, the most encouraging development is that after many years of contradictory approach, the World Bank now considers hydropower of all sizes and configurations to be renewable. At present, hydropower accounts for more than half of the World Bank group's renewable energy portfolio. It is stated by the World Bank that hydropower infrastructure plays a dual role in meeting the climate change challenges. It is the largest source of affordable, renewable energy, and a low carbon fuel plays a critical role in mitigating greenhouse gas (GHG) emission. Increasing the share of hydropower in India's energy mix from the present 24% to around 35% (CEA generally proposed about 40%) will avoid 138 Mt CO_2 per year from alternative coal generation, equal to 8.5% of emission in India in 2015.

Hydropower infrastructure also plays an important role in climate adaptation. Climate change will exacerbate hydrologic variability, the consequence changes in the long-term water balance, and intensification of extreme weather events. In India, the rainfall season is well defined and covers only a period of 4–5 months out of 12 months, and out of 4 months, 70–80% of the rainfall comes in 20–25 days in the full monsoon period (the effect of hydrological variability shall be even more intense); this increases the risk and uncertainty in the hydrological infrastructure management and operation.

In the recent past, hydropower development in India was largely affected due to many issues over and above the financing of the project and are as follows: poorly identified and managed projects, environmental risks, a narrow approach to resettlement issue based on compensation for land, and also most importantly no priority to the socioeconomic development of the people in and around of the project areas. **Table 3** [6] gives a target versus achievement of hydropower capacity addition plan-wise that reflects sluggish development.

6.08.1.1 World Bank Comments

Hydropower being an indigenously available, clean and renewable source of energy, the Government of India is keen to use the largely untapped potential in this area – currently, only 23% of India's hydro potential is being utilized to provide the additional generating capacity it needs [4].

Table 3 Break up of hydropower capacity addition from each 5-year plan

No.	Plans	Target capacity addition (MW) Central	State	Private	Total	Actual capacity addition (MW) Central	State	Private	Total	Achieved (%)
1	5th Plan (1974–79)				4654				3812	82
2	6th Plan (1980–85)				4768				2873	60
3	7th Plan (1985–90)				5541				3828	69
4	8th Plan (1992–97)	3260	5860	162	9282	1464	795	168	2427	26
5	9th Plan (1997–02)	3455	5815	550	9820	540	3912	86	4538	46
6	10th Plan (2002–07)	8742	4421	1170	14393	4495	2691	700	7886	55
7	11th Plan (2007–12)	9685	3605	3263	15627					

Moreover, additional hydropower capacity is desirable in India's generation mix, as it provides the system operator with technically vital flexibility to meet the changes in demand that typically affect a power network like that of India. The high density of household demand in India means that the system can experience a peaking load of anything between 20 000 and 30 000 MW. This sudden spurt in demand can be best met by hydropower plants that have the ability to start-up and shutdown quickly. Other sources of power cannot do this as economically as hydropower plants.

Also, the Government of India is committed to developing world-class companies that are able to design, construct, and maintain hydropower projects to international standards, and has requested the World Bank's support in this endeavor. In addition to helping with financing, the Bank brings extensive experience in developing such projects across the world.

A number of factors are essential for such projects:

- Careful selection of the site and appropriate engineering design
- Solid initial investigations, especially regarding geological conditions
- Strong and competent implementing agencies
- Continued and substantive consultations with stakeholders
- Early attention to social and environmental aspects of projects, in particular, mitigating the negative social and environmental impacts of the projects
- Appropriate financing and tariff design that are critical to the financial sustainability of projects with long gestation periods.

Another important goal that is a little sensitive but important from the Indian context is the major tribal and ethnic groups that live along the Himalayan region. These groups need to be integrated into the mainstream of Indian socioeconomic growth, without imposing change to their basic social and ethnic cultural structure. In the remote areas of the Himalayas where agriculture is limited, hydropower development will probably be the only major driving force for socioeconomic development of these people.

6.08.2 Hydrology and Climate Change

Out of the total precipitation, including snowfall, of around 4000 km^3 in the country, the available surface water and replenishable groundwater is estimated to be 1869 km^3[5]. Due to various constraints of topography and uneven distribution of resources over space and time, it has been estimated that only about 1128 km^3, including 690 km^3 from surface water and 433 km^3 from groundwater resources, can be put to beneficial use. Table 4 shows the water resources in the country at a glance.

Table 4 Water resources in India

Estimated annual precipitation (including snowfall)	4 000 km^3
Average annual potential in rivers	1869 km^3
Estimated utilizable water	1123 km^3
Surface	690 km^3
Ground	433 km^3
Water demand = Utilization (for year 2000) (634 km^3)	
Domestic	42 km^3
Irrigation	541 km^3
Industry, energy, and others	51 km^3

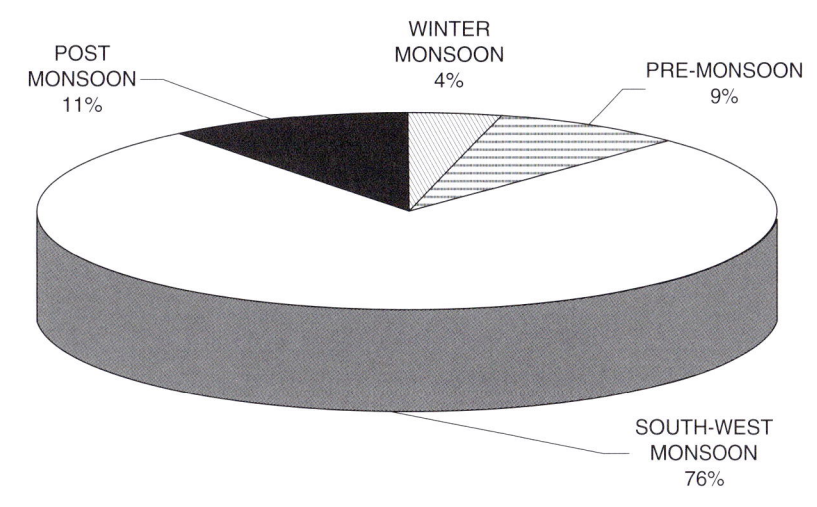

Figure 3 Season-wise rainfall in the country (1.1.2003 to 31.12.2003).

Many Indian rivers are perennial, though few are seasonal. Precipitation over a large part of India is concentrated on the monsoon season during June to September and October. Precipitation varies from 100 mm in the western parts of Rajasthan that has desert characteristics to over 11 660 mm at Cherrapunji in northeastern Himalaya in the state of Meghalaya. **Figure 3** shows the season-wise rainfall in the country as documented by the Central Water Commission in 2003. As already discussed, the monsoon season is between June/July to September/October depending on the region of the country.

There are 12 major river basins with a catchment area of 20 000 sq km and above. The total catchment area of these rivers is 2.53 million sq km, out of which three Himalayan rivers namely Ganges has a catchment area of 861 452 sq km, Brahmaputra and Barak has a catchment area of 236 136 sq km, and Indus up to the Indian border has a catchment area of 321 289 sq km. Other major peninsular rivers are Mahanadi, Godawari, and Krishna. River basin-wise riverine length is given in **Figure 4**.

The distribution of water resources potential in the country shows that as against the national per capita annual availability of water of 1905 m^3, the average availability in Brahmaputra and Barak is as high as 16 589 m^3, while it is as low as 360 m^3 in the Sabarmati basin. The Brahmaputra and Barak basin with 7.3% of geographical area and 4.2% of population of the country has 31% of the annual water resources. The per capita annual availability for the rest of the country, excluding the Brahmaputra and Barak basin, works out to about 1583 m^3. Any situation of availability of less than 1000 m^3 per capita is considered by international agencies as scarcity conditions. Cauvery, Pennar, Sabarmati, east flowing rivers, and west flowing rivers are some of the basins that fall into this category.

The Himalayas is a large mountain system, influencing the interaction between climate hydrology and environment. The total spread of Himalayas between latitude 25° and 35° N and longitude 60° to 105° E covers an area of 844 000 sq km.

All the major north and northeast Indian rivers own their origin to thousands of glaciers in the Himalayas. There are 9575 glaciers in the Indian Himalayas as per the latest update of the glacier inventory maintained by the Geological Survey of India.

The Indian part of the Himalayas above elevation 1060 m covers an area of 350 000 sq km out of which 190 000 sq km form a part of Jammu and Kashmir, Uttarakhand, and Himachal Pradesh, and the rest covered by eastern Himalayas. Distribution of glaciers is controlled by the altitude orientation, slope, and climate zone in which they fall. The Indus basin has 7997 glaciers, the

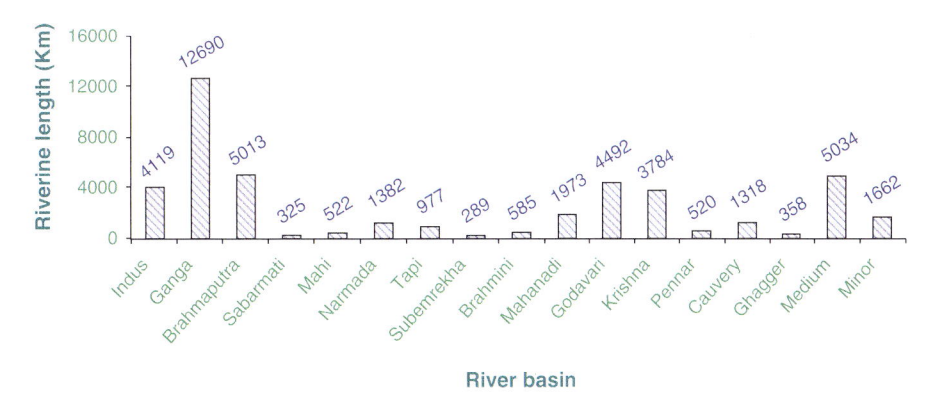

Figure 4 River basin-wise riverine length.

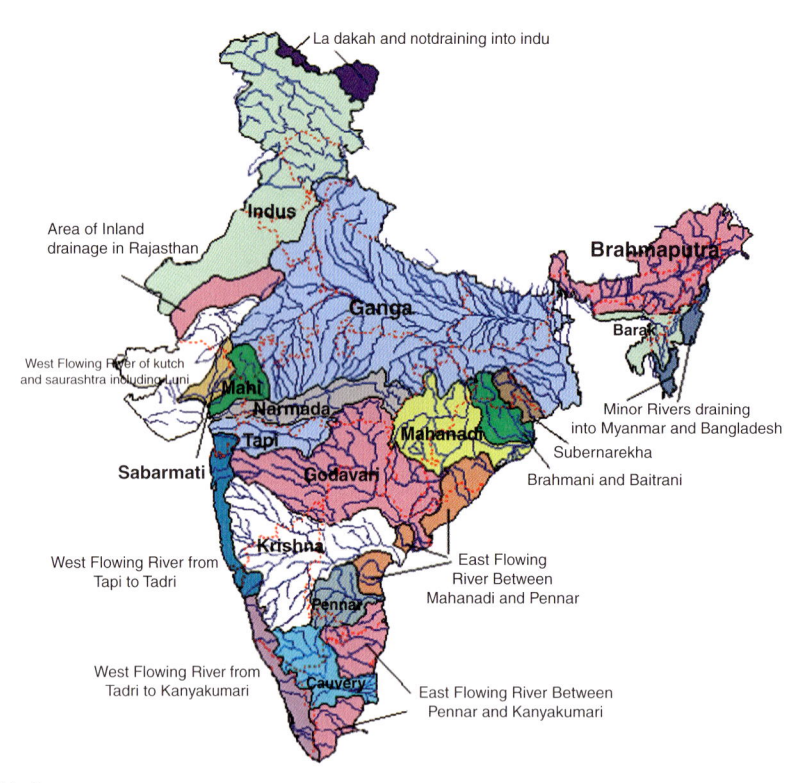

Figure 5 River basins of India.

Ganga basin has 968 and the Brahmaputra along with the Teesta has 610. The Brahmaputra through its major tributary Siang is fed by Tibetian cold desert that keeps its non-monsoon flow quite high.

The total area covered by the Indian glaciers is about 18 054 sq km, whereas the volume is about 1219 km^3. A basin map of India is show in **Figure 5**.

In India, several studies have been carried out to determine the change in temperature and rainfall and its association with climate change. Investigators using different data lengths and studies have been reported using more than a century of data. All such studies have shown warming trends on the country scale. An analysis of the seasonal and annual air temperature from 1881 to 1997 by Parthsarthy and Kumar shows that there has been increased trend in mean annual temperature by the rate of 0.57 °C per 100 years. The trend and magnitude of global warming over India/Indian subcontinent over the last century has been observed to be broadly consistent with the global trend and magnitude. In India, warming is found to be mainly contributed by the postmonsoon and winter season. The monsoon temperature does not show a significant trend in any part of the country, except for a negative trend over northwest India. This temperature anomaly is given in **Figure 6**.

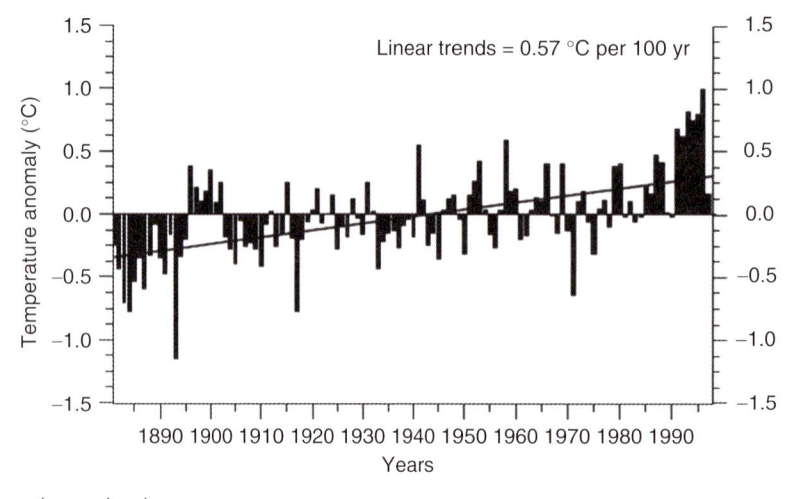

Figure 6 Temperature anomaly over decades.

Even during the twentieth century, an analysis of long-term temperature records (1901–82, 73 stations) has shown an increasing trend of mean annual surface air temperatures over India.

It was observed that about 0.4 °C warming has taken place on a country scale during a period of 80 years. However, studies do not show an increasing trend over the entire country. The temperature shows cooling trends in the northeast and northwest India, that is, along the Himalayas.

Studies related to change in rainfall over India have shown that there is no clear trend of increase or decrease in average annual rainfall over the country. The examination of trend of annual rainfall over India has indicated that 5 year running mean has fluctuated from normal rainfall within ± 1 standard deviation. Summer monsoon rainfall anomalies all over India are shown Figure 7.

Mirza *et al.* carried out trend and persistent analysis for Ganges, Brahmaputra, and Meghna river basins. These have shown that precipitants in the Ganges basin are by and large stable. One of the three divisions of the Brahmaputra basin shows decreasing trends, while another shows increasing trend. As in coming years a major number of hydroelectric and water resources projects are to be built in Brahmaputra and Ganges basins. This information shall have a qualitative contribution in planning, development, and management of water resource in these basins.

Basin-wise flow and storage potential of the major rivers as documented by the Central Water Commission is shown in Figure 8. From the figure it can be seen that when the Brahmaputra and Barak has an average annual flow of 585.6 billion m^3, only 11.68 billion m^3 of live storage capacity has been developed. Similarly for the Ganga that has an annual flow of 525.02

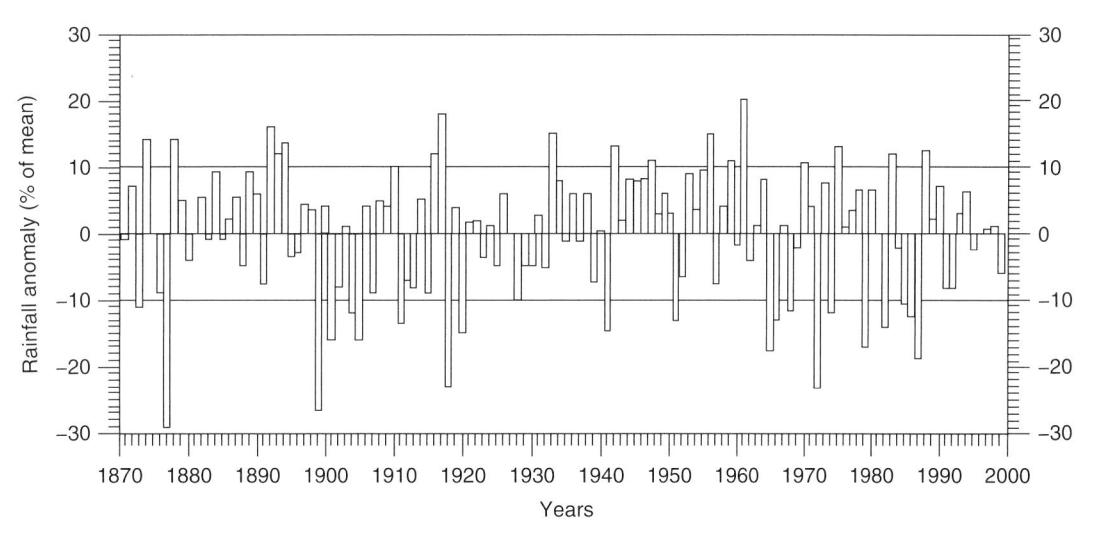

Figure 7 Rainfall anomaly from 1870 to 2000.

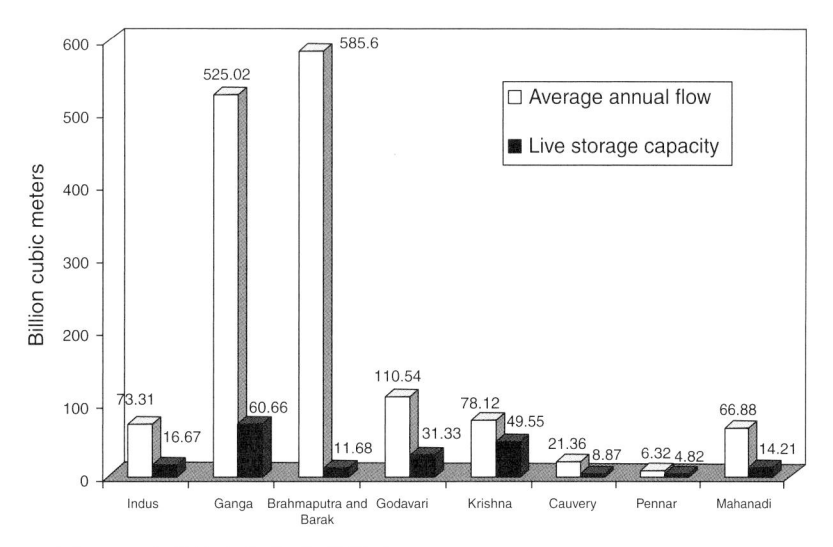

Figure 8 Basin-wise flow and storage potential in India (up to IX Plan).

billion m³, only 60.66 billion m³ has been developed up to 2002. Hence, it is obvious that both in the Brahmaputra and the Ganges the immediate requirement of development of storage capacity is there. Over and above the requirement of hydroelectric projects they will help in flood control, especially in Brahmaputra, and also removing the future uncertainty of water supply and food security.

Development of hydropower in the Himalayas has its own challenges. Many of the proposed projects that are going to be built in the coming years will be in the remotest corners and in the hostile geohydrological environment. Glacier lake out burst flood, cloud burst flood, land slide dam burst, huge sediment movement generated due to described events and also due to bank failure, infrastructure activity, and so on, are some of the additional hydrological hazards with such projects, during construction and also in postconstruction stages.

6.08.3 Environment Study

As per Ministry of Environment and Forest (MOEF) notification of 1994 under the provision of the Environment Protection Act of 1986, environmental clearance is mandatory for river valley projects, including the multipurpose ones. Environmental Impact Assessment (EIA) Notification 2006 requires an application seeking prior environmental clearance in all cases shall be made after the identification of prospective site for the project and/or activities to which the application relates, before commencing any construction activity, or preparation of land, at the site by the developer. The developer shall furnish, along with the application, a copy of the pre-feasibility project report.

The environmental clearance process for new projects will comprise of a maximum of four stages, all of which may not apply to particular cases as set forth in the notification. These four stages in sequential order are:

Stage (1) Screening
Stage (2) Scoping
Stage (3) Public consultation
Stage (4) Appraisal

Screening will entail the scrutiny of an application seeking prior environmental clearance made, for determining whether or not the project or activity requires further environmental studies for preparation of an EIA for its appraisal prior to the grant of environmental clearance depending upon the nature and location specificity of the project. For the majority of hydropower projects prior environmental clearance is required.

Scoping refers to the process by which the Expert Appraisal Committee determine detailed and comprehensive terms of reference (TOR) addressing all relevant environmental concerns for the preparation of an EIA report in respect of the project or activity for which prior environmental clearance is sought. The Expert Appraisal Committee shall determine the TOR on the basis of the information furnished in the prescribed application; TOR proposed by the applicant may or may not be a site visit by a subgroup of the Expert Appraisal Committee or state-level Expert Appraisal Committee.

After the EIA and Environment Management Plan has been submitted by the project authority, subsequent stages start.

Public consultation refers to the process by which the concerns of local affected persons and others who have a plausible stake in the environmental impacts of the project or activity are ascertained with a view to taking into account all the material concerns in the project or activity design as appropriate.

The public consultation shall ordinarily have two components comprising of:

1. A public hearing at the site or in its close proximity, district-wise, to be carried out in the manner prescribed, for ascertaining concerns of local affected persons.
2. Obtain responses in writing from other concerned persons having a plausible stake in the environmental aspects of the project or activity.

The public hearing at, or in close proximity to, the site in all cases shall be conducted by the State Pollution Control Board or the Union Territory Pollution Control Committee concerned in the specified manner and forward the proceedings to the regulatory authority.

For obtaining responses in writing from other concerned persons having a plausible stake in the environmental aspects of the project or activity, the concerned regulatory authority shall invite response in writing. After completion of the public consultation, the applicant shall address all the environmental concerns expressed during this process.

Appraisal means the detailed scrutiny by the Expert Appraisal Committee of the application and other documents like the final EIA report, outcome of the public consultations including public hearing proceedings, submitted by the applicant to the regulatory authority concerned for grant of environmental clearance.

It shall be mandatory for the project management to submit half-yearly compliance reports in respect of the stipulated time prior to environmental clearance terms and conditions to the regulatory authority concerned, on 1 June and 1 December of each calendar year.

A prior environmental clearance granted for a specific project or activity to an applicant may be transferred during its validity to another legal person entitled to undertake the project or activity by following the laid down procedure.

Many factors related to high capital costs, uncertain geology, construction scheduling and construction management, climate change and variable hydrology evolving an uncertain market role, multidisciplinary and cross-sectoral project design, and last but not the least corruption are contributing to the risks. Of particular areas are the risks associated with environment management, inclusion, and appropriate sharing of benefits and rents. The effects of all the above factors are prominently visible and are major bottlenecks in developing the fragile socioeconomic environment in the Himalayas. Like many other countries, one of the biggest challenges in preparing the environment assessment and report is that the TOR the Government issues to guide the study is only general and not site-specific. As a result, when the project comes for examination or for the consent on the socioeconomic issue, a specific factor, which otherwise may turn out to be important for that project, has not been studied because of reference issued by the government. Such study cannot be site-specific as it would require much more specific and elaborate study in advance for identification of such issues by the government. However, to some extent this situation could be avoided by developing site- and region-specific TORs, with each of the agencies involved in the approval process specifying the details that it will require to approve the project.

In one case, the catchment area treatment plan for Chamera Hydroelectric Project Stage III in the state of Himachal Pradesh was prepared based on remote sensing data and the silt yield index method, as per the guidelines of MOEF, Government of India, the approving agency for forest, and environment study. However, the State Forest Department of Himachal Pradesh wanted several additions in the plan expanding its scope and cost. Hence, the cost of the Catchment Area Treatment (CAT) plan went up from US $3.46 million to US$8.51 million. The issue became a point of dispute between the owner of the project and the State Government with MOEF as arbitrating agency. Ultimately, it was finalized at a cost of US$6.34 million and considerable time was lost whose hidden cost is not accounted for [7].

In recent years, the Supreme Court of India has taken over the final clearance authority of diversion of forest land for project construction or any other activity. It also has put a restriction on diversion of declared wild life area and reserved forest. In October 2002, the Supreme Court of India issued an order for Net Present Value (NPV) payable on forest area when directed for non-forest purpose. The NPV is charged in addition to the following compensation costs/expenses paid to concerned State Forest Department by the project developer; in lieu of diversion of forest land:

1. Cost of tree, poles, and so on, falling in the forest area
2. Cost of any other structures constructed within the required forest land
3. Cost of compensatory afforestation for raising plantation over degraded forest area, twice in extent of the required forest land. In case degraded forest land is not available, non-forest land is provided by the user agency and transferred to the State Forest Department for raising compensatory afforestation
4. Cost of implementation of CAT plan.

The NPV rate as approved by the Supreme Court ranges from US$12 000 to US$19 575 per hectare, which is quite substantive. Some of the projects for which environmental cost has been estimated are given in Table 5.

Such high-cost provision for environmental preservation have created problems in two ways:

1. High cost for environmental protection is making the project sometimes unviable.
2. There is no structured mechanism with State Government to spend such money in a proper way.

To take care of the second issue, an authority to be known as the 'State Compensatory Afforestation Fund Management and Planning Authority' (State CAMPA) is intended as an instrument to accelerate activities for preservation of natural forests, management of wildlife, infrastructure development in the sector, and other allied works.

The State CAMPA receive monies collected from user agencies toward compensatory afforestation, additional compensatory afforestation, penal compensatory afforestation, NPV, and all other amounts recovered from such agencies.

State CAMPA shall seek to promote:

1. Conservation, protection, regeneration, and management of existing natural forests
2. Conservation, protection, and management of wildlife and its habitat within and outside protected areas including the consolidation of the protected areas

Table 5 Environmental cost compared with total project cost

No.	Items	Subansiri lower (2000 MW)	Teesta lower dam – III (132 MW)	Siyom (1000 MW)	Tipaimukh (1500 MW)
1	Total project cost Million US$	1 406	166	1 000	1 246
2	Environmental cost Million US$	17.70	4.53	24.79	65.9
3	Environmental cost with NPV Million US$	81.53	8.26	67.05	300
4	Total environmental cost as % of total project cost	5.8	4.96	6.71	24.18

3. Compensatory afforestation
4. Environmental services.

However, this mechanism is yet to be functional and effective.

Baseline information for environment assessment and reports should be prepared by government experts, not only to reduce the cost to the industry or preparing the reports but also to increase confidence in their conclusions. This will expedite the project implementation process.

It is understandable that different agencies of government in both state and central focus on different aspects of a particular project; yet a more holistic approach would enable potential developers to fine-tune their projects from the start. However, today the forest and environmental clearance takes more than a year after submission of EIA study and Detail Forest Land Acquisition Proposal by the developer to the MOEF. This needs to be expedited.

Perhaps a significant improvement could come from creating an independent council that engages all the agencies and the sector's professional involved in the planning and approval process especially on environment and social issues. Such a body would provide a unified presence that would inevitably lead to greater understanding and awareness of the multiple needs that the project must address. An independent council with proper authority granted to it would also be able to remove potential obstacles from the beginning and serve as a forum for resolution of the problem that might occur along the way.

The 412 MW Rampur Hydropower Project, located in the state of Himachal Pradesh is planned as a cascade plant to India's largest hydroelectric power plant, the 1500 MW Nathpa Jhakri. The World Bank is actively involved in this project. A 15 km underground tunnel will carry water emerging from the Nathpa Jhakri plant and bring it downstream to a powerhouse located near Bael village in Kulu district. It uses silt-free water from the Nathpa Jhakri plant; the Rampur Project will neither involve the construction of any dam or reservoir or desilting chamber nor will any land be inundated for the scheme. The project has funding assistance from the World Bank.

The location and design of the Rampur Project has been finalized with the aim of minimizing adverse impacts on local people and their natural environment. Some 79 ha spread across eight panchayats (village elected bodies' jurisdiction) was acquired for the project; of this, 49 ha is forest land (although largely without forest cover) belonging to the Himachal Pradesh state government and some 30 ha is private land belonging to 141 families comprising 167 landowners (**Figure 9**).

The displaced families who lose their houses will each get a plot of 280 sq m at a site of their choice on which they can build their new houses. The families had a choice of opting a developed house or a plot, but all chose to construct their own houses. They will be given monetary help for the construction of 60 sq m of built-up plinth area on which they can construct their new homes, as well as a monthly rental allowance to help them tide over the period of construction (18 months) in a rented house. Each family will also receive a lump sum amount to help them meet the costs of shifting from one house to another.

A special package has been worked out for those 35 families who will be left with less than 5 Bighas (1 Bigha = 809 sq m in the State of Himachal Pradesh) of land after the project has acquired their land it needs. Apart from the compensation for the acquired land, they will also receive a rehabilitation grant, depending on the amount of land left with them after acquisition. In order to help the project-affected persons (PAPs) recover from any loss of livelihood and also in order to help those interested in setting up additional income-generation schemes, the owner will also offer seed money of up to US$640.

The company has also undertaken to give preference to suitably qualified candidates from landless families whenever a job opening comes up. The contractors working on the civil works of the project have also been directed to give preferential employment to people from the project-affected area while hiring labor. All petty contracts on the project up to a value of US$21 275 are also being ear-marked for PAPs. About US$255 320 of such contracts have already been awarded to PAPs and more worth US$2 million have been given to people from other parts of Himachal Pradesh. Children from project-affected families and areas are being offered merit scholarships to acquire technical and vocational skills and the first batch of 35 students, including four girls, are already receiving training in a variety of trades.

Figure 9 Public consultation on the resettlement action plan.

Figure 10 A footpath to village Bakhan constructed under the project's community infrastructure program.

The villages impacted by the project have also been ear-marked for special development assistance (**Figure 10**).

The owner has set aside US$2.66 million to be invested over a period of 5 years in infrastructure and development schemes for these villages. Here again, the people have led the local area development exercise, choosing the infrastructure schemes they would like to see implemented in their villages. From street-lighting, through improved water supply to footpaths and footbridges, the villagers have identified their particular needs that are being funded by the scheme. The company also runs a mobile health van that travels round the project-affected villages taking basic healthcare to the doorstep of people living in remote areas and the project is also setting up a dispensary at the village, near the site of the proposed powerhouse for the Rampur Project.

The owner, who as the developer of the already operational Nathpa Jhakri Project has a long-standing relationship with the region, is also helping improve the quality of people's lives beyond the project-affected villages. The Company is helping finance the renovation of the bus stand at Rampur town; it is also helping build several access roads and bridges and helping improve infrastructure in local schools.

Benefits to Himachal Pradesh is apart from the 12% free power it receives as royalty (worth approximately US$13 million), the host state of Himachal Pradesh will also get an additional 30% of power generated at Rampur Project (109 MW) at cost; this is equivalent to its share of equity percentage in the project. And, as part owner of the developing agency, developing the Rampur Project, Himachal Pradesh will also receive dividends on its investment in the project and also be entitled to a share in the remainder of the power generated from the project.

The state also stands to gain in terms of job creation and income generation. The Rampur Project has already generated some 2500 man-months of work for the people of Himachal Pradesh over the past 1 year, and some US$2.28 million of petty contracts on the project have already gone to people belonging to the state. So far 145 members of the families affected by the project were offered work under contractors.

6.08.4 Reservoir and Downstream Flow

The major projects in the Himalayas are being conceived as classical run-of-the-river (ROR) schemes. Even some of the large projects that are proposed to have considerable storage and can be termed as storage projects are also ROR projects.

For example, some large projects investigated/under construction over river Brahmaputra are shown in **Table 6**.

Though the projects have substantive storage, they have all been designed as ROR projects to maximize the benefit of the power generation. Hence, the role of the storage for such big hydroelectric projects is limited from the point of view of power generation. However, their role to mitigate fluctuation in power generation due to climate change and for flood control, water supply,

Table 6 Gross storage, energy, and MW of some major projects

No.	Projects	MW	Gross head (m)	Energy (MUs)	Gross storage (Mcum) at FRL
1	Dibang	3000	288	11330	3748.21
2	Siyom	1000	188	3641	558.33
3	Siang lower	2000	110	10980.52	1421
4	Subansiri lower	2000	91	7421.59	1365
5	Subansiri middle	1600	171.8	4874.88	1687.7
6	Subansiri upper	2000	199.5	6581.29	1743

downstream environmental flow, and so on, is quite important. All the projects listed in the table and many such projects that are being conceived and designed over the major rivers with a very high monsoon flow are being conceived as storage with the high head generated by constructing dams leading to longer reservoirs but relatively narrow and deep. While conceiving such projects it is being ensured that such reservoir remains in the river channel and in the river flowing valley and does not develop a wide reservoir area submerging agricultural land, household, and reserved forests. As a result, such a reservoir is having a benign effect on the population and environment as already shown in **Figure 2**. For such projects, the powerhouse is almost at the tow of the dam or a few hundred meters downstream if the powerhouse is underground. Hence, the requirement of dedicated environmental flow during the monsoon period is not a necessity.

However, for such Himalayan rivers the daily average flow goes down to 20–30% in the dry season compared to the monsoon season. During this period if the powerhouse is conceived to generate peak power near to its full capacity for a few hours in a day, then it is required to be done by storing the water and releasing by a few hours in a day when it is generating the maximum. In such cases, there is a substantive fluctuation in downstream flow in two ways:

1. By stopping the flow during the longer period of the day
2. Releasing high/very high discharge compared to daily dry flow for limited hour.

In such cases, optimized generation planning during the dry season can take care of the necessity for environmental flow; however, the quantum and quantity requirement of such a flow requires that they be studied in detail for the dry season period. Premonsoon, monsoon, and postmonsoon periods are not really affected by such storage and dam-toe powerhouses as far as downstream environmental flow are concerned.

The majority of the storage projects being constructed in the Himalayas are in the downstream reach, almost to the foothill where the river carries very high discharge and has a relatively flat slope. However, GHG emissions from such a reservoir are likely to be little due to reasons that they are shallow and also located in a temperate/cool weather region.

There is considerable debate at present on how GHG emissions from reservoirs should be determined.

Draft guidelines of the Intergovernmental Panel on Climate Change set up potential methodologies based on decay of flooded vegetation, discounted gross carbon dioxide emissions, and undiscounted gross methane emissions. All potential methodologies overestimate the anthropogenic contribution of hydropower reservoirs.

Using a completely different approach, the Executive Board of the Clean Development Mechanism set qualification parameters for hydropower in March 2006, based entirely on the capacity density of a hydropower scheme:

1. Where the scheme has a density of less than 4 Watts installed capacity per square meter of reservoir, it is deemed to not qualify.
2. Where the scheme has a density of greater than 10 Watts m^{-2} of reservoir, it is deemed to qualify.

In between 4 and 10 Watts m^{-2} of reservoir, it is given a default value of 90 tonnes GWh^{-1}.

This method is even considered unsatisfactory and effectively excludes most storage hydro from the Clean Development Mechanism.

However, following this criteria a few of the reservoirs for hydropower projects in Himalaya capacity density have been calculated as given below (**Table 7**). It can be seen that all of them satisfy the criteria, except Tehri Multipurpose.

The other kind of hydroelectric development in the Himalayas and specially in the upper reaches of the Himalayas where the discharge is not very high but the river is quite steep is conceived by building some diversion structure with very little storage and the generating head by constructing tunnel/channel of considerable length ranging from 4 to 5 km to even 20 km and then mainly underground powerhouse and in some cases surface powerhouse. In such projects design flow for full capacity of generation will be around 50–75% of the regular monsoon flow; hence, the downstream flow during the monsoon period is not really much affected as 25–50% of flow continues in the river up to the powerhouse. Further downstream full flow is revived. However, during

Table 7 Capacity density of some hydropower stations in Himalaya

No.	Name of project	Capacity (MW)	Submergence area (ha)	Capacity density in watt sq m^{-1} of submergence
1	Baira Siul	180	15.2	1184.21
2	Chamera I	540	975	67.29
3	TLDP III	132	172.9	76.34
4	Subansiri Lower	2000	3436	58.21
5	Siyom	1000	1891.44	52.87
6	Uri II	240	75	320.0
7	Dibang Multipurpose	3000	4009	74.83
8	Nimo Bazgo	45	342	13.16
9	Tehri Multipurpose	1000	42000	2.38

Table contains both ROR and reservoir projects.

non-monsoon months if the diversion structure is holding back the daily flow for maximizing the power generation, then the problem is again seen in two ways:

1. During this period in between diversion structure and the powerhouse throughout the day flow is very little, that is, for a long stretch of river.
2. The problem is also in fluctuation of flow from powerhouse to downstream. A careful evaluation of requirement of environmental flow taking into account regenerated flow, flow coming from other streams and rivulets in this reach, and different aspects of environmental requirement required to be accessed and release of downstream flow from the diversion structure, become important factors.

For design and engineering of such projects, a more exhaustive study is required for environmental flow. In many of the cases, such areas are not accessible due to its isolated location, steep mountains, deep forests, and so on. Information for such components are collected more by indirect techniques and sometimes from regional and local information available in macroscale. All these factors increase the uncertainty of downstream flow study.

6.08.5 Rehabilitation and Resettlement

In February 2004, the Ministry of Rural Development adopted a National Policy on Resettlement and Rehabilitation. It was stated in the preamble that there is a need to minimize large-scale displacement and to handle the issues related to resettlement and rehabilitation with utmost care. The intention of the policy is "to impart greater flexibility for interaction and negotiation so that the resultant package gains all round acceptability in the shape of a workable instrument providing satisfaction to all stakeholders/ requiring bodies".

Then again the Ministry came up with a new National Rehabilitation and Resettlement Policy that came into operation on the 31 October 2007. Two major points mentioned in the objectives are:

1. To provide a better standard of living, making concerted efforts for providing sustainable income to the effected families
2. To integrate rehabilitation concerns into development planning and implementation process.

Following are some important provisions:

1. The Act shall apply to the rehabilitation and resettlement of persons affected by acquisition under the Land Acquisition Act, 1894.
2. The definition of the affected family also quite exists. Besides the land holder and tenants and lessees of the acquired land, it includes any agricultural or nonagricultural laborer, landless person (not having homestead land, agricultural land, or either homestead or agricultural land), rural artisan, self-employed person; who has been residing or engaged in any trade, business or occupation or vocation continuously for a period of not less than 5 years in the affected area proceeding the date of declaration of the affected area, and who has been deprived of earning his livelihood or alienated wholly or substantially from the main source of his trade, business, occupation, or vocation because of the acquisition of land in the affected area or being involuntarily displaced for any other reasons.
3. A provision is made in the amending bill to the Land Acquisition Act to ensure that a social impact assessment shall be carried out in cases involving the physical displacement of 400 or more families in plains or 200 or more families in tribal or hilly areas.
4. The bill provides that the social impact assessment clearance shall be granted in such manner and within such time as may be prescribed. It appears that the clearance is to be given by the expert group and that it can be conditional. However, there is no provision that is should be published and made available to the public.
5. It is provided that in case of projects displacing 400 or more families in plains and 200 or more in tribal or hilly regions, the state government shall appoint in respect of that project, an officer for formulation, execution, and monitoring of the R&R Plan.
6. Apart from the notifications under the Land Acquisition Act, the appropriate government is to issue a notification declaring affected areas where displacement affects 400 or more families, and so on.
7. The developer shall contribute to the socioeconomic development of such geographic area on the periphery of the project site as may be defined by the appropriate government, and for this it shall earmark a percentage of its net profits or in case no profits are declared in a particular year such minimum alternative amount determined by the appropriate government in consultation with the requiring body.
8. For such project displacing 400 or more families in plains and 200 or more in the Himalayas, there shall be a committee called the Rehabilitation and Resettlement Committee to monitor and review the progress of the implementation of the rehabilitation scheme and to carry out postimplementation social audits.

However, it is more important how and what way such policy is getting implemented. The political and administrative will to implement such policy in letter and spirit and also continuous course correction of such implementation will be the cornerstone of success of large-scale hydropower development.

6.08.6 Project Planning and Implementation

As per Electricity Act, 2003, any generating company can establish, operate, and maintain a generating station without obtaining a license if it complies with the technical standard relating to connectivity with a grid specified by the CEA. However, certain clearances/approvals are required for taking up hydropower projects, which are as follows:

1. Consent from the respective state government for setting up the project including certificates for land and water availability
2. Techno-Economic Clearance (TEC) from CEA as per Electricity Act of 2003
3. Clearance from MOEF from the point of view of environmental impact including resettlement and revalidation
4. Clearance/Recommendation from State Forest Department for acquiring forest land including river channel and subsequent clearance from MOEF for the same
5. Clearance from Ministry of Water Resource for international rivers, interstate rivers, and also for multipurpose projects including flood control and irrigation
6. Clearance from Ministry of Social Justice and Enforcement/Tribal Affairs in case scheduled tribe (ST) population is likely to be affected
7. Clearance from the Ministry of Defense in case defense issue/land is involved.

The TEC required in the project involves interstate rivers and the cost of the project is more than US$106 million or the project is more than 100 MW. This is with a view to ensure that (1) the proposed river work will not prejudice the prospect of best possible development of the river or its tributary for power generation, consistence with the requirement of drinking water, irrigation navigation, flood control, or other public purposes; (2) adequate studies have been done on the optimum location of the dam and other hydraulic structures; and (3) dam safety requirement is met. However, it is felt that at least for hydrological and geological study for all hydropower projects, such clearance or examination should be applicable irrespective of size and cost of the projects in view of impact of such data/study is there for all the projects in a basin.

In the 12th Plan as already discussed, CEA has identified 20 000 MW to be built between 2012 and 2017, out of which about 18 000 MW will be from the Himalayan rivers. The majority of these projects are ROR projects and few are with storage capacity. Building of around 18 000 MW project in 5 years into the remote areas of the Himalayas will require an extensive preproject planning and substantive quantum of implementation planning. The projects that are having a capacity around and more than 100 MW would require reasonable infrastructure such as roads, standard capacity bridges/culvers built over river/falls. Such development activities in interior parts of the Himalayas without affecting its ecological balance itself are a time-consuming and financially expensive venture. Projects of all sizes that are built into the interior part of the Himalayas are vulnerable to scarce data availability for hydrological and geological study and even the climate effects on hydrology. Though it is possible to a large extent to take care of the geological study by collecting and investigating during the project investigation stage (undoubtedly remoteness and hostile climate hinder such informations collection) but nonavailability of historical hydrological data for such a project site is an impediment with a high risk for viable development of a hydro project. Application of an overall river basin concept management and study of the cluster of such projects in a basin in totality, use of regional hydrological analysis, application of remote sensing for such basin as a whole for development of reliable hydrological model may to some extent help; still such projects will require certain techno-economic flexibility built in to take care of the uncertainty and risks. Provision of storage in such basin/sub-basin can also impart the flexibility in project conception and design.

Geological uncertainty and the risks associated with it have been always talked about especially for the Himalayan projects. Understanding and dealing with such risks and uncertainty in the Himalayas has been much more difficult not only due to the very complex nature of geology of this area but also due to the hostile and inaccessible environment where at the initial stage data collection and information generation have been a very difficult job both from the point of view of physical hardship and financial investment. Again here to mention that a basin/sub-basin-wise study for such geological uncertainty and the risk can really be a great supporting factor for development of such hydro projects. In recent years, applications of state-of-the-art tools such as remote sensing, geophysical investigation over and above conventional geological investigation, have been adopted but mainly concentrated around the medium and large projects.

However, the project that has a capacity less than 100 MW or little more are yet to find a way out for a rational investigation and information collection to reduce the risks of such projects. In recent years government agencies both in central and state are not the dominant players in hydropower development in the Himalayas. One of the fallouts of this is that the private developers are not in a position to invest reasonable resource for geological, hydrological, and environmental studies especially during investigation and planning stage. It has resulted in the increase of risk during the project development and on economic and social optimization of such projects. This problem needs to be looked into and the role of the government for study and investigation of the project that require inputs of fund, knowledge, and frontier technology should be increased. For sustainable, viable, and socially acceptable hydropower development into the tribal and ethnic belt of the Himalayas, it is not only the hydrological and geological study that have to be a priority but social and environment study beyond the normal visible options are to be an important part of the project development. The role of government agencies research groups and multidisciplinary approach region-wise/basin-wise has become important.

Out of the 20 000 MW project proposed to be built in the 12th Plan, that is, 2012–17, about 12 000 MW will be by government agencies, both state and central, and about 8000 MW by private developers. It is a substantive change in the development of

Table 8 Projects categorized by MW range

	No. of projects	Total MW
> 500 MW	11	10 660
200–500 MW	21	6 758
100–199 MW	16	2 257
< 100 MW	39	2 639

hydropower in India whereas private developers contribution is likely to be about 40% compared to almost little in the 11th Plan. The switchover of hydro development with major priority for private developers have opened up new opportunities of development and has also raised many complexities in the process of hydro development; some of which have already been discussed. In brief, we mention them again:

1. Planning and conceiving project
2. Optimization of project
3. Infrastructure development and its availability for the project development
4. Environmental and social issues and its mitigation
5. And most importantly a sustainable development of the area and its surroundings.

Out of the total 20 000 MW proposed for development, the projects having reasonable storage provision is only of about 4000 MW and balance 16 000 MW is the ROR projects. To take care of hydrological uncertainty that is likely to increase more due to climate change, storage requirement for such projects is more important and lack of storage schemes may not be an ideal situation.

Projects proposed for the 12th Plan can be catagorized as given in **Table 8**.

Out of 22 314 MW under execution, about 20 000 MW is likely to be commissioned in the 12th Plan. A major quantum of power shall be from big projects, that is, above 500 MW and some of them are in remote areas that will require infrastructure developments, including roads, bridges, and so on. Projects of < 100 MW capacity are few in number though power contribution may not be so high. Such projects face more problems due to constraint in data and information, remoteness, and communication costs and will require effective power evacuation planning and other assistance. But such projects being spread over larger areas and being more environmentally benign can bring substantive local development if scientifically planned and executed. The same can be stated for projects between 100 and 200 MW also.

The Indian construction industry would play an important role in achieving the goal for the hydropower project in coming years. Today it is not in a position to cope with the huge construction requirement in coming years for these hydropower projects. Though there is large number of construction firms in this country, the firms in the field of heavy constructions are few. It is well known that a large-scale development of construction activity in the field of hydropower projects in remote parts of the country can only be achieved by the country's own construction capabilities and not by depending upon outside agencies. A wide array of organizational issues, policies, and practices that result in inefficiencies and loss of productivity are present.

The large number of construction firms and their size make it difficult to deploy new technologies, best practices, or other innovations effectively across a critical mass of owners, contractors, and subcontractors. The industry is also segmented into at least four distinct sectors – residential, commercial, industrial, and heavy construction. These sectors differ from each other in terms of the following:

1. The characteristics of project owners, their sophistication, and their involvement in the construction process
2. The complexity of the projects
3. The source and magnitude of financial capital
4. Required labor skills
5. The use of specialty equipment and materials
6. Design and engineering processes
7. Knowledge base and training process.

Obstacles to rationalize their functions are:

1. A diverse and fragmented set of stakeholders: Owners, users, designers, builders, suppliers, manufacturers, operators, regulators, manual laborers, and specialty trade contractors, including plumbers, electricians, masons, carpenters, and roofers.
2. Segmented processes: Planning, financing, design, engineering, procurement, construction, operations, and maintenance.
3. The image of the industry: Work that is cyclical, low-tech, physically exhausting, and unsafe, which makes it difficult to attract and retain skilled workers and recent graduates.
4. The one-of-a-kind, built-on-site nature of most construction projects.
5. Variation in the standards, processes, materials, skills, and technologies required by different types of construction projects.
6. The lack of an industry-wide strategy to improve construction efficiency.

7. The lack of effective performance measures for construction-related tasks, projects, and as a whole.
8. The lack of an industry-wide research agenda and inadequate levels of funding for research.

Government is to initiate now an action to facilitate and motivate the modernization and skill development for this industry and more important by applied research agenda. So that by the start of the 12th Plan the industry has available expertise, knowledge, and organizational framework to attract talent. The motivation and incentive to modernize require an immediate push so that the gigantic task of building these difficult projects is achieved. Action on this line has been initiated but requires to be expedited.

Similarly, the Indian manufacturing industry requires developing capacity of manufacturing of hydro turbine and hydro generator and other ancillary equipment to cope with the requirement. Until recently, India had one large manufacturing unit owned by the Government of India to manufacture hydro turbine and hydro generators and related ancillaries. However, some of the international manufacturers have set up shops in recent years to manufacture and or assemble the turbines and generators in this country. The Government of India is already encouraging Indian manufacturing industry to set up facilities for the turbines and generators mainly and also for other components required for hydro plants.

It is obvious that such large and expeditious development of hydropower cannot be achieved without development of internal capacity both for construction and manufacturing.

6.08.7 Storage and ROR Hydroelectric Projects

For designing and engineering projects, preconstruction investigation, that is, during the period by which accessibility to the site components have been developed and actual construction to start, should be more affectively used for further detailed investigation. Even hydrological data collection and information generation during this period should be more effectively implemented. Such information should be incorporated into the design and engineering of the project as a continuous updating of project detail and as and when the data flow in. This process, though a little bit difficult and require inbuilt flexibility in project planning, can take care of many uncertainties in the implementation. Implementation planning includes construction planning, infrastructure planning, construction equipment planning, and so on, along with the risk evaluation, flexibility provision of risks adjustments, and a strong risk mitigation mechanism. Use of modern technology, meticulous quality control, and quality management are the key to the success of such projects. Recently, serious attempts along these lines are being made.

Regarding the engineering and planning of high dam and large surface/underground powerhouses into the relatively difficult geology and very high hydrological uncertainty requires very meticulous investigations, high quantum of hydrological and geological data collection, and use of state-of-the-art technology for design and analysis. For successful execution of such projects, extensive implementation planning that includes construction planning, construction management planning, construction equipment planning, and the scientific evaluation of risk and its mitigation. In recent projects such ideas are being implemented, sometimes not very successfully, but undoubtedly there is a visible attempt for course correction. A major number of such dams and powerhouses are located at very high earthquake intensity zones which increases the risk for them.

Tehri Dam Hydropower Project Stage I is one of the largest reservoir projects built recently in the Himalayas. The project is located over the river Ganges, upstream of the holy city of Haridwar. The dam has a height of 260.5 m above the bed. The width is 1125 m at the river bed. This is a conventional earth and rock-fill dam constructed in a region of high earthquake intensity. **Figure 11** shows the surface spillway of the dam and **Figure 12** shows the top of the dam.

The dam has a gross storage of 3540 million m^3 and life storage of 2615 million m^3. The water spread at full supply level, that is, EL830 m is 42 sq km. This dam has submerged the whole township of Tehri and its replacement in a new and modern township with all infrastructure and amenities have been built over the mountain (see **Figure 13**). The project has an installed capacity of 1000 MW (4 × 250 MW).

Figure 11 Tehri dam surface spillway.

Figure 12 Tehri dam top.

Figure 13 Resettlement town of Tehri.

Nathpa Jhakri Hydroelectric Project has been built over river Satluj as an ROR scheme. It has a capacity of 1500 MW (6 × 250 MW). This has a 62.5 m high concrete dam over the Satluj river and an underground desilting basin comprising four chambers each 525 m long, 16.31 m wide, and 27.5 m deep, which is one of the largest underground complexes for the generation of hydropower in the world. It also comprises a 10.15 m diameter and 27.397 km long headrace tunnel, which is one of the largest hydropower tunnels in the world. An underground powerhouse with a cavern size of 222 m long, 22 m wide, and 49 m deep having six Francis turbine units of 250 MW each are its major components. This project faces a severe problem due to the high quantum and concentration of sediment that flows during the monsoon in the river. Dam and intake with upstream pond is shown in **Figure 14**. The machine hall of the powerhouse is shown in **Figure 15**.

Figure 14 Dam and intake of Nathpa Jhakri Hydroelectric Project.

Figure 15 Inside of powerhouse of 1500 MW Nathpa Jhakri Hydroelectric Project.

Figure 16 Chamera Hydroelectric Stage I dam.

Another project that has a part head created by a 120 m high, 295 m long concrete arch gravity dam is the 540 MW Chamera Hydroelectric Project Stage I (**Figure 16**). To create further head, it has been provided with a 6.4 km long, 9.5 m diameter headrace tunnel and also a 2.4 km long, 9.5 m diameter tailrace tunnel. The dam is located over the river Rabi, a major tributary of the river Indus and an underground powerhouse containing three units of 180 MW each. This is also a ROR project with limited storage. The project also faces the problem of high sediment load coming in the reservoir but is being effectively managed by using a recent concept of sediment removal management. **Figure 16** shows the 120 m high concrete dam and **Figure 17** shows the underground powerhouse of the project.

The 1000 MW Indira Sagar Project is also a storage project. It is located over the river Narmada in peninsular India, and does not have the benefit of snow melt during the dry season; however, having a large catchment area of around 61 642 sq km, it has a large annual flow. Hence, the dam has been built with gross storage of 12.22 billion m^3 and life storage of 9.75 billion m^3. **Figure 18** depicts the dam and **Figure 19** depicts the powerhouse which is a surface powerhouse.

It has an installed capacity of 1000 MW (8 × 125 MW) with Francis turbines.

Another peninsular project over the river Krishna that has been recently commissioned is again a multipurpose reservoir project where the main purpose will be for irrigation. About 408 747 ha of area is proposed to be irrigated. The dam has a gross storage of 3.78 billion cm^3 and life storage of 3.07 billion cm^3; however, it has a dam-toe powerhouse of capacity of 297 MW. **Figure 20** depicts the dam and dam-toe powerhouse of the Upper Krishna Project.

6.08.8 Sediment Transport and Related Issues

All the rivers on the three major river basins in the Himalayas, namely Ganges, Brahmaputra, and Indus, and also the major peninsular rivers carry huge quantities of silt every year. Sediment load in major Indian rivers is given in **Table 9**.

For the river Indus in the Indian part, the annual sediment load shall be of the order of 250 million tonnes and annual discharge of around 150×10^8 m^3 yr^{-1}. Huge sediment load of all these rivers is again transported mainly during the monsoon season, that is,

Figure 17 Chamera Hydroelectric Stage I underground powerhouse.

Figure 18 Dam 1000 MW Indira Sagar hydroelectric project.

Figure 19 Powerhouse 1000 MW Indira Sagar hydroelectric project.

around 4 months of a year, and even during the 4 months, the major quantum of sediment is transported in the early monsoon month and during the latter period of the monsoon. During the months of June–July, the concentration is quite high as the sediment source for the rivers has a large amount of loose materials accumulated during the previous dry season. Similarly, during the end months of the monsoon in September–October, the concentration may be high as during this period of monsoon a new source of sediment becomes activated in the river basin.

Figure 20 Upper Krishna Project.

Table 9 Major rivers, discharge, and sediment load [10]

River	Discharge ($m^3 \times 10^8$ yr^{-1})	Sediment load (m yr^{-1})
Mahanadi	67	142
Krishna	30	251
Godavari	92	310
Cauvery	21	88
Ganges	493	750
Brahmaputra	510	580

In the Himalayas and especially in the larger sub-basins of Ganga, Brahmaputra, and Indus, the huge quantity of the sediment load during the monsoon period causes a considerable problem for social, environmental, engineering, and economical issues. Effects of generation and transportation of large quantum of sediment result in large-scale erosion of the river bed, river bank erosion and instability, engineering instability to structures across the river channel, sediment depositions and rise of river bed in the channel, meandering and braiding of river channel sometimes resulting in change of river course, sedimentation of reservoir, pollution of water supply, choking of drainage channel, and so on. The huge sediment load during the peak monsoon discharge of 4 months for the hydropower projects in the Himalayas has become a major impediment for the development of sustainable hydropower in the region. Such sediment load during this period is affecting the maximization of hydropower generation due to the following reasons:

1. Large-scale sediment deposition, upstream of reservoir and/or diversion structures.
2. Entry of large quantum of sediment-laden water in the intake and water conducting system resulting in deposition of sediment in such system and sometimes choking them.
3. Entry of such high sediment-laden water into the turbine resulting in large-scale erosion and cavitations to the rotating underwater parts and also even the static underwater parts around which high water velocity takes place.
4. The ancillary systems of powerhouse such as cooling water, turbine seals, and so on, either getting eroded severely or choked.
5. The over flow structure such as spillway and energy dissipating system such as ski jump bucket, stilling basin, and roller buckets getting severely damaged.
6. Deep and large scour at the toe of dam/diversion structures.

Though the erosion and cavitation damage largely depend on the size of sediment particles and also on the hardness of such material, it has been observed that in general almost all Himalayan rivers are inflicting large-scale damage on the hydro turbine systems and also to the majority of the hydraulic structures, located across the river. Extensive research and prototype experiment both by the turbine manufacturer, hydraulic laboratory, and hydraulic designers for the projects have been going on for many years. The use of long desilting basins wherein the water conductor system a large expansion of the conductor system with a considerable length is introduced to bring down the water velocity in that basin length to a substantive low value of around 0.2–0.3 m s^{-1} so that bigger particles settle down in the basin to a large extent and only the smaller and finer particles gets transported further. The basic parameter for the design being that around 90% of the particles having a size of greater than 0.2 mm/0.15 mm will get deposited in the basin and flushed out from the bottom and the particles which are smaller than 0.2 mm/0.15 mm will get transported further. The basic premise being that damaging capacity for particle sizes less than 0.2 or 0.15 mm is much less. Generally, such basins

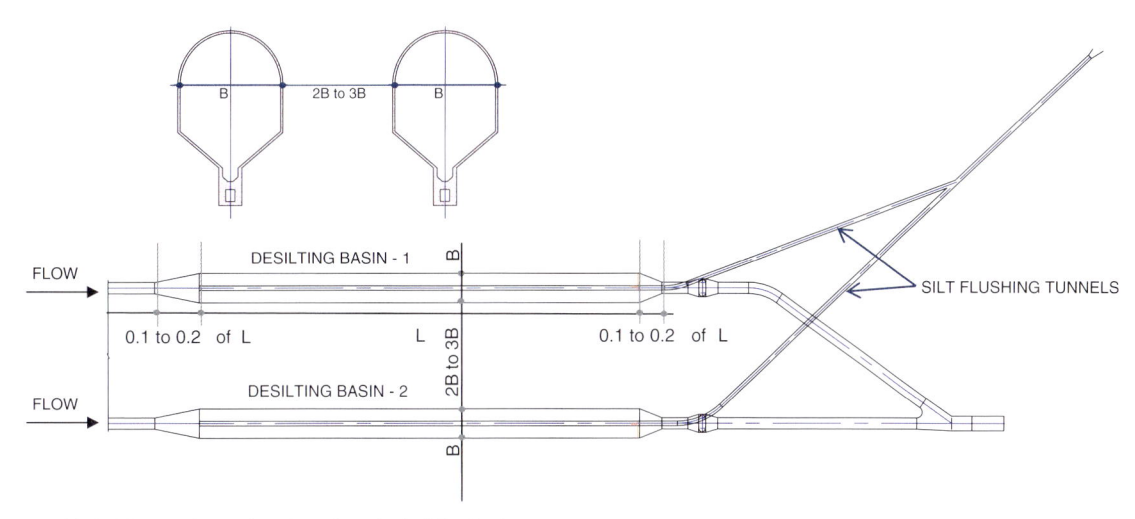

Figure 21 Typical layout plan and section of a twin desilting basin.

Figure 22 Nathpa Jhakri desilting basin.

depending upon the discharge will have a width of 15–20 m and around 300–400 m length and more than one in number. A typical plan of a twin desilting basin is shown in **Figure 21**.

Real efficiency of such basins in controlling the damage due to erosion and cavitations and also damages and hindrance to powerhouse ancillary structures are not very well established. **Figure 22** shows a huge desilting basin of Nathpa Jhakri Hydropower Station.

However, constructing such a desilting basin specially by excavating a large open area or by large underground caverns are very costly and time-consuming propositions. Hence, the search for more reliable and effective arrangements for silt elimination entering into the intake and water conductor system are going on. One of the concepts of design for sediment elimination in India is to build the dam of reasonable height around 35–50 m with some amount of storage space with a range of 5–10 million cm^3 and then provide a big spillway as near as possible to the river bed keeping in view the hydraulic, geological, and morphological character of the river and providing this spillway opening with big size radial gates that are submerged. The power intakes are provided at least 15–20 m above the spillway crest and in case such intake can be provided on a much higher level, the benefit of providing lower level spillway can be tremendous. During the high flood and sediment load season, these spillways are kept partly or fully open depending upon the discharging capacity requirement and operation procedure finalized. A major quantum of sediment of bigger and also smaller particles passes through this spillway without entering the intake; hence, through the intake only the fine particles travel to the powerhouse. Such an arrangement has been found to contribute substantively in sediment load reduction from the point of view of damage potential for a turbine and its components in some of the projects such as Chamera Hydroelectric Project Stage I – 510 MW, Chamera Hydroelectric Project II – 300 MW, Teesta Stage V – 540 MW, and so on. This concept has been implemented and results are being studied.

Regarding the damage of hydraulic civil structures and its severity, it is been noticed that the damage and its frequency are much higher in its intensity in the hydropower stations in the Himalayas compared to the peninsular rivers. Some of the peninsular rivers like Godavari though with high sediment content do not show much damage in the hydraulic structures. It appears that the majority of such projects in peninsular rivers are storage schemes and the reservoir to date has not been filled up to its dead storage. Hence, sediment deposition especially of coarse particles are taking place in the reservoir and finer particles are passing through the structures and turbine. It is also felt that hardness/density of particles transported over the structure and through the turbine is much less than that transported through Himalayan rivers.

The importance of provision of a reasonable size of reservoir having a capacity of 10 million m^3 or more is established in the Himalayas at least from the point of view of better sediment handling. However, the associated problems with lower level spillways is of two kinds. First, the spillway is put near the river bed and they are submerged, so that very high velocity of the flow is generated right from the spillway crest to downstream. This high velocity flow with high sediment concentration during the monsoon season severely damages the concrete in the spillway and energy dissipating system. The problem is increased due to the bigger particle sizes and also in a majority of cases the presence of quartz with a high abrasion value.

Second, as such reservoirs are flushed regularly during the monsoon months, at least once a month by depletion of reservoir, very high concentration of silt as deposited in the reservoir gets flushed out through the spillway. Though in this process velocity is not that high, but quantum of sediment and size of particles damage the spillway and energy dissipating system. The reservoir is flushed by lowering the full reservoir level and by opening the spillway gates fully. During the monsoon months in small and medium reservoirs this is done once a month. **Figure 23** shows the reservoir flushing of Rangit reservoir.

A prototype study on the erosion and cavitation damage of concrete by increase in strength and higher performance level are going on in the projects such as Salal Hydroelectric Project, Chamera Hydroelectric Project Stage I, Dhauliganga Hydroelectric Project, and so on. Use of high strength concrete and increasing its performance level by using microsilica, steel fibers, and so on, is being tested. However, to date results in controlling erosion and cavitation damage of concrete have not been found satisfactory. It is observed that even the use of a synthetic material like Alag concrete, ceramic tiles, coating by epoxy-resin, and so on, have not been able to help much; however, use of high performance concrete has to some extent extended the cycle of damage repair. Hence, today, more stress is put upon repairing the spillway and stilling basins more easily, in a cycle of 3–5 years, and this forms part of the design concept. However, this is the area that requires further study for future sustainability of hydropower projects and specially the dams in India. **Figure 24** shows the damage to the energy dissipation system of Rangit Dam.

Figure 25 shows the damage to the Salal dam spillway bucket. Repair by different materials has been going on at Salal since 1996 and lot of information and data on damage characteristic and repair methodology have been generated.

High head radial gates with bigger size opening have certain limitations. To date the gates have been generally designed up to a head of 60–70 m with an opening of about 60–70 sq m, the constraint being the huge hydrostatic and hydrodynamic pressure and

Figure 23 Rangit reservoir flushing by lowering the reservoir.

Figure 24 Damage to stilling basin of Rangit dam.

Figure 25 Damage of Salal dam spillway bucket.

also related operational uncertainties. But keeping in view of the requirement of such large gates in the future. further study and knowledge development for designing and planning bigger gates are very much required.

Regarding the damage to hydro turbines, there are different points of views on the mechanism of damage in silt-laden water. One view is that damage is caused by cavitation erosion inflicted by the solid particles; another view is considered as a result of combined action of cavitation and abrasion. Extensive study of such damage has been done in Europe, China, and India.

In one of the studies on the effect of particle size, its concentration, and its density, the author states that tests on curved hydraulic conduits show that on the outside curve of the conduit even at a relatively low velocity particles above 1 mm in diameter will not follow the hydraulic contour and will impact upon and damage the hydraulic surface. Particles with diameter between 0.1 and 1 mm will tend to be channelled along the outer hydraulic contour, and their capability for damage will be progressively less. For particles below 1 mm, the surface damage increases considerably. This is because small particles become entrained in the turbulent boundary layer, which encases all hydraulic surfaces and results in a sand blasting of the surface [9].

It is opined that overall erosion from fine particles if in sufficient quantity can be as great as that from large particles. It has also been expressed by many that the inside bend surface experiences steady increase in damage as particle size decreases. However, this has not been established but with the general understanding that the finer particles have also a reasonable contribution to the erosion of the turbine blade, the contribution of the desilting basin as discussed in previous paragraphs requires it to be seriously reviewed, keeping in view that provision of such long, wide, and deep basins in the mountainous region is a costly and time-consuming concept.

The effect of particle density is similar to that of size. A particle of greater density will have a greater momentum and thus be more inclined to reach the surface in the case of larger particles and layers inclined to be entrained in the boundary layer in the case of smaller particles. It is stated by Gummer that a particle can only appreciably damage a softer surface and particle with hardness of 5 Mohs is generally considered as cutoff value for hydraulic turbines.

In the Himalayas, silt has substantive quantity of particles of hardness 7 Mohs or even more. It is believed that the damage rate of the abrasion is generally proportional to the cube of flow velocity. The higher the velocity, the more severe the damages. This fact has been established in many turbines erected in India. So for high head power stations, desilting basins are designed with particle elimination up to 0.15 mm.

However, all the above discussions only indicate that the mechanism of erosion and damage of the turbine and other components such as guide vanes with respect to the effect of sediment movement have not been well understood to date.

In such a situation, as in the case of the spillway, ease of repair is the design concept of the powerhouse. Hence, the arrangement of easy runner removal during the annual maintenance period, which is generally from November to March, when the discharge in the river is less and the powerhouse runs in its reduced capacity is part of the design criteria. For the runner removal, a bottom gallery with the turbine pit is being provided through which the runner is taken out and lifted by an electrically operated travel (EOT) crane for the purpose of repair/replacement. For reduction of commercial losses there is a provision of extra runner and guide vanes which are immediately replaced in place of a damaged runner. Undoubtedly, this process has increased the space requirement, width-wise in the powerhouse and also in the service bay. Cost of extra runner and also annual repair and maintenance of runner and other parts are quite high. Such damages also affect the efficiency of the turbine especially the turbines designed for the higher efficiency range. Repair of runner and other components are being experimented by using various coatings, both at the initial stage of installation and subsequently in the repair and maintenance stage. **Figure 26** shows the damage in the scroll case and runner blade of the Kaplan turbine of Tanakpur Hydroelectric Project. **Figure 27** shows the damage of the runner of the Baira Siul Hydroelectric Project. Various coatings on the runner which is being used today for repair and also to enhance the life cycle can be categorized in two coatings.

Protective coatings fall into two categories: 'hard' coatings, such as welded Stellite and thermally applied ceramic and tungsten carbide, and 'soft' coatings, which are typically a brush, trowel, or spray-on polymer. Variants of the pure hard coatings are the thermally applied systems of hard particles in a softer matrix. These hybrid systems bridge the gap between hard and soft coatings while maintaining the potentially superior bonding strength of the thermal application process when compared with the brush or spray-on application of soft coatings. Conversely, the resistance of soft coatings against particle erosion depends on the type of

Figure 26 Tanakpur Hydroelectric Project damaged runner blade of Kaplan turbine.

Figure 27 Damaged runner taken out for repair in 2005–06 Baira Siul Hydroelectric Project, NHPC.

polymer, the surface quality, and the bond efficiency. Given the correct composition and bond for the particular application, a soft coating can be every bit as effective against particle erosion as a hard coating.

Coating has also its problems as many of them cannot be executed at site due to the constraint of technology, cost, and time; it has also the problem when recoating requires to be done. However, extensive field experiments with a different coating are going on in many of the power plants such as Nathpa Jhakri, Salal, Chamera-II, Dhauliganga, and so on. Nathpa Jhakri that faces the problem with severity has set up a workshop for hard coating at the project site (see **Figure 28**).

Figure 28 Workshop for hard coating at the project site.

Sustainability of the dam and reservoir from the point of view of sediment deposition, damages to the hydraulic structures, and also to the turbine and its components are the major issues that require handling in coming years for economical and social viability of such projects. Control of the sediment transport through such Himalayan rivers is a difficult proposition. Socioeconomic development, infrastructure development, and development of new habitat into the Himalayas along with land-use change pattern shall influence sediment generation for such rivers. Another important issue is that as more and more projects are being built into the remote and upper reaches of the Himalayas, the sediment particles size, its geological characteristics, and mineralogical composition are apparently more aggressive than for the downstream projects, which has also resulted in more severe damages.

6.08.9 Socioeconomic Development and Hydropower in the Himalaya Northeast Region

Eight political units of the union of India, namely, Arunachal Pradesh, Assam, Manipur, Meghalaya, Mizoram, Nagaland, Sikkim, and Tripura constitute Northeast India. They are together commonly known as North Eastern Region (NER).

- The hallmark of the eight political units is the diversity on account of terrain, climate, ethnicity, culture, institution, land system, language, food habits, dresses, and so on.
- These states have evolved in different time and function under different provisions of the constitution of India.
- The regional identity of eight states as NER is a concept based on extreme intraregional diversity.

The NER of the country forms an area of low per capita income and major growth requirements. Growth in social infrastructure through national program must be complemented by development of physical and economic infrastructure. In this context, the development efforts of the states have to be supplemented in order to minimize certain distinct geophysical and historical constraints.

The process of development had been slow in the NER for many reasons. The traditional system of self-governance and social customs of livelihood in NER remained virtually untouched during British rule. The creation of a rail network for linking tea-growing areas for commercial interests was the only major economic activities taken up in the region during this period. The partition of the country in 1947 further isolated the region.

This has also disturbed the socioeconomic equations in many parts of the region resulting in the demand for autonomy by the relatively more backward areas. While development efforts over the years have made some impact, the region is deficit in physical infrastructure which has a multiplier effect on economic development.

It is the home of over 140 major tribes out of 573 in the country besides nontribal with diverse ethnic origin and cultural diversity. The ST as defined in Indian constitution population (2001 census) is 12.41% of India's ST total. It is 26.93% of NER's total population and is dominant in Arunachal Pradesh (64.22%), Meghalaya (85.94%), Mizoram (94.46%), and Nagaland (89.15%). The group is quite large also in Manipur (34.20%), Tripura (31.05%), Sikkim (20.60%), and Assam (12.41%). Scheduled cast (SC) as defined in Indian constitution population is 1.49% of India's total. It is 6.40% of NER's total population. Maximum concentration is in Tripura (17.37%) followed by Assam (6.85%) and Sikkim (5.02%).

NER has a hydropower potential of 63 257 MW (42.54%), including Sikkim. Sikkim was later added to NER from ER, against the all-India potential of 148 701 MW. Arunachal Pradesh alone has the potential of 50 328 MW, which is 80% of the total hydropower potential of the NER and 34% of the total potential of the country. Despite recognizing this potential, the desired thrust is not there as hydropower development requires huge investments. The sectoral summit on power suggested a two-pronged strategy for power generation with focus on small/localized hydropower and thermal power projects for local needs and high-capacity hydropower and thermal power projects with associated transmission lines for meeting the demands of the region and also supply to the rest of the country. Transmission, sub-transmission, and distribution system improvements have been identified as one to the thrust areas for the 11th Plan.

Two broad types of land tenure systems operate in the region:

1. Revenue administration under government operates in the plains and valleys of Assam, Tripura, Manipur, and in the hilly state of Sikkim and
2. Customary land tenure system under village level authority operates in the hilly states of Arunachal Pradesh, Meghalaya, Mizoram, and Nagaland, and in the hilly parts of Assam, Manipur, and Tripura.
    - Cadastral survey is not done in these areas.
    - Land is held almost by all. Landless people are negligible. Marginal (< 1 ha) and small farmers (1.0–2.0 ha) are the two dominant categories (78.92%).
    - Distribution is largely egalitarian rooted in the principle of community way of living and sharing.
    - Operational availability of land is a small fraction of total availability in the hills.

Land acquisition for hydropower projects is a major hindrance, due to land tenure system prevalent in the region. The Planning Commission constituted a Task Force on Connectivity and Promotion of Trade and Investment in NER. The main recommendations of the Task Force are to take up the Trans Arunachal Highway on priority; road links in Manipur; construction of a bridge at Sadia-Dholaghat over the Brahmaputra River; completion of ongoing railway projects; priority funding for identified projects;

construction of three greenfield airports at Pakyong in Sikkim, Itanagar in Arunachal Pradesh, and Cheithu in Kohima; modernization of airports of NER; and harnessing of the maximum potential of inland water as a mode of transport.

All these infrastructure developments will go a long way in developing hydropower projects in the North East and especially expeditious development of big hydropower projects. Such developments shall have a cascading effect in socioeconomic development by building local roads, institutions, schools, health centers, and local markets backed by a new communication system. There would also be in some projects development of local navigation route, and if planned properly, can be connected with the national waterway of Brahmaputra–Ganga route.

The Ministry of Power's Policy on Hydropower projects provide the following and the project cost shall include the following:

– cost of the approved R&R Plan of the project that shall be in conformity with the following:
  1. The National Rehabilitation and Resettlement Policy currently in force
  2. The R&R package as provided in the policy notification by the Ministry of Power, Government of India.
– the cost of project developers' 10% contribution toward rural electrification project in the affected area as per the project report sanctioned by the Ministry of Power.
– 1% of power for contribution toward the Local Area Development Fund as constituted by the state government.
– Energy corresponding to 100 units of electricity to be provided free of cost every month to every project-affected family notified by the state government to be offered through the concerned distribution licensee in the designated resettlement area/projects area for a period of 10 years from the date of commissioning. As displacement-related issues for hydropower in this area is small, other issues related to maintaining their sociocultural identity and fear of encroachment by outsiders in their life are an important hurdle. Such issues need to be resolved in the best interest of the project-affected people and others in the periphery.

In conjunction with the policy, developmental activity in the form of development of horticulture and fruits, pisciculture, poultry, animal farming, handicraft, weaving, rural industry in the form of cottage industry for which NER is well known and suitable marketing arrangement for such products and goods centered on such projects utilizing better infrastructures shall bring new economic development of the area. Benefit sharing is a key parameter for such development.

6.08.10 Conclusion

The Indian economy is growing fast. The requirement of power and water are also growing at a fast pace. Necessity of development of sustainable and renewable energy will be a cornerstone of power generation growth in this country. Today hydropower is the most important source of commercially viable renewable energy in India. Development of hydropower in fast pace is been targeted by policy-makers and regulators in this country. Governments, both central and state, are encouraging private developers to enter into hydropower development in a big way to achieve such fast development. Advanced planning, introduction of frontier technology, advanced concept of project management, environmentally sustainable, and socially acceptable development are the instruments to achieve this goal. To achieve all these it is most important that such development is oriented primarily toward the growth of economy and society of the people around the projects. Their prosperity and well-being shall automatically expedite the growth of hydropower. It is not only the commercial development of hydropower but it is also the commercially viable model of social development which is to be perused.

References

[1] Ramanahan K and Abeygunawardena P (2007) *Hydropower Development in India: A Sectoral Assessment*. Asian Development Bank, Manila, Philippines.
[2] *International Conclave on Key Inputs for Accelerated Development of Indian Power Sector for 12th Plan and Beyond: Base Paper*. 18–19 August 2009, Ministry of Power and Central Electricity Authority.
[3] *Hydro Development Plan for 12th Five Year Plan Approach Paper: Central Electricity Authority, Hydro Planning and Investigation Division*. September 2008.
[4] The World Bank (2009) *Direction in Hydropower: Scaling Up for the Development: Note No 21*. June 2009, The International Bank for Reconstruction and Development.
[5] Preliminary consolidated report on effect of climate change on water resources. June 2008, New Delhi, India: Ministry of Water Resources, Government of India. http://cwc.gov.in/main/downloads/Preliminary_Report_final.pdf (last accessed 5 July 2011).
[6] Sen SP (2008) Hydropower potential and its development in India: Implementation and delays. *Book Seminar on Water Resources Management, Role for Water Sector in India*. New Delhi, India: Indian National Academy of Engineering, 21–22 February.
[7] Bhat U (2008) Environmental issues concerning development of hydropower projects. *Book Seminar on Water Resources Management, Role for Water Sector in India*. New Delhi, India: Indian National Academy of Engineering, 21–22 February.
[8] World Bank (2005) *India's Water Economy: Bracing for a Turbulent Future*. Report No. 34750-IN, 22 December 2005. World Bank, Washington, DC.
[9] Gummer JH (2009) Combating silt erosion in hydraulic turbines: Hydroworld.com. http://www.hydroworld.com/index/display/article-display/354757/articles/hydro-review-worldwide/volume-17/issue-1/articles/combating-silt-erosion-in-hydraulic-turbines.html (last accessed 5 July 2011).
[10] Subramanian V The sediment load of Indian rivers – an update – school of environmental studies, Jawaharlal Nehru University, New Delhi 110 067, India. http://iahs.info/redbooks/a236/iahs_236_0183.pdf (last accessed 5 July 2011).

6.09 Hydropower Development in Iran: Vision and Strategy

E Bozorgzadeh, Iran Water and Power Resources Development Company (IWPCO), Tehran, Iran

© 2012 Elsevier Ltd. All rights reserved.

6.09.1	Introduction	253
6.09.2	Energy Generation in Iran	253
6.09.2.1	Energy Flow in Iran	253
6.09.2.2	Electricity Generation in Iran	254
6.09.3	Considerations and Requirements for Hydropower Developments	256
6.09.3.1	Requirements	256
6.09.3.2	Restrictions and Limitations	256
6.09.3.2.1	Geographic issue	256
6.09.3.2.2	Technical issues	256
6.09.3.2.3	Organizational issues	257
6.09.3.2.4	Economic and financial issues	257
6.09.4	Potentiality of Hydropower Projects	257
6.09.4.1	Under Operation Projects	257
6.09.4.2	Under Construction Projects	258
6.09.4.3	Under Study Projects	258
6.09.4.3.1	Storage hydropower projects	258
6.09.4.3.2	Large run-off-river hydropower plants	258
6.09.4.3.3	Medium run-off-river power plants	258
6.09.4.3.4	Small run-off-river power plants	258
6.09.4.3.5	Pumped-storage power plants	258
6.09.4.3.6	Synthetic and conclusion	260
References		263

6.09.1 Introduction

The long-term average precipitation of Iran is around 250 mm, which is nearly one-quarter of the world's average amount, so Iran is classified as an arid and semi-arid country. In addition, the precipitation is not evenly distributed all over the country, so about two-thirds of total run-off flows in one-third of the country at 16 major rivers. The total precipitation and renewable water amounts – including surface and groundwater – are 413 and 130 bcm, respectively. The total surface water amount is 92 bcm of which around 27 BCM flow into three major basins, namely the Karoun, Dez, and Karkheh river basins, which are located in the south-west of the country over the Zagros mountain chains where the major hydropower projects are located. Northern and Northwestern regions have relevant precipitation and pertinent topography to develop medium- and small-sized hydropower plants too. So, Iran has been attempting to develop hydropower stations in these areas.

6.09.2 Energy Generation in Iran

6.09.2.1 Energy Flow in Iran

The latest energy flow diagram was prepared by the Tavanir Company. According to this, the total energy sources were equal to 2583.5 million barrels of oil equivalent (MBOE) of which 1052.7 MBOE were consumed in the country at 2008.

In addition, the consumption of petroleum and its products reached 85.5 MBOE for energy generation in thermal power plants. This means that nearly 75% of petroleum import and its products were consumed for electricity generation in thermal power plants. Also, the amount of natural gas consumption to generate electricity was equal to 232.9 MBOE, which was 29% of the country's natural gas production. In addition, the ratio of fossil energy sources, which was consumed in thermal power plants, to the total consumption of energy was about 0.3, whereas this ratio was about 0.01 for hydropower and renewable energy. In other words, the contribution of hydropower and renewable energy to generate electricity reached 10.7 MBOE, which was 0.14% of the total productions in 2008.

Energy price is one of the outstanding features of the energy section of the country. The energy cost is very cheap due to subsidies which are paid by the government. Based on the analysis of the annual budget approved by the parliament, the total subsidy paid in energy sections by the government was about 3.6 times the total budget appropriated in civil engineering activities in the country in the financial year of 2005. It seems that more hydro stations should be developed to reduce the consumption of fossil fuel sources and to improve budgetary appropriations to better the situation as well.

6.09.2.2 Electricity Generation in Iran

The total nominal capacity of power plants reached 52 586 MW in 2008. This shows an increasing rate of 7.1% in comparison with last year. At the same time, the total installed capacity of hydropower plants has met 7672 MW. **Figures 1** and **2** display the contribution of different types of power plants in terms of total installed capacity and energy generation, respectively. As demonstrated, the contribution of steam power plant, gaseous stations, and hydropower plants are 45.4%, 43.1%, and 14.5%, respectively. In other words, the hydropower plants are in the fourth step. Recently, because of drought phenomena, hydropower plants contribution was only 2.2% in term of energy generation, whereas their contribution was 9% during 2006–08. The maximum contribution of hydropower plant had been 13% during the last 5 years.

In order to generate electricity, the thermal power plants have totally consumed 43 412 million cubic meters of gas, 4398 million liters of gas oil, and 8911 million liters of oil in 2008. So their percentages were 76, 16.4, and 7.6 for gas, oil, and gas oil, respectively. In **Table 1**, the pertinent indexes for electrical energy generation have been depicted.

Figures 3 and 4 show how the nominal capacity and energy generation varied during 1992–2008 and 1966–2008, respectively. According to these figures, the diesel power plants' nominal capacity and energy generation decreased at these periods.

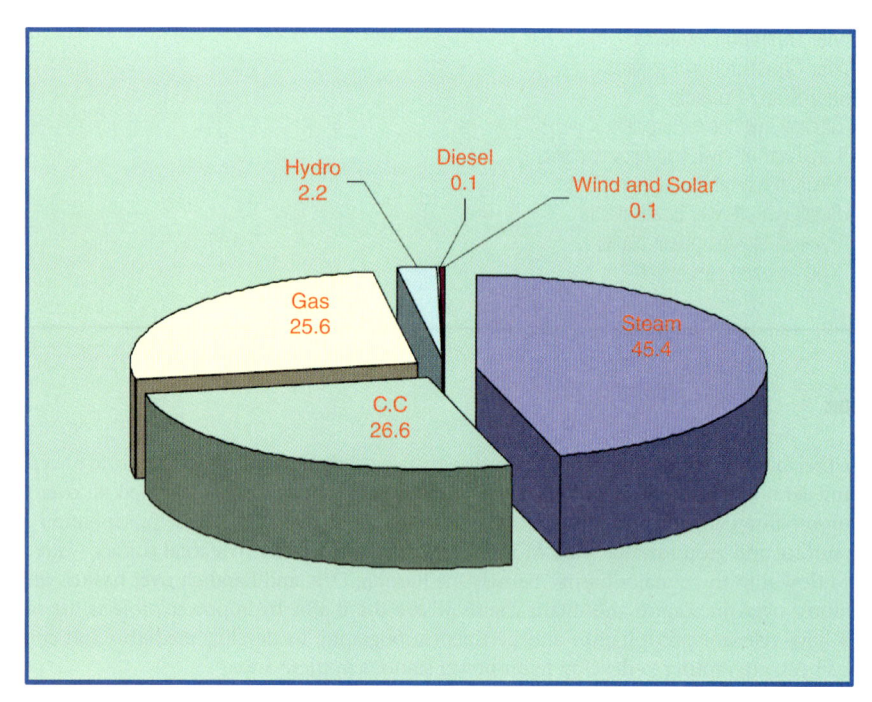

Figure 1 Contribution of all types of power plants – energy generation (2008).

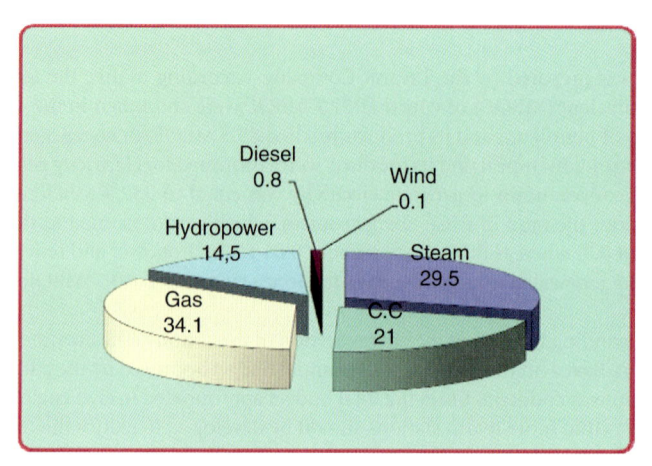

Figure 2 Contribution of all types of power plants – capacity (2008).

Table 1 Pertinent indexes of electricity networks in 2008

Index	Amount	Increasing rate in comparison to last year
Nominal capacity	52 944 MW	7.1
Average actual capacity	47 589 MW	6.7
Maximum concurrent power capacity	34 270 MW	−0.9
Maximum load	37 651 MW	7.6
Power per capita	738 W	5.9
Generation per capita	2987 KWh	3.8

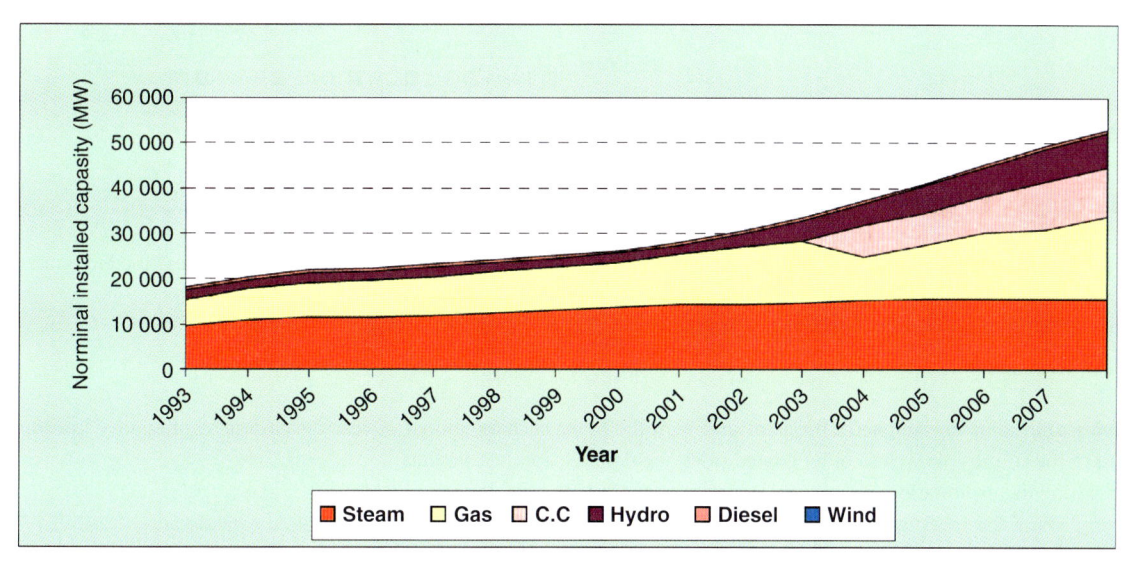

Figure 3 Nominal capacity trends of all types of power plants.

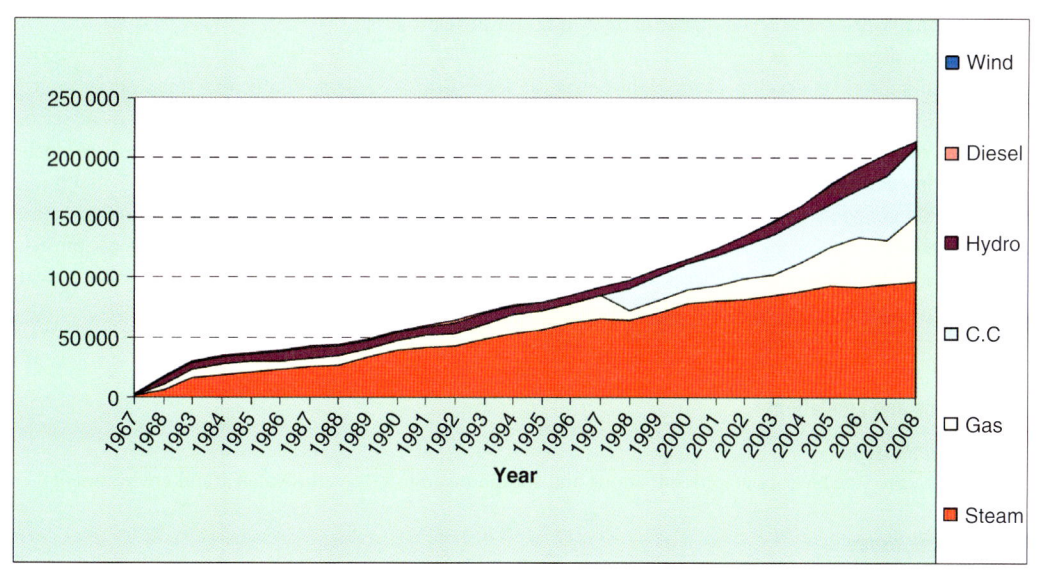

Figure 4 Electricity generation trends of all types of power plants.

The trends of different sources for electricity generation have been depicted in **Figure 5** during 1966–2008. The contribution of hydropower plants has changed from 35.8% to 3.1% in 1966 till 1999, respectively. In other words, the hydropower plants contribution has not increased for a long time due to imposed war (Iran–Iraq war) and has enhanced since 2000 so that it reaches 14.5% at present.

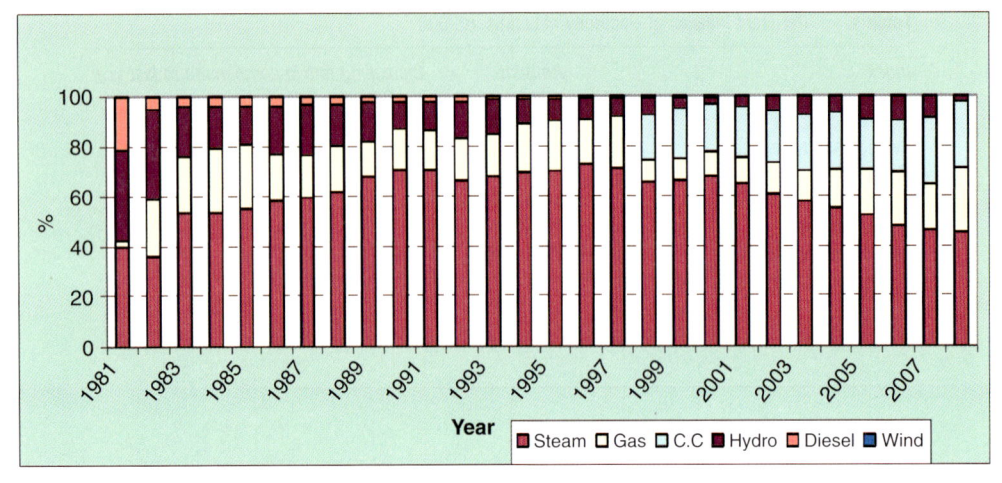

Figure 5 Trends of different power plants for electricity generation (percentage).

6.09.3 Considerations and Requirements for Hydropower Developments

In this section, the considerations, limitations, and requirements for hydropower plant developments are described in point of view of diverse effects such as technical, legal, and organizational issues, and so on.

6.09.3.1 Requirements

- Obeying and adjusting long-term program and periodic plans such as annual, 5-year development programs and long-term vision of the country, as well as other related rules, regulations, and instructions.
- Accelerating the exploitation of renewable sources of energies to save opportunity costs.
- Accomplishing the investigation phases of hydropower projects including master planning, reconnaissance, feasibility studies, and the detailed design stage as soon as possible.
- Perfect appropriating of actual incomes gained from hydro generation to develop more hydropower projects.
- Designing hydropower plants to operate at peak hours so that the best combination of diverse power sources is created.
- Real assessment of all kinds of benefits of hydropower stations including energy generation, frequency control, flood control, irrigation, recreational aspects, and so on, particularly in multipurpose projects.
- Actualizing the price of energy.
- Enhancing financial sources to develop more hydropower stations such as securing loans and financial support from donor international institutions such as the World Bank, Islamic Development Bank, CDM, as well as from the private sector.
- More attention to structural aspects of water and energy resources management in design, construction, and operation stages. In this regard, the integrated water resources management issue should be implemented.
- Concurrent completion of hydro projects and other related projects, such as watershed management to protect those projects against sedimentation problems, executing the irrigation network to gain agricultural benefits in the case of multipurpose projects, and so on.
- Revising the master planning projects of energy and water resources planning in all major basins to be adjusted with social and environmental constraints and sustainable development goals.
- Education, documentation, and establishing a robust database.

6.09.3.2 Restrictions and Limitations

6.09.3.2.1 Geographic issue
Hydropower plants require suitable topographic situations and should be adjusted to atmospheric and environmental conditions.

6.09.3.2.2 Technical issues

- Since Iran has been classified as an arid and semi-arid country owing to the low amount of precipitation, there are usually inconsistencies and conflicts between hydropower projects and inter-basin water transfer projects in particular.
- Lengthening the construction period of the product.
- There are no abundant suitable and feasible sites to develop hydropower projects in most parts of the country, particularly conventional ones.

- Competing with alternative methods to generate the required energy.
- Adverse environmental impacts and archeological problems.
- Lack of suitable database.

6.09.3.2.3 Organizational issues

- There are a lot of stakeholders and institutes which are involved in water issue with diverse interests so that decision-making becomes sometimes either exhausting or impossible.
- The organization which is chiefly responsible to develop hydropower projects sometimes has not been authorized as much as necessary.
- Experts and software insufficiency in consultancy companies, clients, and contractors.
- Existence complexities and ambiguities in water laws and identifying stakeholders needs.

6.09.3.2.4 Economic and financial issues

- Providing the required investments are sometimes difficult, for example, the large hydropower projects need high amounts of investment cost.
- Intensive dependency to governmental budgets. The private sector has not been involved much due to lack of efficient strategies to encourage them to invest.
- Existence ambiguities in economic assessment of these projects.

6.09.4 Potentiality of Hydropower Projects

In recent decades, many projects have been carried out to identify suitable sites to develop hydropower plants all over the country by different consultancy companies. Many of the projects identified are either under construction or in operation, and the rest are studied at different phases. In this part, the results are reviewed comprehensively and described in the following sections.

6.09.4.1 Under Operation Projects

The total installed capacity of hydropower projects in the country reached 7733 MW at 2008. **Map 1** shows the spatial distribution of large hydropower projects in the country.

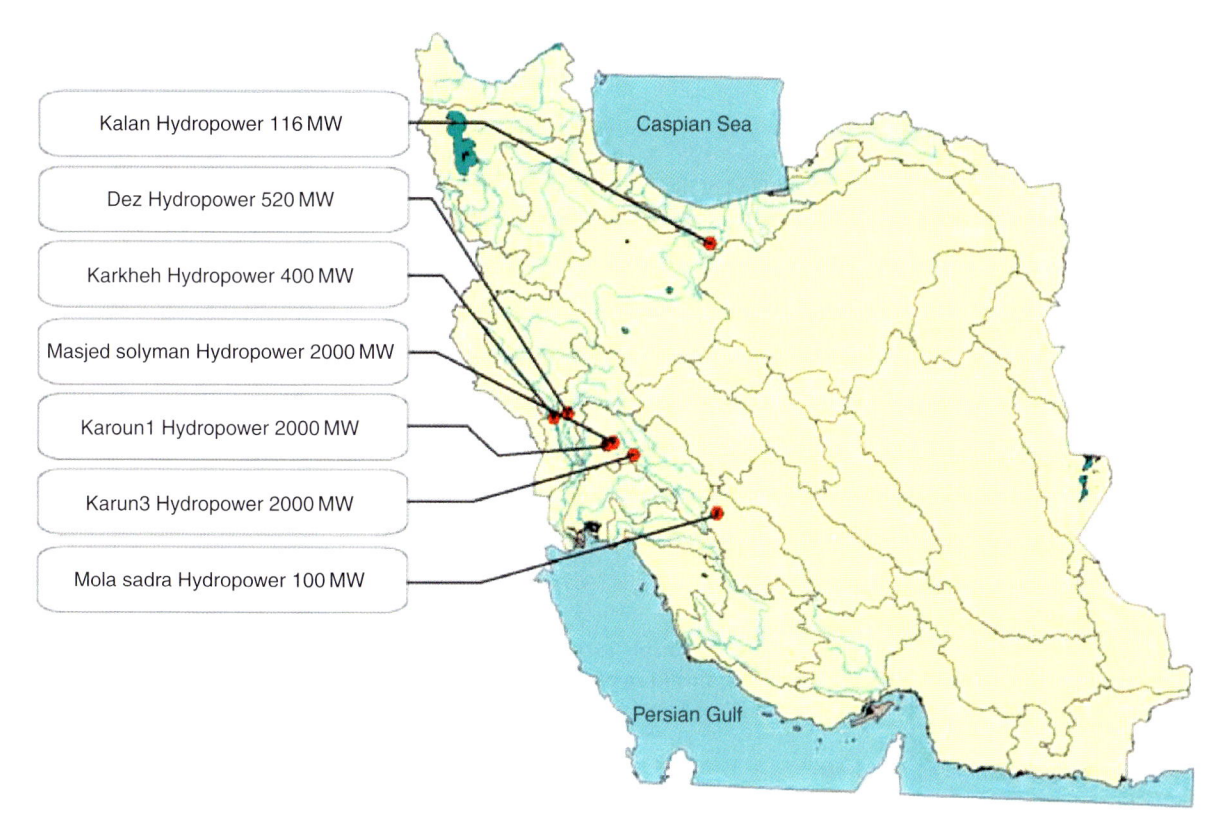

Map 1 Spatial distribution of large hydropower projects.

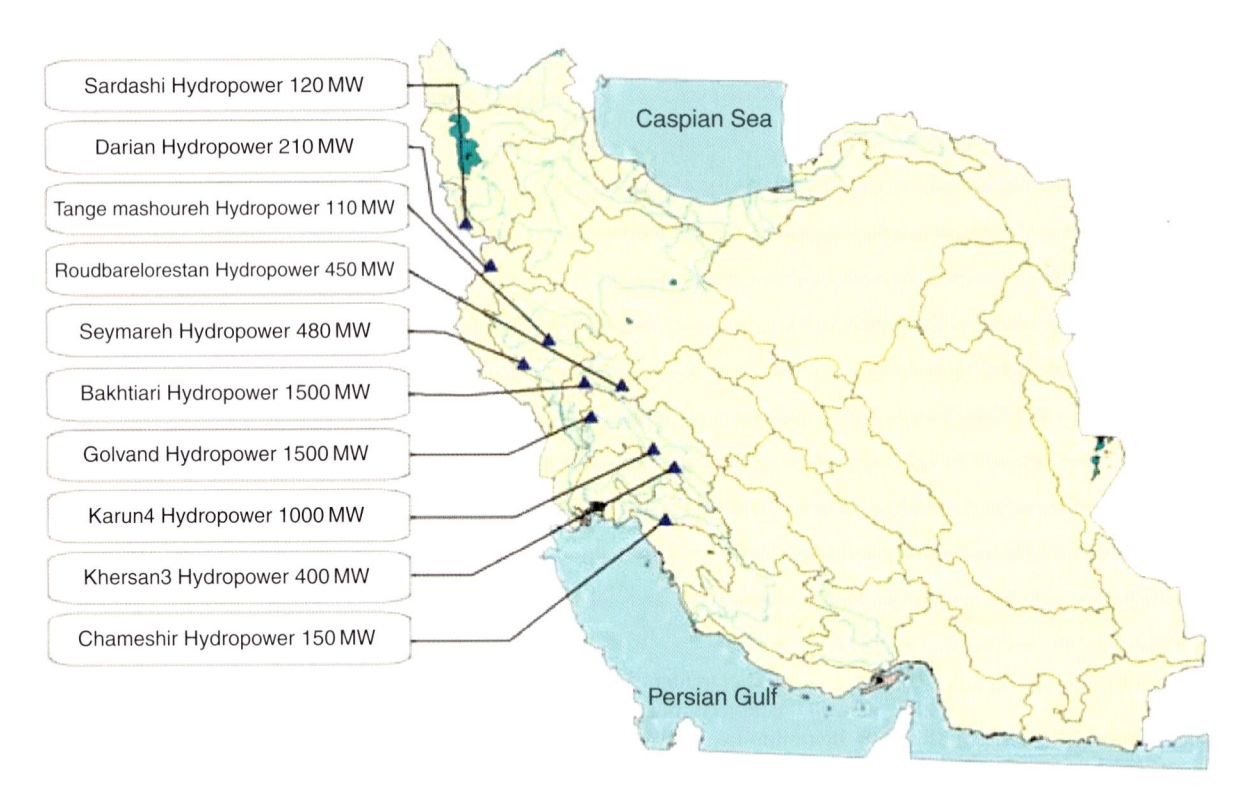

Map 2 Location of large high-pressure processings.

6.09.4.2 Under Construction Projects

The total capacities of hydropower projects which are under construction are 6037 MW conventional projects and 1040 MW nonconventional projects. All of them would be launched until 2019. **Map 2** shows the location of large hydropower projects in the country.

6.09.4.3 Under Study Projects

6.09.4.3.1 Storage hydropower projects
Due to climatic and topographic conditions, this kind of hydropower project has been located mainly in the mountainous regions of the Zagros mountain chains and partly in the Alborz mountain chains. In these areas, there are 14 power plants with a total capacity of 4200 MW (**Map 3**).

6.09.4.3.2 Large run-off-river hydropower plants
In total, five large run-off-river power plants with a capacity of 2800 MW were studied in the Dez, Karoon, Karkheh, and Aras basins (**Map 4**).

6.09.4.3.3 Medium run-off-river power plants
According to the studies which were carried out in most parts of the country, this kind of hydropower project could be developed in the mountainous regions of Alborz and Zagros. In this case, a number of projects which are feasible, with a total capacity of 1500 MW, were identified over nine main basins (**Map 5**).

6.09.4.3.4 Small run-off-river power plants
Over the Zagros and Alborz mountain chains, there are suitable locations to develop small run-off river. In these areas, there are suitable heads and discharges at the rivers. All feasible small run-off-river projects have been classified according to their locations and other considerations so that they can be set at 17 packages with a total capacity of 460 MW (**Map 6**).

6.09.4.3.5 Pumped-storage power plants
In order to perform the frequency control and to balance loads and demands in electrical network, the Siah Bisheh 1000 MW pumped-storage project has been constructed. In addition, a number of pumped-storage projects have been investigated that have a

Zalaky: Installed Capacity 466 MW Dez Basin

Lirou: Installed Capacity 324 MW Dez Basin

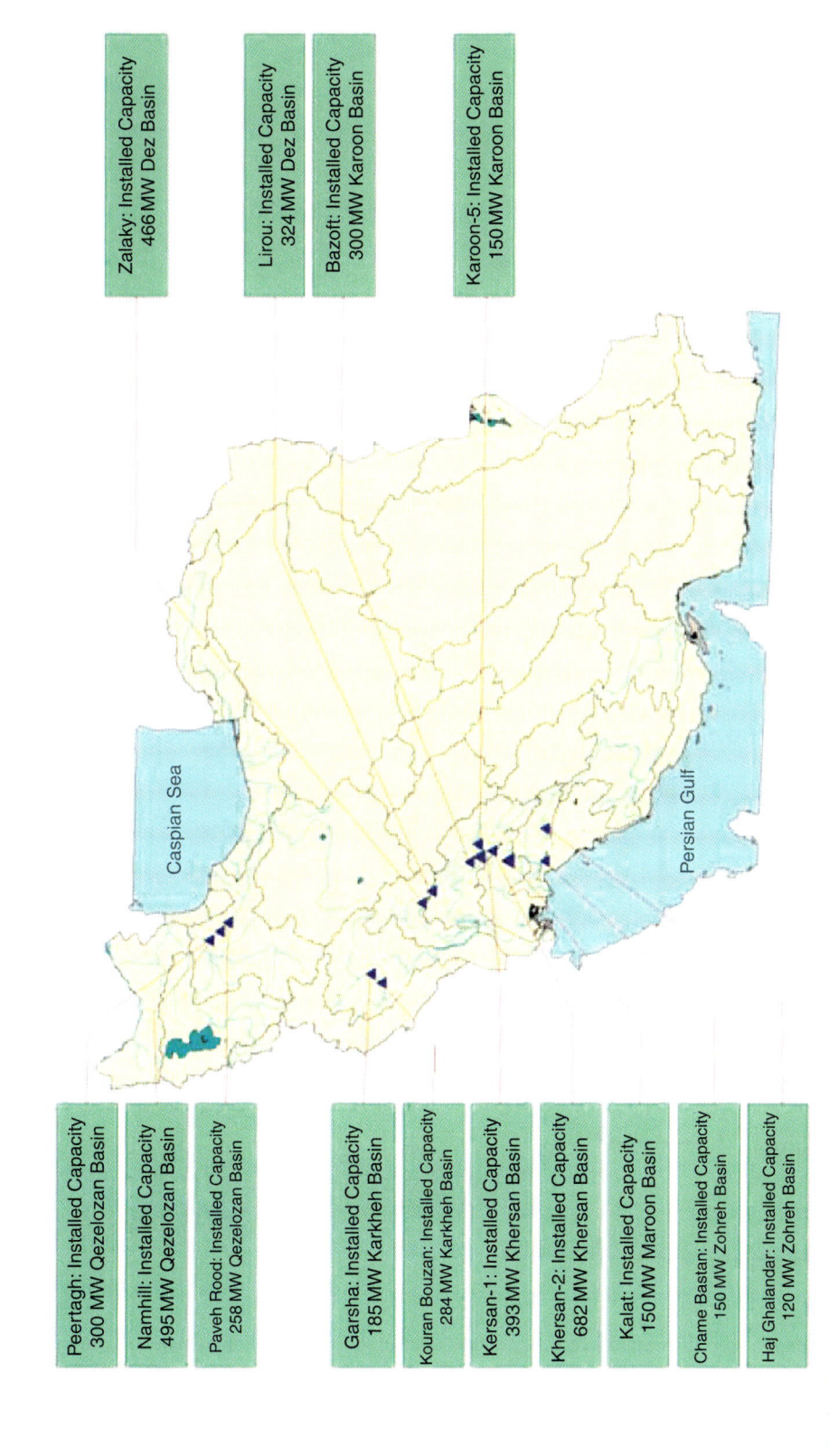

Bazoft: Installed Capacity 300 MW Karoon Basin

Karoon-5: Installed Capacity 150 MW Karoon Basin

Peertagh: Installed Capacity 300 MW Qezelozan Basin

Namhill: Installed Capacity 495 MW Qezelozan Basin

Paveh Rood: Installed Capacity 258 MW Qezelozan Basin

Garsha: Installed Capacity 185 MW Karkheh Basin

Kouran Bouzan: Installed Capacity 284 MW Karkheh Basin

Kersan-1: Installed Capacity 393 MW Khersan Basin

Khersan-2: Installed Capacity 682 MW Khersan Basin

Kalat: Installed Capacity 150 MW Maroon Basin

Chame Bastan: Installed Capacity 150 MW Zohreh Basin

Haj Ghalandar: Installed Capacity 120 MW Zohreh Basin

Map 3 Spatial distribution of storage high-pressure processings.

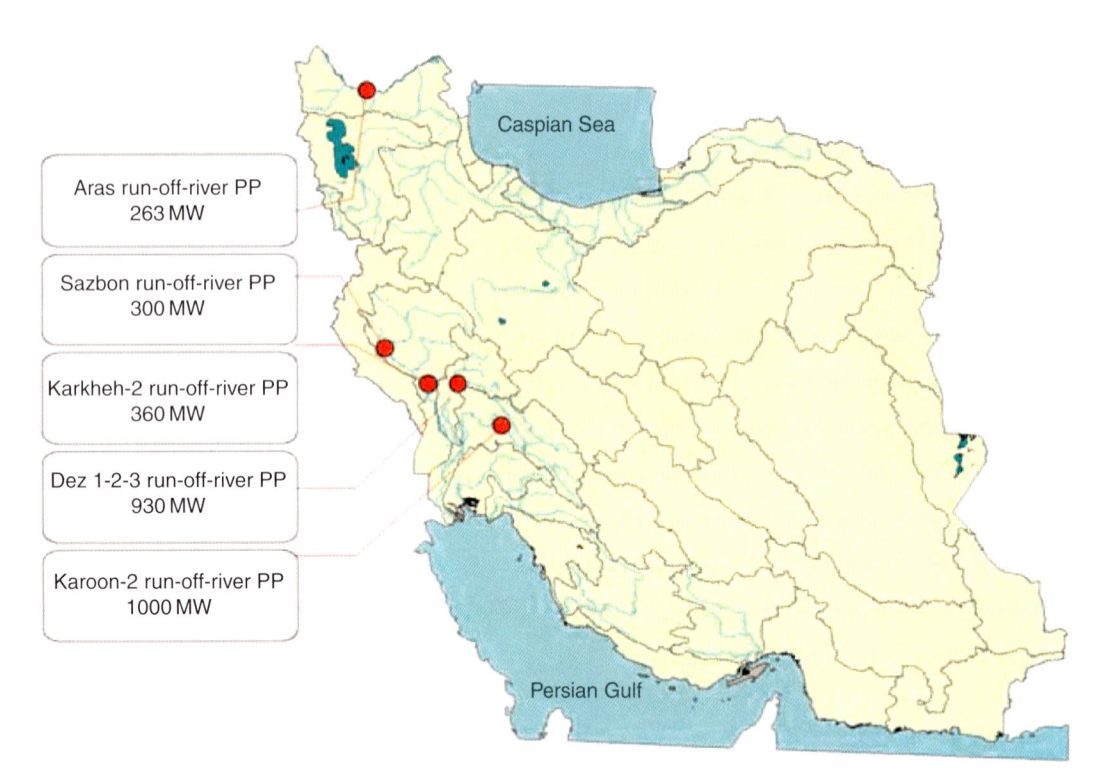

Map 4 Spatial distribution of large run-off-river projects.

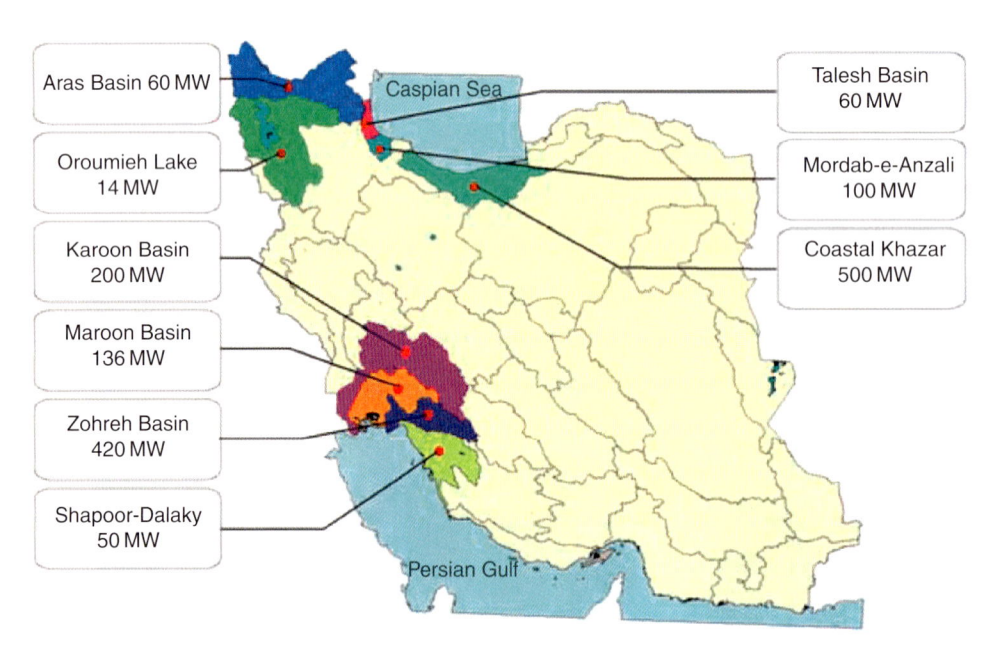

Map 5 Spatial distribution of medium run-off-river projects.

capacity of 7000 MW. **Map 7** shows the regions which are under investigation for the development of nonconventional hydropower projects.

6.09.4.3.6 Synthetic and conclusion

The last situation of under study/construction hydropower is shown in **Table 2** in brief, and the planned trend to develop hydropower projects in the future is depicted in **Figure 6**.

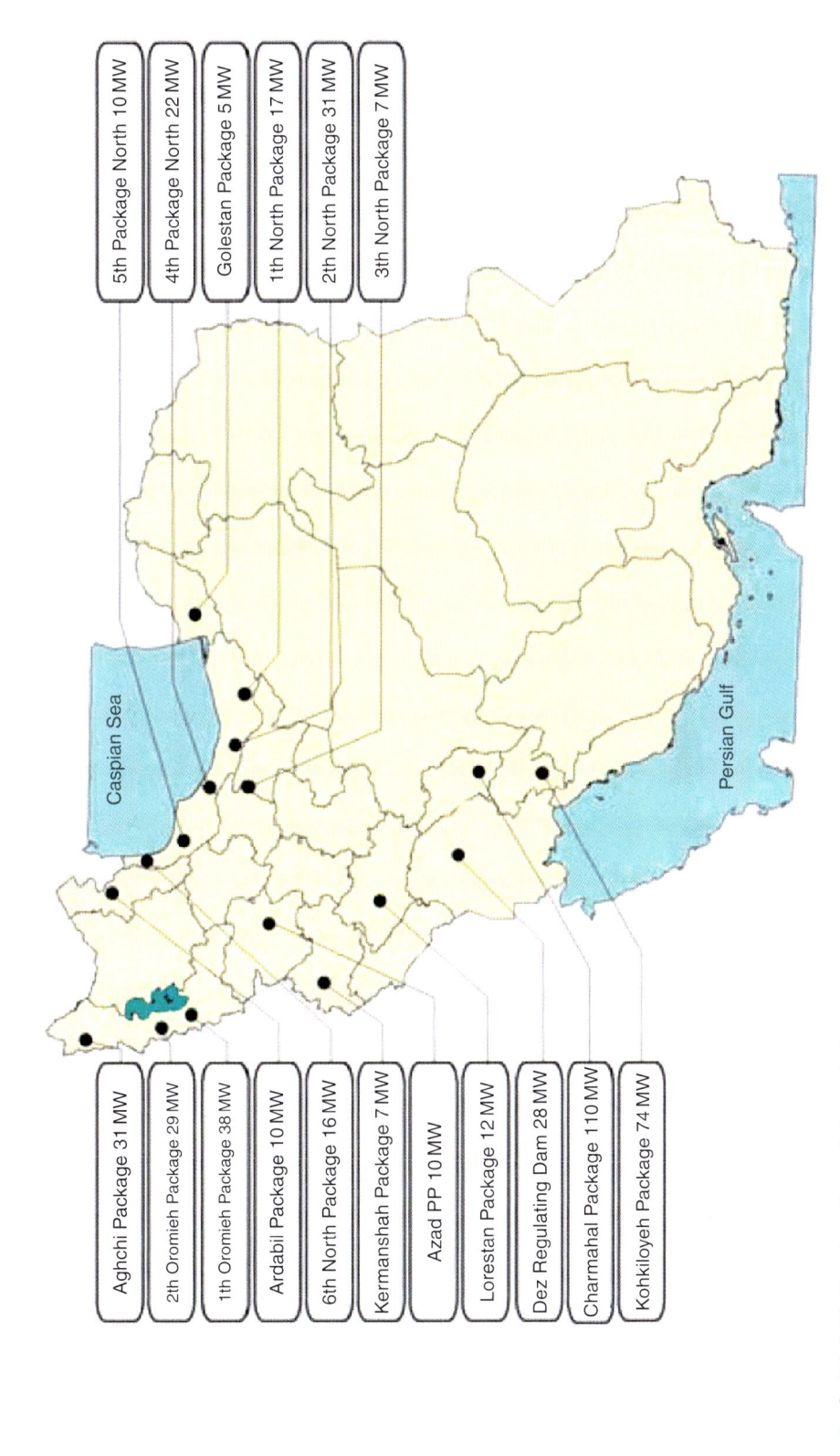

Map 6 Spatial distribution of small run-off-river projects.

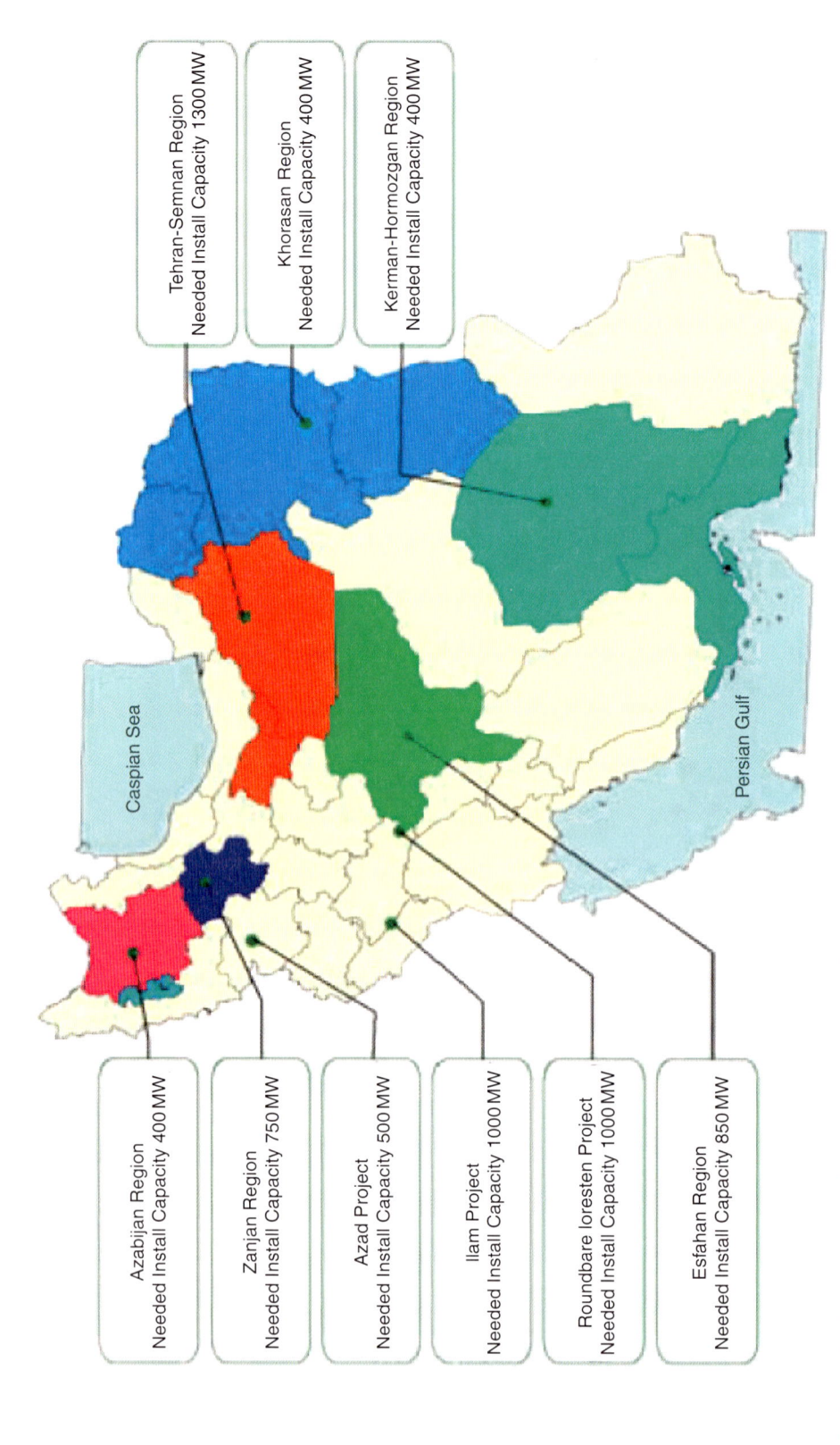

Map 7 Spatial distribution of regions which are under investigation for pumped-storage projects.

Table 2 Last situation of under study/construction high-pressure processings (HPPs)

Type of hydropower	Capacity (MW)
Storage HPP	4200
Large run-off-river HPP	2800
Medium run-off-river HPP	1500
Small run-off-river HPP	460
Under construction HPP	7077
Pumped-storage HPP	6950
Sum	Around 23 000

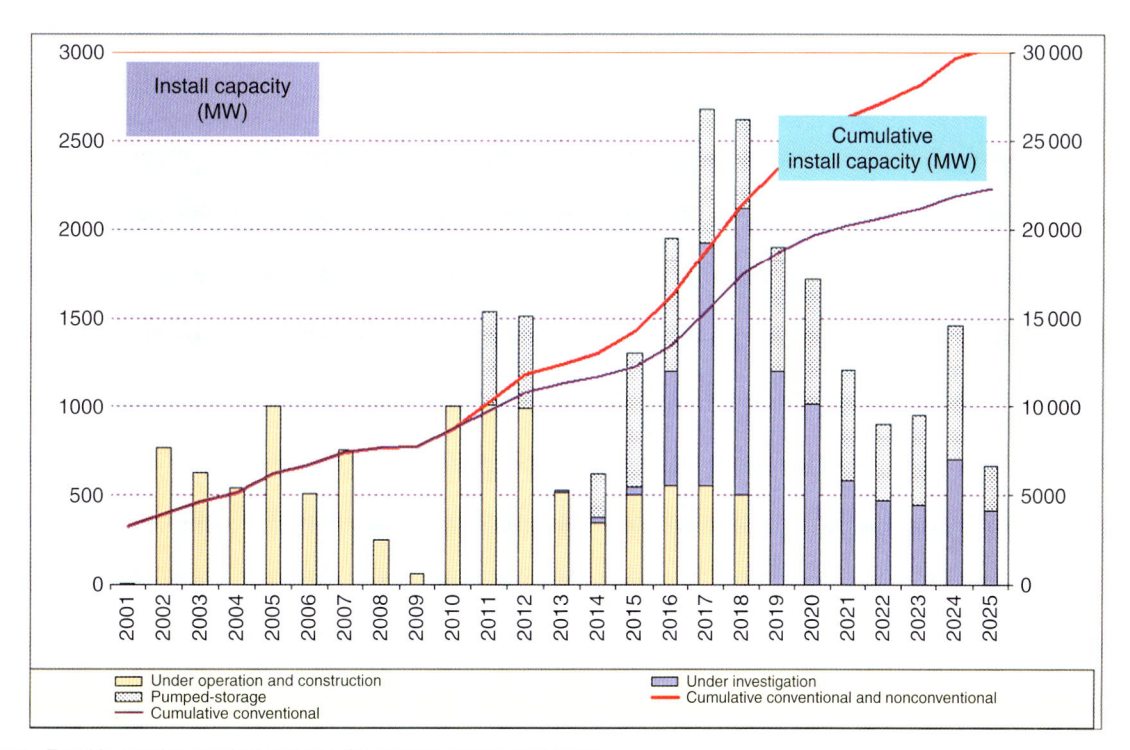

Figure 6 Trend in growing installed capacity of hydropower plants until 2025.

At present, the total capacities of under construction and study high-pressure processings including conventional and nonconventional projects are around 7000 and 16 000 MW, respectively. It is anticipated that the total capacity of hydropower projects would reach 30 000 MW by 2025, if the hydropower development strategies and actual plans were executed in compliance with **Figure 6**. In other words, the total capacities of conventional and nonconventional hydropower projects would increase 900 and 400 MW yearly. It is estimated, based on 2009 unit prices, that the total investment for developing of new projects would reach more than 34 BUS$, that is, 2.13 BUS$ per year.

References

[1] Vision of Islamic Republic of Iran by 2025, IREC, 2003.
[2] Fourth 5-year program on economic, social, and cultural development of Islamic Republic of Iran, Government of IR of Iran, 2004.
[3] Draft of fifth 5-year program on economic, social, and cultural development of Islamic Republic of Iran, Government of IR of Iran, 2009.
[4] Long-term strategy on Water Resources Development, Government of IR of Iran, 2003.
[5] Collection of rules, regulations, and instructions submitted by government on water and energy issues.
[6] Energy balance sheets, Tavanir Company, up to 2008.
[7] Dam's and hydropower's data bank, Iran Water and Power Resources Development Company (IWPCO).
[8] Technical reports and documents of different hydropower projects submitted by domestic and international consultancies and contractors.

6.10 Hydropower Development in Japan

T Hino, CTI Engineering International Co., Ltd., Chu-o-Ku, Japan

© 2012 Elsevier Ltd.

6.10.1	Outline of the History of Hydropower Development in Japan	265
6.10.1.1	The Start of Hydropower Production	266
6.10.1.2	The Start of Long-Distance Transmission of Electric Power and Large Hydropower Dams	266
6.10.1.3	The Development of Dams and Conduit-Type High-Capacity Hydropower Production	267
6.10.1.4	The Increased Use of River Water as an Energy Source	268
6.10.1.5	Electric Power Shortages and the Postwar Reorganization of Electric Power	269
6.10.1.6	Development of Large-Scale Dam-Type Hydropower Plants	270
6.10.1.7	Hydropower Dams from the Rapid Economic Growth Period to the Stable Growth Period	270
6.10.1.7.1	Electric power demand and the roles of hydropower dams during the rapid economic growth period	270
6.10.1.7.2	The redevelopment of hydropower by consistent hydropower development in a river system	271
6.10.1.7.3	Hydropower development centered on pumped-storage-type hydropower	273
6.10.2	Current State of Hydropower in Japan	274
6.10.2.1	Primary Energy in Japan	274
6.10.2.2	Development of Hydroelectric Power in Japan	276
6.10.2.3	Hydroelectric Power Development	276
6.10.2.4	Development of Pumped-Storage Power Plant	277
6.10.3	Hydropower in Japan and Future Challenges	280
6.10.3.1	Energy Situation in Japan and Hydropower	280
6.10.3.2	Hydropower in Japan and Future Challenges	280
6.10.4	Successful Efforts in Japan	281
6.10.4.1	Large-Scale Pumped-Storage Power Plants in Tokyo Electric Power Company	281
6.10.4.1.1	Outline of the project	281
6.10.4.1.2	Features of the project area	282
6.10.4.1.3	Benefits	284
6.10.4.1.4	Effects of the benefits	286
6.10.4.1.5	Reasons for success	286
6.10.4.2	Sediment Flushing of Reservoir by Large-Scale Flashing Facilities in the Kansai Electric Power Company	288
6.10.4.2.1	Outline of the project	290
6.10.4.2.2	Features of the project area	290
6.10.4.2.3	Major impacts	291
6.10.4.2.4	Mitigation measures	291
6.10.4.2.5	Results of the mitigation measures	292
6.10.4.2.6	Reasons for success	293
6.10.4.3	Reservoir Bypass of Sediment and Turbid Water during Flood in the Kansai Electric Power Company	294
6.10.4.3.1	Outline of the project	294
6.10.4.3.2	Features of the project area	294
6.10.4.3.3	Major impacts	294
6.10.4.3.4	Mitigation measures	296
6.10.4.3.5	Results of the mitigation measures	296
6.10.4.3.6	Reasons for success	298
6.10.4.4	Measures for Ecosystems	298
6.10.4.4.1	Outline of the project	299
6.10.4.4.2	Features of the project area	299
6.10.4.4.3	Major impacts	299
6.10.4.4.4	Mitigating measures	299
6.10.4.4.5	Results of the mitigation measures	305
6.10.4.4.6	Reasons for success	306
Relevant Websites		307

6.10.1 Outline of the History of Hydropower Development in Japan

Hydropower production that began with waterwheels on small rivers has expanded to include the run-of-river type, conduit type, dam and conduit type, and dam type.

During the last half of the 1880s in Japan, hydropower production appeared as an economical power production method to replace coal-fired thermal power production, meeting the growing electric demand. The first hydropower was run-of-river type with small-scale intake weirs installed to stabilize the intake water level. In about 1900, the construction of large-scale hydropower plants in mountainous regions far from demand regions began in response to progress in long-distance electric power transmission technology. From about 1910, the hydro-first/thermal-second stage arrived, and the construction of hydroelectric stations as part of dam regulation pond construction began. In the 1920s, dam–conduit-type hydropower plants appeared, providing a base load supply in response to soaring demand for industrial electric power.

Although the end of Second World War was followed by a temporary surplus of electric power, its demand soared because of shortages of power source, and postwar rehabilitation. Such circumstances triggered demand for the immediate start of work to establish the postwar electric power development system. Advanced thermal power stations were being constructed to provide electricity to meet rising demand, and provide base load. On the other hand, large-scale hydropower plants that were intended to meet the peak demand for electric power, increased in importance, spurring their construction.

The hydro-first/thermal-second electric power structure continued until 1962 and was followed by the advance of thermal power and nuclear power, but even after 1960, reservoir and regulating pond-type hydropower plants continued to be developed as valuable peak supply power. The concept of river hydropower development is to construct groups of hydropower plants appropriately from upstream to downstream to efficiently produce hydropower from the overall river perspective, and is called Consistent Hydropower Development in a River System. The oil shock of 1973 was followed by large-scale pumped-storage electric power production as part of valuable clean energy and as power to respond to peak electric power production.

6.10.1.1 The Start of Hydropower Production

Hydropower production was first developed for in-house use by the spinning and mining industries. The first electric power station developed to provide commercial electric power was constructed in Kyoto: the Keage Power Station (1892) that used water drained from Lake Biwa. Its power was used to operate the first electric street cars in Japan.

Demand for electric power for lighting began in 1887 and records of electric power demand for factories appeared in 1903, when Japanese industry finally modernized.

Early electric power projects were primarily intended to supply electric power for lighting from thermal power stations. During this period, transportation within Japan was inconvenient and transporting coal was costly, so it was difficult to produce thermal power in inland regions of Japan. Therefore, most power produced in such regions was hydropower. In other words, hydropower development began in regional cities close to hydropower zones.

Table 1 is a table of the oldest hydropower plants in various regions.

6.10.1.2 The Start of Long-Distance Transmission of Electric Power and Large Hydropower Dams

After the Russo-Japan war, the Japanese economy underwent rapid growth. Because electric power demand was also expanded rapidly by the Russo-Japan war, the electric power industry acquired an important position in Japanese industry. This growth of electric power demand grew in two areas: spreading electric lighting in homes and the electrification of power provision in factories.

The earliest hydropower plants in Japan were extremely close to their demand regions, and their generator output and transmission voltage were both low. However, in 1899, the transmission of 11 kV for 26 km and the transmission of 11 kV for 22 km were achieved in the Chugoku and Tohoku Regions, respectively, permitting longer distances between hydropower plants and consumption regions, thereby contributing greatly to electric power production projects in Japan. Later, electric power companies worked to increase transmission voltages, lengthen transmission distances, and to develop high-capacity hydropower plants.

Table 1 Oldest hydropower plants in each region

Region	Name of power plant	River system	Effective head (m)	Maximum discharge ($m^3\,s^{-1}$)	Maximum output (kW)	Beginning of operation	Classification	Current state
Tohoku	Sankyozawa	Natori	26.67	5.57	5	1888.7	In-house use	1000 kW operating
Kanto	Simotsuke Asa Bouseki (owner)	Tone			17	1890.7	In-house use	Abolition
Chubu	Iwazu	Yahagi	53.94	0.37	50	1897.7	Project use	130 kW operating
Kansai	Keage	Yodo	33.74	16.7	80 × 2	1891.11	Project use	4500 kW operating

Source: Electric Power Civil Engineering Association.

Table 2 Large-scale hydropower plants constructed by the beginning of the twentieth century

Name of power plant	Name of river system	Dam or water resource	Beginning of operation	Maximum output (kW)	Voltage (V)	Distance (km)
Komabashi	Sagami	Lake Yamanaka	1907.12	15 000	55 000	75
Yaotsu	Kiso	Lake Maruyama Sosui	1911.12	7 500	66 000	34
Yatsuzawa	Sagami	Oono Desanding Basin	1912.7	35 000	55 000	75
Shimotaki	Tone	Kurobe Dam	1912.12	31 000	66 000	125
Inawashiro No. 1	Agano	Lake Inawashiro intake weir	1914.10	37 500	115 000	225

Source: Electric Power Civil Engineering Association.

During this period, intake facilities used to generate electric power also changed as low fixed water intake weirs that could take in the flow rate in the dry season were replaced by dams with gates, and these were expanded to include dams with regulating ponds. Large-scale hydropower plants developed in this way are shown in **Table 2**.

Of these, the Shimotaki Power Station in the northern Kanto Region supplied power to Tokyo at that time, providing almost the entire demand (~40 million to 80 million kWh yr^{-1}) to run trams in Tokyo.

In addition, the Yatsuzawa Power Station (Tokyo Electric Power Company, Inc. (TEPCO), 1912) in western Kanto was not only a high-capacity dam but also a conduit type with a large regulating pond (effective capacity: 467 000 m^3). It was an epoch-making type of dam at that time.

The development of large-scale hydropower plants had a number of important impacts on the management of the electric power industry. First, it allowed a drop in the price of electricity, because hydropower could be produced more cheaply than thermal power. Second, it permitted companies to meet the daytime demand for power for industry, in addition to the nighttime demand for lighting power.

Because most hydropower plants were the conduit type at that time, it was impossible to control daytime and nighttime flow rates. This means that when an appropriate customer could not be found, it was impossible to produce power using the daytime flow rate when demand for electric power was lower than at night, and this encouraged a rise in electric power production costs. Electric power companies attempted to obtain daytime demand by lowering their daytime electricity charges, contributing to the profitability of industries that used this cheap electric power (**Figure 1**).

During this phase, the structure of electric power production facilities underwent a sharp change, from thermal-first/hydro-second to hydro-first/thermal-second. Hydro surpassed thermal power in 1911, ushering in the age of hydro-first/thermal-second in the Japanese electric power industry: for about a half century from 1911 to 1960.

Figure 2 shows changes of electric power production until the late 1930s.

6.10.1.3 The Development of Dams and Conduit-Type High-Capacity Hydropower Production

During the Taisho Period (1912–26), the Japanese economy was affected by worldwide economic growth, resulting in lively growth of about 5% per annum of Japan's manufacturing industries, until the start of the Second World War. The growth of the electrochemical industries and the machinery and iron and steel sectors after the First World War was remarkable. To support production, these industries required an abundant and low-priced supply of electricity. The electric power supply grew explosively at a rate in excess of 20% per annum, and the first shortage of electric power since the establishment of the industry occurred during a drought in 1918.

Other reasons for the rapid development of hydropower were the successful introduction of long-distance power transmission, permitting the development of large hydropower production in mountainous regions, and the flourishing of industries that used

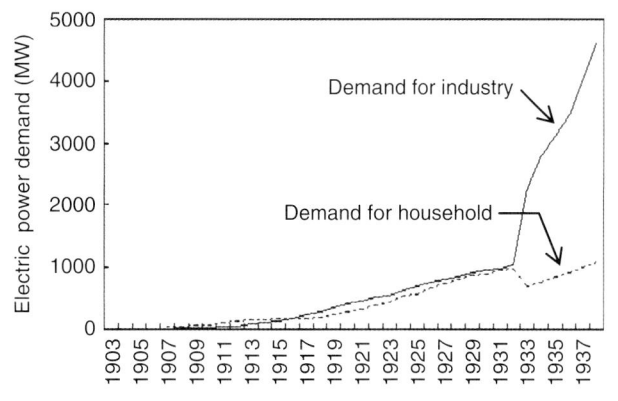

Figure 1 Transition of maximum electric power demand before the Second World War. Source: Electric Power Civil Engineering Association.

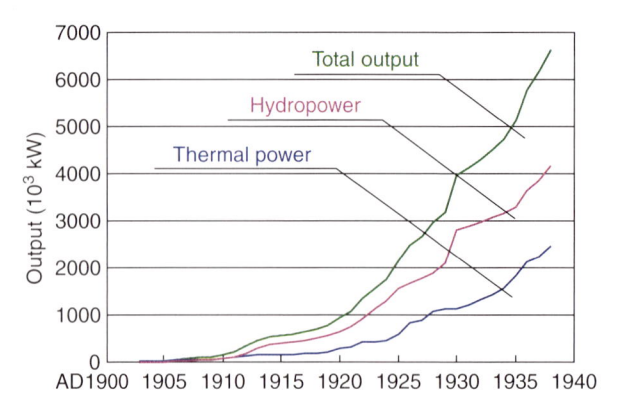

Figure 2 Changes of electric power production at the early stage. Source: Electric Power Civil Engineering Association.

low-priced electric power available at times when electric power demand was low. Operating thermal power stations in parallel to supplement hydropower production during dry seasons ensured stable electric power and encouraged the expansion of the electric power industry. This gave the industry the idea of effectively using a quantity of water in excess of the flow rate in the dry season by creating complementary hydropower–thermal power systems, and increased the maximum intake to approximately the average water flow rate.

Under these circumstances, companies that planned and conducted large-scale hydropower development were established, one after another. They developed the rich hydropower of mountainous areas and constructed high-voltage transmission lines to supply electric power to cities. One example was the successful transmission of 154 kV for 238 km, from the Suhara Power Station (Kansai Electric Power Co., Inc. (KEPCO), 1922) in Chubu to the Osaka Substation in Osaka (now the Furukawabashi Substation), in 1923. The achievement and spread of long-distance electric power transmission, by increasing voltage, spurred hydropower development in mountainous regions, with particularly remarkable development of high-capacity hydropower beginning in the late Taisho Period (1912–26).

An example is the Oi Electric Power Station that includes the Oi Dam (PG, 53.4 m) in Chubu Region. The Oi Electric Power Station, a dam–conduit-type power station developed on the Kiso River, was completed in 1924. It was originally planned as a conduit type, but it was converted to a dam type that can respond to peak demand, making it the first power station to include a large-scale dam constructed in Japan. The maximum output of this power station was equivalent to half of the entire electric power demand in Aichi Prefecture at that time.

The Shizugawa Power Station (KEPCO), which includes the Shizugawa Dam in Kyoto Prefecture, developed in 1924, provided more than 10% of all electric power used in Osaka Prefecture at that time. When it was developed, the Osaka–Kyoto–Kobe Metropolitan Region was particularly short of electric power, so its development made a big contribution to the supply of electric power in the region.

Table 3 shows typical dam-type and dam–conduit-type hydropower plants that were constructed in various districts during the Taisho Period.

6.10.1.4 The Increased Use of River Water as an Energy Source

During the early Showa Period (1926–45), large-capacity hydropower development continued in response to the results of economic evaluations of two approaches that began to spread in the late Taisho Period, making the maximum intake quantity approximately the average water flow rate and using thermal electric power as supplementary power during dry seasons. At the same time as national government control of industry strengthened, mining and manufacturing industry production soared, and metal,

Table 3 Hydropower plants representing each region after the First World War

Present owner	Name of power plant	Name of river system	Name of dam	Height of dam (m)	Maximum output (kW)	Beginning of operation
HEPCO	Nokanan	Ishikari	Nokanan	30	5 100	1918
CEPCO	Kamiasou	Kiso	Kamiasou Hosobidaani	22.4	24 300	1926
KEPCO	Shizugawa	Yodo	Shizugawa	35.2	32 000	1924
KEPCO	Oi	Kiso	Oi	53.4	42 900	1924
ENERGIA	Taishakugawa	Takahashi	Taishakugawa	62.1	3 706	1924

Source: Electric Power Civil Engineering Association.

chemical, and machinery industries grew at a particularly rapid rate. Electric power companies responded to trends in the manufacturing industry by devising and implementing the concept of successively developing hydropower plants mainly from the downstream reaches of large-scale rivers.

Of these, the development of hydropower on the Kurobe River in Hokuriku Region began with the completion of the Yanagawara Power Station (1927), and moving upstream, was followed by the Kurobegawa No. 2 Power Station (1936) supplied by the Koyadaira Dam (PG, 54.5 m), then the Kurobegawa No. 3 Power Station (1940) supplied by the Sennindani Dam (PG, 43.5 m).

When the national government took control of electric power, continued surveys moved upstream, but because it was followed shortly by the Second World War, hydropower development ended with the construction of the Kuronagi No. 2 Power Station (1947) on a tributary. The later successive development of the Kurobe River is described later.

As successive developments were carried out along the river, dams used exclusively to produce hydropower were constructed separately at locations in the river basin where topographical conditions suited hydropower development, advancing the use of river water as an energy source.

In the 1920s, on the Oi River System in Chubu Region, the water intake dam, the Tashiro Dam (PG, 17.3 m), was constructed as the furthest upstream dam located 160 km from the river mouth, and hydropower plants (Tashiro River No. 1 and No. 2 Power Stations) were developed, carrying the water into the Hayakawa River on the Fuji River System. This hydropower was transmitted to the Metropolitan Tokyo area. The power produced by these power stations was equivalent to about three times the demand by Tokyo at that time.

In the middle reaches of the Oi River and on the Tenryu River, dam-type hydropower plants were constructed, forming the core electric power development of each river system at that time. The maximum output of the Oigawa Power Station was so massive that it equaled approximately half of the contract kilowatts for all electric power in Shizuoka Prefecture at that time. The Yasuoka Dam (PG, 50.0 m), the first dam constructed on the Tenryu River to produce electric power, was also completed during that period.

Table 4 shows representative dam-type and dam–conduit-type hydropower plants that were constructed in various regions during that period.

6.10.1.5 Electric Power Shortages and the Postwar Reorganization of Electric Power

The end of the Second World War was followed by a temporary surplus of electric power, because electric power consumption was halved from its former level by stagnation of manufacturing activities caused by the wartime destruction of manufacturing plants. As a consequence of the spread of electric heaters to heat people's homes in response to shortages and soaring prices of coal, petroleum, and gas, and of the spreading use of electric power that could be obtained easily and cheaply as a power source to restore manufacturing, electric power demand soared. Annual energy supply that was down to 19.5 billion kWh in 1945 had leaped to 29.4 billion kWh in 1947.

However, new electric power sources were not developed, as little work was done to restore electric power systems damaged by the war and to continue projects initiated before the war. Later, at the end of 1949, approval was given for hydropower development at 33 locations, with an intended production of 1180 MW, as hydropower development funded by the US Economic Rehabilitation Fund.

The system of state control of the electric industry that had been implemented through Japan Power Generation and Transmission Co. Inc., during the war, ended with the 1951 breakup of the electric power industry into the current nine regional companies as an occupation policy of moving away from Japan's overcentralized economy. That year, the Korean War that boosted electric power demand was accompanied by an extremely severe drought in the autumn, resulting in an unprecedented electric power crisis. At that time, frequent power failures made candles a standard form of emergency lighting in homes.

Such circumstances triggered demand for the immediate start of work to develop a large-scale hydroelectric source. In 1952, the Electric Power Development Promotion Act was enacted. Under this law, the Electric Power Development Co., Ltd. (J-POWER) was

Table 4 Hydropower plants representing each region before the Second World War

Present owner	Name of power plant	Name of river system	Name of dam	Height of dam (m)	Maximum output (kW)	Beginning of operation
KEPCO	Komaki	Sho	Komaki	79.2	72 000	1930
KEPCO	Kurobegawa No. 3	Kurobe	Sennindani	43.5	81 000	1940
TEPCO	Tashirogawa No. 2	Oi	Tashiro	17.3	20 362	1928
CEPCO	Yasuoka	Tenryu	Yasuoka	50.0	52 500	1936
CEPCO	Oigawa	Oi	Oigawa	33.5	62 200	1936
			Sumatagawa	34.8		

Source: Electric Power Civil Engineering Association.

founded with government funding, to establish a power source development system with the primary task of directly investing government funds in regions where development was difficult. This law also stipulated that the Electric Power Development Coordination Council would prepare long-term basic plans for electric power and annual implementation plans, including all electric power development projects conducted by electric power companies and public bodies. In these ways, the postwar electric power development system was established.

6.10.1.6 Development of Large-Scale Dam-Type Hydropower Plants

When the growth of hydropower production in Japan began, it was centered on run-of-river-type hydropower plants that required relatively little initial funding, and until the 1950s, the decline in hydropower production during the drought season was supplemented by thermal power production. In the late 1950s, of the hydropower plants at approximately 1460 locations, only about 40 were equipped with reservoirs that could regulate their flow.

Thermal power stations operating after the war were powered by coal, but their electric power production efficiency and profitability both fell remarkably because of delayed supplies of coal, a decline in its quality, and a rise in its price. As a result, the construction of dam-type hydropower plants was reemphasized as a way of increasing water usage and overcome the seasonal imbalance.

More advanced thermal power stations were being constructed to provide electricity to meet rising demand in response to the postwar rehabilitation of industry, and with these stations providing base load, large dam-type hydropower plants that were intended to meet the peak demand for electric power, increased in importance, spurring their construction. Table 5 shows the major hydropower dams that were completed during this period.

An example is the Sakuma Power Station that became the key to promoting electric power development.

The Tenryu River carries a large volume of water as a result of the heavy snow and rain that fall in its mountainous middle reaches as it flows through the Chubu Region. As a result, the region had sought development for many years dating back to the Taisho Period. Following a severe drought in 1951, electric power had to be developed very quickly, so J-POWER, which was founded in 1952, decided to develop electric power at Sakuma.

The Sakuma Power Station was designed to handle peak loads and J-POWER also developed electric power resources at the reregulating reservoir, further downstream at Akiha.

The Sakuma Dam is a concrete gravity dam with a height of 155.5 m and reservoir capacity of about 330 million m^3. The maximum output of the Sakuma Power Station is 350 000 kW, equivalent to 2.3% of the total electric power output in Japan at that time. Its annual electric power production of about 1.5 billion kWh has been the largest in Japan from the time it was completed until now, and a pioneer in large-capacity reservoir-type hydropower plants. The power it produces has been shared by Chubu Electric Power Co. (CEPCO) and TEPCO, thus making a major contribution to stabilizing supply and demand in the Tokyo–Yokohama area and Nagoya area and to the economical operation of advanced thermal power plants.

6.10.1.7 Hydropower Dams from the Rapid Economic Growth Period to the Stable Growth Period

6.10.1.7.1 *Electric power demand and the roles of hydropower dams during the rapid economic growth period*

The development and spread of household electrical appliances was a particularly remarkable aspect of the process of postwar economic growth in Japan, as washing machines, television sets, and refrigerators spread rapidly in homes during the late 1950s. This was followed by a revolution in consumption of the so-called three Cs: cars, coolers (home air conditioners), and color televisions.

Economic growth was accompanied by a rapid growth in energy demand. Beginning about 1948, a series of large-scale oil fields were discovered in the Middle East and technological progress encouraged a switch from coal to petroleum in the industrial world. Demand for electric power also increased so that, as shown in Figure 3, from the 1950s to 1965 the quantity of electric power consumed soared almost as rapidly as the annual 10% increase in the mining and manufacturing industry's production index. During this period, electric power companies were compelled to ensure energy supplies by means of large-scale electric power

Table 5   Large-scale hydropower dams in the 1950s

Name of dam	Owner	River	Dam height (m)	Type	Name of power station	Output (MW)	Commencement of operation
Maruyama (PG, 98.2 m)	KEPCO	Kiso	96	PG	Maruyama	125	1955
Kamishiiba (VA, 110.0 m)	Kyusyu EPCO	Mimi	110	VA	Kamishiiba	90	1955
Sakuma	J-POWER	Tenryu	155.5	PG	Sakuma	350	1956
Ikawa	CEPCO	Oi	103.6	HG	Ikawa	62	1957

Source: Electric Power Civil Engineering Association.

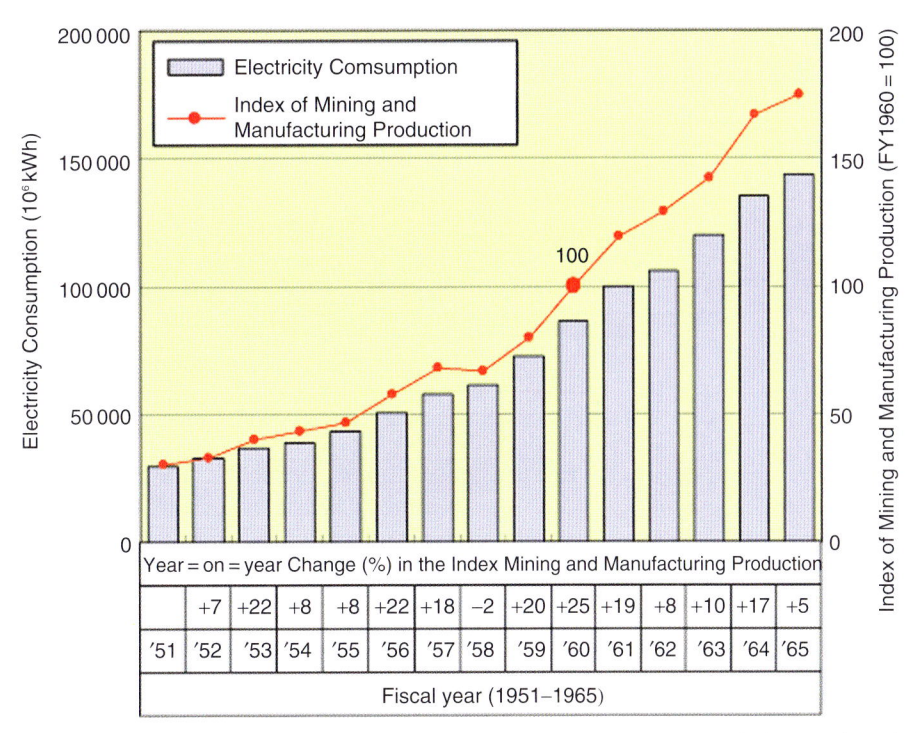

Figure 3 Change in the index of mining and manufacturing production and the quantity of electricity consumption between the end of the war and 1965. Source: Ministry of Economy, Trade and Industry (METI).

source development. **Figure 4** shows changes in the state of hydropower output excluding pumped-storage-type hydropower increased significantly, and especially during a period of more than 10 years beginning in about 1955.

Table 6 shows examples of dams for large-scale reservoir-type electric power stations following the Sakuma Dam, which was constructed prior to the rapid economic growth period. These were the dams that created the golden age of hydropower development.

The hydro-first/thermal-second electric power structure continued until 1962 and was followed by the advance of thermal power and nuclear power, but even after 1960, reservoir and regulating pond-type hydropower plants continued to be developed as valuable peak supply power. The concept of river hydropower development is to construct groups of hydropower plants appropriately from upstream to downstream to efficiently produce hydropower from the overall river perspective, and is, accordingly, called Consistent Hydropower Development in a River System. The oil shock of 1973 was followed by large-scale pumped-storage electric power production as part of valuable clean energy and as power to respond to peak electric power production.

6.10.1.7.2 The redevelopment of hydropower by consistent hydropower development in a river system

On major river systems in Japan, hydropower development was started in the 1920s by constructing dams in the central and downstream reaches of rivers, where it was easy to construct electric power stations. Thereafter, large-scale hydropower development shifted upstream. From about 1960, the construction of hydropower plants resumed from the upstream reaches to the central and downstream reaches of each river system, and electric power plants were developed or redeveloped in the central and downstream reaches of rivers to efficiently utilize the head drop and water quantity. Good examples are the Kiso River, Hida River, Oi River, Agano River, Sho River, and Kurobe River.

Below, the Kurobe River, which is the location of the Kurobe Dam (VA, 186.0 m), the highest dam in Japan, is introduced as an example of Consistent Hydropower Development in a River System.

As mentioned above, until the 1940s, electric power stations were constructed on the Kurobe River in a series of steps, thus taking advantage of the head drop of river water from the old Yanagawara Power Station to the Kurobegawa No. 3 Electric Power Station (**Table 7** and **Figure 5**).

After the Second World War, an age when electric power production had shifted to advanced thermal power while hydropower played a role as a large reservoir-type peak load supply, the KEPCO responded by preparing a plan to construct a dam to form a large reservoir at the furthest upstream part of the Kurobe River, which was to play a pivotal role in the Consistent Hydropower Development in a River System. The Kurobe Dam, which attracted attention as one of the century's giant projects, was completed in 1963 and was the product of the finest civil engineering technology in Japan at that time. The Kurobegawa No. 4 Hydropower Station that takes water from the reservoir at the Kurobe Dam began operating in 1961, prior to the completion of the Kurobe Dam.

The completion of the Kurobe Dam with its total reservoir capacity of approximately 200 million m³ improved the downstream flow regime remarkably. To use its capacity effectively, the new power stations located downstream were constructed in succession, completing the entire Consistent Hydropower Development on the River (**Figure 5**). In this way, the Kurobe River became a

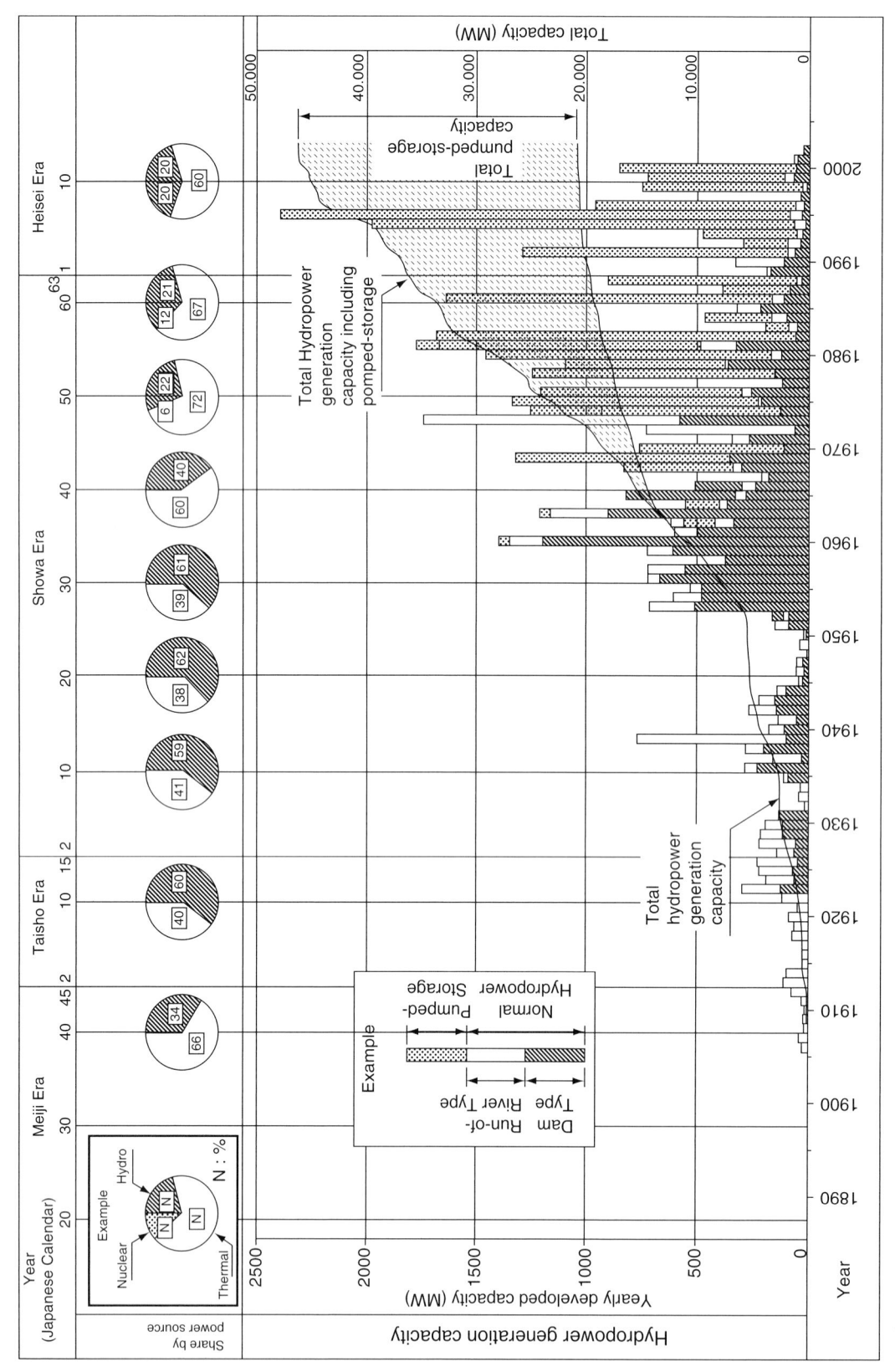

Figure 4 Change in hydropower generation capacity and its share in the whole generation capacity. Source: Electric Power Civil Engineering Association.

Table 6 Hydropower dams completed around 1960

Name of dam	Owner	River system	Dam height type	Name of power station	Output power (MW)	Start of operation
Tagokura	J-POWER	Agano	145 m PG	Tagokura	380	1959
Okutadami	J-POWER	Agano	157 m PG	Okutadami	360	1969
Miboro	J-POWER	Sho	131 m ER	Miboro	215	1961
Kurobe	KEPCO	Kurobe	186 m VA	Kurobegawa No. 4	335	1960

Source: Electric Power Civil Engineering Association.

Table 7 Consistent hydropower development on the Kurobe River

Type of development	Completion	Power station (dam)
Development in the lower reach	1927–47	Yanagawara P.S.
		Aimoto P.S.
		Kurobegawa No. 2 P.S. (Koyadaira Dam)
		Kurobegawa No. 3 P.S. (Sennindani Dam)
		Kuronagi No. 2 P.S.
Large-scale reservoir development in the upper reach	1961	Kurobegawa No. 4 P.S. (Kurobe Dam)
Redevelopment in the lower reach	1963–85	Shin-Kurobegawa No. 2 P.S. (Koyadaira Dam)
		Shin-Kurobegawa No. 3 P.S. (Sennindani Dam)
		Otozawa P.S. (Dashidaira Dam)
	1993–2000	Shin-Yanagawara P.S. (Dashidaira Dam)
		Unazuki P.S. (Unazuki Dam)

Source: Electric Power Civil Engineering Association.

power-source river with a series of peak power stations that took full advantage of the head drop of more than 1300 m from the Kurobe Dam reservoir water level (elevation 1448 m) to the Otozawa Power Station (elevation 131.1 m).

6.10.1.7.3 Hydropower development centered on pumped-storage-type hydropower

In the late 1950s, high-capacity, advanced thermal power stations took over the base load supply of electric power, with peak adjustment handled by large-scale reservoir-type hydropower plants.

During the 1960s, rapid urbanization and a rise in the people's standard of living driven by rapid economic growth resulted in a remarkable increase in office and home electricity demand for air conditioners. This trend shifted the annual maximum power demand, which had formerly been on winter evenings, to the daytime during summer. The summer peak exceeded the winter peak in 1968.

This summer peak created a new demand pattern, marked by a sharpened peak during the day, a pattern that was beyond the adjustment capacity of reservoir-type power stations, thereby creating a need for pumped-storage power stations that are better suited to adjusting the gap in daytime and nighttime electric power demand.

In 1960, the Resources Council of the Science and Technology Agency of the Prime Minister's Office issued a policy statement calling for the diversification of energy supply sources. In the recommendation concerning the survey of pumped-storage electric power production, it called for hydropower surveys of pumped-storage power station locations in order to establish a power source development approach that treats thermal, nuclear, and pumped-storage power stations as harmonized sources.

Under these circumstances, the electric power companies also studied policies to promote hydropower development from a new perspective, thus establishing large-scale redevelopment plans for pumped-storage power stations, both standalone and as part of comprehensive development projects.

A pumped-storage power station requires two reservoirs: an upper and a lower reservoir. Until about 1970, many were constructed as mixed pumped-storage power stations that could also produce ordinary hydropower where the inflow of river water to the upper reservoir was sufficient. However, as the number of available economical locations declined, the development of pure pumped-storage power stations at locations where either no water or extremely little water flows into the upper reservoir began to flourish, beginning with the station at Numappara (J-POWER, 1973) in the early 1970s. This was made possible by the development of new technologies: a steel penstock with a head drop in the 500 m class and high-capacity reversible pump-turbines.

Many efforts to reduce energy dependency on petroleum were initiated following the first and second oil shocks in 1973 and 1979, then in 1980, the Act on the Promotion of Development and Introduction of Alternative Energy was enacted, shifting priority to the construction of nuclear power stations.

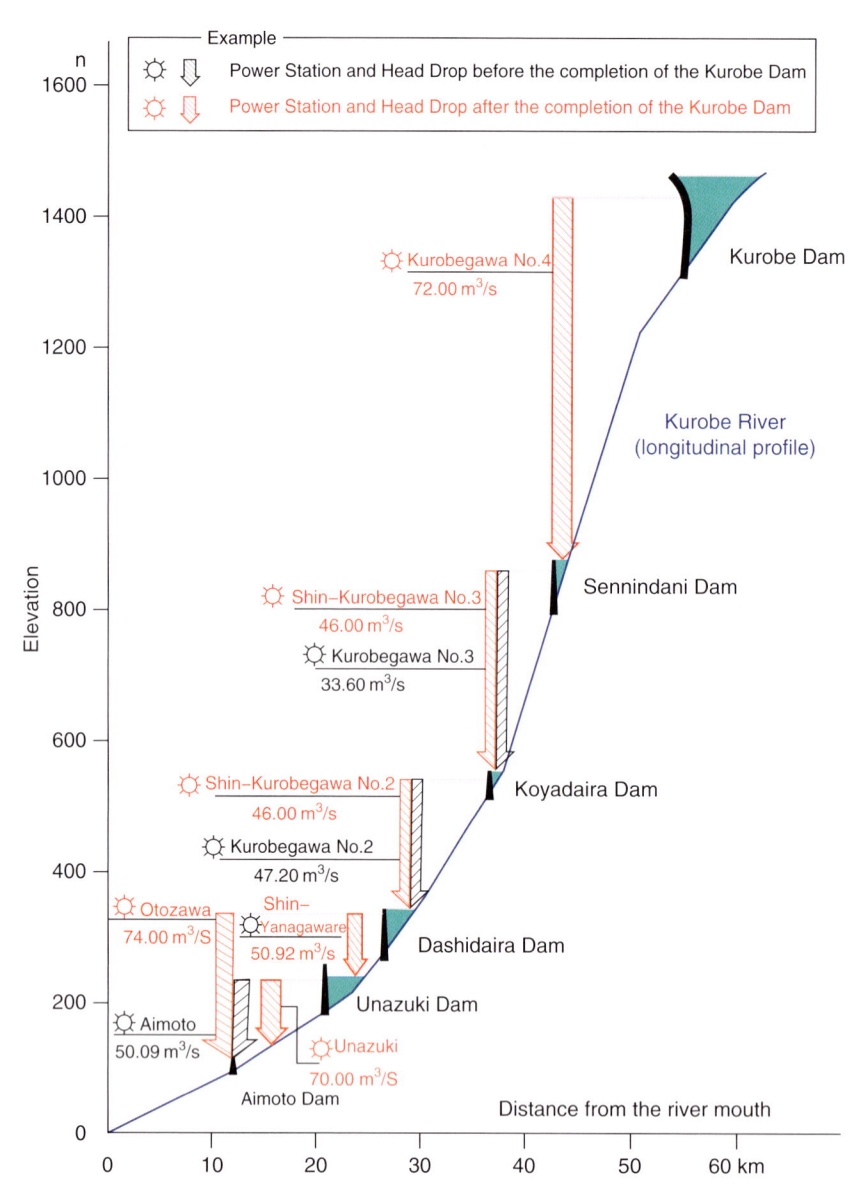

Figure 5 Schematic view of hydropower generation on the Kurobe River. Source: KEPCO.

Because both total electric power demand and the annual maximum demand stopped rising, almost all plans for new pumped-storage power plants have either been postponed or cancelled since the 1990s.

Output from pumped-storage hydropower plants at the end of 1960 was only 58 MW (0.3% of total electric power production output), but it had grown to 3390 MW (5.8% of total electric power output) by 1970. Then, its output increased by about 20 000 MW from 1970 to 2001, as its share of all power production facilities rose from about 6–11% (**Figure 6**).

The structure of power supply by power source is shown in **Figure 7**, revealing that in recent years, the base load supply has been provided by run-of-river hydropower, nuclear power, and coal-fired thermal power, load fluctuations during the daytime are handled by liquefied natural gas (LNG) and by LPG thermal power plants, and short-term peaks are supplied by dam-type hydropower and pumped-storage power.

6.10.2 Current State of Hydropower in Japan

6.10.2.1 Primary Energy in Japan

Resource-poor Japan is dependent on imports for 96% of its primary energy supply; even if nuclear energy is included in domestic energy, dependency is still at 81%. Thus, Japan's energy supply structure is extremely vulnerable. Following the two oil crises in the 1970s, Japan has diversified its energy sources through increased use of nuclear energy, natural gas, and coal, as well as the promotion of energy efficiency and conservation (**Figure 8**).

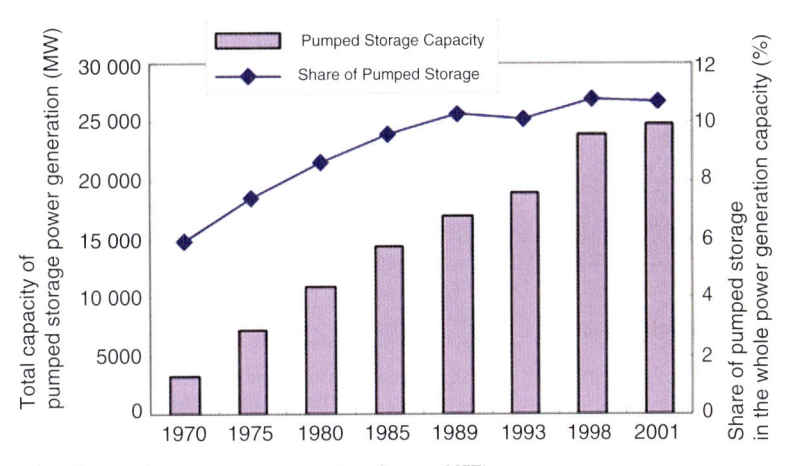

Figure 6 Change in the capacity of pumped-storage power generation. Source: METI.

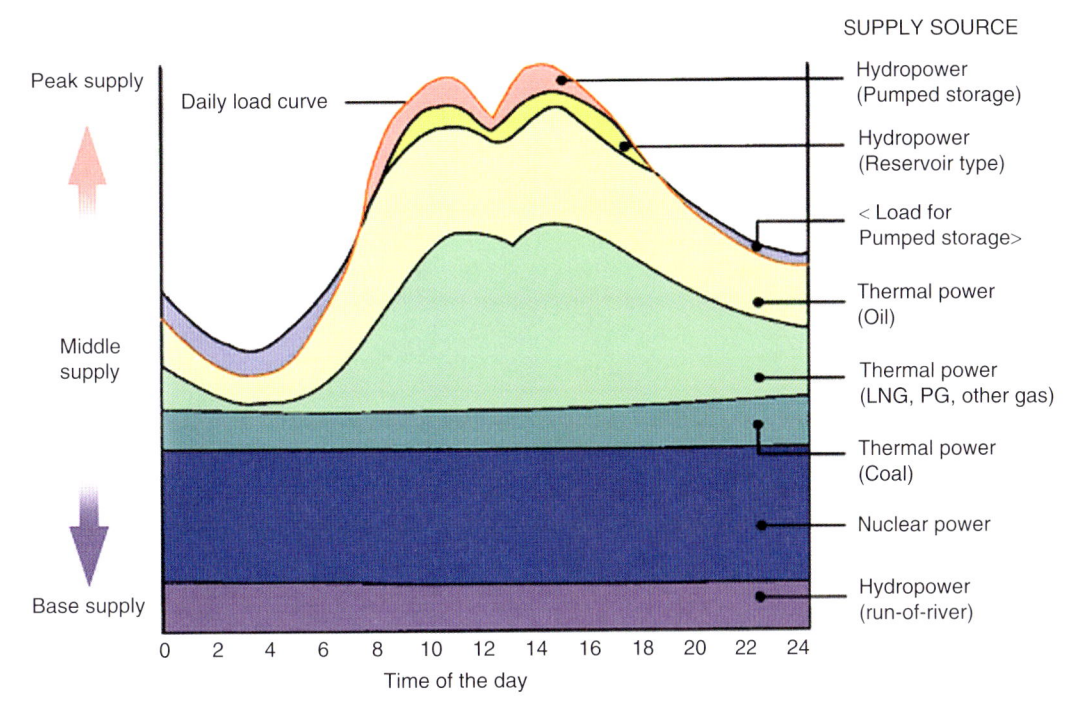

Figure 7 Combinations of electric power supplies by the time of day. Source: FEPC

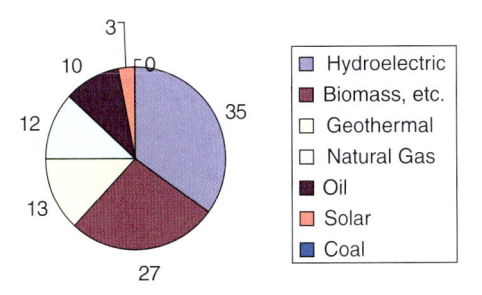

Figure 8 Share of primary energy %. Source: IEA/Energy Balances of OECD Countries 2003–04 (2006 Edition).

Hydroelectric power is one of the few self-sufficient energy resources in resource-poor Japan. Hydroelectric power is an excellent source in terms of stable supply and generation cost over the long term. Hydroelectric power saw a rebirth in development following the oil crises of the 1970s. Although steady development of hydroelectric power plants is desired, Japan has used nearly all available sites for the construction of large-scale hydroelectric facilities, and so recent developments have been on a smaller scale.

6.10.2.2 Development of Hydroelectric Power in Japan

Table 8 shows the developed hydropower, ongoing hydropower, and potential hydropower. As for annual hydroelectric power generation, two-thirds of the total was already developed.

6.10.2.3 Hydroelectric Power Development

Figures 9 and 10 show the history of electric power development and annual electric power generation in Japan, respectively. The hydro-first/thermal-second electric power structure continued in early 1960s and was followed by advanced thermal and nuclear power, but even after 1960, reservoir- and regulating pond-type hydropower plants continued to be developed as valuable peak supply power (**Figures 11** and **12**).

Table 8 Potential hydroelectric power

		Number of sites	Maximum output (MW)	Annual power generation (GWh)
Developed	Conventional	1 888	21 852	91 995
	Pumped-storage (mix)	19	5 710	2 572
Under construction	Conventional	32	750	2 043
		−5	−48	−235
	Pumped-storage (mix)			
Not developed	Conventional	2 713	12 128	45 877
		−257	−1 003	−6 877
	Pumped-storage (mix)	18	6 916	1 651
		−10	−98	−647
Conventional total		4 622	33 683	132 804
		−262		
Pumped-storage (mix)		37	12 528	3 576
		−10		
Total				136 381

Source: RPS (Renewable Portfolio Standard), http://www.enecho.meti.go.jp/hydraulic/data/stock/top.html.

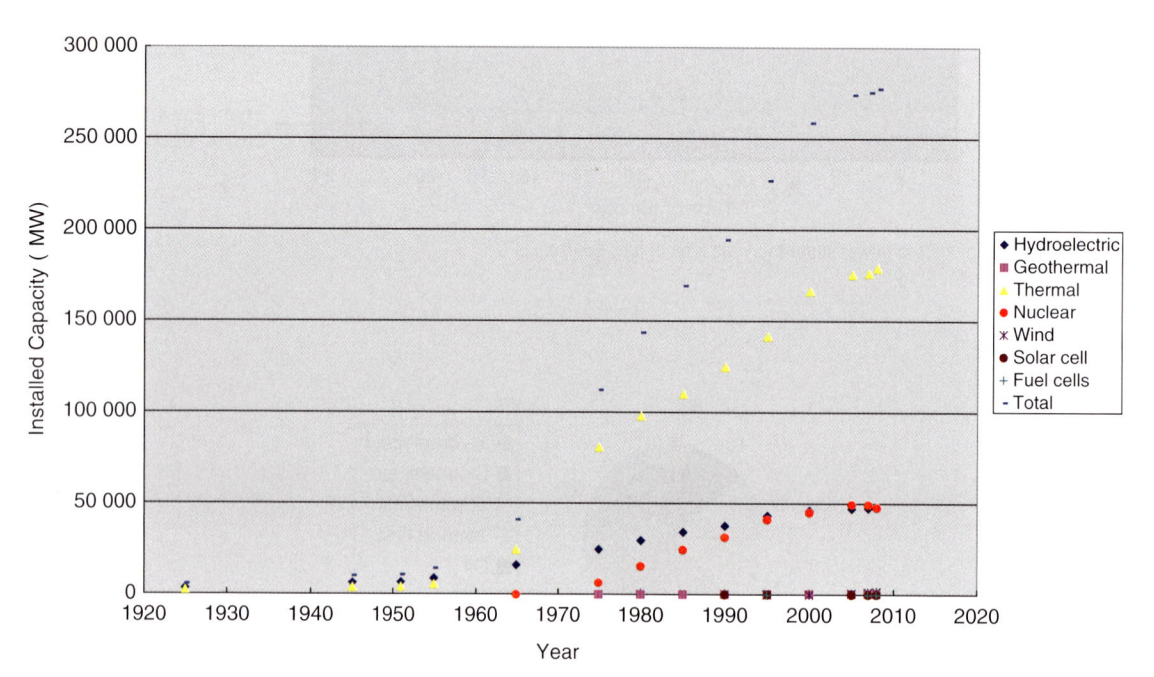

Figure 9 History of electric power development.

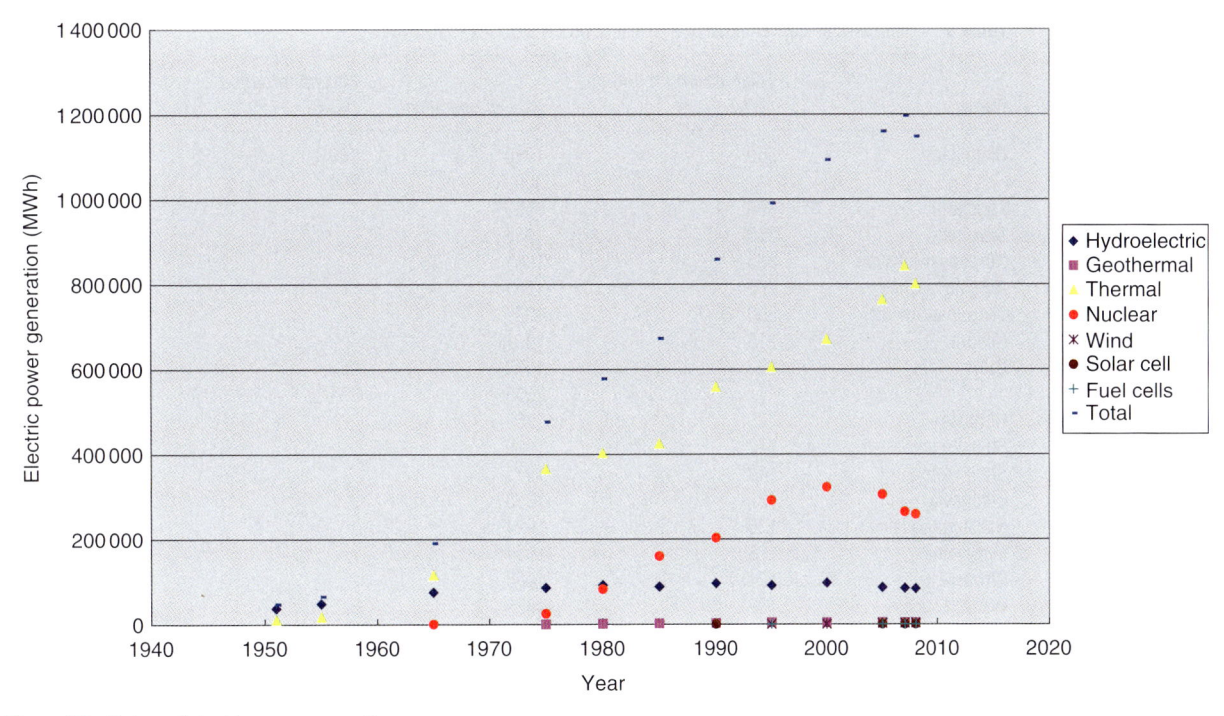

Figure 10 History of electric power generation.

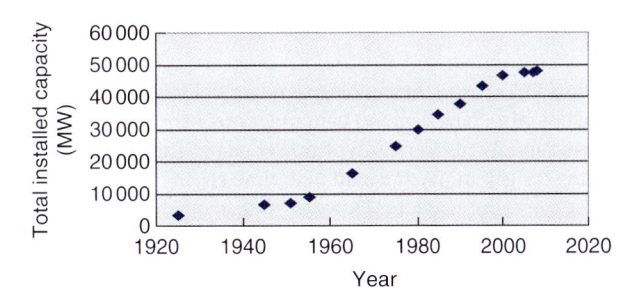

Figure 11 Hydroelectric power.

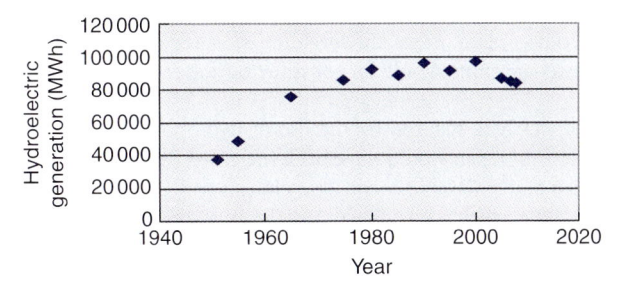

Figure 12 Hydroelectric generation. Source: Hand Book of Electric Power Industry, 2009, Japan Electric Association.

Most of large-scale conventional hydropower stations shown in **Table 9**, with maximum capacity excess 100 MW, were developed from 1950 to 1980. These hydropower stations have scale merit; therefore, they were given priority to smaller ones.

6.10.2.4 Development of Pumped-Storage Power Plant

As the gap in demand between daytime and nighttime continues to grow, electric power companies are also developing pumped-storage power generation plants to meet peak demand. The share of pumped-storage generation facilities of the total hydroelectric power capacity in Japan is growing year by year.

Table 9 Large-scale conventional hydroelectric power stations (>100 MW)

Name	Total maximum output (MW)	Year of operation	Maximum output (MW)
Okutadami	560	1960	360
		2003	200
Tagokura	390	1959	
Sakuma	350	1956	
Kurobegawadaiyon	335	1961	
Ariminedaiichi	265	1981	
Tedorigawadaiichi	250	1979	
Miboro	215	1961	
Otori	182	1963	95
		2003	87
Hitotuse	180	1963	
Shinanogawa	177	1929	
Shimokotori	142	1973	
Kinugawa	127	1963	
Nakatugawadaiichi	126	1924	
Maruyama	125	1954	
Otozawa	124	1985	
Wadagawadaini	122	1959	
Ariminedaini	120	1981	
Yomikaki	117	1923	
Kiso	116	1968	
Akimoto	108	1940	
Shinkurobegawadaisan	107	1973	
Hiraoka	101	1952	

During the 1960s, rapid urbanization and a rise in the people's standard of living driven by rapid economic growth resulted in a remarkable increase in electricity demand. Also, new demand pattern was created, marked by a sharpened peak during daytime, a pattern that was beyond the adjustment capacity of reservoir-type power plants, thereby creating a need for pumped-storage power plants that are better suited to adjusting the gap in daytime and nighttime electric power demand.

Electricity is normally supplied at a constant frequency. However, this frequency is not constant since it declines when supply capacity falls short of demand and increases in excess of demand. The adjustment of generation output in response to demand fluctuations is thus an important way of ensuring the supply of high-quality power of a stable frequency.

Pumped-storage power plants reach maximum output within 3–5 min of start-up and their output can be adjusted in a matter of seconds (**Figures 13** and **14** and **Table 10**).

Though both the share of installed capacity of hydroelectric power in total installed electric power and the share of electric generation are only 17% and 7%, respectively, hydroelectric power is one of the few self-sufficient energy resources in resource-poor Japan. So, even smaller hydroelectric power plant, its steady development is required in Japan (**Figure 15**).

The CO_2 emissions from hydropower are emitted only for constructing and repairing the facilities. Hydropower stations do not emit CO_2 during operation.

Approximately 70 million tons of CO_2 was reduced by the use of hydropower in fiscal 2006. Without power supply from hydropower stations, the CO_2 emissions in Japan would have been increased by about 6% compared to the level in fiscal 1990. Today, hydropower is viewed as a clean and renewable energy that emits zero CO_2 and is effective for preventing global warming.

The benefits of hydropower production, which is one type of recyclable energy, are that it is purely domestic, recyclable resource that can be recovered, produces low emissions of CO_2, has a long service life, contributes to regional development, and its technologies are established.

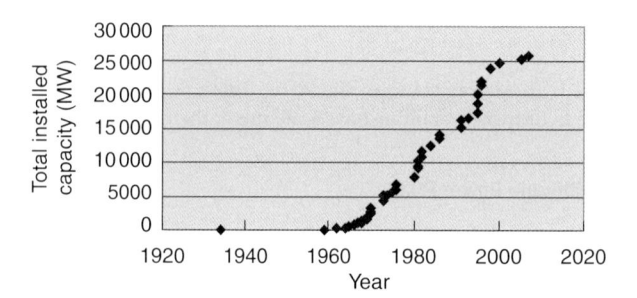

Figure 13 Total installed capacity: Pumped-storage power plant. Source: Hand Book of Electric Power Industry, 2009, Japan Electric Association.

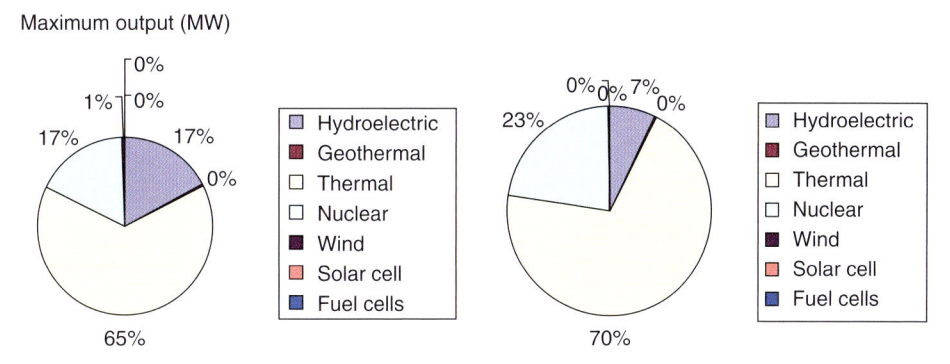

Figure 14 The share of installed capacity of hydroelectric power. Source: Hand Book of Electric Power Industry, 2009, Japan Electric Association.

Table 10 Pumped-storage power stations (>500 MW)

Plant name	Maximum output (MW)	Year of operation
Okutataragi	1932	1998
Okumino	1500	1995
Sintakasegawa	1280	1981
Ookochi	1280	1995
Okuyoshino	1206	1980
Tamahara	1200	1986
Matanogawa	1200	1996
Sintoyone	1125	1973
Imaichi	1050	1991
Okukiyotu	1000	1982
Shimogo	1000	1991
Shiobara	900	1995
Kazunogawa	800	2000
Okuyahagi–Daini	780	1981
Numappara	675	1973
Azumi	623	1970
Nabara	620	1976
Honkawa	615	1984
Tenzan	600	1986
Okukiyotu-Daini	600	1996
Omarugawa	600	2007
Oobera	500	1975

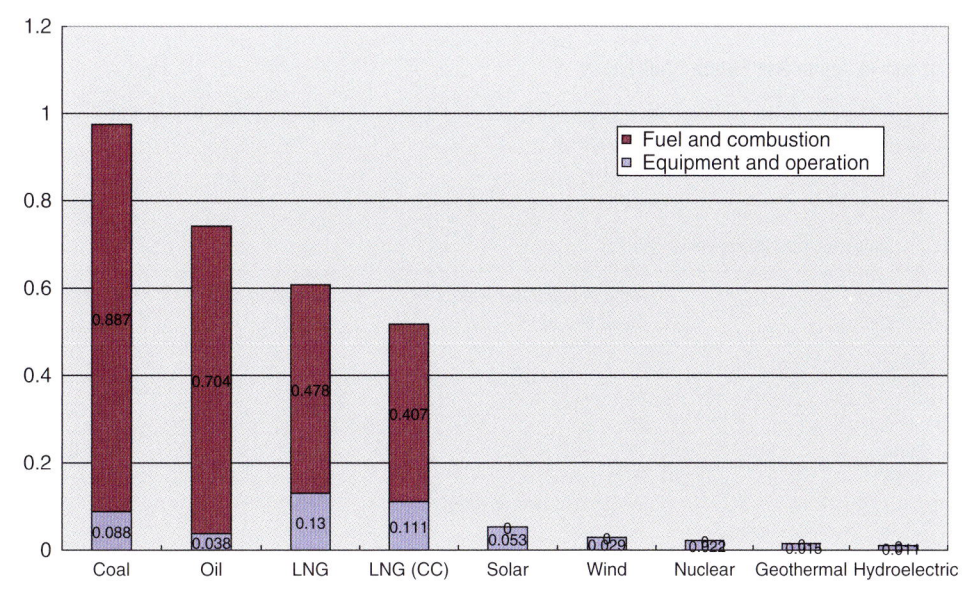

Figure 15 The CO_2 emissions.

Medium and small hydropower development will make a great contribution to ensuring valuable domestically produced energy, and will go beyond merely producing power to create regional industries centered on hydropower generation, and so on, by providing other functions that contribute to autonomous development of the region.

6.10.3 Hydropower in Japan and Future Challenges

6.10.3.1 Energy Situation in Japan and Hydropower

Just before the first oil shock, oil provided the highest percentage of Japan's primary energy, accounting for 77% of all energy. Later, the oil shocks led to the introduction of nuclear power, LNG, coal, and so on, so that by 1998, Japan's dependency on oil was down to about 52%.

The percentage of primary energy provided as electric power was 41% in 2000, while other forms were sent directly to consumers as fuel. With the exception of the small supplies provided by new energy sources (about 1%), energy, other than electricity, is produced almost entirely from fossil fuels. To prepare for the depletion of fossil fuels and to stop global warming, hydrogen and other secondary energy media should be developed. If hydrogen energy becomes a replacement fuel for oil energy in the near future, electric energy will be needed to produce hydrogen, and in order for this to be as independent of fossil fuels as possible, nuclear power, new energies, and hydropower must be developed.

Next, an examination of the breakdown of electric power sources shows that it used to be mainly hydropower, but from about 1962, hydropower was surpassed by thermal power. Fuels used to produce thermal power are oil, coal, LNG, and so on. At peak production times, more than 60% of all electric power was produced from oil. The oil shocks were followed by the development of electric power sources such as nuclear power, coal, LNG, and so on, as substitutes for the oil that is high-priced and its supply is unstable.

As shown in Table 11, nuclear power now accounts for 31% of annual electric power production. Incidentally, hydropower accounts for 9% of annual electric power production (Table 11).

The share of oil used for electric power production is low at 9%. Its electric power production cost is high and it generates a lot of CO_2. Therefore, the use of oil will continue to decline in the future.

Nuclear power, which is the largest source, generates almost no CO_2 and its production cost is the lowest, but recent accidents at nuclear power plants have made it difficult to boost nuclear power, and ensuring safety and back-end measures are other challenges.

The second-largest source, LNG, provides relatively low-cost power and is considered the cleanest among fossil fuel sources, but it produces a lot of CO_2: between 500 and 600 g kWh^{-1}.

Coal, which is the third largest source of electric power, ensures superior fuel supply stability and cost, but it produces 975 g kWh^{-1} of CO_2, the largest of any energy source.

The fourth source is hydropower, which produces extremely small quantities of CO_2, even less than new energy sources, is clean, and as energy produced entirely in Japan, offers a very stable supply. Although hydropower's initial investment costs are relatively high, its long-term cost is low. To develop future hydropower plants, it is necessary to reduce costs and protect the environment. In addition, it is important that efficient maintenance be performed continuously to extend the equipment's lifetime in existing hydropower systems, which offer long-term cost superiority.

New energies still provide less than 1% of electric power generation, but under the Renewable Portfolio Standard Act, electric power companies are now legally required to provide a certain percentage of their power by wind power, waste material power, or other reusable energies, and this is expected to promote the development and spread of these technologies. Nevertheless, to develop these in the future, their costs must be reduced and they must supply electricity more stably.

6.10.3.2 Hydropower in Japan and Future Challenges

The following will be important as policies to restrict emissions of CO_2, thereby lowering dependency on fossil fuels while responding to the anticipated increase in energy consumption:

Table 11 Comparison of electric power sources

Electric power sources		Ratio of generated energy (as of end of FY 2002)	Generation cost (Yen kWh^{-1})	Unit CO_2 discharge (g-CO_2 kWh^{-1})
Hydro		9	13.6	11
Nuclear		31	5.9	22
Coal-fired thermal		22	6.5	975
LNG-fired thermal		27	6.4	500–600
Oil-fired thermal		9	10.2	742
New energy	Wind	<1 (total of new energy)	10–24	29
	Waste material		9–12	

Source: METI.

1. Restricting energy consumption and saving energy.
2. Developing and adopting recyclable energy (hydropower, wind power, geothermal, photovoltaic, hydrogen, wave power, seawater temperature difference power, biofuel, etc.).

The benefits of hydropower production, which is one type of recyclable energy, are that it is a purely domestic recyclable resource that can be recovered, produces low emissions of CO_2, has a long service life, contributes to regional development, and its technologies are established. Its negative impacts are that it is expensive and that it may affect the natural environment.

Japan produces 1076×10^6 MWh of electric power, of which 94×10^6 MWh is hydropower. Including existing hydropower, Japan's potential hydroelectricity equals 135×10^6 MWh.

It is presumed that hydropower will be implemented as described below.

At this time, construction at almost all locations suitable for large-scale development has been completed, and in this century, it will be necessary to maintain, manage, and prolong the life of dams and hydropower plants that have already been constructed.

Future development will presumably be done by introducing power production technologies that are kind to the environment and suited to locations with short falls and low flow rates, thereby developing hydropower plants that take advantage of unused small falls, including that at dams other than hydropower dams, while reducing production costs. Hydropower is a clean 100% domestically produced recyclable energy, and will naturally be passed on to future generations as a valuable asset.

Medium and small hydropower development will make a great contribution to ensuring valuable, domestically produced energy, and will go beyond merely producing power to create regional industries centered on hydropower generation, and so on, by providing other functions that contribute to autonomous development of the region. A trend toward using small falls effectively, mainly on agricultural channels and rivers at a scale ranging from a few tens of kilowatts to a few hundred kilowatts, has emerged.

Hydropower generation during this century will probably meet the demands of the times from various perspectives, including resolving global environmental problems.

6.10.4 Successful Efforts in Japan

6.10.4.1 Large-Scale Pumped-Storage Power Plants in Tokyo Electric Power Company

Pumped-storage-type power plants have been developed in Japan since 1930. Tokyo Electric Power Co., Inc. (TEPCO) has nine pumped-storage power plants with approximately 10 000 MW in total, including two under construction. They have contributed to stable operation of a huge power network in Kanto district including Tokyo metropolitan area, functioning as peak load power sources, storage of electric power, spinning reserve, voltage support ability to control reactive power, and black start capability for power network recovery.

6.10.4.1.1 Outline of the project

Pumped-storage power generation uses two adjustment reservoirs that are located at different elevations and are connected together by conduits together with reversible pump-turbines, to utilize surplus electricity generated during the low-demand small hours and weekends to pump water from the lower adjustment reservoir up to the upper adjustment reservoir so that the water can be used to generate electricity during the daytime peak demand hours and/or in the event of an emergency. Tokyo Electric Power Company (TEPCO) currently owns a total of nine pumped-storage power plants (including two under construction), which are being operated by TEPCO to meet the daytime peak electricity demand. **Table 12** and **Figure 16** show a list of TEPCO's pumped-storage power plants and their locations, respectively (**Table 13**).

Table 12 TEPCO's pumped-storage power plants

Name	Output (MW)	Operational since	Type
Yagisawa	240	1965	Mixed
Azumi	623	1969	Mixed
Midono	245	1969	Mixed
Shin-Takasegawa	1280	1981	Mixed
Tamahara	1200	1982	Pure
Imaichi	1050	1988	Pure
Shiobara	900	1994	Pure
Kazunogawa	1600	1999 Partially commissioned	Pure
Kannagawa	2820	2007 Partially commissioned	Pure

Figure 16 Locations of TEPCO's pumped-storage power plants.

Table 13 Main facilities of Tokyo Electric Power Company's pumped-storage power plants

	Upper dam and adjustment reservoir			Lower dam and adjustment reservoir			
Name of power plant	Type	Dam height (m)	Dam volume (10 000 m^3)	Total storage capacity (10 000 m^3)	Type	Dam height (m)	Dam volume (10 000 m^3)	Total storage capacity (10 000 m^3)
Yagisawa	Concrete arch	131	57	204	Concrete gravity	72	19.8	2 850
Azumi	Concrete arch	155.5	66	12 300	Concrete arch	95.5	30	1 510
Midono	Concrete arch	95.5	30	1 510	Concrete arch	60	7	1 070
Shin-Takasegawa	Rock-fill	176	1 159	7 620	Rock-fill	125	738	3 250
Tamahara	Rock-fill	116	544	1 480	Concrete gravity	95	41.5	5 249
Imaichi	Rock-fill	97.5	252	707	Concrete gravity	75.5	19	910
Shiobara	Rock-fill	90.5	211	1 190	Concrete gravity	104	59	1 050
Kazunogawa	Rock-fill	87	406	1 120	Concrete gravity	105.2	62	1 150
Kannagawa	Rock-fill	136	722	1 836	Concrete gravity	120	72	1 910

6.10.4.1.2 Features of the project area

6.10.4.1.2(i) Supply and demand in TEPCO's service area

TEPCO is supplying electricity to approximately 42.8 million people in its service area that covers most of the Kanto Region including the Tokyo metropolitan area. The total area of the service area is approximately 39 500 km^2. The total amount of electricity sales and the peak demand in fiscal year 2009 were about 280 billion kWh and 60 million kW, respectively. To ensure that TEPCO will be able to supply electricity in a stable, uninterrupted manner for the years to come, it is striving to achieve, taking into consideration the anticipated global energy demand trends, the most efficient mix of energy resources which best accommodates the hourly, daily, and seasonal fluctuations of electricity demand and is best from the standpoints of economics, environmental protection, and securing stable sources of fuel procurement. The pattern of daily electricity usage in the summertime in TEPCO's service area is as follows (**Figure 17**): The demand starts increasing sharply at around 6.00 a.m. and continues to increase up until the lunch hour when it dips slightly. It starts increasing again at 1.00 p.m. and continues to increase up until around 2.00 p.m. when

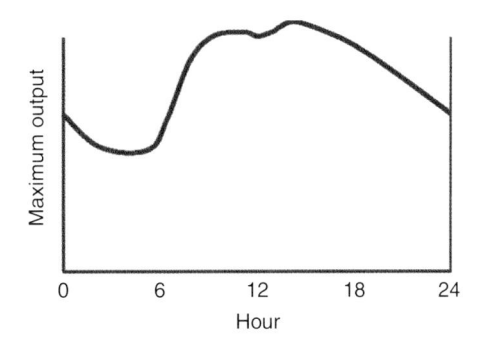

Figure 17 Pattern of daily electricity usage in TEPCO's service area.

Table 14 Characteristics of the individual components of electricity demand and the requirements for suitable power sources

Demand component	Characteristics of demand component	Power source requirements — Operational requirements	Power source requirements — (cost)	Suitable power sources
Peak	Sharp fluctuations. The duration of power generation operation is short	Load adjustment capability. Hot reserve and frequent start/stop capability	Low fixed costs. Relatively high variable costs can be tolerated as long as this requirement is satisfied[a]	Pumped-storage- and pondage-type hydropower, gas turbine. Oil and LNG-fired thermal power
Middle	Large daily fluctuations. The duration of power generation operation is relatively long	Capable of being activated and deactivated at a relatively high frequency during the day or of otherwise being adjusted so that similar effects can be achieved	Both variable and fixed costs are relatively low	Coal-fired thermal power
Base	Negligible fluctuations. Power is generated all day	Capable of continuous 24 h operation	Low variable costs. Relatively high fixed costs can be tolerated as long as this requirement is satisfied	Run-of-river-type hydropower. Nuclear power

[a] 'Variable costs' mainly refers to fuel costs and 'fixed costs' mainly refers to depreciations and interests on the construction cost.

it peaks. It then decreases gradually up until around 6.00 p.m. when it starts to decrease sharply. The demand continues to decrease until it reaches the bottom at around 4.00 a.m. when it starts increasing again. To accommodate this variation in an economical and efficient manner, it is necessary to develop dedicated power sources for each of the peak, middle, and base demand portions explained in Table 14 and use them in combination.

6.10.4.1.2(ii) History of pumped-storage power plant development in TEPCO

After the Second World War, Japan's electricity demand increased sharply as the Japanese economy developed rapidly into an autonomous economy, but thermal power plants were used as the primary means to accommodate the sharp increases in electricity demand, with pondage- and reservoir-type hydroelectric power plants (which have high adjustment capabilities) developed as peak load power sources. At the time, thermal power plants were improving their thermal efficiencies thanks to the development of high-temperature, high-pressure equipments, which were considered to be optimal power sources to meet the electricity demand which was increasing at a rate of more than 10% per year, because of their large capacities and short construction periods. As a result of the accelerated development of thermal power plants, the share of thermal power relative to the total amount of electricity generated surpassed the share of hydropower by the early 1960s, signaling the advent of the so-called 'era of thermal electricity'. Hydroelectric power plants continued to be developed and used as important peak load power sources, but as the number of sites suitable for hydroelectric power plant development decreased as a result of progressive exploitation of economical sites, mixed pumped-storage hydroelectric power stations started to be developed. Mixed pumped-storage hydroelectric power plants are pondage-type hydroelectric power plants added with pumped-storage power generation systems to enable them to make large-scale daily adjustments to meet peak demand. Examples include the Yagisawa Power Plant (Tone River, 240 MW, operational since 1967) in Gunma Prefecture, the Azumi Power Plant (Shinano River, 623 MW, operational since 1970) in Nagano Prefecture, and the Shin-Takasegawa Power Plant (Shinano River, 1280 MW, operational since 1969) in Nagano Prefecture. From around the second half of the 1970s, the need for mixed pumped-storage hydroelectric power plants started to increase as the summertime peak electricity demand increased sharply due to sharp increases in the cooling- and air-conditioning-related consumption of electricity. However, because the number of suitable sites for mixed pumped-storage power plant development had decreased as a result of progressive exploitation of sites where natural river flows can be utilized effectively, pure pumped-storage hydroelectric power plants started to be developed. Because pure pumped-storage hydroelectric power plants essentially have no river water inflow into their upper adjustment reservoirs and generate power using water pumped up from their lower adjustment reservoirs only, they can be sited without the need to consider river system conditions as long as the heads are sufficiently large. The scales of pumped-storage power plant development projects and the proportion of the pumped-storage capacity as a percentage of the total capacity of the entire power network are determined based on the results of a power network system analysis that aims to minimize the power generation cost of the entire power network taking into consideration the above-mentioned pattern of daily electricity usage in TEPCO's service area. The current optimal proportion of the pumped-storage capacity as a percentage of the total capacity of the entire power network in TEPCO's service area is estimated to be about 10–15% (**Figure 18**). In line with the increases in electricity demand in recent years, the Tamahara Power Plant in Gunma Prefecture (1200 MW, head = 518 m, operational since 1982), the Imaichi Power Station in Tochigi Prefecture (1050 MW, head = 524 m), the Shiobara Power Station in Tochigi Prefecture (900 MW, head = 338 m), the Kazunogawa Power Station in Yamanashi Prefecture (1600 MW, head = 714 m), and the Kannagawa Power Station in Gunma Prefecture (2820 MW, head = 653 m, currently under construction) were planned and constructed sequentially to maintain the proportion at the optimal level.

6.10.4.1.3 Benefits

6.10.4.1.3(i) Functions of pumped-storage power plants

Pumped-storage power plants play a wide range of roles in power network system, including such functions as peak supply source, storage of electricity, hot reserve capacity, phase modification function, and power source for black start for power network system recovery.

6.10.4.1.3(i)(a) Peak load power source For the peak portion of the demand, it is desirable to use a power source whose fixed costs are low even if it means relatively high variable costs, because the duration of power generation operation is short. Pumped-storage power plants are lowest-cost power plants in terms of fixed costs because they can be constructed at a low unit construction cost per kilowatt and comprise long-life structures such as dams and conduits. In terms of fuel costs, which make up the bulk of the total variable costs of a power plant, approximately 30% of the fuel consumed to run a pumped-storage power plant is wasted in the form of losses due to the upward and downward transport of water in the waterway and losses of reversible pump-turbines and generator-motors, but the pumped-storage power plant can be run at a lower total fuel cost by using low-cost electricity generated by nuclear power as a power for pumping water than that for a coal-fired thermal power plant. **Figure 19** shows the relationship between the annual operating hours and energy costs (i.e., fixed costs plus variable costs) by power plant type. As is

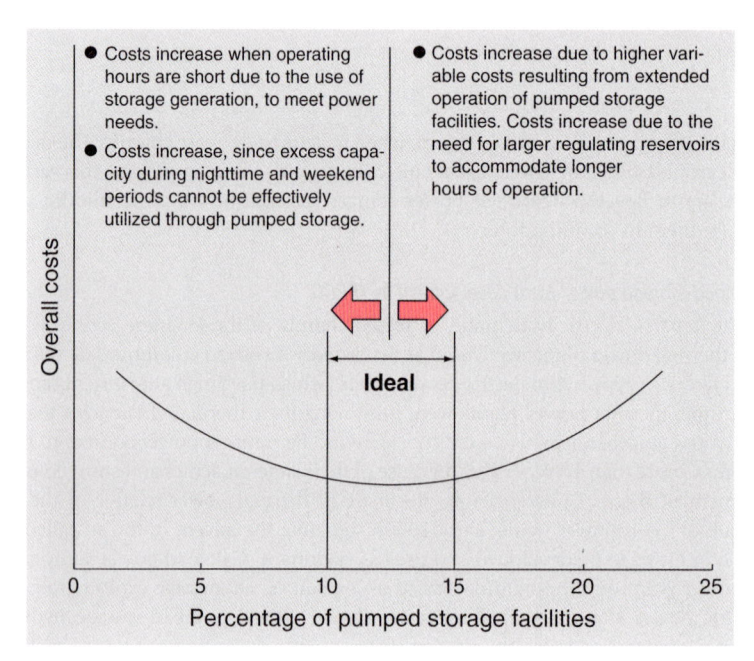

Figure 18 Optimal proportion of the pumped-storage capacity as a percentage of the total capacity of the entire power network.

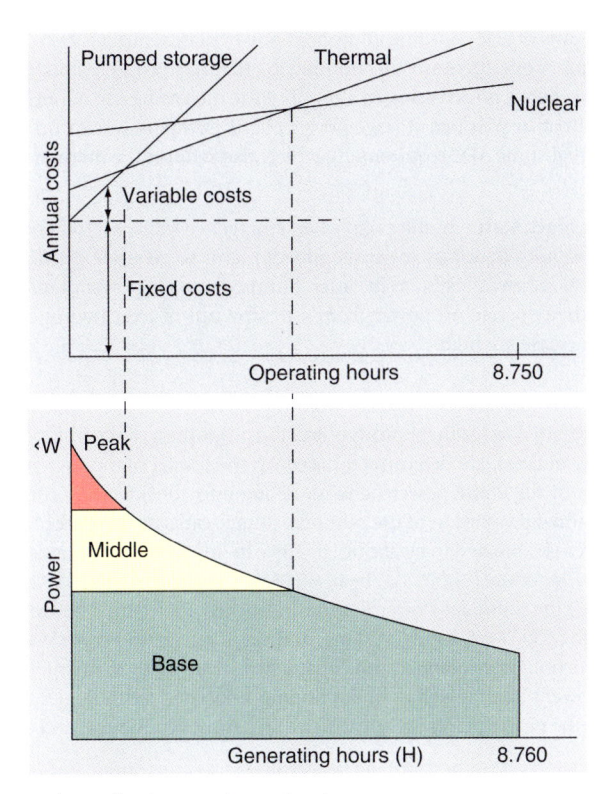

Figure 19 Relationship between the annual operating hours and annual costs.

clear from the figure, the economical approach is to use nuclear power and thermal power for the base and middle demand portions, respectively, because the annual operating hours for these portions are long. For the peak demand portion, however, pumped-storage hydropower generation is the lowest cost, because the annual operating hours is short. In addition, the peak portion of the electricity demand is characterized by sharp load fluctuations and thus requires a power source that has an excellent load adjustment capability and is also capable of frequent start/stop. These operational requirements can only be met by pumped-storage hydroelectric power plants, which can adjust their outputs quickly and can start/stop in a matter of minutes. Because of these economic and operational characteristics, pumped-storage hydroelectric power plants have been developed and used as peak load power sources.

6.10.4.1.3(i)(b) Storage of electricity Because electricity demand changes daily, weekly, and seasonally, it is convenient to utilize the cheaply available electricity generated by nuclear and coal-fired thermal power plants (whose variable costs are low) during the low-demand hours such as midnights and weekends to operate pumped-storage power plants, so that low-cost electricity can be stored in the form of water in upper adjustment reservoirs, and it can be used as a generator during peak load hours of weekdays to reduce the overall electricity supply cost, saving the use of power sources that are higher cost in terms of fuel costs (such as oil-fired thermal power plants).

6.10.4.1.3(i)(c) Hot reserve capacity To ensure stable, uninterrupted supply of electricity, it is necessary to provide for unexpected demand increases and unscheduled power source outages, as well as output reductions by having sufficient reserve capacities in place. In general, it is desirable to achieve this with power sources whose fixed costs are low even if it means relatively higher variable costs, because the operating hours is much less than that in the case of ordinary power generation. In addition, a reserve capacity should be capable of being activated instantly in the event of a power source failure or other emergency to make up for the lost capacity to ensure that the supply of electricity is not disrupted or reduced. Because these requirements, which are similar to those for peak load power sources, are best satisfied by pumped-storage power plants, it is best to use pumped-storage power plants as hot reserve capacities.

6.10.4.1.3(i)(d) Phase modification and frequency Control Functions Because of parallel capacitance increases in power networks due to the increase of long overhead and underground transmission lines, voltage rises and drops occur at receiving ends when the load is low and high, respectively. These phenomena are usually controlled by means of shunt reactors and power condensers installed in substations, but electricity utilities can also use pumped-storage power plants as synchronous phase modifiers that adjust power network voltages by operating generator-motors without load and adjusting magnetic field currents to provide or absorb reactive power. In addition, the recent development of variable-speed pumps has enabled pumped-storage power plants to be used as a means of power network frequency control (AFC, automatic frequency control) during nighttime hours. AFC is usually achieved by means of extra burning at thermal power plants, but it is becoming increasingly difficult to do this with operating thermal power plants alone, because the number of thermal power plants that have to shut down during the low-demand nighttime hours and weekends is increasing as a result of the increasing use of nuclear power stations as power sources for weekends and nighttime hours. Because pumped-storage power plants pump up water during the night and weekends, they can also be used as a means to meet the nighttime AFC requirement, which also reduces the thermal power generation fuel consumption and hence the overall cost.

6.10.4.1.3(i)(e) Power source for black start In the event of an emergency such as a total outage of an entire power system due to a major accident, and so on, it becomes necessary for some power plants to generate electricity with black start, and charge the transmission lines and restore the power network system in order. Pumped-storage power plants are very suitable to be used as such emergency power sources because they operate on power from a nearby run-of-river hydropower plant, they can be activated in 3–5 min and their rates of output increase are high.

6.10.4.1.3(ii) Planning and development of pumped-storage power plants that is in line with increases in electricity demand

The scales of pumped-storage power plant development projects and the proportion of the pumped-storage capacity as a percentage of the total capacity of the entire power network are determined based on the results of a power network system analysis that aims to minimize the power generation cost of the entire power network taking into consideration the pattern of daily electricity usage in TEPCO's service area. The current optimal proportion of the pumped-storage capacity as a percentage of the total capacity of the entire power network in TEPCO's service area is estimated to be about 10–15%. In line with the increases in electricity demand in recent years, the Tamahara Power Plant in Gunma Prefecture (1200 MW, head = 518 m, operational since 1982), the Imaichi Power Plant in Tochigi Prefecture (1050 MW, head = 524 m), the Shiobara Power Station in Tochigi Prefecture (900 MW, head = 338 m), the Kazunogawa Power Plant in Yamanashi Prefecture (1600 MW, head = 714 m), and the Kannagawa Power Plant in Gunma Prefecture (2820 MW, head = 653 m, currently under construction) were planned and constructed sequentially to maintain the proportion at the optimal level. In locating and constructing these power plants as well as in developing associated technologies and techniques, TEPCO has used the following four criteria to ensure that the power plants are constructed in a most economical and efficient manner.

6.10.4.1.3(ii)(a) High storage capacity Pumped-storage power plants require upper and lower dams. Sitting requirements for the dams include a topography that will enable large reservoirs to be created behind small dams, as well as a geological structure strong enough to hold the weight of the dams and the pressure of the water.

6.10.4.1.3(ii)(b) Good access to power supply network Power plants must be built as close as possible to demand areas in order to minimize power loss and transmission costs, such as the cost of building transmission lines and substations. In addition, they must be built in locations that provide good access to electric power from thermal and nuclear power plants, since power is required for pumping operations.

6.10.4.1.3(ii)(c) Suitability for excavation of large-scale underground caverns Since it is most economical to link the upper and lower dams by the shortest route, most powerhouses are built underground. The powerhouse cannot be constructed economically unless the subterranean rock mass is hard and extensive enough to allow the excavation of a large cavern. In recent years, however, it has become possible to reduce the size of caverns thanks to improvements in output capacity per generator. **Figure 20** shows the chronological changes in the maximum output capacity per pump-turbine.

6.10.4.1.3(ii)(d) High head with short waterway The output capacity of a pumped-storage power plant is determined by the volume of water used and its effective head. Efficiency can therefore be optimized by minimizing the distance between the upper and lower reservoirs and maximizing the head. These requirements are also reflected in efforts to improve the maximum head of pump-turbines (**Figure 21**).

6.10.4.1.4 Effects of the benefits

Figure 22 shows the operation record of the pumped-storage power plants for 24 July 2001, on which day the highest peak demand was recorded in TEPCO's service area. **Figure 23** shows the combination of energy sources to meet changing demand. As can be seen from these figures, pumped-storage power plants are being fully utilized as peak load power sources to help meet the electricity demand during the summer, which is the peak load season in Japan. **Table 15** shows the average cost for pumping up water at the pumped-storage power plants per unit electricity (kWh) generated versus the average cost of extra burning at the oil-fired thermal power plants per unit electricity (kWh) generated as of 2001. TEPCO is minimizing the overall power generation cost of the power network as a whole by utilizing its pumped-storage power plants, whose unit cost of power generation is lower than the unit cost of extra burning at oil-fired power plants during daytime, and achieving an 'electricity storage' effect.

6.10.4.1.5 Reasons for success

As mentioned above, pumped-storage power plants have been successfully used in Japan as useful power sources that contribute to stabilizing the operation of power systems. This is partly attributable to the presence of the following conditions in the Tokyo Electric Power Company's service area that includes Tokyo, the center of Japan's economic activities.

6.10.4.1.5(i) Sharp increases in the peak electricity demand

As mentioned above, the electricity demand in TEPCO's service area starts increasing sharply at around 6.00 a.m., continues to increase up until the lunch hour when it dips slightly, starts increasing again at 1.00 p.m. and continues to increase up

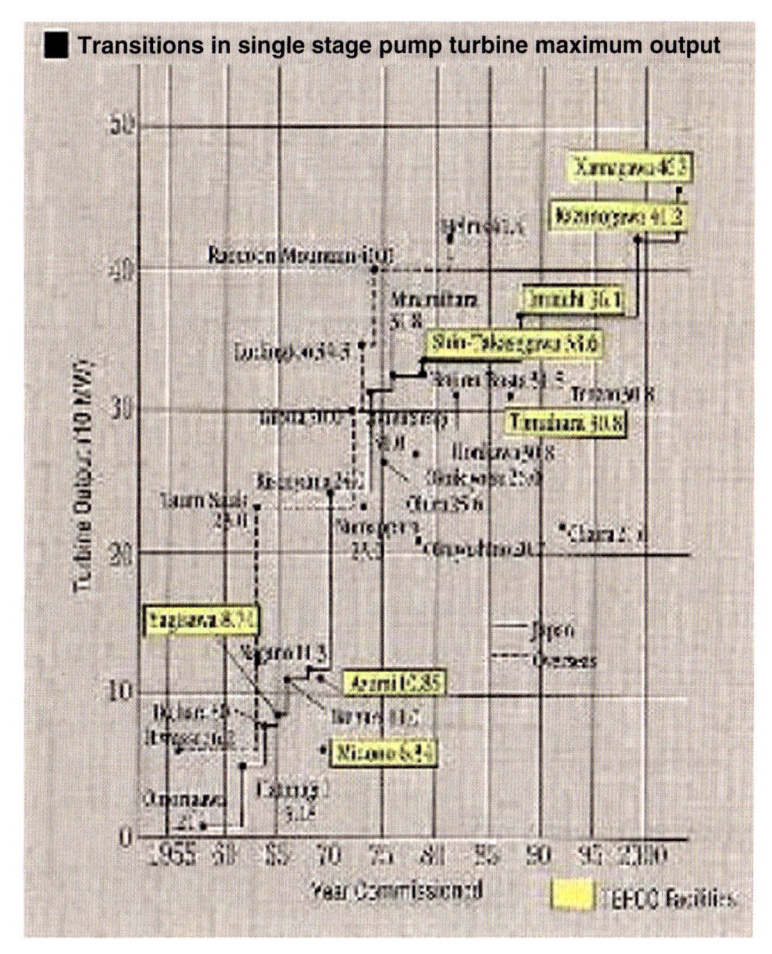

Figure 20 Chronological changes in the maximum output capacity per pump-turbine.

until around 2.00 p.m. when it peaks. This peak demand, which is largely due to the consumption of electricity by factories and office buildings, as well as the summertime electricity consumption for air conditioning, has been increasing sharply over the years as a result of the growth of the Japanese economy, with the current load factor standing at around 55%. Thus, pumped-storage power plants in this region are an important means to accommodate the increasingly sharpening peak portion of the daily load curve.

6.10.4.1.5(ii)   Spatial expansion of power networks
TEPCO's service area includes the Tokyo metropolitan area and the surrounding cities, which together comprise one of Japan's heaviest electricity-consuming regions, but these days new power plants tend to be sited in areas remote from this region because of the scarcity of sites suitable for power plant development. With the increase of longer-distance overhead transmission lines and underground transmission lines, the need for power network voltage control has been increasing.

6.10.4.1.5(iii)   Improved economic efficiencies of pumped-storage power plants
Because the two oil crises which Japan experienced in the 1970s revealed the vulnerability of the energy supply system of the nation which was heavily dependent on oil, Japan started an all-out effort to develop new nonoil power sources, which has resulted in a sharp increase in the share of nuclear power relative to the total amount of electricity generated. As a result, Japanese electricity utilities have gradually been increasing the proportions of nuclear power relative to those of thermal power as a source of electricity to run their pumped-storage power plants, because nuclear power is less expensive than thermal power in terms of fuel cost. This has significantly improved pumped-storage hydroelectric power plants' economic efficiencies.

6.10.4.1.5(iv)   Social responsibility as a public utility
As a public utility serving a region that includes Tokyo, which is the center of Japan's economic activities, TEPCO must ensure that stable, uninterrupted supply of electricity is maintained 24 h a day, 7 days a week, including during the peak hours explained above. For this reason, it is vital that TEPCO develop peak load power sources even if they require higher costs than ordinary power sources.

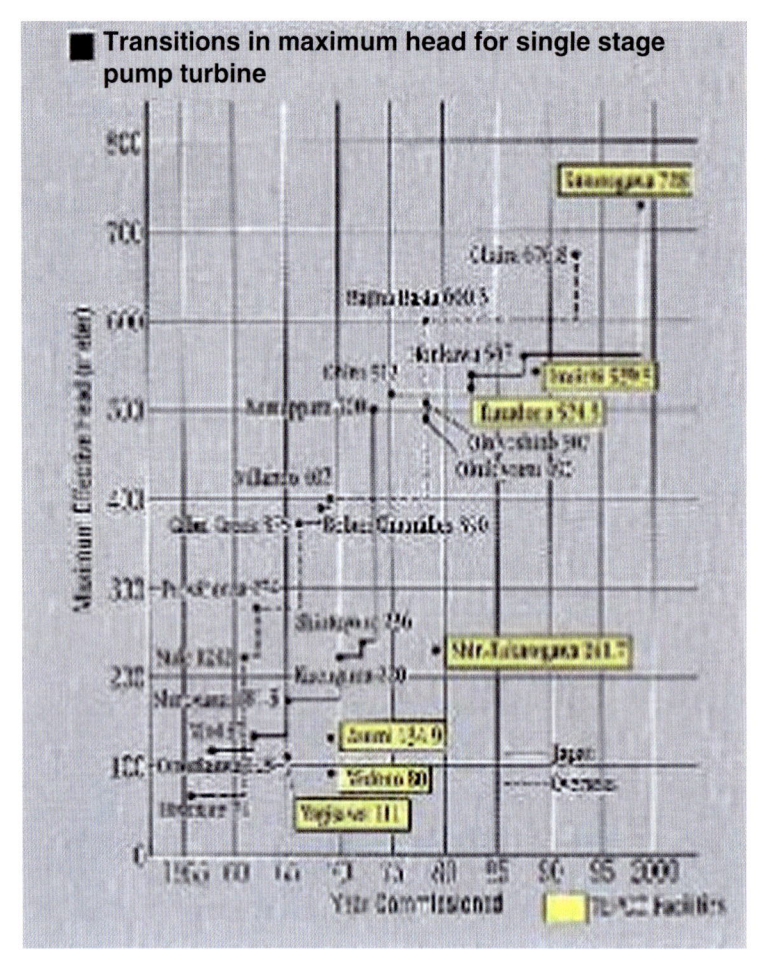

Figure 21 Chronological changes in the per-pump maximum head.

6.10.4.2 Sediment Flushing of Reservoir by Large-Scale Flashing Facilities in the Kansai Electric Power Company

The watershed of the Kurobe River Basin yields a lot of sediment and the Dashidaira Dam constructed in 1985 is equipped with sediment flushing gates. Through the precise prediction of environmental impact and the meeting with stakeholders including local residents and technical experts, appropriate mitigations for environmental impact has been established.

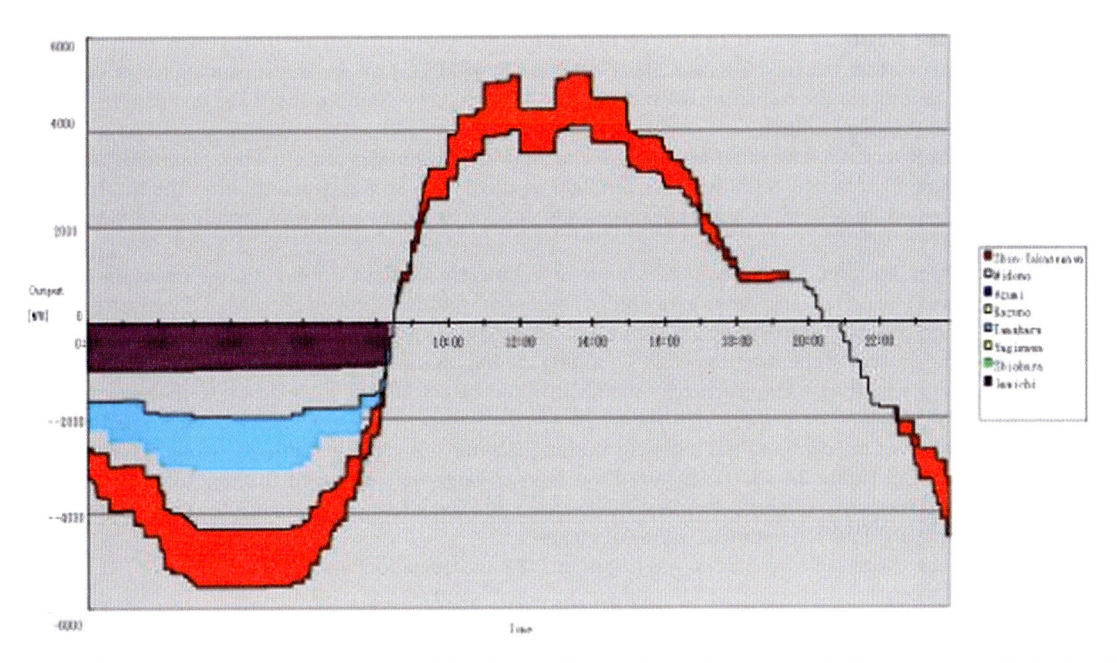

Figure 22 Operation record of pumped-storage power plants for 24 July 2001, on which day the highest electricity consumption of 64.3 million kW was recorded in the Tokyo Electric Power Company's service area.

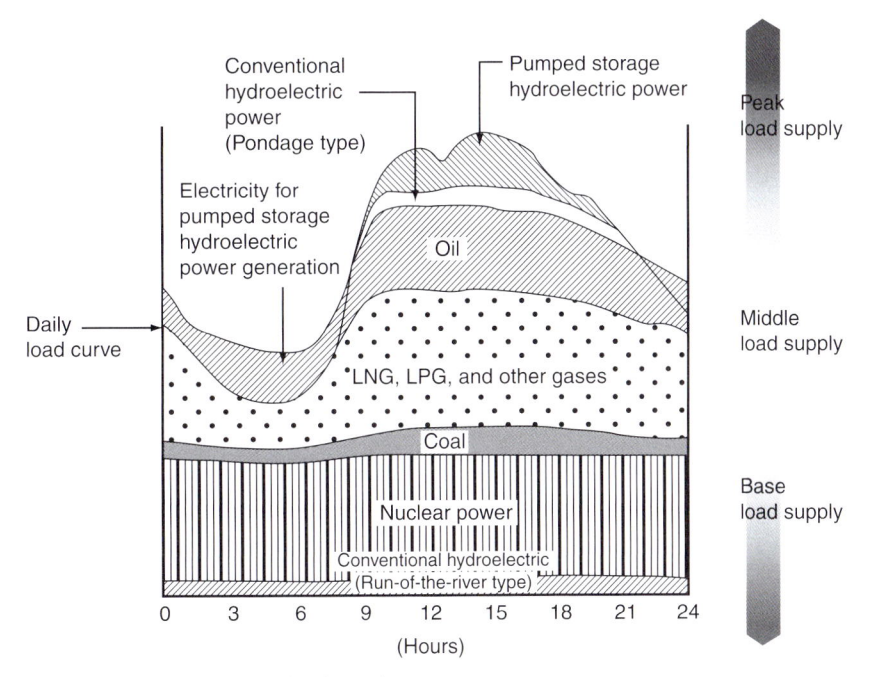

Figure 23 Combining of energy sources to meet changing demand.

Table 15 Comparison of the average unit cost for pumping up water at TEPCO's pumped-storage power plants and the average unit cost of extra burning at TEPCO's oil-based thermal power stations

	Unit power generation cost	Remark
Average unit cost for pumping up water at pumped-storage power plants	4–5 yen kWh^{-1}	Pumping efficiency = 66%
Average cost of extra burning at oil-fired thermal power stations	6–7 yen kWh^{-1}	Unit fuel cost under operation with a utilization factor of 30%

6.10.4.2.1 Outline of the project

The Dashidaira Dam was constructed by the Kansai Electric Power Co. (KEPCO) on the midstream stretch of the Kurobe River (~26 km from the river mouth) for the regulating reservoir of the Otozawa Power Station, and is the first dam in Japan equipped with large-scale sediment flushing facilities.

Streams of the Kurobe River Basin have extremely heavy sediment loads and a major concern when planning construction of the Dashidaira Dam was how to solve the sedimentation problem as there were strong demands from the local community for prevention of coastal erosion. In general, measures against sedimentation at a dam consist of installing sediment storage weirs, dredging, and so on.

At the Dashidaira Dam, however, (1) a very large volume of sediment is brought down from the upstream catchment area, (2) even if dredging were to be done, transportation of dredged sediment would be difficult because of the constraints imposed by the site consisting of a gorge, and (3) conventional methods such as dredging would be unable to solve problems of degradation in the downstream area and erosion of the coastline. KEPCO, noting the importance of mitigating riverbed degradation and coastal erosion, decided to adopt a flushing method whereby sediment would be discharged downstream to the same extent as before construction of the dam.

Specifications in outline of the dam and the sediment flushing facilities are given in **Table 16**, while an outline view of the Dashidaira Dam is shown in **Figure 24** The Dashidaira Dam has two large-scale sediment flushing tunnels in its body. Their structures are that when it is desired to release sediment downstream, the water level at the dam is lowered for free flow of water inside the reservoir so that accumulated sediment will be discharged.

The construction of the Otozawa Power Station was begun in 1982 and operation was started in 1985.

6.10.4.2.2 Features of the project area

The Kurobe River springs from the Northern Alps Mountain Range in the Chubu Sangaku National Park runs down from mountainland of elevation from 2000 to 3000 m cutting steeply graded, deep gorges before dropping into the Sea of Japan. The catchment area is 682.5 km^2, and with a length of 86.0 km, that is, the Kurobe is one of the swiftest rivers even in Japan. The river basin has an annual mean precipitation of approximately 4000 mm to make it as one of the most rainy and snowy areas in the country, and there is an abundant flow of water throughout the year. The entire catchment of the Kurobe River consists of new and old granites that are low in water retention capacity, and the runoff ratio is extremely high. There are approximately 7000 collapse areas totaling 31 km^2 out of the 667 km^2 of mountainland in its catchment area. It means that this area is characterized by extremely heavy sediment load.

The mountainland part of the Kurobe Basin is designated a special area of the Chubu Sangaku National Park and a pure natural state is maintained. The Kurobe River, with its abundant water is looked upon as a rich source of electric power and a stable fountain of domestic and agricultural water supply. It also contributes greatly to the local economy as a tourism resource (**Figure 25**).

Table 16 Specifications of Dashidaira Dam

Item	Specification	
River system	Kurobe River, Kurobe River System	
Catchment area	461.18 km^2	
Power station	Name	Otozawa Power Station
	Maximum output	124 MW
	Maximum discharge	74.0 m^3 s^{-1}
	Effective head	193.5 m
Dam	Type	Concrete gravity
	Height	76.7 m
	Crest length	136.0 m
	Volume	203 000 m^3
Reservoir	Gross storage capacity	9.01 × 10^6 m^{3a}
	Effective storage capacity	1.66 × 10^6 m^{3a}
	Available depth	18 m
Flushing channel	Quantity	2 lines (steel-lined)
	Dimensions	5.0 × 5.0 m
Sediment flushing gate	Upstream side	Slide gate
	Intermediate	Roller gate
	Downstream side	Radial gate

[a]When constructed.

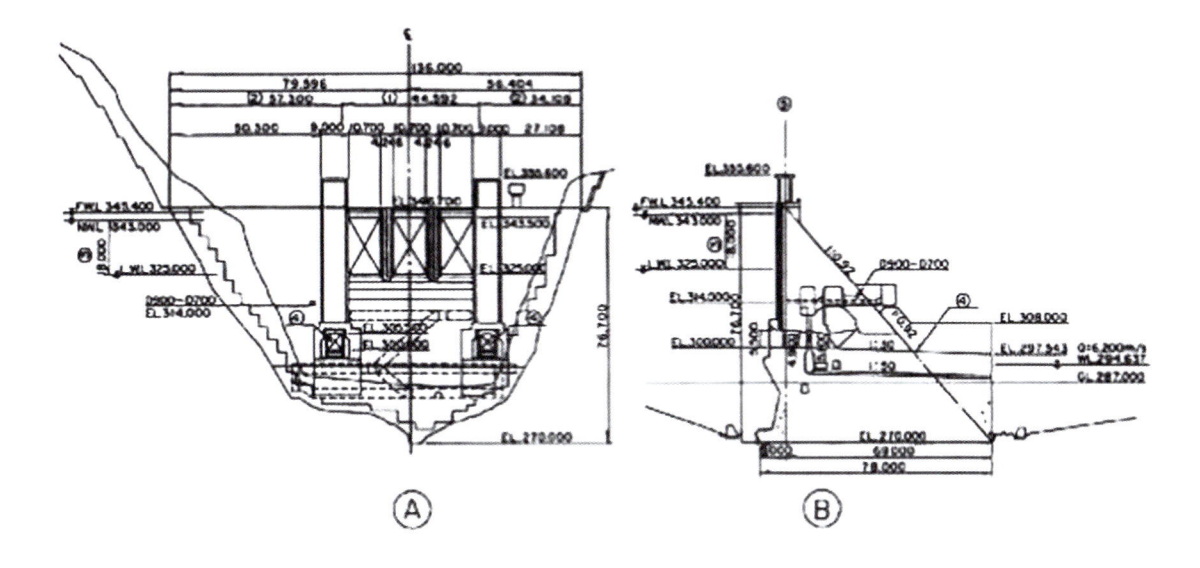

Figure 24 Outline view of Dashidaira Dam. Source: Federation of Electric Power Companies of Japan (FEPC), www.fepc.or.jp/english/index.html.

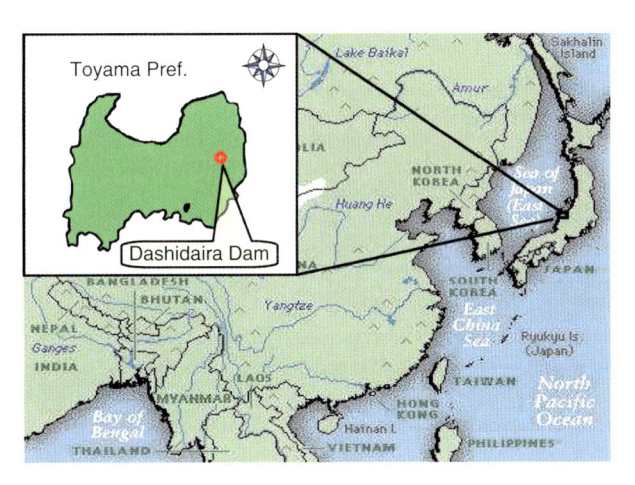

①Overflow section ②Non-overflow section ③Drawdown range ④Flushing channel ⑤Dam Axis

Figure 25 Dashidaira Dam location.

6.10.4.2.3 Major impacts

As of June 2000, sediment flushing had been carried out a total of eight times at the Dashidaira Dam. The annual sediment discharges and the cumulative discharge are shown in **Figure 26**.

In December 1991, 6 years after completion of the dam in 1985, the sediment accumulated had reached approximately 3 million m^3, making it possible for discharge to be done from the gates, and the first flushing was carried out. The result, contrary to expectations, was that turbid water of a dark gray color and with a putrid odor was discharged, and moreover, this turbid water spread out into the sea area and discharge was discontinued at the request of the local community.

A committee including local representatives and knowledgeable persons was organized in order to study the impacts and suitable methods of the sediment discharge. The committee carried out investigations of the impacts on fisheries and agriculture, and of the impacts on the environment, and also conducted sediment flushing tests (February 1993). It was learned that organic matter had become degenerated by long-term deposition of sediment in the dam reservoir, and this had affected the downstream environment when the sediment was discharged. **Table 17** gives the environmental impact items investigated.

6.10.4.2.4 Mitigation measures

The method of discharging sediment from the Dashidaira Dam is to temporarily lower the water level of the reservoir and wash out deposited sediment by free flow of river water. The previously mentioned committee recommended that in order to minimize

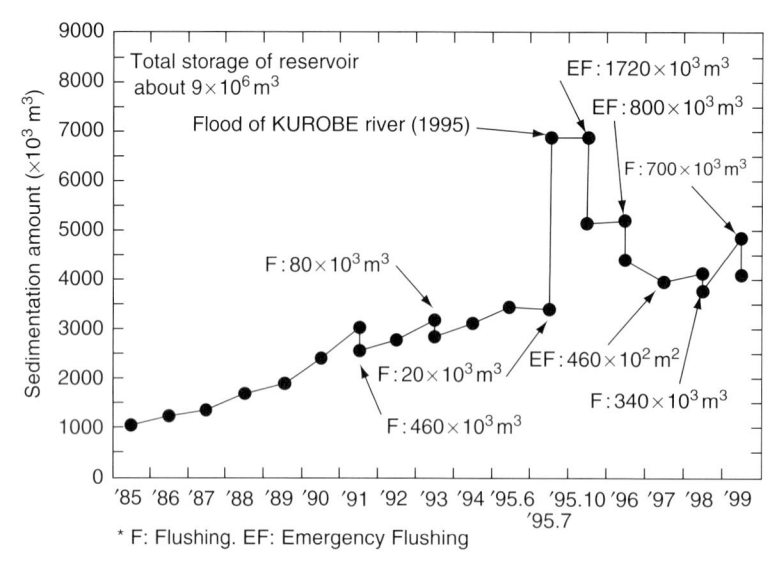

Figure 26 Temporal change of reservoir sedimentation.

Table 17 Environmental impacts investigated

Items investigated	Site investigated			Contents of investigation
	Dam	River	Sea	
Water quality	●	●	●	Water temperature, pH, SS, Turbidity, BOD, COD, T-N, T-P, etc.
Bottom material (sediment)	●	●	●	Appearance, mud temperature, smell, pH, COD, ignition loss, T-N, T-P, grain size distribution, etc.
Aquatic organism	–	●	●	Fish, attached algae, chlorophyll-a, benthic organisms, zoo/phytoplankton, etc.
Sedimentation condition	●	●	–	Cross-sectioning

impacts on the downstream environment, discharge should be done during floods when the volume of river flow is large so that the discharge of sediment from the dam would be close to natural conditions, and sediment flushing was done during floods according to the committee recommendation.

Approximately 3.4 million m³ of sediment were newly deposited at the Dashidaira Dam by severe local rain in 1995 so that the stability of the dam was endangered, so that it was decided that emergency flushing for disaster recovery should be carried out over 3 years (1995–97). These emergency discharges were also carried out during floods in accordance with recommendations of the committee.

6.10.4.2.5 Results of the mitigation measures

As of June 2000, sediment flushing had been carried out a total of eight times at the Dashidaira Dam. The annual sediment discharges and the cumulative discharge are shown in **Figure 26**.

The sediment flushing facilities of the Dashidaira Dam produced results as expected in the aspect of flushing sediment out of the reservoir. As a result of adopting the procedure of releasing sediment to coincide with flood discharge, it became possible to carry out sustainable sediment flushing without causing any great problem in the downstream area. It was found on investigating sediment flushing records and the environment when flushing that (1) the sediment flushing facility of the Dashidaira Dam is effective as a means of discharging sediment from within the reservoir, (2) the sediment flushing carried out to coincide with flooding is effective as a measure for mitigating the impact on the downstream environment, and (3) if a proper sediment flushing method is adopted, the environment is not greatly affected even when large-volume sediment flushing is hurriedly done. Flushing has since been done annually at the same time as flooding in accordance with the advice of the committee. Meanwhile, in order to ascertain the environmental impacts, KEPCO has carried out investigations on the river and sea area downstream of the dam.

The results of the investigations may be summarized as follows:

1. In water quality investigations of the river, turbidity conditions of the downstream river are showing improvement year by year, and prominent impacts due to sediment flushing are not seen.
2. In water quality investigations of the sea area, indices of turbidity and organic matter were temporarily high in the vicinity of the river's mouth during sediment flushing, but in investigations 1 day after flushing, the conditions were seen to have returned more or less to normal.

Table 18 Results of sediment flushing impact investigations (water quality–SS measurements)

			Immediately below dam	Shimokurobe bridge (near estuary)	Point C (sea area)	Point A (sea area)
1995 Emergency flushing	Before flushing		23	230	490	4
	During flushing	Maximum observation	103 500	26 000	1 000	31
		Average	18 000	7 500		
	After flushing	1 Day after	30	193	6	3
1996 Emergency flushing	Before flushing		764	1 520	1 500	21
	During flushing	Maximum observation	56 800	6 770	1 200	52
		Average	10 000	2 900		
	After flushing	1 Day after	194	879	76	7
		1 Month after	8	6	5	3
1997 Emergency flushing	Before flushing		4	8	3	1
	During flushing	Maximum observation	93 200	4 330	3 500	24
		Average	10 000	2 200		
	After flushing	1 Day after	108	757	86	14
		1 Month after	35	22	6	6

The maximum observations of turbidity and SS at point C in 1997 emergency flushing are higher than for previous two emergency flushings because in 1996 emergency flushing rough weather during peak of turbidity of river made sea area investigations impossible. Unit: mg l^{-1}.

3. In bottom material investigations, changes in conditions before and after sediment flushing were not seen and impacts on bottom materials were not detected.
4. Regarding aquatic organisms (benthic animals), as a whole, there were reductions in populations immediately after sediment flushing, but in investigations 1 month later, the situation had returned more or less to the condition before sediment flushing and close to a natural flood condition.

As examples of investigation results, **Table 18** shows values of SS measurements in the river and sea area, and **Figure 27** the transitions in the population of benthic organisms. This population was calculated taking six locations downstream of the dam, counting the population of organisms living in an area of 0.5 m^2 at each location, and totaling the counts of the six.

6.10.4.2.6 Reasons for success
The following may be cited as reasons for success:

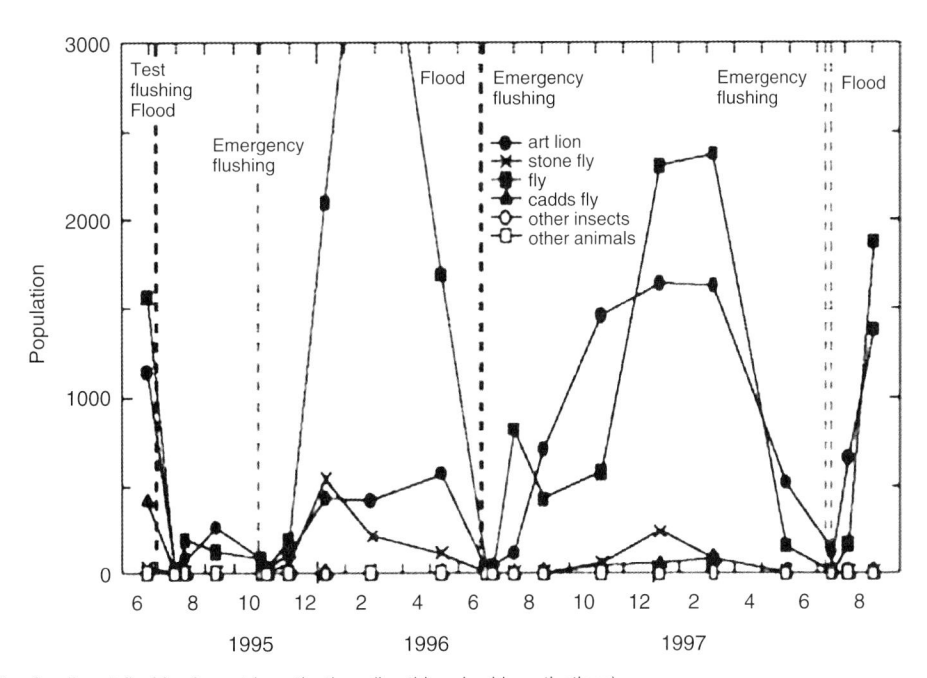

Figure 27 Results of sediment flushing impact investigations (benthic animal investigations).

6.10.4.2.6(i) Study of sediment flushing operation scheme considering environmental impacts
In order to minimize impacts on the downstream environment, a mode of operation was adopted in which sediment flushing was made to coincide with flooding so that sediment would be discharged in a condition close to natural floods.

6.10.4.2.6(ii) Consultation with scientists, experts, stakeholders, and so on
A committee to study the impacts of sediment flushing composed of knowledgeable persons and representatives of local government, fisheries, and agricultural organizations was formed; measures to deal with sedimentation at the Dashidaira Dam were examined from various angles; a study of the possibility of sediment flushing from the dam giving consideration to environmental aspects was made; and a consensus was reached with the local community.

6.10.4.2.6(iii) Establishment of prediction method for environmental impact by sediment flushing
Numerical simulations were made of items such as SS and DO in the downstream part of the river and in the sea area when flushing sediment from the dam, enabling prediction to some extent of impacts on the environment when sediment flushing was done, and this made it possible to plan a better method of operation through utilization of the prediction results.

6.10.4.3 Reservoir Bypass of Sediment and Turbid Water during Flood in the Kansai Electric Power Company

The Asahi Dam had been suffering from the turbid water persistence. The KEPCO installed a bypass tunnel connecting between the upstream end of the reservoir and the downstream of the dam. The bypass tunnel helps to restore the downstream environment as well as to resolve the turbid water persistence.

6.10.4.3.1 Outline of the project
The Oku-yoshino Power Plant is the third pumped-storage-type hydropower for the KEPCO with a maximum output of 1206 MW, following the Kisenyama Power Plant (466 MW) and the Oku-tataragi Power Plant (1212 MW).

The investigation for the construction started in 1971 and the construction started in 1975 and ended in 1980.

The specifications are shown in **Table 19** and the location is shown in **Figure 28**.

6.10.4.3.2 Features of the project area
The Asahi Dam is situated in the Shingu River System rising from the Omine Mountains in the southern part of Kii Peninsula, the rainiest area of Japan, and the site has an annual precipitation in excess of 2000 mm. Precipitation is heavy during the period from the rainy season in June to the typhoon season in September with the past maximum discharge of 662 m^3 s^{-1} recorded in September of 1990.

Mature, rugged mountainland of elevation from 1000 to 1800 m is developed in the watershed. River valleys are V-shaped and river gradients are steep, from 1/6 to 1/7. Conifers such as cedar and Japanese cypress have been planted on the steep mountain slopes while there are also mixed stands of oak and red pine. Locations where collapses have occurred have been increasing in the catchment ever since construction and a comparison of survey results for 1966 and 1990 shows that collapsed areas have increased by 12 times.

6.10.4.3.3 Major impacts
At the Asahi Dam Reservoir, the lower pond of the Oku-yoshino Power Plant, preventive measures against turbidity such as operation of selective intake, installation of a filtering weir immediately downstream of the dam, and protective works against slope collapses around the regulating reservoir had been carried out since the completion of the construction. However, due to changes in the watershed caused by activities upstream such as logging, and especially because of mountainside collapses resulting from large-scale floods brought by typhoons in 1989 and 1990, the problem of turbid water persistence has become prominent. In

Table 19 Asahi Dam specifications

Item	Specifications	
River system	Asahi River, Shingu River System	
Catchment area	39.2 km^2	
Power plant (stand-alone pumped storage)	Name	Oku-yoshino power plant
	Maximum output	201 MW/unit × 6 units
	Maximum discharge	288.0 m^3 s^{-1}
	Effective head	505.0 m
Dam	Type	Arch
	Height	86.1 m
	Crest length	199.41 m
	Volume	147 300 m^3
Reservoir	Gross storage capacity	15.47 × 106 m^{3a}
	Effective storage capacity	12.63 × 106 m^{3a}
	Available depth	32 m

[a] When constructed.

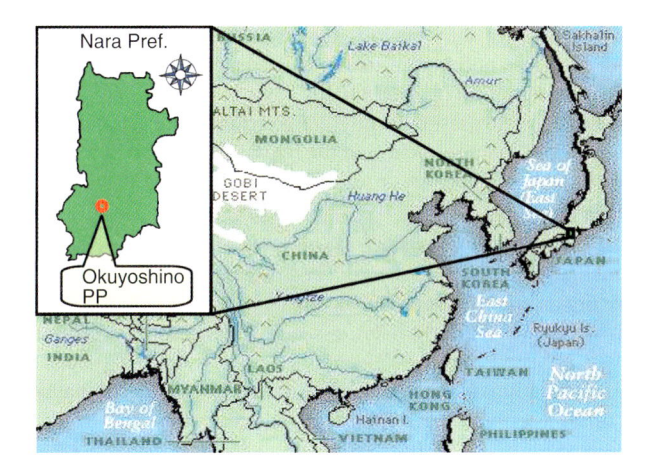

Figure 28 Asahi Dam location.

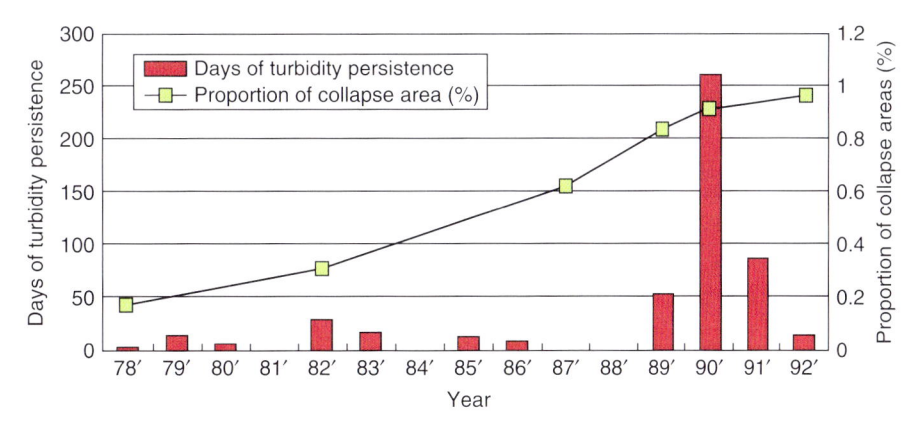

Figure 29 Number of days of turbidity persistence downstream of Asahi Dam and ratio of upstream collapse areas.

addition, sedimentation far in excess of original estimates has become a matter of great concern, and radical countermeasures have become necessitated.

Figure 29 shows the number of days turbidity persisted downstream of the Asahi Dam and the transition in the collapsed area ratio upstream of the dam. According to the results, collapsed areas gradually increased after operation of the dam, specifically triggered by the large-scale typhoons of 1989 and 1990. It caused that huge quantities of sediment were washed down from collapse areas and carried into the regulating reservoir to cause extremely turbid water persistence.

6.10.4.3.4 Mitigation measures

Various countermeasures to turbid water persistence were carried out since the start of operation of the dam, but satisfactory results were not obtained against lasting turbidity caused by very large floods. With strong requests for improvement from the local community, proposals of mitigation measures were studied from 1991. Improvements on selective intake operations, protection of collapse areas, gravel filtration in the downstream channel, forcible settling through use of coagulants, filtering with turbidity-preventing membranes, and sediment bypassing were some of the steps contemplated, and installation of a bypass, the first in Japan, which would be a radical measure resolving the problem of sedimentation at the same time, was chosen. To elaborate, there is no need to store water from the flow of the river since the plant is a standalone pumped-storage type, while the catchment area is comparatively small. The sediment bypassing facility would consist of a bypass tunnel to route turbid water and sediment load around the reservoir and into the downstream river channel.

In planning and designing facilities, the fundamental layout was first selected based on characteristics of the site such as the river channel configuration, and not only wash load but also suspended and traction loads were considered from the points of view of lessening turbid water persistence and of reducing sedimentation. Technical problems such as determination of the optimum tunnel discharge capacity and avoidance of tunnel blockage by sediment were addressed carrying out model hydraulic tests and numerical simulations. Furthermore, various examinations were made concerning predictions of riverbed changes upstream and downstream of the bypass, hydraulic stability, and problems of maintenance such as abrasion among others. Since the start of operation in 1998, the bypass has basically been used only during floods to detour water and sediment through the tunnel, clear water in normal times being allowed to enter the reservoir. This is because the Asahi Dam is for the regulating pond of a pumped-storage power station and thus does not require inflow of water for storage, but inflow would improve circulation of water inside the reservoir and prevent deterioration of water quality.

The particulars of the dam and sediment bypassing facility are given in **Table 20** and a sketch of the waterway is given in **Figure 30** (hereafter referred to as 'bypass'). The construction of the bypass was started in 1994 and its operation in April of 1998.

6.10.4.3.5 Results of the mitigation measures

In order to ascertain the effectiveness of the bypass since starting its operation, investigations of water quality (turbidity persistence, eutrophication), sedimentation inside the reservoir, sedimentation in the river (river cross-section), riverbed gradation, shoals and pools, and aquatic organisms have been carried out as indicated in **Table 21**. These investigations are for seeing how turbid water persistence and sedimentation have been lessened and what impacts there have been on the downstream riverine environment.

According to the results of these investigations and measurements, it may be considered that sediment bypassing has been highly effective in mitigating persistent turbidity, inhibiting buildup of sedimentation, and restoring the environment of the river downstream.

First, as an example of the effects concerning the problem of turbid water persistence, the results comparing turbidity conditions upstream and downstream of the dam and in the reservoir before and after starting operation of the bypass are shown in **Figures 31–33**. The floods used in comparison were of approximately the same scales. Even for floods from which turbidity had lasted close to 1 month before operation of the bypass (BO in the figures), after starting operation (AO), only 3 days after flooding had ended, the turbidity had become the same as that upstream with the condition back to normal, and the effectiveness was clearly confirmed. The turbidity inside the reservoir was at a fairly low level compared with that before operation, while it was found that sedimentation was held to approximately one-tenth compared with before operation.

Table 20 Specifications of bypass facilities

Sediment bypassing facility		
Weir	Height × crest length	13.5 × 45.0 m
	Structure	Steel
Intake	Height × width	14.5 × 3.8 m
	Length	18.50 m
	Structure	Reinforced concrete, steel-lined
	Gate	1
Bypass tunnel	Height × width	3.8 × 3.8 m (hood shape)
	Length	2350 m
	Gradient	~1/35
	Maximum discharge capacity	140 m^3 s^{-1}
	Structure	Reinforced concrete lined
Outlet	Width × length	8.0–5.0 × 15.0 m
	Structure	Reinforced concrete

Hydropower Development in Japan

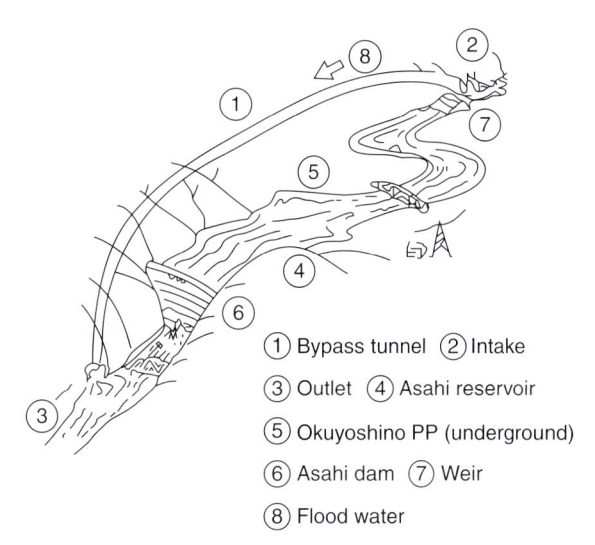

Figure 30 Outline of Asahi Dam sediment bypassing facilities. Source: Federation of Electric Power Companies of Japan (FEPC), www.fepc.or.jp/english/index.html.

① Bypass tunnel ② Intake
③ Outlet ④ Asahi reservoir
⑤ Okuyoshino PP (underground)
⑥ Asahi dam ⑦ Weir
⑧ Flood water

Table 21 Items of environmental impact investigation concerning bypass operation

Items investigated	Dam	DS river[a]	Contents of investigations
Water quality (turbidity persistence)	●	●	Water temperature, turbidity
Water quality (eutrophication)	●	–	Water temperature, turbidity, BOD, COD, T-N, T-P, etc.
Sedimentation condition	●	●	Cross-sectioning
Shoal, pool conditions	–	●	Distribution survey, cross-sectioning
Aquatic organisms	–	●	Habitat environment, attached algae, benthos, fish surveys

[a] DS, Downstream.

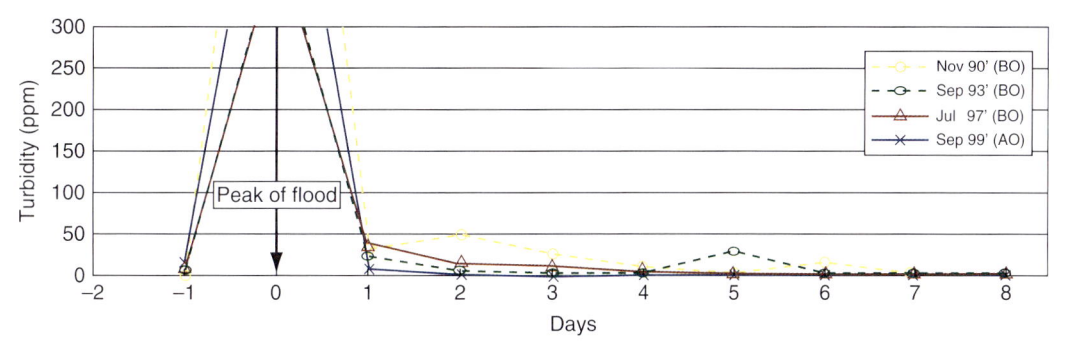

Figure 31 Turbidity conditions upstream of dam.

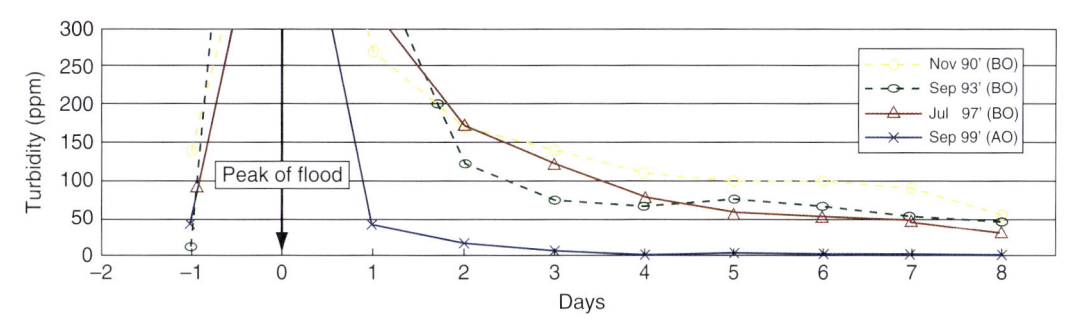

Figure 32 Turbidity conditions downstream of dam.

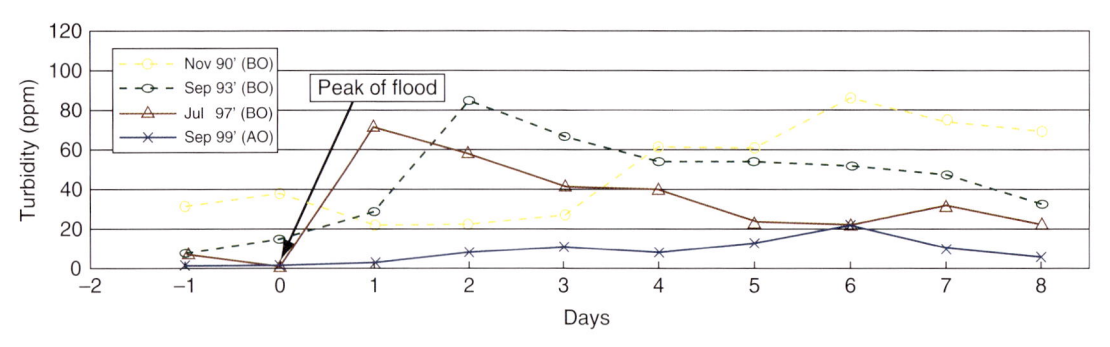

Figure 33 Turbidity conditions in Asahi Dam regulating reservoir.

Next, as the impact on the downstream environment of the river, it was made possible for sediment that had been stopped before by the dam to go downstream unobstructed via the bypass and this is thought to have had the effects of preventing degradation and armor coating of the downstream riverbed. In fact, it was confirmed in investigations of shoals and pools and of riverbed gradation that the river profile had changed (recovered), and it was commented by local people that "Whities (pretty white pebbles specific to the upstream area) had been getting scarcer and scarcer since the dam was built, but now they've come back again. The river is returning to its old self."

6.10.4.3.6 Reasons for success
The following may be cited as reasons for success.

6.10.4.3.6(i) Planning and implementation of sediment bypassing as the most effective mitigation measure
Various mitigation schemes, including examples in foreign countries, were compared and studied.
Features of the site were taken into consideration and sediment bypassing of the reservoir was planned and implemented as a radical solution measure.

6.10.4.3.6(ii) Detailed investigations, analyses, and studies at planning and designing stages
Leading authorities on the subjects were consulted in detailed investigations and analyses of hydrology, meteorology, and topography at the planning and designing stage, and in hydraulic design of structures, large-scale hydraulic model experiments, and numerical simulations were carried out, and the results were reflected in design.

6.10.4.4 Measures for Ecosystems

The Okutadami and Ohtori Power Stations became two of the largest conventional projects following the expansion that increased the power output (combined) of 455 MW by 287 MW to a total of 742 MW. The expansion started in full swing in July 1997 and was completed in June 2003. The expansion was carried out in a rich natural environment within a natural park – the habitation of large predatory birds in danger of extinction such as golden eagles and Hodgson's hawk eagles. This, therefore, necessitated efforts to minimize environmental loads during the planning and construction stages. With particular attention to ensuring no interference to the habitation and breeding of golden eagles and Hodgson's hawk eagles, environmental conservation measures were taken to protect their nesting places and minimize interference with the ecosystem supporting the survival of these birds. Moreover, to facilitate social consensus building on the expansion, extra effort was put into information disclosure to promote accountability. Thanks to such efforts, the expansion is considered an example of successful coexistence of natural protection and development in Japan.

6.10.4.4.1 Outline of the project

Hydropower generation attracts interest as a clean, recyclable energy source free from CO_2 emissions. However, the sites for economically feasible hydropower development are few in Japan. This circumstance led to the redevelopment of existing conventional hydropower. For the purpose of improving the peak supply capacity, the expansion of power generating facilities was planned using existing dams and reservoirs at the Okutadami Dam (normal water surface level of elevation 750 m and total storing capacity of ~600 tons, completed in 1961) located on the border between Fukushima Prefecture and Niigata Prefecture and at the Ohtori Dam (normal water surface level of elevation 557 m and total storing capacity of ~16 million tons, completed in 1963). The expansion of the power stations (hereafter referred to as the 'expansion'), carried out by Electric Power Development Co. Ltd. (J-POWER), started in full swing in July 1997 and the operation of the new facilities started in June 2003. With the completion of the expansion, the Okutadami Power Station has a combined (original and additional) power output of 560 000 kW and the Ohtori Power Station has a combined power output of 182 000 kW. The Okutadami Power Station in particular became the largest conventional hydropower in Japan.

Table 22 shows the output and specifications of original power generating facilities at the Okutadami and the Ohtori Power Stations, as well as those of additional power generating facilities built by Electric Power Development Co. Ltd. (J-POWER).

Figures 34 and 35, on the other hand, respectively show a plane view and aerial view of the Okutadami Power Station and Figure 36 shows an aerial view of the Ohtori Power Station.

6.10.4.4.2 Features of the project area

The area around the project site forms a valley along the Tadamigawa River, which originates from Oze, and is surrounded with mountains in the range of 1200–1500 m. The area is climatically one of the snowiest areas in Japan, with the depth of snow accumulation near the Okutadami Dam sometimes exceeding 5 m.

The vegetation in the area comprises natural forests of Japanese beeches and is classified as natural vegetation level 9. (According to the natural environmental conservation survey report – 1976 by the Environment Agency – the natural vegetation level is classified into 10 levels depending on the degree of human interference, and the highest natural vegetation level is 10.)

The project site is located in such a rich natural environment inhabited by rare predatory birds such as golden eagles designated as precious natural product and is specified as first-class special zone in the Echigo Sanzan Tadami Quasi-National Park (**Figure 37**).

6.10.4.4.3 Major impacts

During the environmental impact survey, two pairs of golden eagles and one pair of Hodgson's hawk eagles were found nesting in the area surrounding the expansion site. Both pairs of golden eagles were found nesting on ledges of steep rock walls and the Hodgson's hawk eagles were found nesting in a Japanese beech in a forest of deciduous broadleaf trees.

Golden eagles and Hodgson's hawk eagles are very small in number and as shown in **Table 23** are designated as rare species under the Law for the Conservation of Endangered Species of Wild Fauna and Flora. Since predatory birds such as golden eagles and Hodgson's hawk eagles are positioned at the top of the ecosystem (food chain) and their survival depends on the availability of prey animals, they are considered indicator organisms that represent the level of natural richness and diversity.

Two goals were therefore set during the expansion: first, to protect the two pairs of golden eagles and the one pair of Hodgson's hawk eagles found nesting in the area, and second, to conserve the natural environment that supports the survival of these predatory birds (**Table 23**).

Critically endangered (I) – Species in danger of extinction (IA and IB in order of increasing risk)
Vulnerable (II) – Species in increasing danger of extinction
Near threatened – Species whose survival is jeopardized
Data deficient – Species for which available data are too fragmentary for assessment

6.10.4.4.4 Mitigating measures

6.10.4.4.4(i) Life cycle of golden eagles and Hodgson's hawk eagles
The annual life cycle of the golden eagle and Hodgson's hawk eagle comprises the nest building period and nonnest building period and they are believed to be more susceptible to external disturbances during the nest building period (courting and nest building period, egg laying and incubation period, and hatching and nest breeding period).

6.10.4.4.4(ii) Fundamental policies relating to the protection of golden eagles and Hodgson's hawk eagles
Based on the observation (hereafter referred to as the 'territorial zone survey') of flying routes and resting places of golden eagles and Hodgson's hawk eagles nesting around the construction site, the geographically important zone for the survival of these birds was determined in consideration of the life cycle of each species (**Figure 38**). Then, a construction plan as explained below was

Table 22 Specifications of the Okutadami and Ohtori Power Stations (original and additional facilities)

Items		Okutadami		Ohtori	
	Existing	Expansion	Existing	Expansion	
River		Tadami River in Agano River System			
Dam and reservoir	Name	Okutadami Dam, Okutadami Reservoir (existing)		Ohtori Dam, Ohtori Reservoir (existing)	
	HWL	Elevation 750 (m)		Elevation 557 (m)	
	LWL	Elevation 690 (m)		Elevation 551 (m)	
	Effective depth	60 (m)		6 (m)	
	Reservoir area	11.5 (km^2)		0.89 (km^2)	
	Catchment area	595.1 (km^2)		656.9 (km^2)	
	Dam type	Concrete gravity-type dam		Concrete gravity arch-type dam	
	Height	157 (m)		83 (m)	
	Length of dam crest	480 (m)		187.9 (m)	
	Volume	1 636.3 × 10^3 (m^3)		160.0 × 10^3 (m^3)	
	Gross reservoir capacity	601 × 10^6 (m^3)		15.8 × 10^6 (m^3)	
	Effective reservoir capacity	458 × 10^6 (m^3)		5.0 × 10^6 (m^3)	
	Design flood discharge	1 500 (m^3 s^{-1})		2 200 (m^3 s^{-1})	
Location		Aza Komagatake Hinoematamura Minamiaizugun Fukushima Prefecture		Aza Iriyama Oaza Tagokura Tadamichou Minamiaizugun Fukushima Prefecture	
Generation type		Dam–conduit type	Dam–conduit type	Dam type	Dam type
Maximum output		120 000 kW × 3	200 000 kW × 1	95 000 kW × 1	87 000 kW × 1
Maximum discharge		249 (m^3 s^{-1})	138 (m^3 s^{-1})	220 (m^3 s^{-1})	207 (m^3 s^{-1})
Maximum effective head		170.0 (m)	164.2 (m)	50.8 (m)	48.1 (m)
Power house		Type – Underground	Type – Underground	Type – Semi-underground	Type – Underground
		H – 37.80 (m)	H – 39.20 (m)	H – 50.80 (m)	H – 48.20 (m)
		W – 18.50 (m)	W – 17.90 (m)	W – 37.20 (m)	W – 22.00 (m)
		L – 87.60 (m)	L – 45.00 (m)	L – 28.45 (m)	L – 44.50 (m)
Turbine	Type	Vertical Francis	Vertical Francis	Vertical Kaplan	Vertical Kaplan
	Maximum output	137 000 kW × 3	205 000 kW × 1	100 000 kW × 1	89 500 kW × 1
Generator	Type	3 Phases vertical	3 Phases vertical	3 Phases vertical	3 Phases vertical
	Maximum output	133 000 kVA × 3	223 000 kVA × 1	100 000 kVA × 1	97 000 kVA × 1
Tailrace length		3 048 (m)	3 444.67 (m)	109.25 (m)	
Penstock	Length	No. 1, 185.9 (m)	280.41 (m) × 1	69.2 (m) × 1	93.39 (m) × 1
		No. 2, 3, 189.5 (m)			
	Diameter	4.3–3.8 (m)	6.5–4.0 (m)	7.5–6.35 (m)	6.8–6.2 (m)
Operation service date		2 December 1960	7 June 2003	20 November 1963	7 June 2003

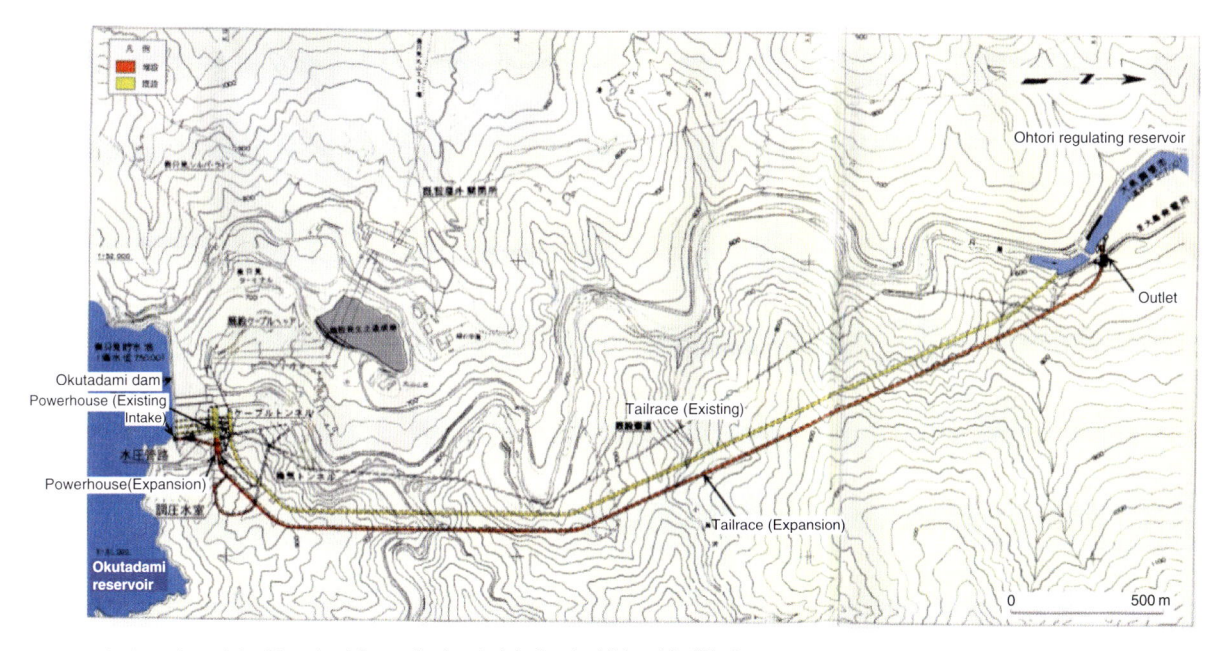

Figure 34 A plane view of the Okutadami Power Station (original and additional facilities).

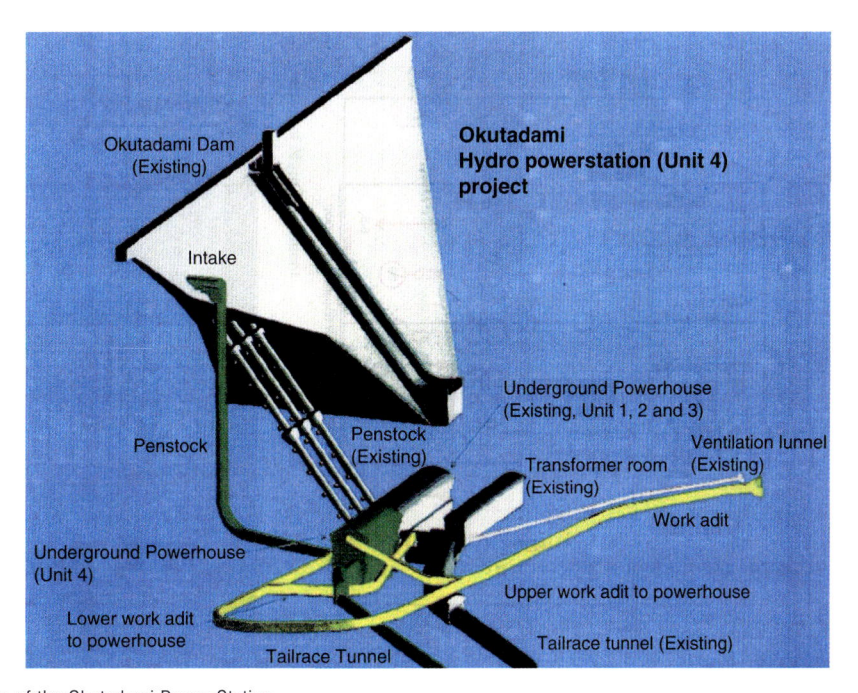

Figure 35 An air view of the Okutadami Power Station.

formulated with attention to minimizing interference with nest building by golden eagles and Hodgson's hawk eagles. Monitoring was also performed during the construction period.

6.10.4.4.4(iii) Protection measures for golden eagles

6.10.4.4.4(iii)(a) **Restriction of the construction period** During the nest building period of golden eagles (generally considered to be between November and June of the following year), no above-ground construction was planned within the important nest building zone (hereafter referred to as the 'core area') determined from the territorial zone survey. The above-ground construction within the core area, therefore, was limited to the 4-month, nonnest building period of July to October.

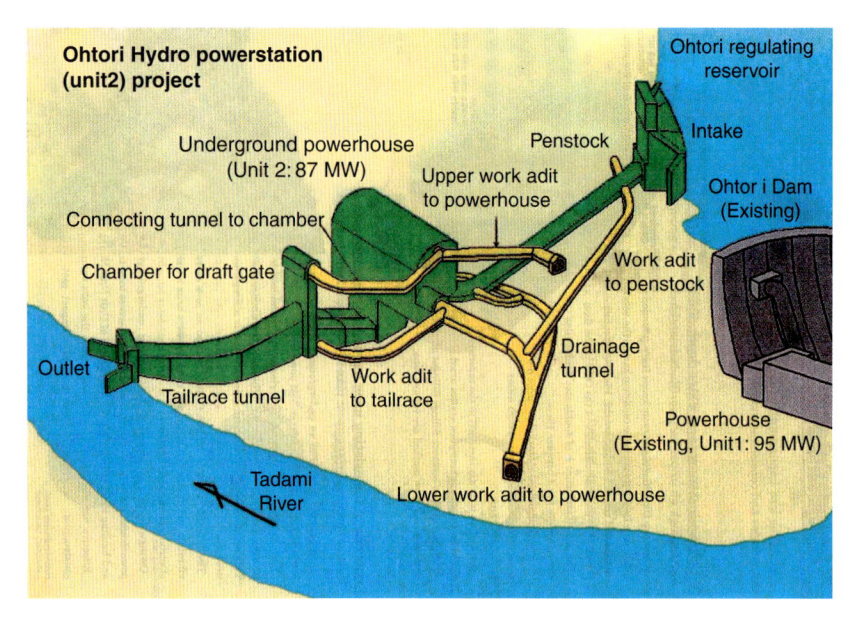

Figure 36 An aerial view of the Ohtori Power Station.

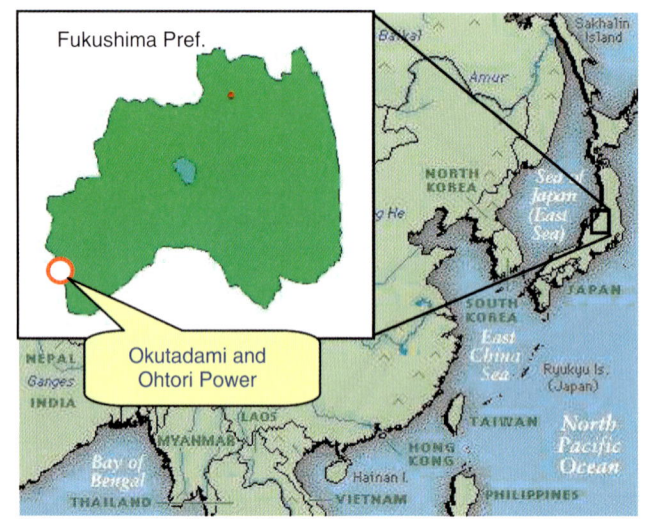

Figure 37 Location map for the Okutadami and Ohtori power plants.

6.10.4.4.4(iii)(b) **Restriction on the construction in consideration of the fledging of young birds** With regard to the resumption of the construction in July following successful breeding by golden eagles, the measures below were taken based on the concept of adaptive management in consideration of newly fledged young birds (because they can only cover a small part of the territorial zone for about a month after fledging and they are still fed by a parent bird).

1. Within the core area and the zone considered to be inhabited by newly fledged young birds in which they are still fed by a parent bird, no construction was planned for implementation for about a month after young birds fledging and the expansion was resumed after it was confirmed that the zone covered by young birds had expanded sufficiently.
2. Outside the core area, the construction started, initially on a small scale, and gradually increased in scale, after it was confirmed by monitoring that the construction had no adverse effects on young birds.
3. The construction was temporarily halted when it was found from the monitoring of young birds and their parents that the construction may have adverse effects. For example, when a young bird flew into the construction site, the construction was immediately brought to a temporary halt when requested to do so by survey staff, and when it was confirmed that the young bird left the site, the construction was resumed.
4. In consideration of where young birds are staying and the extent of their territorial zones, the period and extent of the construction mentioned above were set with flexibility.

Table 23 Precious animals and plants observed around the expansion site of the Okutadami and Ohtori power plants

Category	Species	Precious natural product	Rare domestic species	Red list
Mammals	Japanese serow	○		
	Japanese small flying squirrel			
Birds	White-tailed sea eagle	○	○	Endangered (IB)
	Hodgson's hawk eagle		○	Endangered (IB)
	Golden eagle	○	○	Endangered (IB)
	Steller's sea eagle	○	○	Vulnerable (II)
	Fish hawk			Near threatened
	Goshawk		○	Vulnerable (II)
	Peregrine falcon		○	Vulnerable (II)
	Honey buzzard			Near threatened
	Sparrow hawk			Near threatened
Plants	Iris gracilipes			Near threatened
	Agrostis hideoi Ohwi			Data deficient

Precious natural product: Precious natural product designated under the law for the protection of cultural properties. Rare domestic species: Rare domestic wild animal and plant species designated under the law for the conservation of endangered species of wild fauna and flora. Red list: List of threatened wild animal and plant species in Japan (Wildlife Protection Division, Nature Conservation Bureau, Environment Agency 1998). Each category is defined as follows.

Figure 38 A young golden eagle (238 days old).

6.10.4.4.4(iv) Protection measures for Hodgson's hawk eagles

With regard to Hodgson's hawk eagles found nesting around the construction site, it was confirmed that the construction site and the construction roads were not included in the geographically important zone for the survival of the birds. There was, however, an overlap of about 2 months (July and August) between the above-ground construction period (July–October) set with attention to protecting golden eagles and the nest building period of Hodgson's hawk eagles (January–August) (Figure 39).

Nonetheless, restricting conditions imposed by the heavy snowfall in winter and the luxuriant growth of broadleaf trees in summer made it difficult to directly observe the breeding of Hodgson's hawk eagles (see Table 26). Under this circumstance, it was assumed from the perspective of protection and conservation that the susceptibility of Hodgson's hawk eagles increased with the progress of breeding, and measures to reduce noise, for example, by maintaining long intervals between construction vehicles, were used on some sections of construction roads relatively close to the nesting place.

6.10.4.4.4(v) Natural environmental conservation measures

The environmental conservation measures indicated below were carried out not only to protect golden eagles and Hodgson's hawk eagles but also to conserve the natural environment inhabited by prey animals supporting the survival of these predatory birds.

6.10.4.4.4(v)(a) Reduction of the renovation area

1. The plan included the installation of a head gate in the existing Okutadami Dam using the dam drilling method and underground construction of a large part of power station facilities.

Figure 39 A Hodgson's hawk eagle (female).

6.10.4.4.4(v)(b) Measures against noise and vibration

1. To reduce blasting noise accompanying underground tunneling and excavation for the construction of an underground power station, a soundproof door was installed in the pitmouth. A method that allows delay blasting control was used to reduce blasting vibration.
2. Concrete and aggregate production facilities that cause large noises were housed in a building to reduce the outdoor noise level (**Figure 40**).
3. Low noise construction machines were used.
4. A speed limit of 30 km h^{-1} was applied to construction vehicles and a stop to idling was encouraged when vehicles were at a stop.

6.10.4.4.4(v)(c) Water quality conservation measures

1. A double layer of pollution prevention membranes were installed for the underground construction (of a head gate) in order to prevent the spreading of polluted water.

6.10.4.4.4(v)(d) Measures relating to lighting and coloring

1. Minimum nighttime lighting necessary for construction safety was used. The high-voltage sodium lamp was used because it has only minor effects on insects and plants.
2. Blinds were hung from the windows of temporary buildings so as not to allow interior lighting to leak to the outside and headlights of vehicles were turned off when vehicles were at a stop.
3. The use of colors disliked by birds (yellow and red) was restricted as the exterior colors of temporary facilities and construction machines (**Figure 41**).

Figure 40 A concrete production facility. Housed in a building (within the temporary facility site at the Okutadama outlet).

Figure 41 Entire view of the temporary facility site in Ohtori.

6.10.4.4.4(v)(e) Measures to compensate for the use of the marshland as the reclamation site of excavated rocks

1. Since restricting conditions made it difficult to transport excavated rocks generated from the underground construction outside the construction site, a plan was made to reclaim the marshland (hereafter referred to as the 'existing marshland') within the construction site using excavated rocks.

    However, since the existing marshland was inhabited by aquatic plants and dragonfly species, it was considered essential to take measures to compensate for the use of the marshland environment. These measures comprised the construction of a new marshland as a replacement within the reclamation site and the restoration of the original marshland environment. More specifically, a site next to the existing marshland was reclaimed to construct a replacement and the existing and replacing marshland were used until the existing marshland was completely reclaimed. These measures allowed animals inhabiting the marshland including dragonflies to freely travel between the two marshlands and thus made natural changes of generations possible (**Figure 42**).

6.10.4.4.4(v)(f) Protection of other animals and plants

1. When small animals were found in the construction site or on the construction roads, vehicles were brought to a temporary stop until the animals left the area at their own will.
2. Precious plants found in the renovation area were transplanted according to specialist advice.

6.10.4.4.5 Results of the mitigation measures
Tables 24–26 show the state of breeding by two pairs of golden eagles and one pair of Hodgson's hawk eagles whose nesting places were located.

Figure 42 Restoration of the Marshland environment in Yasaki reclamation site (restored Marshland – in the right of the photo).

Table 24 State of breeding by one pair of golden eagles in Okutadami

Year	State of breeding	Breeding result
1994	Located the nesting place and witnessed the fledging of young birds in June	○
1995	Eggs laid (2/24) and incubation abandoned (4/6)	×
1996	Eggs laid (2/20) and eggs hatched (4/6). Confirmed the death of the chicks (4/7)	×
1997	Eggs laid (3/5) and eggs hatched (4/16). Confirmed the death of the chicks by the attack of a crow (4/30)	×
1998	Eggs laid (2/27) and incubation abandoned (3/28)	×
1999	Eggs laid (2/24) and incubation abandoned (3/22)	×
2000	Eggs laid (3/1) and eggs hatched (4/13). Witnessed the fledging of young birds (7/2)	○
2001	No eggs laid due to the interference by the young birds born in the previous year	×
2002	Eggs laid (3/6 and 3/7) and eggs hatched (4/18). Witnessed the fledging of young birds (7/4)	○

Table 25 State of breeding by one pair of golden eagles in Ohtori

Year	State of breeding	Breeding result
1995	Located the nesting place in May and witnessed the fledging of young birds in June	○
1996	Witnessed the fledging of young birds in June	○
1997	Breeding failed (the course of breeding unknown)	×
1998	Eggs laid (3/2) and eggs hatched (26/13). Witnessed the fledging of young birds (6/18)	○
1999	Eggs laid (2/15), eggs hatched (3/28), and the breeding of the chicks abandoned (4/11)	×
2000	Eggs laid (2/18) and eggs hatched (3/30). Confirmed the death of the chicks (between 4/2 and 4/27)	×
2001	Eggs laid (2/25) and incubation abandoned (end of March)	×
2002	Nest building discontinued	×

Table 26 State of breeding by one pair of Hodgson's hawk eagles whose nesting tree was located

Year	State of breeding	Breeding result
1998	Located the nesting place in October (the course of breeding unknown)	?
1999	Found that eggs were laid. But, the later course of breeding unknown. Breeding may have succeeded since the presence of young birds was witnessed	○ ?
2000	The course of breeding unknown	×
2001	The course of breeding unknown	×
2002	The course of breeding unknown. Breeding may have succeeded since the presence of young birds was witnessed	○ ?

The rate of successful breeding of golden eagles in Japan is estimated to be between 20% and 30%. One of the two pairs of the golden eagles had successful breeding twice, first in 2000, immediately after the start of the construction, and second in 2002. The other pair had unsuccessful breeding soon after the start of the construction. However, around the nesting place of the pair, another pair of young golden eagles with no breeding experience was frequently observed.

As already explained, since the site conditions make the observation of the nesting place difficult, the state of breeding by Hodgson's hawk eagles remained unknown. However, the presence of the birds inhabiting the area was continuously witnessed throughout the construction period.

These facts appear to lead to the conclusion that the expansion had no adverse effects on the golden eagles and Hodgson's hawk eagles nesting around the construction site.

Since the marshland constructed as a replacement in the reclamation site was found inhabited by animals and plants that previously inhabited the existing marshland, the compensating measures appear to have succeeded.

Moreover, noise and vibration caused by the expansion as well as the quality of construction drainage were monitored and made to fall within the criteria voluntarily adopted. Although there were times when they temporarily fell outside the criteria, the cause was investigated to prevent reoccurrence. These measures, monitoring, and corrective measures were carried out in accordance with ISO 14001 (environmental management system).

6.10.4.4.6 Reasons for success

The expansion was carried out while pursuing cost-effectiveness as well as with great emphasis on environmental conservation. The success of environmental conservation was specifically reflected in the successful breeding of golden eagles immediately after the start of the construction. The expansion of the Okutadami–Ohtori Power Stations will be regarded as a successful example of achieving the coexistence of natural protection and development.

The following factors are considered to have contributed to the success:

1. Attention was directed toward minimizing changes to the land from the planning stage.
2. A construction plan with attention to environmental conservation was formulated with the advice of specialists on predatory birds and natural ecosystems.
3. With regard to the fledging of young golden eagles, the concept of adaptive management was followed. This more specifically means that the construction was carried out while monitoring these young birds and ensuring no effects on these birds and that flexible solutions were provided, including immediate implementation of corrective measures, in the event that something unexpected happened.
4. The environmental management system certified to ISO 14001 was used to make sure systematic and proper implementation of various environmental conservation measures.

The four points listed above comprised the technical solutions we used. It is also worthy of mention that a major effort was placed on information disclosure in order to fulfill our accountability to the society.

To win the understanding of a wide spectrum of general public about development activities in areas such as this construction site with an intact, rich natural environment, information sharing and mutual communication were considered most important. The expansion of the Okutadami–Ohtori Power Stations, therefore, entailed the implementation of technical solutions for environmental conservation as well as other various efforts to win social acceptance. These specifically included the creation of a home page to provide information including the pattern of habitation of golden eagles and the progress of the construction, and the issuing of environmental reports that included the results of environmental conservation measures taken. Efforts were also directed to promoting greater understating of the construction plan through public relations activities that included information provision to mass media whenever necessary and press conferences. Moreover, extra efforts were placed on information disclosure and dialog with certain environmental protection organizations that raised objections to the expansion.

We believe, as the entity responsible for the expansion project, that various efforts explained above proved a success, as reflected not only in successful breeding of golden eagles but also in successful building of social consensus. Two important factors considered to have contributed to the success were technical solutions for environmental conservation and information disclosure to fulfill our accountability obligations.

Relevant Websites

http://www.chuden.co.jp/english/CHUBU Electric Power Co., Inc.
http://www.ci.nii.ac.jp CiNii Sediment characteristics of Dashidaira Dam Reservoir at Kurobe River and Toyama Bay, and flushed suspension impacts on fishes.
http://www.jpower.co.jp/english/index.html Electric Power Development Company (EPDC).
http://www.wds.iea.org/Energy Balances of OECD Countries.
http://www.fepc.or.jp/english/index.html Federation of Electric Power Companies of Japan (FEPC).
http://www.iso14000-iso14001-environmental-management.com ISO 14000 environment management systems and standards, including ISO 14001, ISO 14004, ISO 14010, ISO 14011 and ISO 14012.
http://www.jepoc.or.jp/english/english01.html Japan Electric Power Civil Engineering Association.
http://www.meti.go.jp/english/Ministry of Economy, Trade and Industry (METI).
http://www.enecho.meti.go.jp/hydraulic/data/stock/top.html Renewable Portfolio Standard (RPS).
http://www.kepco.co.jp/english/KEPCO – The Kansai Electric Power Co., Inc. Electric generation company involved in nuclear, thermal, and hydroelectric power, transmission and distribution, environmental protection.
http://www.tepco.co.jp/ The Tokyo Electric Power Company.

6.11 Evolution of Hydropower in Spain

A Gil, Hydropower Generation Division of Iberdrola, Salamanca, Spain
F Bueno, University of Burgos, Burgos, Spain

© 2012 Elsevier Ltd. All rights reserved.

6.11.1	Hydroelectric Power in Spain	309
6.11.1.1	Electric Power and Hydroelectric Power	309
6.11.1.2	The Strategic Importance of Hydroelectric Power	310
6.11.1.3	Hydrology, River Network, and Hydroelectric Development	311
6.11.1.4	Power Plants and Main Developments	313
6.11.1.5	Producing Companies	315
6.11.2	Evolution of Schemes and First Developments	315
6.11.2.1	Periods in the Evolution of Development	315
6.11.2.2	The 1890–1940 Period	317
6.11.2.2.1	First steps of electricity in Spain	317
6.11.2.2.2	The electricity sector in the first decades of the twentieth century	318
6.11.2.2.3	Main hydropower developments	318
6.11.2.3	The 1940–60 Period	323
6.11.2.3.1	The electricity after the civil war	323
6.11.2.3.2	Main hydropower developments	324
6.11.2.4	The 1960–75 Period	327
6.11.2.4.1	The electricity sector	327
6.11.2.4.2	The golden age of dam engineering in Spain	328
6.11.2.4.3	Main hydropower developments	329
6.11.2.5	The Last Three Decades	333
6.11.2.5.1	The electricity sector	333
6.11.2.5.2	Main hydropower developments	333
6.11.3	A Representative Case: The Duero System and Its Evolution	336
6.11.4	The Future of Hydroelectric Power in Spain	339
References		341

6.11.1 Hydroelectric Power in Spain

6.11.1.1 Electric Power and Hydroelectric Power

In the first years of hydroelectric power development in Spain, at the end of the nineteenth century, it was the thermal plants that covered most of the electric power demand. With the general use of alternating current and transformer stations this changed, and in the first four decades of the twentieth century hydraulic power increasingly became the main source of supply, reaching 93% of the total supply in 1936.

With slightly lower values, this relevance was maintained until, from the first years of the 1960s, a large number of classic thermal power plants started operating, and nuclear power plants started from the beginning of the 1970s, which meant that in 1975 hydroelectric production was only 35% of the total. In this century, the construction of combined cycle power plants and wind farms has led to the current situation, in which the installed hydroelectric power is 20% of the total and the coverage of demand is around 12% in an average year.

Thus, the installed hydroelectric power at the end of 2008 was 18 700 MW, from which 16 700 corresponded to the ordinary production system and 2000 to mini power plants under the special production system, over an installed total of all types of energy of 96 000 MW. Combined cycle power plants are those that provide the highest installed power to the group, whereas wind power is practically the same as hydroelectric power (**Figure 1**).

As for energy produced, hydroelectric power accounted for 26 000 GWh in 2008, from which 21 500 corresponded to the ordinary system and 4500 to the special system, compared to the nearly 295 000 GWh of the system's total net generation (**Figure 2**). These values are below average, the average being 35 000 GWh, as 2008 was a dry year.

The pluviometric irregularity that characterizes the Spanish territory results in irregularity of superficial runoff and, as a consequence, affects hydroelectric production. In 1979, good hydraulicity resulted in attaining an absolute maximum hydroelectric production of 47 473 GWh, which meant 45% of the total. On the contrary, the drought in 1992 resulted in the energy produced only reaching 20 750 GWh, which meant 13% of the total.

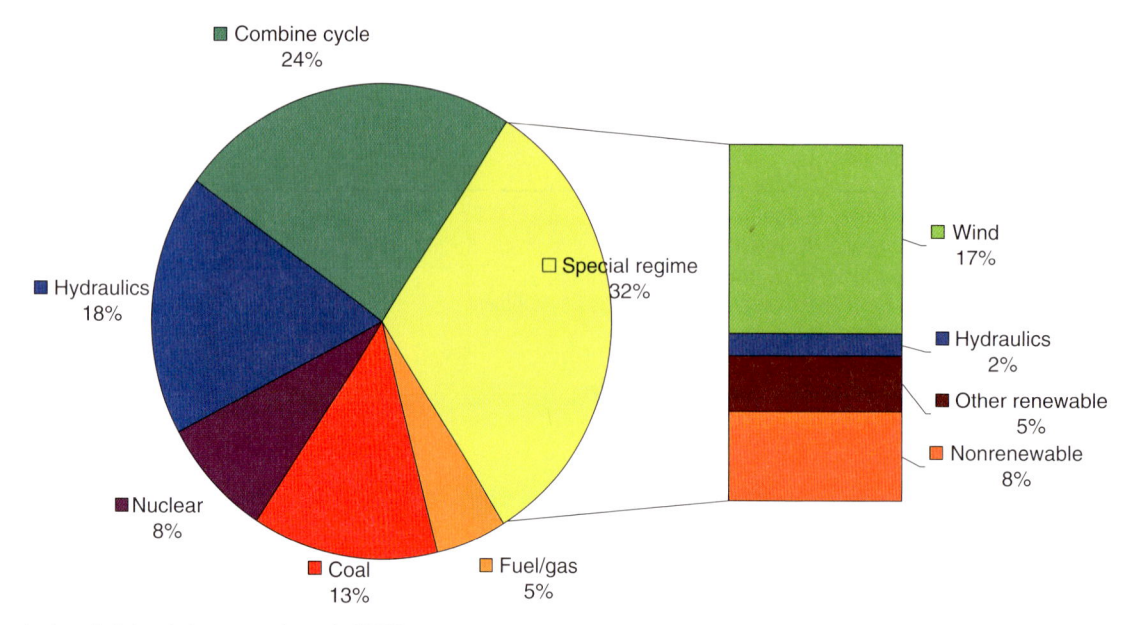

Figure 1 Installed electrical power at the end of 2008.

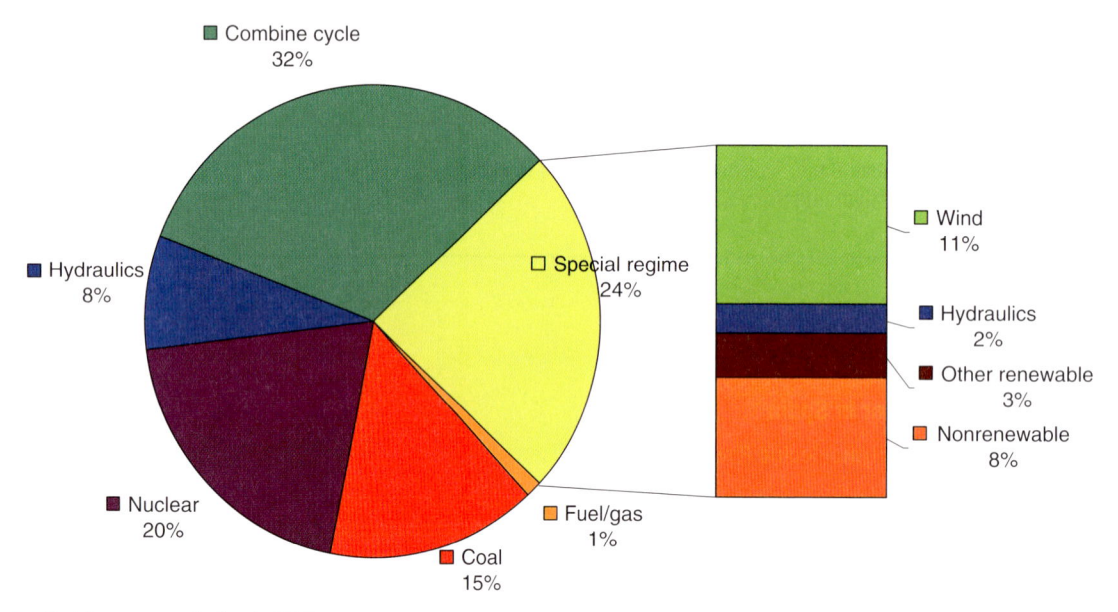

Figure 2 Electric power production in 2008.

6.11.1.2 The Strategic Importance of Hydroelectric Power

Hydroelectric power has a series of important qualities that make it one of the most strategically important energies from the technical, economic, and environmental points of view. From a technical point of view, due to its high degree of use in comparison to its potential, as the high efficiency of the turbines and alternators must be added to the low load losses in intake and return pipes, achieving a global efficiency of the plants between 85% and 90%, which has never been achieved in any other type of power plant. From the economic point of view, the cost of the raw material is very low or nil, which affects the total generation costs very favorably. From an environmental point of view, its main characteristic is in the nonemission of greenhouse gases. Each hydroelectric kWh avoids the emission of up to 1 kg of CO_2, 7 g of SO_2, and 3 g of NO_x. The average production in Spain is equivalent to not emitting 35 million tons of CO_2.

In addition, the developments related to regulation reservoirs and pumping provide a high quantity and guarantee electrical energy supply, facilitating load curve management and the regulation of frequency and voltage. They are also an installed power reserve in view of possible unavailability of other types of generation.

Besides, hydroelectricity is a source of energy in itself, an important fact in a country and high energy dependence. National hydroelectric production in an average year is equivalent to that obtained with 6 billion cubic meter of natural gas, 13.2 million tons of coal from abroad, or 9.3 million tons of fuel in plants that consume these fuels. The cost of avoided imports may amount to nearly €1100 million in the case of gas, €680 million in that of coal, or €1900 million in that of fuel.

6.11.1.3 Hydrology, River Network, and Hydroelectric Development

The average annual precipitation in Spain is around 650 mm and is characterized by its irregularity, both spatial and temporal. Spatial irregularity results in two differentiated areas: Wet Spain and Dry Spain (**Figure 3**). Temporal irregularity of precipitations results in that for any considered period – multiannual, annual, or seasonal – the gap between the maximum and minimum values is very big. To this we need to add a high evapotranspiration, which makes the average value of runoffs around one-third of precipitation. From the 330 000 hm^3 of precipitation, only around 110 000 become runoff.

All this results in the natural regulation level in Spain being close to 6–8%. The current regulation level is around 40–42%, for which it has been necessary to build more than 1300 large dams. Without them, economic and social development in Spain throughout the twentieth century would have been impossible (**Table 1**).

The Spanish hydrographic network is characterized, in a first approach, by the existence of rivers with two types of structure, some with a well-developed river network (considerably long tree-shaped riverbeds with a large number of tributaries) and others with rather parallel riverbeds and short in length.

Among the first we find the Miño, Duero, Tajo, Guadiana, and Guadalquivir that flow into the Atlantic Ocean, and the Ter, Llobregat, Mijares, Ebro, Júcar, Turia, and Segura that flow into the Mediterranean Sea. The second type are characterized for flowing in a perpendicular direction between the Cantabrian mountain ranges and the Cantabrian Sea in the north of the peninsula, and between the Andalusian mountain ranges and the Mediterranean Sea in the south. The proximity of these mountain ranges with the coast give these rivers characteristics of short lengths, steep slopes, perpendicularity to the sea, and the nonconnection between them despite being close to each other (**Figure 4**).

From the river network structuring point of view, in the first type not only has the full use of the main rivers with their tributaries been possible, but also in some cases full use of both has been possible. This layout of the river network has favored a higher use of

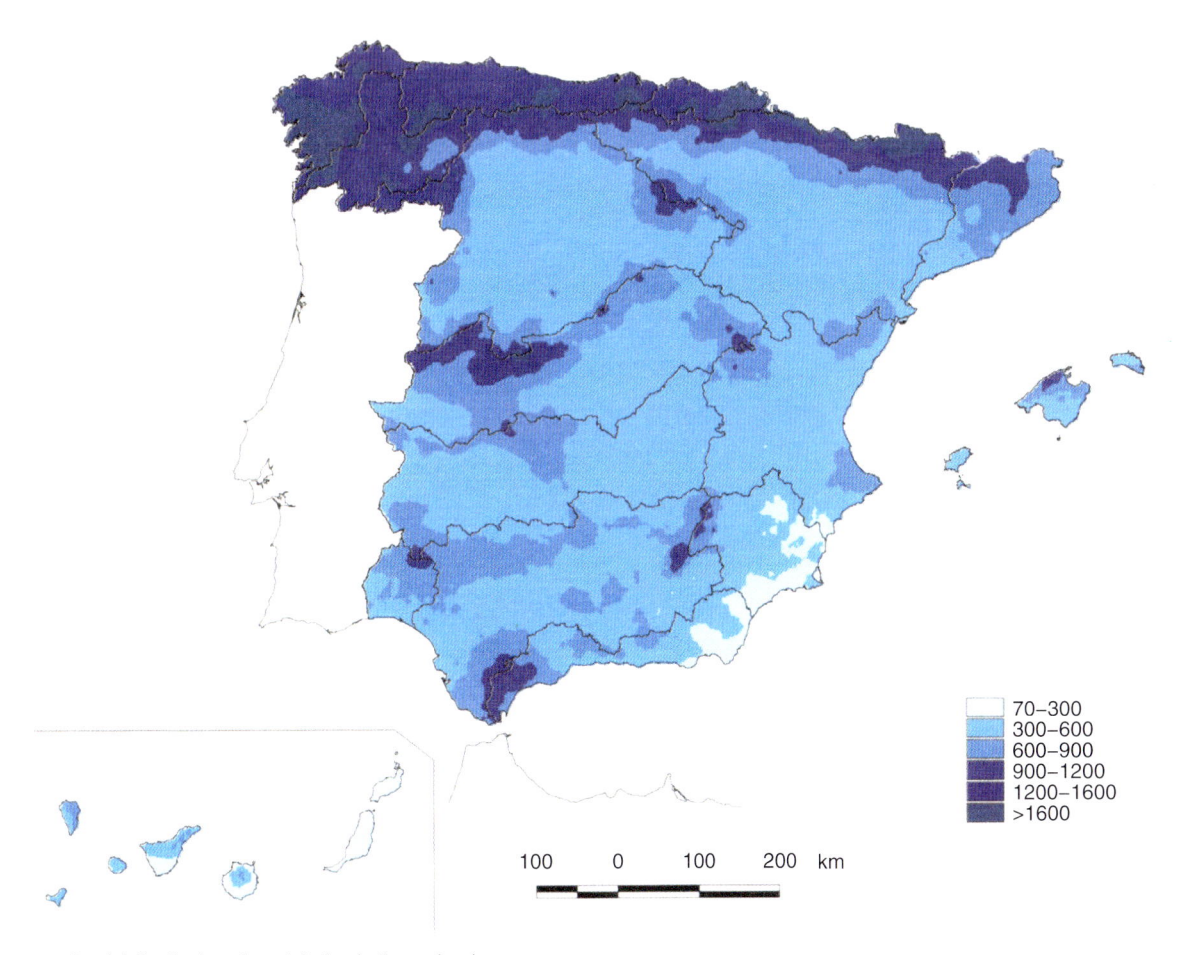

Figure 3 Spatial distribution of precipitation in the peninsula.

Table 1 Natural regulation and artificial regulation by hydrographic basins

Basin	Natural resources ($hm^3\ yr^{-1}$)	Natural regulation ($hm^3\ yr^{-1}$)	Natural regulation (%)	Reservoir capacity (hm^3)	Available resources (hm^3)	Available resources (%)
Norte I	12.689	916	7	3.040	3.937	31
Norte II	13.881	1.146	10	559	1.837	16
Norte III	5.337	251	6	122	353	8
Duero	13.660	742	6	7.667	6.095	49
Tajo	10.883	490	5	11.135	5.845	54
Guadiana I	4.414	44	1	8.843	1.922	47
Guadiana II	1.061	7	1	776	228	23
Guadalquivir	8.601	208	3	8.867	2.819	35
Sur	2.351	18	1	1.319	359	26
Segura	803	192	25	1.223	626	83
Júcar	3.432	771	28	3.349	2.095	76
Ebro	17.967	1.819	11	7.702	11.012	64
C. I. Cataluña	2.787	190	11	772	791	46
Galicia Costa	12.250	426	6	688	1.223	18
Total	110.116	7.219	8	56.063	39.175	41

Figure 4 Spanish hydrographic basins.

some of the tributaries, fed by high and medium-height mountains, than those of the main rivers, whose middle sections are less steep and whose use has been destined to irrigation. In the rivers of the second group, hydroelectric use has followed classic steep development schemes.

Hydroelectric development of the rivers in Spain has been conditioned by competition with other uses: that of supply and especially that of irrigation. Currently, 70–75% of the consumptive uses of water are destined to irrigation, which occupy the center of the Atlantic river basins and the lower sections of the Mediterranean rivers, with the consequential need of regulation reservoirs at the headwaters of its tributaries, on mountain fringes. The development of irrigation began in the early twentieth century, at the same time as the origins of hydroelectric development, being direct competitors in some lands.

6.11.1.4 Power Plants and Main Developments

There is a great variety of hydroelectric plants, both regarding the size and the facilities characteristics. In 2004, there were more than 1500 plants, including mini power plants under the special system. There were nearly 900 power generation units under the ordinary system.

There are five power stations with more than 500 MW and 21 with more than 200 MW, which represent more than half the installed power. Another 14 power stations exceed 100 MW and represent 12% of the power; those that exceed 50 MW represent 14% and those with less than 50 MW, including mini power plants, the rest (Table 2 and Figure 5).

The largest plants are those of Aldeadávila I and II, with 1243 MW, Jose M. de Oriol with 933 MW, Cortes-La Muela with 915 MW, Villarino I and II with 810 MW, and Saucelle I and II with 520 MW. The first, fourth, and fifth are located in the Duero System, the second in the Tajo river, and the third in the Júcar river.

Table 2 Hydroelectric power plants in Spain with an installed capacity of more than 100 MW

Hydro plant	Turbining capacity	River	Pumping capacity (MW)
Aldeadávila I and II	1.243	Duero	Pure 435
José María Oriol	934	Tajo	
Cortes – La Muela	915	Júcar	Pure 635
Villarino	825	Tormes	Mixed 825
Saucelle I and II	520	Duero	
Estangento – Sallente	451	Flamisell	Pure 451
Cedillo	500	Tajo	
Tajo de la Encantada	360	Guadalhorce	Pure 360
Aguayo	362	Torina	Pure 362
Mequinenza	324	Ebro	
Puente Bibey	316	Bibey	Mixed 316
San Esteban	265	Sil	
Ribarroja	263	Ebro	
Conso	270	Camba	Mixed 270
Belesar	258	Miño	
Valdecañas	250	Tajo	
Moralets	221	Nog. Ribagorzana	Pure 221
Guillena	210	Ribera de Huelva	Pure 210
Bolarque I and II	246	Tajo	
Villalcampo I and II	227	Duero	
Castro I and II	194	Duero	
Azután	200	Tajo	
Los Peares	168	Miño	
Ricobayo I and II	328	Esla	
Tanes	126	Nalón	Mixed 126
Frieira	154	Miño	
Torrejón	133	Tajo – Tiétar	Mixed 133
Salime	160	Navia	
Cofrentes	124	Júcar	
Cornatel	132	Sil	
Tavascán Superior	120	Lladore-Tabascán	
Castrelo	130	Miño	
Gabriel y Galán	111	Alagón	Mixed 111
Canelles	108	Nog. Ribagorzana	
Cíjara I and II	102	Guadiana	

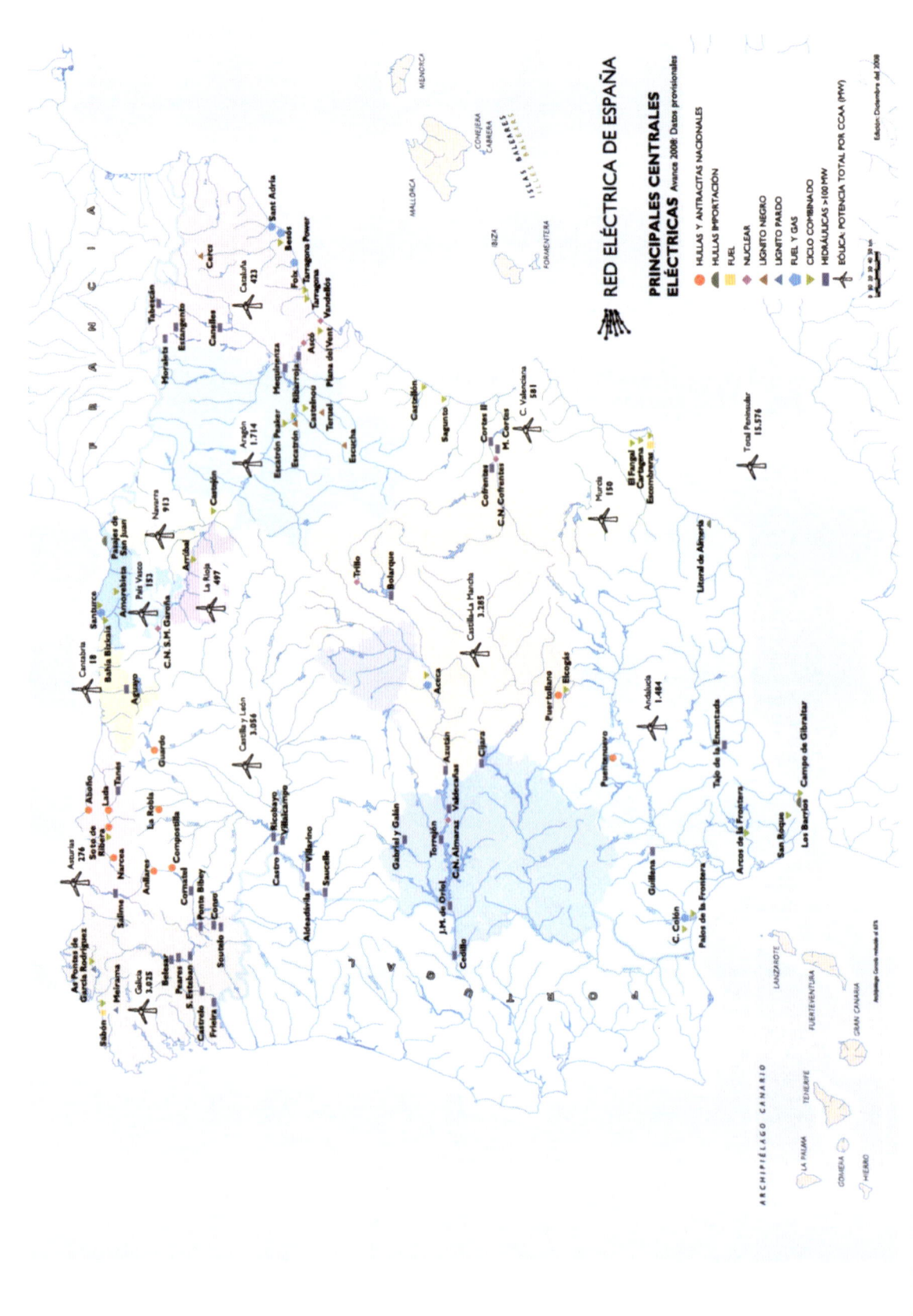

Figure 5 Location of Spain's main hydroelectric power plants.

Approximately 10 000 MW have a high seasonal regulation, of which 2500 MW are equipped with pumping. There are around 2350 MW in important power systems but with scarce regulation, and around 1300 MW in developments at the base of a dam. The rest of the hydroelectric facilities consist of small power plants, many of run of rivers.

Spain's hydroelectric potential is estimated at around 162 000 GWh yr^{-1}, of which a little over 64 000 GWh are technically usable. Taking into account that average yearly power production is around 35 000 GWh yr^{-1}, there is still technical margin available. The economically viable potential is estimated at 37 000 GWh, in accordance with the most recent data, not including the pumping plants. This means that Spain is close to the economically acceptable ceiling. However, some clarifications must be made. On the one hand, this ceiling is moveable, as it depends on the economic conditions not only of the jumps themselves but also, and especially, of the power production strategies at national level, which depend on the degree of dependence, on power vulnerability, on petrol and gas prices, or on the consideration of other types of plants' environmental costs, among others. On the other hand, power production is far from this level in the last few years, partly due to low rainfall. Exploitation during these dry years may be increased by building new facilities or improving the existing ones.

6.11.1.5 Producing Companies

The large electric utilities that have hydroelectric power stations in Spain are Iberdrola, Endesa, Gas Natural SDG, Acciona, E.ON España, and HC Energía. The last two concentrate their hydroelectric activity mainly in the north part of the peninsula, while the others distribute their facilities over greater areas of the national territory. A high number of other small companies must be added to these large ones, including those that have mini power plants with less than 50 and 10 MW and that are subject to the special electric production system. The Administration is also the owner of a high number of toe of dam schemes, most of these in dams intended for regulation for irrigation or integral regulation of rivers.

Iberdrola is the result of a merger in 1991 between Iberdrola and Hidroeléctrica Española and owns 9187 MW. Iberduero had its origins in Hidroeléctrica Ibérica, founded in 1901, and in Saltos del Duero, founded in 1918 with the purpose of exploiting the great hydroelectric potential of the Duero System. They were merged in 1944, soon after joined by Saltos del Sil, born to exploit the great hydroelectric site of the Sil river and its tributaries. Hidroeléctrica Española was founded in 1907, with its origins also being some of the schemes and concessions of Hidroeléctrica Ibérica. The main hydroelectric development of Hidroeléctica Española took place in the Júcar and Tajo basins.

Endesa, with a current installed hydroelectric power of 4511 MW, was created with public funding in 1944 with the purpose of helping the private sector in hydroelectric development. In 1983 the Endesa group was created, with the acquisition of some electricity companies such as Enher or Gesa, among others, from the National Institute of Industry. In the 1990s it acquired Electra del Viesgo, the historical Sevillana de Electricidad, Hidroeléctrica de Cataluña, and Fuerzas Eléctricas de Cataluña.

Unión Fenosa was the result of the merger between Unión Eléctrica and Fuerzas Eléctricas del Noroeste (FENOSA) in the year 1982. The first had its origins in 1889, with the creation of the Compañía General Madrileña de Electricidad, which after several groupings became Unión Eléctrica Madrileña in 1912. The second was created in 1943 to exploit several hydroelectric schemes in Galicia, in the northeast of Spain. Recently, Unión Fenosa has merged into Gas Natural as Gas Natural SDG has a hydroelectric power of 1860 MW.

Acciona acquired Energía Hidroeléctrica de Navarra and assets from Endesa, Saltos del Nansa among them, to achieve the 857 hydroelectric MW.

The presence of E.ON. is more recent, as it dates back to 2007 through its renewable energies affiliate and to 2008 as a market unit and as E.ON España, with 668 MW. This presence is due to the acquisition of assets from Ente Nazionale per L'Energia Elettrica (ENEL), who in turn had acquired the old Electra de Viesgo from Endesa, one of the Spain's historical companies created in 1906.

The historical Hidroeléctrica del Cantábrico has merged into the EDP group (Electricidade do Portugal) in the last few years under the name HC Energía. It has 433 MW of hydroelectric power.

6.11.2 Evolution of Schemes and First Developments

6.11.2.1 Periods in the Evolution of Development

The demand of electric energy has been increasing from the first days until today, with variable rates according to economic growth in general and the industrialization level in particular. The relationship between the industrialization and electric energy demand has always been similar. This demand has been met throughout the twentieth century with different types of power plants, among which hydroelectric power plants have had varied importance.

The distribution of generation to satisfy such demand between the different stations has depended on several factors, among which we must mention the hydroelectric potential and its level of exploitation, the cost of produced power and the environmental problems of the different power plants, or the strategic decisions to protect different sectors of the national economy, just to mention some of the main ones. The political and economic situation was also important in Spain in certain periods as it conditions the availability of equipment and building technology, as we will see later on. From a technical point of view, the main factors were those corresponding to the state of the art of the technologies available for those elements that affect hydroelectric developments: turbines, turbo pumps, or generators, as well as hydraulic engineering, dam engineering, and tunnel engineering, essential elements in Spain's hydroelectric development.

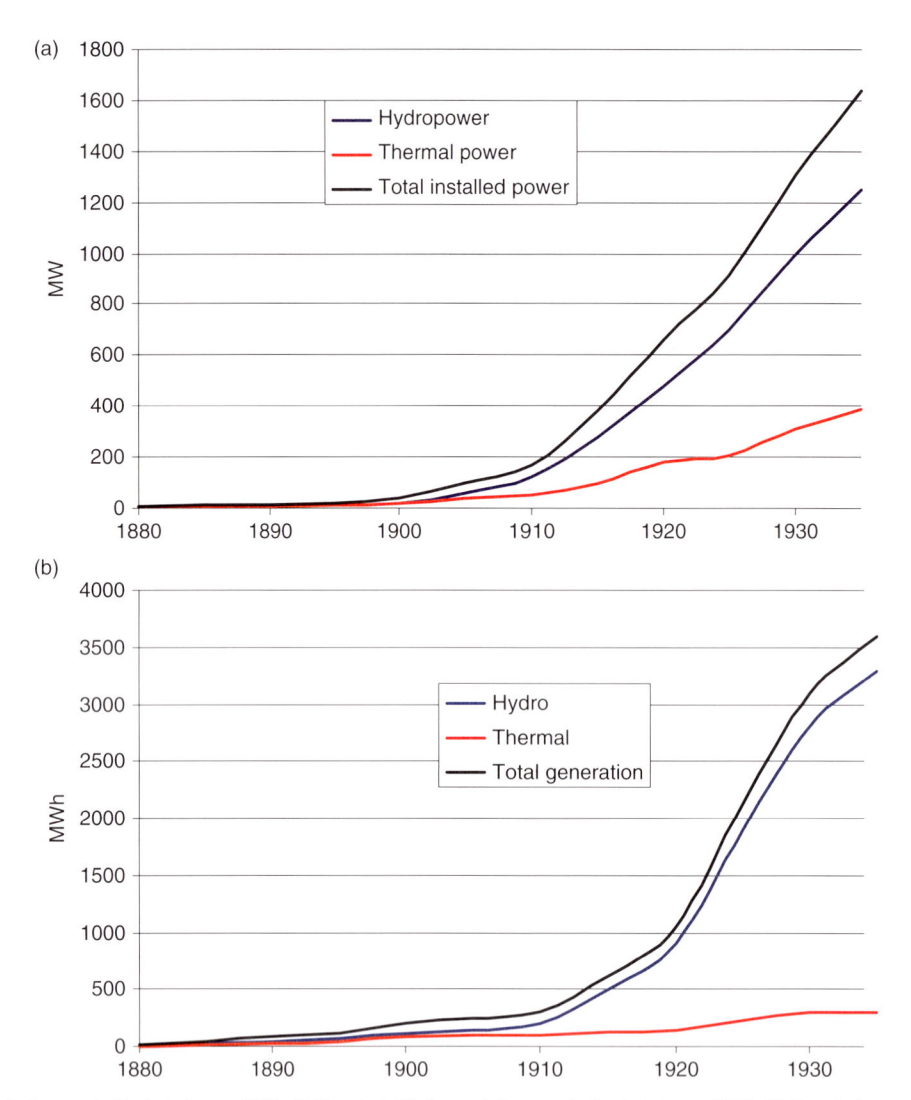

Figure 6 (a) Installed power in Spain between 1880–1940 period. (b) Generated power in Spain between 1880–1940 period.

All these factors have contributed to the variation of the development and the hydroelectric use systems and their importance in the electricity sector in Spain throughout the twentieth century, with clearly different characteristic periods.

The first period goes from the first steps of electrical energy in Spain to the civil war (1890–1940), characterized by an important increase of installed power and of produced energy, especially since 1910. After some years during which thermal power plants supplied most of the energy, with the new century it was the hydroelectric power plants that began to absorb most of the demand. In 1940, the installed hydroelectric power was 78% of the total and the generated power was 93% (**Figures 6(a)** and **6(b)**).

The 1940–73 period is characterized by the importance of hydroelectric production, which represented more than half the total electric production. From 1973, the increase of thermal production in classic and nuclear power plants made hydroelectric power stop being the main source, and has from then on lost relative importance in terms of energy (**Figure 7**).

If we take other factors into account, we can clearly distinguish two different phases within this period. One is from 1940 to the second half of the 1950s, characterized by the use of building equipment already used before the civil war, as a result of the autarchic politic and of Spanish isolation, which meant a certain continuation of previous productions.

The second period begins with the change of decade from the 1950s to the 1960s, during which the confluence of several factors allowed for the great development of hydroelectric power. The great development of dam engineering, the development of reversible power units, and the cost reduction in the construction of all types of works – hydraulic, underground, mechanical, and so on – due to the availability of new equipment, are the reason for the quick evolution in the development schemes. On the other hand, the more constructive and economic facility in underground works allowed for a greater flexibility in hydroelectric schemes, making it possible to build developments unthinkable just a few years before. From then, the construction of underground power plants and reversible power plants were commonplace.

From the middle 1970s, the increase of installed power continued increasing, but less than the classic thermal or nuclear power plants, despite of which the produced power stagnated to previous levels, with the exception of very favorable years in hydrological

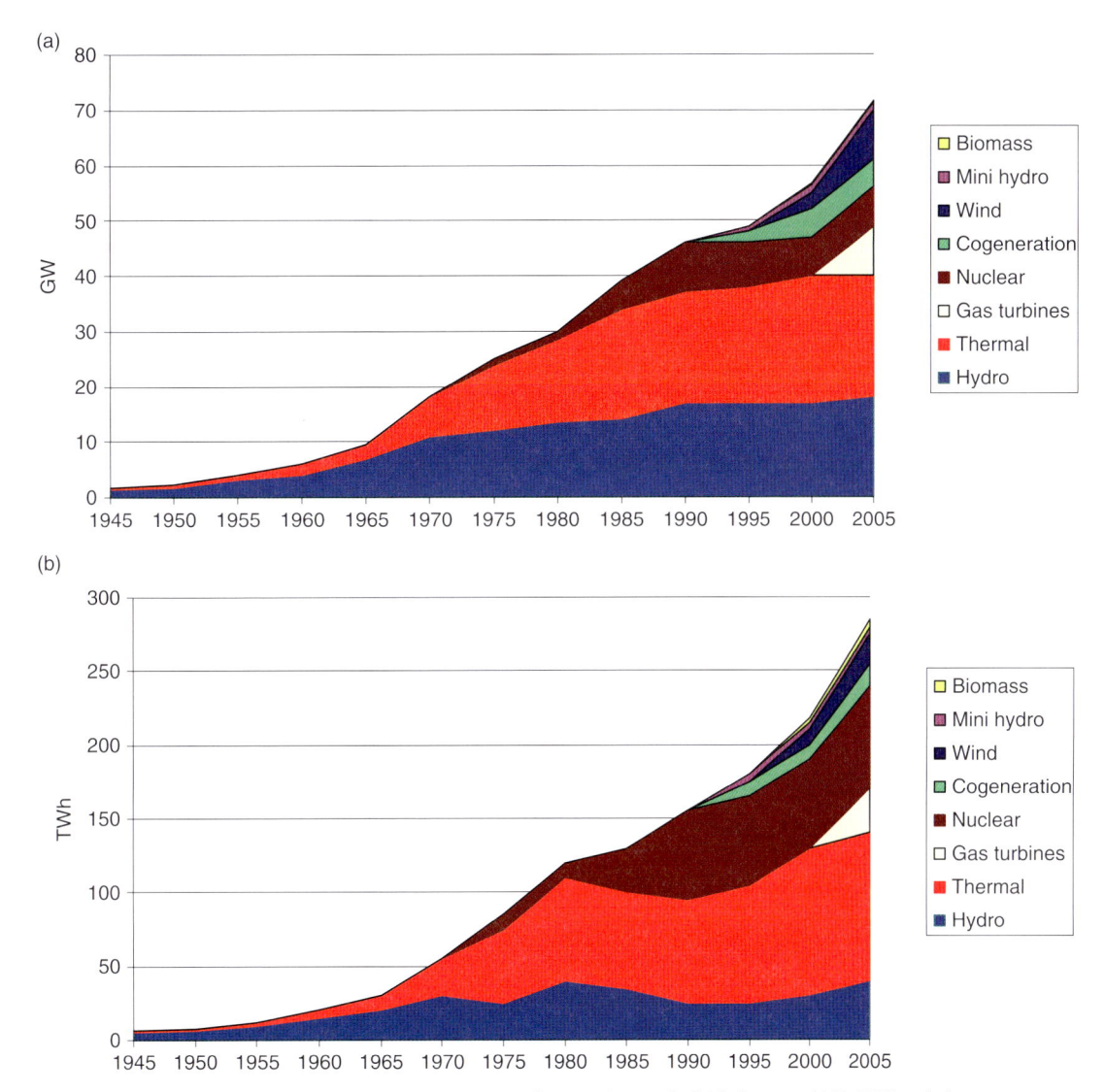

Figure 7 (a) Installed power in Spain between 1940–2008 period. (b) Generated power in Spain between 1940–2008 period.

terms. From 1990, the increase of installed power has been very little mainly due to electric system regulation installations and power increases in already existing plants (**Figures** 7(**a**) and 7(**b**)).

In the last decade of the past century, a construction process of small plants and the renovation of others that were not in use began. In the origins of this process we find, on the one hand, the development of more and more reliable power units and the remote control and/or centralization of operations, with the corresponding reduction of maintenance and operation costs and, on the other hand, the inclusion from 1997 of these plants in the special electricity production system, whose purpose is the promotion of renewable energy.

In each one of these periods, hydroelectric power and hydroelectric developments have had clearly different characteristics.

6.11.2.2 The 1890–1940 Period

6.11.2.2.1 First steps of electricity in Spain

The first reference to the practical application of electricity in Spain dates back to 1852, when the pharmacist Domenech lit up his premises in Barcelona with a method invented by him. That same year lighting tests were carried out in several public spaces using galvanic cells. From then on, and with a higher intensity in the 1870s, certain areas of some cities began to be lit up, generally using dynamos powered by steam engines.

The production of electrical energy in important quantities for that time began in the mid-1870s, using thermal power units powered by coal and low-quality gas. The first electricity supply contract dates back to 1876, Sociedad Española de Electricidad was the first Spanish electricity company. In 1878, several squares, streets, and important buildings in Madrid were lit up, and at the beginning of the next decade in cities such as Valencia and Bilbao. Increasing demand in the last two decades of the nineteenth century resulted in several companies being established with the only purpose of supplying electricity, both for public and private use.

In the last few years of the nineteenth century and in the beginning of the twentieth century it was supplied as direct current, which forced the power plants to be located near the consumption centers, which in turn limited the building of hydroelectric schemes and favored the use of thermal power plants. With the first use of alternating current in the last few years of the century and its great development in the first decade of the twentieth century, the limitations of the location of hydroelectric power plants disappeared, making the boom that took place during the second decade of the century for this type of energy possible.

In 1901 there were 861 electric power plants in Spain, with a total installed capacity of nearly 100 MW. Around 65% of them were thermal and 35% were hydroelectric, 650 mostly dedicated to public services, and the rest for private use. More than half of the total, around 510, produced in direct current and the rest in alternating current. The uses of this energy were around 87 000 incandescent lamps and 1500 arch lamps for public lighting, and 1 240 000 incandescent lamps, 2800 arch lamps, and 2036 engines for private use.

6.11.2.2.2 The electricity sector in the first decades of the twentieth century

In the 1901–30 period, the total electric power was multiplied by 12, reaching 1200 MW, which became 1600 MW in 1936, a number that slightly dropped at the end of 1939 as a result of the civil war destructions, which paralyzed the development of the Spanish economy. This big increase of installed power resulted in there being a small excess of installed production over consumption during this period. Increase in demand was variable: 8% between 1901 and 1922, 10% up to 1930, and 5% up to 1936, very similar values to those of economic and industrial growth.

If in 1890 installed hydroelectric power in Spain was 30% of the total electric power, this percentage went up to 69% in 1910 and 77% in 1920. This percentage would become stable until 1936, the year the civil war paralyzed electric development. As far as production is concerned, these percentages were even more favorable for hydroelectric power plants due to a higher amount of operation hours than those of thermal power plants. So, in 1929, 81% of production was hydroelectric. This percentage rose to 93% in 1936 (**Figure 6**).

Apart from the general use of alternating current, the origins of this development lie in the fewer total costs of hydroelectric power despite a higher initial investment, this was not only due to the zero cost of water but also to the increase in the cost of coal, gas, and petroleum products throughout these first few decades.

The construction of hydroelectric schemes of considerable size was generalized from the last years of the century's first decade to cope with the demand, which involved high economic investments. This was only possible due to the creation of a large number of electricity companies intended for production and distribution, which was led by private initiatives and with important participation of the banking sector.

6.11.2.2.3 Main hydropower developments

Between 1901 and 1902, the first large (for that time) hydroelectric plants began operating, among which we must mention those of Molino de San Carlos, near Zaragoza, Navallar, on the Manzanares river, the first to supply this type of power to Madrid, and San Román, on the Duero river near Zamora.

The Navallar power plant, with 1750 CV installed in four power units was the first of series of facilities that supplied power to the capital city from the rivers located to the north of it (**Figure 8**). The source of the scheme was the first dam of Manzanares el Real, with a height of 10 m and which was raised in 1906. A canal flowed from the dam which, apart from feeding the plant's surge chamber, served as a reservoir from which water was pumped to the city of Colmenar. It was one of the first important dams destined for hydroelectric and multiuse purposes.

The San Román power plant, which had an installed capacity of 5000 CV and took advantage of a long meander of the Duero river, with a 6 m-high dam and a 15 m head meant an important leap in terms of installed capacity, superior to those built until then.

Until 1910, a large number of similar schemes were built, most of them with installed capacities of 1500–2000 kW and located in medium river watercourses. Most of them were characterized by their diversion schemes, with a weir, a canal which was not too long and flowing parallel to the river, a small surge chamber and one or more horizontal configuration power units with Francis turbines.

In 1909, the Salto de Molinar, on the Júcar river, was inaugurated. It was the most remarkable hydropower project built in Spain until then. The development has been historically considered as one of the best and most profitable in Spain. It was projected for a production of 70 GWh, when Madrid's consumption was 30 GWh. Three 4500 kW power units were initially installed, in the following year an additional power unit was installed, achieving an installed capacity of 22 500 kW. If the development was

Figure 8 Navallar power plant (1902). Cross-section and inside the power plant.

Figure 9 Molinar power plant (1909).

important for its singular technical characteristics, it was also important for the transport line, 250 km long with a voltage of 60 kV, which supplied power to Madrid. This line was Europe's most important one at that time (**Figure 9**).

For similar purposes, the Villora project meant another leap in hydroelectric development in Spain. Located on a tributary of the Júcar river, the Cabriel river, it had an installed capacity of 12 000 kW. The scheme comprised a dam, a tunnel that crossed the watershed with a tributary of the Cabriel, and a power plant that hosted two horizontal axis power units. A second dam that regularized the turbined waters completed the system. In 1925, the increase in demand in the Madrid area made it necessary to install a new unit that doubled the installed capacity, which was the first vertical axis power unit installed in Spain. A new power unit, similar to the previous one, was installed in 1945, after the war, with the purpose of collaborating in relaunching the battered Spanish economy (**Figure 10**).

Figure 10 Villora power plant. In the foreground are the two vertical power units, one of them the first of this type in Spain. Behind and at a lower level are the two initial horizontal axis power units.

In the same Júcar–Cabriel system as the previous ones, in 1928 the Millares project works began, inaugurated in 1933 with two 20 000 kW vertical power units, those with the highest installed capacity in Spain until then, to which others were added in 1933 and 1942, achieving an installed capacity of 80 MW in 1942, which were very important at that time.

The power supply to Madrid had, apart from those already mentioned and several others, the Bolarque project, which was put into operation in 1910. Its scheme, even though it was also a bypass, was different from the others as it had a more regulated reservoir and a short high capacity canal of 60 m^3 s^{-1}. The connection line with Madrid was important, to which it supplied electric power, with a length of 78 km and a voltage of 50 kV. After completing in 1952, a rise with which the dam went from being 26 m above the riverbed to being 37 m above it, a power plant was built at the base of the dam, with pressure intakes, taking better advantage of the regulation and height. A pumping station was also installed to safeguard the watershed between the Tajo basin and the headwaters of the Tajo-Segura water diversion canal, with a north to south course crossing a large area of the eastern part of the peninsula, taking water for supply and irrigation purposes to an important part of the Mediterranean basins (**Figure 11**).

The first important hydroelectric development in Catalonia was that of Salto de Capdella, built from 1910 to 1914 in the Pyrenees, which supply power to the city of Barcelona. It uses the waters of the upper watercourse of the Flamisell river, a tributary of the Noguera Pallaresa river, in turn one of the main tributaries of the Ebro river. Collection and regulation takes place in 28 glacial reservoirs interconnected by means of canals and tunnels, all of them located between 2140 and 2534 m above sea level. Nine of these reservoirs were raised with 10–20 m-high dams. The lower reservoir, which collects the waters of all the others, is that of Estany Gento, from which nearly 5 km canal flows to feed two pipes that supply the water to the turbines. Four power units with Pelton turbines of 8500 CV each were initially installed.

Along with the singularity of collection and regulation, due to its height, we must mention the head of 836 m, the highest in Spain and one of the highest in Europe at that time (**Figure 12**).

Figure 11 The Bolarque power station and dam are in the foreground, the power plant is at the base of the dam. At the back is the pumping station for the Tajo-Segura diversion.

Figure 12 Capdella power plant: Diagram of the scheme and photograph of the power plant.

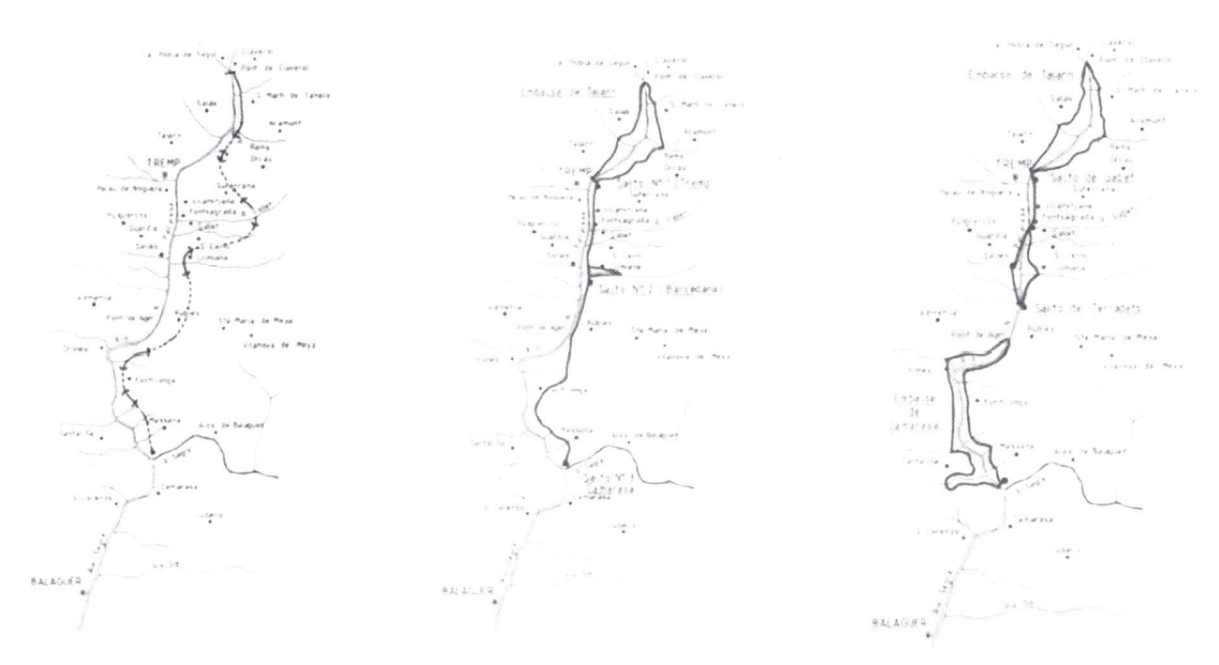

Figure 13 Evolution of development schemes in the middle watercourse of the Noguera Pallaresa.

Downstream from the Capdella power plant, three other power plants turbined the waters coming from it and from its own basins, one of them, that of Molinos, with a head of 273 m and an installed capacity of 13 500 kW. The system of these schemes was the same as interconnected bypass jumps, so they did not need a dam.

The next step in the exploitation of the Noguera Pallaresa did not follow this scheme. For the middle section of the river, between the confluence of the Flamisell and its confluence with the Segre, the first phase of the studies planned the construction of two diversion schemes with parallel canals near to the riverbed. The next phase of the studies was carried out after the unification of both concessions, both from the administrative and from the technical points of view. This solution seemed to be a complex one and it was decided that the first section was to be exploited with a dam and a station at its base: the Talarn power plant. The advantage of this solution was the achieved regulation, which would exploit more efficiently downstream (**Figure 13**). This type of evolution was commonplace during the first two decades of the century.

In 1916 the construction of the Talarn dam and power plant ended, which acted as the primary regulation of the Noguera Pallaresa river. It was the first large reservoir of hydroelectric use in Spain, with more than 200 hm³ achieved, thanks to the 86 m-high dam, which was the highest at that time and one of Spain's dam engineering milestones. The power house is located half way down the slope on the left margin and initially had four Francis power units with a total power of 30 000 kW (**Figure 14**). With this record, in 1920 the construction of the Camarasa dam ended. It was 103 m high and with spillway capacity for 2000 m³ s^{-1}. Like its predecessor, this dam also established a height record in Spain and was a new milestone in our country's dam engineering. The power station was equipped with four Francis power units, with a total power of 56 000 kW (**Figure 15**).

These two schemes of the Noguera Pallaresa were the beginning, wherever possible, of new way of conceiving hydroelectric exploitation of the rivers by means of their integral exploitation, which included the regulation with dams and reservoirs and steep

Figure 14 Talarn dam and power plant.

Figure 15 Camarasa dam and power plant.

schemes in large river sections by means of power plants at the base of the dams. In rivers with medium flow rates and in winding orography, this solution proved economically competitive in comparison to the classic diversion schemes, which, on the other hand, carried on being viable and were the best solution in many other cases.

Integral exploitation studies of basins and rivers began to be carried out from the 1920s. The most representative example is the one carried out by José Orbegozo for the exploitation of the lower watercourse of the Duero river and its tributaries in Spain, from the Esla to the Agueda (see Section 6.11.3). The construction of the 97 m-high Ricobayo was finished in 1935, with a reservoir, that exceeded 1000 hm^3 and an installed capacity of 133 MW. All these numbers were new milestones in Spain's hydroelectric development.

In the western Pyrenees, important projects were also built in the 1920s and 1930s. The most important ones had the same scheme: a medium regulation dam and high heads. Among these we must mention Lafortunada-Cinca-Pineta, with a 475 m head; Lafortunada-Cinqueta, with a 375 m head; Barrosa-Avellaneda, with a 205 m head; and Urdiceto, with a 426 m head. All these jumps were equipped with Pelton turbines.

In the Guadalquivir river basin and some of its tributaries, more projects stand out, more due to the characteristics of the dams than for the importance of the power plants. That is the case of the Cala power plant, which began being operational in 1927, with a 53 m-high dam, a maximum power of 13 000 kW and a 193 m head. One of the most beautiful dams built in Spain was built on the Jándula river in 1932. With a height of 90 m, its aesthetic characteristics are due to its neat plant and the location and design of the hydroelectric power station, which adhered to the downstream wall, anticipating designs that in the future the great André Coyne would use in his Dordogne dams and that would be adopted in the Grandas de Salime and Contreras dams in Spain (**Figure 16**).

On the Guadalquivir itself and during this period, three projects were built. These were Mengibar, El Carpio, and Alcalá del Río, in 1916, 1922, and 1930, respectively, with head created by barrages (gated dams) in each one. The meticulous designs and execution of the plants were a common characteristic of these projects. Mengibar was the first gated dam built in Spain (**Figure 17**). They were all part of the 'Channelling and exploitation of the Guadalquivir's power between Córdoba and Seville Project', and the ambitious project included 11 installations between both cities, each one with a power plant, a lock and a gated dam. Finally, navigability was discarded, and from the 11 installations, only four dams with their respective power plants were built.

In the Sur river basins, the El Chorro power plant and dam were built on the Guadalhorce river, in an abrupt location ending in an impressive gorge. Soon after, in 1927 the construction of the Gaitanejo dam and power station ended, a little downstream from the previous reservoir. It supplied power to the city of Málaga from the time after it began operating. The dams were the main elements of both developments: the first, 80 m high with a large regulation reservoir, and the second, of smaller dimensions was 20 m high, and very singular in design, with the power station included in it and the spillway above it (**Figure 18**).

Figure 16 Jándula Dam. View during construction and cross-section.

Figure 17 Mengibar power station and dam.

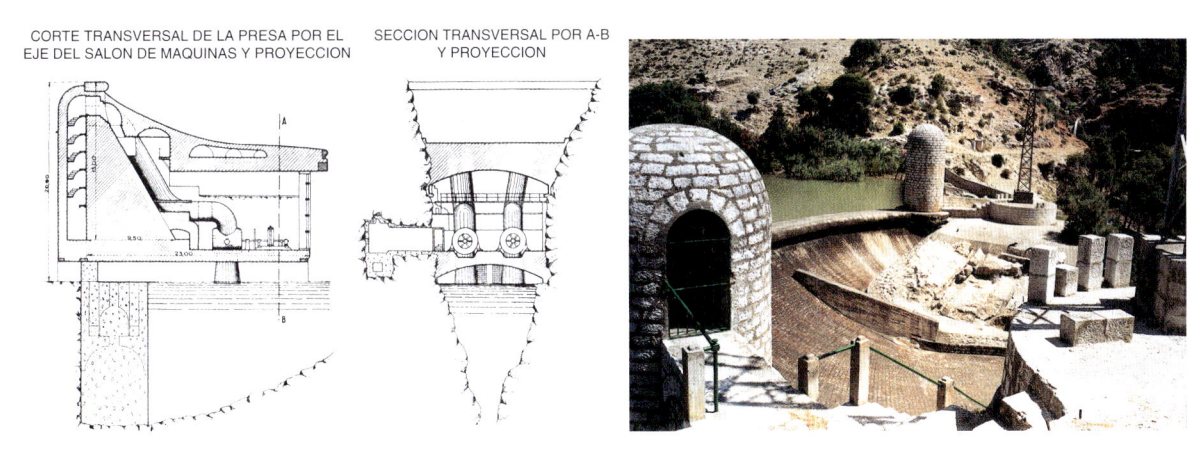

Figure 18 Gaitanejo dam: Cross-section and photo.

Next to the previous one and near the city of Ronda we find the Montejaque dam, which was a 84 m-high first modern arch dam built in Spain. It was finished in 1924, and was the top of a hydroelectric scheme which had to be abandoned soon after, as the reservoir could not be filled due to the high leakage through the calcareous basin.

6.11.2.3 The 1940–60 Period

6.11.2.3.1 *The electricity after the civil war*

After the civil war, the situation in Spain was characterized by the economic isolation from abroad, in turn a consequence of two situations that fed each other. On the one hand, the international embargo and, on the other hand, the desire of the regime to become a self-sufficient autarchy, aside from the international economy. To this we need to add the state of the economy after the physical destruction of industries and facilities, and the disappearance of the emerging industrial infrastructure and, in the first years, the consequences of the Second World War resulting in generalized poverty in Europe.

All of this resulted in having to use the same building equipment as that used in the previous decades, a lot of which was in poor condition and needed to be repaired; despite this they carried on using them and achieved good results.

During the years of war (1936–39), the energy production capacity stalled, although some power plants continued to be operative, others were damaged or destroyed. After this period there was a severe drought, from 1944 to 1946, which resulted in a considerable decrease of hydroelectric production. In this way the production capacity excesses of the previous decade turned into important production deficits versus demand, whose growth values in that decade were very high due to the logical reconstruction needs.

The installed capacity went from 1731 to 2553 MW in the 1940–50 period, due to the new hydroelectric facilities. In the 1950–60 period, it rose to 4600 MW. In this case from the 2000 MW installed, 1300 belonged to thermal plants, which especially occurred in the second half of the decade. The application of a new rate system, which allowed companies to initiate new investments, influenced the higher increase of this decade. With this, the usual restrictions began to disappear, and totally did so from 1958 (**Figure 7**).

The end of the international embargo and a certain economic liberalization gave way to the 1959 Stabilization Plan, which was enforced by the international economic institutions to provide assistance. With this assistance, the two development plans established in the following decade allowed for the Spanish economy's big take off.

6.11.2.3.2 *Main hydropower developments*

As a result of what has been previously mentioned, Spanish hydroelectric development can be considered as a continuation of the previous periods. However, this continuation was optimized due to the experience acquired in previous decades and to the Spanish engineers' know how. With few means they were able to build remarkable constructions for that time. As an example we must mention Villalcampo, Castro, and Saucelle projects, the second phase of the building of the Duero System, that were put into operation between 1949 and 1956 with capacities of 206, 190, and 240 MW, respectively. Those projects were built in difficult execution conditions, both in economic and in technical terms (see Section 6.11.3).

Important hydroelectric systems began to be developed during this period, which in some cases meant the integral exploitation of basins. This is the case of the Sil System, whose first steps were taken during these years and which was completed from 1960 with more complex and advanced installations. Between 1952 and 1960, nine power plants were built in it, with seven dams (**Figure 19**) and a total installed power of around 450 MW, from which more than half corresponded to the power station at the base of the San Esteban dam, with 266 MW.

The dams were an important part of these first power plants of the Sil System. Seven of them have heights between 22 and 42 m and functions of forebay, with small or medium regulation for diversion schemes. The other two are more important, that of San Esteban and that of Chandreja, with relatively small reservoir capacities for height, and whose main purpose is to create height head in order to locate the power plants at their toe. That of San Esteban is a 115 m-high arch-gravity dam, and the most modern means began to be used. We could say that it is a transition dam between this period and the next. That of Chandreja is an 85 m-high buttress dam. This type of dam was used frequently in Spain during that period, due to the lesser need of the scarce concrete and the higher manpower, which were abundant and cheap in those years (**Figure 20**).

There are numerous power plants and dams of both types from this period in Spain especially with small and medium height dams, and diversion schemes of all types. Those located in high plateaus in the Pyrenees stand out. This is the case of the Baños power plant, fed by several dams built between 1942 and 1961 in plateaus located between 2100 and 2550 m above sea level, or that of Moncabril, fed by several reservoirs raised by means of dams in glacial lakes in Sanabria (**Figure 21**) with a net head of 514 m.

The first important dam in that period was the Grandas de Salime, located on the Navia river in the north of Spain. It was built between 1946 and 1955 in abrupt terrain, and is 128 m high, creating an important scheme of 160 MW. This dam was a milestone in Spanish dam engineering at that time due to how quickly it was built despite the rather inefficient construction means they had, reaching a European record for daily and weekly concrete laying. The location of the power plant stands out, at its base and under

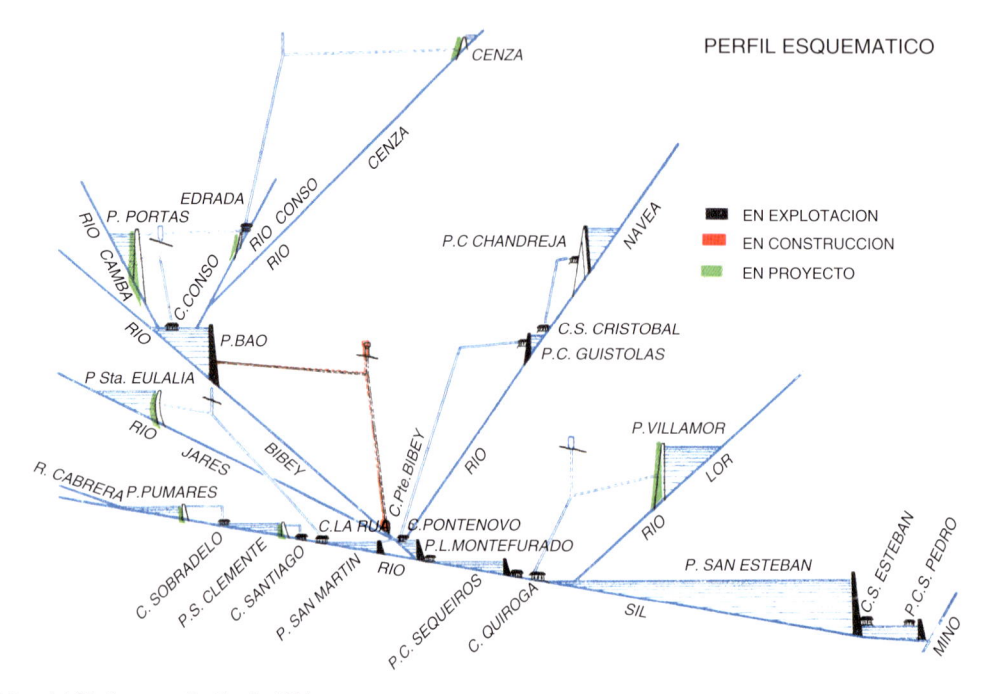

Figure 19 Saltos del Sil diagram, situation in 1961.

Figure 20 San Esteban and Chandreja dams.

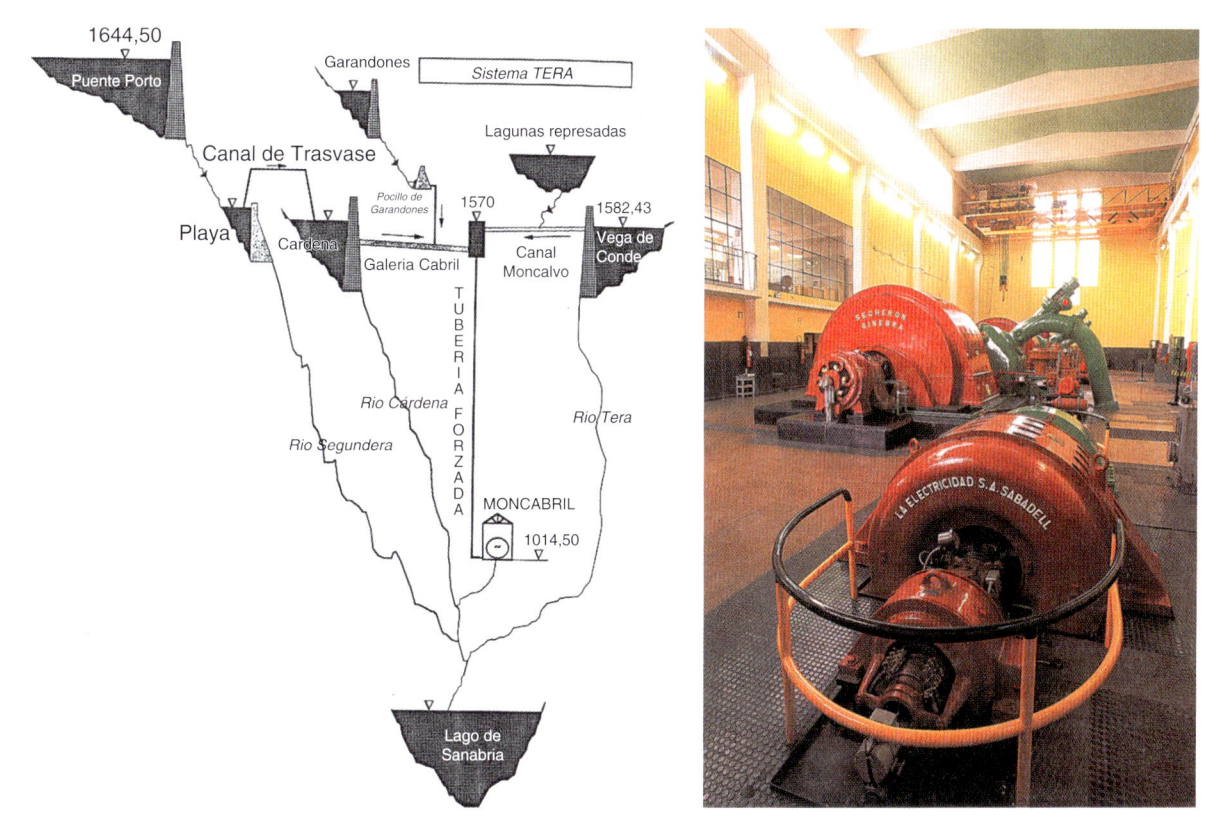

Figure 21 Diagram of Salto de Moncabril in Sanabria and inside the power plant.

the spillway designed for a flow of 2000 m^3 s^{-1}. This arrangement, contrary to the hydraulic interests of flood outflow, required several model tests, achieving an effective design (**Figure 22**).

In the Pyrenean Noguera Ribagorzana, parallel to the Noguera Pallaresa, we find, among others, the power plants of Escales and Canelles, both with an underground power house near them, the first important ones of this type. The first is a 125 m-high gravity dam with a 118 m head. The turbined waters of this power plant also make use of the Puente Montañana power plant, which, in turn, makes use of the necessary compensating reservoir located downstream. Downstream we find the 150 m-high Canelles dam, built between 1958 and 1961. It is a peculiar arch dam due to its irregular shape, located in a narrow calcareous area and practically entirely designed by means of structural tests on models. Once it was finished, important injection screens had to be carried out on the left hillside in order to stop the important water leaks from the basin (**Figure 23**), as well as a great reinforcement on the right side as a result of new studies carried out due to the breaking of the Malpasset dam. The power plant is a cavern on the left side and has an installed power of 108 MW. With this one and the other two previous power plants, the Noguera Ribagorzana river is totally regulated from the hydroelectric point of view.

Figure 22 Grandas de Salime power plant and dam.

Figure 23 Canelles jump power plant and Canelles dam.

There are two other arch dams that we must mention: La Cohilla, 116 m high, whose purpose is the general regulation of the Nansa river for the power plants located downstream (**Figure 24**), and Eume, the first modern Spanish arch dam, with a 103 m height and a 55 MW power plant downstream with a 245 m head. The first was built with very scarce means. The second stands out for its design and for being the first arch projected with modern criteria and parameters (**Figure 24**).

The hydroelectric development of the middle-lower section of the Ebro, the main Spanish river, dates back to this period, although the last of those projects was built in the 1960s. The Flix power plant was built in 1948, next to the barrage (gated dam) of the same name, which is 26 m high. The plant has a power of 40 MW, despite the small 12 m head, thanks to a maximum flow of 400 m^3 s^{-1}, exploited with four vertical axis Kaplan power units.

The study of the section located upstream of the latter was performed over the first few years of the 1950s, with the building of the Mequinenza dam at the end of the decade and that of Ribarroja at the beginning of the next decade, with the object of staggering the construction of the complex. The conception of the schemes can be considered as classic, far from the criteria that would have been followed in the following decade.

These are two gravity dams in two wide enclosures that allowed for a 'comfortable' design, as the power plant could be placed at the toe of the dam, as well as the large spillways suitable for draining floods of 11 000 and 8500 m^3 s^{-1}. In both cases, the transformer stations are located on the vast banks. The Ebro's large flow rate allowed for high power plants, 324 and 256 MW, respectively, with Francis turbines in the first and Kaplan turbines in the second (**Figure 25**).

The construction of supply systems to towns and cities was important in these two decades. This is the case of the improvement of Bilbao's water supply by means of the interbasin diversion from the Zadorra, a tributary of the Ebro, and the construction of two

Figure 24 La Cohilla dam during construction and Eume dam.

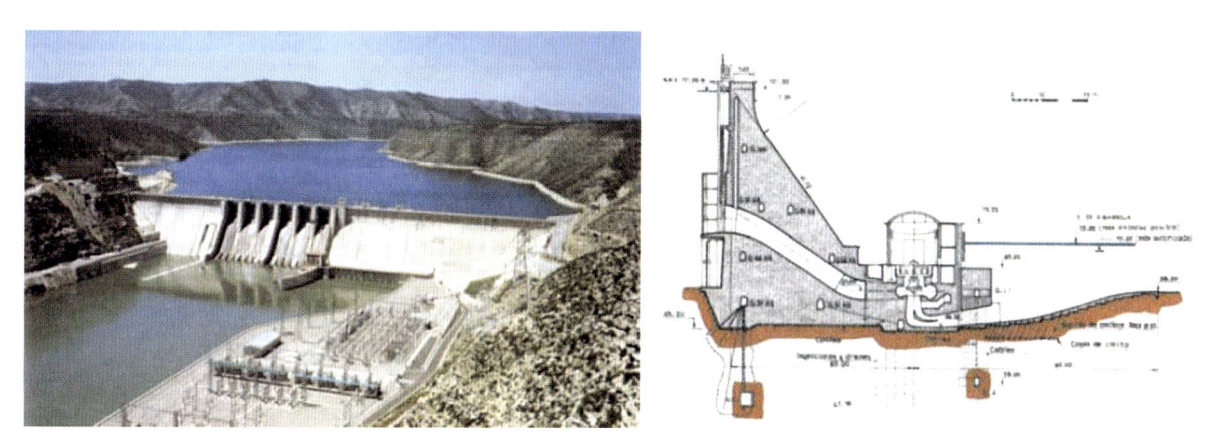

Figure 25 Mequinenza power plant and dam.

dams. The diversion of the waters, with a very constant flow, is turbined to the Barázar power plant, of 83 MW, taking advantage of a 330 m height variation levels, resulting in an average power production of 170 GWh yr^{-1}.

In many of the rivers where dams and power plants were set up before the war, they continued building during this period by making use of the existing regulation and infrastructures.

6.11.2.4 The 1960–75 Period

6.11.2.4.1 The electricity sector

The 1959 Stabilization Plan, which established a stable framework for growth, the First Development Plan of 1964 and the opening to the 'outside world', which favored incoming currencies and commercial exchange, was the source of the great Spanish economic development from 1960. The electricity sector was an active collaborator in this development, quickly adapting to the demand. The previous existence of a fairly large and comprehensive supply network and its extension throughout these years was a decisive factor. In these years they achieved to electrify practically the entire national territory.

The installed power rose from 6600 MW in 1960 to 18 000 MW in 1970 and to 25 500 in 1975, hence four times as much in 15 years. Production also rose from 18 600 GWh in 1960 to 56 500 in 1970 and to 82 515 in 1975, five times as much in the same period of time.

During these years, hydroelectric power units were greatly developed, from 4600 MW installed to 12 000. However, the high increase was due to the building of new thermal plants, both conventional fuel oil plants in a context of low oil prices, and nuclear power plants, with the start up of José Cabreras in 1968, of 160 MW, Garoña in 1971, with 466 MW, and Vandellós I in 1972, with 500 MW (**Figure 7**).

In 1973, the first oil crisis took place, which multiplied the source price by six, despite that the important fleet of thermal power plants under construction used this fuel. Spain only reacted to this in 1975, when the National Energy Plan was passed, but effective measures were not taken to change the energy model, which was indeed done one decade later, in 1983, with the II National Energy Plan.

6.11.2.4.2 The golden age of dam engineering in Spain

Since the Roman times the construction of dams in Spain has been common, due to the semiarid nature of a good part of the territory and the temporal irregularity of precipitation and runoffs. Since then, and throughout all the periods, dams have been essential in social and economic development. Dam engineering has always been level with the world's best with important and numerous constructions, even during unfavorable periods, as previously mentioned. This important and continuous experience reached its climax from 1960 with the conjunction of all the mentioned factors, which favored the building of important dams with different purposes. We can therefore affirm that the 1960s and 1970s were the golden age of dam engineering in Spain. From all the dams of this golden age, those built for hydroelectric purposes stand out, most of them among the highest dams built in Spain (Table 3). This golden age of dam engineering was so especially in the construction of arch dams in general and double-curvature dams in particular, but also with a few important examples of buttress dams and embankment dams.

In the Duero System the arch-gravity dam of Aldeadávila, 140 m high, and the double-curvature dam of Almendra, 202 m high, were built in this period. The first was finished in 1963 and is located in an impressive granite canyon formed by the Duero river. A double-curvature dam was the first selection, but the high floods made them opt for the arch-gravity type. It is one of the most beautiful Spanish dams, not only for the surroundings but also for the dam itself. In 1970, the Almendra double-curvature dam was finished, the highest in Spain, which rises beyond the closed topography, thanks to two gravity abutments on which it is supported. For the closure of the lateral troughs, a buttress dam and an embankment dam with a bituminous concrete face were built (see Section 6.11.3).

In the Sil Sytem, the construction of the Santa Eulalia ended in 1967. It had the slimmest and most curved double-curvature dam ever built in Spain (Figure 26). Most of the building equipment that was later used to build the great Almendra dome was fine tuned here. In the same system, the Las Portas double-curvature dam was finished in 1975, with a height of 141 m, which was the third highest hydroelectric dam (Figure 26).

The Belesar dam, on the Miño river and 132 m high, and the Valdecañas dam, on the Tajo river, 78 m high with a singular arrangement, with the power station at the base of the dam, protected by a small arch cofferdam and with spillways in tunnels on both sides (Figure 27), were built in 1963 and 1964. Other double-curvature dams built during these years included La Jocica (1964), 87 m high and very narrow, and that of La Barca (1966), 74 m high, over the Cantabrian Narcea river. The Susqueda dam (1968) was built over the Ter river, with a height of 135 m.

Among the buttress dams we must mention that of José María de Oriol (1969), associated to the Alcántara reservoir, with a height of 130 m and a double-buttress or 'Marcello' type, which was a world record for its height in this type of dam until the Itaipu dam was built. This dam changed the single-buttress dam type used in profusion throughout the 1950s and 1960s by hydroelectric companies. It was also the last important dam of this type built in Spain (Figure 28).

Gravity dams continued to be the most used for lower heights, while embankment dams were used to a lesser extent, of which that of Portodemouros (1967), 91 m high, stands out. Among the gated dams, we must mention those of Velle, Castrelo, and Frieira, all over the Miño river, with heights in the 25–35 m range and built in the 1960s.

Table 3 Main hydroelectric dams in Spain

Dam	Height (m)	Typology	Year	River	Location	Reservoir capacity (hm^3)
Almendra	202	VA	1970	Tormes	Salamanca	2.649
Canelles	151	VA	1960	Noguera Ribagorzana	Huesca	687
Portas, Las	141	VA	1974	Camba	Ourense	535
Aldeadavila	140	VA-PG	1963	Duero	Salamanca	114
Susqueda	135	VA	1968	Ter	Gerona	233
Belesar	132	VA	1963	Miño	Lugo	654
José María de Oriol	130	CB	1969	Tajo	Caceres	3.162
Escales	125	PG	1955	Noguera Ribagorzana	Huesca	153
Salime	125	PG	1956	Navia	Oviedo	266
Cohilla, LA	116	VA	1950	Nansa	Santander	12
Cortes II	116	VA-PG	1988	Jucar	Valencia	118
Matalavilla	115	VA	1967	Valseco	Leon	65
San Esteban	115	VA-PG	1955	Sil	Ourense	213
Bao	107	PG	1960	Bibey	Ourense	238
Eume	103	VA	1960	Eume	Coruña, A	123
Ricobayo	99	PG	1934	Esla	Zamora	1.150
Doiras	95	PG	1934	Navia	Oviedo	96
Tanes	95	PG	1978	Nalon	Oviedo	33
Peares, Los	94	PG	1955	Miño	Lugo	182
Portodemouros	91	ER	1967	Ulla	Coruña, A	297

Figure 26 Santa Eulalia and Las Portas dams, in the Sil System.

Figure 27 Valdecañas dam.

6.11.2.4.3 Main hydropower developments

From a technical point of view, the important hydroelectric development in this period was based on three cornerstones: the construction of large dams, the construction of underground power plants, and the development of pumping installations. The first enabled the creation of higher heads and a higher regulation. The second, due to the development and availability of building equipment and building techniques, enabled a higher flexibility in the arrangement of development schemes without the technical or environmental conditionings of surface locations. The third enabled for the development of electric power storage schemes in the way of potential hydraulic power. This was an important factor in an electricity system such as the Spanish one which was beginning to generate more through thermal plants. The pumping installations of that time continued to be used for decades and are actually still in use, although with different criteria.

At present, and as explained further on, we need pure pumping facilities that allow us to supply very concentrated peaks and with a relatively low number of operative hours. In the 1960s and 1970s pumping was understood as a storage means, by means of pressure diversions, in lateral basins of the main riverbeds, where the reservoir capacity and the flow availability did not coincide geographically. With the change of generating model, this concept changed toward the pure pumping facilities in the 1980s with the purpose of serving as system regulation and to be able to supply the strong peaks. This does not mean that both approaches are different, but complementary, when not coincidental.

The main pumping developments carried out in this period are connected to three of the main Spanish development schemes: the Duero System, the Sil System, and the Tajo Inferior Development. The first development was carried out from the first decades of the twentieth century, and was continued during the 1940s and 1950s with classic schemes, while the second began its development

Figure 28 JM de Oriol Dam.

after the war, as we have already pointed out. But the definite boost during this period in both of these was, to a great extent, due to the application of pumping. The third, that of the Tajo, was developed in a concentrated manner between 1960 and 1975, and had pumping as its main element from the start.

The Sil System takes advantage of the waters of the Sil river, as well as of most of its tributaries. It currently has an installed power in generation of 1270 MW and in pumping of 400 MW, with 19 power plants, 45 power units, and 17 large dams, all built between 1952 and 1994. It has the highest concentration of hydroelectric developments in Spain (**Figure 29**).

There are three pumping stations in the system: Camba-Conso, Bao-Puente Bibey, and Santiago-Jares. The first (1975) is the main one, with an installed power of 230 MW, a flow of $120 \, m^3 \, s^{-1}$ and a 230 m head between the power house and the large multiannual Las Portas reservoir, which confers a great power reserve for all the power plants located downstream, among which is that of Bao (1964), with one of the four pumping units with a ternary arrangement. This station, along with that of Aldeadávila I, built at the same time, were the first step in the application of the modern underground excavation techniques applied in hydroelectric power plants and galleries in Spain (**Figure 30**).

The analysis of the study phases of the Tajo Inferior Development is interesting, as it highlights the change in the design of hydroelectric systems that took place in only a few years. Downstream from Toledo, the Hidroeléctrica Española concession allowed for the exploitation of the river itself, with a total height variation of 280 m (**Figure 31**). The top 40 m were exploited with a conventional power plant in accordance with the existing easements, fundamentally agricultural and for town use.

For the lower 240 m there were three solutions. The first was a conventional scheme, with four development steps by means of dams and power plants at their base. Subsequently, the inclusion of the exploitation of one of its tributaries, the Tiétar, was

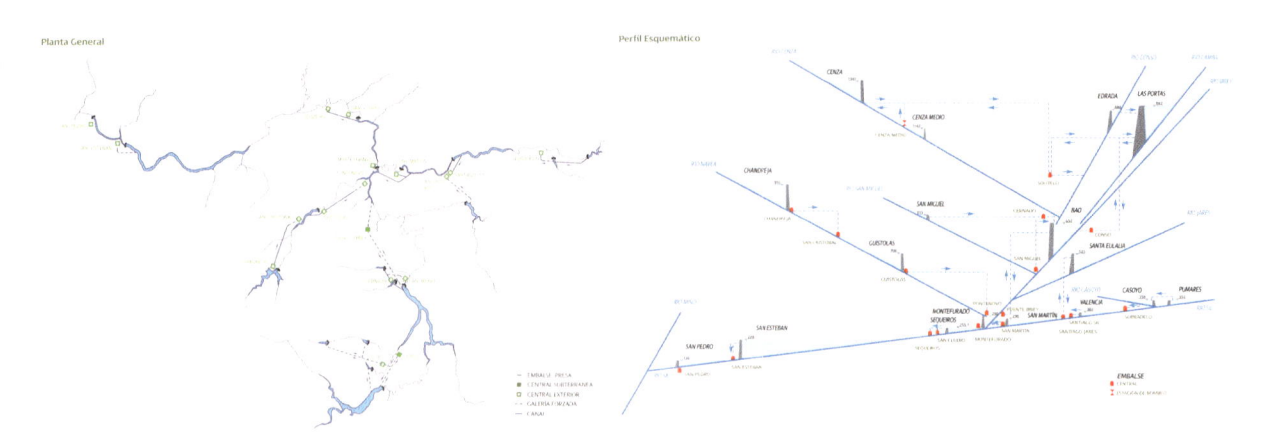

Figure 29 Current schematic plan and profile of the Sil System.

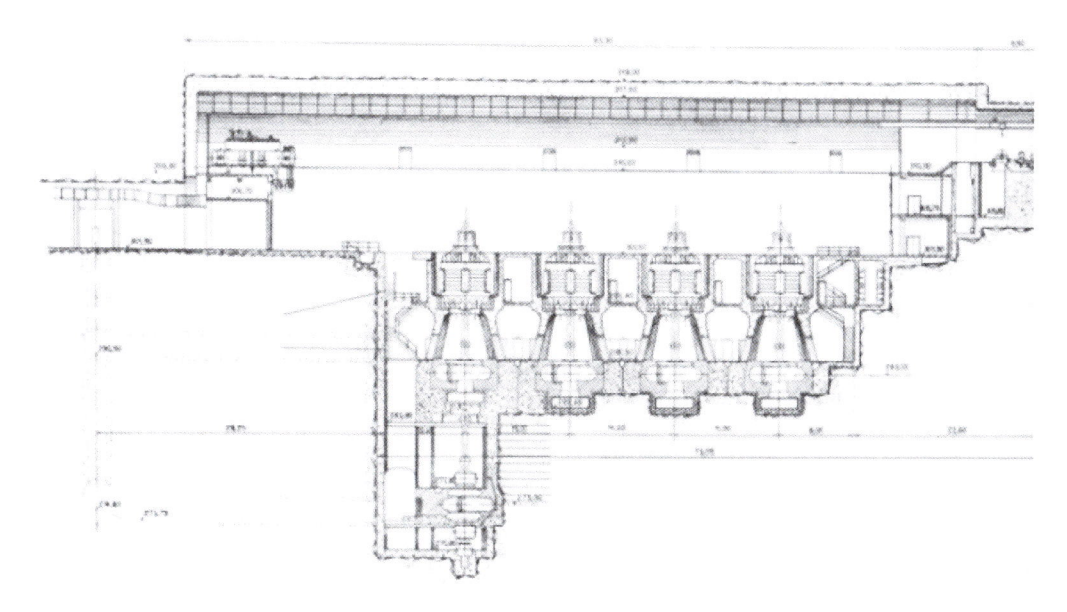

Figure 30 Bao power plant. Cross-section, in which the difference between the three generation units and the pumping-generation unit are shown.

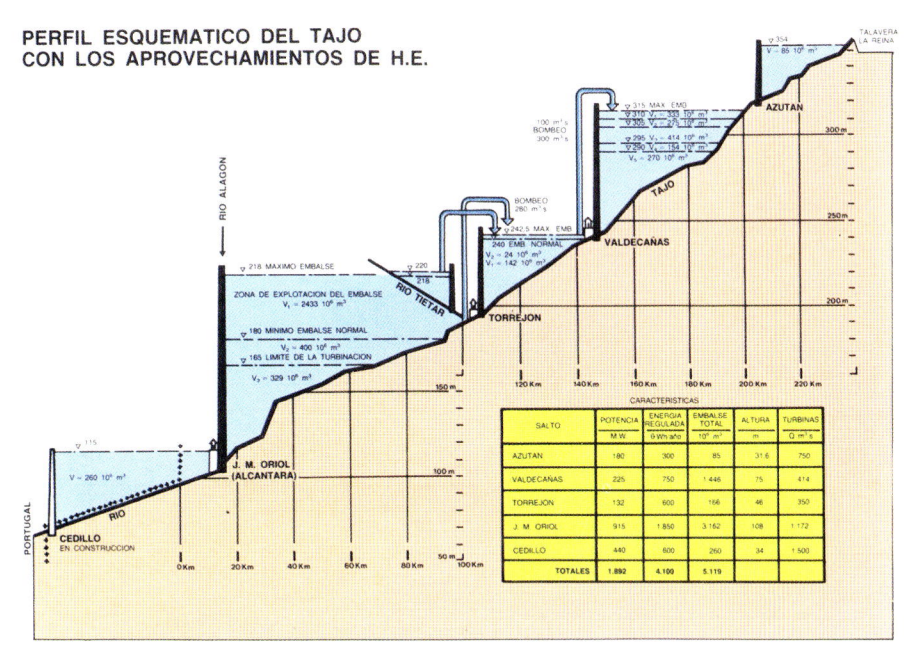

Figure 31 Scheme of the Tajo project, situation in 1964.

considered. And finally, soon after the previous one, the possibility of including pumping was considered, which was indeed done in two of the power plants.

On the basis of the need to create a large regulation reservoir, it was situated as far upstream as possible, resulting in the Valdecañas dam. In order to create a minimum head of 50 m, a volume of 270 hm^3 was sacrificed, while the oscillation between a height of 50 and 75 m created a regulated water reserve of 1275 hm^3, not excessive if we take into account that, although it is one of the main ones in Spain, it has annual irregularities of 1–6 and monthly of 1–140. As downstream the Tajo river did not receive important contributions, the construction of more regulation reservoirs was not considered, using the following step, of 46 m, with a second dam, that of Torrejón, with a power plant at the toe of the dam.

This scheme was considered as satisfactory until the mid-1950s, with two power plants, one of 225 MW in Valdecañas and the other of 130 MW in Torrejón, plus another two to be exploited later on. The expected productivity of this scheme in the first two power plants was 550 GWh in the first and 345 GWh in the second, that is, a total of 895 GWh, but with strong oscillations, from 1 to 4 in annual values.

On the other hand, a group of power plants at the base of dams fed by an important regulating reservoir such as that of Valdecañas was considered as ideal to supply the connection load peaks. In this way, first the partial exploitation of the Tiétar river's resources was considered, being a tributary of the Tajo downstream from the projected Torrejón jump, and second the large-scale adoption of pumping with reversible power units in both sites.

The final result was the construction of the Valdecañas power plant, at the toe of the dam, with a power of 250 MW and pumping from the lower reservoir, that of Torrejón. The incorporation of the Tiétar to the scheme immediately presented the convenience of also using its contribution, for which the possible pumping from it to the Tajo in the Torrejón reservoir was considered. For this purpose a power house was built to serve both reservoirs, that of Torrejón and a new one on the Tiétar, in an area in which they are both very close to each other, upstream from their confluence. Both reservoirs have a difference in height of 20 m. This power plant was designed to have a great operational flexibility between both rivers, in such a way that Tajo-Tajo and Tiétar-Tajo turbination is possible, as well as Tajo-Tajo, Tiétar-Tajo, and Tajo-Tiétar pumping. In this way, the expected productivity was 710 and 420 GWh, respectively, with a total of 1130 GWh, with a minimum improvement of 26% and only a 1–2 variation (**Figure 32**).

The scheme continued with the construction of the José M. de Oriol power plant, which was initially expected for a power of 600 MW and finally achieved 935 MW, being Spain's second largest today. The last phase of the development was the construction of the Cedillo power plant, located at the point where the Tajo enters Portuguese territory, in the confluence of the Tajo and Sever rivers. The power house is located in the dam with the same name, between two large spillways, one per river. Its installed power is 500 MW (**Figure 33**).

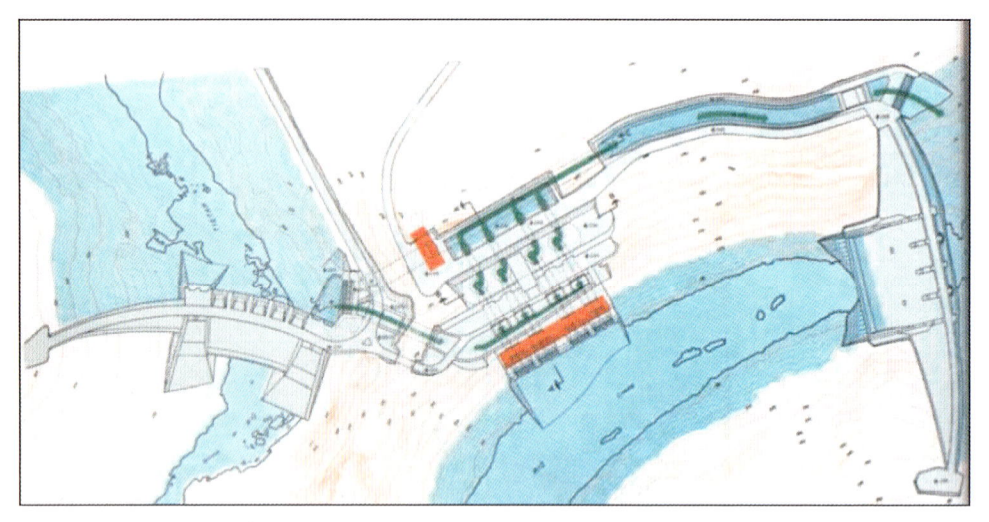

Figure 32 José M. de Oriol power plant. Tiétar-Tajo pump operation.

Figure 33 Cedillo power plant and dam.

Many other developments were built over this period of strong hydroelectric expansion. The schemes used in most of them were those with power plants at the toe of dams and those with diversion schemes, both in areas with strong height variations such as in the century's first decades, and in lesser height variations and more important flows in more regulated river sections.

6.11.2.5 The Last Three Decades

6.11.2.5.1 *The electricity sector*

At the end of the 1970s and beginning of the 1980s, several coal (both national and imported) power plants began to operate. In addition, the nuclear program continued and between 1983 and 1988 seven power plants began operating, with a total installed power of more than 7000 MW.

The quick fuel plant replacement process and the development of the nuclear program had a double effect on the electricity sector: The strong indebtedness of the electricity companies and the overcapacity made it necessary for a Legal and Stable Framework to be established in 1988, which stabilized the sector and in the 1990s enabled the electricity companies to reorganize, mainly in two large groups: Endesa and Iberdrola, which also began to expand internationally.

In 1995, the Law for the Regulation of the National Electricity System was enacted, and in 1996, the EU Council passed the Directive concerning common rules for the internal market in electricity. As a result, in 1997 the Electricity Sector Law was enacted, introducing the most important regulatory changes in this sector's history, in the line of a liberalization of the electricity market and of the separation of the generation, transport, distribution, and marketing activities. Subsequently, the figures of the Market Operator, whose purpose was the market's economic management, and that of the System Operator, whose purposes were to the system's technical management and the management of the transport network, were created.

At the beginning of this century, the Spanish electricity sector was characterized by a low reserve of installed power, as a transport network with congestion problems in certain areas, and an important increase in demand, a result of the strong economic growth. All this was the cause for an important development in the construction of new power plants: on the one hand, combined cycle plants, and on the other hand, renewable energy plants in general and wind farms in particular, under the shelter of favorable laws for this type of energy and of the social concern in the ambit of environmental protection.

6.11.2.5.2 *Main hydropower developments*

From 1985 the economically exploitable hydroelectric potential was practically used, so from then on the developments have been of four types. Among the medium and great power developments, the pumping plants stands out, especially pure pumping plants, and those built as an extension of relatively important power plants, under the shelter of the gradually greater regulation upstream with reservoirs of all uses throughout time. Two types stand out among the small power developments: small power plants built under the special production system and those that exploit the existing regulation hydraulic installations, such as the irrigation or supply regulation dams.

Among the pure pumped storage we must mention those of Aguayo, with an installed power of 362 MW; Estangento-Sallente, with 450 MW; Guillena, with 210 MW; La Muela, with 635 MW; Moralets, with 222 MW; and Tajo de la Encantada, with 360 MW; 2240 MW in total, built in these last few decades, to which we must add the 112 MW from the Gabriel y Galán and the 126 MW of the Tanes mixed pumped storages.

To complete the pumping power plant overview in Spain, we would have to add the over 2600 MW of installed power of mixed pumping plants built before 1975, in the aforementioned Saltos del Sil, Tajo System, Duero System, and others with less power. Total power installed in reversible power units is, in terms of turbination power, of the order of 5100 MW. This capacity will be increased in the coming years with the installations being built and those being projected.

The first pure pumping plant built in Spain was that of Guillena, in the Guadalquivir basin near the city of Seville in 1970. Between 1983 and 1984, the reversible power plant of Aguayo became operative in the Cantabrian mountain range, whose lower reservoir was formed by raising the existing Alsa dam (**Figure 34**). This reservoir is also part of the Ebro-Besaya interbasin diversion.

Between 1986 and 1989, the Moralets pumping plant was built at the headwaters of the Noguera Ribagorzana river. The location of the Tajo de la Encantada pure pumping plant is rather singular, with the building inside the reservoir (**Figure 35**).

The pure pumped storage of Estangento-Sallente, Spain's second largest in terms of power, is located at the highest watercourse of the Capdella river, and uses as its upper water tank the lake which was raised with the Estangento dam, which has been mentioned previously (**Figure 11**). The initial dam has the same purpose as that of the beginning of the twentieth century, which is to collect the waters for the power station, while the rise is used to regulate the pumping. The lower tank was formed by building a 90 m-high embankment dam (**Figure 36**).

The most important pure pumped-storage plant in Spain is that of La Muela, located near the Cofrentes Nuclear plant, on the Júcar river. This pumping plant is currently being enlarged, which will make it Spain's most powerful plant (**Figure 37**).

The schemes of the two mixed power plants built in this period are interesting. That of Tanes is part of the supply system to the central region of Asturias and is located between two reservoirs, Tanes and Rioseco, and has the particularity that the power plant is located near the middle point of the hydraulic circuit. The Gabriel y Galán mixed pumping plant is part of a development built in the 1980s and of which the Guijo de Granadilla dam is also part. Both the power plant and the dam are located between two dams built in the 1960s for irrigation purposes; the upper dam, Gabriel y Galán, is a regulation dam, and the lower dam, Valdeobispo, acts

Figure 34 Alsa dam and Aguayo reversible power plant.

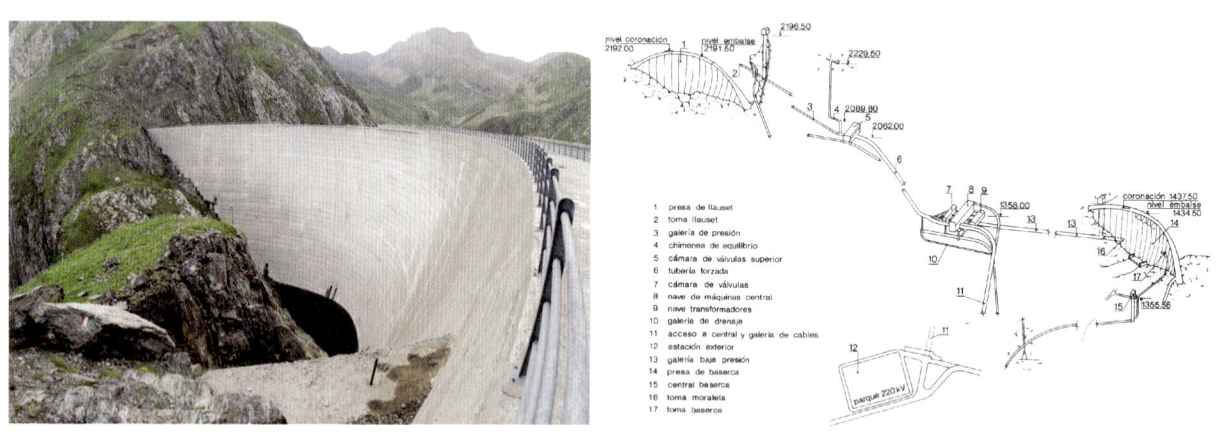

Figure 35 Llauset dam and diagram of the Moralets jump.

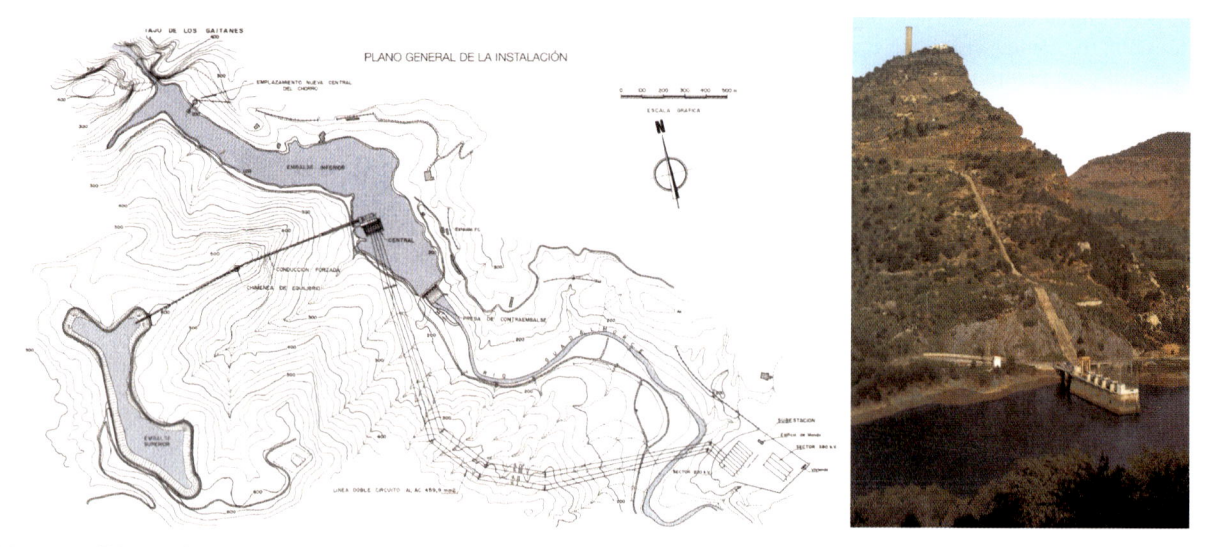

Figure 36 Tajo de la Encantada reversible pumping plant: Diagram and plant.

Figure 37 Aerial view of La Muela pumping plant.

as a diversion dam. In between there is a height variation that was taken advantage of by building the Guijo de Granadilla dam, which enabled the overlapping of the three reservoirs and allowed for the installation of two pumping plants, the aforementioned Gabriel y Galán, of 112 MW, and that of the intermediate plant, of 54 MW (**Figure 38**).

The Canal de Isabel II Hydroelectric Power Plant System is an example of one of the developments with less installed power in existing hydraulic installations. It is a public entity that has supplied water to Madrid since 1850. The entity took advantage of the regulation dams and pipes, and in the first phase in the 1990s seven new power units were installed and some of the existing units were modernized.

With regards to the small hydro, some have been built from scratch and others by using old power plants built in the first years of the nineteenth century and which were abandoned in the second half of that century due to their scarce profitability and competition from large power plants of all types and sizes. The promotion of renewable energy over the last decades, the existence of usable infrastructure, and the general use of automation and remote control have made them be profitable once again (**Figure 39**).

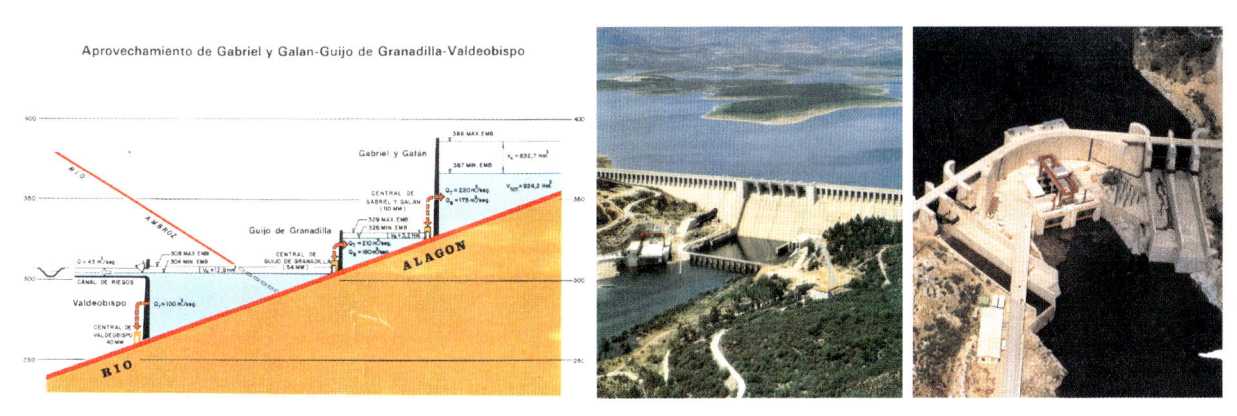

Figure 38 Scheme of the Alagón river hydroproject. Gabriel y Galán and Guijo de Granadilla dams and power plants.

Figure 39 Canal of the Tranco del Diablo small hydro.

6.11.3 A Representative Case: The Duero System and Its Evolution

The hydroelectric production Duero System is exploited by Iberdrola, the heir and continuation of the companies that studied and built the scheme projects from the first years of the twentieth century, first Saltos del Duero and then Iberduero.

It is the most comprehensive and most complex hydroelectric system with the most installed power in Spain. In addition, it has the first and third largest individual facilities in terms of power (Aldeadávila and Villarino), the highest dam (Almendra), and one of the most beautiful (Aldeadávila), among other important ones (**Figure 40**).

The Duero river, after crossing the large sedimentary basin with rather gentle height variations, narrows in the Arribes del Duero granitic massif, where the height variations between the top peneplain and the riverbed can exceed 400 m in some points. In the section where this narrowing takes place, it has three sections from the administrative point of view: a top section in Spanish territory, a second section on the Portuguese border exploited by Portugal, and a third section also on the border and exploited by Spain. All of this was a result of an agreement reached between both countries in 1927.

Spain's main jumps are located on the Duero itself, on the two previously mentioned sections, and on its tributaries: the Esla, which at its confluence nearly has as much flow rate as the main river and the Tormes, which joins with it in the separation between the two international sections, where the Spanish one begins. We also have to add those of the Tera river, a tributary of the Esla (**Figure 41** and **Table 4**).

The general scheme of the Duero, Esla, and Tormes development was already planned in the second decade of the twentieth century. It consisted of a large regulation reservoir in the Esla river, as it could not be built on the Duero, which is the headwaters of the entire system. Downstream, in the Spanish Duero, these regulated flow rates are exploited in the Villalcampo and Castro power plants, built in the 1940s and beginning of the 1950s. Further downstream, on the International Duero exploited by Portugal, we find the Portuguese Miranda, Picote, and Bemposta power plants. The Tormes river discharges its waters at the end of this section.

Figure 40 Almendra and Aldeadávila dams.

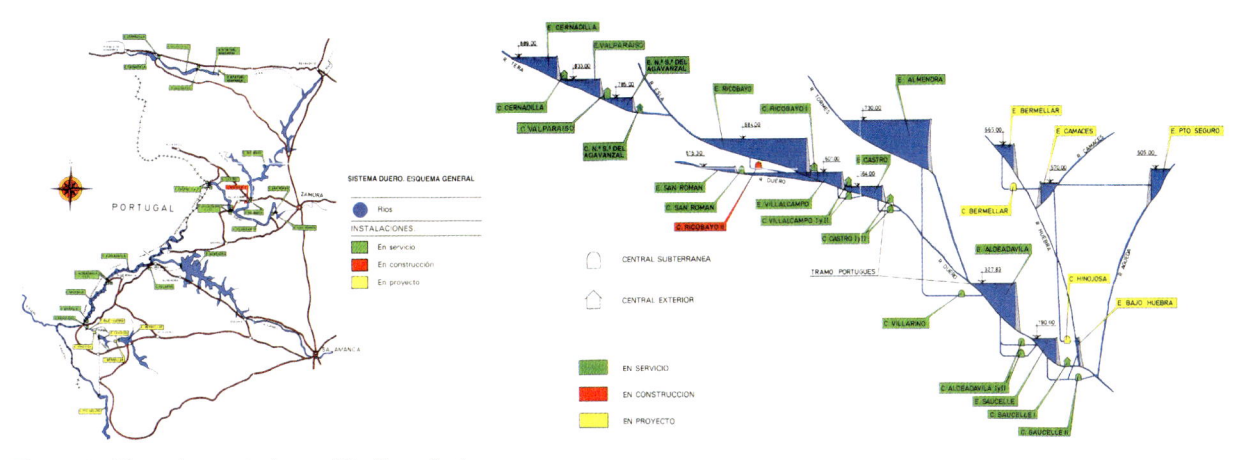

Figure 41 Plan and current scheme of the Duero System.

Table 4 Duero System power plants: Main characteristics

Power plant	Dam type/height	MW	Head	Flow	Number of units	Pumped-storage type	Production
Subsystem Tera							
Cernadilla	PG – 69	30	56	60	1		82
Valparaiso	PG – 67	62	48	160	2		75
Santa María Agavanzal	PG – 43	23	36	72	3		64
Esla river							
Ricobayo I	PG – 100	133	83	240	4		638
Ricobayo II		135	83	210	1		46 + 264 Sist
Duero national							
San Román	PG – 6	6	15	43	1		35
Villalcampo I	PG – 52	96	37	303	3		744
Villalcampo II		110	37	340	1		
Castro I	PG – 53	80	42	270	2		755
Castro II		110	42	340	1		
Duero International							
Aldeadávila I	AG – 140	718	138	617	6		3807
Aldeadávila II		420	138	350	2	Mixto	
Saucelle I	PG – 83	240	63	468	4		1302
Saucelle II		250	63	480	2		321
Tormes							
Villarino	BV – 202	810	402	232	6	Mixto	1376
Total		3223			39		

They are exploited downstream by the Aldeadávila I and II power plants and subsequently by those of Saucelle. On the other hand, the waters of this last reservoir can be pumped to the Aldeadávila reservoir and from it to the great Almendra reservoir built on the Tormes (**Figure 41**).

From the 1970s, the Tera development was initiated, a tributary of the Esla, with the construction of the three new dams: Cernadilla, Valparaíso, and Santa María de Agavanzal, and their corresponding power plants at their base, which in turn increased the regulation of the system downstream. This fact, together with the regulation increase at the headwaters of the Esla by means of dams for irrigation purposes, enabled the construction of the new previously mentioned Ricobayo II power plant.

From the first studies that date back to the first decade of the nineteenth century and the first execution, the Ricobayo dam (1932), until the last execution, the construction of the Ricobayo II power plant (1998), nearly one century has elapsed, during which the social, economic, technical, and technological conditions have been changing. This meant that the projects and constructions were addressed differently, with different schemes and different dams and power stations. And all of this without changing the general exploitation scheme proposed by José Orbegozo.

The first period goes from 1902, the date of the first concession in the International Duero, to 1927, date in which the joint exploitation with Portugal was signed. From a technical point of view, the first developed schemes were successive diversion channel schemes over the international Duero, with canals on the Spanish side seeing the problems posed by Portugal to reach an agreement (**Figure 42**).

In 1924, seeing the proximity of the agreement being signed, Orbegozo presented the project in an almost definite manner. On the Spanish side, a regulation dam on the Esla, the current Ricobayo, and a dam on the national Duero, that of Villardiegua, took advantage of the entire section of the current Villalcampo and Castro reservoirs. The first section of the international Duero, and awaiting the treaty, it was solved based on a diversion waters by means of a canal that led to a power house, also on the Spanish margin, located just before the confluence of the Tormes. The second section of the International Duero exploited that of Aldeadávila by means of a dam and another diversion scheme, that of Saucelle (**Figure 43**). On the Tormes, exploitation was to be by means of two dams and their corresponding power houses and a diversion scheme with the corresponding power house, between them.

This 'visionary' solution was the one that was developed later on, in accordance with the logical technical advances. Thus, in the second period, from 1927 to 1936, the Ricobayo dam was built, with a height of 100 m instead of the 70 initially conceived, as a result of the greater experience in dam building. The location did not change at all. It was the highest in Spain at that time and one of the most important in Europe (**Figure 44**).

The third phase took place during the 1940–60 period, in which due to the circumstances, the construction of only one dam on the Spanish Duero was discarded and two smaller dams were built: those of Villalcampo and Castro and the development of the Saucelle project, built as a dam and not as a diversion scheme (**Figure 45**).

The fourth phase is that of the development of the Aldeadávila and Almendra-Villarino dams and power plants, more technically advanced installations than the previous ones, which included underground power plants and mixed pumped storages. The Villarino power plant, with a large 202 m-high dam, replaced the previous three Tormes steeps, something unthinkable 50 years

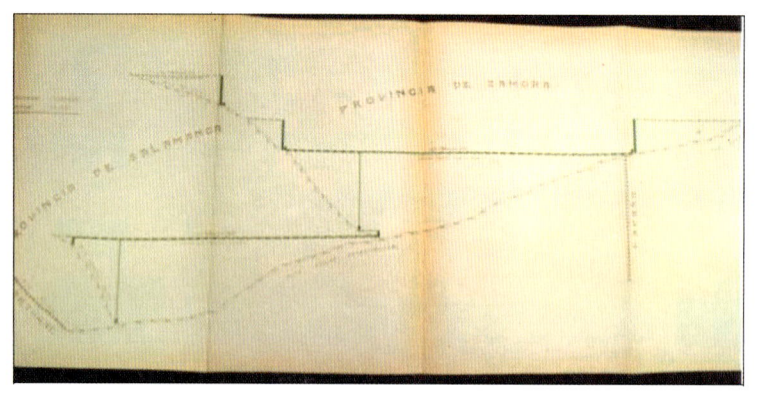

Figure 42 Ugarte Solution (1919).

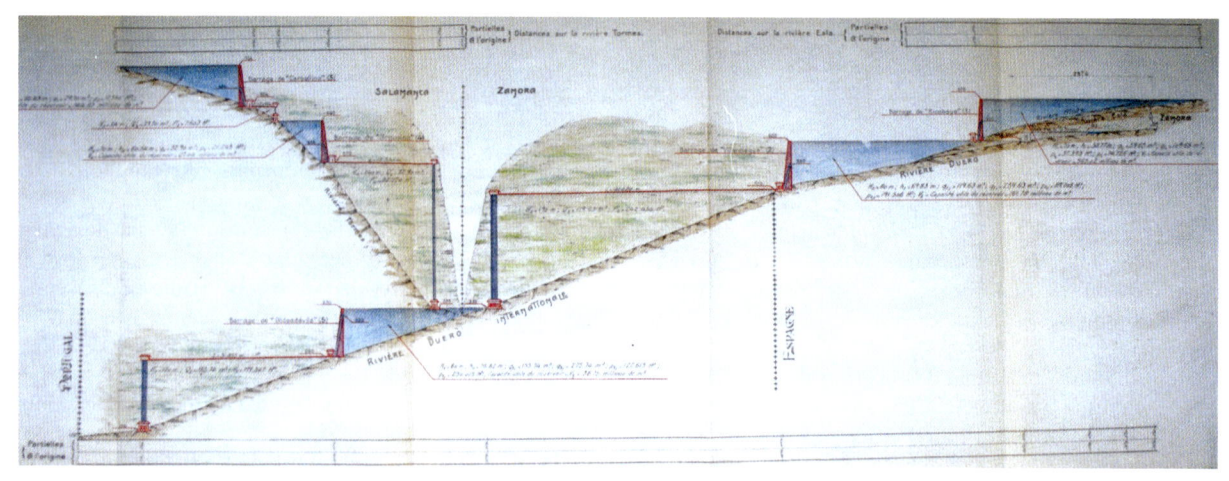

Figure 43 1924 Orbegozo Solution.

Figure 44 Ricobayo dam.

Figure 45 Castro and Saucelle dams.

before. Modern underground excavation techniques were used for the first time in the Aldeadávila I power plant. Technical advance allowed to substantially improve the installed power and system's total power with the same initial concept (**Figures 46** and **47**).

In the fifth phase, from 1975 to today, executions have taken two directions: the exploitation of the Tera by means of the aforementioned power plants and the power increase of the existing power plants, those of Villalcampo and Castro in 1977, Saucelle II in 1989, and Ricobayo in 1988, by building new power units in all of them.

6.11.4 The Future of Hydroelectric Power in Spain

The need of new hydroelectric power in the Spanish Electricity System is mainly determined by three factors. First, the increase in demand of electricity, which was 27% in the 2001–07 period, a big necessary investment in new power generation plants. Environmental conditions and technological advances have made the technologies developed throughout this decade to be renewable energy, especially wind energy, and combined cycles. These conditions make the investments in new hydroelectric equipment along with other renewable energy having to decisively contribute by achieving the long-term objectives of greenhouse gas emissions.

Second, the increase in peak demand, which in the same previous period was nearly 29%, rising from 35 000 to 45 000 MW and which is expected to be nearly 59 000 MW in the year 2016. In order to cover this demand, there needs to be fast-response technologies that operate few hours per year. Hydraulic power plants with high regulation can fulfill this mission.

340 Hydropower Schemes Around the World

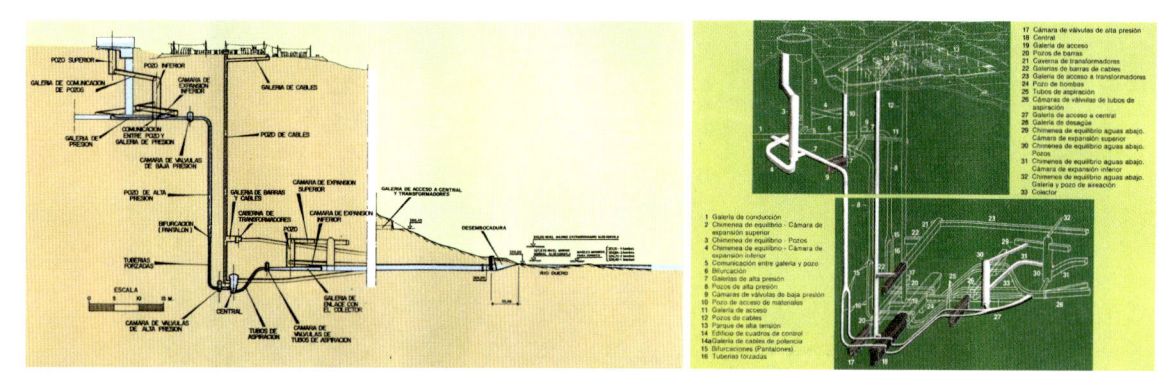

Figure 46   Villarino power plant: Diagram.

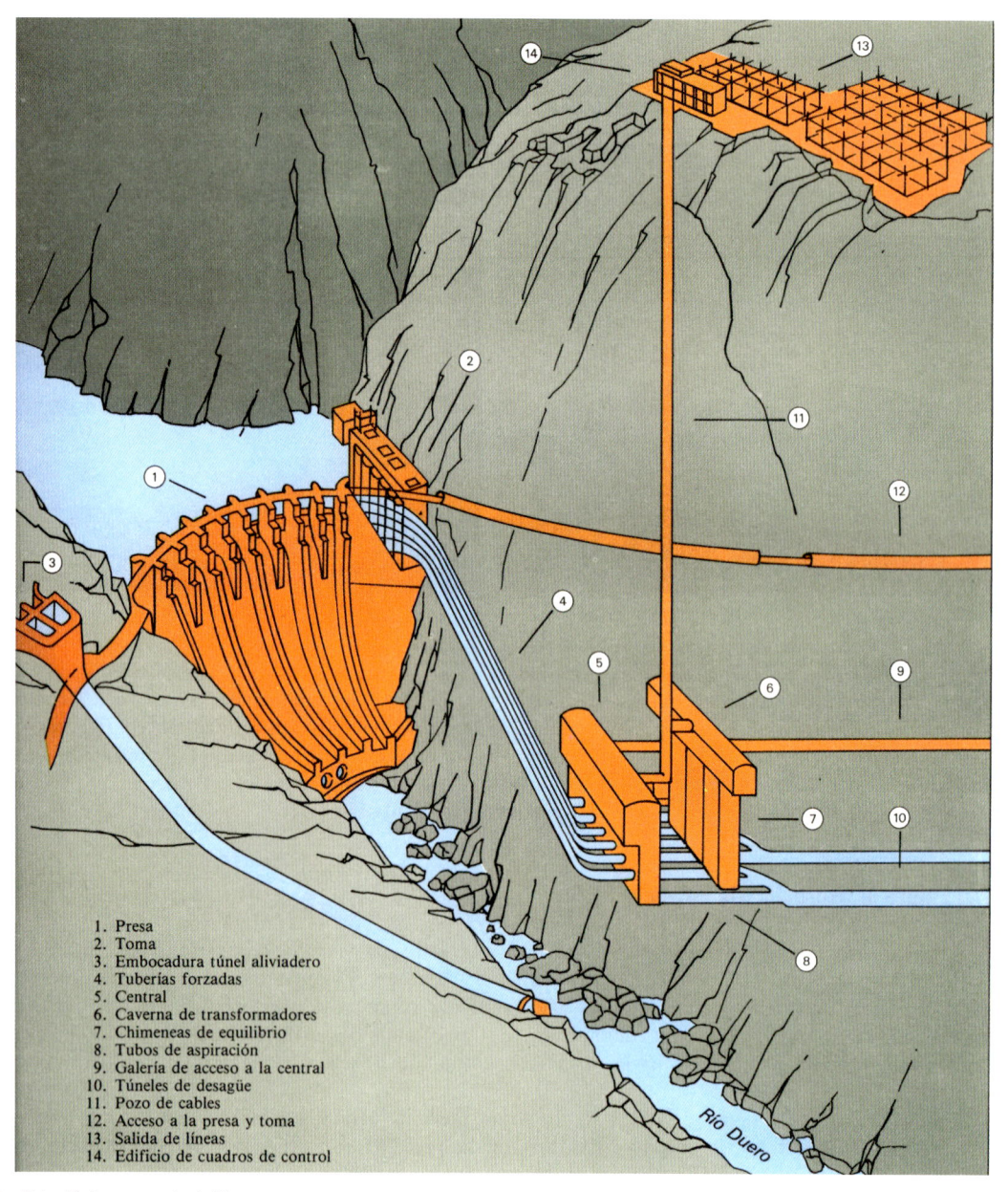

1. Presa
2. Toma
3. Embocadura túnel aliviadero
4. Tuberías forzadas
5. Central
6. Caverna de transformadores
7. Chimeneas de equilibrio
8. Tubos de aspiración
9. Galería de acceso a la central
10. Túneles de desagüe
11. Pozo de cables
12. Acceso a la presa y toma
13. Salida de líneas
14. Edificio de cuadros de control

Figure 47   Aldeadávila power plant: Diagram.

And third, the uncertainty and volatility of wind power production. In Spain, the current wind power is 17 000 MW, which is expected to increase to 29 000 by 2016, 25% of the total. Great variations are already taking place in produced wind power. Thus, in November 2008, in two days wind power went from covering 43% of the demand to only 1.2%. In this way, couplings to the network can go from 1000 to 6000 MW in 8 h or to 11 000 MW in 24 h.

This volatility of wind power requires the system to have quick-start or quick-stop high power technologies that replace it. Pure pumping hydropower plants and those with regulation reservoirs play a vital role to fulfill this mission.

As an example, and with these purposes, Iberdrola expects to increase power by nearly 2000 MW between pure pumping plants and the increase of power in existing power plants:

1. *La Muela II*. Pure pumping with 850 MW of turbination power. It is the second part of the current La Muela I pumping plant, with a power of 630 MW. If to these we add the power plant at the base of the Cortes II dam, with 280 MW, the complex will achieve 1760 MW of installed power, the largest in Spain.
2. *Santa Cristina*. Pure pumping with 750 MW of turbination power. Located in the San Esteban reservoir.
3. *San Esteban II*. Underground power plant with 175 MW that will be added to the current 264 MW.
4. *San Pedro II*. New 25 MW power plant designed parallel to the current one.

In the case of the Canary Islands, the problem gets worse due to the higher rigidity of the electricity production by thermal power units, the development of wind power, and the nonexistence of hydroelectric power plants. In order to minimize this problem, projects are being implemented. These are small projects but important ones due to the problems they solve. Among these we must mention the Soria pumping plant, located on the island of Gran Canaria, which will use the existing reservoir of the same name, created by a 130 m-high double-curvature dam, and which will have an installed turbination power of 150 MW, of which 50 MW will only come from turbination and 100 MW from pure pumped storage.

References

[1] Buil Sanz JM and García AG (2006) Hydropower. *Dams in Spain*. Madrid, Spain: Colegio de Ingenieros de Caminos, Canales y Puertos y SPANCOLD.
[2] Perán F and Pérez JJ (2009) Pumped storage to support wind. *HRW* 17(3).
[3] Iberdrola (2006) *Large Dams*. Salamanca, Spain: Iberdrola.
[4] García AG and Buil Sanz JM (2006) The role of hydro and future pumped-storage plans in Spain. *International Hydropower and Dams* 3.
[5] Villalba J (2000) The value of water for energy generation. *Water Economics Conference*, Valencia, Spain.
[6] Marcos Fano JM (2004) History and current overview of the Spanish electrical system. *Energía y Sociedad (Physics and Society) Magazine* Number 13, Madrid, Spain.
[7] Bueno Hernández F *History of Electrical Energy in Spain*.
[8] Díez-Cascón Sagrado J and Bueno Hernández F (2000) *Dams Engineering: Concrete Dams*. Universidad de Cantabria, Santander.
[9] Endesa Worldwide (2006) *Large Dams*. Endesa, Madrid.
[10] Unesa (2004) *The Electricity Sector through Unesa. 1944–2004*. Unesta, Madrid.

6.12 Hydropower in Switzerland

B Hagin, Ingénieur-Conseil, Lutry, Switzerland

© 2012 Elsevier Ltd. All rights reserved.

6.12.1	Short Recall of Switzerland's Characteristics	343
6.12.2	The Drainage Basins of Switzerland	343
6.12.3	Electricity Production in Switzerland	343
6.12.3.1	In General	343
6.12.3.2	Large-Scale Hydropower Plants	345
6.12.3.3	Small-Scale Hydropower Plants	346
6.12.3.4	Dams	346
6.12.4	List of the Dams in Switzerland	347
6.12.5	New Developments	349
6.12.6	Dixence, Grande-Dixence, and Cleuson-Dixence Schemes as an Example of Capacity Increase	350
6.12.6.1	In General	350
6.12.6.2	First Stage: The Dixence Scheme	350
6.12.6.3	Second Stage: The Grande-Dixence Scheme	350
6.12.6.4	Third Stage: The Cleuson-Dixence Scheme	351
6.12.7	New Hydroelectric Schemes Presently under Construction in Switzerland	352
6.12.7.1	The Nant de Dranse Scheme	352
6.12.7.2	The Linthal 2015 Project	353
6.12.7.3	The Hongrin-Léman Plus Project	354
Relevant Websites		354

6.12.1 Short Recall of Switzerland's Characteristics

Switzerland is a landlocked alpine country in Central Europe. The country borders Germany to the north, France to the west, Italy to the south, and Austria and Liechtenstein to the east. The area of Switzerland is 41 285 km² with a population of 7 783 000 inhabitants in 2010 (**Figure 1**).

The Alps cover 65% of Switzerland's surface area, making it one of the most alpine countries. Among this, 9788 km² or 24% of the territory is above 2000 m elevation and 936 km² or 2.3% are above 3000 m elevation. The highest summit is 'La Pointe Dufour', with an altitude reaching 4634 m.

The glaciers of the Swiss Alps cover an area of 1230 km² (3% of the Swiss territory), representing 44% of the total glaciated area in the Alps (2800 km²).

The average annual rainfall is 1456 mm (European annual average is around 790 mm). The area around Sion in Valais, in the Rhône Valley in the south, being the driest zone with an average value of 600 mm yr^{-1} and the wettest part with 2900 mm yr^{-1}, is around the Säntis mountain in Appenzell, north-east of the country (**Figure 2**).

6.12.2 The Drainage Basins of Switzerland

The south-west side of the Swiss Alps is drained by the Rhône River, flowing to France and then to the Mediterranean Sea. The north is drained by the Aar and the Rhine River, flowing to Germany and then to the North Sea. The south-east is drained by the Ticino River, flowing to the Pô, through Italy, and then to the Adriatic Sea. The extreme east is drained by the Inn River, flowing to the Danube River and then to the Black Sea (**Figure 3**).

Switzerland possesses 6% of Europe's freshwater and is sometimes referred to as the 'water tower of Europe'.

6.12.3 Electricity Production in Switzerland

6.12.3.1 In General

Thanks to its topography and high levels of annual rainfall, Switzerland has ideal conditions for the utilization of hydropower. Toward the end of the nineteenth century, hydropower underwent an initial period of expansion, and between 1945 and 1970, it experienced a genuine boom during which numerous new power plants were opened in the lowlands, together with large-scale storage plants in the Alps.

Figure 1 Map of Switzerland with the main cities and the 26 cantons.

Figure 2 Topographical map of Switzerland.

Based on the estimated mean production level, hydropower still accounted for almost 90% of domestic electricity production at the beginning of the 1970s, but this figure fell to around 60% by 1985 following the commissioning of Switzerland's nuclear power plants, and is now around 56%. Hydropower, therefore, remains Switzerland's most important domestic source of renewable energy.

In a European comparison, Switzerland ranks fourth in terms of contribution of hydropower toward electricity production, behind Norway, Austria, and Iceland.

The total number of power plants of a capacity higher than 300 kW is presently equal to 543, representing an installed capacity of 13 480 MW and an annual production of 35 601 GWh.

The statistics of the hydropower plants are placing them in four categories:

The run-of-river power plants with an installed capacity of 3707 MW and a production of 16 611 GWh.
The storage power plants with an installed capacity of 8073 MW and a production of 17 397 GWh.
The combined power turbine–pump plants with an installed capacity of 1383 MW and a production of 1594 GWh.
The pure turbine–pump power plants with an installed capacity of 316 MW.

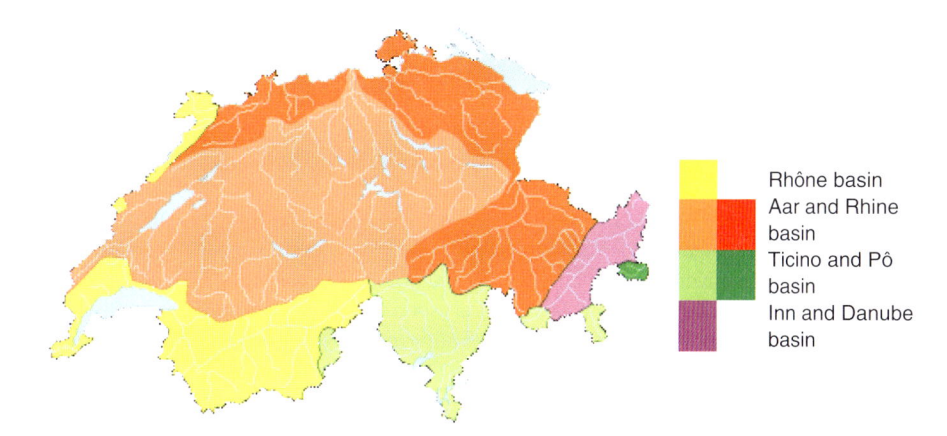

Figure 3 The drainage basins of Switzerland.

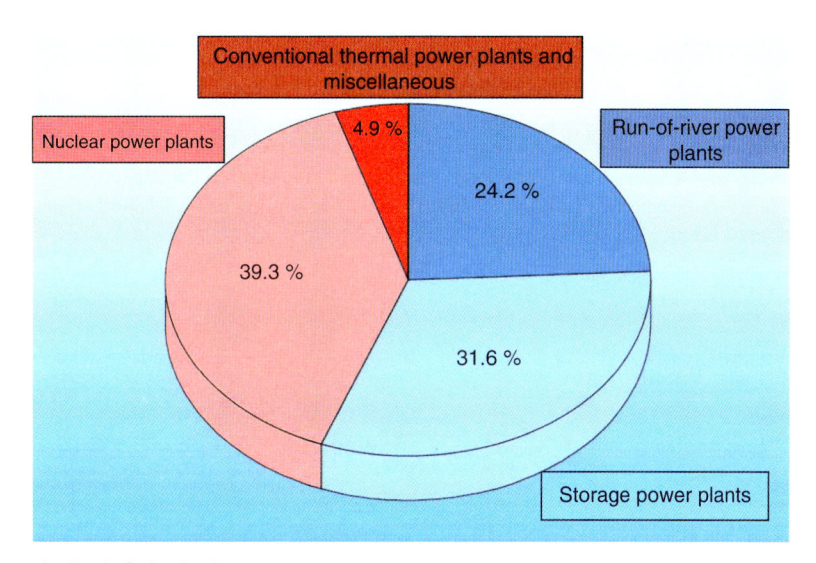

Figure 4 The electricity production in Switzerland.

The big hydraulic plants with an installed capacity over 10 MW, produce about 90% of the hydroelectricity of Switzerland (**Figure** 4).

Hydropower plays a major role in Switzerland's energy production, with a share of around 56%. In addition, storage plants are an important factor for power production at short notice and for the changeover of production from summer to winter. Thanks to its storage capabilities, Switzerland plays a central role as an electricity supplier in the European networks. Hydropower is the most important, CO_2-free energy source.

6.12.3.2 Large-Scale Hydropower Plants

Today, there are 543 hydropower plants in Switzerland, each having a capacity of at least 300 kW, and these produce an average of around 35 600 GWh per annum, 47% of which is produced in run-of-river power plants, 49% in storage power plants, and approximately 4% in pumped-storage power plants. The main water sources (feeding 476 power plants) are the Rhine (into which the Aar, Reuss, and Limmat flow) and the Rhône.

Two-thirds of hydroelectricity are generated in the mountain cantons of Uri, Grisons, Ticino, and Valais, while Aargau and Bern also generate significant quantities. Roughly 10% of Switzerland's hydropower generation comes from facilities situated on bodies of water along the country's borders.

In Switzerland's hydropower plant statistics, a distinction is made between four types of plants: run-of-river plants (3707 MW, 16 611 GWh), storage plants (8073 MW, 17 397 GWh), pumped-storage plants (1383 MW, 1594 GWh), and basic water flow plants (316 MW).

Large-scale hydropower plants (capacity >10 MW) account for around 90% of Switzerland's total hydropower production.

6.12.3.3 Small-Scale Hydropower Plants

In Switzerland, the term small-scale hydropower plant refers to facilities that produce a mean mechanical gross capacity of up to 10 MW.

Small-scale hydropower plants have been around for a long time in Switzerland. At the beginning of the twentieth century, there were already around 7000 in operation. But with the advent of low-cost electricity from large-scale power plants, many of these ceased production.

Today, there are more than 1000 small-scale hydropower plants in operation, with an installed capacity of approximately 760 MW and an output of 3400 GWh per annum.

Electricity production in small-scale hydropower plants is attractive from both an economical and an ecological point of view, and an expansion of output is perfectly feasible, as long as ecological aspects are duly taken into account. The potential is estimated at around 2200 GWh per annum. Technological innovations and measures to lessen environmental impacts make small-scale hydropower plants inexpensive energy sources that provide renewable energy on an independent basis and help protect the environment.

In addition to small-scale hydropower plants in rivers and streams, it is now possible to utilize other sources, for example, excess pressure in drinking water systems.

6.12.3.4 Dams

The Swiss Federal Office of Energy (SFOE) is the highest supervisory authority for all dams in Switzerland. In practice, however, the SFOE delegates responsibility for the supervision of several hundred small dams to the relevant cantonal authorities, so that it can focus on the country's larger facilities (195 reservoirs with 217 dams). Eighty-four percent of these are for the production of hydropower. One hundred and thirty-four dams take the form of concrete walls (78 gravity dams, 52 curved dams, 2 multiple-curve dams, and 2 pierhead dams), 78 are soil and stonefill constructions, and five are in the form of river weirs. Twenty-five are higher than 100 m and four are over 200 m high, namely the Grande-Dixence gravity dam (285 m) and the Mauvoisin (250 m), Luzzone (225 m), and Contra (220 m) curved dams. Most of the large-scale dams are located in the Alps.

The oldest dams date from the nineteenth century, though most of the biggest dams in Switzerland were constructed in the period between 1950 and 1970 (**Figure 5**).

Figure 5 Location of the main dams and power plants in Switzerland.

6.12.4 List of the Dams in Switzerland

The list of the main Swiss dams is given hereunder in alphabetic order.

Name	Height (m)	Type	Year	Reservoir volume (mio m^3)	Location canton
Airolo	20	PG/TE	1968	0.37	Ticino
Albigna	115	PG	1959	71.00	Graubünden/Grigioni
Alp Dado	24	TE	1995	0.06	Graubünden/Grigioni
Arnensee	17	TE	1942	10.50	Bern
Arniboden Süd	15	TE	1910	0.24	Uri
Bagnes GD	31	TE	1957	0.30	Valais/Wallis
Bannalp	37	TE	1937	1.70	Nidwalden
Barberine	79	PG	1925	40.00	Valais/Wallis
Barcuns	29	PG	1947	0.12	Graubünden/Grigioni
Bärenburg	64	PG	1960	1.00	Graubünden/Grigioni
Bortelsee	20	ER	1989	3.66	Valais/Wallis
Bremgarten-Zufikon	19	PG	1975	2.20	Aargau
Brigels	18	TE	1960	0.30	Graubünden/Grigioni
Buchholz	19	PG	1892	0.25	St. Gallen
Burvagn	20	PG	1949	0.20	Graubünden/Grigioni
Carassina	39	VA	1963	0.31	Ticino
Carmena	40	VA	1969	0.30	Ticino
Cavagnoli	111	VA	1968	29.00	Ticino
Chapfensee Nord	20	PG	1948	0.43	St. Gallen
Châtelard CFF	33	TE	1975	0.25	Valais/Wallis
Châtelot	74	VA	1953	20.60	Neuchâtel/France
Cleuson	87	PG	1950	20.00	Valais/Wallis
Contra	220	VA	1965	105.00	Ticino
Croix	15	TE	1955	0.09	Valais/Wallis
Curnera	53	VA	1966	41.10	Graubünden/Grigioni
Darbola	22	PG	1958	0.11	Graubünden/Grigioni
Dixence	87	CB	1935	50.00	Valais/Wallis
Egschi	40	PG	1949	0.40	Graubünden/Grigioni
Emosson	180	VA	1974	227.00	Valais/Wallis
Esslingen	16	TE	1988	0.10	Zürich
Ferden	67	VA	1975	1.89	Valais/Wallis
Ferpècle	28	VA	1964	0.10	Valais/Wallis
Garichte	42	PG	1931	3.29	Glarus
Garichte Nebenmauer	18	PG	1931	3.29	Glarus
Gebidem	122	VA	1967	9.20	Valais/Wallis
Gelmer	35	PG	1929	14.00	Bern
Giétroz-du-Fond	15	PG	1965	0.02	Valais/Wallis
Gigerwald	147	VA	1976	35.60	St. Gallen
Godey	35	TE	1974	0.93	Valais/Wallis
Göscheneralp	155	ER	1960	76.00	Uri
Göschenerreuss	36	PG	1949	0.10	Uri
Grande-Dixence	285	PG	1961	401.00	Valais/Wallis
Greuel	17	TE	1984	0.11	Aargau
Gries	60	PG	1965	18.60	Valais/Wallis
Gübsensee Ost	24	PG	1900	1.50	St. Gallen
Gübsensee West	17	TE	1900	1.50	St. Gallen
Hongrin Nord	125	VA	1969	53.20	Vaud
Hongrin Sud	90	VA	1969	53.20	Vaud
Hospitalet	21	PG	1962	0.01	Valais/Wallis
Hühnermatt	17	TE	1937	96.50	Schwyz
Icogne	17	TE	1962	0.04	Valais/Wallis
Illgraben	50	PG	1970	0.00	Valais/Wallis
Illsee	25	PG	1923	6.60	Valais/Wallis
In den Schlagen	33	PG	1936	96.50	Schwyz
Innerferrera	28	PG	1961	0.23	Graubünden/Grigioni
Isenthal	20	PG	1955	0.03	Uri

(Continued)

(Continued)

Name	Height (m)	Type	Year	Reservoir volume (mio m^3)	Location canton
Isola	45	VA	1960	6.50	Graubünden/Grigioni
Jougnenaz	21	PG	1937	0.01	Vaud
Käppelistutz	18	PG	1945	0.06	Nidwalden
La Fouly	18	PG	1972	0.02	Valais/Wallis
La Luette	15	PG	1918	0.40	Valais/Wallis
Lago Bianco Nord	15	PG	1912	18.60	Graubünden/Grigioni
Lago Bianco Süd	26	PG	1912	18.60	Graubünden/Grigioni
Le Chalet	15	PG	1894	0.10	Vaud
Le Pontet	22	PG	1970	0.07	Vaud
Les Clées	32	PG	1955	0.74	Vaud
Les Esserts	20	TE	1973	0.26	Valais/Wallis
Les Marécottes	19	MV	1925	0.05	Valais/Wallis
Les Toules	86	VA	1963	20.15	Valais/Wallis
Lessoc	33	CB	1976	1.50	Fribourg/Freiburg
Limmern	146	VA	1963	93.00	Glarus
List	17	PG	1901	0.04	Appenzell A.-Rh.
Löbbia	26	PG	1959	0.20	Graubünden/Grigioni
Loré	21	TE	1996	0.07	Ticino
Lucendro	73	CB	1947	25.00	Ticino
Luzzone	225	VA	1963	108.00	Ticino
Maigrauge	24	PG	1872	0.40	Fribourg/Freiburg
Malvaglia	92	VA	1959	4.60	Ticino
Mapragg	75	PG	1976	5.30	St. Gallen
Marmorera (Castiletto)	91	TE	1954	60.00	Graubünden/Grigioni
Mattenalp	27	PG/TE	1950	2.10	Bern
Mattmark	120	TE	1967	101.00	Valais/Wallis
Mauvoisin	250	VA	1957	211.50	Valais/Wallis
Moiry	148	VA	1958	78.00	Valais/Wallis
Molina	54	PG	1951	0.81	Graubünden/Grigioni
Montsalvens	55	VA	1920	12.60	Fribourg/Freiburg
Mühleberg	29	PG	1920	25.00	Bern
Muslen	29	PG	1908	0.08	St. Gallen
Nalps	127	VA	1962	45.00	Graubünden/Grigioni
Naret I	80	VA	1970	31.60	Ticino
Naret II	45	PG	1970	31.60	Ticino
Oberaar	100	PG	1953	61.00	Bern
Orden	42	VA	1971	1.67	Graubünden/Grigioni
Othmarhang	20	TE	2000	0.06	Valais/Wallis
Ova Spin	73	VA	1968	7.40	Graubünden/Grigioni
Palagnedra	72	PG/TE	1952	4.26	Ticino
Panix	53	PG	1989	7.30	Graubünden/Grigioni
Pfaffensprung	32	VA	1921	0.15	Uri
Pilgersteg	17	PG	1920	0.07	Zürich
Piora	27	PG	1920	53.90	Ticino
Plan-Dessous	17	PG	1957	0.10	Vaud
Plans Mayens	20	TE	1971	0.13	Valais/Wallis
Prä	20	PG	1961	0.00	Graubünden/Grigioni
Preda	20	VA	1961	0.27	Graubünden/Grigioni
Proz-Riond	20	TE	1957	51.00	Valais/Wallis
Punt dal Gall	130	VA	1968	164.60	Graubünden/Italia
Räterichsboden	94	PG	1950	27.00	Bern
Rempen	32	PG	1924	0.50	Schwyz
Rhodannenberg	30	TE	1910	56.40	Glarus
Robiei	68	PG	1967	6.70	Ticino
Roggiasca	68	VA	1965	0.52	Graubünden/Grigioni
Rossens	83	VA	1947	220.00	Fribourg/Freiburg
Rossinière	30	PG	1972	2.90	Vaud
Runcahez	33	PG	1961	0.48	Graubünden/Grigioni
Rütiweiher	22	TE	1836	0.20	St. Gallen

(Continued)

(Continued)

Name	Height (m)	Type	Year	Reservoir volume (mio m³)	Location canton
Safien-Platz	15	TE	1957	0.24	Graubünden/Grigioni
Salanfe	52	PG	1952	40.00	Valais/Wallis
Sambuco	130	VA	1956	63.00	Ticino
Sanetsch	42	PG	1965	2.80	Valais/Wallis
Santa Maria	117	VA	1968	67.30	Graubünden/Grigioni
Schiffenen	47	VA	1963	65.00	Fribourg/Freiburg
Schlattli	25	PG	1965	0.36	Schwyz
Schlundbach	22	VA/PG	2000	0.01	Luzern
Schöni	17	PG	1961	0.02	Uri
Schräh	111	PG	1924	150.00	Schwyz
Schwänberg	15	VA	1916	0.08	Appenzell AR/St. Gallen
Secada	21	VA	1982		Ticino
Seeuferegg	42	PG	1932	101.00	Bern
Sella	36	PG	1947	9.20	Ticino
Serra	22	VA	1952	0.20	Valais/Wallis
Sihl-Höfe	19	PG	1961	0.08	Schwyz
Simmenporte	20	PG	1908	0.25	Bern
Solis	61	VA	1986	4.07	Graubünden/Grigioni
Sosto	20	VA	1963		Ticino
Spitallamm	114	VA	1932	101.00	Bern
St-Barthélemy B	45	VA	1975	0.15	Valais/Wallis
St-Barthélemy C	51	VA	1984	0.50	Valais/Wallis
Steinibach	28	PG	2000	0.02	Luzern
Sternenweiher	17	TE	1874	0.10	Zürich
Sufers	58	VA	1962	17.50	Graubünden/Grigioni
Tannensee	25	TE	1958	3.80	Obwalden
Teufenbachweiher	16	TE	1895	0.23	Zürich
Tobel	30	VA	1989	0.10	Uri
Totensee	20	PG	1950	2.60	Valais/Wallis
Turtmann	32	VA	1958	0.80	Valais/Wallis
Ual da Mulin	16	TE	1962	0.06	Graubünden/Grigioni
Val d'Ambra	32	TE	1965	0.40	Ticino
Valle di Lei	141	VA	1961	197.00	Graubünden/Grigioni
Vasasca	69	VA	1967	0.40	Ticino
Verbois	34	PG	1943	12.00	Genève
Vieux-Emosson	45	VA	1955	13.80	Valais/Wallis
Vordersee	15	TE	1986	0.56	Valais/Wallis
Waldialp Nord	15	TE	1961	0.25	Schwyz
Wettingen	29	PG	1933	3.35	Aargau
Zen Binnen	22	VA	1953	0.17	Valais/Wallis
Zervreila	151	VA	1957	100.50	Graubünden/Grigioni
Zervreila A'becken	44	ER	1957	0.14	Graubünden/Grigioni
Zeuzier	156	VA	1957	51.00	Valais/Wallis
Z'Mutt	74	VA	1964	0.85	Valais/Wallis
Zöt	36	VA	1967	1.65	Ticino

PG/TE, Concrete gravity and fill dam; PG, concrete gravity dam; TE, earth dam; ER, rockfill dam; VA, arch dam; CB, butress dam; MV, multiple arch dam.
Source: For more details on each dam, see http://www.swissdams.ch/Dams/damList/

6.12.5 New Developments

The federal government wants to promote the use of hydropower to a greater extent through a variety of measures. In order to exploit the realizable potential, existing power plants are to be renovated and expanded while taking the related ecological requirements into account. The instruments to be used here include cost-covering remuneration for feed-in to the electricity grid for hydropower plants with a capacity up to 10 MW and the measures aimed at promoting hydropower included in the 'Renewable energy' action plan. In terms of quantity, the goal is to increase the mean estimated production level by at least 2000 GWh versus the level recorded in the year 2000 by renovating existing hydropower plants and constructing new ones.

6.12.6 Dixence, Grande-Dixence, and Cleuson-Dixence Schemes as an Example of Capacity Increase

6.12.6.1 In General

Situated in the Canton of Valais in the Val des Dix, a contributory of the Rhône River, the Grande-Dixence scheme is well known for its concrete gravity dam, the highest concrete dam in the world, at 285 m high. It is an example of several extensions to improve its capacity and especially the production of peak energy.

This scheme was developed in three stages.

6.12.6.2 First Stage: The Dixence Scheme

The first stage is the Dixence scheme with the Dixence dam, a 80 m high vault-gravity concrete dam with a storage capacity of 50 000 000 m³ built in the period from 1930 to 1935. It collects the water of the Val des Dix and later on receives the water pumped from the Cleuson buttress dam, which has a storage capacity of 20 000 000 m³ built in the period from 1948 to 1950 in a lateral valley. The water head was 1747 m (a world record at that time) with a discharge of 10 m³ s^{-1} to the 120 MW Chandoline Power House located in Sion, in the Rhône Valley (**Figure 6**).

6.12.6.3 Second Stage: The Grande-Dixence Scheme

The second stage is the construction of the Grande-Dixence gravity concrete dam, 285 m high, downstream of the Dixence dam. It has a storage capacity of 400 000 000 m³ and it is filled by collecting the water on a drainage basin of 350 km², located above 2000 m in altitude and extending to the East, up to Zermatt and the Matterhorn.

The drainage basin contains 35 glaciers, and the water is collected through 75 water intakes and 100 km of galleries, four pumping stations, with altogether 18 pumps. There are four compensating reservoirs, made by the Z'Mutt arch dam, the Stafel Basin, both above Zermatt, and the reservoir of the Ferpècle arch dam and the underground storage basin of the Arolla pumping station in the Val d'Hérens (**Figures 7–9**).

Figure 6 The Dixence dam, now inside the Grande-Dixence lake.

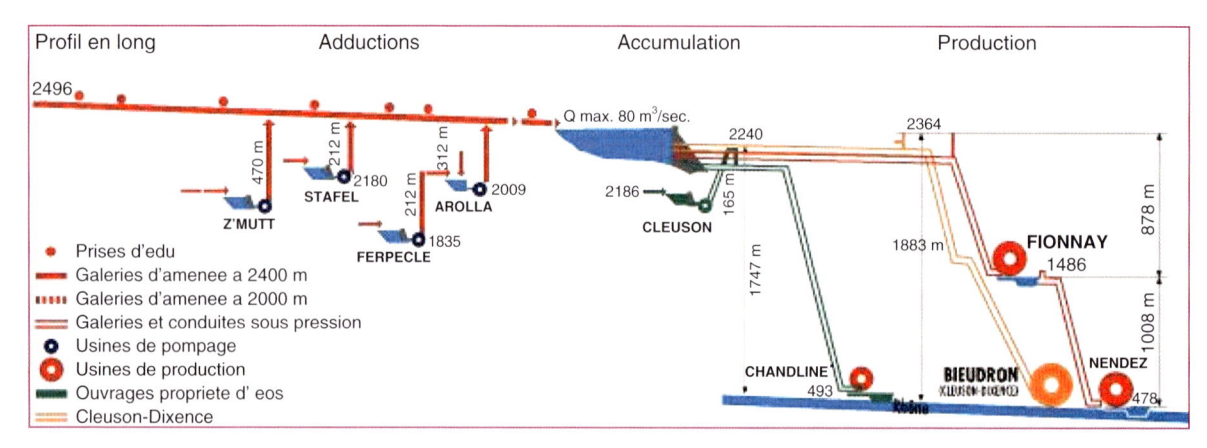

Figure 7 Schematic view of the Grande-Dixence scheme.

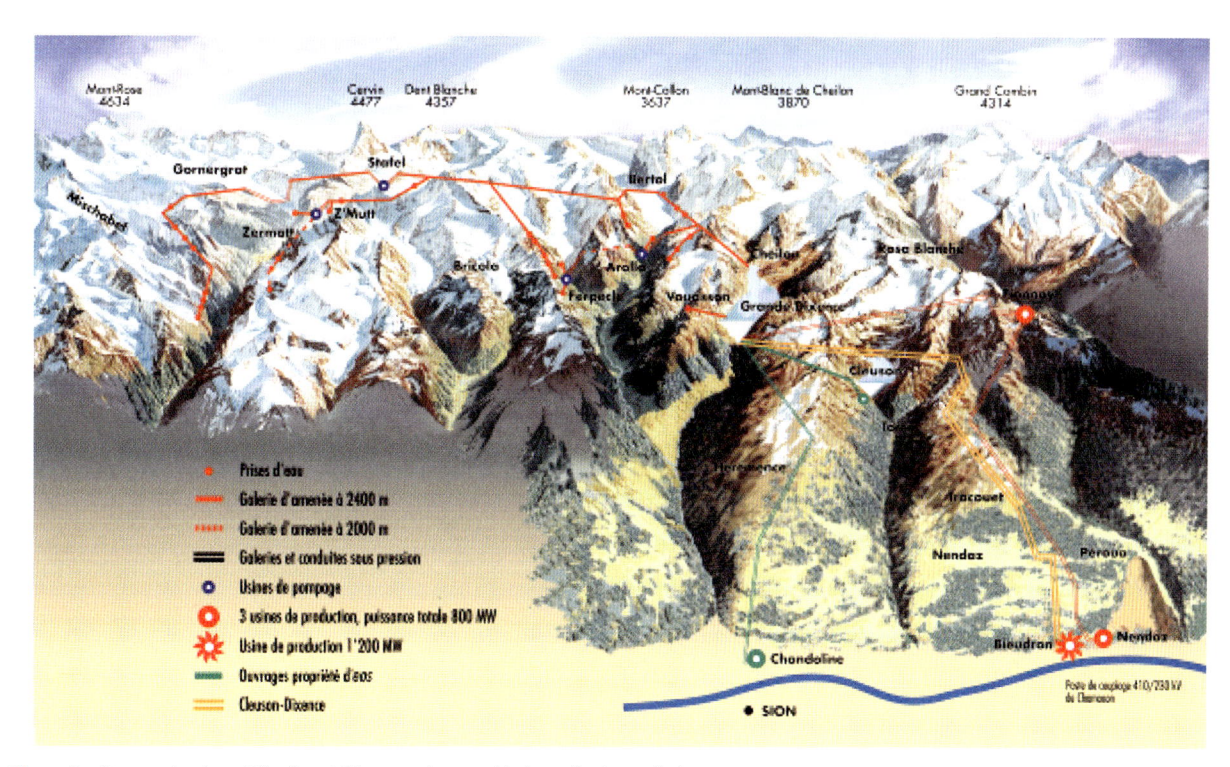

Figure 8 Panoramic view of the Grand-Dixence scheme with the collecting galleries.

Figure 9 The Dixence dam and the Grande-Dixence dam in April at low reservoir.

The total water head of the Grande-Dixence scheme is 1886 m, with water discharge of 45 m^3 s^{-1}. The water is turbined in two steps: at first in the intermediate power house of Fionnay, with a water head of 878 m, and then in the lower step in the Nendaz power house, in the Rhône valley, with a water head of 1008 m. The total capacity of the two power houses is 680 MW (**Figure 10**).

The total installed capacity of the Grande-Dixence scheme together with Chandoline is therefore 800 MW, with a total water discharge of 55 m^3 s^{-1}.

With a storage capacity of 400 000 000 m^3 and an annual inflow of about 520 000 000 m^3, the whole Grande-Dixence scheme produces energy for about 2000–2200 h a year, mainly in the winter time and during the strong demand periods.

6.12.6.4 Third Stage: The Cleuson-Dixence Scheme

The Cleuson-Dixence scheme was built during the period from 1993 to 1998, with the view to increase the installed capacity from 800 to 2000 MW in order to produce super peak energy on the base of 1000 h a year, to be injected on demand on the Swiss and the European electrical network.

Figure 10 The Grande-Dixence dam, 285 m high.

It consists of a new water intake in the Grande-Dixence dam, a new pressure gallery, a new penstock, and a new power house in the Rhône Valley, the Bieudron power house.

The water head is 1883 m (a world record) with 75 m^3 s^{-1}, and the Bieudron power house is equipped with three Pelton turbines of 400 MW each (also a world record) (**Figures 11** and **12**).

6.12.7 New Hydroelectric Schemes Presently under Construction in Switzerland

Several projects to increase the installed capacity, and especially for producing peak energy, are presently under construction in Switzerland. Among them, the three main ones are the following.

6.12.7.1 The Nant de Dranse Scheme

The Nant de Dranse scheme is a development of the Emosson high head storage scheme.

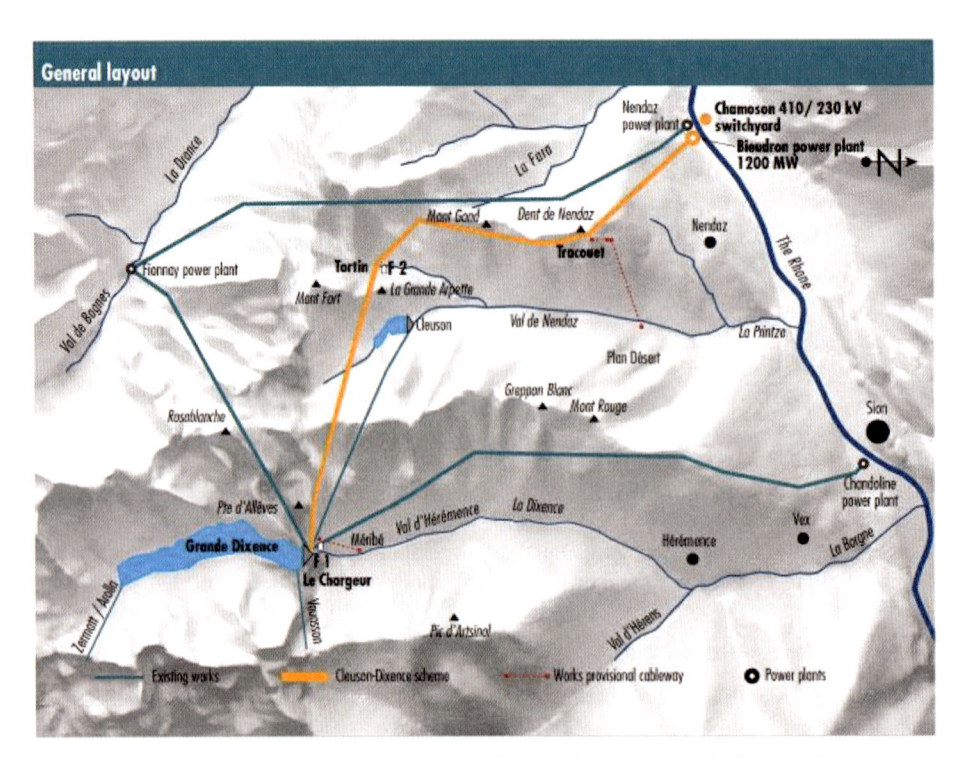

Figure 11 The general layout of the Dixence, the Grande-Dixence, and the Cleuson-Dixence schemes, high head galleries, and penstocks.

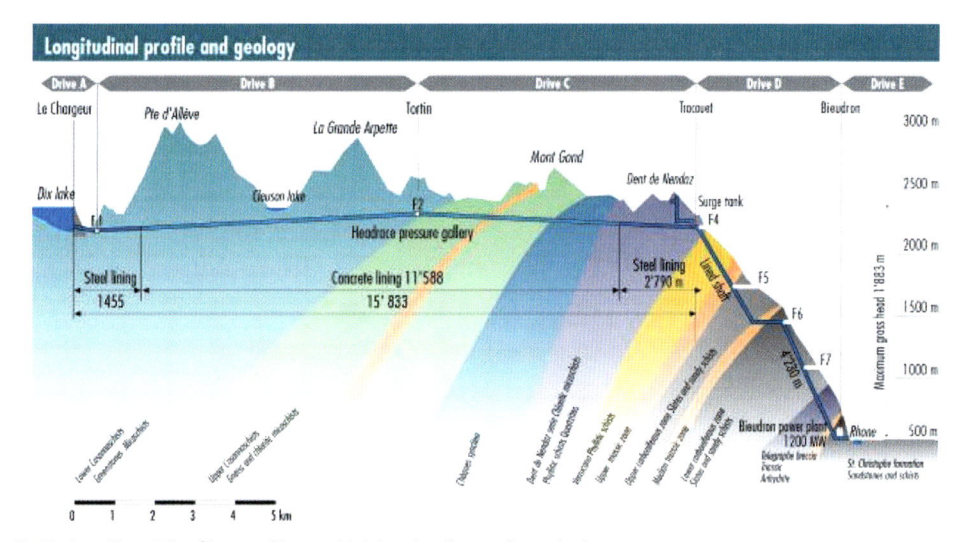

Figure 12 Longitudinal section of the Cleuson-Dixence high head gallery and penstock.

Emosson arch dam is located on the Swiss-French border in the Canton du Valais, between Martigny and Chamonix. Its extension is called the 'Nant de Dranse 600 MW Project'.

It connects the reservoir of the 'Vieil Emosson', a concrete gravity dam which has a reservoir capacity of 11 400 000 m^3 situated at elevation 2205 m, with the reservoir of the Emosson arch dam, having a reservoir capacity of 210 000 000 m^3 at elevation 1930 and with adding an underground power house equipped with four Francis pump–turbine of 150 MW each, to produce peak energy by pumping during the low-demand energy and then to turbine during the peak hours (**Figure 13**).

6.12.7.2 The Linthal 2015 Project

The Linthal 2015 project is a development of the Linth-Limmern high head storage scheme by extending with a pump–turbine power house with a capacity of 1000 MW.

The storage capacity of the natural lake of Mutt will be increased from 12 000 000 to 25 000 000 m^3, from elevation 2446 to 2474 m, by building a gravity dam which is 35 m high.

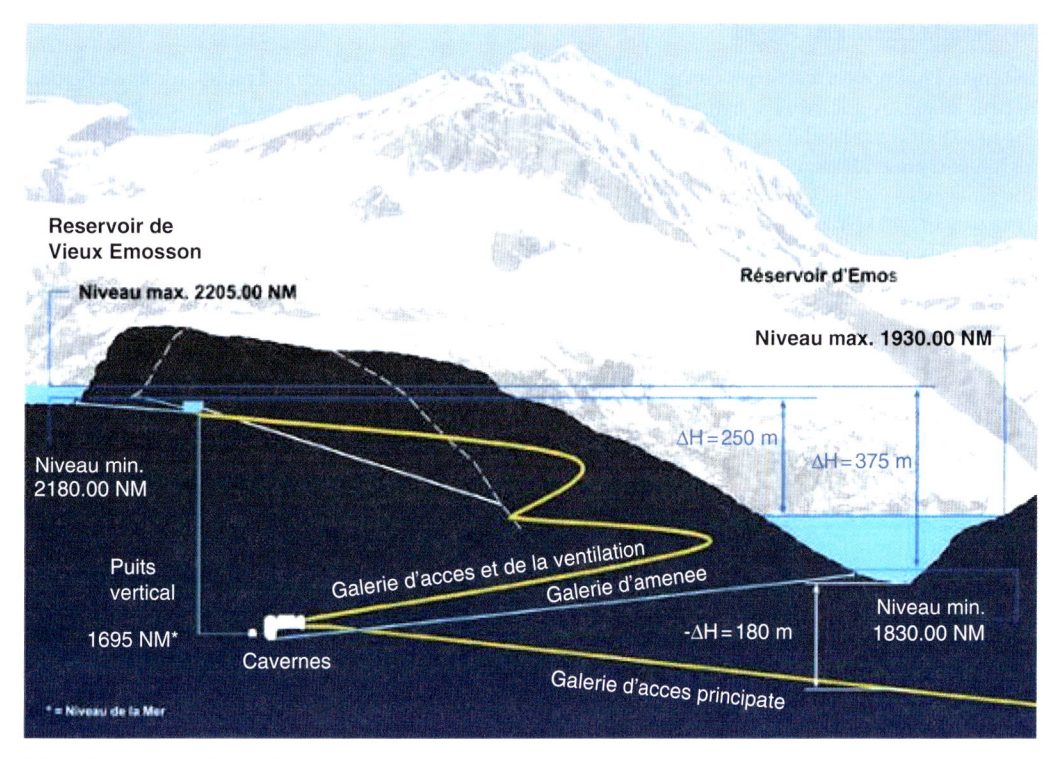

Figure 13 Schematic view of the Nant de Dranse project.

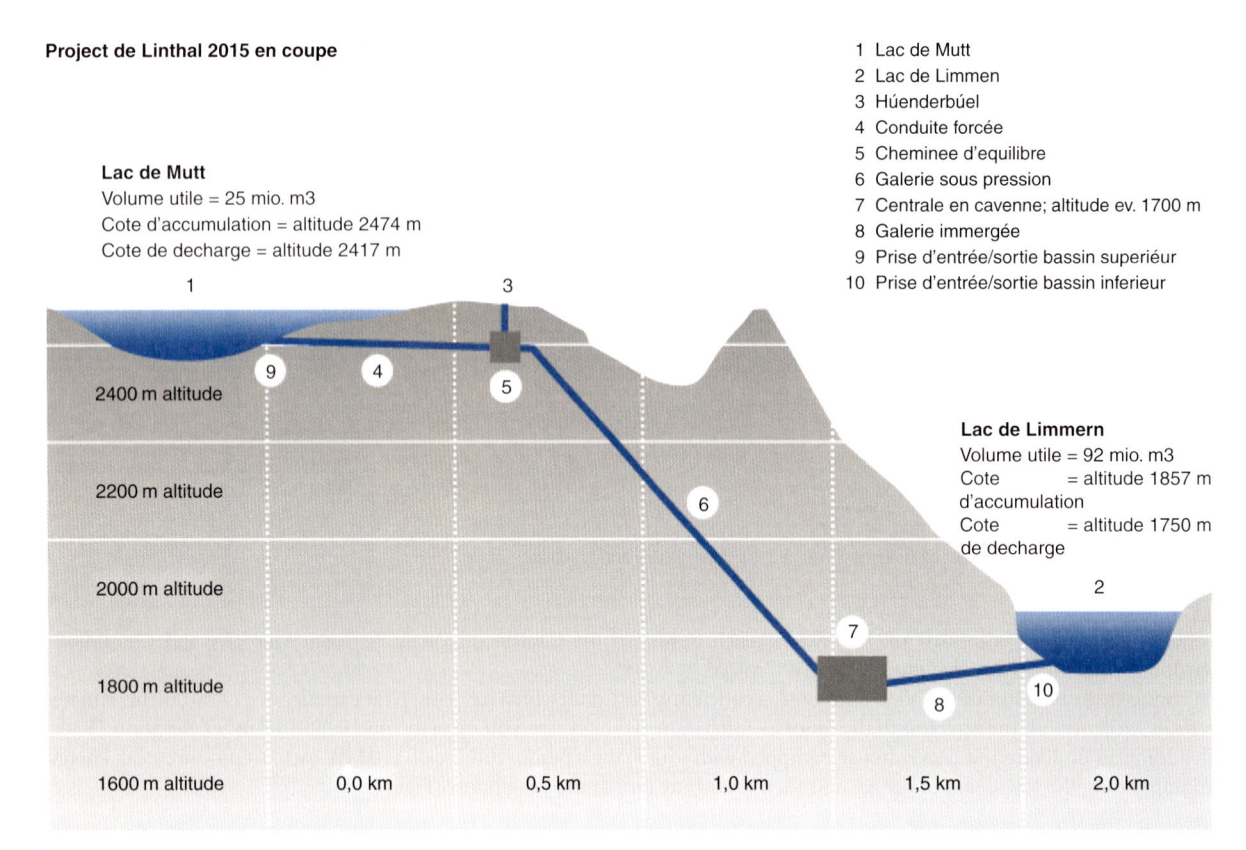

Figure 14 Schematic view of the Linthal 2015 scheme.

The water of the lower reservoir of the Linth-Limmern arch dam at elevation 1857 will therefore be pumped during the low-demand hours to the upper reservoir, 600 m higher, and turbined during the peak hours. The installed pump capacity, as well as the turbine installed capacity, will be 1000 MW. The work should be completed by 2015 (**Figure 14**).

6.12.7.3 The Hongrin-Léman Plus Project

The 'Hongrin Léman Plus' project is an extension of the Hongrin-Léman high head storage and pump–turbine scheme between the Hongrin reservoir made by the two Hongrin arch dams at elevation 1255 m and the lake of Geneva at elevation 377 m, corresponding to 878 of water head in one step.

The power house is presently equipped with four Pelton turbines, 4 × 60 MW, and four pumps, also 4 × 60 MW, with pump and turbine on the same axis.

The total capacity of 240 MW will be increased by 180 to 420 MW by adding a new underground power house with pumps and turbines, connected to the existing penstock. The work should be completed by 2013.

Relevant Websites

http://www.alpiq.com – Alpiq; Forces Motrices Hongrin-Léman S.A. (FMHL).
http://www.axpo.ch – Axpo Hydro Energy.
http://www.hydro-exploitation.ch – Cleuson-Dixence.
http://www.grande-dixence.ch – Grande Dixence.
http://www.swissdams.ch – Swiss Committee on Dams (CSB in French, STK in German).
http://www.bfe.admin.ch – Swiss Federal Office of Energy (OFEN in French, BFE in German).

6.13 Long-Term Sediment Management for Sustainable Hydropower

F Rulot, BJ Dewals, S Erpicum, P Archambeau, and M Pirotton, University of Liège, Liège, Belgium

© 2012 Elsevier Ltd. All rights reserved.

6.13.1	Introduction	355
6.13.1.1	General	355
6.13.1.2	The DPSIR Framework	357
6.13.2	**Driving Forces**	357
6.13.3	**Pressures**	357
6.13.3.1	Variability in Pressures	358
6.13.3.1.1	Variation in space	358
6.13.3.1.2	Variation in time	359
6.13.3.2	Measurement Techniques	360
6.13.3.2.1	Reservoir survey	360
6.13.3.2.2	Fluvial data	361
6.13.3.2.3	Modeling	361
6.13.4	**State**	362
6.13.4.1	Capacity Loss	362
6.13.4.2	Sedimentation Pattern	362
6.13.5	**Impact**	363
6.13.6	**Responses**	365
6.13.6.1	Measures Addressing Driving Forces	365
6.13.6.2	Measures Addressing Pressures	365
6.13.6.2.1	Sedimentation basins	365
6.13.6.2.2	Sediments bypass	366
6.13.6.2.3	Sediment routing	366
6.13.6.3	Measures Addressing the State	366
6.13.6.3.1	Dredging	366
6.13.6.3.2	Flushing	366
6.13.6.4	Measures Addressing Impacts	367
6.13.6.5	Numerical Modeling for Sustainable Sediment Management	367
6.13.6.5.1	Modeling scales and approaches	367
6.13.6.5.2	The WOLF modeling system	368
6.13.6.5.3	Case study: Alpine shallow reservoir	368
6.13.6.6	Assessing Sustainability of Hydropower Projects	370
6.13.7	**Conclusion**	376
References		376

6.13.1 Introduction

6.13.1.1 General

Hydraulic constructions have a significant influence not only on river flows but also on sediment fluxes. Therefore, they may cause long-term morphological changes in the watercourses. In particular, due to reduced flow velocity in reservoirs, transported sediments tend to settle down on the bottom of the reservoir, whereas erosion takes place downstream of the dam.

This reservoir sedimentation process in turn has a number of important consequences. The reduced available reservoir capacity undermines water supply and hydropower production. Flood control effectiveness is also decreased, and conditions may ultimately be reached in which the dam would be overtopped during an extreme flood. Operation of low-level outlets, gates, and valves is disturbed, while the extra pressure acting on the dam as a consequence of sediment deposition may affect dam stability. The abrasive action of sediment particles can roughen the surfaces of release facilities and cause cavitation as well as vibrations. Downstream of the dam, degradation can undermine the foundations and also deteriorate dam stability. Sedimentation also affects water quality.

As the life span of a dam is determined by the net sedimentation rate and since many existing major reservoirs are approaching a stage in which sediments clog low-level outlets, it is a key priority to take sedimentation into better consideration in the planning, design, operation, and maintenance of dams and reservoirs.

One way of preserving reservoir storage is to remove sediments out of the reservoir. For example, under favorable conditions, it is possible to flush sediments through outlet works within the dam. This technique can be applied both to existing dams (with adaptations) and to new dams. However, the technique is only effective depending on site-specific conditions and is not applicable in all geographical areas. An alternative consists in building more dams to replace the depleting storage of the existing stock. However, there are less and less suitable dam sites available, and many new dam projects are considered as leading to serious environmental and social consequences. Moreover, between half and one percent of the worldwide storage capacity of dams is lost annually as a result of reservoir sedimentation, resulting in the need to build approximately 400 dams every year just to compensate for lost storage capacity [1]. In addition, the global demand for water is increasing at a rate even higher than the rate of population growth. In contrast, the commissioning of large dams tends to decrease with time, as shown in **Figure 1**.

Worldwide storage in reservoirs reaches almost 6815 km^3. Seventy percent of the existing world stock of reservoir storage is situated in America, northern Europe, and China. Sedimentation rate can be expressed as the percentage of total original reservoir volume lost each year; this rate depends on the geographic region as shown in **Table 1**. Biggest annual loss of storage occurs in China with 2.3% of storage lost by sedimentation. There are also significant differences between the regional averaged rates and the rates for individual reservoirs, showing high spatial variability. For example, data gathered from 16 reservoirs in Turkey give a mean annual rate of storage loss of 1.2%, but the rates for individual reservoirs ranged from 0.2% to 2.4% [3], confirming that the problem is indeed highly site-specific.

As the industrialization of nations increases worldwide, power consumption is growing. Hydropower turns out to be an increasingly attractive alternative ever to generate electricity. The percentage of hydroelectricity actually exploited in 2005 as

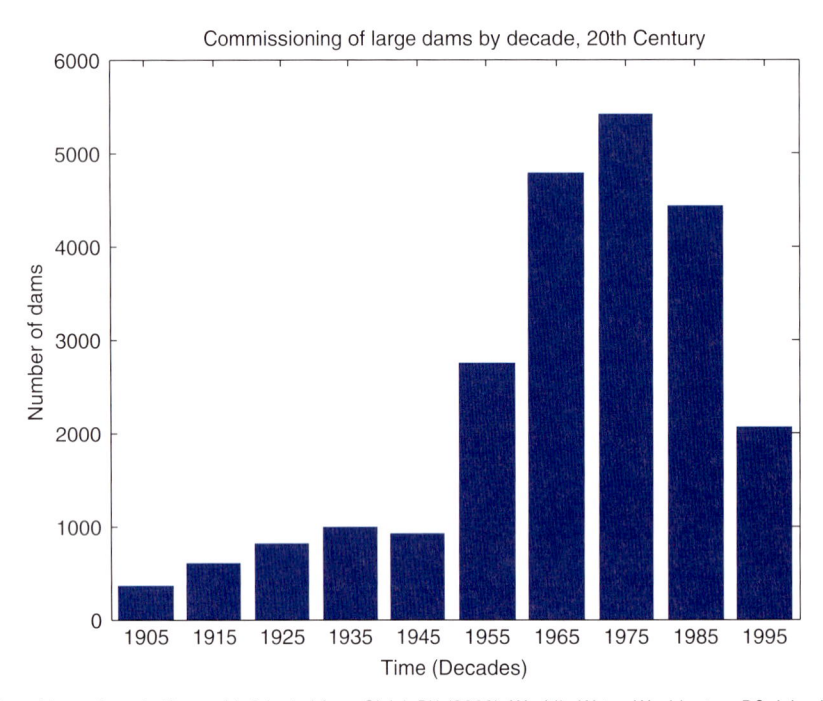

Figure 1 Commissioning of large dams in the world. Adapted from Gleick PH (2002) *World's Water*. Washington, DC: Island Press [2].

Table 1 Worldwide rate of reservoir sedimentation

Region	Inventoried large dams	Storage (km^3)	Annual percentage storage loss by sedimentation	Hydroelectricity produced with respect to potential for hydroelectricity
China	22 000	510	2.3	
North America	7 205	1 845	0.2	~2/3
Europe	5 497	1 083	0.17–0.2	~2/3
Africa	1 246	763	0.08–1.5	<1/10
Worldwide	45 571	6 325	0.5–1.0	~3/10

Adapted from White R (2001) *Evacuation of Sediments from Reservoirs*. London, UK: Thomas Telford [1]; Morris GL, Annandale G, and Hotchkiss R (2008) Reservoir sedimentation. In: Marcelo HG (ed.) Sedimentation Engineering: Processes, Measurements, Modeling, and Practice. *American Society of Civil Engineers* 110: 579–612 [3].

compared to the existing potential, as shown in **Table 1**, reveals that the potential of hydroelectric energy available is largely used in Europe and North America whereas it is far from being totally exploited in Africa.

6.13.1.2 The DPSIR Framework

It is very important to understand the complex problem of sedimentation management along with its causes and consequences. For the assessment of such environmental problems, the European Environment Agency (EEA) recommended the use of a specific framework, developed by the National Institute of Public Health and Environment (RIVM), which distinguishes driving forces, pressures, states, impacts, and responses. It is known as the Driver, Pressure, State, Impact, Response (DPSIR) framework. **Figure 2** shows the DPSIR model [4]. According to the DPSIR framework, there is a chain of causal links starting with 'driving forces' that exert 'pressures' on the environment and, as a consequence, the 'state' of the environment changes. This leads to 'impacts' that may elicit a societal 'response'. The response provides feedback to the driving forces, pressures, state, and/or impacts through adaptation of curative action. A driving force is an anthropogenic activity that may have an environmental effect, like agricultural or industrial human activities. The pressures account for the direct effects of the driving forces. As an example, industrial human activities can cause pressures like gas emissions or waste generation. The state is the condition of the water body resulting from both natural and anthropogenic factors. It is the physical, chemical, and/or biological state of the water. For example, the state of the water becomes acidic due to industrial wastes and emissions. The impacts are the environmental and/or human health effects of the pressure(s). Finally, the responses are the measures taken to improve the state of the water body. For example, water acidity could be reduced if agriculture is managed in a more environmentally friendly manner. The DPSIR framework is applied in the following sections with regard to the specific issue of long-term sediment management for sustainable hydropower generation.

6.13.2 Driving Forces

Driving forces can be seen as independent, autonomous, 'outside' forces directly or indirectly affecting a dependent system. For dams or reservoirs principal natural driving forces are geology, slope, climate, and/or vegetative cover. Main human driving forces are modifications in land use like urban development, deforestation, and agriculture, as well as drainage density. The features of the soil change, and therefore the behavior of erosion and deposition also changes. For example, urban development in the catchment area of a dam often reduces the sediments supply. Driving forces depend on the surveyed problem and may not be the same for all hydropower dams. A good example is deforestation. Deforestation may play an important part in flood generation because when rain falls in a geographic site where deforestation has happened, water is no longer absorbed and the runoff seriously erodes the soil. Deforestation occurs mainly in Amazonia, south Asia, Indonesia, and central Africa; for example, the 183 km² catchment of Ringlet reservoir in Malaysia has been gradually changed from forests to plantations and holiday facilities, which has resulted in a dramatic increase of the amount of sediment since the mid-1960s (**Figure 3**).

6.13.3 Pressures

Pressures are direct stresses deriving from the anthropogenic system (i.e., caused by humans, like deforestation) and natural systems, and affecting the natural environment. Particle input and transport, bottom and bank erosion, and resuspension are the principal pressures. Sedimentation is a more general term used to describe these pressures. Distribution, frequency,

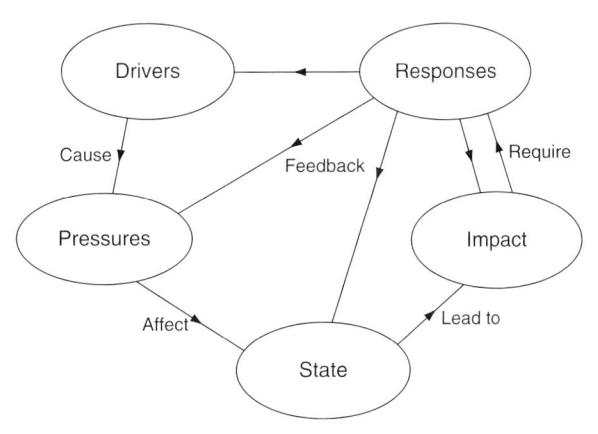

Figure 2 The DPSIR assessment framework.

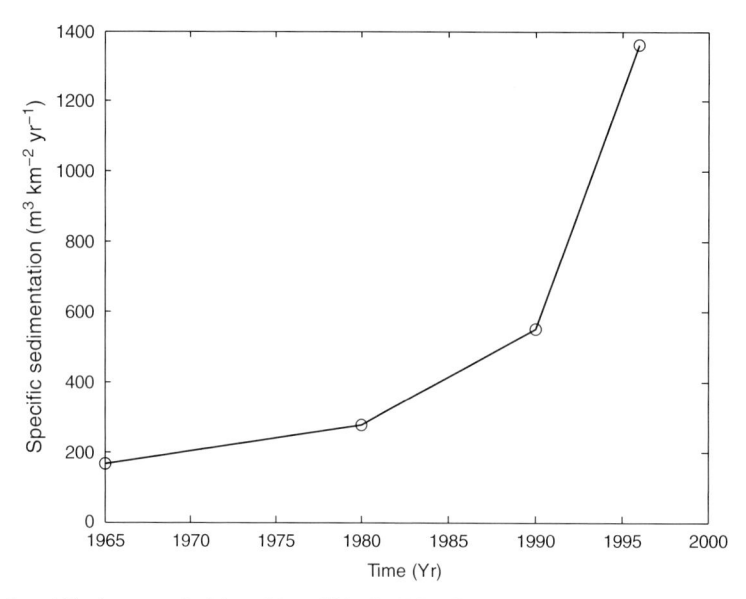

Figure 3 Specific sedimentation of Ringlet reservoir. Adapted from White R (2001) *Evacuation of Sediments from Reservoirs.* London, UK: Thomas Telford [1].

and intensity are the characteristics (or magnitude) of the pressures, which are explained first in this section. Next we discuss spatial and temporal variations in sediment yield. Finally, techniques to measure the magnitude of these pressures are discussed.

6.13.3.1 Variability in Pressures

The amount of sediment exported by a basin (drainage network) over a period of time is referred to as 'sediment yield'. Obviously, it is always less than the amount of sediment eroded within a watershed, owing to redeposition prior to reaching reservoirs. 'Sediment delivery ratio' is the ratio of delivered sediment to eroded sediment. 'Specific sediment yield' is the sediment yield per unit area.

6.13.3.1.1 Variation in space

Sediment yield is highly variable over space. In some cases, even a small part of the landscape unit contributes a disproportionate amount of the total sediment yield. For instance, intensive local logging leads to a substantial increase in sediment yield. Sometimes, the yield ratio between a logging zone and a 'normal' one can reach several hundreds. Hence, knowledge of the spatial variation in yield is required to focus yield reduction efforts on the landscape units that deliver the maximum amount of sediments to the reservoir.

Jansson [5] analyzed the data from 1358 gauging stations worldwide with watersheds of various sizes from 350 to 100 000 km^2. These data are shown in **Table 2**, where the stations are classified into four categories. Only 8.8% of the land area accounts for 69.1% of the sediment load, whereas 58.8% of the total land area contributes only 4.2% of the total sediment yield. Therefore, watershed areas that produce the maximum amount of sediment must be identified and controlled. Obviously, sediment yield depends on the human and natural driving forces as detailed in Section 6.13.2.

The wide variation in specific sediment yield in the global data set is also reflected at all levels of analysis: national, regional, and within-watershed. The phenomenon is highly site specific. Specific sediment yields typically vary by up to 3 orders of magnitude depending on the geographic region.

Table 2 Sediment yield from gauging stations worldwide

Yield class ($t\,km^{-2}\,yr^{-1}$)	Number of gauging stations	Gauged land area (%)	Total gauged sediment load (%)
0–100	687	58.8	4.2
101–500	426	25.6	14.7
501–1000	145	6.9	12.0
>1000	179	8.8	69.1

Adapted from Jansson MB (1988) A global survey of sediment yield. *Geografiska Annaler Series A, Physical Geography* 70(1/2): 81–98 [5].

6.13.3.1.2 Variation in time

Estimates of long-term sediment yield have been used for many decades to size the sediment storage pool and estimate reservoir life. However, the models that estimate long-term sediment yield are not accurate for floods, as most of the sediment is exported from watersheds during this relatively short period of the year. For instance, Santa Clara river basin in Southern California is reported to have discharged 50×10^6 tons of sediments during a single flood event, which represents more than 700 times the measured average annual load. In the United States, more than half of the annual sediment load is discharged in only 1% of time. Thus temporal variation is also a key factor to study.

Techniques for evaluating sediment yield depend on the choice of time horizon. Very long-term trends in sediment yields appear after decades and can usually be correlated with human activities in the watershed. As an example, it has been reported [3] that the Piedmont area of the eastern United States was completely deforested by the mid-1800s, leading to increased erosion rates and sediment yield. After 1920, erosion rates declined because hillside farms were abandoned and revegetated naturally, while soil-conservation methods were implemented in the remaining farms. Despite the high erosion rate of soil over a 150-year period, the sediment delivery ratio remained as low as approximately 5% because eroded sediments were deposited further downstream in channels and on floodplains.

Long-term trends can be visualized by constructing a cumulative-mass curve for water and sediment. **Figure** 4 gives a better idea of trends than a timewise plot because it helps compensate for runoff variability. The dotted curve accounts for an equivalent system in which flushing is employed, thus decreasing the cumulative suspended sediment discharge. In applying regional curves to a particular study site, care must be taken to consider local features such as upstream reservoirs, land use, and topographic or geological conditions that may depart from regional norms.

A cyclic seasonal variation can be observed in specific regions of the world. For example, seasonal variations of erodibility were observed in Nepal because of monsoon and vegetation cycles. In some other regions, wind plays an important part, transporting sediment from ridges into depressions where it becomes available for fluvial transport.

Short-term trends are attractive for describing a flood or storm situation. Usually, a suspended solid concentration (C) versus discharge (Q) plot is used to represent the short-term phenomenon. As shown in **Figure** 5 and **Table** 3, storm or flood events can be divided into three categories. The characteristics correspond to the forms of the C–Q graphs observed in **Figure** 5 [6].

Class I represents cases for which sediment concentration responds immediately to a variation in discharge. In the graph, discharge is just a scaling of the sediment concentration. This implies that sediment supply through the flood is uninterrupted and sediment concentration should be directly related to hydraulic factors alone. Class I occurs not very often, but these curves are widely used for their simplicity.

In contrast, Class II occurs commonly. This pattern is usually observed when sediment concentration reaches a peak value before discharge as shown in **Figure** 5. Under certain conditions, it also happens that sediment concentration and discharge peak simultaneously. Three causes can lead to clockwise C–Q loops:

- The sediment accumulated or the easily erodible material in the watershed is washed out when water discharge increases a little bit, and sediment load supply decreases over the duration of the event because sediment becomes less readily erodible.
- During the event, prior to the peak of water discharge, sediment supply from the bed becomes limited because of the development of an armor layer.
- Spatial variations in rainfall and erodibility across the watershed can concentrate sediment discharge from areas of high sediment production near the catchment area outlet before the peak of discharge.

Class III occurs when soil erodibility is high and erosion is prolonged during flood or as a result of specific rainfall and erodibility distributions across the watershed. In such cases, occurrence of the peak of the sediment concentration curve is delayed.

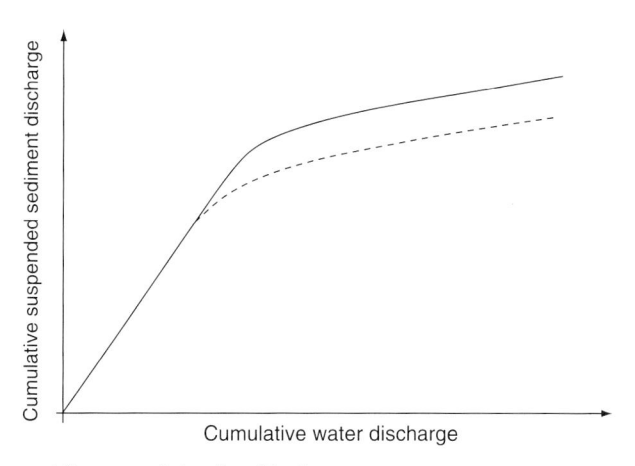

Figure 4 Cumulative discharge vs. cumulative suspended sediment load.

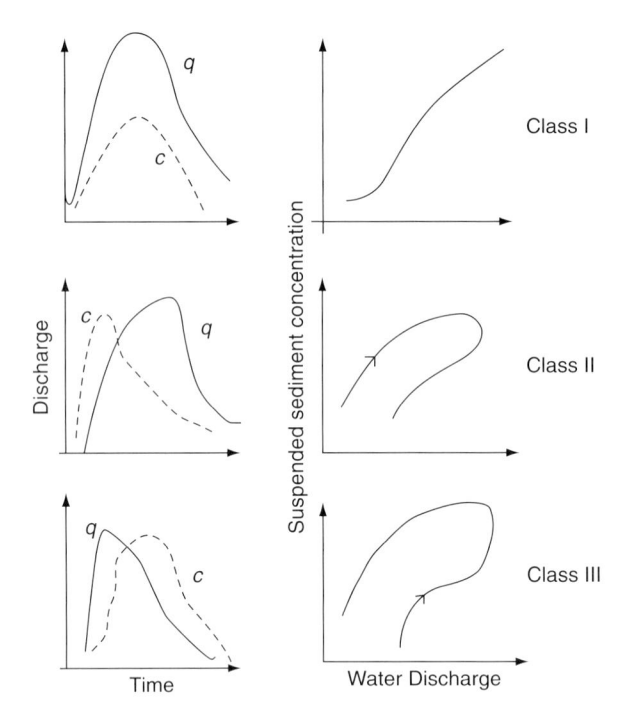

Figure 5 C–Q graphs showing typical relationships between discharge and sediment concentration observed during floods.

Table 3 Classification of C–Q graphs

Name	Characteristic	Occurrence
Class I	Single-value curve	Rare
Class II	Clockwise loop	Common
Class III	Counterclockwise loop	Common

Unfortunately, the C–Q relationship is not fixed for a given watershed but it can vary from one flood to another because the factors of intensity and areal distribution of the rainfall and sediment supply always keep changing. Nevertheless, discharge–concentration data pairs from many events can be combined to develop a fairly reliable relationship.

6.13.3.2 Measurement Techniques

There are two approaches for measuring sediment yield:

- inspection of the sediments volume deposited in the reservoir
- continuous monitoring of fluvial sediment discharge.

The main advantage of the first method is its accuracy because the construction of a reservoir eliminates problems of missed or underreported events at fluvial gauge stations. The main advantage of the second method is the good description of spatial and temporal patterns; it is thus easy to identify and diminish the sediment yield. These two strategies are detailed below.

6.13.3.2.1 Reservoir survey

The first reservoir survey method is bathymetric mapping, which is often combined with local surveys to determine the grain size of the deposits and enables verification of computational or mathematical models. Generally, reservoir measurements may be performed at intervals of 5–20 years; it depends essentially on budgetary constraints, rate of storage decrease, and management requirements. However, unscheduled surveys may be called for after a major flood or other phenomena that lead to a surge in sediment yield. Surveys should also be conducted downstream of the dam. More than 20 years of surveys may be needed to get a reliable trend in long-term sediment accumulation. There are two main techniques to compute reservoir volume: the range line and contour surveys. The original volume of reservoir is often computed using the contour method based on preimpoundment topographic mapping.

The widely used range line method uses a system of cross sections where depths are measured. Each range line is tied to the initial elevation–capacity relationship of the reservoir reach corresponding to that range and provides the base against which all future surveys will be compared. This sequence is repeated at regular time intervals, and an elevation–capacity relationship with respect to time can be plotted. The range line method is less accurate than the contour method because the latter entails a complete survey of the reservoir. The use of global positioning system (GPS) facilitates these measurements today.

An alternative method for drawing a contour map of the reservoir is used when the pool is often drawn down or emptied. When the water level decreases, photographs of the reservoir may be taken from an aircraft or a satellite at different stages, gradually drawing the contour map of the reservoir.

The rate of sediment accumulation can also be determined by measuring the depth of deposition above an identifiable and datable layer of ^{137}Cs. This is a toxic radioactive isotope of cesium. It is water-soluble but penetrates only a short distance into clayed soil. As the half-life of ^{137}Cs is approximately 30 years, it can be used as a dating tool. Nevertheless, ^{137}Cs is a toxic element, so its production is forbidden. Therefore, if this method is to be employed, the watershed must be impacted by uncontrolled events such as large fires, volcanic eruption, or Chernobyl-like radioactivity that produce or have produced ^{137}Cs. The datable horizon is limited in time. Indeed, before the year 1954, which corresponds to the first atmospheric nuclear testing, ^{137}Cs never appeared in measurable amounts [7]. Another limitation is that the procedure does not work when the reservoir is drawn down. Moreover, it is necessary to take several samples from a number of locations to reliably map deposition thickness because of the uneven deposition in reservoirs. An isotope of lead ^{210}Pb is another radioactive element sometimes used to measure sediment deposition.

6.13.3.2.2 Fluvial data

Sediment-rating curves are one of the widely used tools for the estimation of sediment discharge in a river based on fluvial data. Over a given period, data (obtained by gauging stations) are plotted in an instantaneous discharge–concentration relationship. These graphs are often in log–log scale. A sediment-rating curve from several years of field data that include sampling of flood events can be applied to a long-term discharge data set to estimate long-term sediment yield. There are different procedures to be considered for the development of accurate rating curves. Regression techniques very often incorporate bias if data are too widely scattered. In such cases, data of a particular kind of runoff event should be gathered (e.g., seasonal runoff). It is thus important to back test a rating relationship by applying it to the original stream flow data set to ensure that it correctly computes the total load. Sometimes a multiple slope is also necessary to have accurate values at high discharge.

If sediment concentration data can be measured frequently, the use of sediment-rating curve becomes redundant; sediment load can be computed directly as the product of discharge and concentration at short intervals. The main advantage of this method is that time variation can be accurately represented; for instance, looped rating curves can be detected. In highly variable hydrographs, for example, rivers, short sampling intervals are required to accurately track sediment yield.

Turbidity measurement can give an idea of the suspended sediment concentration. Turbidity is the term used to describe the reduction in water clarity due to particulate matter suspended in solution. The attenuation (reduction in strength) of light passing through a sample column of water gives a measure of its turbidity. An automatic pumping sampler is used to take samples at short intervals, but laboratory costs are high. Moreover, it is possible that the sample bottles are filled before the end of the peak event, resulting in high undercounting errors. When pumped samples and turbidity measurement are analyzed together, it represents a viable strategy for improving the quality of sediment discharge data. There is no direct correlation between turbidity and suspended sediment concentration; hence errors are unavoidable. However, turbidity data can be recorded every few seconds and averaged. As a matter of fact, errors can be reduced if local sensors take into account the suspended sediment concentration at several points over a cross section of the river.

The method discussed above can measure the amount of suspended sediment. For bed load, the method is obviously different because riverbed particles are usually bigger than suspended ones. Bed load samplers directly measure the load of particles moving along the bed. The main difficulty in measuring bed load is the highly irregular rate of bed load transport even at a constant discharge. Another challenge is that the bed load transport is multidirectional; it can also occur in the transverse direction of the flow. Hence, sampler efficiency is defined in such a way that sampled transport rate divided by sampler efficiency determines the true transport rate. Sampler efficiency is determined by calibration in a hydraulic flume in the laboratory and varies as a function of grain size and transport rate. The method widely used for measuring the transport distance and transport rate of sediment is the marking of stones (painting, embedding magnets) in one section of the stream and relocating them repeatedly during a certain period of time. This method can prove appropriate to determine the condition of initiation of motion in different areas of the streambed. Methods for collecting the grains depend on their size and the depth of water. Coarse grains (gravels, cobbles) are collected by hand if not too heavy, whereas smaller grains (sand) can be sampled with a mechanical system if the river is shallow and the velocity remains moderate.

6.13.3.2.3 Modeling

Neural network models are numerical models rather than experimental ones. They are nonlinear black boxes that establish a link between input data (stream flow, rainfall, temperature, and other parameters from gauging stations) and output data (sediment concentration) by training their internal algorithms and their weighting scheme. The main advantage of this type of modeling as compared to sediment-rating curves is that sediment concentration can be correlated with several inputs. It is thus easier to show the effects of any one parameter on sediment concentration. Several approaches exist for this method. One approach is to use these models to develop rating relationships based on channel hydraulic characteristics. Another approach is to predict suspended sediment concentration or discharge based on channel with the help of watershed and hydraulic parameters or watershed parameters alone.

Another way to compute sediment yield is to use spatial modeling. Spatially distributed data may be analyzed to compute the yield of both water and sediment from the watershed based on observation of the soil, hydrologic input parameters, and land use, and the output data (sediment load and runoff) are routed to the watershed exit. Thus, the main disadvantage of this method is that the problem highly depends on watershed data. The next step consists in coupling the empirical erosion prediction model with a sediment delivery module to simulate sediment yield. Alternately, models that simulate both sediment detachment and transport processes may be coupled with fluvial routing procedures to simulate sediment yield. As an advantage, several land use scenarios can be compared to identify areas where erosion control would provide the highest benefit.

6.13.4 State

The state accounts for the environmental conditions of the system. It corresponds to a description of the system subjected to pressures and driving forces. Here, the amount of sediment trapped in the reservoir, the reservoir sediment deposition, and its geometry describe the state of the system. Another important point developed below is the expected future evolution of the reservoir.

6.13.4.1 Capacity Loss

When a tributary enters an impounded reach, flow velocity decreases and the sediment load begins to deposit. The volume of the sediment deposited in a reservoir depends on the trap efficiency of the reservoir and the density of the deposited sediment. Trap efficiency is the percentage of sediment load that stays in the reservoir over a given period of time. It depends highly on the fall velocity of sediment particles, the shape and size of the reservoir, and the variation of flow through the reservoir. This parameter is computed with the help of graphics. There are two evaluation methods widely used. The first one was developed by Brune [8] for large-storage reservoirs. The trap efficiency is given as a function of the ratio of reservoir capacity to average annual inflow. The capacity of the reservoir is taken at the mean operating pool level for the period to be analyzed. For smaller reservoirs, Churchill [9] developed a specific trap efficiency curve.

Though it is possible to estimate the sediment deposition in the reservoir based on these empirical formulae, if the anticipated sediment accumulation is larger than one-fourth of the reservoir capacity, trap efficiency should be determined for incremental periods of the reservoir life because trap efficiency generally decreases with time.

Periodic reservoir surveys are often considered as one of the most suitable methods for the determination of sediment yield from an upstream watershed. The volume of sediment trapped in a reservoir during a period between two surveys is simply the difference in reservoir volume between these two surveys. The difference between the original capacity (water volume) and the actual gives a global estimation of the loss of storage in a reservoir (**Figure 6**). In association with the difference in area, **Figure 6** also gives some insight into the distribution of sediments at a given elevation.

The average percentage values of annual loss of storage due to sedimentation vary gradually between 0.5% and 1%. Except for China where the mean loss of storage per year reaches 2.3% (**Table 1**), storage loss generally tends to grow faster in smaller reservoirs than in larger ones. Today, the number of dams and reservoirs commissioned worldwide tends to decrease, and the rate of loss of storage is not counterbalanced by the newly available storage.

6.13.4.2 Sedimentation Pattern

A highly conceptual sketch of the sedimentation processes is presented in **Figure 7**. The coarse fraction of the sediments entering the reservoir (typically cobbles) creates a delta. This part is called the 'topset bed.' Downstream limit of the delta is characterized by an

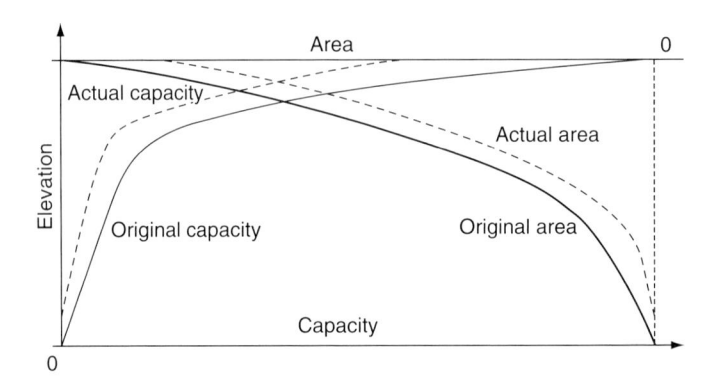

Figure 6 Area–capacity curve.

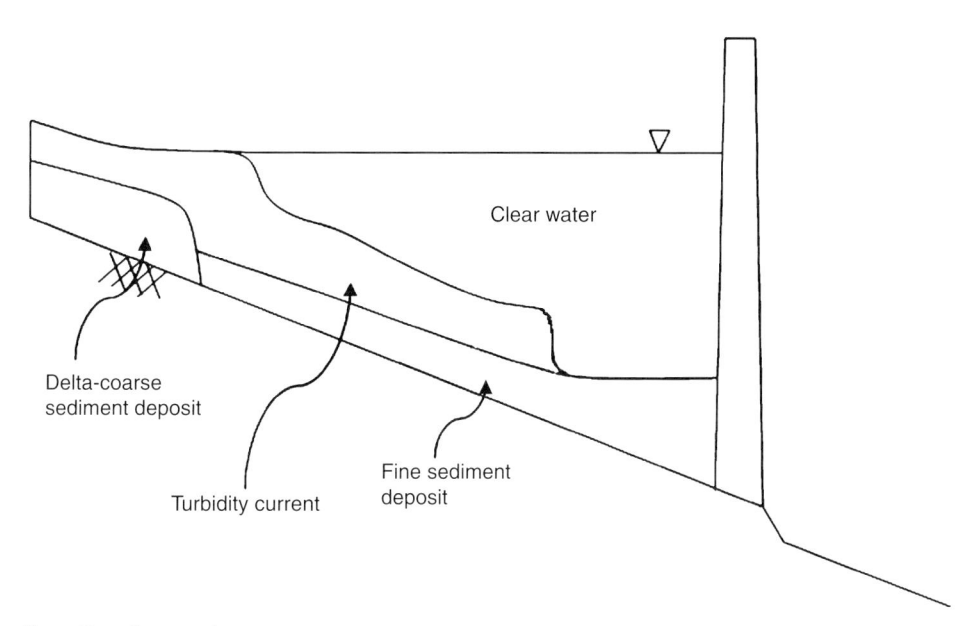

Figure 7 Deposition patterns in reservoirs.

abrupt reduction in grain size; it also corresponds to the downstream limit of bed material transport in the reservoir. Upstream limit is not well defined, and sediment deposits extend into the river. After the delta section, there may be a plunge point if turbidity currents take place in the reservoir. Turbidity currents are flows of water and very fine sediments (< 100 μm) driven by the difference in density between clear water and sediment-laden water. The 'bottomset bed' consists of fine sediments, which are deposited beyond the delta by suspension and turbidity currents. Under specific conditions, such as floods, reservoir drawdowns, and slope failures, coarser sediments may be transported further downstream into the reservoir. It is thus possible to observe several layers of different grain sizes near the dam. Although longitudinal deposition patterns can have different shapes, depending on pool geometry, discharge, grain size characteristics of the inflowing load, and reservoir operations, the most typical pattern is well represented by **Figure 7**.

Regarding lateral depositional pattern, the deposition in a cross section of the reservoir occurs first in its deepest part and subsequently spreads out across the submerged floodplain to create broad flat sediment deposits (**Figure 8**; stage I).

Sedimentation rates may be alternatively expressed by means of reservoir half-life, which is the time required to lose half of the original capacity of the reservoir. In contrast, 'reservoir life' is defined as the time between the construction of the dam and the total filling of the usable storage pool preceding the abandonment of the structure. Three successive stages may be distinguished in (**Figure 8**):

Stage 1: *Continuous sediment trapping.* This is the period when sediments fill the deepest parts of the cross section of the reservoir. During this first stage, sediment inflow is not counterbalanced by sediment outflow.

Stage 2: *Partial sediment balance.* This stage represents a mixed regime between deposition and removal of sediments. If sedimentation proceeds uninterruptedly, the former pool area looks like a channel–floodplain configuration. The inflow and discharge of fine sediment may be nearly balanced, whereas the coarse bed materials continue to accumulate. Sediment-balancing techniques (e.g., flushing) can produce a partial sediment balance to help preserve useful reservoir capacity.

Stage 3: *Full sediment balance.* This is the stage when long-term sediment inflow counterbalances long-term sediment outflow. This balance is obtained when sediments can be transported beyond the dam or artificially removed (e.g., flush).

Most of the dams are designed to work in the continuous sediment trapping mode, but some reservoirs have been designed to achieve sediment balance, such as Three Gorges reservoir on China's Yangtze River designed to reach full sediment balance after about 100 years.

Sediment management can postpone the filling of the reservoir. It is also possible to increase the capacity of the reservoir. Capacity history curves may be drawn to visualize historical and anticipated changes in usable storage volume under different management strategies.

6.13.5 Impact

Sedimentation impacts not only the reservoir but also a short distance upstream of the reservoir and areas far downstream of the dam. Table 4 reviews the main impacts with respect to these three areas.

The primary impact due to sedimentation in a reservoir is the storage loss that makes water control, hydropower generation, and navigation difficult. It also affects water supply. Coarse material can abrade hydromechanical equipment, and sediment that

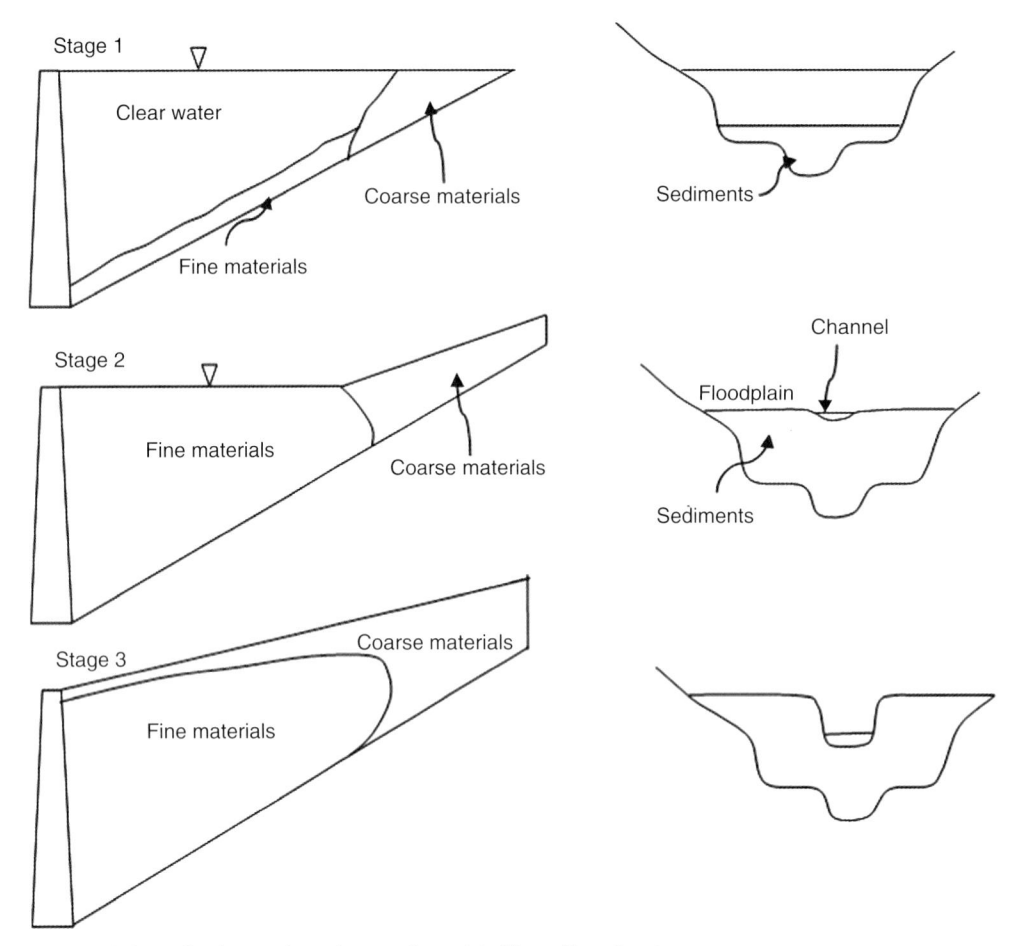

Figure 8 Long-term evolutions of an impoundment in case of complete filling with sediment.

Table 4 Sedimentation impacts

Locations	Above the normal pool	Pool area	Below the dam
Impacts	Bed aggradation	Reduced conservation and flood control pool volumes	Channel incision
	Higher level flooding	Clogging of intakes	Bank erosion
	Higher groundwater levels	Abrasion of structural equipment	Lower groundwater level
	Impaired navigation	Environmental impacts Increased static load on the dam	Scouring below the dam

Adapted from Morris GL, Annandale G, and Hotchkiss R (2008) Reservoir sedimentation. In: Marcelo HG (ed.) Sedimentation Engineering: Processes, Measurements, Modeling, and Practice. *American Society of Civil Engineers* 110: 579–612 [3].

deposits on the dam increases the static loading on the structure. The presence of contaminants in sediments can compromise the feasibility of procedures to remove the sediments (dredging, flushing, etc.).

Deltas create an increase in the upstream water level and bed aggradation upstream of the reservoir. The general increase in water level can cause an increase in the frequency and severity of floods and also reduce the clearance beneath bridges. Aggradation will also increase groundwater level, leading to ecological impacts such as alteration of habitats.

Below the dam, trapping of sediments leads to an incision in the channel. A general decrease in water level is thus observed and the following impacts are noticed: tributaries' degradation, destabilization and undercutting of streambanks, undermining of structures like bridge piers. Net erosion below the dam occurs only if there is a sediment deposit below the dam. Other environmental effects are also observed in tributaries below the dam, as a result of lower groundwater levels, such as dewatering of wetlands. During the first decade of reservoir operation, erosion of the river below the dam will be limited by the formation of an armor layer, preventing the erosion of finer sediments by clogging them with coarse ones.

These environmental problems are at the root of public's and environmental organizations' opposition to the construction of new reservoirs. Nevertheless, if long-term impacts of the reservoir are taken into account from the early stages of dam design, they can be mitigated to a great extent by means of a proper management scheme.

6.13.6 Responses

In man-made reservoirs, constructed for water supply, irrigation, flood and low-flow control, or hydropower generation, both the loss in storage capacity and the location of deposits are a concern. These reservoir sedimentation issues would be solved if watershed erosion could be stopped, or at least controlled, and sediment yields drastically reduced. This may, however, turn out to be economically nonfeasible and would create other problems such as upstream river bed degradation and scouring.

In contrast, authorities in charge of reservoir sustainability may implement responses which, in line with the DPSIR framework, may be targeted toward any component of the causal chain, between driving forces and impacts.

Therefore, possible responses may be classified into a number of categories, depending on which stages of the DPSIR chain they affect. Among other possibilities, sediment-control measures are related to driving forces, sediment bypass is linked to pressures, whereas sediment dredging or flushing are oriented toward the state of the reservoir.

As a result of the complexity and natural variability of the involved sedimentation processes (such as the influence of turbulence or grain sorting) and site-specific parameters, there is no single measure generally suitable for solving sediment management concerns. Therefore, an optimal site-specific strategy needs to be developed. For this purpose, a comprehensive understanding of the fundamentals of sediment transport, erosion, and deposition is a prerequisite. Very valuable is also a thorough quantitative knowledge of the sedimentation processes that take place on the site, which requires suitable measurement devices and monitoring programs. For practical purposes, a wide range of possible responses needs to be reviewed to lead to a cost-effective and sustainable sediment management strategy, usually involving a combination of several carefully selected measures. The optimal combination of sediment management measures may vary in time during the life of the reservoir and depends mainly on the purposes of the reservoir, its hydrological size (capacity vs. inflow), and site-specific environmental challenges.

This section provides an overview of responses for mitigating sedimentation and its impacts, including both standard practice approaches and more advanced techniques. These responses are classified depending on the component of the DPSIR chain they address.

6.13.6.1 Measures Addressing Driving Forces

From the perspective of mitigating sedimentation in reservoirs, reducing soil erosion in the catchment may appear as the ideal solution, though difficult to successfully implement in practice. It typically involves measures such as terracing or suitable agricultural practices, as well as structural measures such as bank protection and slope reduction using sills in thalwegs.

Experience shows that it may be particularly effective in small catchments or catchments with confined intensive erosion-producing areas, while being economically unrealistic if the reservoir has a large drainage area.

Furthermore, implementation of such measures requires cooperation of a potentially large number of landowners throughout the basin. Obtaining the commitment of all the influencing parties may be difficult, because reduced reservoir sedimentation usually benefits other parties than those who own and exploit upstream land. Nevertheless, involvement of stakeholders may be facilitated by the numerous side-benefits of reduced soil erosion, including enhanced soil fertility, water quality, and state of the environment.

In some specific areas, wind erosion may play a part and thus needs to be addressed by appropriate measures including increasing vegetative cover and construction of wind barriers.

6.13.6.2 Measures Addressing Pressures

For a given set of driving forces, measures addressing pressures tend to reduce net sediment inflow into the reservoir, by means of upstream sediment retention, sediment bypass, or sediment routing.

6.13.6.2.1 Sedimentation basins

Sedimentation basins constructed upstream of the reservoir may prove efficient to catch the coarse sediments. They need to be periodically dredged by mechanical means, or sediments need to be flushed through the main reservoir.

Designing of cost-effective sedimentation basins remains a challenge. Indeed, models currently used in practice are still relatively unable to predict the pattern of deposition as a function of the geometry of the reservoir, the hydraulic conditions, and the sediment characteristics. Empirical approaches developed during the last few decades focus on predicting the amount of deposits [10, 11], whereas they fail to provide predictions of the spatial distribution of deposits, which is a prerequisite for developing an optimal sediment removal strategy and for implementing proper basin maintenance. To obtain this additional information, knowledge of the flow pattern is necessary, as detailed by Dewals *et al.* [12] and Dufresne *et al.* [13]. The inaccuracy of the current empirical methods could also result from the fact that they disregard the flow pattern when estimating trapping efficiency of the basin.

6.13.6.2.2 Sediments bypass

An ideal way of managing sediments is to prevent them from entering the reservoir by diverting flows with high sediment concentrations. Sediments can be bypassed from reservoirs, for instance, by installing the dam in a meander and diverting the sediment-laden flows through the inner floodplain.

Such sediment bypass has been used since 1998 at Asahi reservoir in Japan, where a diversion channel successfully transports most of the bed load and suspended load material from upstream of the reservoir directly toward downstream of the dam [14]. Such structures are generally costly, and care must be taken to control abrasion of the diversion tunnel, through which flood flows with high sediment rates are conveyed.

Sediments bypass becomes straightforward if the reservoir is constructed off-stream and supplied through a diversion channel. The water intake from the main course of the river should be designed in such a way as to prevent large amounts of sediments from reaching the reservoir. Side-benefits of such a measure include essentially unaltered natural valley and undisturbed fish migrations, as well as navigation in the main course of the river. A key drawback of off-stream reservoirs is their inherent inability to store the entire water yield of the river.

6.13.6.2.3 Sediment routing

Sediment routing enables the passing of sediment-laden floods through the reservoir by maintaining a high velocity in the impoundment by means of reservoir level drawdown. Specific outlets may be constructed in the dam for the purpose of routing (fine) sediments as well as turbidity currents.

Transportation toward downstream of previously deposited sediments may happen, without being the objective of sediment routing operations. Compared to hydraulic flushing, sediment routing thus induces lower concentrations and less deposition in the downstream reach.

Depending on the size of the reservoir and its catchment, pool drawdown may be scheduled either on a seasonal basis, based on flood hydrograph predictions, or based simply on a rule curve depending on inflow discharge.

In wide reservoirs, a channel–floodplain configuration may form and hence limit the storage capacity maintained by sediment routing.

Routing turbidity currents may be particularly beneficial since turbidity currents are likely to cause sedimentation of very fine material focused in critical areas such as near the outlets, while no significant deposition occurs elsewhere. Fortunately, due to their ability to run along the bottom of the reservoir and reach the dam, density currents may to a certain extent be successfully routed through the reservoir.

6.13.6.3 Measures Addressing the State

Dams may be designed with a so-called dead storage, situated below the lowest outlets or water intake, where sediments may be stored for a long period of time without disturbing normal operation of the dam and reservoir. Sediment storage has also been achieved using (successive) heightening of the dam and/or heightening of bottom outlets and water intakes. Such modifications in the structure inevitably raise environmental concerns regarding downstream impacts, particularly turbidity during the work period, as, for instance, at Mauvoisin dam in the Swiss Alps [15].

For a given net inflow of sediments, measures addressing the state of the reservoir are supposed to preserve or restore a maximum storage capacity in the reservoir. As dead storage and dam heightening are in essence not sustainable indefinitely, preserving the reservoir storage capacity may, after a certain stage, only be achieved by sediment removal techniques, such as dredging or hydraulic flushing.

6.13.6.3.1 Dredging

Being often technically hardly feasible and economically ineffective, dredging sediments from large reservoirs is rather an unusual operation. In addition, disposal of significant amounts of dredged sediments is generally costly, partly as a result of the environmental concerns it may legitimately cause. The cost of transportation of dredged material to available disposal sites may exceed the cost of the dredging operation itself. Moreover, as far as fine sediments are concerned, the volume required for disposal is higher than the recovered storage capacity in the reservoir. When separated from finer material, sand and gravel may be beneficially exploited as construction material, while the finer particles may prove suitable for agriculture. Sustainability of the strategy nonetheless remains highly questionable.

In contrast, focused dredging with the single purpose of removing sediments from the vicinity of water intakes or outlets may prove effective. In such a case, sediments may even be relocated elsewhere in the pool, where they do not disrupt the operation of the dam and reservoir. This approach fails, however, in cases where the dredged area refills quickly during subsequently occurring floods.

6.13.6.3.2 Flushing

Flushing consists in opening the outlets to lower the impoundment level and create a stream flow in the reservoir for a long enough period of time, so that a certain amount of bottom material gets eroded. Typical duration of flushing ranges from a couple of days to 1 week, except for particularly large reservoirs where flushing may last longer. Given the sediment properties, successful flushing requires low-level outlets characterized by a sufficiently high discharge capacity so as to induce an erosive flow in the reservoir.

In contrast to sediment routing, the aim of hydraulic flushing is to transport downstream sediments that have previously settled down at the bottom of the reservoir, and not just pass-through incoming sediments. Sediment concentration downstream during flushing operations significantly exceeds the inflow concentration. Therefore, environmental impacts on the downstream valley should not be undermined and require a thorough analysis according to local regulations, especially as far as large amounts of polluting sediments are concerned.

Sediment management by flushing may be suitable for a specific site only if the flow regime of the river and geometry of the valley offer appropriate conditions for the flushed sediments to be transported further downstream and not clog the valley immediately downstream of the dam. Besides, water used for flushing is usually lost, except if it can be exploited for hydropower production. Consequently, in most cases, rapid refilling of the reservoir is an important issue, which makes flushing operations effective mainly for hydrologically small reservoirs (i.e., storage smaller than about one-third of the mean annual inflow).

Rivers with a regular and predictable high-flow season (due to rainfall or snowmelt) favor flushing operations. Sound hydrological knowledge of the basin, either based on discharge time series recorded over long periods or resulting from rainfall runoff modeling, enhances the predictability of high-flow periods suitable for refilling the reservoir. Cost–benefit analysis enables verification of whether the cost of sacrificed water is balanced by the benefits of flushing sediment, so that the operation is indeed effective. In addition, flushing schemes may be optimized based on such cost–benefit analysis, as described, for instance, by Bouchard [16].

The usual result of a flushing operation is the scouring of a relatively narrow flushing channel in the reservoir, whereas lateral deposits, the so-called submerged floodplains, remain essentially unaltered. After flushing, fine sediments will first settle down in the flushing channel, thus facilitating their evacuation during a subsequent flushing operation.

On the contrary, flushing often turns out to be inadequate and ineffective for controlling deltaic deposition of coarser material and preventing propagation of the delta within the reservoir. As a consequence, the ability of flushing operations to move the coarser sediments will critically influence the sustainability of reservoirs where sediments are managed by hydraulic flushing.

Flushing with partial instead of complete drawdown of the pool level is called 'pressurized flushing'. It is known to be far less effective than full drawdown flushing, because it mainly leads to the movement of deposits from the upstream part of the reservoir toward the downstream part, where the water level remains too high for the sediments to be transported further downstream. An erosion hole is also generally created nearby the outlets.

The success and efficiency of flushing operations is highly dependent on the reservoir shape and does not apply universally. The narrower the reservoir and the steeper the longitudinal slope, the higher the chance of recovering significant storage capacity by means of flushing. Sediment grain size also strongly affects the efficiency of flushing. While coarse sediments may be difficult to mobilize and fine silt or clay tends to settle down on the submerged floodplains and consolidate, sand and coarse silt are the sediments most conducive to efficient flushing. A risk when hydraulic flushing is performed is the creation of a deep and narrow scouring hole or channel nearby the reservoir outlets, which results in a failure to recover a large part of the storage capacity. As compared to reservoirs located in the lower parts of watercourses those situated in the upper reaches have more chance to enable efficient flushing, due to the generally smaller reservoir capacity and steeper slopes facilitating high sediment transport rates.

Besides a thorough knowledge of the basin hydrology and sediment characteristics, designing a flushing scheme requires specific data collection, possibly including bathymetry survey and measurement of the sequence of sediment inflow. In addition, reliable hydraulic modeling coupled with suitable sediment transport modeling is recognized as a cornerstone of any detailed assessment of flushing efficiency [1]. Numerical modeling for sustainable sediment management is discussed in subsection 6.13.6.5.

6.13.6.4 Measures Addressing Impacts

Using focused dredging or other techniques to rearrange deposits, the location of deposits may be controlled to some extent, in order to prevent their accumulation in places where they most critically affect the functioning of components such as water intakes or outlets. This can also be achieved by means of variations in the reservoir elevation or by judiciously breaching self-formed channels in the reservoir. Apart from such sediment-focusing techniques and ultimate measures such as decommissioning of the reservoir, measures aiming at reducing the impacts of a given decline in storage capacity remain limited.

6.13.6.5 Numerical Modeling for Sustainable Sediment Management

As a result of the complexity of the governing physical processes and the significant uncertainties affecting input data, numerical modeling tools with a genuine predictive capacity, such as comprehensively validated models for flow and sediment transport, constitute the key elements to provide quantitative decision support in the design and planning of sediment management schemes.

6.13.6.5.1 Modeling scales and approaches
The modeling problem can be divided into four segments:

1. Water and sediment yield from the watershed;
2. Rate and pattern of sediment transport, deposition, or scour before the dam;
3. Localized pattern of deposition and scour near the hydraulic structures (in the reservoir, for example); and
4. Scour, transport, and deposition of sediment in the river below the dam.

Numerical sediment transport models simulate flows in one, two, and three dimensions. One-dimensional models are widely used because of their robustness and short computation time. Moreover, many reservoir and river systems have a highly elongated geometry, which is well suited for one-dimensional analysis.

Every model needs inputs like geometry data (cross section, width, depth, slope, grain size, and distribution). The grain size and load must also be known at the upper limit of the model. Hydraulic and sediment transport equations are solved through a series of time steps.

To date, depth-averaged simulations of flow and sediment transport remain an appealing approach, due to the difficulties in collecting the required input and validation data for three-dimensional modeling. Provided sufficient mixing occurs over the water depth, depth-averaged modeling definitely offers sufficiently accurate results for practical engineering purposes, and particularly as a support for sustainable management of sediments.

A challenging issue in numerical modeling of sediment transport is the need to handle accurately and efficiently the wide range of time scales involved in the relevant phenomena. Indeed the time scales of interest extend from a few seconds or minutes (e.g., rapid scouring, slope failures, or bank collapse during flushing) to periods as long as years or decades (long-term sedimentation). Therefore, specific numerical modeling tools must be combined to handle reliably and at an acceptable CPU cost the processes characterized by time scales spanning such a wide range.

For this purpose, a number of modeling systems have been developed over the past few years. Such numerical models for flow and sedimentation in reservoirs were recently reviewed by ICOLD [17], while fundamentals of flow and sediment transport modeling are presented by Wu [18]. As an example, details of the modeling system WOLF, developed by the authors, are presented below.

6.13.6.5.2 The WOLF modeling system

The modeling system WOLF, developed in about a decade at the University of Liège, is based on a series of complementary numerical tools designed to be combined for covering the widest possible range of relevant time scales in sedimentation processes. It includes the following components, all based on similar finite-volume schemes [19]:

1. Steady flow and sediment transport model, computing bed equilibrium profile;
2. Unsteady model loosely coupling sediment transport and flow computation (quasi-steady); and
3. Unsteady model tightly coupling sediment transport and flow computation (fully transient).

Besides, in cases where a direct coupling between sediment transport and flow computations turns out to be unnecessary, several postprocessing tools (including a Lagrangian-type tracking of sediment particles) are available to analyze the results of the hydrodynamic depth-averaged simulations in terms of transport capacity or erosion risk.

An original treatment of locally rigid beds has been developed, enabling a very general applicability of the models to real cases, often involving nonerodible areas. Besides, several turbulent closures are implemented, such as Smagorinsky type or k-ε, which play an important part since turbulence modeling directly affects predictions of sediment transport.

6.13.6.5.3 Case study: Alpine shallow reservoir

Numerical modeling of flow and sediment transport is presented below for the case of a hydropower project involving a shallow reservoir, with a focus on reservoir sedimentation issues and on the long-term management of the sediments by means of periodic flushing.

The hydropower project consists in a diversion dam (9 m high, 40 m wide) and a water intake, which diverts the flow through a penstock directly to a power plant located almost 10 km downstream. The total available head exceeds 200 m. The reservoir is about 2 km long, and its storage capacity is approximately 200 000 m³.

The upper part of the catchment of the river is situated in mountainous areas, and several of its tributaries are highly torrential. Therefore, high sediment inflows are expected in the reservoir especially during the flood season, which takes place in summer as a result of snowmelt in the basin. As a consequence, there was a need to assess the reservoir sedimentation process and to evaluate the feasibility and efficiency of flushing operations. To achieve those goals, a three-step procedure has been applied.

6.13.6.5.3(i) Step 1: Assessment of short-term sedimentation in the reservoir

First, the sediments likely to reach the water intake at the beginning of exploitation are characterized. For this purpose, by means of the above-mentioned two-dimensional shallow-water model, the flow field has been simulated in the reservoir with a grid resolution of 1 m × 1 m and an adequate k-ε turbulence closure (**Figure 9**). Next, this flow field is analyzed in terms of sediment transport capacity in order to predict the maximum grain size able to reach the water intake and thus contribute to damaging the turbines (**Figure 10**).

6.13.6.5.3(ii) Step 2: Assessment of long-term sedimentation in the reservoir

Second, the long-term equilibrium bathymetry of the reservoir is computed with a module of WOLF 2D, handling mobile beds (**Figure 11**). Due to the long time scale of the process, the computation is based on a quasi-steady approach (iterative steady state hydrodynamic simulations). The sensitivity of the results has been verified to remain reasonably low with respect to variations in the main assumptions such as sediment yield and grain size.

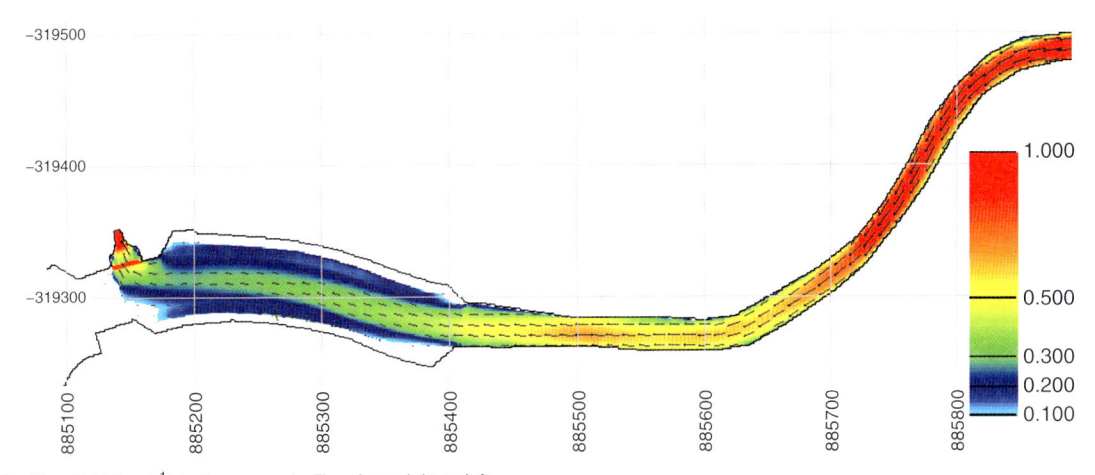

Figure 9 Flow field (m s^{-1}) in the reservoir. Flow from right to left.

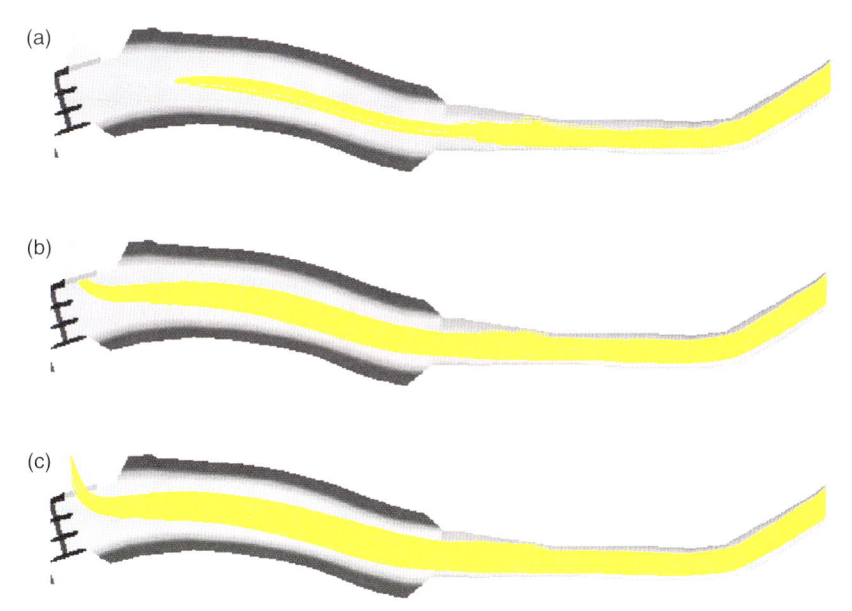

Figure 10 Particles trajectories (in yellow), as predicted by the Lagrangian sediment particle tracking in the downstream part of the reservoir, revealing that grains of 425 and 300 μm, respectively, are trapped in the reservoir and in the vicinity of the water intake, whereas grains of 250 μm enter the water intake.

6.13.6.5.3(iii) Step 3: Evaluation of the efficiency of flushing operations

Finally, the rapidly transient flow with high erosive capacity during a flushing operation has been simulated with a fully unsteady module of WOLF 2D, tightly coupling the computation of flow and sediment transport. As a result, this numerical study enables evaluation of the effect of a given flushing scenario in terms of changes in bathymetry in the downstream part of the reservoir as well as in terms of released discharge. The overall efficiency of the flushing operation can then be evaluated with respect to the recovered storage capacity and its extension in space.

The sequence of reservoir bathymetry during the flushing operation is given in **Figure 12** for the first 1.5 h, whereas further results of computed bathymetry after 3, 6, 12, and 24 h are given in **Figure 13**. Simulation results reveal that scouring initially takes place mainly immediately upstream of the dam, as confirmed in **Figure 12** for the first 90 min of flushing, whereas erosion occurs much further upstream at a later time, especially from 3 h after flushing has started. At such later times, bathymetry in the vicinity of the dam becomes almost stable, with hardly any ongoing erosion. **Figures 12** and **13** also show that significant amounts of deposits, reaching a depth as high as 1.5 m, are predicted in the reach downstream of the dam.

Flow velocity fields and Froude numbers are represented in **Figures 14** and **15**. The latter shows that transcritical flow takes place as far as approximately 100 m upstream and downstream of the dam, thus confirming that only the fully transient model tightly coupling sediment transport and flow computation applies for simulating flushing operations.

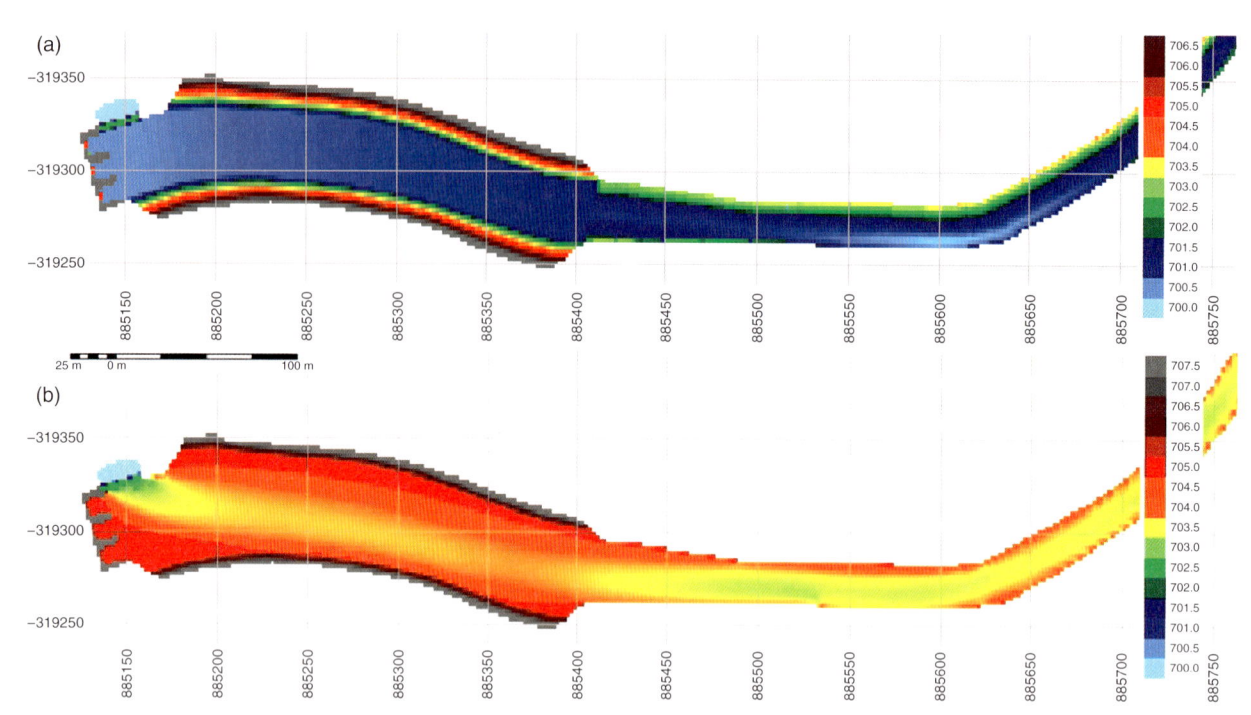

Figure 11 (a) Initial reservoir bathymetry (m); (b) Long-term equilibrium profile.

6.13.6.5.3(iv) Step 4: Surge downstream

Additional analysis was conducted to evaluate the feasibility of cleaning the flushing-induced deposits from the reach located downstream of the dam. For this, simulations were undertaken to represent the erosion induced downstream as a result of a surge of relatively clear water released by the spillway of the dam during 36 h. The model fully coupling flow and sediment transport computations was used again, together with the depth-averaged k–ε turbulence model. The initial condition of the present simulation corresponds to the ultimate stage of the bathymetry, that is, after 24 h of flushing in the reservoir.

Bathymetry changes resulting from the surge are displayed in **Figure 16**, showing a complex pattern in the lateral direction of the river. The highest deposits were located in the middle of the main riverbed and along the left bank, and simulation results predict that they get successfully cleared as a result of the surge. The efficiency of the surge is found to be strongly affected by the sediment concentration in the released water: the lower the concentration, the higher the surge efficiency.

6.13.6.6 Assessing Sustainability of Hydropower Projects

Traditionally, hydraulic constructions like dams or reservoirs were not designed as sustainable infrastructures. For the past few years the concept of sustainable development has proliferated and corresponding design principles have been developed. New infrastructures should not compromise the ability of future generations to access the same resources. Biological diversity and environmental integrity must be maintained. The potential for catastrophic events resulting from infrastructure collapse or obsolescence must be minimized. Finally, activities or infrastructures that imply environmental restoration or infrastructure rehabilitation obligations for future generations must be avoided.

Economic analysis of a project is based on cost–benefit analysis, thus disregarding noneconomical parameters that may play an important part. As the benefits decrease with time (discount rate), new dams are often considered as economically nonviable in the long term and are not designed for long-term management. For example, large low-level flushing outlets are rarely built. Consequently, short-term economic gain tends to override long-term sustainability and ecological considerations. Cost–benefit analysis also faces other limitations, such as limited knowledge of the cost of impacts and uncertainties on future market trends.

To compensate for these limitations in the assessment of economical and ecological viability, the RESCON approach has been developed [20]. The methodology proceeds in three stages:

1. determining which method of sediment management is technically feasible;
2. determining which alternative is more desirable based on economic analysis; and
3. incorporating environmental and social factors to select the best course of action for sediment management.

The methodology may also be expressed as:

$$\text{Maximize} \sum_{t=0}^{T} NB_t.d^t - C + V.d^T \qquad [1]$$

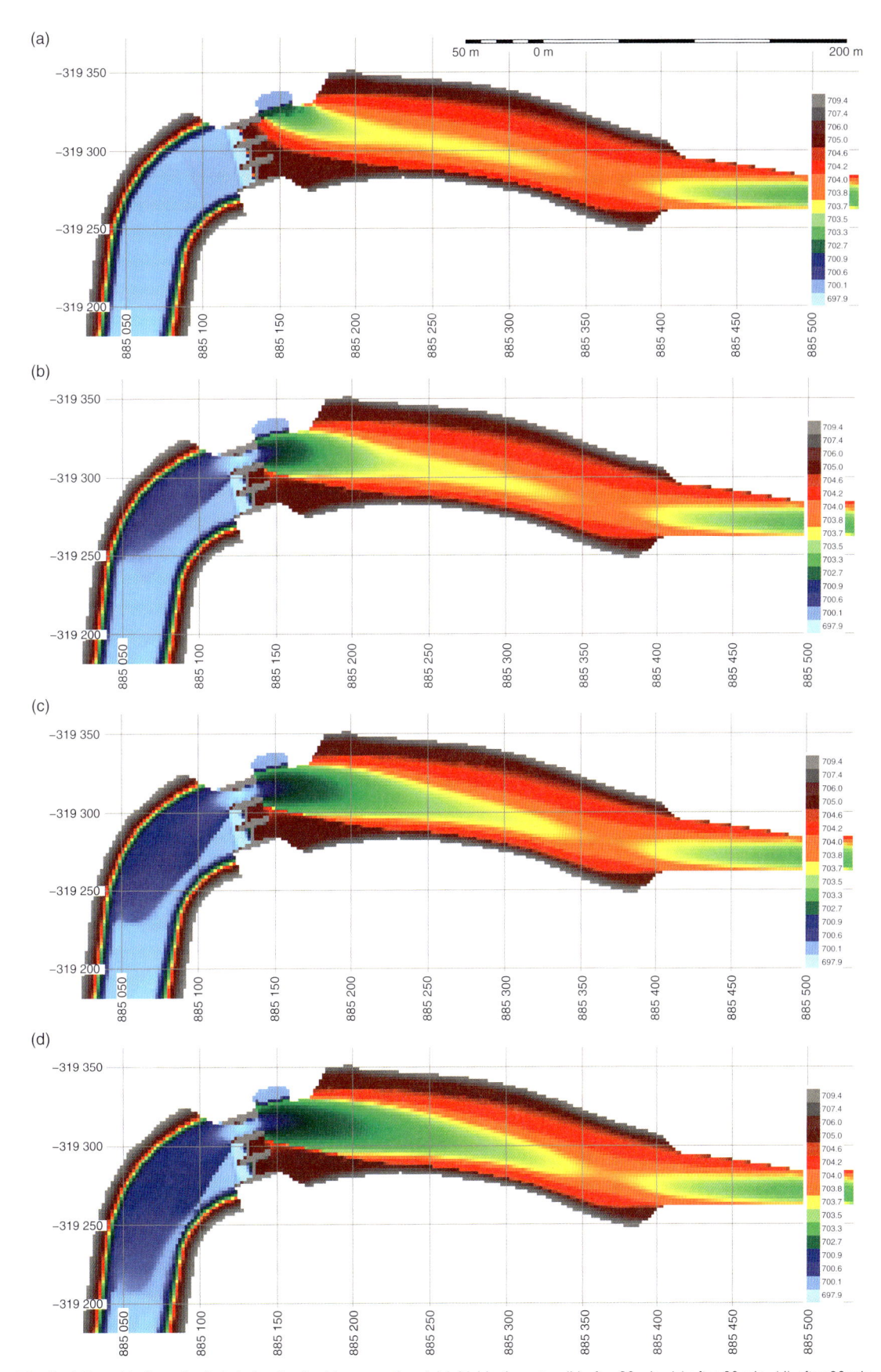

Figure 12 Evolution of bathymetry (m) during the flushing operation: (a) initial bathymetry; (b) after 30 min; (c) after 60 min; (d) after 90 min.

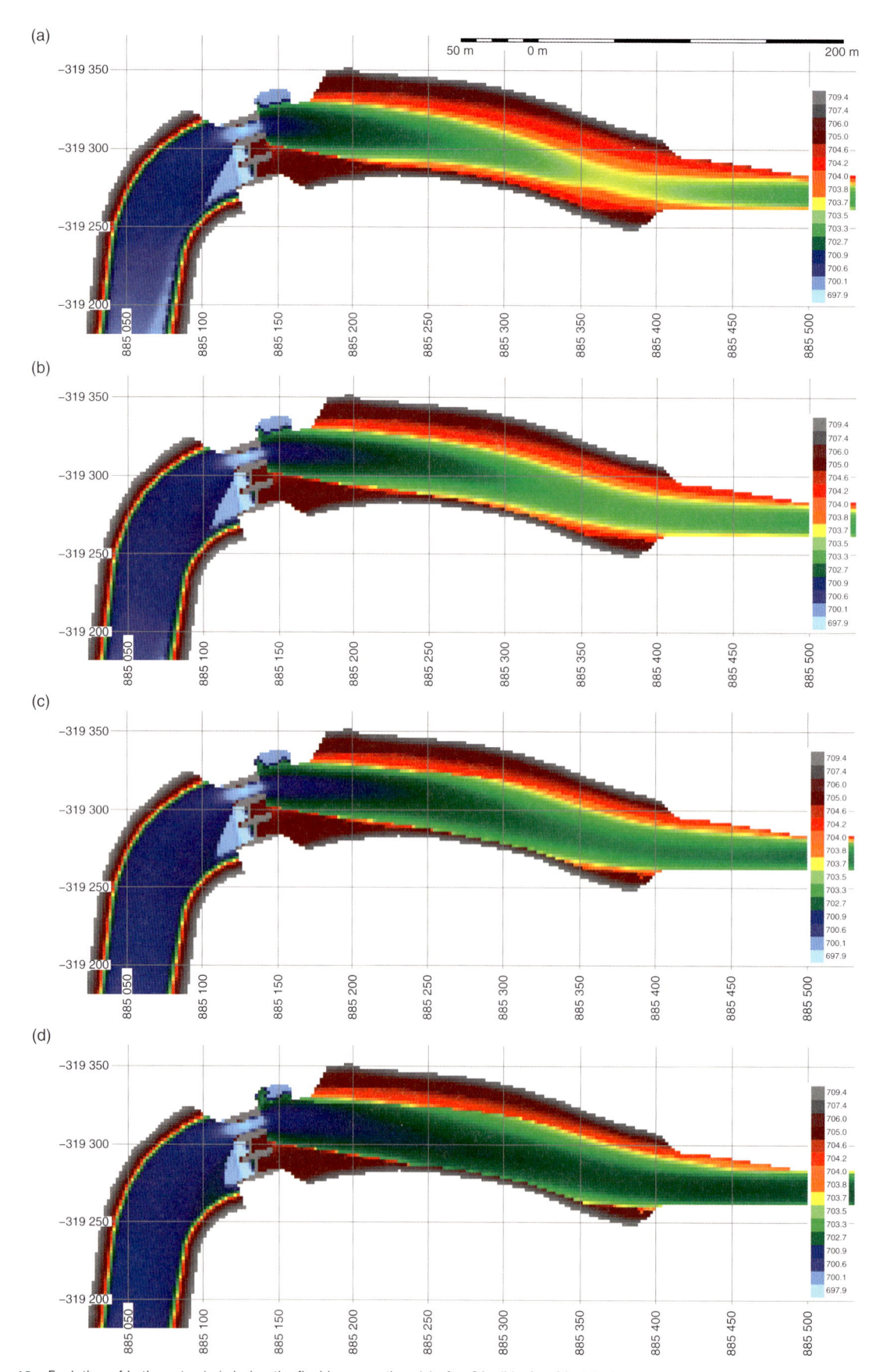

Figure 13 Evolution of bathymetry (m) during the flushing operation: (a) after 3 h; (b) after 6 h; (c) after 12 h; (d) after 24 h.

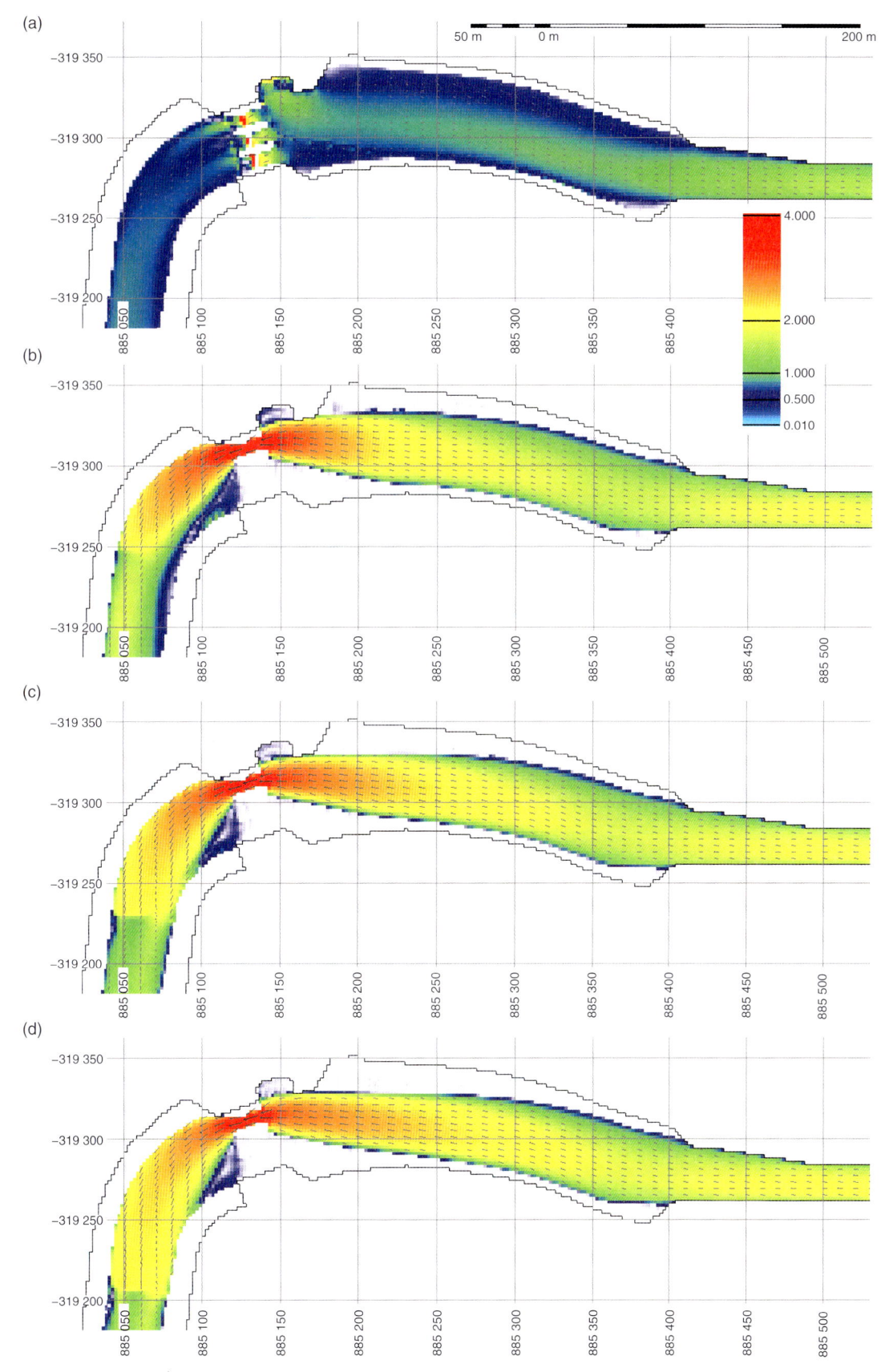

Long-Term Sediment Management for Sustainable Hydropower 373

Figure 14 Flow velocity (m s^{-1}) during the flushing operation: (a) initial bathymetry; (b) after 30 min; (c) after 60 min; (d) after 90 min.

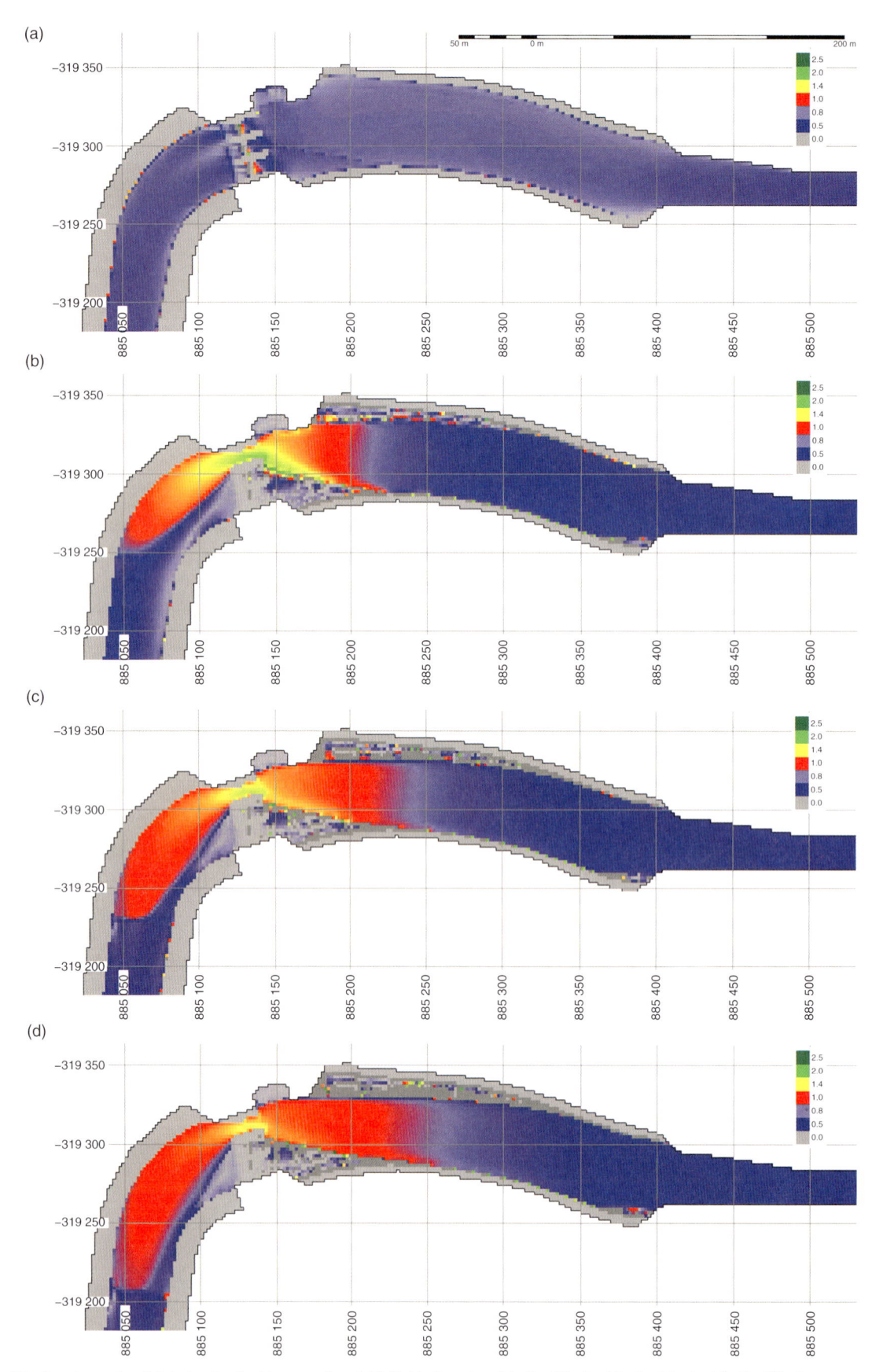

Figure 15 Froude number (–) during the flushing operation: (a) initial bathymetry; (b) after 30 min; (c) after 60 min; (d) after 90 min.

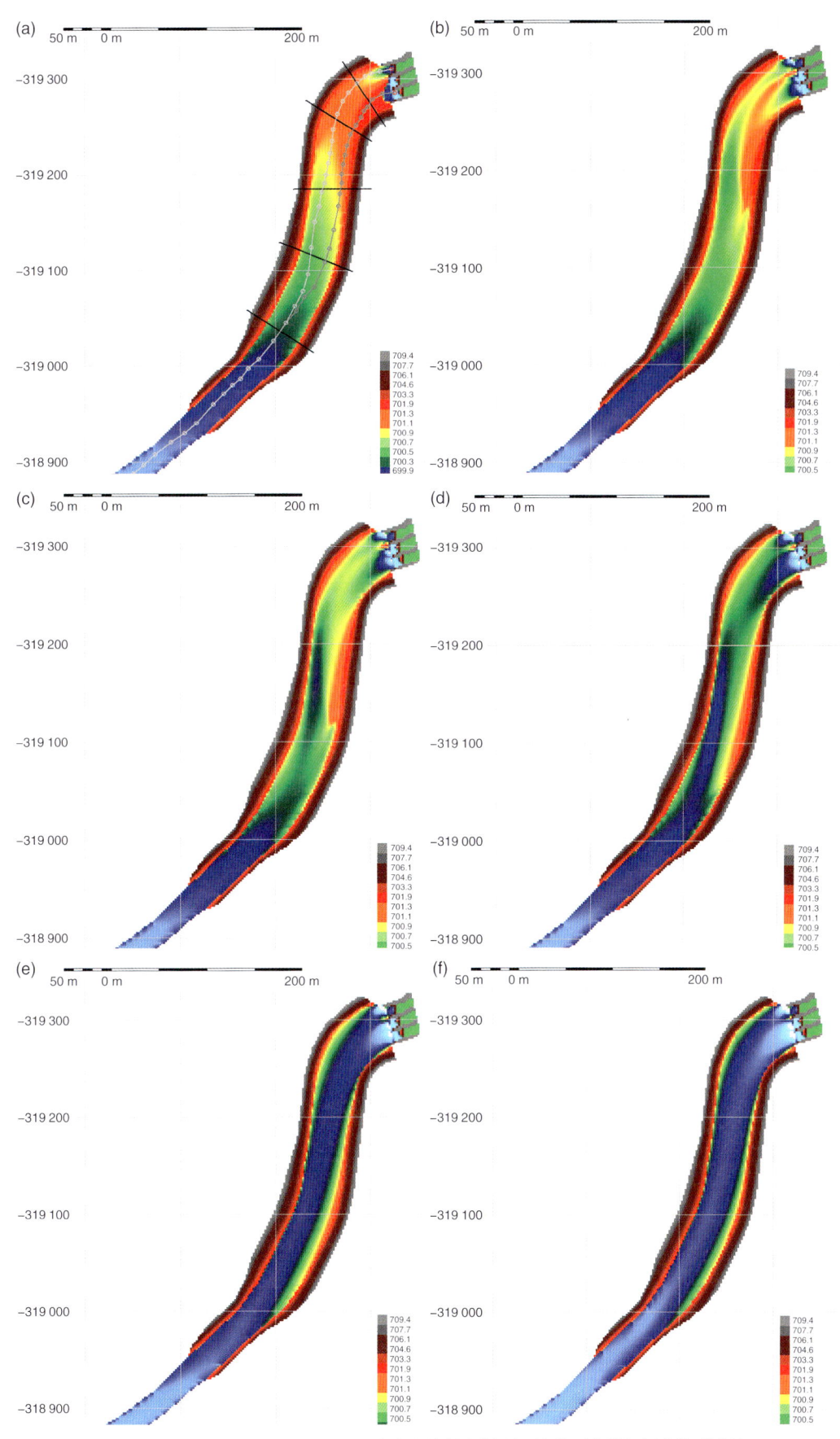

Long-Term Sediment Management for Sustainable Hydropower 375

Figure 16 Bathymetry of the downstream reach during the surge (m): (a) initial; (b) 3 h; (c) 6 h; (d) 12 h; (e) 24 h; (f) 36 h.

subject to

$$S_{t+1} = S_t - M + X_t \qquad [2]$$

This expression maximizes the algebraic sum of net benefits, capital cost, and salvage value, given the initial capacity and other physical and technical constraints. In this expression NB_t is the net benefit in year t; d^t is the discount rate factor in year t defined as $1/(1+r)$_ with the discount rate r; C is the initial capital cost of construction; V is the salvage value; T is the terminal year; S_t is the remaining reservoir capacity in year t; M is the trapped annual incoming sediment; and X_t is the sediment removed in year t.

The salvage value V of a reservoir is usually negative, because it represents the cost of decommissioning at terminal year T. In view of sustainable development, a part of the yearly benefit must be saved for intergenerational equity. This could be achieved by means of an annual investment, which will be available for the coming generations to decommission the facility. This investment must be subtracted from the annual net benefits.

6.13.7 Conclusion

Sediments eroded from the catchment cause various damages and disruptions to reservoirs, dams, and hydropower plants. These issues have been discussed in this chapter and possible mitigation actions reviewed.

At a time when appropriate dam sites are becoming more difficult to find, sustainability of hydropower may be preserved only if long-term sediment management is regarded as a key concern and objective from the early stages of dam design, dimensioning, and construction, as well as during the whole life span of the reservoir through proper maintenance and operation rules.

There is no universal strategy to achieve these goals, but an optimal combination of structural and nonstructural measures need to be identified on a site-specific basis. Although highly specialized expertise (in hydrology, sedimentology, hydraulic engineering, etc.) is absolutely necessary to understand, quantify, and master the complex turbulent flow and sedimentation processes involved, skilled technical experts also need to account for the more global scenario incorporating all other issues of a successful integrated management of the river basin. The impact of the impoundment on the long-term geomorphology of the water course should also be accounted for, considering even possible ultimate dam decommissioning.

There is definitely a need for further research and a more comprehensive understanding both in terms of fundamental sedimentation processes and regarding integrated assessment of sustainable sediment management.

References

[1] White R (2001) *Evacuation of Sediments from Reservoirs*. London, UK: Thomas Telford.
[2] Gleick PH (2002) *World's Water*. Washington, DC: Island Press.
[3] Morris GL, Annandale G, and Hotchkiss R (2008) Reservoir sedimentation. In: Marcelo HG (ed.) *Sedimentation Engineering: Processes, Measurements, Modeling, and Practice*. American Society of Civil Engineers 110: 579–612.
[4] European Commission (2003) Analysis of pressures and impacts. Guidance note No 3: 157.
[5] Jansson MB (1988) A global survey of sediment yield. *Geografiska Annaler Series A, Physical Geography* 70(1/2): 81–98.
[6] Morris GL (1997) *Reservoir Sedimentation Handbook: Design and Management of Dams, Reservoirs, and Watersheds for Sustainable Use*. New York: McGraw-Hill.
[7] Williams HFL (1995) Assessing the impact of weir construction on recent sedimentation using cesium-137. *Environmental Geology* 26(3): 166–171.
[8] Brune (1953) Trap efficiency of reservoirs. *Transactions American Geophysical Union* 34(3): 407–418.
[9] Churchill MA (1948) Discussion of 'Analysis and Use of Reservoir Sedimentation Data', *Proceedings of Federal Interagency Sedimentation Conference* (edited by Gottschalk LC), Denver, Colorado, pp. 139–140.
[10] Garde RJ, Raju KGR, and Sujudi AWR (1990) Design of settling basins. *Journal of Hydraulic Research* 28(1): 81–91.
[11] Ranga Raju KG, Kothyari UC, Srivastav S and Saxena M (1999) Sediment removal efficiency of settling basins. *Journal of Irrigation and Drainage Engineering* 125: 308–314.
[12] Dewals BJ, Kantoush SA, Erpicum S, et al. (2008) Experimental and numerical analysis of flow instabilities in rectangular shallow basins. *Environmental Fluid Mechanics* 8: 31–54.
[13] Dufresne M, Dewals BJ, Erpicum S, et al. (2010) Classification of flow patterns in rectangular shallow reservoirs. *Journal of Hydraulic Research* 48(2): 197–204.
[14] Harada M, Morimoto H, and Kokubo T (2000) Operational results and effects of sediment bypass system. *Transactions of the XXth Congress on Large Dams*, Beijing, China, ICOLD 2 (Question 77): 967–984.
[15] Durand P (2001) Barrage de Mauvoisin Projet de surélévationpumping de la prise d'eau et vidange, impacts sur l'environnement. *La Houille Blanche* 6–7: 44–48.
[16] Bouchard J-P (2001) La gestion des sédiments par chasse: Outils d'optimisation et de prévision d'impact. *Houille Blanche-Revue Internationale* 6–7: 62–66.
[17] ICOLD (2007) *Mathematical Modelling of Sediment Transport and Deposition in Reservoirs: Guidelines and Case Studies*. International Commission of Large dams. Paris, France.
[18] Wu W (2007) *Computational River Dynamics*. London: Taylor & Francis.
[19] Dewals BJ, Erpicum S, Archambeau P, et al. (2008) Hétérogénéité des échelles spatio-temporelles d'écoulements hydrosédimentaires et modélisation numérique. *Houille Blanche-Revue Internationale* 5: 109–114.
[20] Palmieri A, Shah F, Annandale GW and Dinar A (2003) *Reservoir Conservation: The RESCON Approach*. Washington DC: The World Bank.

6.14 Durability Design of Concrete Hydropower Structures

S Jianxia, Design and Research Institute, Yangzhou City, Jiangsu Province, China

© 2012 Elsevier Ltd. All rights reserved.

6.14.1	Introduction	378
6.14.1.1	General	378
6.14.1.1.1	Background of concrete hydropower structures	378
6.14.1.1.2	The inadequacy of concrete hydropower structures	379
6.14.2	**Early Cracking**	380
6.14.2.1	General	380
6.14.2.2	Definition of Early Cracking	380
6.14.2.3	The Causes of Early Cracking	380
6.14.2.3.1	Thermal stress	380
6.14.2.3.2	Autogenous shrinkage	381
6.14.2.3.3	Drying shrinkage	381
6.14.2.3.4	Uneven settlement of the foundations	381
6.14.2.3.5	Foundation friction	382
6.14.2.4	Controlling the Cracking of Concrete Hydropower Structures	382
6.14.2.4.1	Different design concepts on crack width and monolith size	382
6.14.2.4.2	Experience of limiting crack width and monolith size	382
6.14.2.4.3	The influence of reinforcement on concrete cracking	382
6.14.2.5	State-of-the-Art Joints	382
6.14.2.5.1	General	382
6.14.2.5.2	Joint type	383
6.14.2.5.3	Joint filling materials	383
6.14.2.5.4	Present condition of joints	384
6.14.3	**Durability Problems**	385
6.14.3.1	General	385
6.14.3.2	How Concrete Hydropower Structures Become Damaged	386
6.14.3.3	Carbonation and Reinforcement Corrosion	386
6.14.3.3.1	General	386
6.14.3.3.2	Working mechanism	386
6.14.3.3.3	Influential factors	386
6.14.3.3.4	Prevention methods	387
6.14.3.4	Freezing and Thawing	387
6.14.3.4.1	General	387
6.14.3.4.2	Influential factors	387
6.14.3.4.3	Prevention methods	387
6.14.3.5	Concrete Expansion and Contraction	387
6.14.3.5.1	General	387
6.14.3.5.2	Influential factors and prevention methods	388
6.14.3.6	Chloride and Sulfate Attack	388
6.14.3.6.1	General	388
6.14.3.6.2	Chloride attack	388
6.14.3.6.3	Sulfate attack	388
6.14.3.7	Alkali–Aggregate Reaction	389
6.14.3.7.1	General	389
6.14.3.7.2	Sodium equivalent	389
6.14.3.7.3	AAR in the world	389
6.14.3.7.4	Influential factors	390
6.14.3.7.5	Prevention methods	390
6.14.3.8	Seepage Scouring	390
6.14.3.8.1	General	390
6.14.3.8.2	Influential factors	390
6.14.3.8.3	Prevention methods	390
6.14.3.9	Abrasion and Cavitation	390
6.14.3.9.1	General	390

6.14.3.9.2	Abrasion- and cavitation-induced deterioration in China	391
6.14.3.9.3	Prevention methods	391
6.14.3.10	Other Less Important Factors	391
6.14.3.10.1	Structural design	391
6.14.3.10.2	Construction defects	391
6.14.3.10.3	Biological process	391
6.14.3.10.4	Fire damage	391
6.14.4	**Durability Design**	392
6.14.4.1	Performance-Based Durability Design	392
6.14.4.2	Expert System	392
6.14.4.3	The Direction of Future Research	392
6.14.5	**How to Maintain a Durable Concrete**	393
6.14.5.1	Material Selection	393
6.14.5.1.1	General	393
6.14.5.1.2	Mineral admixtures	393
6.14.5.1.3	Chemical admixtures	394
6.14.5.1.4	Polymer fibers	394
6.14.5.1.5	Geomembranes	394
6.14.5.1.6	Self-healing concrete	394
6.14.5.1.7	Self-compacting concrete	394
6.14.5.1.8	High-performance concrete	395
6.14.5.2	Design	395
6.14.5.2.1	Mix design	395
6.14.5.2.2	Structural design	395
6.14.5.3	Construction	396
6.14.5.3.1	Construction sequence	396
6.14.5.3.2	Reinforcement placement	396
6.14.5.3.3	Concrete casting	396
6.14.5.3.4	Curing	396
6.14.5.3.5	Quality evaluation	396
6.14.5.4	Inspection and Assessment	397
6.14.5.4.1	General	397
6.14.5.4.2	Visual inspection	397
6.14.5.4.3	Instrumental inspection	397
6.14.5.5	Rehabilitation	397
6.14.5.5.1	General	397
6.14.5.5.2	Joint repairing	397
6.14.5.5.3	Crack repairing	398
6.14.6	**Case Histories for Durable Concrete**	399
6.14.6.1	Shi Lianghe Key Project in China	399
6.14.6.2	Fengman Hydropower Structure in China	399
6.14.6.3	Haikou Key Project in China	399
6.14.6.4	Butgenbach Dam in Belgium	400
6.14.6.5	Baoying Key Project in China	401
6.14.7	**Conclusion**	403
References		403

6.14.1 Introduction

6.14.1.1 General

6.14.1.1.1 Background of concrete hydropower structures

Concrete is one of the oldest and most durable building materials. Its earliest known use was for a hut floor in the former Yugoslavia, dating from 5600 BC. Later, more notable examples of the use of concrete include the Great Pyramid of Giza, Egypt, and the Pantheon in Rome, Italy (**Figure 1**). Portland cement came into use in 1854 and quickly became one of the most versatile and frequently applied construction materials in the world, owing to its availability, low cost, suitability for making simple shapes, and fire-resistant properties.

Concrete, however, is a brittle material with very low tensile stress, which limits its application. To solve this problem, reinforced concrete started to be used in the mid-nineteenth century. Nowadays, most concrete hydropower structures are applied in combination with steel reinforcement, which brings several advantages.

Figure 1 The Pantheon, Rome, Italy [1].

- Steel can withstand a lot of tension but is not able to resist large compressive loads. The opposite is true of concrete; therefore, the mass of concrete provides stability and protects against buckling.
- Unprotected steel is subject to corrosion under normal atmospheric exposure, whereas concrete remains stable. The alkalinity of concrete provides a passive environment where steel is less likely to corrode.
- Steel and concrete possess similar coefficients of thermal expansion (CTEs). The CTE of steel is 1.2×10^{-5} and that of concrete is $1.0-1.5 \times 10^{-5}$. When steel and concrete are combined, they do not exert undue strains on one another when subjected to wide temperature extremes.

Although concrete can be reinforced or prestressed with steel, which can enhance its tensile and flexural strength, it is still a material that cracks easily. Many concrete structures suffer cracking, spalling, loss of strength, or steel corrosion. These faults are not only expensive to repair, but they also limit the life of structures.

Most of the concrete hydropower structures in China have a life of only 50 years. In Europe and the United States, the structures have a longer life but not as long as one might expect of such structures. Research is needed to shed more light on the durability of concrete hydropower structures. The aim is to at least double the life of concrete hydropower structures in the next century.

6.14.1.1.2 The inadequacy of concrete hydropower structures

Generally, research on the inadequacy of concrete hydropower structures is divided into two branches, with one focusing on early cracking and the other on durability (see **Figure 2**).

The main causes of the early cracking of concrete include thermal effects, cement shrinkage, and foundation conditions. These are the main factors that limit the monolithic dimension of concrete hydropower structures. The main factors that influence the durability of concrete hydropower structures include carbonation and reinforcement corrosion, freezing and thawing, concrete expansion and contraction, chloride and sulfate attacks, alkali–aggregate reaction (AAR), seepage water scouring, abrasion, and cavitation.

The early cracking of concrete damages a whole structure's integrity and makes it vulnerable to other faults due to an increase in permeability and a loss of mechanical resistance in the concrete. Therefore, early cracking is actually a reason for concrete deterioration as well. In this chapter, early cracking is treated separately from the durability problem because of the following:

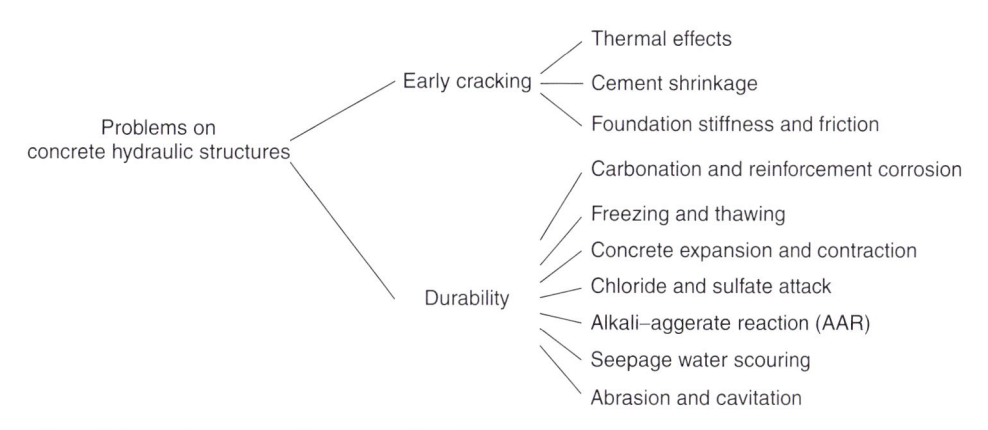

Figure 2 Inadequate performance of concrete hydropower structures.

- Controlling early cracking is very important for the correct performance of concrete hydropower structures during their lifetime. It is cheaper to repair early cracks than it is to fully restore a structure that has fallen into further disrepair.
- The conditions that cause concrete to crack early generally happen over a short period in contrast to the conditions that cause long-term durability problems.

This chapter will focus on concrete hydropower structures. Concrete hydropower structures have all the characteristics of concrete structures, but since they are generally situated in water, they are more likely to experience the problems associated with water. These problems include freezing and thawing, chloride and sulfate attacks, and AAR.

6.14.2 Early Cracking

This section will introduce the concept of the early cracking of concrete and will focus on the characteristics of young concrete and the main factors that influence early cracking and the general condition of joints.

6.14.2.1 General

Concrete is a material vulnerable to cracking. New concrete hydropower structures crack whenever the tensile stresses exerted on the concrete exceed the concrete's tensile strength. It is important to deal with the early signs of cracking to ensure that the structure remains safe and functional for as much of its anticipated life as possible. Partial or full-depth cracking often requires expensive repairs to be carried out and sometimes complete reconstruction.

To prevent concrete from cracking, transverse or longitudinal contraction joints are generally placed at the site where cracks are expected to occur. However, each transverse and longitudinal joint induces a point of weakness where the joint filling material may age or be flushed away by seepage water or become permeable (when the joints are required to be impermeable). Therefore, making reliable stress estimations to determine the onset of cracking and the width of the cracks for young concrete is the basis for the dimension design of concrete hydropower structures.

6.14.2.2 Definition of Early Cracking

According to the hydration degree of cement paste, concrete formation is generally divided into three stages: plastic stage, early stage, and matured stage [2].

1. Plastic stage
   The plastic stage is the time from when the concrete is cast to when the concrete turns from a plastic to a solid state. The end of this stage is called set time. For plain concrete, the set time is reached about 6–12 h after concrete casting. At the plastic stage, concrete is in a plastic state and under no stress. Cement hydration is quite strong and the physical and chemical properties are very unstable. The volume changes tremendously.
2. Early stage
   The early stage refers to the period lasting between 12 h and 90 days after concrete casting. It is generally divided into two parts. The first part is from 12 to 72 h after concrete placement. At this stage, the cement hydration process is half complete and the microstructure of concrete is basically formed. The stress and elastic modulus develop very fast. The second part refers to the 72 h to 90 days after concrete placement. Cement hydration is almost finished at the end of this stage, and the development of the stress and elastic modulus slows down. The concrete is nearly matured.
3. Matured stage
   The matured stage refers to the 90 days after concrete casting. Slight hydration is still taking place inside the concrete. The physical and chemical characteristics of the concrete are still developing but very slowly. Research on the early cracking of concrete tends to be focusing on this stage presently.

In this chapter, the term 'early cracking' refers to cracks that happen during the 12-h to 90-day period after concrete casting. The main factors that cause early cracking will now be examined.

6.14.2.3 The Causes of Early Cracking

6.14.2.3.1 Thermal stress

The ability to expand or contract in response to changes in temperature (thermal stress) is one of the main characteristics of concrete. Thermal stress as a cause of early cracking is considered to be serious only for construction using mass concrete.

'Mass concrete' is any large volume of concrete with dimensions large enough to require measures to be taken to cope with the generation of heat from the hydration of cement and volume change to minimize cracking.

When a concrete structure is relatively thick, excessive heat from cement hydration occurs. The heat at the surface of the concrete is easily released at the ambient temperature, while the center of the concrete remains at a high temperature. This results in a large temperature difference in mass concrete, especially when the ambient temperature changes abruptly. This then causes the center of the structure to expand at a different rate than the sides, and when the shrinkage or expansion is limited by the structure's foundations (external restraint) or the structure itself (internal restraint), cracks may happen. The main factors that influence the occurrence of thermal stress are:

- the cement content of the concrete – the more cement present in the concrete, the greater the chance of cracking occurring;
- ambient conditions – these include the air and subbase temperatures, and the friction of the subbase; and
- the size of the monolith – the greater the size of the monolith, the more difficult it will be to transmit hydration heat to an ambient environment and the greater the limitation of the subbase. To prevent the influence of those effects, a hydropower structure is generally divided into several parts with joints.

There are many ways to reduce concrete thermal stress. The generally accepted strategies are to decrease the cement content or use low-heat cement to reduce the amount of heat produced, to apply low thermal expansion and crushed aggregate, to slow down the hydration process through the use of various admixtures, to reinforce the concrete with steel or fibers, to precool the concrete constituent materials, to decrease the internal temperature, to protect the exposed surfaces and formwork from environmental extremes, to section off or split concrete for a single structure with temporary or permanent joints, and to add concrete in several lifts or pours.

The casting and curing for mass concrete is very important. The general thermal controlling process for a hydropower structure during construction is as follows: temperature anticipation → thermal stress controlling scheme → concrete casting and curing → temperature monitoring → scheme adjusting.

6.14.2.3.2 Autogenous shrinkage

Autogenous shrinkage is the uniform reduction of internal moisture due to cement hydration, which is typical of high-strength concrete. Autogenous shrinkage contributes significantly to concrete cracking when the water–cement (w/c) ratio is less than 0.4 [3]. The use of concrete with a somewhat higher w/c ratio can mitigate this problem. However, the strength and impermeability of concrete will be decreased if the w/c ratio is increased.

The main factors that influence the autogenous shrinkage of concrete are w/c ratio and concrete maturity. Concrete maturity is mainly influenced by ambient temperature and the type of cement. There is no effective way to mitigate the autogenous shrinkage for high-performance concrete (HPC).

6.14.2.3.3 Drying shrinkage

Drying shrinkage is caused by nonuniform drying of concrete after curing and the removal of forms. It increases with a greater w/c ratio [4]. Cracks caused by cement drying shrinkage generally happen 1 week after concrete casting or shortly after concrete finishes curing.

Ambient conditions such as air temperature, wind, relative humidity, or sunlight may draw moisture from exposed concrete structures, which will cause water content to gradient from the inner core to the surface of the concrete. As the moisture is lost in the small pores of the surface concrete, the surface tension of the remaining water tends to pull the pores together, which results in a loss of volume over time. However, this shrinkage will be limited by ambient conditions such as the inner concrete or the foundations. If the tensile stress produced surpasses the tensile strength of the surface concrete, cracks will appear.

Hydropower structures such as stilling basins, flashboards, and working bridges, whose surface areas are large relative to their volume, are more likely to have cracks caused by drying shrinkage.

Two factors influence concrete drying shrinkage: the concrete characteristics and the ambient conditions. Concrete characteristics include w/c ratio, cement type, dosage of cement, aggregate characteristics and dosage, admixtures, and so on. Ambient conditions include air temperature, relative humility, and so on, which may draw moisture from the concrete.

Techniques for reducing concrete drying shrinkage are as follows: to apply low shrinkage cement or cement with low hydration heat, to decrease the quantity of cement (as long as the concrete satisfies the strength requirements and can obtain adequate placement), to decrease the w/c ratio, to maintain at a minimum the quantity of fine aggregate that will just produce adequate workability and finishing characteristics, to prolong the moisture curing time of concrete, to set contraction joints (temporary or permanent), and to apply admixtures such as a water reducer or a shrinkage reducing admixture.

6.14.2.3.4 Uneven settlement of the foundations

There are two possible reasons for the uneven settlement of the foundations: one is the unevenness of the foundation soil and the other is the unevenness of the stress works on the foundations. Uneven settlement of the foundations can cause often deep or even transverse cracks to occur. The direction of the cracks is influenced by the condition of uneven settlement. Cracks are generally perpendicular or 30–45° to the vertical direction. When settlement is finished, the width and length of the cracks will remain stable.

In China, national design codes for hydraulic structures dictate that the mean settlement for a monolith is not allowed to be larger than 10 cm, and the uneven settlement for a single structure is not allowed to be larger than 5 cm.

6.14.2.3.5 Foundation friction

Besides foundation settlement, foundation friction is another factor that may induce uncontrolled cracking because of the high friction and, in some cases bonding, between the foundations and the concrete slab. The friction or bonding restrains the concrete's volume change (shrinkage or contraction), inducing higher stresses than might occur in concrete that directly comes into contact with the foundations. The higher the foundation friction, the more likely that cracks will occur. For concrete sat on rock foundations, the problem caused by foundation friction is more prominent.

Friction-initiated cracks are likely to happen from the bottom of the slab and travel toward the slab surface. Cracks from high friction can be erratic in orientation but follow zones of restraint between the concrete and the subbase.

To decrease the friction between the concrete and the subbase, a thin layer (generally 10–20 cm) of cementitious material (such as cement, asphalt, sand, or clay) is often applied at the interception. However, this method can only be applied when the horizontal force working on the structures is relatively small and the structures are stable under the decreased friction force.

6.14.2.4 Controlling the Cracking of Concrete Hydropower Structures

6.14.2.4.1 Different design concepts on crack width and monolith size

The process of avoiding the onset and further development of cracks in concrete is very complicated, and accurate solutions are not available. At present, countries fall into one of two groups in the way they deal with the cracking of concrete:

- For countries in the first group, there are no definite rules on the maximum crack width or the maximum size of the monolith. In national design codes for hydraulic structures, calculations concerning cracking are only for reference and engineers design structures based on their own experience. Many hydropower structures have no contraction/expansion or settlement joints. If cracks that do occur cannot be blocked, other complementary repair methods will be undertaken, such as drainage. Japan, the United Kingdom, and the United States apply this design concept.
- For countries in the second group, there are formulae in national design codes that must be obeyed in estimating cracks caused due to loading. There are no specific methods for estimating cracks caused by deformation. However, there are limitations on the maximum monolith size. It is supposed that if the size of the monolith is limited, the problems of cracking will not exist. The Soviet Union, some European countries, and China adhere to this design concept.

6.14.2.4.2 Experience of limiting crack width and monolith size

Cracks that can be seen with the naked eye are generally wider than 0.02–0.05 mm. Cracks with a width less than 0.05 mm are not harmful to the structure. No-cracking concrete refers to concrete whose crack width is less than 0.05 mm [5].

The criteria applied by engineers all over the world to limit the width of cracks are more or less the same, as follows:

- If no harmful ions are present in the environment and there are no seepage control requirements, the maximum allowable crack width is 0.3–0.4 mm.
- If the environment contains slightly harmful ions and there are no seepage control requirements, the allowable crack width is 0.2–0.3 mm.
- If the environment contains very harmful ions and the structure has seepage control requirements, the allowable crack width is 0.1–0.2 mm.

Generally, the size of the monolith of a hydropower structure is limited to 25 m on rock foundations and 35 m on soil foundations. Monoliths are becoming larger due to improvements in concrete construction techniques such as thorough cooling, consolidation, curing, and appropriate admixtures. When a structure is taller than the above values, joints are generally needed to separate the monolith into several parts.

6.14.2.4.3 The influence of reinforcement on concrete cracking

Thanks to research, the potential for cracks to appear due to thermal stress, foundation condition, and structural load can be estimated accurately with software on the basis of finite element methods. However, how the cracks will develop, especially under the influence of reinforcement, and the exact interrelationship between concrete drying shrinkage and steel reinforcement are topics for future research.

It is widely agreed that steel reinforcement can decrease the width and depth of a crack once a crack emerges. Cracks in reinforced concrete will be denser and smaller compared with cracks in non-reinforced concrete.

6.14.2.5 State-of-the-Art Joints

6.14.2.5.1 General

In order to prevent concrete cracking caused by uneven settlement, thermal effects, autonomous shrinkage, drying shrinkage, or other effects, different parts of hydropower structures are often connected with transverse joints. These joints are generally set at the site where the internal force is low or the elevation of the base changes abruptly. The normal width for the joints is about 1.0–2.5 cm. The joints permit minor differential movements between adjacent blocks, and in their absence major transverse cracks may develop.

6.14.2.5.2 Joint type

Nearly every concrete hydropower structure has joints that must be sealed to ensure the integrity and durability of the structure. Joints are divided into two categories: temporary joints and permanent joints.

Temporary joints: Temporary joints refer to construction joints – the actual surfaces where two successive placements of concrete meet. To assist in the construction and the placement of concrete, temporary joints are designed and positioned at certain locations during the placement of mass concrete. The concrete at the point of stoppage becomes a construction joint when concrete placement continues. Size of placement and time are contributory factors in the use of construction joints. In monolithic placements, temporary joints are required to be fully bonded across the construction joint for structural integrity. If the concrete construction is not too large, temporary joints may be designed to correspond with permanent joints.

Permanent joints: Permanent joints are generally set for concrete expansion, contraction, or uneven settlement. Permanent joints can be made to be cross-sectional or just shallow grooves on the concrete's surface. Shallow grooves are generally formed by a sawing machine. They are applied only as an auxiliary method to facilitate concrete expansion or contraction. Cross-sectional joints divide a structure into two or more monoliths. To avoid the abutting concrete elements from touching each other, they are often filled with soft and elastic materials.

Joints are further divided into two types according to their direction: vertical joints and horizontal joints. Vertical joints are normally flexible, allowing for uneven settlement between different parts (**Figure 3**). Horizontal joints (**Figure 4**) are generally rigid, allowing for only slight horizontal movements.

6.14.2.5.3 Joint filling materials

For hydropower structures, joints are often required to endure both minor structural movements and high hydrostatic pressures. Therefore, the joints need to be robust, flexible, and sometimes waterproof. To meet those requirements, the space of the joint needs to be filled with some premolded, nonabsorbent, nonreactive, nonextruding materials that are either rigid or flexible. Rigid materials are metallic materials: steel, copper, and, occasionally, lead. Flexible materials include polyvinyl chloride (PVC) mats, rubber, asphalt-impregnated fiber sheets (or rolls), or compressible foam strips. **Figure 5** illustrates the application of compressible foam mats in a concrete structure.

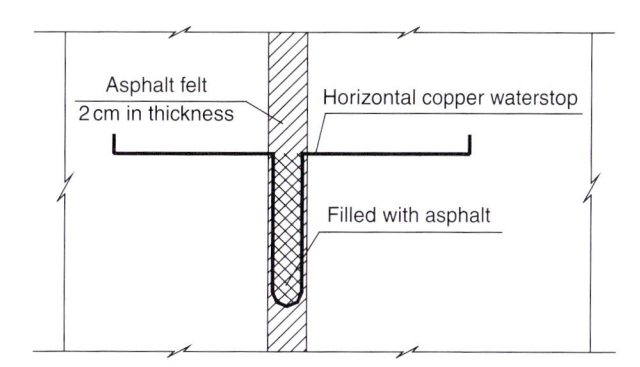

Figure 3 Vertical joints.

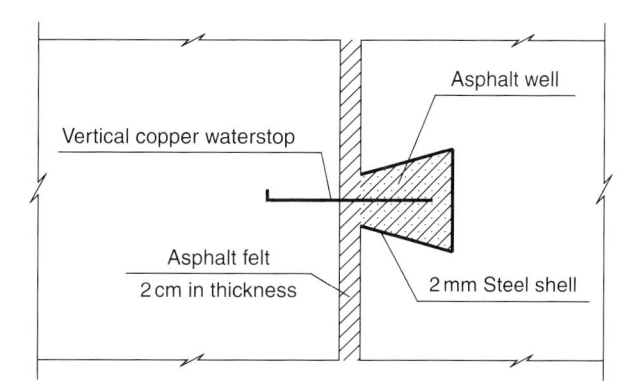

Figure 4 Horizontal joints.

Figure 5 Foam mats.

Joint filling materials must be resistant to harsh weather conditions, alkalis, fungi, musts, oils, greases, and other agents, as well as silicone or polyurethane-based mastics. They are installed across the joints. One end of the joint filling material is often embedded in concrete, while the other end is free in the adjoining monolith to permit slight movement of the concrete without any restriction.

Copper is an expensive joint filling material, yet it is very durable and waterproof (**Figures 3** and **4**). Copper joints are generally applied where water stoppage is very critical. For very important projects, two lines of copper waterstop are often required. The general thickness for copper waterstop is 1.2–1.4 mm.

The joint filling materials such as PVC or rubber are more affordable than copper, but they age quickly, especially in sunlight. Nevertheless, they are common types of waterstop. In China, PVC or rubber is mainly used as the joint-filled material for small hydropower structures or as the remedy of the main waterstop.

6.14.2.5.4 Present condition of joints

The selection of joint-filling material differs from country to country, often depending on a country's economic conditions and the availability of materials. In China, copper joints are widely used as joint-filling material for important hydropower structures. However, copper is expensive and working with it is labor intensive, which limits its application in developed countries. In Europe, rubber joints are more frequently used, but their quality is yet to match that of copper joints.

Following the expansion of the rubber and plastics industry, some new shapes of rubber or PVC joints have started to be used in hydropower structures in China (**Figures 6** and **7**). **Figure 8** is rolls of rubber waterstop applied in a hydropower structure. However, these types of joints are still at an experimental stage since it is not known how long their life will be.

Figure 6 PVC waterstop.

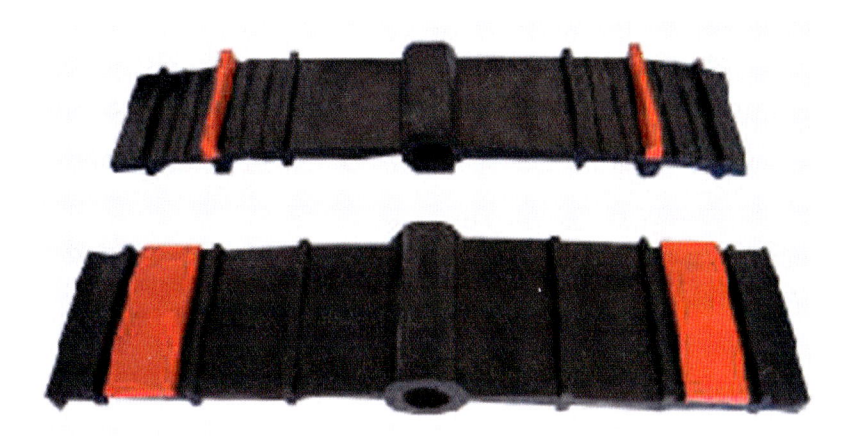

Figure 7 Rubber waterstop.

Figure 8 Rolls of rubber waterstop.

Although joints are being widely used and presenting operational results, the repair or recovery of the joints is very difficult should an accident or localized rupture in the joints occur after they have been installed. Because of the durability problem and the difficulty of repairing joints, the joints are often sited at the weakest point of a hydropower structure. Therefore, it is preferable to construct a concrete hydropower structure without joints if cracks are to be prevented.

6.14.3 Durability Problems

This section will examine the factors that influence the durability of concrete hydropower structures.

6.14.3.1 General

With the public sector investing in more and larger infrastructures, and part of those infrastructures being built underground or underwater where maintenance and repair is often impossible or extremely expensive, the development of durable concrete hydropower structures with a long life is vital.

In the quest to build environmentally sustainable concrete hydropower structures, it is clear that durability is a more important consideration than strength [6]. Durable concrete can be defined as concrete that is designed, constructed, and maintained to perform satisfactorily in the expected environment for the specified designed life. It is the ability of the concrete to resist any process of deterioration.

In the early part of this century, research into the durability of concrete is focused on long-term performance standards, materials, optimization techniques, the science of chemical admixtures, and advancing the understanding of aggregate–paste transition zones [7]. The latter part of the century could offer opportunities to research affordable defect-free concrete that is able to stand the test of time in any predicable environment.

6.14.3.2 How Concrete Hydropower Structures Become Damaged

The main cause of concrete damage is concrete permeability resulting from the environment [8]. If concrete is constructed correctly, the microcracks and voids inside it will not form interconnection pass ways to the surface of the concrete; the concrete is basically waterproof. After a certain period of time, however, the voids and microcracks in the concrete will enlarge due to environmental factors such as freezing–thawing cycles, loading, and chemical attacks. Once an unconnected pass way becomes interconnected, reactive ions or water can enter the concrete, causing the pass way to enlarge further and the concrete to gradually fall into a state of disrepair due to inflation, cracking, weight loss, and decreased strength.

The factors that influence the durability of concrete hydropower structures include carbonation and reinforcement corrosion, freezing and thawing, expansion and contraction, chemical attacks, AAR, seepage water scouring, abrasion, and cavitation. Other aspects such as structural design, construction defects, biological process, and fire damage can also influence the durability of concrete hydropower structures but to a lesser extent.

6.14.3.3 Carbonation and Reinforcement Corrosion

6.14.3.3.1 *General*

In fresh concrete, the presence of abundant amounts of calcium hydroxide gives concrete a very high alkalinity with a pH of 12–14. However, when carbon dioxide, which is present in the air, dissolves in the pore solution of the surface concrete and reacts with calcium hydroxide according to the following equation, the pH value will drop:

$$Ca(OH)_2 + CO_2 \rightarrow CaCO_3 + H_2O$$

This is called carbonation. The pH value of a fully carbonated concrete is about 7 (neutral). During carbonation, the outer part of concrete is affected first, but with the passage of time, the inner mass also becomes affected as carbon dioxide diffuses inward from the surface. If all influencing factors remain constant, the depth of penetration of the carbonation front is thought to be proportional to the square root of the time of exposure. Concrete carbonation and reinforcement corrosion are major factors that affect the durability of concrete hydropower structures in North America, Europe, the Middle East, and other parts of the world. The problem is more serious in places where the temperature and levels of humidity are relatively high.

6.14.3.3.2 *Working mechanism*

Normal carbonation results in a decrease of the porosity of concrete, making the carbonated paste stronger. Carbonation is, therefore, an advantage in non-reinforced concrete but a disadvantage in reinforced concrete. When concrete is at an early age, the high alkalinity results in the formation of a passive film on the surface of the embedded steel in reinforced concrete. This layer is durable and self-repairing, and as long as the film is undisturbed, it will protect the steel from corrosion. When carbonation happens, the alkalinity of concrete reduces. If the pH value of the carbonated concrete drops below 10.5, the passive layer will decay, exposing the steel to moisture and oxygen and making it susceptible to corrosion. The solid corrosion product (rust) occupies a larger volume (2–2.5 times of the previous volume) than that of the steel destroyed and exerts a pressure on surrounding concrete, causing cracking and spalling.

For a hydropower structure that is subject to concrete carbonation and reinforcement corrosion, the concrete appears to be sound with relatively little macroscopic cracking initially. Over time the macroscopic cracks enlarge and the concrete surface is stained by reddish corrosion products. Furthermore, spalling of the concrete covering the reinforcing steel is visible due to the formation of voluminous corrosion products. Eventually, severe spalling of the concrete covering the reinforcing steel is evident, leaving the reinforcing steel bars directly exposed to the atmosphere. Besides carbon dioxide (CO_2), hydrochloric acid (HCl), sulfur dioxide (SO_2), and chlorine (Cl_2) in the environment can also decrease the high alkalinity of concrete and break the passive film over steel. However, this is not very common generally.

6.14.3.3.3 *Influential factors*

The speed of carbonation is influenced by two factors: the quality of the concrete and the environmental conditions.

Quality of the concrete: For good quality concrete that is properly strengthened and has no cracks and little porosity, the expected rate of carbonation is very low. The quality of concrete is influenced by the *w/c* ratio, cement content, type and dosage of admixture, cement grade, type of curing aid, the water curing period, and the temperature when the concrete is initially cast and exposed. A decrease in the *w/c* ratio and an increase in cement content and the water curing period will result in a decrease in concrete carbonation.

Generally, the exposed surfaces of concrete should be kept wet for at least 7 days from the date of placing. However, longer curing periods – up to 28 days – are recommended for blended cement. Concrete that is initially cast and exposed in the winter exhibits lower carbonation than concrete that is cast in the summer. Compared with plain concrete, concrete that incorporates mineral admixtures such as fly ash generally shows higher resistance to carbonation.

Environmental conditions: The speed of carbonation is also affected by environmental conditions such as relative humidity, carbon dioxide density, and ambient temperature. Carbonation occurs fastest when the relative humidity reaches 55% [9]. Above this value, with an increase in the relative humidity, the speed of carbonation decreases. Although the carbonation speed slows down for concretes when the relative humidity is 90–95%, the rate of corrosion is at its highest. Concrete carbonation increases with rising carbon dioxide levels, and this fact goes some way to explaining why some concrete structures located close to highways or factories show little resistance to carbonation. The ambient temperature also influences the rate of chemical reactions. Carbonation speed will increase with an increase in temperature. With an increase in temperature of 10 °C, the rate of reaction is approximately doubled.

6.14.3.3.4 Prevention methods

Having examined the factors that affect the speed of carbonation (the quality of the concrete and the environmental conditions), it is possible to deduce two ways to prevent concrete carbonation from occurring. One way is to make concrete stronger, for example, by decreasing the w/c ratio while increasing the amount of concrete covering the steel, thus preventing concrete from cracking. However, this method can only be used for newly built concrete structures. A second way is suitable for old concrete structures. It involves the application of specialized surface coatings if the depth of carbonation is less than the depth of the concrete covering the steel.

6.14.3.4 Freezing and Thawing

6.14.3.4.1 General

Hardened cement paste, like sand and stone aggregates, is a porous solid and will absorb water. If the temperature drops to freezing point, the water in the porous concrete will freeze. The transition of water from a liquid state to a solid state involves an increase in volume by about 9%, which can damage the cement paste by pushing the capillary walls and generating hydraulic pressures. Concrete subjected to continuous or frequent wetting is susceptible to damage by freezing and thawing cycles. Several cycles of freezing and thawing of water may result in the spalling of concrete.

The freezing and thawing of plastic or green concrete is serious and usually results in permanent damage, even with a single cycle. The volume change that accompanies the freezing tends to increase the space between particles of cement and aggregate so that later bridging by hydration products can only be partly achieved. Such concretes, when hardened, have sharply reduced strengths and much higher porosity.

6.14.3.4.2 Influential factors

Concrete's resistance to freezing and thawing depends upon several parameters, such as pore size, the distribution of pores and capillaries, the age of the concrete, aggregate type, the degree of saturation, and freezing–thawing cycles. Deterioration of concrete caused by freezing and thawing may occur when approximately 91% of the pores are filled with water. The freezing–thawing cycle will start from the first freezing and thawing and will continue throughout successive winters, resulting in repeated loss of concrete surface.

Concrete with a high water content and a high w/c ratio is less resistant to freezing and thawing than concrete with a low water content. Because concrete is workable, the level of water in the mix is normally much greater than that needed for hydration. Excess water in the mix is undesirable because spaces filled with water in the original mix become voids in the concrete when the water not used in hydration evaporates.

Following the curing period, normal drying of dense concrete is vital in aiding resistance to damage caused by freezing and thawing because once the original water in the concrete has been largely used by hydration or lost by evaporation, the process of reabsorption through normal rewetting is very slow. In dry concrete, it is difficult to achieve a near-saturation state with exposure to periodic wetting. Therefore, adequate drainage to provide rapid water runoff is an important way to prevent damage caused by freezing and thawing.

6.14.3.4.3 Prevention methods

Concrete hydropower structures can be protected from freezing and thawing damage through air entrainment or by improving the quality of concrete.

Air entrainment: Air entrainment is an effective way of protecting concrete from freezing and thawing damage. It is achieved by adding a surface active agent in very small dosages to the concrete mixture. This action creates a large number of small, closely spaced air bubbles in the hardened concrete. The air bubbles relieve the buildup of pressure caused by ice formation by acting as expansion chambers.

Improving the quality of concrete: Concrete that is dense is difficult to permeate and has a low w/c ratio is relatively resistant to the effects of freezing and thawing. In actual fact, high grade concrete (grade C50 or above) is generally more resistant to freezing and thawing because the higher cement content needed for producing the higher strength concrete effectively lowers the w/c ratio.

6.14.3.5 Concrete Expansion and Contraction

6.14.3.5.1 General

Temperature-related expansion and contraction of surface moisture exerts a mechanical action and results in the gradual wearing of concrete's surface. Concrete expansion and contraction happens in those areas where the humidity and temperature changes periodically.

As already mentioned, hardened concrete is a porous solid that can absorb water. If the ambient conditions such as air temperature, wind, relative humidity, and sunlight change, the moisture (or water content) or temperature in concrete will change and, hence, cause structures to expand or contract. If those volume changes are limited by aggregates, foundations, or neighboring structures, cracks may occur.

Thin hydropower structures, whose surface areas are large relative to their volume, are subject to the strains of concrete expansion and contraction. Cracks caused by concrete expansion and contraction can occur on upper structures such as flashboards or bridges – the splash zone of the piers, the stilling basin, and the apron being frequently exposed to air.

6.14.3.5.2 Influential factors and prevention methods

The deterioration of concrete resulting from expansion and contraction is influenced by two main factors: the geometric condition of the structure and the ambient conditions. For hydropower structures exposed to the air or in the splash or tidal zone, irregular cracks are very likely to occur. However, the ambient conditions are generally very difficult to change. The only way to relieve the deterioration of concrete through expansion and contraction is to decrease the limitation of neighboring structures or to make the structures more crack resistant. The use of aggregates with low coefficient of expansion may also prevent cracks from happening but in a less effective way.

6.14.3.6 Chloride and Sulfate Attack

6.14.3.6.1 General

Chemical processes are the most important factor influencing the durability of concrete structures. Among them, chloride and sulfate attacks are the most common aggressive actions leading to the deterioration of concrete.

6.14.3.6.2 Chloride attack

Chloride-induced rebar corrosion results mainly from the use of deicing salts in cold climates and/or exposure to marine environments. When concrete undergoes alternate wetting and drying, the water, which contains chlorides, can penetrate through concrete and cause concrete deterioration. This is called a chloride attack. Chloride ions can enter concrete in two ways: they may be added during mixing either deliberately as an admixture or as a contaminant in the original constituents. Calcium chloride has been used as an admixture in the past because it is relatively cheap, accelerates setting, and provides early strength in concrete. The corrosive effects of chloride were not observed until 20 years after the application of calcium chloride, when it had already caused some damage to hydropower structures. Today, non-chloride-containing admixtures exist, and it is required that only these admixtures be added into concrete mix. Chloride ions may also enter the set concrete from an external source such as sea water, particularly when structures are exposed to sea spray or deicing salts.

Once chloride ions have reached the reinforcement in sufficient quantities, they will depassivate the embedded steel by breaking down the protective oxide layer normally maintained by the alkaline environment, and hence cause the reinforcement to corrode. The rust formed has a larger volume than that of the steel consumed, which will cause concrete to crack.

The speed of chloride diffusion is influenced by two main factors: the quality of the concrete and the ambient conditions. For dense concrete, the diffusion speed of the chloride ions will be slow, and hence more time will be taken for chloride ions to reach the steel surface. The environment to which the reinforced concrete structure is exposed affects the extent of a chloride attack as well. The speed of chloride penetration is greatly influenced by the degree of saturation.

There are generally two ways to prevent a chloride attack. The first method involves enhancing the quality of concrete (e.g., by choosing proper materials, having adequate cover over reinforcements, paying attention to the environmental changes during construction) and concrete coating. The second method can be very effective for newly built structures or for old structures whose chloride quantity at the depth of the reinforcement is below the chloride threshold.

6.14.3.6.3 Sulfate attack

A sulfate attack occurs in hardened concrete when sulfates, found in sea water, in some soils or in wastewater, react with the tricalcium aluminates (C_3A) in Portland cement paste. The reaction causes a material called ettringite to form. The ettringite produced occupies a greater volume within the concrete than the calcium aluminate hydrates, which results in concrete expansion and irregular cracking. The cracking of concrete provides further access to penetrating substances and to progressive deterioration. The higher the C_3A content of the cement, the more chance there will be of a sulfate attack.

Gypsum (calcium sulfate) is present in some of the clay soils in the south of England and also occurs in desert soils at locations where the water table is close to the surface. The more soluble sulfate salts – sodium sulfate (Glauber's salts) and magnesium sulfate (Epsom salts) – may be present in some rock formations in the United Kingdom but are more extensive in the alkali soils of North America. In northeast China, a lot of hydropower structures are experiencing sulfate-induced deterioration.

In addition to the two methods that can help to prevent a chloride attack, concrete can also be protected against a sulfate attack by limiting the aluminates to a level between 3% and 8%. According to research, blended cements perform better than ordinary Portland cement when subjected to a sulfate attack, and pozzolanic materials such as fly ash, silica fume, and rice husk ash provide moderate resistance.

6.14.3.7 Alkali–Aggregate Reaction

6.14.3.7.1 General

As already mentioned, the high pH value within the concrete pore structure provides a protective coating of oxides and hydroxides (passive film) on the surface of the steel reinforcement. However, the alkalinity leads to AAR.

AAR is a chemical reaction where the aggregate reacts with the high pH pore solution in the hardened concrete. The most common form of AAR is alkali-silica reaction (ASR): the reaction occurs between the alkali ions in the concrete pore water and the reactive silica in the aggregate. ASR can continue indefinitely if sufficient alkalis are available. ASR procedure can be expressed as

$$\text{alkali} + \text{reactive silica} + \text{water} \rightarrow \text{ASR} \rightarrow \text{expansive gel}$$

The gel produced by the reaction has a very strong affinity for water, and thus has a tendency to swell. Once the gel is formed, it may migrate through the porous structure of the concrete, following preexisting voids and fractures. If it meets with water, the gel may absorb the water and expand, causing microfracturing of the material and, ultimately, a characteristic map-cracking pattern on exposed surfaces.

Besides ASR, carbonate minerals may also cause deterioration of concrete due to an alkali attack but in an uncommon way (alkali–carbonate reaction (ACR)). Until recently, ACR was found only in North America in some concretes containing dolomite aggregates. The vast majority of sources of dolomitic limestone aggregates have not been found to be deleteriously reactive, and have been used successfully for the production of concrete [10].

The reaction time of AAR ranges from 2–3 to 40–50 years depending on the different aggregates in the concrete. For ACR, cracks will happen only after 2–3 years, while for ASR, the reaction time is generally around 10–20 years. For some ASR, the reaction time can be very long, with cracks happening after 40–50 years. However, once cracks appear, the compressive strength and the modulus of elasticity of the concrete will decrease tremendously, which is very difficult to rectify.

Often, the first external signs of alkali reaction in a structure are short fine cracks on the surface radiating from a point. The cracks occur adjacent to fragments of reacting aggregates and are caused by the outward swelling pressure. As time passes, the cracks propagate and eventually join up to form maplike patterns (see **Figure 9**). The macrocracks, which appear on the surface, are generally found to be 25–50 mm deep and to occur roughly at right angles to the surface.

6.14.3.7.2 Sodium equivalent

In order to make simple comparisons between different cements, the alkali content is usually expressed as total alkali content or 'sodium equivalent'. To obtain the equivalent sodium oxide content, the potassium oxide content is factored by the ratio of the molecular weights of sodium oxide and potassium oxide, and added to the sodium oxide content [10].

$$\text{Total alkali (as equivalent Na}_2\text{O)} = \text{Na}_2\text{O content} + 0.658 \times (\text{K}_2\text{O content})$$

The alkali content of cement depends on the materials from which it is manufactured and also to a certain extent on the manufacturing process, but it is usually in the range of 0.4–1.6%.

In 1941, the United States declared that, in order to prevent AAR, the aggregates used in concrete construction must be tested and the alkali content of cement must be less than 0.6%. More than 20 countries, including China, have accepted and added this criterion to their national standards. In many countries such as New Zealand, England, and Japan, cement is produced with the alkali content lower than 0.6%.

6.14.3.7.3 AAR in the world

Damage due to AAR in concrete was first recognized in the United States in 1940 and has been observed in many countries since then. The maintenance costs of concrete structures suffering from AAR in Western Europe accounts for approximately 10% of the

Figure 9 ASR attack, Ontario, USA.

total maintenance costs. The reduction in the life of structures that are damaged due to AAR cannot be estimated, but it is probably significant.

A wide variety of aggregate rock types in structures throughout the world have been reported as being alkali-silica reactive. It has also been discovered that the great majority of concrete structures reported to be deteriorating due to ASR have been constructed using a high-alkali Portland cement, which is the most widely used cement in the world.

6.14.3.7.4 Influential factors

If the concrete is kept in a consistent environment, AAR damage develops progressively over the years [11]. The basic ingredients for an AAR attack are reactive aggregates, pore solution alkalinity, and moisture (usually 85% relative humidity) or water. Besides these three main ingredients, air content and temperature are also important variables that increase the expansion of concrete affected by AAR. Studies indicate that even HPC is susceptible to AAR if reactive aggregates are used.

6.14.3.7.5 Prevention methods

To prevent AAR from occurring, one or more of the ingredients for an attack needs to be substituted. If there are no alternative ingredients, then it is necessary to prevent the reaction or the expansion of the existing AAR gel by:

- decreasing the alkali content of the pore solution – this can be achieved by using low alkali cement or replacing part of the cement with low alkali mineral admixtures;
- applying nonreactive aggregates – aggregates that have no reaction with the pore solution alkalinity will prevent AAR from happening; and
- decreasing the water content of the hardened concrete – ASR will be minimized if concrete is kept dry. This can be achieved by using a low w/c ratio when making concrete mix, by applying admixtures to minimize water content, or by putting a layer of waterproofing material outside of the structure.

Air-entraining admixtures applied in concrete can provide some voids in concrete and thus mitigate the expansion of gel produced.

6.14.3.8 Seepage Scouring

6.14.3.8.1 General

Seepage scouring commonly affects the durability of hydropower structures. Cement paste includes calcium silicate hydrate (CSH), calcium hydroxide ($Ca(OH)_2$), and calcium sulfoaluminate hydrates. A sufficient quantity of calcium hydroxide is necessary for the existence of other hydrates. If a hydropower structure is not water resistant, calcium hydroxide will be taken away by seepage water and white calcium carbonate ($CaCO_3$) crystals will form on the downstream of the structure. This action will break the pH equilibrium of the concrete and cause the decomposition of other components dissolved in the pore solution. According to statistics published by the Soviet Union, when 25% of calcium hydroxide is taken away from the concrete, the compressive strength decreases to 50% of the previous value. And when the percentage of calcium hydroxide increases to 33%, concrete will lose its strength and be damaged [12].

Seepage scouring can also cause other problems. For example, seepage water in concrete can accelerate the freezing and thawing break, and thus increase the corrosion of reinforcements.

6.14.3.8.2 Influential factors

Seepage scouring is influenced by many factors such as water head, water level, the presence of a crack and its characteristics (width, depth, and distribution), temperature, humidity, the quality of concrete, and the condition of waterstop.

6.14.3.8.3 Prevention methods

Seepage scouring can be prevented in two ways:

- By making the concrete water resistant
- By adding a layer of waterproofing material outside of the structure. This is the same method as for AAR prevention.

6.14.3.9 Abrasion and Cavitation

6.14.3.9.1 General

For common hydropower structures, the surface may be weathered by wind, rain, snow, or mechanical actions such as people and traffic. This is called abrasion. Abrasion is generally caused by flowing water due to grinding, rolling, and impacting effects of suspended particles.

Cavitation is a characteristic type of deterioration for hydropower structures but works in a different way to abrasion. All hydraulic surfaces contain some construction defects. This irregularity causes small areas of flow separation and in these regions the pressure will be lowered. If the velocities are high enough, the pressure may fall below the local vapor pressure of the water and vapor bubbles will form. When these bubbles are carried downstream into the high-pressure region, the bubbles will collapse, giving rise to high impact and possible damage. Experimental investigation shows that the damage can start at clear water velocities between 12 and 15 m s^{-1} and up to velocities of 20 m s^{-1} [13].

6.14.3.9.2 Abrasion- and cavitation-induced deterioration in China
According to China's national investigation on hydraulic structures in 1985, abrasion and cavitation accounted for the damage in 70% of the structures investigated. Some hydraulic structures were damaged by cavitation so frequently that rehabilitation had to be undertaken every year. Therefore, China has paid much attention to research on materials that are resistant to abrasion.

6.14.3.9.3 Prevention methods
Both abrasion and cavitation start at the surface, so special attention should be given to the quality of concrete surfaces.

- Use HPC at the surface to withstand the abrasion of high-velocity water (sometimes with suspended particles). Recently, polymer fibers have been applied in many hydropower structures in China to increase resistance to abrasion. However, the long-term performance of these fibers is still the subject of research.
- Make the surface of the structure as smooth as possible, and thus make the concrete more resistant to cavitation.

6.14.3.10 Other Less Important Factors

In addition to the factors that have already been mentioned, structural design, construction defects, biological process, and fire damage can also influence the durability of a concrete hydropower structure to some extent.

6.14.3.10.1 Structural design
The structural design aspect includes strength failure, geometric design, and accessibility of a suitable replacement. Hydropower structures are generally designed under the guidance of a national design code. Therefore, if structures are operated under the designed static and dynamic loads, structural strength failure can be prevented.

For geometric design, the exposed surface of a concrete hydropower structure should be of a simple nature to avoid local deterioration. Complexities often lead to maintenance problems later. Furthermore, differential settlement and thermal effects should be considered in the design to avoid inexplicable cracking.

Structural components, such as joints, seals, the drainage system, and waterproofing treatments, should be planned for easy replacement later on, if necessary, without damaging the adjacent structural components.

6.14.3.10.2 Construction defects
Construction defects arise from incomplete consolidation in tamping, inadequate curing, removal of the forms before full curing has taken place, settlement of forms, movement of support, and the structure being subjected to the load ahead of time, especially during the first few hours following the setting of the concrete, before the concrete has had a chance to develop to its full strength. Construction defects can be prevented if structures are built under the guidance of a construction code. However, a constructor should be aware of the constructional aspects in order to foresee any problems. Buried components of structures (such as footing and piles) cannot be reached or inspected after construction. Such inaccessible components require greater attention and care at the construction stage than other components.

6.14.3.10.3 Biological process
Biological process may cause both mechanical and chemical deterioration of concrete hydropower structures.

- *Mechanical process*: Plant roots may penetrate cracks and other weak spots of the concrete. The resulting bursting forces may widen the existing cracks and cause spalling of concrete.
- *Chemical process*: In the presence of sewers and biogas plants, the hydrogen sulfide produced in the anaerobic conditions may be oxidized in the aerobic conditions and form sulfuric acid, which attacks concrete above the water level.

6.14.3.10.4 Fire damage
Concrete structures are relatively good at resisting the effects of damage caused by fire. Concrete is a poor conductor of heat and thermal conductivity is reduced as the temperature increases.

However, if concrete is exposed to a rise in temperature, the water contained in its pores and capillaries is first driven off. Should the temperature increase to above 100 °C, some of the water, combined with the calcium silicate hydrates in the cement paste, is also lost. However, this desiccation has little influence on the strength of concrete.

Should the temperature rise to between 300 and 400 °C, there will be a more pronounced chemical change in the cement paste. The calcium silicate will be converted into calcium oxide and silica. The cement shrinkage resulting from water desiccation and the aggregate expansion resulting from a temperature increase will cause concrete to crack. Considerable damage may also occur because of the thermal shock experienced by the concrete as water is sprayed onto the structure during firefighting. This damage may be in the form of cracks or further spalling. At this stage, the concrete will have experienced a tremendous loss of strength. If the aggregate in concrete, which is exposed to temperatures in this range, has silica or limestone in it, there will have been a distinct color change. Generally, if the concrete has changed to pink, the strength may have already decreased. If the color has turned to gray, the concrete will have become brittle and porous.

6.14.4 Durability Design

The deterioration of concrete hydropower structures is usually a medium- to long-term process. The onset of deterioration and its acceleration may be stimulated by many factors such as material properties, construction technique, environmental conditions, foundation characteristics, and the presence of cracks. But to what extent can these factors affect the life of hydropower structures? Is there a relationship between these factors and the life of a hydropower structure?

6.14.4.1 Performance-Based Durability Design

The *European Standard for the Basis of Design* states that works shall meet essential requirements for their economically reasonable 'working life'. This introduces a time-dependent element into performance specifications called 'performance-based durability design'. When referring to performance-based durability design, the term 'performance' means that with physically correct mathematical models it can be shown that the essential functions of the structures are fulfilled with reference to a period of time. To do this, a performance limit is defined and it is then shown that the probability of falling below this limit is acceptably low, demonstrating that the structure is reliable. The performance (such as loads or material properties) with reference to a time period, the performance limits, the service life, and the reliability level (or the accepted probability of failure) can be given by codes or stated by the owner of the structure, or they can be derived from an economic optimization.

Performance-based specifications may drive many of the durability improvements in concrete hydropower structures. The value of the concrete will be defined in terms of maturity, permeability, air-void structure qualification, sulfate resistance, chloride penetration, strength, and *in situ* performance [7]. This method offers the advantage of a seamless transition between structural design and durability design. In the last decade, Australia, North America, Japan, and countries in Europe have put much effort into the development of performance-based durability design of concrete structures. However, such an approach is not yet available in practice since performance-based durability design needs qualified knowledge of material science, engineering, environmental actions, construction techniques, assessment and repair, and probabilistic approaches.

6.14.4.2 Expert System

Based on the idea of 'performance-based durability design', the 'expert system' is emerging.

The first expert system came into being in the late 1980s. Called the DURCON system (DURurable CONcrete), it was codesigned by the American bureau of standards and an American concrete durability committee. The system can supply durability design on concrete structures prone to steel corrosion, freezing and thawing, deicing salts, AAR, and sulfate attack. The Finland technical research center has also designed an expert system. This system can design prestressed concrete mix. Research into expert systems is being carried out in other countries, too. However, the construction industry awaits the availability of a uniform expert system for concrete hydropower structures. Much needs to be done before the application of an expert system. For example, all the efforts made so far have the same weak points; the environmental factors are sorted into single factors; and the result of the durability design is the simple add up of all those factors. In reality, the damage caused by many of those factors is complicated and the effects severe. However, those synergistic effects are difficult to express and, therefore, much more investigation is needed in this field.

Setting up an expert system involves information collection, analysis, and sorting. Information collection is the most critical point that decides the accuracy of the system. The system should have the capability of solving problems in the event of uncertain or incomplete data. An expert system may be designed for a whole nation or just for one certain district.

6.14.4.3 The Direction of Future Research

In the future, when more information is available, it will be possible to set up an expert system that can be applied in engineering practice, where when the initial data of a project are known (data input), the possible results can be estimated (output) with theories, practical knowledge, and experiences. Future expert systems should have the following functions:

- Durability design on concrete hydropower structures. Durability design includes concrete mix design, casting, and curing.
- The ability to anticipate the life of a hydropower structure within certain environments.
- The ability to apply technical support for operation, maintenance, and rehabilitation of concrete hydropower structures.

The expert system of the future needs to limit the scope between rehabilitation and reconstruction, and hence comes up with a standard. At present, when engineers complete cost analysis to decide whether a structure should be reconstructed or rehabilitated, the further life of the old structure if it is rehabilitated is not considered. Therefore, rehabilitation is often applied when the cost is relatively low. However, it may not be wise to think that the life of a rehabilitated structure will be greatly shortened compared with a newly built structure. In Jiangsu province, China, about 70% of the hydraulic structures are rehabilitated instead of rebuilt. In some cases, only the base and the piers are left; the other auxiliary structures have to be rebuilt. Consideration needs to be given to how cost-effective this is in the long run. On the other hand, demolition and reconstruction can be impractical once resource intensity, social disruption, and environmental effects have been taken into consideration. This is especially true in developed countries. Therefore, there is great demand on the expert system to come up with a standard to make a rational judgment about whether a structure should be rehabilitated or reconstructed.

6.14.5 How to Maintain a Durable Concrete

The main task facing durability design of concrete hydropower structures is maximizing their life, taking into consideration cost and the environment in which they will be placed and used. This is a task that requires the efforts of not only designers but also constructors and administrators. The requirements that must be met to ensure a durable hydropower structure exist at the stages of material selection, design, construction, inspection, and rehabilitation.

6.14.5.1 Material Selection

6.14.5.1.1 General
Material selection, and proportion, will play a very important role in the improved concrete of the new century. Every concrete mix should be proportioned taking into account the conditions it is most likely to be exposed to (such as freezing and thawing, sulfates, deicing chemicals, acids, varying moisture conditions, and abrasive loadings), construction considerations, and structural criteria.

Proper selection of aggregate or cement is the fundamental factor in creating durable concrete. Both materials need to be checked to prevent excessive expansion due to AAR or thermal gradients. Furthermore, judicious selection of mineral admixtures (fly ash, blast furnace slag, silica fume) or chemical admixtures (air-entraining admixture, water-reducing admixture, etc.) can reduce the possibility of shrinkage cracking or the permeability of concrete and, therefore, extends the life of a hydropower structure as well.

Innovative concrete materials are continually being developed to enhance performance, improve construction, and reduce waste, for example, mineral admixtures, chemical admixtures, polymer fibers, geomembranes, and self-healing, self-compacting, and HPC. As the materials industry develops, durable concrete should become available in the not-too-distant future.

6.14.5.1.2 Mineral admixtures
Concrete admixtures are used to improve the behavior of concrete under a variety of conditions and are of two main types: mineral and chemical.

Mineral materials include fly ash, silica fume, and slag. The use of these materials in blended cement is becoming more important in the construction industry for countering durability problems.

Fly ash: Fly ash is derived from burning coal. Research indicates that the application of fly ash can enhance the strength of concrete, make concrete easier to work with and decrease bleeding, reduce drying shrinkage, enhance the impermeability of concrete, decrease the heat of hydration; reduce corrosion, and mitigate AAR, sulfate, and salt attacks.

Silica fume: Silica fume, or microsilica, is a by-product of the electric arc furnace reduction of quartz into silicon and ferrosilicon alloys used in the electronics industry. The use of silica fume can enhance concrete's early strength, impermeability, and resistance to corrosion. One pound of silica fume produces about the same amount of heat as a pound of Portland cement, and yields about 3–5 times as much compressive strength. The amount of silica fume required is in the range of 8–15% (by the weight of cement). It is typically added to, and does not replace, the existing Portland cement. It is worth mentioning that the addition of silica fume will decrease the workability of concrete. Therefore, the higher the percentage of silica fume used, the greater the amount of superplasticizer that will be needed.

Ground granulated blast furnace slag: Iron is produced in blast furnaces that are charged with the raw materials of iron ore, coke, and limestone. In the smelting process, iron is produced in molten form and slag forms on its surface. The slag results from the fusion of limestone with ash from the coke, and aluminates and silicates from the ore. The compounds that the slag contains are similar to those in Portland cement. To produce a material suitable for use as cement, it is necessary to cool it quickly so that it is solidified in a glassy state. This kind of material is called ground granulated blast furnace slag (GGBS). GGBS may be interground with Portland cement or added as a separate ingredient. The application of GGBS in concrete has the following effects: high strength, low permeability, low potential of chemical attack, and low hydration heat.

6.14.5.1.3 Chemical admixtures

Chemical admixtures are usually added as liquids or powders in relatively small quantities and may be used to modify the properties during the plastic or hardened state of concrete. Chemical admixtures can be divided into five types: accelerating, retarding, water reducing/plasticizing, air entraining, and waterproofers.

Chemical admixtures should be used judiciously because the addition of wrong quantities can affect the long-term performance of concrete in many ways. For example, the use of calcium chloride as an accelerator can lead to reinforcement corrosion; overdosage of air-entraining admixtures can lead to reductions in strength, which in turn could lead to structural problems; and overdosage of plasticizers may lead to segregation or bleeding.

With the development of the chemical industry, new kinds of admixtures have started to come into use. They include self-curing admixture, shrinkage reducing/compensating admixture, corrosion inhibitors, alkali-silica reactivity inhibitors, and so on.

6.14.5.1.4 Polymer fibers

The corrosion problem associated with steel rebar is the most important factor in limiting the life expectancy of reinforced concrete structures. In some cases, the repair costs can be twice as high as the original costs. To increase the life span of reinforced concrete structures, government organizations, private industry, and university researchers are seeking ways to avoid the corrosion problems and thereby mitigate the burden of never-ending repair costs. One preferred solution, which has assumed the status of cutting-edge research in many industrialized countries, is the use of fiber reinforced polymer (FRP) in concrete. These FRP materials can be used for new structures (reinforced and prestressed concrete) as well as for the rehabilitation of existing structures (external FRP sheet bonding and FRP external posttensioning).

FRP materials include two types: metallic and nonmetallic. Metallic materials include carbon, steel, and stainless steel. Nonmetallic materials include carbon fiber, aramid fiber, and glass fiber.

Nonmetallic FRP materials are advanced composites made of continuous synthetic or organic fibers with high strength and stiffness embedded in a resin matrix. They generally offer many advantages over the conventional steel or metallic FRP materials. For example, they

- are one-quarter to one-fifth of the density of steel;
- do not corrode even in harsh chemical environments;
- are neutral to electrical and magnetic disturbances; and
- have a higher tensile strength than steel.

Extensive research on the use of FRP in concrete structures started in Europe about 25 years ago. Since then, there has been great interest in FRP reinforcement in Europe, and some pioneering work in this field has been done. As a result, the world's first highway bridge using FRP posttensioning cables was built in Germany in 1986. Commercial use of externally bonded FRP reinforcement started mainly in Switzerland around 1993 and soon followed in other European countries.

As these early developments were commercially less successful than it was hoped, today the engineering application of FRP reinforcement in Europe is less exploited than in North America and Japan. However, given the advantages, interests, and efforts, FRP reinforcement is becoming more widespread. Today, the use of FRP is becoming a standard technique. FRP reinforcement is available in different European countries by local suppliers and a few manufacturers [14]. It is believed that the new materials will become economically available in structural engineering in the not-too-distant future.

6.14.5.1.5 Geomembranes

Geomembranes are thin, flexible materials that are manufactured in factories in a controlled environment. Geomembranes can be permeable or impermeable. Impermeable geomembranes are often used as a water barrier in hydropower structures, while permeable geomembranes are applied for the seepage water to pass by without taking away the soil. If a geomembrane is associated with a geotextile, it is called 'geocomposite'.

Geomembranes have been used in dams since 1959 and their use in hydropower structures continues to grow. Geomembranes have been used, exposed, and covered, for rehabilitation and new construction. The life of a geomembrane can be as short as 2 months or as long as 50 years; it depends on the quality and the environmental conditions.

6.14.5.1.6 Self-healing concrete

Self-healing concrete is a new type of concrete. It imitates the automatic healing of body wounds by the secretion of some kind of material. To create self-healing concrete, some special materials (such as fibers or capsules), which contain some adhesive liquids, are dispensed into the concrete mix. When cracks happen, the fibers or capsules will break and the liquid contained in them will then heal the crack at once. However, self-healing concrete is only at the research stage. Its application in the concrete industry is still some way off.

6.14.5.1.7 Self-compacting concrete

Self-compacting concrete (SCC) is a special type of concrete mix that can be cast without compaction or vibration. It is applied at sites where vibration is very difficult or even impossible (such as underwater). SCC is quick to apply, leading to savings in staff costs. Overall costs savings are realized even when the addition of superplasticizer is taken into account.

Although very fluid, SCC is also very cohesive and has a low tendency of experiencing segregation and bleeding. The important basic principle for SCC is the use of superplasticizer combined with a relatively high content of power materials such as Portland cement, mineral additions (fly ash, silica fume, etc.), or very fine sand, and with some limits on the maximum size of the coarse aggregates (< 25 mm).

SCC and cohesive concrete were first studied in 1975–76 [15, 16] following the advent of superplasticizers. At that time, the maximum slump level permitted by the American Concrete Institute (ACI) was 175 mm. It was not until the 1990s that the term self-compacting concrete was used. At present, SCC is considered to be the most promising material for concrete works [17]. To make SCC more applicable in construction, future research will shed more light on two aspects:

- More effective superplasticizers with lower slump loss
- Viscosity-modifying admixtures, which can reduce bleeding and increase the resistance to segregation, particularly in SCC with a low content of cement.

6.14.5.1.8 High-performance concrete

The 1970s saw the advent of HPC. Any concrete which satisfies certain criteria proposed to overcome the limitations of conventional concretes may be called high-performance concrete (HPC). It can be made through selected mix design, and proper mixing, transporting, placing, consolidation, and curing.

HPC may include concrete that provides either substantially improved resistance to environmental influences or substantially increased structural capacity while maintaining adequate durability. It may also include concrete that significantly reduces construction time without compromising long-term serviceability. Therefore, it is not possible to provide a unique definition of HPC without considering the performance requirements of the intended use of the concrete. HPC is often of high strength, but high strength concrete may not necessarily be of high performance.

The Federal Highway Administration (FHA) has proposed criteria for four different performance grades of HPC. The criteria are expressed in terms of eight performance characteristics including strength, elasticity, freezing/thawing durability, chloride permeability, abrasion resistance, scaling resistance, shrinkage, and creep. Depending on a specific application, a given HPC may require a different grade of performance for each performance characteristic. However, the above-mentioned criteria only suit highways. For hydropower structures that are situated in water, the water-related characteristics, such as AAR and seepage water scouring, may be more prominent. Therefore, further work is needed on the criteria of HPC on concrete hydropower structures.

6.14.5.2 Design

Durability is influenced by many factors such as the materials used, the construction technique, the structural layout, and the region where the structure will be built. In modern constructions, durability of concrete has become the most important factor that governs the life of the structures. Some researchers [18] state that the design of a hydropower structure should be 'double controlling'. That is, the weight given to strength control is 25%, while the importance of durability should occupy 75%. To maintain a durable concrete, structural design and durability should be considered together rather than as separate entities. The design of hydropower structures includes both mix design and structural design.

6.14.5.2.1 Mix design

Mix design refers mainly to w/c ratio and cement content. The w/c ratio influences the permeability of concrete and should be decreased with increasing harsh environmental conditions. The w/c ratio for a hydropower structure is generally in the range of 0.4–0.55, depending on the harshness of the environment. The cement content of concrete is of less significance than the w/c ratio for structural durability, provided the mix is of adequate workability. The cement content for a hydropower structure is normally between 220 and 375 $kg\,m^{-3}$.

Concrete mix design is decided by two factors: strength and durability. Mix design based on strength is the more advanced at present, and plenty of codes and theories are available.

6.14.5.2.2 Structural design

Structural design mainly refers to the rational distribution of the structure. Besides the strength requirements, there are some points that must be stressed for durable concrete:

- In order to resist the ingress of deleterious substances, the concrete cover over steel should be dense, strong, and impermeable. The thickness of concrete cover is generally in the range of 15–75 mm.
- The exposed surface of a hydropower structure should be of a simple nature to avoid local deterioration. Complex details often lead to maintenance problems later. Sudden changes in cross section should be avoided to prevent stress concentration within concrete.

- Differential settlement and thermal effects should be considered in the design to avoid inexplicable cracking.
- Designers should consider the accessibility, reparability, and replaceability of various structural components such as joints, seals, and drainages systems. Those components should be planned for easy placement without damaging the adjacent structural components.

6.14.5.3 Construction

Appropriate construction techniques are essential to improve concrete durability. They include construction sequence, reinforcement placement, concrete casting, curing, and quality evaluation.

6.14.5.3.1 Construction sequence

Construction sequence is often the most important factor, since it can reduce costs and help to prevent harmful cracks if applied correctly.

To mitigate the effects of side loads, it is preferable to construct the heavy part of hydropower structures first (such as the main part of the structure and the retaining wall) and then the light part (such as the stilling basin). The general construction sequence of a hydropower structure is as follows: The main part → the retaining wall → back fill the soil to at least 70% of the designed elevation → the stilling basin, river bank, river bed protection → back fill the soil to the designed elevation. A monolith can also be constructed with temporary joints, and then the joints filled with concrete. Obviously, this technique decreases the internal stress, and hence decreases costs greatly.

6.14.5.3.2 Reinforcement placement

Proper attention should be paid at the time of positioning the reinforcement so that its usability is at an optimum level. A too thin protection layer over steel may induce reinforcement corrosion, while a concrete layer that is too thick may induce concrete cracking.

6.14.5.3.3 Concrete casting

During construction, the concrete should be well mixed and placed as near as possible to its final position. It should not be placed in large quantities at a given point and then allowed to run over a long distance in the forms. This practice results in segregation, because the mortar tends to flow out ahead of the coarser material. During casting, the consolidated mass should be uniform without rock pockets or honeycombed areas.

During concrete casting, the arrangement of temporary joints and the methods for bonding successive lifts of concrete are also important details that can affect the performance of the structure even though the concrete itself is durable. Drainage should be made at the point where the structure is susceptible to constant saturation in order to avoid damage caused by freezing.

6.14.5.3.4 Curing

Curing is very important in the construction of durable concrete. This includes protection against extremes of temperature as well as provision of moisture during the critical early period. No other element of concrete construction offers such possibilities for increased strength and durability at such a low cost than better curing.

Moisture curing: For proper curing, the exposed surfaces of concrete should be kept continuously wet for at least 7 days from the date of placing. However, longer curing periods, up to 28 days, are recommended for blended cement.

Curing temperature: Curing temperature is also very critical to concrete quality. Freezing of the concrete should be avoided for the first few days because it may reduce the strength of the concrete tremendously. Conversely, increasing the curing temperature will accelerate the chemical reaction within concrete and enhance its early strength. However, a curing temperature that is too high may negatively affect the strength of the concrete after 7 days. Research shows [5] that a too fast chemical reaction (which is caused by high curing temperature) may cause the formation of some voids and uneven materials within concrete. Those unevenly distributed materials will form some weak points, which will decrease the strength of the concrete.

Commonly used methods for curing include covering the concrete with a polyethylene sheet, spraying a liquid curing membrane on the concrete, and continuously wetting the concrete with a soaker hose. For mass concrete, correct curing is especially important as large temperature gradients may be developed between the center and the surface of the structure, inducing cracks.

6.14.5.3.5 Quality evaluation

Specimens for strength tests in compression (or in flexure) should be made from all trial batches after the mixture has been established to determine if the strengths are within the range intended. Furthermore, if concrete is exposed to unfavorable environmental conditions, tests for chloride penetration, shrinkage, permeability, and the air-void system or resistance to freezing–thawing cycles would be desirable.

6.14.5.4 Inspection and Assessment

6.14.5.4.1 General

In order to extend the life of existing concrete hydropower structures, they must be inspected periodically. A detailed investigation of deteriorated structures is essential before planning remedial measures. Concrete inspections are divided into two types: visual inspection and instrumental inspection. General investigation procedure includes initial inspection, surveying for cracks and other defects, monitoring, sampling and laboratory testing, measuring of the concrete cover, assessing the material strength, reviewing history documents, and so on. This procedure includes both visual inspection and instrumental inspection. As part of the procedure, a document review is a very important tool for analysis. This is a review of historic records such as original drawings, engineering reports, field data, and photographs.

Analysis will be carried out based on the investigations mentioned above. This procedure, which requires thorough knowledge of structures and materials, is particularly important where historic concrete is involved since improper repairs can cause additional deterioration. Generally, the analysis considers whether the structure behaves normally; if damage is present, the location and intensity of that damage; the residual life span of the structure; and the cause of deterioration.

6.14.5.4.2 Visual inspection

Visual examination includes the inspection of weathering, unusual or extreme stresses, alkali or other chemical attacks, erosion, cavitation, vandalism, and other destructive forces. General signs of deterioration include concrete cracking, spalling, deflection, exposure of reinforcing bars, large areas of broken-out concrete, misalignment at joints, undermining and settlement in the structure, rust stains on the surface of concrete, clogging of drains and drainage paths, inadequate growing of vegetation, and so on.

6.14.5.4.3 Instrumental inspection

To confirm that the design will remain strong over time and to predict the service life of hydropower structures, cost-effective and reliable inspection instruments are required. Inspection instruments form the basis for the owners' and operators' measurements of the quality of a structure and allow them to rate the repair items and estimate their repair budgets.

Common instrumental monitoring and assessment methods include the use of sensors, examination of extracted core samples, measurements of carbonation depths, tests for the strength and permeability of concretes, mix composition analysis by weight and volume, chemical reaction analysis for alkalinity, carbonation, chloride and other components, deflection measurements, and so on.

However, some methods are complicated or involve breaking the natural structure (to extract samples from the structure). Recently, research has focused on nondestructive evaluation methods, and some new techniques are starting to be applied. Those instruments generally inspect the temperature and strength of concrete, the onset of cracking, and the development of cracking. There is no reliable, nondestructive measurement for testing the degree of corrosion.

6.14.5.5 Rehabilitation

6.14.5.5.1 General

With the information gained through inspection and assessment, structures can be repaired and strengthened by applying appropriate rehabilitation methods.

Common durability problems for old hydropower structures are concrete aging, joint leakage, and cracking. If the strength of the concrete decreases to such an extent that it cannot carry the stress caused by the design load, the structures will have to be rebuilt. While for joint leakage and cracking, different methods will be taken corresponding to different conditions.

6.14.5.5.2 Joint repairing

The joint-filling materials used in hydropower structures are also called waterstops. Their main function is to prevent water or soil from passing through the structural body. Nearly all hydropower structures need waterstops.

Waterstops are susceptible to damage in hydropower structures. Typical causes of waterstop failure include excessive movement of the joint, which ruptures the waterstop; a honeycombed concrete area adjacent to the waterstop; contamination of the waterstop surface, which prevents bonding to the concrete; and material aging under sun radiation or other environmental factors. This last cause is the most common durability problem for joints. Rigid joint-filled materials such as copper, steel, or lead tend not to experience aging problems. Meanwhile, flexible materials such as PVC mats, rubbers, asphalt-impregnated fiber sheets (or rolls), or compressible foam strips will lose function over a certain period of time. Agreement has been reached about the life of flexible joint-filling materials. The material manufacturers say that they can function for 40–50 years. However, it seems not true for many structures.

Since it is usually impossible to replace an embedded waterstop, grouting or installation of the remedial waterstop is the most common repair method. Remedial measures are generally grouped into three types: surface waterstop, caulked waterstop, and drilled holes filled with elastic material. The type used depends on a number of factors, including joint width and the degree of movement, hydraulic pressure in the joint, environment, type of structure, economics, available construction time, and access to the joint face.

Surface waterstop: A surface waterstop is generally a few layers of waterproofing material that span and attach to the surface with bolts.

The general type of waterstop consists of a rigid plate (normally stainless steel) with a crimp and a layer of deformable rubber. The plate, together with the deformable rubber, is anchored with bolts on both sides of the monolith, which provides initial pressure on the rubber and hence waterproofing.

Recently, the use of geomembranes as external waterstops has drawn much attention. This kind of surface waterstop consists of a waterproofing geomembrane, a drainage layer (in case there is still some water leakage), a supporting layer, and an anti-puncturing layer. The waterproofing geomembrane layer is anchored with bolts on each side of the monolith.

The use of surface waterstops as remedial waterstops is particularly useful when the repair site is freely accessible. Repairs can be carried out simply and cheaply. This method can also be applied underwater when dewatering is difficult. Potential problems with this type of repair include loosening of the anchor bolts, barge tearing, ice pressure from moisture trapped behind the waterstop, and aging of the flexible layer (rubber or geomembrane) under the exposed condition, especially under the sun.

The methods mentioned above can also be applied in crack repairing and the crimps in the middle of the rigid plate can be canceled.

Caulked waterstop: Caulked waterstop is a simple and economical waterstop-repairing method that is achieved by sawing the joints along the leaking part and then filling the cut with an elastic waterproofing material. If the monoliths are chamfered, the 'V' formed by the chamfers can be used for the caulking if the concrete is in good shape. The saw cut should be wide and deep enough to fill the elastic material and penetrate any unsound, cracked, or deteriorated materials.

Since the repair materials are essentially on the surface of the structure, they can easily be handled and removed if the repair is unsuccessful. This technique is also economical as the materials are easy to install if dewatering of the structure is not necessary.

However, this technique is limited if large movements between monoliths are expected, which can stress the joint filling material beyond its elastic limit. Another disadvantage is that, in some circumstances of joint closure, the caulk has a tendency to extrude from the joint if the joint is not closed with a surface membrane. In many of these cases, a surface plate or a surface membrane is applied in conjunction with the chalked joint to resist extrusion of the joint-filled material as the monolith expands.

Drilled holes filled with elastic material: This approach consists of drilling a large diameter hole from the top of the structure, along the vertical joint between monoliths, and filling the hole with an elastic material to create a seal against water penetration. The hole is typically drilled by a 'down-the-hole' hammer or core drill.

This procedure is typically used under conditions in which the site to install the remedial waterstop is not accessible from the face of the structure. The grout filler will form a continuous elastic bulb within the drilled hole and press tightly to the downstream side of the hole under water pressure. However, if the filler material, which is often in liquid form (such as cement, acrylamide, and hydrophilic polyurethane gel), travels out from the drilled hole during placement and into the joint before it sets, the repair may fail. Another disadvantage of this method is that, with the liquid grouting system, large movements of the joint may cause the material to break.

In some cases, continuous tube-type, flexible liners are inserted into the drill hole to contain the filler material. Liner materials include reinforced plastic fire hose, rubber, elastomer-coated fabric, neoprene, and felt tubes. A variety of materials have been used as fillers inside the liners, including water, bentonite slurry, and various formulations of chemical grout.

Materials experts are working on the invention of new types of joints and the improvement of the properties of rubber or PVC waterstops. The requirements of future joints are as follows:

- Simple construction
- Resistance to harsh weather conditions and other deteriorating agents
- Elastically deformable together with the structure
- The ability to sustain high hydrostatic pressures
- Easy *in situ* repairing or replacement if the joints are damaged.

6.14.5.5.3 Crack repairing

Crack repairing can only be done when the reason for the crack formation is known and the crack stops developing. If the reason is overloading, the extra load must be removed before crack repairing. If it is because of uneven settlement, the repair can only be undertaken when the settlement is stable.

Concrete cracking is a complex interaction of a variety of seemingly unrelated factors. According to the working mechanism, cracks may be divided into two branches: loading cracks and deformation cracks. In the case of loading cracks, the exertion of internal forces, the development of cracks, and even ultimate failure generally happen within a short period. Loading cracks are often very serious and immediate action needs to be taken to repair them. Deformation cracks are caused by concrete deformation due to foundation settlement and friction, thermal effects, temperature changes, drying shrinkage, autogenous shrinkage, and so on. For this type of crack, the onset and enlargement of cracks needs an internal force accumulation and transmission procedure. According to statistics, deformation cracks (or cracks that are mainly caused by concrete deformation) account for 80% of the cracks in engineering. Loading cracks (or cracks that are mainly induced by overloading) account for 20% of the cracks in engineering practice.

Cracks in concrete are divided into two types according to severity: active and dormant cracks. Dormant cracks are often hairlike and irregular, and may be caused by weathering or poor construction. Minor surface cracking does not affect the structural integrity and performance of the concrete structure. However, moisture infiltration through cracks may accelerate concrete deterioration and hence reduce the life of the structure and increase maintenance costs. Dormant cracks may be induced by thermal effects, autogenous shrinkage, or drying shrinkage. They usually require observation and limited corrective action to prevent moisture infiltration. Active cracks are more serious than dormant cracks and indicate severe problems. They are generally identified by long, single or multiple diagonal cracks with accompanying displacements and misalignment. Those cracks are generally caused by structural loads. Active cracks may be induced by structural overloading, foundation settlement, inherent design flaws, or other deleterious conditions. They require careful monitoring and possible corrective action; otherwise they may propagate to ultimate failure. Active cracks can also be temporary or continuous, or become dormant.

General crack-repairing methods include grouts, epoxy, and coatings. Surface or narrow cracks are generally not structural and therefore not dangerous. They can be patched with 'neat cement' mortar (a Portland cement and water mixture) or filled with nonshrinkage plastic grouts by injection (grouting). The nonshrinkage grouts usually contain silica fume or other stable aggregates. If the cracking is not harmful, it can only be brushed with a thin layer of epoxy. If severe cracking has occurred and extends through a structural member and shows signs of movement, extensive repair is required. In this condition, insertion of dowels and/or grouting may be required. For thin and deep cracks that extend through a structural member, grouting is often very difficult. These cracks can be cut into a 2–5 cm wide slot, then the slot is filled with nonshrinkage materials (generally cement or cement–epoxy mixture). For vertical and overhead conditions, epoxy adhesives or forming may be required for proper installation.

The application of flexible surface coatings is an effective corrosion control measure for the ingress of chlorides, sulfates, carbon dioxide, oxygen, and moisture. However, coatings should be applied before structural deterioration occurs, and not afterward, to be effective.

A common crack-repairing material is nonshrinkage cement, or a small amount of fine sand with neat cement. Recently, some new types of chemical materials such as epoxy or plastic have been applied. Epoxy can be applied or added to the cement to repair the cracks. The epoxy liquid or the cement–epoxy mixture has the advantage of strong cohesion, less shrinkage, fast solidification, and less aging. It is widely applied for the repairing of concrete cracks now.

6.14.6 Case Histories for Durable Concrete

This section includes case histories on the durability problems associated with concrete and looks at ways to make cost-effective and durable hydropower structures. The case histories are divided into two categories: lessons and experience.

6.14.6.1 Shi Lianghe Key Project in China

Shi Lianghe key project lies in Lianyungang city, Jiangsu province, China. The reservoir storage is 0.531 m^3 and the reservoir surface area is 85 km^2. This project was built in 1958, but because it was poorly constructed, it had to be rebuilt after only 40 years. The main problems experienced were concrete peeling off, carbonation, and concrete cracking.

Concrete carbonation and losing: **Figure 10** is the pier of Shi Lianghe key project. From the photograph, it is possible to see that a huge block of concrete was peeled off from the pier. Lianyungang city is situated in east China. In this area, winter time lasts for about 4–5 months, with the temperatures around 0 °C. Because of frequent freezing and thawing cycles and poor quality construction, the piers were greatly damaged.

Concrete cracking: **Figure 11** is the weir surface of Shi Lianghe key project. This photograph shows a transverse crack through the weir body. The foundation of the sluice is rock. When the ambient conditions changed, the weir contracted or expanded. For this reason, and the poor quality construction, a transverse crack came up.

6.14.6.2 Fengman Hydropower Structure in China

Fengman complex (**Figure 12**) lies on Songhua river, Jilin province, north China. It includes a hydropower plant and a gravity dam (with the height 90.5 m). The winter temperature in this area is around −30 °C. This project was built in 1937. The rock cracks in the dam foundations were not given any treatment during construction and the quality of the concrete was very bad. As a result, the seepage in the corridor is very serious and the upstream face has deteriorated severely due to water freezing. In 1986, during the spillway spillage period, huge blocks of the spillway surface were flushed away and rehabilitation had to be implemented.

6.14.6.3 Haikou Key Project in China

Haikou key project is the last control project on the Huai watercourse, China. Construction began in 2004 and finished in 2006.

In April 2006, just 1 month after its completion, part of the wave preventing board on top of the retaining wall was peeled off. The abutment connected to the retaining wall was also slightly damaged. The reason is that although there were joints between the retaining walls, the wave preventing board, which was on top of the retaining wall, was connected together. When the temperature rose, the concrete started to expand. Since there was no joint in the wave-preventing wall, cracks emerged in the concrete. The expansion of concrete was so severe, that even waterstop between the retaining walls was squeezed out (see **Figure 13**).

To relieve the expansion of concrete, permanent cross-sectional joints have been set on the wave-preventing board.

Figure 10 Pier of Shi Lianghe sluice.

Figure 11 Weir surface of Shi Lianghe sluice.

6.14.6.4 Butgenbach Dam in Belgium

Butgenbach dam lies in Belgium at the border near Germany. It was built in 1929–32. The project consists of a dam, a bottom outlet, a hydropower plant (one unit), a two-bay, ogee-weir-type sluice spillway, and a siphon spillway. Butgenbach dam is a

Figure 12 Fengman hydropower structure.

Figure 13 Squeezing out of waterstop due to concrete expansion.

multiple arch dam. The reservoir surface area is 125 ha with the dam height 23 m. The main function of this dam is hydropower, water supply, and flood control.

After 70 years' operation, the strength of the concrete in the Butgenbach dam decreased tremendously due to AAR, and the surface concrete also became very brittle due to concrete carbonation.

To prevent AAR, a layer of waterproof membrane was installed on the upstream face of the dam to prevent seepage water from entering. Between the membrane and the dam surface, there are gutters to guide the seepage (there may still be some seepage water even if the membrane is waterproof) downstream. **Figure 14** shows the rehabilitation of this dam.

Since the concrete carbonation is not very severe, only a layer of epoxy is brushed on the surface of concrete.

6.14.6.5 Baoying Key Project in China

Baoying key project is the starting point of a national project in China – the transfer of water from the south to the north. The size of the base of this project is 33.4 × 24 × 1.2 m. The foundation is sandy clay. Because this is an important project and high durability is required, some measures were taken during construction:

- Stone filling inside mass concrete – to decrease the hydration heat caused by cement hydration, blocks of stones were filled in the center of the concrete base and side wall (see **Figure 15**). It was required that concrete with a minimum thickness of 0.4 m be kept on the surface of the concrete to satisfy the strength requirements. This method, which has been applied in many projects that have mass concrete, proved to be successful.
- Adding fibers – fibers were applied in this project to enhance the tensile strength of concrete and hence decrease the risks of cracking (see **Figure 16**). Since research on the long-time performance of fiber reinforcement is at an early stage, the fibers in this project were added very cautiously and were only small in number.

Construction of this project began in 2003 and finished in 2005. No cracks have been detected so far.

Figure 14 Butgenbach multiple arch dam, Belgium.

Figure 15 Stone filling inside mass concrete.

Figure 16 Fiber adding in concrete.

At present, the principles for producing concrete and the laws of concrete behavior are well established through long experience and extensive research. This makes it possible for structural design to meet the recognized requirements of engineering practice and safety. However, with new requirements and challenges arising, there is a great need for continued research into new methods, materials and machines for construction, new monitoring instruments, and so on.

6.14.7 Conclusion

The most cost-effective way to maximize the life of a hydraulic structure is to produce concrete that is fit for purpose during the construction stage. This, and careful consideration of the selection of materials and the design, construction, servicing, and repair of concrete hydropower structures, should ensure that the life of the structures is 80–100 years. In the future, it is possible that the life of a hydraulic structure could be 150 years or even longer.

Looking ahead, it is predicted that expert systems with a life of 150 years or longer will be able to be designed to meet the requirements of specific construction projects. When this happens, it will be possible to estimate the anticipated life of a structure and thus aid in decision-making about the value of repairing a structure.

Setting up an expert system involves information collection, analysis, and sorting. Information collection is the most critical point that decides the accuracy of the system. The system should have the capability of solving problems in the event of uncertain or incomplete data. More importantly, the system is needed to limit the scope between rehabilitation and reconstruction, and hence come up with a standard. For this standard, different weight and consideration (concerning rehabilitation and reconstruction) will be given to the cost, the future service life, resource availability, social favorability, and environmental compatibility.

To prolong the life of hydraulic structures, a standard is needed for HPC. This standard could be expressed in terms of performance characteristics including strength, elasticity, permeability, shrinkage, creep, freezing–thawing durability, abrasion resistance, AAR, water head, and so on. Depending on a specific application, a given HPC may require a different grade of performance for each performance characteristic.

References

[1] http://www.dl.ket.org/humanities/arch/pantheon.fwx.
[2] Yong Y (2003) *Controlling of Early Cracking on Concrete Structures*. Beijing: Science Press.
[3] Tazawa E and Miyazawa S (1993) Autogenous shrinkage of concrete and its importance in concrete technology. In: Bazant ZP and Carol I (eds.) *Creep and Shrinkage of Concrete*, pp. 159–168. London: E & FN Spon.
[4] Le Roy R, De Larrard F, and Pons G (1996) *The AFREM Code Type Model for Creep and Shrinkage of High-Performance Concrete, 4th Internal Symposium on Utilization of High-Strength/High-Performance Concrete*, pp. 387–396. Paris, France, 29–31 May.
[5] Tiemeng W (2000) *Cracking Controlling of Engineering Works*, p. 6. China: China Construction Industry Press.
[6] Mehta PK and Burrows RW (2001) Building durable structures in the 21st century. *Concrete International* 23(3): 57–63.
[7] Tikalsky PJ, Mather B, and Olek J (2006) A2E01. Committee on durability of concrete. http://onlinepubs.trb.org/onlinepubs/millennium(accessed July 2006).
[8] Jianquan Z, Huiqiang L, and Xiaogen S (2005) Study of reinforced concrete structures' durability based on holistic view-point. *Journal of Central China Technology University (Urban Science Edition)* 22: 35–37.
[9] Weiliang J and Yuxi Z (2002) *Durability of Concrete Structures*. Beijing: Science Press.
[10] Kay T (1992) *Assessment and Renovation of Concrete Structures*, p. 1. New York, NY: John Wiley & Sons, Inc.
[11] Wood JGM and Johnson RA (1993) The appraisal and maintenance of structures with alkali–silicon reaction. *The Structural Engineer* 71(2): 19–23.
[12] Jinyu L and Jianguo C (2004) *Research and Application of Durability in Hydraulic Engineering Concrete*. Beijing: Hydropower Press.
[13] Wood IR (1985) Air water flows. *Proceedings of 21st IAHR Congress*. Melbourne, Australia, 19–23 August.
[14] Taerwe L and Matthys S (1999) FRP for concrete construction: Activities in Europe. *Concrete International* 21: 33–36.
[15] Collepardi M (1975) Rheoplastic concrete. *Il Cemento* 1975: 195–204.
[16] Collepardi M (1976) Assessment of the "rheoplasticity" of concretes. *Cement and Concrete Research* 6(3): 401–408.
[17] Collepardi M (1982) The influence of admixtures on concrete rheological properties. *Il Cemento* 289–316.
[18] Zhong-wei W (2000) High performance concrete – Green concrete. *Concrete and Cement Products* 2: 1.

6.15 Pumped Storage Hydropower Developments

T Hino, CTI Engineering International Co., Ltd., Chu-o-Ku, Japan
A Lejeune, University of Liège, Liège, Belgium

© 2012 Elsevier Ltd. All rights reserved.

6.15.1	Inroduction	406
6.15.2	Electrical Energy Storage	406
6.15.2.1	General Issues	406
6.15.2.1.1	Benefits of storage	406
6.15.2.1.2	Barriers to the deployment of electrical energy storage	406
6.15.2.1.3	Location of storage systems	407
6.15.2.2	Applications	407
6.15.2.2.1	Load management	408
6.15.2.2.2	Spinning reserve	408
6.15.2.2.3	Transmission and distribution stabilization and voltage regulation	408
6.15.2.2.4	Transmission upgrades deferral	408
6.15.2.2.5	Distributed generation	408
6.15.2.2.6	Renewable energy applications	409
6.15.2.2.7	End use applications	410
6.15.2.2.8	Miscellany	410
6.15.2.3	Storage Technologies	410
6.15.3	Pumped Storage Hydropower Plant	412
6.15.3.1	Characteristics	412
6.15.3.2	History	412
6.15.3.3	Characteristics of Pump–Turbines	413
6.15.3.3.1	Elements of pump–turbine hydraulic design	413
6.15.3.3.2	Elements of pump–turbine mechanical design	416
6.15.4	Examples of Remarkable Pumped Storage Power Plants	418
6.15.4.1	Okinawa Seawater Pumped Storage Power Plant	418
6.15.4.1.1	Outline of the project	419
6.15.4.1.2	Features of the project area	420
6.15.4.1.3	Major impacts	421
6.15.4.1.4	Mitigation measures	422
6.15.4.1.5	Measures during construction	422
6.15.4.1.6	Permanent measures	423
6.15.4.1.7	Results of the mitigation measures	424
6.15.4.1.8	Reasons for success	426
6.15.4.2	Goldisthal Pumped Storage Power Plant	426
6.15.4.2.1	Introduction	426
6.15.4.2.2	Developing the Goldisthal project	426
6.15.4.2.3	Main features of the project	427
6.15.4.2.4	Choosing variable-speed machines	427
6.15.4.2.5	Operation to date	428
6.15.4.3	Tianhuangping Pumped Storage Power Plant	429
6.15.4.3.1	Introduction	429
6.15.4.3.2	Two pump storage reservoirs	429
6.15.4.3.3	Benefits of the project	430
6.15.4.4	Coo-Trois Ponts Pumped Storage Power Plant	431
6.15.4.4.1	Introduction	431
6.15.4.4.2	Main features of the project	431
6.15.4.4.3	Generating equipment	432
6.15.4.4.4	Special features of interest	433
References		434

6.15.1 Inrodution

Pumped storage hydroelectric projects differ from conventional hydroelectric projects. They store electrical energy normally by pumping water from a lower reservoir to an upper reservoir when demand for electricity is low. Water is stored in the upper reservoir for release to generate power during periods of peak demand. These projects are uniquely suited for generating power when demand for electricity is high and for supplying reserve capacity to complement the output of large fossil-fueled and nuclear steam electric plants. Start-up of this type of plant is almost immediate, thus serving peak demand for power better than fossil-fueled plants do, which require significantly more start-up time. Like conventional projects, they use falling water to generate power, but they use reversible turbines to pump the water back to the upper reservoir. This type of project is particularly effective at sites having high heads (large difference in elevation between the upper and the lower reservoir) [1].

6.15.2 Electrical Energy Storage

6.15.2.1 General Issues

The concept of electrical energy storage has become a controversial issue in recent years. Many questions are raised in the electricity sector: Why is energy storage needed? What are the alternatives? How much do storage systems cost, and how much added value does a storage system provide? Will storage contribute to the increased utilization of renewable sources?

The storage issue must be viewed in the frame of a changing electricity sector characterized by

- restructuring of the electricity market;
- growth in new/renewable energy sources;
- increasing reliance on electricity and demand for higher-quality power;
- move toward distributed generation; and
- more stringent environmental requirements.

As part of these changes, there are growing pressures to operate the electrical network more efficiently while still maintaining high standards of reliability and power quality. Accommodation of renewable generation and ever more stringent environmental requirements are combining strongly to further influence electricity companies' decisions on how they should be developing their future network designs. With these driving forces as a backdrop, the rapidly accelerating rate of technological development in many of the emerging electrical energy storage technologies, with anticipated system cost reductions, now makes their practical application look attractive.

Energy storage is not a new concept in the electricity sector. Utilities across the world have built a number of pumped hydro facilities in the last few decades, resulting in a storage component of roughly 5% of the power generation capacity of all the European countries, which is 3% in the United States, and 10% in Japan. These pumped hydro plants, and to a lesser extent compressed air storage systems, have been used for load leveling, frequency response, and voltage/reactive control. Likewise, storage facilities based on other technologies such as lead–acid batteries have been installed by a number of utilities to fulfill a variety of functions. At a different scale, energy storage is also commonly used at the user level to ensure reliability and power quality for customers with sensitive equipment. Another traditional application is the electrification of off-grid networks and remote telecommunications stations, mostly in connection with renewable sources. The market penetration achieved by electrical energy storage to date has been heavily constrained by its cost and the limited operational experience, resulting in high technical and commercial risk. However, the presence of storage systems is growing fast owing to the circumstances mentioned above.

6.15.2.1.1 Benefits of storage

Storage contributes to optimizing the use of existing generation and transmission infrastructure, reducing or deferring capital investment costs. It contributes to integrating renewable sources (and in general distributed sources) into the system, enhancing their availability and market value. The environmental benefits must be highlighted, in terms of both reduction of emissions from conventional power plants and increase of renewable sources' penetration. Energy storage facilities can also help maintain transmission grid stability by providing ancillary services, including black start capability, spinning reserve, and reactive power.

At the consumer level, storage improves power quality and reliability, and can provide capability to control or reduce costs. Energy storage is of growing importance as it enables the smoothing of transient and/or intermittent loads, and downsizing of base load capacity with consequent substantial potential for energy and cost savings.

However, it is acknowledged that energy storage systems will have to compete within the context of present overcapacity of power stations and power generators with short start-up times, such as open-cycle gas turbines and gas or diesel motors with appropriate emission controls.

6.15.2.1.2 Barriers to the deployment of electrical energy storage

Electrical energy storage involves significant investment and energy losses, which must be weighed against the benefits and compared to other nonstorage solutions. There are a number of key barriers to a more widespread use of storage systems:

- *Immaturity of some technologies and lack of operating experience.* More demonstration projects are needed to gain customers' confidence. Further research and development is necessary in some aspects such as the implementation of power conditioning and control process for a multi-application energy storage system.
- *High initial capital costs.* Technological advances and large manufacturing volumes will bring these costs down.
- *Uncertainty over the quantified benefits.* This is true especially when, as usually happens, there are multiple different benefits associated with a storage system.
- *Uncertainty over the regulatory environment.* The future shape of the electricity market, in relation to not only energy trading but also ancillary services trading, will affect decisively the viability of electrical energy storage. The use of storage systems for the provision of ancillary services currently provided by the system operator will depend on the deregulatory process.

6.15.2.1.3 Location of storage systems

Utility-scale energy storage systems are envisaged as forming an integral part of the future energy system. Depending on the application, they can be implemented in any of the different segments of the electricity supply system (**Figure 1**). In a liberalized market, the different segments of the electricity sector are increasingly being separated. Each segment offers different potential opportunities to energy storage applications. Correct location of the storage systems is important to maximizing the benefits. Large-scale, that is, multimegawatt, centralized storage could improve generation and transmission load factors and system stability. Smaller-scale, localized, or distributed storage could deliver energy management and peak-shaving services, as well as improving power quality and reliability. Distributed storage would be an ideal complement to distributed generation, especially on account of the increasing levels of renewable energy generation.

One of the axioms of energy storage is that storage units should be located as close as possible to the end consumer of electricity. This is because the storage device can improve the utilization of all components in the network. In order to place a storage device close to the end consumer, the device would need to be matched for both power and energy storage capacity to the requirements of the consumer. Since the specific capital cost increases as the system becomes smaller, the optimum position for a storage device in the network tends to move closer to the generation source. For this reason, Price [2] maintains that many storage systems can, and should be, located near to substations or grid distribution points. When storage systems are utilized to facilitate renewable energy source integration, the picture changes, however, since the fluctuations in the generated power are usually greater than those in the load. As a result, the optimum location is likely to be close to the generation points, thus maximizing the capacity of the transmission and distribution lines.

6.15.2.2 Applications

Applications of electrical energy storage are numerous and varied, covering a wide spectrum, from larger-scale generation- and transmission-related systems to smaller-scale applications at the distribution network and the customer/end use site. This chapter deals specifically with the application of storage for renewable energy source integration; however, this is closely connected to other applications. Storage systems usually provide multiple benefits, and thus it is necessary to review all their possible functions. Interesting reviews of the applications can be found in Schoenung [3], Herr [4], and Butler [5]. Ultimately, the purposes of all these applications come down to

- improved load management
- provision of spinning reserve
- transmission and distribution stabilization and voltage regulation
- transmissions system upgrade deferral
- facilitating distributed generation
- facilitating renewable energy deployment
- end use applications
- miscellaneous (including ancillary services)

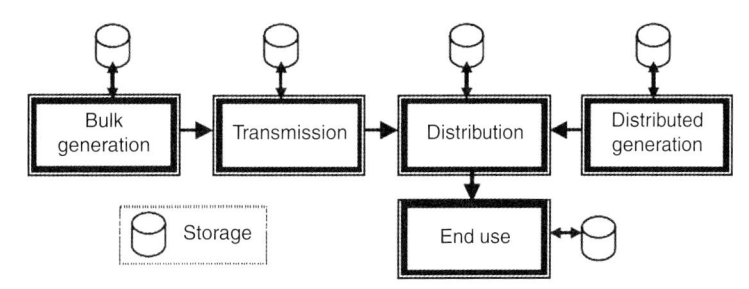

Figure 1 Storage locations in the electricity supply system.

6.15.2.2.1 Load management

Load management includes the traditional 'load leveling', a widespread application for large energy storages, in which cheap electricity is used during off-peak hours for charging, while discharging takes place during peak hours, providing cost savings to the operator. In addition, load leveling can lead to more uniform load factors for the generation, transmission, and distribution systems. Although load leveling was the first application of energy storage that utilities recognized, the difference in the marginal cost of generation during peak and off-peak periods for many utilities is moderate. Therefore, Butler [5] concludes that load leveling is likely to be a secondary benefit derived from an energy storage system installed for other applications that offer greater economic benefits. Load leveling requires energy storage systems on the order of at least 1 MW and up to hundreds of megawatts, and with several hours of storage capacity (2–8 h). For utilities without a strong seasonal demand variation, a system used for load leveling would operate on weekdays (250 days per year). Other types of load management are 'ramping and load following', in which energy storage is used to assist generation to follow the load changes. Instantaneous match between generation and load is necessary to maintain the generators' rotating speed and in turn the frequency of the system. Storage systems serving this application should be able to deliver on the order of 10–100 MW to absorb and deliver power as demand fluctuates. The system would have to be able to dispatch energy continuously, especially during peak-load times, in frequent, shallow charge–discharge cycles that would occur. This service is usually provided by conventional generation.

6.15.2.2.2 Spinning reserve

The category 'fast-response spinning reserve' corresponds to the fast-responding generation capacity that is in a state of 'hot standby'. Utilities hold it back to be put in use in case of a failure of generation units. Thus, the required power output for this application is typically determined by the power output of the largest unit operating on-grid. The conventional spinning reserve requires a less quick response. Storage systems can provide this application in competition with standard generation facilities. Since the power plants that they would temporarily replace may have power ratings on the order of 10–400 MW, storage systems for reserve must be in this same range. Generation outages requiring rapid reserve typically may occur about 20–50 times a year. Therefore, storage facilities for rapid reserve must be able to address up to 50 significant discharges that occur randomly through the year.

6.15.2.2.3 Transmission and distribution stabilization and voltage regulation

Transmission and distribution stabilization are applications that require very high power ratings for short durations in order to keep all components of a transmission or distribution line in synchronous operation. This includes phase angle control, and voltage and frequency regulation. In the event of a fault, generators may lose synchronism (due to difference in phase angle) if the system is not stabilized, making the systems collapse. Energy storage devices can stabilize the system after a fault by absorbing/delivering power from/to the generators as needed to keep them turning at the same speed. Fast action is essential for quick stabilization. Response time limitations demand an appropriate power-conditioning interface designed to ensure a reliable mitigation of short-duration electrical disturbances, which can range from a couple of cycles to 2 min. The portability of the storage systems might be an important factor in many cases. Some applications are temporary in nature, and Boyce [6] points out that for a storage system the ability to be transferred from site to site can significantly increase its overall value. With the liberalization of the electricity market there will be an increasing need to maintain and to improve the stability of the electrical grid. The risk of voltage instability, being the source of failures in automatic production centers and the base of cascading outages, will become more and more serious. Many of the utility grids cannot properly react to transient events with the limited transmission capacity they have. In cases of fast-changing load flow patterns or changes in the distribution of the loads or power plants on the grid, the risk of voltage instability increases. To offset the effect of impedance in transmission lines, utilities inject reactive power and maintain the same voltage at all locations on the line. Traditionally, fixed and switched capacitors have provided the reactive power necessary for 'voltage regulation'. Storage systems deployed by transmission or distribution network operators for any other primary application can provide reactive power to the system to augment the existing capacitors and replace capacitors planned for future installation. An energy storage system for voltage regulation should provide reactive power on the order of 1–10 MVAR for several minutes, mainly during daily load peaks.

6.15.2.2.4 Transmission upgrades deferral

When growing demand for electricity approaches the capacity of the transmission system, utilities add new lines and transformers. Because load grows gradually, new facilities are designed to be larger than necessary at the time of their installation, and utilities under-use them during their first several years of operation. To defer a new line installation or transformer purchase, a utility can employ an energy storage system until the load demand will make better use of a new line or transformer. The power requirement for this application would be on the order of hundreds of kilowatts to several hundred megawatts. Butler [5] states that the energy storage system should allow for 1–3 h of storage to provide support to the constrained transmission facility.

6.15.2.2.5 Distributed generation

The growing presence of distributed sources opens a new market for storage systems, which can assist during transient conditions of generation units such as microturbines and diesel engines, with a slower dynamic response and thus limited capability to adjust to load changes. In this way, storage can increase the distributed generation capacity that can be embedded on a distribution network and avoid cost-intensive reinforcements. A less-demanding application of storage technologies in distributed generation is 'peaking generation', which can also avoid reinforcement of distribution lines. Areas with temporary high demands, for example, at daytime, could be equipped with storage that supply power at peak times and are recharged through off-peak hours. These applications are often referred to as 'distribution capacity deferral'. An energy storage system deployed to defer installation of new distribution capacity requires power on the order of tens of kilowatts to a few megawatts, and must provide 1–3 h of storage.

6.15.2.2.6 Renewable energy applications

Electrical energy storage is very promising as a means of tackling the problems associated with the intermittency of renewable sources such as wind and solar energy. This application will cover a wide range of power capacity and discharge duration. With the increasing market penetration of renewable sources, these applications are more and more likely to gather momentum within future energy systems, as conventional generation utilities' ability to even out the intermittent renewable energy production is limited. There is a variety of denominations in the technical literature for the use of storage in connection with renewable energy-related applications. Butler [5] states that some authors call it 'renewable integration' or 'renewable energy management'. Schoenung [3] identifies only one utility-scale application under the term 'renewable matching', referring to the use of storage to match renewable generation to any load profile, making it more reliable and predictable and hence more valuable. This does not seem to be applicable to the storage of renewable energy at off-peak times to be delivered at peak times. Herr [4], however, broadens the scope of renewable matching, by referring to applications making renewable electricity production more predictable throughout the day and bringing renewables closer to demand profiles, especially providing high power outputs at peak hours. Baxter and Makansi [7] identify four categories within renewable energy-related storage: distributed generation support, dispatchable wind, base load wind, and off-grid applications. Storage systems with a longer discharge duration can cover longer mismatches (up to several hours). In the longer term, a utility with a significant percentage of renewable power may require storage capacity of days to ride through periods with windless days. In Table 1, a number of short- and long-discharge renewable matching applications are included. Both will be referred to later as 'renewable integration'. Indeed, a broader scope can be given to renewable integration, including short-time applications that also contribute to tackling the problems associated with intermittent sources. The storage system required for either application would need to provide from 10 kW to 100 MW. According to Butler [5], the storage system would need response time on the order of fractions of a second if transient fluctuations are to be addressed. The cycling of the storage

Table 1 Applications of storage systems with different discharge times

	Application	Power rating	Discharge duration	Storage capacity	Response time	System location
Fast discharge	Transit and end use ride-through	<1 MW	Seconds	~2 kWh	<1/4 cycle	End use
	Transmission and distribution stabilization	up to 100 s MVA	Seconds	20–50 kV Ah	<1/4 cycle	Transmission and distribution
Short to long discharge	Voltage regulation	up to 10 MVAR	Minutes	250–2 500 kV Arh	<1/4 cycle	Transmission
	Fast response spinning reserve	10–100 MW	<30 m	5 000–500 000 kWh	<3 s	Generation
	Conventional spinning reserve	10–100 MW	<30 m	5 000–500 000 kWh	<10 min	Generation
	Uninterruptible power supply	<2 MW	~2 h	100–4 000 kWh	Seconds	End use
	End use and transmission peak shaving	<5 MW	1–3 h	1 000–150 000 kWh	Seconds	End use and distribution
	Transmission upgrade deferral	up to 100 s MW	1–3 h	1 000–500 000 kWh	Seconds	Transmission
	Renewable matching (short discharge)	<100 MW	Min–1 h	10–100 000 kWh	<1 cycle	Generation
Long discharge	Renewable matching (long discharge)	<100 MW	1–10 h	1 000–100 000 kWh	Seconds	Generation
	Load levelling	100 s MW	6–10 h	100–10 000 MWh	Minutes	Generation
	Load following	10–100 s MW	Several hours	10–1 000 MWh	< cycle	Generation and distribution
	Emergency back-up	<1 MW	24 h	24 MWh	Seconds	End use
	Renewables back-up	100 kW–1 MW	Days	20–200 MWh	Seconds–Minutes	Generation and end-use

systems coupled with wind energy will be rather unpredictable and could range from 100 to 1000 cycles per year or more. In remote locations not connected to the grid, it may be useful to include energy storage to minimize the generation capacity. This is especially attractive in renewable-based supplies. Renewable backup applications should be capable of substituting renewable production when this is not available for time lengths that could go up to a week. The power rating would depend on the corresponding power output of the renewable system.

6.15.2.2.7 End use applications

The primary end use application of energy storage is in maintaining/improving power quality. Outages and poor power quality phenomena are important concerns for many business sectors – a survey estimated losses between $119 billion and $189 billion in the US economy alone. Energy storage systems are being successfully deployed to provide reliable and high-quality power to sensitive loads. 'Transit and end use ride-through' are applications requiring very short durations combined with very quick response times. They cover electric transit systems with remarkable load fluctuations and customer power services like voltage stabilization and frequency regulation to prevent the events that can affect sensitive processing equipment and can cause data and production losses. The demand for quality power is growing within industry and is becoming a matter of concern also for electricity suppliers, who may also install systems at the distribution level to improve the power quality. Uninterruptible Power Supply (UPS) devices provide protection against electricity supply downtimes. Primarily they prevent production losses; however, if the serving systems have very short response times they can also be used for power quality assignments (protection against voltage sags and power surges, frequency regulation, etc.). UPS systems often consist of a storage device that usually acts for a short duration of time until a generation set takes over. Although the provision of UPS is usually taken up at the user level, generation facilities can also use storage systems to remove particularly the short-term fluctuations from their supply. The attractiveness of the investment will depend on any penalties imposed on generating units that fail to provide a quality supply. There are other customer uses such as 'end use peak shaving', which can avoid additional demand charges by reducing demand peaks. 'Emergency backup' at customer site requires power ratings of approximately 1 MW for durations up to 1 day. Presently, most of these applications are served by reciprocating engines.

6.15.2.2.8 Miscellany

The provision of ancillary services by storage systems can also include black start capability, which consists of the supply of electricity for the start-up of generators after a network failure. This is usually performed by relatively expensive diesel engines. Some storage options also need an auxiliary electricity supply, but several can start without an electricity source. Once again, this service can be provided as an addition to the other applications listed. A number of different lists and different descriptions of the storage applications can also be found in the literature. Some authors include deferment of new capital equipment as a separate category. This is in fact simply an aggregation of some of the applications already mentioned that can be performed by conventional equipment such as peaking plants, new lines, substations, and so on. Installation of storage systems on the transmission and distribution grid in order to expand the grid capacity, decouple generation and load, and thus reduce congestion has already been included as a separate application (transmission upgrade deferral). The use of storage systems to improve transmission stability also reduces or defers the need for transmission upgrades. Likewise, storage units installed for load leveling, spinning reserve, or peak shaving delay the need for new generation capacity. Other authors also refer to the improvement of power plant efficiency as a category, but this is rather a driving force for applications such as load management and spinning reserve. Environmental benefits are also sometimes quoted as an application, but it is rather a consequence derived from the application of storage systems as spinning reserve, peak shaving, and others that results in a cut in emissions that conventional technologies cause. Energy storage can enhance the environmental performance of a network in a number of ways:

- Conventional generating units used to provide spinning reserve and other ancillary services could be replaced by energy storage.
- Generators, which operate best at constant load, can be combined to provide ramping and peaking duties.
- Grid upgrades can be avoided.
- System control issues arising from intermittent RE sources can be mitigated, thus increasing the proportion of renewable energy generation that the system can absorb.

6.15.2.3 Storage Technologies

Storage systems generally comprise three key elements, namely, storage subsystems, power conversion systems (PCS), and balance of plant systems (BOP), as illustrated in **Figure 2**. Depending on the storage system, certain elements within the scheme may be unnecessary; for example, pumped hydro and compressed air energy storage do not need a rectifier and inverter, as pumps and compressors operate using alternating current (AC).

There is a wide range of energy storage technologies at utility scale that are at various stages of development. Each technology has different features that make it more or less desirable for the various applications. **Table 2** provides an overview of possible selection criteria. The relevance of the different features varies largely depending on the application to be served. Fundamental criteria for any technology will be power capacity (including the reactive power capacity for some applications), energy capacity/discharge time,

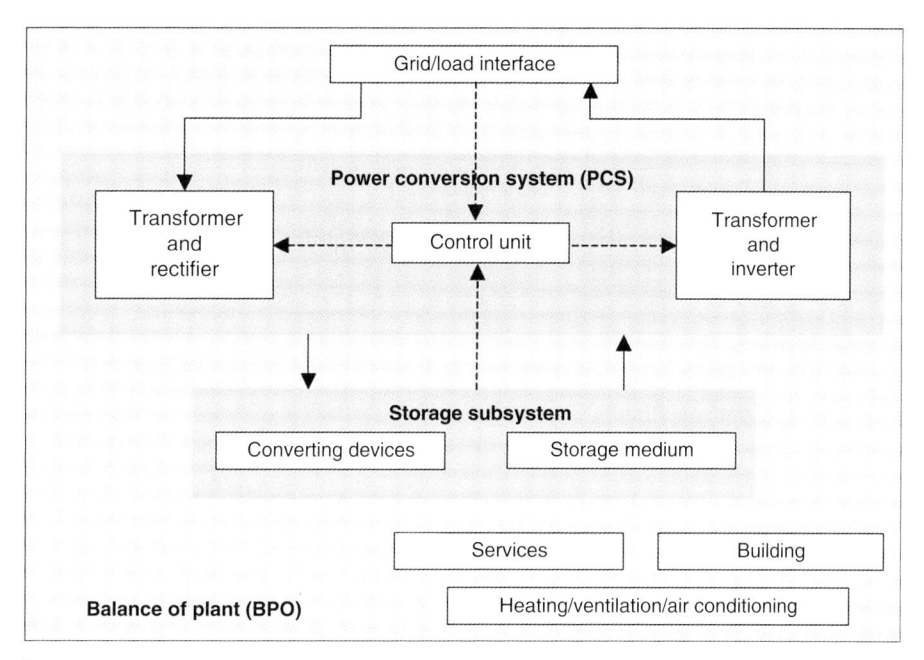

Figure 2 Scheme of a storage system.

Table 2 Criteria for the selection of a storage system

Design	Operating	Financial	Others
Power rating	Overall cycle efficiency	Capital cost per energy stored	Health and safety aspects
Storage capacity/discharge duration	Lifetime/maximum	Capital cost per power rating	Environmental impacts
Response time	number of	Fixed O&M cost	Synergies with other sectors
Energy density per unit area (foodprint)	charge–discharge cycles	Variable O&M cost	
Energy density per unit volume and weight	Parasitic losses	Replacement cost	
Maturity of technology		Disposal cost	
Reliability		Commercial risk	
Modularity			
Siting requirements			
Portability			
Synergies with other energy applications			

and reaction time. Some applications, like grid support, require discharge to commence less than a second after starting; others, like power sales, can be scheduled allowing for a reaction time of a few minutes.

Ideally, energy storage technologies should

- entail low capital, operating, and maintenance cost
- have a long lifetime
- be flexible in operation
- have high efficiency
- have a fast response
- be environmentally sustainable

There is a notable absence of detailed technical information on storage technologies. This is surprising given the growing need and opportunities for storage technologies. Somewhat superficial reviews can be found in Gandy [8], Schoenung [9], Swaminathan [10], Ter-Gazarian [11], and Herr [4]. Electricity storage systems can be technically categorized by their inherent physical principles into mechanical, electromagnetic, and electrochemical storage devices (**Figure 3**).

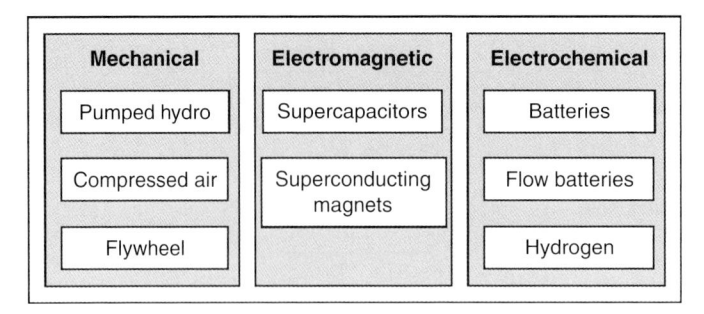

Figure 3 Storage technology categories.

6.15.3 Pumped Storage Hydropower Plant

6.15.3.1 Characteristics

Pumped storage hydroelectricity works on a very simple principle. Two reservoirs at different altitudes are required. When the water is released from the upper reservoir, energy is generated by the down flow, which is directed through high-pressure shafts, linked to turbines. In turn, the turbines power the generators to produce electricity. Water is pumped back to the upper reservoir by linking a pump shaft to the turbine shaft, using a motor to drive the pump.

This kind of plant generates energy for peak load, and at off-peak periods water is pumped back for future use. During off-peak periods, excess power available from some other plants in the system (often a run-of-river, thermal, or tidal plant) is used for pumping the water from the lower reservoir. A typical layout of a pumped storage plant is shown in **Figure 4**.

A pumped storage plant is an economical addition to a system, which increases the load factor of other systems and also provides additional capacity to meet the peak loads. These have been used widely in Europe, and about 50 plants are operational in the United States since the year 1990, reaching a total installed capacity of about 6700 MW.

6.15.3.2 History

Shoenung [12] acknowledges the role of pumped hydro energy storage (PHES) as the most widespread energy storage system currently in use on power networks, operating at power ratings up to 4000 MW and capacities up to 15 GWh. PHES uses the potential energy of water, swapped by pumps (charging mode) and turbines (discharge mode) between two reservoirs located at different altitudes. Currently the overall efficiency is in the 70–85% range, although variable-speed machines are now being used to improve this. Overall efficiency is limited by the efficiency of the deployed pumps and turbines (neglecting frictional losses in pipes and water losses due to evaporation). Plants are characterized by long construction times and high capital costs. One of the major problems related to building new plants is of an ecological/environmental nature. At least two water reservoirs are needed. Some high-dam hydro plants have a storage capability and can be operated as a pumped hydro. A relatively new concept of pumped hydro employs a lower reservoir buried deep in the ground. A good example of underground pumped storage is Dinorwig plant in the United Kingdom, commissioned in 1982, which includes Europe's largest man-made cavern under the hills of North Wales. Open sea can also be used as the lower reservoir – a seawater pumped hydro plant was first built in Japan in 1999. Pumped hydro facilities are available at almost any scale with discharge times ranging from several hours to a few days. PHES can be designed for fast loading and ramping, allowing frequent and rapid (< 15 s) changes between the pumping, generating, and standby spinning modes [13]. Dinorwig plant can go from 0 to 1890 MW (full capacity) in only 16 s. PHES is best suited to load leveling, storing energy

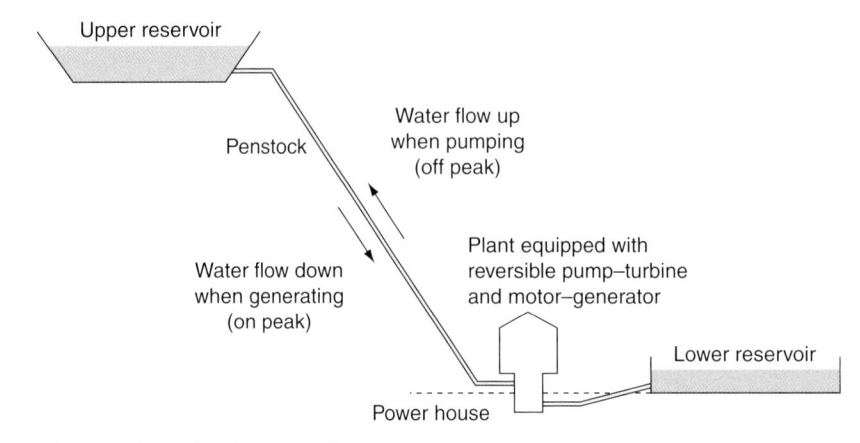

Figure 4 Typical layout of a storage hydroelectricity power plant.

during off-peak hours for use during peak hours, and spinning reserve. They can provide energy to meet peak demands, and, in the pumping mode, they serve as the source of load for base load during off-peak periods, helping to avoid cycling of the regular generating units, thus improving their operating efficiency. PHES systems can provide other benefits, including black start capability (they can begin generation without an external power source) and frequency regulation. There is over 90 GW of pumped storage in operation worldwide in nearly 300 plants, which is about 3% of global generation capacity. In 1998, 10% of Japan's total instantaneous energy requirement was met by pumped hydropower generation [14]. **Table 3** contains some of the most representative pumped hydro plants in the world larger than 1000 MW in current net capacity, which are currently operational or under construction. The 292 MW Turlough Hill Pumped Storage station (representing approx. 5% of installed capacity) is the only bulk electrical energy storage facility in Ireland. Its construction was completed in 1974 and involved the building of a huge cavern in the heart of the mountain, in which the generation plant and controls are housed. A pumped storage system allows for the use of excess electricity capacity during non-peak hours to pump water from the lower to the upper lake at Turlough Hill, and then the release of water in the reverse direction to create electricity in times of maximum demand. Unit prices for pump/turbines have leveled out as the technology has matured. Thus, costs are typically around $600 kW^{-1}. Reservoir costs can vary from almost nothing to more than $20 kWh^{-1}, according to Gordon [13]. Schoenung places that cost at $12 kWh^{-1}.

6.15.3.3 Characteristics of Pump–Turbines

Reversible pump–turbines or separate generating and pumping equipment may be installed in pumped storage plants. Some are equipped with a fixed or adjustable distributor. In the case of separate machines, a clutch operable at standstill, a starting turbine, or a synchronizing torque converter facilitating extremely short changeover times can be provided. Whether a reversible pump–turbine or a turbine and pump combination, these machines are extremely durable. **Figure 5** gives the application range of pump–turbines, and **Figure 6** shows for selected pump–turbines the operation range in the pump mode [15]

6.15.3.3.1 Elements of pump–turbine hydraulic design

Designing a reversible pump–turbine is still a complex and challenging task for the reasons and the characteristics previously stated. This is why the designing turns out to be an iterative process where all steps are interdependent [16]. For all components (spiral case, stay vanes, guide vanes, runner, and draft tube), geometry definition is the first step of the hydraulic design process.

Capabilities of the latest computer systems and computational fluid dynamic (CFD) codes allow the intricate hydraulic computations of all the parts of the pump–turbine with increasing accuracy. The objective is to check the behavior of flows in the entire operating range, in both pump and turbine modes. It also allows verification of whether the hydraulic criteria are satisfied or not. For each component of the machine, even if one of the hydraulic criteria is not satisfied, the geometry of the failing component is modified to attain the required criterion. When the design of the runner and other components is hydraulically validated by CFD, a mechanical check is performed, based on a structural analysis by means of a finite element model (FEM). The objective is to check the mechanical behavior of the components. It has been pointed out that the runner blade thickness, and the stay vanes and the guide vanes thickness have a strong influence on the static stress level.

These points are considered during the design development [17] through hydraulic and mechanical optimization.

Net heads of up to 817 m in the turbine mode and 832 m in the pump mode are considered as a limit for conventional single-stage pump–turbines.

Even though all hydraulic components are important and can affect the pump–turbine performance as a whole, the runner remains, of course, the key element to match the performances. This is the reason why special attention is paid to runner design.

6.15.3.3.1(i)   Runner hydraulic design

Runner design is optimized using Navier–Stokes computations to deliver the best characteristics in terms of efficiency level and cavitation behavior in both pump and turbine modes. The hydraulic criteria to be followed are mainly based on the energy results (head vs. discharge and head losses) as well as the flow field at the runner outlet and cavitation behavior with specific reference to its incidence at the blades' leading edge. For cavitation behavior, specifically in the pump mode, the pressure level at the blades' leading edge is calculated in order to guarantee a free cavitation pitting area over the operating range. A large net head variation can be added to a large frequency variation, thus increasing the operating range.

For various discharge values covering the whole operating range, calculation results can be compared with experimental observations on the model for which the critical cavitation limit is determined by decreasing the net positive suction head (NPSH) until efficiency drops (**Figure 7**).

6.15.3.3.1(ii)   Pressure fluctuation level

Among all the parameters to be mastered in the pump–turbine design, vibratory behavior has a particular importance. Such vibrations result from the mechanical response of the structure to hydraulic solicitations due, both to the operating conditions and to hydraulic interaction between the rotor and the stator.

Since 2000, and especially for the latest PSP designs, a direct application of this R&D program has been performed by decreasing the pressure fluctuation level especially between the runner and the guide vanes as well as in the casing. A comparison is made between the pressure fluctuation levels measured on the model (**Figure 8**) for two hydraulic designs (Nq 32). The first one was developed 6 years ago, whereas the second one has been developed recently.

Table 3 Some of the largest pumped hydro facilities in the world

Station	Country	Capacity (MW)
Bad Creek Hydroelectric Station	United States	1065
Bailianhe Hydroelectric Station	China	1224
Baoquan Pumped Hydroelectric Station	China	1200
Bath County Pumped Storage Station	United States	2772
Blenheim-Gilboa Hydroelectric Power Station	United States	1057
Castaic Dam	United States	1566
Chiotas Dam	Italy	1184
Coo Hydroelectric Power Station	Belgium	1164
Dinorwig Power Station	United Kingdom	1728
Dniester Hydroelectric Power Station	Ukraine	2268
Drakensberg Pumped Storage Scheme	South Africa	1000
Goldisthal Hydroelectric Power Station	Germany	1060
Grand Maison Dam	France	1070
Grande Dixence Dam	Switzerland	2069
Guangzhou Pumped Storage Power Station	China	2400
Heimifeng Pumped Storage Power Station	China	1200
Helms Dam	United States	1200
Huhhot Dam	China	1200
Huizhou Hydroelectric Power Station	China	2400
Ingula Pumped Storage Scheme	South Africa	1332
Imaichi Dam	Japan	1050
Kannagawa Pumped Storage Power Station	Japan	2700
Kazunogawa Dam	Japan	1600
Kruonis Pumped Storage Plant	Lithuania	1600
Lago Delio Hydroelectric Station	Italy	1040
Liyang Hydroelectric Power Station	China	1000
Ludington Pumped Storage Power Plant	United States	1872
Markersbach Dam	Germany	1050
Matanoagawa Pumped Storage Station	Japan	1200
Minghu Dam	Taiwan	1000
Mingtan Dam	Taiwan	1602
Mount Elbert	United States	1412
Mt. Hope Dam	United States	2000
Muddy Run Pumped Storage Facility	United States	1071
Northfield Mountain	United States	1080
Okutataragi Pumped Storage Power Station	Japan	1932
Okuyoshino Pumped Storage Power Station	Japan	1206

(Continued)

Table 3 (Continued)

Station	Country	Capacity (MW)
Piastra Edolo Pumped Storage Station	Italy	1020
Presenzano Pumped Storage Power Station	Italy	1000
Pushihe Pumped Storage Power Station	China	1200
Pyramid Lake	United States	1495
Raccoon Mountain Pumped-Storage Plant	United States	1530
Robert Moses Niagara Power Plant	United States	2515
Shin Takasegawa Pumped Storage Station	Japan	1101
Shintoyone Dam	Japan	1125
Siah Bisheh Dam	Iran	1140
Sir Adam Beck Hydroelectric Power Stations	Canada	1600
Taian Pumped Storage Power Station	China	1000
Tamahara Pumped Storage Power Station	Japan	1200
Tashlyk Hydro-Accumulating Power Station	Ukraine	1494
Tehri Pumped Storage Power Station	India	1000
Tianhuangping Pumped Storage Power Station	China	1800
Tongbai Pumped Storage Station	China	1200
Tumut-3	Australia	1500
Vianden Pumped Storage Plant	Luxembourg	1100
Xiangshuijian Pumped Storage Station	China	1000
Xianyou Pumped-storage Power Station	China	1200
Xilongchi Pumped Storage Power Station	China	1200
Yangyang Pumped Storage Power Station	South Korea	1000
Yixing Pumped Storage Power Station	China	1000
Zagorsk Pumped Storage Station	Russia	1200/1320
Zhanghewan Pumped Storage Station	China	1000
Zhuhai Pumped Storage Station	China	1800

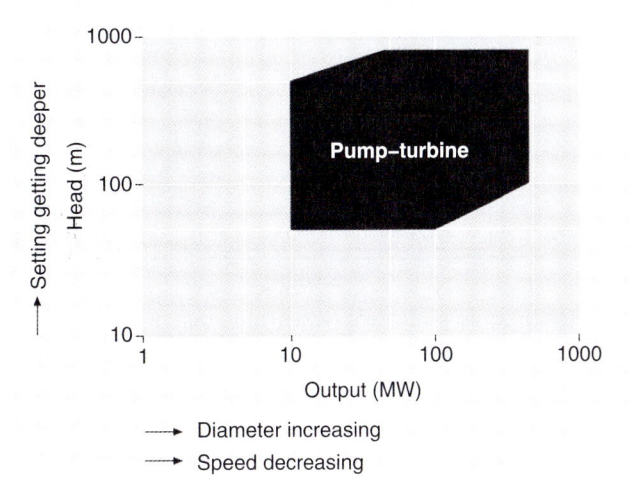

Figure 5 Application range of pump–turbines.

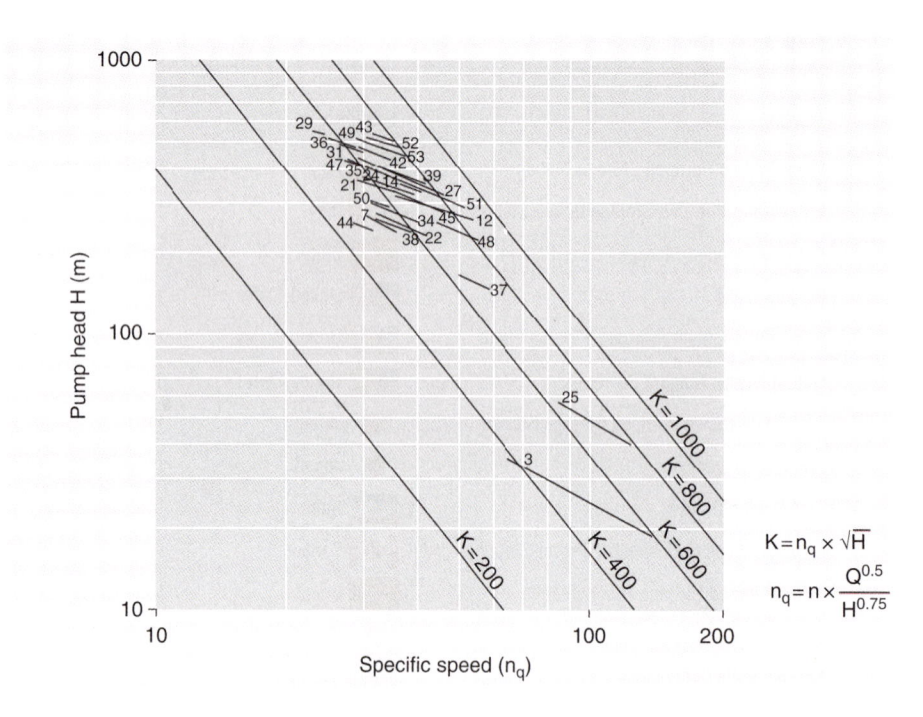

416 Design Concepts

Figure 6 Selected pump–turbines: Operation range in the pump mode.

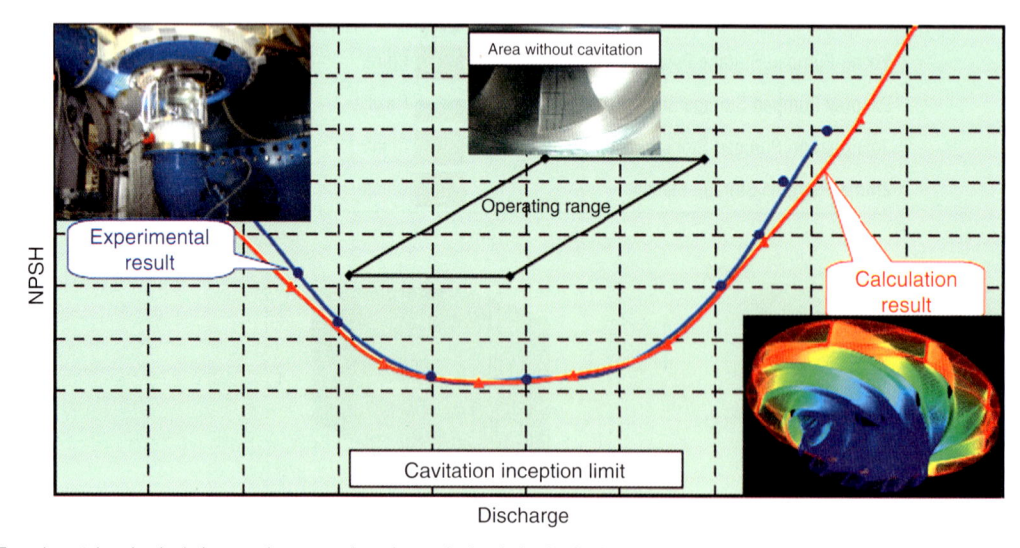

$$K = n_q \times \sqrt{H}$$

$$n_q = n \times \frac{Q^{0.5}}{H^{0.75}}$$

3 Lewiston	7 Coo I	112 Raccoon
14 Rodund II	21 Bath County	22 Coo II
24 Estangento	25 Gabriely Galan	27 Kühtai
29 Obrovac	31 Presenzano	34 Palmiet
35 Bad Creek	36 La Muela	37 Herdecke
38 Chiotas (4stage)	39 Mingtan	42 Shisanling
43 Guangzhou II	44 Edolo (5 stage)	45 Goldisthal (two variable speed)
47 Venda Nova	48 Tai An	49 Siah Bishe
50 Waldeck I	51 Limberg II	52 La Muela II
53 Ingula		

Figure 7 Experimental and calculation results comparison for cavitation behavior in the pump mode.

6.15.3.3.2 Elements of pump–turbine mechanical design

Mechanical design of these reversible pump–turbines is based on both experience and customer's specifications.

The prominent characteristics of main projects are as follows (**Figure 9**):

- Shaft line is resting on three bearings: two for the generator–motor and one for the pump–turbine.
- Thrust bearing is supported by the lower bracket. The inlet valve is of the spherical type.

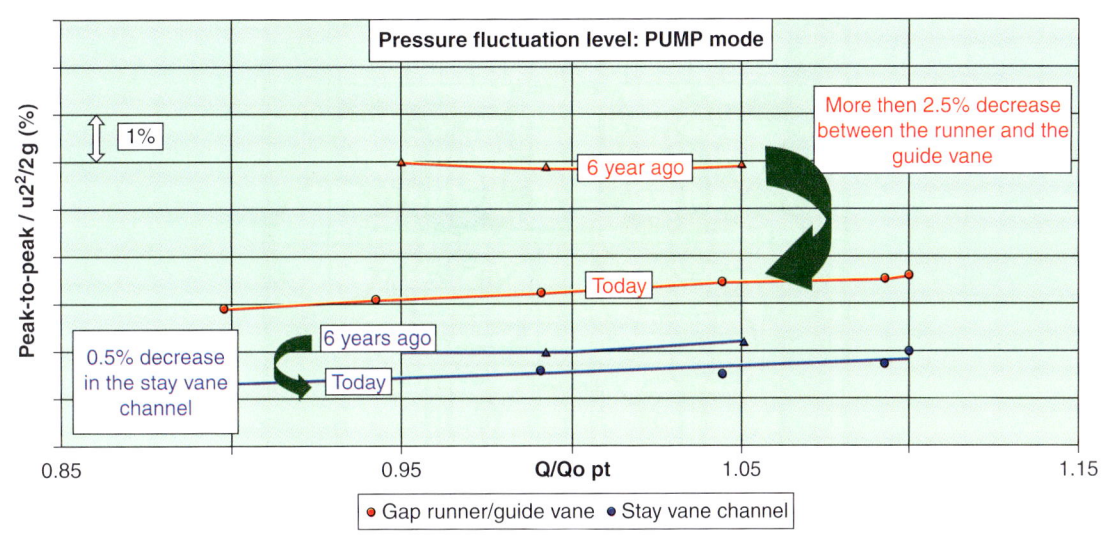

Figure 8 Pump model test result: Decreasing of the pressure fluctuation levels for 6 years.

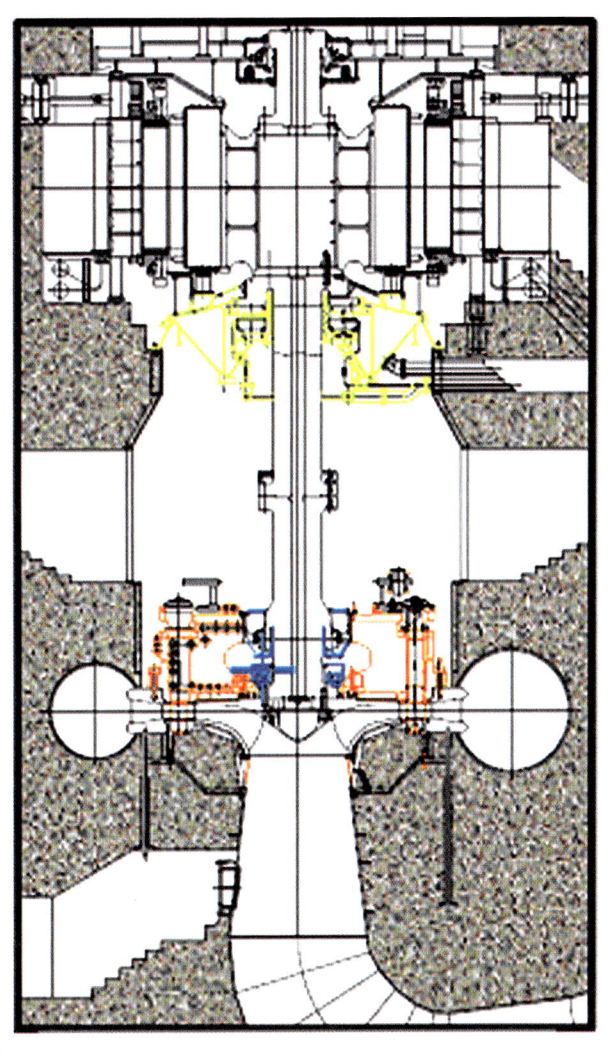

Figure 9 Single-stage pump–turbine.

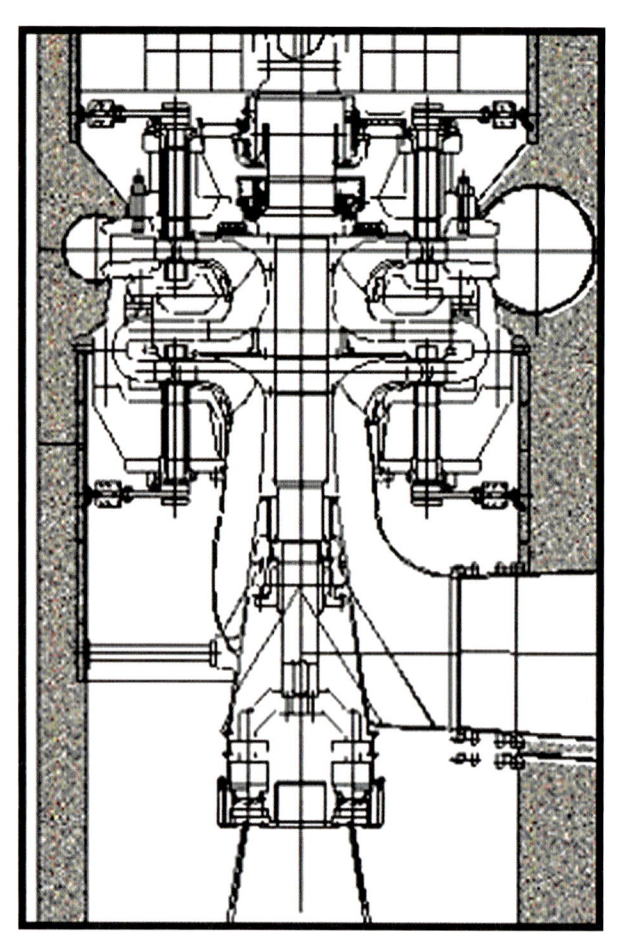

Figure 10 Double-stage pump–turbine.

- The pump–turbine is to be dismantled through the generator–motor stator or at the pump–turbine pit level.
- The spiral case is pressure tested before concreting and is concreted with a partial water pressure inside.
- The bottom ring is embedded into the concrete.
- Guide vanes and distributor mechanism are guided by self-lubricated bushes.
- Guide vanes are operated either by two servomotors connected to one operating ring (mechanical synchronization) or by individual servomotors (electronic synchronization).
- Runner is fabricated and the blades are machined by a five-axis numerically controlled milling machine.
- Pump–turbine guide bearing is pad or shell type.
- Shaft seal is of the hydrostatic axial type.
- Draft tube cone is embedded.
- Starting in pumping mode with dewatered runner can be performed either by a static frequency converter or back to back.

For a double-stage pump–turbine, some characteristics are specific (**Figure 10**).

- Thrust bearing can be located below the pump–turbine
- Draft tube cone is not embedded
- Dismantling of the lower runner by bottom

6.15.4 Examples of Remarkable Pumped Storage Power Plants

6.15.4.1 Okinawa Seawater Pumped Storage Power Plant

This project is located in a unique area in Japan where many original species of flora and fauna indigenous to the Okinawa Islands have been preserved until then. Right from the start of construction activity, several measures were undertaken to mitigate its environmental impacts. These included protection of habitat area, prevention of muddy water outflow, reduction of noise and vibration of heavy construction equipment, installation of slope-type side ditch, restoration and revegetation of the disposal area, and creation of a biotope. These mitigation measures were decided upon based on a study conducted by a special committee formed by well-known local environmental specialists and the local community (**Figure 11**).

Figure 11 Okinawa seawater pumped storage power plant.

6.15.4.1.1 Outline of the project

The project is a demonstration plant for seawater pumped storage power generation located at the northern part of Okinawa Island. In making the concept of seawater pumped storage power generation practical, there was the necessity to find concrete solutions to the technical problems arising from the use of seawater and to the problems of environmental impacts. There was no case of seawater pumped storage power generation actually implemented anywhere in the world till that time, and this pilot plant constituted a first example.

The Electric Power Development Co. Ltd. (J-POWER) [18] undertook implementation of the project on consignment from the Ministry of International Trade and Industry (MITI) [19]. Surveys and research regarding the technical and environmental aspects of using seawater were carried out for 6 years from 1981.

The feasibility of a seawater pumped storage power plant was studied, and, having gained a favorable outlook, construction was performed from 1991 until 1999. Since 1999, a 5-year program of tests has been going on, its aim being to demonstrate the practicality of seawater pumped storage power generation technology.

In Okinawa, where the project site is located, daily load fluctuations are being met with regulated operation of thermal and gas turbine, and the need for pumped storage power generation is well recognized. Water resources are precious on Okinawa Island and construction of a conventional pumped storage power plant using river water would not be appropriate. As for the northern part of the island, it is mostly mountainous, and there are many locations that would be suitable for setting up a seawater pumped storage power plant.

With these in mind, sites meeting the requirements were picked out from all over Okinawa Island, and comparison studies of the impacts on the natural and social environments were made, the result being the selection of the present site.

Table 4 gives the specifications, and **Figures 12** and **13** the general plan and profile, respectively, of the plant.

Table 4 Okinawa Seawater pumped storage power plant specifications

Item		Specification
River system		
Catchment area		
Power plant	Name	Okinawa Yanbaru Power Plant
	Max. output	30 MW
	Max. discharge	26 m^3 s^{-1}
	Effective head	136 m
Upper regulating pond	Type	Excavated type, Rubber sheet-lined
	Max. embankment height	25 m
	Crest circumference	848 m
	Max. width	251.5 m
	Total storage capacity	0.59 × 106 m^3
	Max. depth	22.8 m
Waterway	Penstock	Inside dia. 24 m Length 314 m
	Tailrace	Inside dia. 27 m Length 205 m

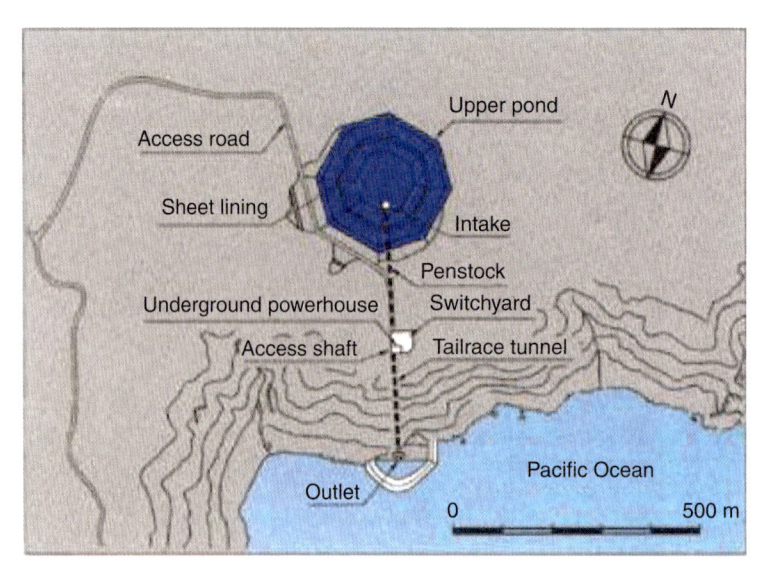

Figure 12 Okinawa project: General plan.

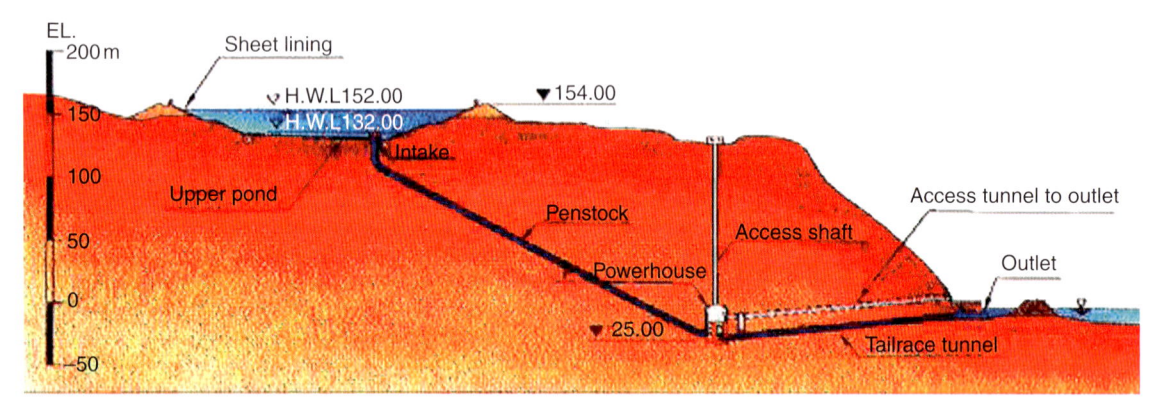

Figure 13 Okinawa project: Profile of waterways.

6.15.4.1.2 Features of the project area

It is said that the Ryukyu Island chain of which Okinawa is a part was connected by land to China on the Asian continent approximately 1.5 million years ago, but due to land upheavals and rising of sea level, a considerable area became submerged, the Ryukyu Islands remaining as a result.

Consequently, fauna and flora that had existed in the area occupied by the islands have lived in an isolated state since then so that original species have been preserved, and there are many species and subspecies indigenous to the Ryukyus and not found anywhere else in the world.

The climate is the closest to tropical in Japan, and it is warm throughout the year with annual mean air temperature being 23 C. Precipitation is approximately 2400 mm, 40% rainier than the national average, short bursts of heavy squalls occurring frequently. Many typhoons strike the islands from July to September.

The regulating reservoir is located at the Pacific Ocean side of Kunigami Village, approximately 600 m from the shore and on a tableland roughly 150 m asl. This area comprises a gently sloped plateau of elevation 150–170 m, gradually declining toward the sea (east), and on nearing the sea, drops into it by a cliff of specific height from 130 to 140 m. Gullies 20–30 m deep and with gentle gradients have developed in dendritic form on this plateau. These gullies gradually join together as the sea is approached, and on reaching the cliff, feed into the sea as a continuous chain of waterfalls. The seashore immediately below the cliff consists of both large boulders and sandy beaches (**Figure 14**).

A red soil classified as Kunigami maaji, a special type of soil in geological terms, covers the greater part of Okinawa Island. Kunigami maaji is a soil from weathered sedimentary rocks, the parent rocks being of various kinds such as phyllite, andesite, and sandstone. Molecular binding is weak and coefficient of permeability is low, from 10^{-5} to 10^{-6} cm s^{-1}, so that surface flows readily occur when it rains, and particles of this soil are washed out. Further features of Kunigami maaji include the following:

- It is highly weathered.
- It contains Kaolinite, a clay mineral.

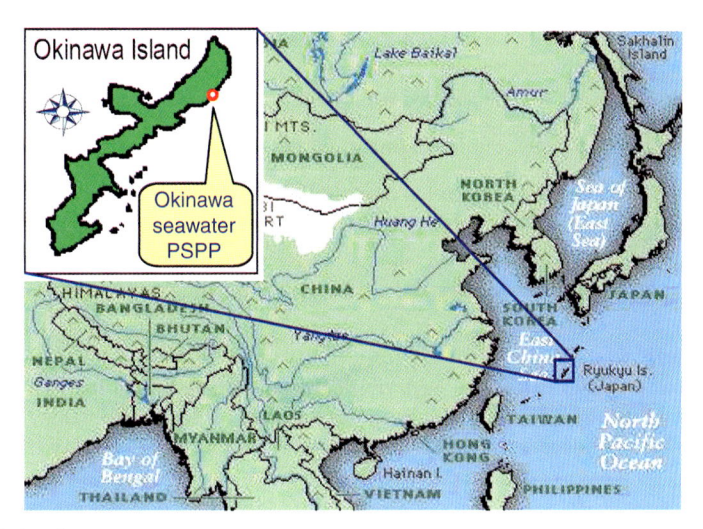

Figure 14 Okinawa project: Site location.

- Its specific gravity is high (2.6–2.8).
- When it is washed out to the sea, it tends to not go out to the open sea, and causes serious damage to the coral reefs along the coast, constituting a social problem for the local community.

The vicinity of the construction site is surrounded by state-owned forestland, and the area had been designated a US Marine Corps training ground under the Japan-US Status-of-Forces Agreement, but a portion was released as demonstration tests were to be conducted, while a separate portion is now leased and is being jointly used. The project site is not in an area designated as a natural park, and there are no cultural assets or recreational facilities.

6.15.4.1.3 Major impacts

J-POWER under consignment from MITI to carry out environmental impact surveys and studies of seawater pumped storage power generation, began field surveys in 1982, and compiled the environmental impact assessment (EIA) report in 1989. In **Table 5** the EIA items are shown.

The environmental problem of greatest concern was the potential effects of salt spray on the surrounding environment due to pump-up of seawater. Wind tunnel water tank tests and simulations using numerical models were carried out in this regard, and it was ascertained that there would be little difference from salt spray flying directly from the sea.

With regard to animal life, it was confirmed that there were 16 species of rare animals inhabiting the area of which five were listed as endangered and seven as threatened in the Red Data Book. As for plant life, there were many indigenous species (Okinawa Island-indigenous: three species, Okinawa Island northern limits-indigenous: five species, Ryukyu Islands-indigenous: nine species), while in the sea area reef-building coral was found.

Under these circumstances, a study committee for the protection of invaluable assets was organized in 1989, comprising mainly of local environmental specialists. Using the findings of the study committee as reference, fundamental principles of environmental conservation were set up as follows:

1. Organisms inhabiting the project area are native to the area and, as such, are to be given consideration with a modest attitude.
2. The development area is to be held to a minimum in order that the ecosystem will be disrupted as little as possible.

Table 5 Okinawa project: Environmental impact assessment items

Environmental impact assessment items		
Meteorology, weather, air quality, water quality, noise, vibration, offensive odor, soil contamination		
Ground settlement, topography, geology, sea current, marine phenomenon		
Salt spray, seawater seepage		
Plant	Vegetation, rare plant, soil profile, etc.	
Animal	Terrestrial animal	Mammals, birds, reptiles, amphibians, insects, soil fauna
	Aquatic organism	Gully animals
	Marine organism	Coral, fishes, benthic organisms, plankton, eggs and fry, tideland organisms, seaweed, grasses

3. The scope of implementation of the measures for natural environment protection is not to be limited to the construction area, and is to be applied to the surrounding area as well.
4. Any damage caused to the environment is to be rectified without delay taking advantage of natural self-healing powers.

Based on these fundamental principles, the environmental impact factors listed below were extracted:

1. Outflow of muddy water (red water) produced from the construction area into the gullies and sea area near the river mouth;
2. Reduction of habitat area due to changes made in land;
3. Noise and vibration from heavy equipment; and
4. Damage to small animals from construction vehicles and accidents due to falling down into roadside gutters.

6.15.4.1.4 Mitigation measures

The environmental conservation measures in **Table 6** were considered and implemented for each environmental impact factor. Meetings were held with the local community to explain these measures and consent was obtained.

Items thought especially to be of unique nature (underlined portions in **Table 6**) are discussed in detail below.

6.15.4.1.5 Measures during construction

6.15.4.1.5(i) Capture and transfer of fauna and flora in the construction area

Since the construction area during the construction period would become an uninhabitable environment for rare animals due to changes made in land and traffic of large heavy equipment, small animals of limited mobility such as frogs, turtles, and newts were captured and moved to favorable environments outside the construction area. Prior to carrying out capture and removal, tests to see whether the animals would survive resettling were conducted and it was confirmed that resettling would be possible. (Resettling survival tests consisted of capturing several individuals of a rare animal species and marking them before release into an optimum environment outside the construction area and confirming several weeks later that they were active and that no dead bodies were to be found, by which it was concluded that resettling could be carried out.)

6.15.4.1.5(ii) Installation of intrusion prevention nets

In order to prevent small animals such as turtles from entering the construction area from outside and being harmed by construction vehicles, polyethylene nets approximately 30 cm high were installed along 8 km of the outer perimeter of the construction area (**Figure 15**).

Table 6 Okinawa project: Countermeasures for environmental impact factors

Environmental impact factor		Countermeasure
Outflow of muddy water produced from the construction area into the gullies and sea area near the river mouth	Construction water	Chemical treatment by turbid water plant
	Turbid water from red soil	Chemical treatment by turbid water plant
		Reduction of turbid water by separation into red water and clear water
		Reduction of red water by spraying asphalt emulsion or seeds on bare ground
		Install a gabion weir downstream of the gully
Reduction in habitat area due to changes in land	Reduction of area changed	Layout of powerhouse and waterways underground
		Omit access road and work adit to outlet and powerhouse
		Reduce construction area by balancing cuts and embankments as much as possible
	Protection, restoration of vegetation	Landscape and green the construction site without delay
		Protect the existing forest by planting low-height trees
Noise, vibration from heavy equipment		Prohibit night time work in surface construction
		Use low-noise machinery
		Drive at low speed inside the construction area
Harm to small animals from construction vehicles and accidents due to falling down into the roadside ditches		Capture and remove animals and plants in the construction area
		Install facilities (intrusion prevention nets) to prevent the entry of rare animals
		PR activities using posters, lecture meetings, pamphlets, etc.
		Prevent accidents of small animals falling down into roadside ditches and getting killed, through the construction of sloping side wall gutters

Figure 15 Okinawa project: Intrusion prevention net.

6.15.4.1.5(iii) PR activities aimed at construction personnel concerned
In order that as many construction personnel concerned as possible may be interested in rare animals, pocket booklets and pamphlets containing photographs of rare animals, precautions for environmental conservation to be taken while executing work, and what to do when a rare animal is encountered were distributed; posters calling attention to protection of rare animals were put up; and lecture meetings concerning rare animals were held. The photographs of the rare animal species – birds, amphibians, and reptiles – contained in the pocket booklets were 16 in number and were in color. A protected-animal monitor was appointed, who, carrying out patrols, checked whether the rare animal protection measures were functioning effectively and whether handling of rare animals was appropriate. At the end of the pamphlet was included a rare animal sighting report form for mandatory reporting in case of encountering a rare animal.

6.15.4.1.6 Permanent measures
6.15.4.1.6(i) Construction of sloped-wall roadside ditches
Ordinarily, roadside ditches are U-shaped and if a small animal were to fall in it, it would have trouble climbing out, and in the hot sun of Okinawa, the animal would be fried to death by the heat absorbed by concrete. There had been a number of cases in the prefecture of chicks falling into U-shaped ditches and being killed by the heat.

Therefore, ditches with sloping sidewalls enabling animals to climb out by themselves were constructed. The ditches have sloping walls on their mountain side and vertical walls on their road side so that animals will be guided toward natural ground, the ditches thus playing a role in promoting traffic safety (**Figure 16**).

6.15.4.1.6(ii) Restoration and revegetation of the disposal yard, and biotope creation
Regarding the construction site, it was considered necessary to regain as quickly as possible a natural environment in which rare animals could live, through the restoration of the site to its original state. Particularly, regarding the disposal yard provided adjacent to the regulating pond where 210 000 m^3 of water had been accommodated in the beginning, it was thought that the conventional concept of a disposal yard could be abandoned, positioning it as a place (biotope) for creating an environment in which different species of wildlife (birds, insects, plants, etc.) can coexist (**Figure 17**).

Figure 16 Okinawa project: Slope-type side ditch.

Figure 17 Okinawa project: Central pond in disposal yard.

In order to recreate vegetation similar to the Itajii (the village tree of Kunigami Village, *Lithocarpus edulis*) forest in the surroundings, about 20 varieties of trees, beginning with priority-status trees such as Itajii and Adeku (*Syzygium buxifolium*), fruit-bearing trees such as Shimaguwa (*Morus australis*) and Sharinbai (*Umbellata*), and pioneer trees such as Akamegashiwa (*Mollotus japonicus*) and sendan, which grow rapidly to form windbreaks for Itajii and others and aid in their growth, were selected, and around 30 000 trees were planted intermixed in the disposal yard of approximately 45 000 m³. Saplings of the trees planted are almost nonexistent commercially, and seeds collected from standing trees were sown in pots and grown for 2–3 years and then transplanted. If left as bare ground after planting, there would have been outflow of red soil whenever it rained; so the surface of red soil was covered with bark and wood chips at flat areas and coconut fiber mats at slopes, which, at the same time, held down evaporation and improved the environment for the growth of trees.

Before development, there were gullies at this construction site and a riverine environment existed. In order to restore this environment, ponds and waterways of various sizes were arranged in a well-balanced manner to create habitats for small animals (aquatic organisms, insects, etc.) thereby aiming for biodiversity. The ponds and waterways were lined, so that small animals could inhabit gaps between stones. The stones used for lining were of different sizes, large and small, for variation. Coconut fiber mats were used to cover the surfaces of small ponds and waterways to prevent outflow of red soil (**Figure 18**).

6.15.4.1.7 *Results of the mitigation measures*

Environmental impact surveys have been conducted regularly since 1990 to assess the impacts of construction work on the surrounding environment. The outline of the environmental monitoring in 1999 is given in **Table 7**.

The condition of growth of the principal varieties of trees at the disposal yard, a part of the results, is shown in **Figure 19**. The numerical values used were obtained from trees growing in areas of 10 × 10 m at ten different locations around the regulating pond. (The data for 1997 were to a great extent affected by a slowly moving typhoon resulting in very long hours of strong wind and rain, and, therefore, were considered unsuitable for comparisons and omitted.)

At the disposal yard biotope, saplings have grown satisfactorily, as can be seen in **Figure 20**. Tadpoles and newts can be seen swimming, and dragonflies flirting about in and around water. Because the yield of fruits from trees and the population of insects, on which animals can feed, have increased, rare birds and boars began to appear. As an unexpected secondary effect, aquatic animals were found to be laying eggs on the coconut fiber mats, which had been spread out at the bottom of water bodies; thus the mats were fulfilling the role of aquatic plants.

Figure 18 Okinawa project: Cover by coconut fiber mats.

Table 7 Okinawa project: Outline of environmental monitoring

Environmental monitoring item			Monitoring frequency
Terrestrial monitoring			
Vegetation (before, after typhoon)		10 points around the plant	Twice a year
Animals	Mammals, birds, reptiles, amphibians, Insects, soil fauna, aquatic organisms	Line census, fixed point survey	
Noise, vibration		4 points around plant	
Regulating pond water quality	Water quality, electrical conductivity	1 point in regulating pond	Continuous
	Transparency, pH, DO, COD, SS, n-hexane extract		Once a year
Marine monitoring			
Water quality	Water temperature, salt content, transparency, pH, DO, COD, SS, n-hexane extract, coliform group number, T-n, T-P	4 points in frontal sea area	Twice a year
Bottom sediment	COD, total sulfides, particle-size distribution, loss on ignition	5 points in frontal sea area	
Organisms	Tideland organisms, Eggs–fry, Zoo/Phytoplankton	2–3 points in frontal sea area	
	Marine algae and grasses	3 traverse lines in frontal sea area	
	Coral	3 traverse lines, 5 points in frontal sea area	
	Fishes and other nekton	1 traverse line in frontal sea area	

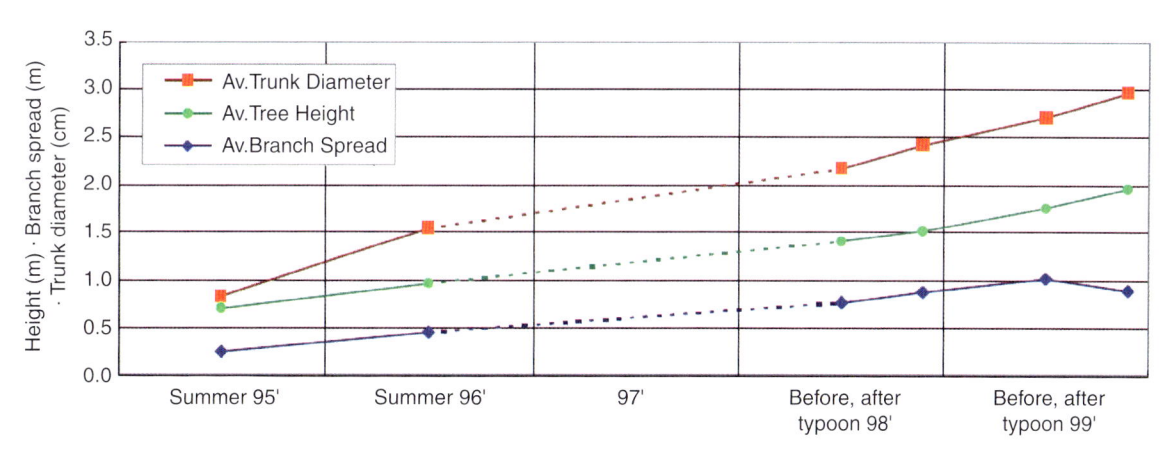

Figure 19 Okinawa project: Environmental impact assessment results.

Figure 20 Okinawa project: Central part of the disposal yard.

Regarding the measures for preventing red soil from running out, there were no outflows even when typhoons had occurred, and it was confirmed that muddy water had not been produced.

Because rare species of flora and fauna were discovered at the site and this was shown on television, many people from the general public started arriving to visit the site. The fact that rare animals and plants can be seen within a short period of time, without having the need to go deep into the surrounding mountains and forests, has won attention. Studies are being conducted at the

Kunigami village office to see whether this feature can be taken advantage of and the site can be used as a place for education of children and for general environmental education.

Figure 20 is a photograph showing the condition at the central part of the disposal yard.

6.15.4.1.8 Reasons for success

The Study Committee for the Protection of Invaluable Assets was formed by local environmental specialists before the start of construction; studies were conducted on environmental conservation measures and their effects; and construction was performed taking these into consideration. These are thought to be the primary reasons for the success of the environmental protection measures.

In particular, the creation of a biotope at the disposal yard with a view of recovering natural habitats can be given high marks. The soil at the disposal yard was inorganic, and there prevailed conditions of severe wind and salt damage; so studies regarding tree planting had been taken up from an early stage of planning. Based on such studies, surveys of salt damage and a survey of the types of trees favorable to nesting of birds were carried out. These were useful in deciding the types of trees to be planted and the method of planting.

6.15.4.2 Goldisthal Pumped Storage Power Plant

6.15.4.2.1 Introduction

The 1060 MW Goldisthal pumped storage plant on the Schwarza River is the biggest hydroelectric project in Germany and the most modern one in Europe. Construction of the project began in September 1997, and the plant started commercial operation in October 2004 [20].

The Goldisthal project is unique in that two of the four vertical Francis pump–turbine units feature variable-speed (asynchronous) motor–generators. This arrangement provides several benefits including power regulation during pumping operation, improved efficiency at partial-load conditions, and enhanced dynamic control of the power delivered, for stabilization of the grid.

6.15.4.2.2 Developing the Goldisthal project

In 1965, scientists within the German Democratic Republic performed a ranking study, which identified the current site on the Schwarza River in Thuringia State as the best location for a large pumped storage hydroelectric project. Beginning in 1972 geologists with Baugrund Dresden of Germany performed intensive geological investigations and began work on infrastructure such as the transformer substation and bypass road. In 1975, Schachtbau Nordhausen GmbH of Germany dug several 4 km-long investigation tunnels to explore the site geology (**Figure 21**).

Work on the Goldisthal project was suspended in 1980, primarily for economic reasons. The increase in energy demand in the country was not as high as expected, and financial problems delayed the construction schedule for the plant.

Nine years later, as a result of the political unification of Germany and restructuring of the East German power supply, investigations to develop Goldisthal resumed. In 1990, Vereinigte Energiewerke AG (VEAG) began the planning and permission procedures to develop Goldisthal. The utility proposed to install four turbine–generator units with a total capacity of about

Figure 21 Goldisthal project.

1000 MW. VEAG determined that the most advantageous arrangement for this plant involved combining conventional pumped storage (synchronous) technology (represented by two single-speed motor–generators) with variable-speed (asynchronous) technology (represented by two variable-speed motor–generators).

Variable-speed motor–generators allow operation of the pump–turbine unit over a wider range of head and flow, making them economically advantageous for a pumped storage facility. VEAG conducted considerable investigations to determine the number of variable-speed units to be installed. The company eventually decided to install two variable-speed units and two conventional units.

The decision was based on a variety of reasons. First, calculations indicated that the company will need about 200 MW of controlled power for the pumping operation, which was within the control range of two variable-speed units. Second, asynchronous machines are not capable of restarting the system during a power outage – they need support from the grid to begin operation. Consequently, conventional units would be required in the event of a power outage. Third, because no company in Europe had any experience in operating large variable-speed units, VEAG considered installing variable-speed units alone as too risky.

VEAG received the building permit for Goldisthal in 1996. Construction began in September 1997 with the work on the access tunnel to the underground powerhouse and transformer caverns, as well as such surface infrastructure work as providing site access.

6.15.4.2.3 Main features of the project

- Upper reservoir with a usable capacity of 12 million m^3. This is the largest artificial reservoir in Germany, covering 55 ha under full storage conditions. The rockfill dam impounding this reservoir is 3370 m long.
- Two 6.2 m-diameter headrace tunnels with steel-armored lining, with a total length of about 870 m.
- Underground powerhouse cavern containing four 265 MW vertical Francis pump–turbines, four motor–generators, and auxiliary systems. This cavern is accessed via a 1 km-long tunnel from the operation building complex.
- Underground transformer cavern containing four unit transformers, the 10 kV transformers of the internal electrical supply, switchgear, and starting frequency converters.
- Two 8.2 m-diameter tailrace tunnels with concrete-armored lining, each 275 m long.
- Lower reservoir with a capacity of 18.9 million m^3, impounded by a rockfill dam (**Figure 22**).

6.15.4.2.4 Choosing variable-speed machines

The most important innovation at the Goldisthal project is the first ever application of variable-speed motor–generators of this size in a hydro plant in Europe. Essentially, turbines have one optimum operating point in terms of head, flow, unit size, and speed. But

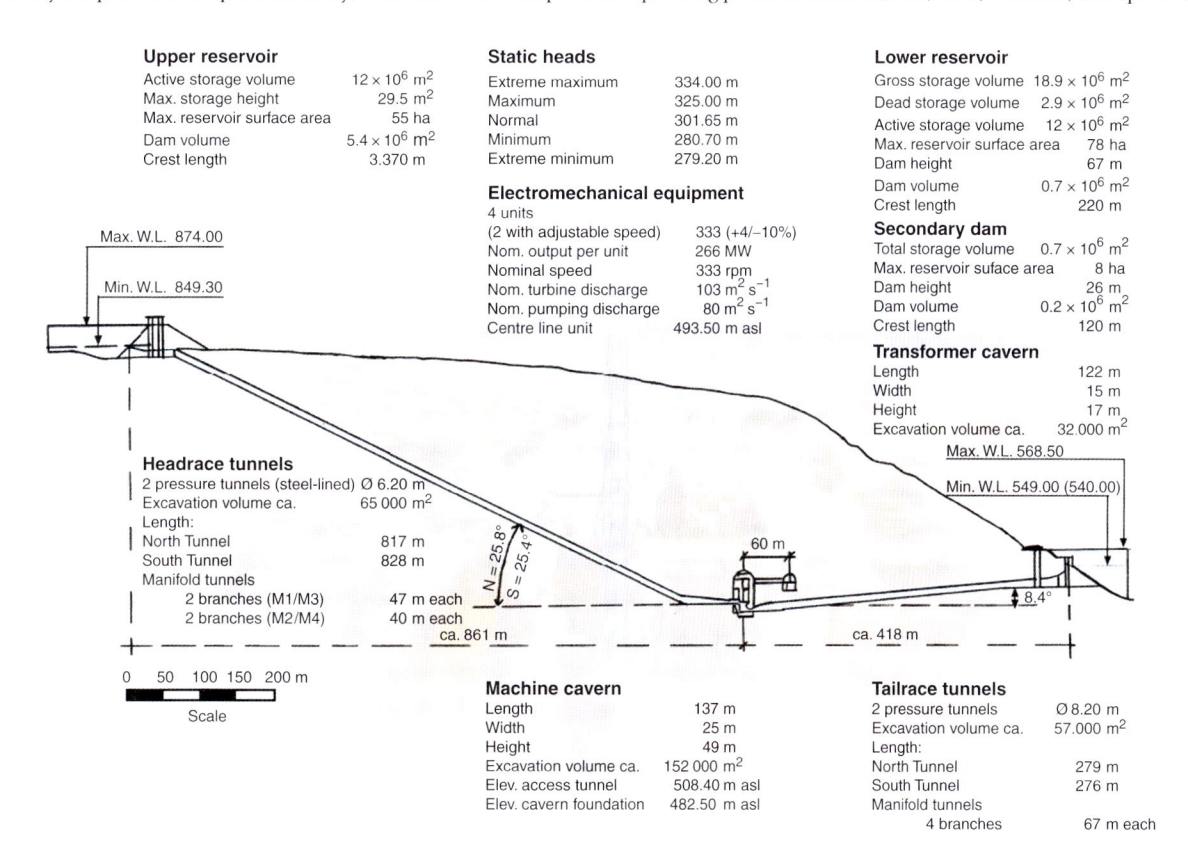

Figure 22 Schemes of Goldisthal project.

Figure 23 Goldisthal project: Variable-speed unit (at left) and one synchronous unit.

when these units are coupled with a variable-speed motor–generator, operating speed can be varied over a certain range of the nominal synchronous speed of the turbine–generator unit. As head and flow vary, the unit is able to increase or decrease its speed to operate closer to its peak efficiency for that unique set of conditions.

Each half of the 1060 MW Goldisthal pumped storage project features one variable-speed unit (**Figure 23**) and one synchronous unit. The variable-speed units allow for efficient operation at a wider head range, during both turbine and pumping operations.

The difference between synchronous and asynchronous machines lies in the construction of the rotor. While classical synchronous generators have salient poles, variable-speed generators have a three-phase winding on the rotor. And whereas the synchronous rotor is energized by a direct current (DC) to create a rotating magnetic field, the asynchronous rotor is energized by a low-frequency AC. A direct frequency converter in the rotor circuit is used to control the frequency. If the frequency is changed, so too is the speed of the unit.

The rotor can be retarded or accelerated, opposite the stator field, from 90% to 104% of the synchronous speed. The variable frequency of the asynchronous generators at Goldisthal ranges from 5 Hz opposite to the stator field of 333 rpm (which provides 300 rpm) to 0.01 Hz (which is nearly the rated speed of the unit) to 2 Hz in addition to the stator field (which provides 340 rpm).

Asynchronous motor–generators provide several advantages, including

- More flexibility in their operation;
- Higher efficiency over a wide range of operation at partial-load conditions;
- A wide range of controllable and optimized power consumption in pump operation;
- Additional and faster features for grid control, such as fast power outlet regulation;
- Better use of the reservoir because higher water level variations can be allowed; and
- Better contribution to grid stability because of the high moment of inertia of the rotating masses.

Asynchronous machines make it possible to regulate power not only in the turbine mode but also in the pumping mode. The range of control at Goldisthal amounts to 190–290 MW.

The power plant at Goldisthal is arranged to be split in half, with each side being a mirror image of the other. Each half of the plant contains one synchronous and one asynchronous machine working together at one headrace tunnel. This allows operators to take half of the plant off line at any one time for maintenance while the other half continues operating.

6.15.4.2.5 Operation to date

The Goldisthal plant was commissioned in October 2004. The variable-speed machines operate an average of 19 h a day in both primary and secondary regulation. When operating under conditions of partial load, these units have an efficiency advantage of about 10% as compared with the single-speed units.

An automatic controller on the two variable-speed machines constantly calculates and adjusts the units for optimal production, based on the momentary head and the power output required.

The asynchronous machines can be started more quickly than the synchronous units because no fixed rotation speed is necessary for synchronization of the variable-speed units. Starting from 95% of full synchronous speed, the frequency converter regulates its parameters to the current speed and releases the unit to synchronization.

The operator expects overall maintenance expenditures to be lower for variable-speed units because of the smaller starting and brake load operation, which is helped by the starting frequency converter. However, periods between maintenance of the variable-speed machines are expected to be somewhat shorter than those of the conventional machines. These units are inspected every 4 weeks, as against every 6 weeks for the synchronous units. Moreover, inspection of the variable-speed units requires 10 h, as compared with 8 h for the synchronous machines.

The higher frequency of inspection and the longer time required result to a large extent from the larger number and size of the auxiliary systems associated with the asynchronous generator. For example, because the rotor of the asynchronous machine needs more power and the voltage and current are much higher than in a synchronous rotor, the slip ring system is much bigger.

With regard to ancillary services, the asynchronous units at Goldisthal have been quite valuable. Because of the large capacity of the units, a large regulation range is available. This is used daily for grid frequency control. The asynchronous machines can be regulated from 40 MW up to 265 MW, while the synchronous machines can only be regulated from 100 to 265 MW. Thus, the asynchronous machines provide 60 MW more for regulation. This allows the operator to take advantage of the lower basic power output of 40 MW, saving water to be used for later generation.

In addition, the asynchronous machines have the ability to respond very quickly. If fast power is needed in the grid, the asynchronous machines can retard their speed and supply additional braking energy to the grid (for a few seconds). In early November 2006, parts of Europe experienced a long blackout. In the eastern part of Germany, frequency on the grid was 50.6 Hz, whereas it normally is 50 Hz. The operator used the Goldisthal units to take energy out of the grid, and the asynchronous units were used for regulation in pumping operation.

On average, about 70% of the working ability of the power plant is used each day. That results in daily production of 5500–7500 MWh during turbine operation.

6.15.4.3 Tianhuangping Pumped Storage Power Plant

6.15.4.3.1 Introduction
East China Electric Power's Tianhuangping pumped storage hydroelectric project is the biggest of its type in Asia. It provides valuable cover for demand surges in the central coastal region, including fast growing Shanghai. It is located in Anji County in Zhejiang, about 175 km from Shanghai, and has a total installed capacity of 1800 MW [21] (**Figure 24**).

6.15.4.3.2 Two pump storage reservoirs
There are two pump storage reservoirs about 1 km apart, which have a difference in elevation of 590 m. Both have a storage volume of 8 million m^3. The lower reservoir is located on the Daxi Creek branch of the Xitiao river and the upper reservoir is an artificial basin cut deep into the mountains. Two 7 m-diameter conduits from the upper reservoir each branch into three 3.2 m-diameter pipes to power the six 306 MW turbines in the underground powerhouse. The maximum generating head is 607.5 m.

The lower reservoir is formed by a CFRD dam 95 m high, with an overall storage capacity of 8.6 million m^3, a catchment area of 24.2 km^2, and an average annual runoff of 27.6 million m^3. The designed maximum pool level of the lower reservoir is 344.5 m. The construction of the CFRD dam commenced in October 1993 and was completed in 1997. The upper reservoir is approximately 600 m higher, in a natural valley basin where the only opening is closed by a 72 m high rockfill dam, so as to maintain the 2315 m long crest constantly at an elevation of 908 m. Four smaller saddle dams have been constructed to form the reservoir. The maximum depth of the reservoir is 50 m. The slope and bottom of the upper reservoir have an asphalt concrete lining, which is best suited to

Figure 24 Tianhuangping project view.

Figure 25 Tianhuangping project: Upper reservoir.

absorb the settlements and deformations of the dams without becoming permeable to water. The lining area amounts to 104 000 m² for the reservoir bottom and 182 000 m² for the reservoir slope, with a slope inclination of 1V:2–2.4H (**Figure 25**).

The substructure consists of a drainage layer made of crushed rock, the thickness of which is 90 cm on the slope and 60 cm on the bottom. Bituminous emulsion was sprayed to stabilize the substructure surface and to achieve a better bond with the asphaltic binder layer. The asphaltic binder layer is 10 cm thick on the slope and 8 cm on the bottom. The thickness of the impervious asphalt concrete layer is 10 cm in both cases. In order to protect the asphalt concrete against aging as a result of ultraviolet radiation associated with oxygen in the air, the slope and the bottom are provided with an asphalt mastic seal coating. In the curve at the junction of the slope and the bottom and at the connections to the concrete structures, a 5 cm-thick protective layer of asphalt concrete is applied, together with a polyester mesh reinforcement. Domestic bitumen products, of which several types were carefully examined, had been ruled out due to their excessive paraffin content, which impairs the bonding characteristics of bitumen (**Figure 26**)

6.15.4.3.3 Benefits of the project

The new plant plays a vital role in stabilizing the entire east China power grid, improving the quality of the power supply in east China and ensuring the safe operation of the nuclear power stations in the surrounding areas (**Figure 27**).

From September 1998, when the first unit was put into operation, to the end of August 2008, the Tianhuangping power station had generated 20.767 billion kWh electricity during peak times and absorbed 25.865 billion kWh during off-peak times. The overall efficiency was 79.2%, which had exceeded the value of designed efficiency (74%); and the station occupied the world's leading position among all the pumped storage stations of its kind. The power station played an important role in peak shaving, frequency modulation, and emergency duty in the network system, and effectively improved energy efficiency and overall efficiency of the network system. As of August 2008, the power station had modulated system frequency during emergencies on more than 50 occasions and had played an important role in its mission of supplying power during the peak periods of every year, for instance, during summer time and the time of major meetings.

Figure 26 Tianhuangping project: Lower reservoir.

Figure 27 Tianhuangping project: Underground powerhouse.

6.15.4.4 Coo-Trois Ponts Pumped Storage Power Plant

6.15.4.4.1 Introduction
The Coo-Trois Ponts (Coo) Pumped Storage Plant is an underground facility located in the Ardennes section of Belgium. The total rated capacity of the plant is 1164 MW provided by six single-stage, reversible pump–turbines. Coo was constructed in two stages. The initial three units totaling 474 MW (Coo I) were put online in 1971. An additional three units totaling 690 MW (Coo II) were put online in 1980. The overall head for the facility is 275 m. The unit hydrograph (UH) ratio varies between Coo I and Coo II from slightly under 2 to slightly over 2. The initial stage of Coo included construction of an upper reservoir, power tunnel, underground powerhouse, and tailrace tunnel. At the first stage itself the powerhouse was constructed to the full size required to accommodate the second-stage expansion also. The second stage included construction of an additional separate upper reservoir, power tunnel, and tailrace tunnel. The original design envisaged the three pump–turbine units installed during the second stage to be identical in size to those in the first stage. Revised turbine designs, however, allowed for larger-capacity units to be installed within the powerhouse, which originally had been designed for the smaller Stage 1 units. The overall layout of the project features is shown in **Figure 28** [22].

6.15.4.4.2 Main features of the project
The upper reservoirs consist of two asphaltic-lined systems with a total storage volume of 8 450 000 m^3. This represents approximately 5 h of generation at full output. The intake structures at both Stage 1 and Stage 2 upper reservoirs are vertical and include a vortex suppression plate (**Figures 29** and **30**).

The underground structures at Coo are located in the massif of Coo, a Paleozoic formation composed of phyllites, quartzo-phyllites, and quartzites. These formations posed numerous problems during construction of the underground facilities. The rock formation revealed expansive characteristics and required special treatment during construction. This treatment included extremely long rock bolts, a temporary concrete tinning, and a final steel lining for all high-pressure water passages. The first- and second-stage waterways consisted of a 6.6 m-diameter, steel-lined, vertical shaft, which connected to the power tunnel section with a 70% slope ending in a manifold upstream of the underground powerhouse. The first stage high-pressure section had an overall length of 183 m. The second-stage tunnel was entirely separate from the initial-stage pressure tunnel and had larger waterway openings to accommodate the larger-capacity units. This tunnel started with a vertical shaft, which extended to a sloping power tunnel. The vertical-shaft tunnel has a diameter of 8 m and an overall length of 354 m. Flow to the units is controlled by spherical valves with a diameter of 2.65 m. The transformers are sited outdoors in a valley adjacent to the powerhouse. The powerhouse cavern is lined with concrete. Due to the expansive character of the rock and numerous fractures, the entire arch of the cavern required a concrete lining. A sequenced excavation plan for the arch and the cavern area was instituted to allow for safe construction. The multiple-stage construction resulted in two concrete-lined tailrace tunnels, each 245 m long with a 9 m diameter. The submergence of the runners is approximately 18 m. A 210 m-long tunnel provides access to the downstream area.

Two embankments were constructed at either end of an ancient oxbow in the Ambleve River to create the lower reservoir. The embankments were constructed of earth and rockfill with an asphaltic-concrete lining. The construction of these two embankments created an essentially closed system, although the lower reservoir area can be dewatered by means of a dewatering conduit located at

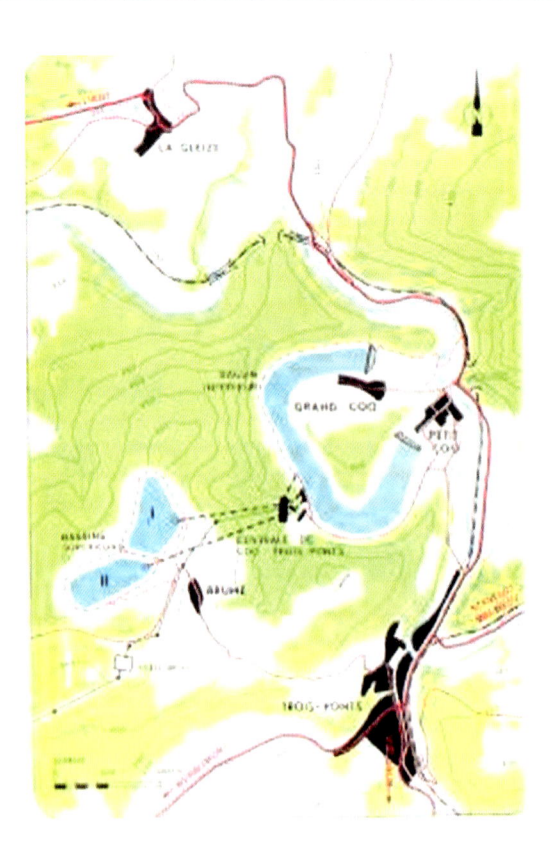

Figure 28 Layout of Coo project.

Figure 29 Intake structure of Coo 1.

the downstream end of the oxbow. The two lower-reservoir embankments are approximately 10 and 30 m high, respectively (**Figure 31**).

6.15.4.4.3 Generating equipment

The generating and pumping equipment for the facility was designed to go from standstill to full generating load in 2.5 min and from standstill to pumping in 6.5–7 min. Thanks to upgrades, those initiation times have been substantially reduced. The units are started in the pumping mode using pony motors (**Figure 32**).

The project currently produces approximately 800 000–1 000 000 MWh per year in the generating mode and utilizes 1 000 000–1 333 000 MWh in the pumping mode. The facility is also utilized for synchronous condensing and spinning reserve (**Table 8**).

Figure 30 Intake structure of Coo 2.

Figure 31 Coo project: Upper and lower reservoirs view.

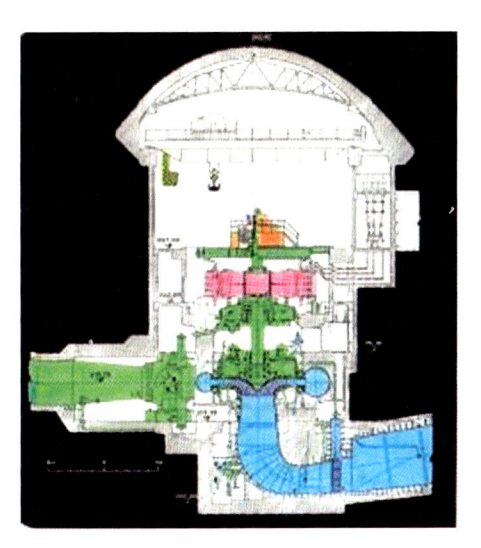

Figure 32 Coo project: Turbine view.

6.15.4.4.4 Special features of interest

The overall Coo-Trois-Ponts project was a staged development. The original planning allowed for the use of two upper reservoir areas with separate high-pressure and tailrace tunnels. The initial construction of the powerhouse included excavation of the total cavern area with sufficient placement of concrete to provide support for the second-stage development. A refined design

Table 8 Coo Project: Performance of turbines

			Max head & power		Min head & power	
			Coo 1	Coo 2	Coo 1	Coo 2
Head	Generating	m	275.35	273.56	235.05	230.85
	Pumping	m	275.35	273.56	235.05	230.85
Capacity	Generating	MW	158	230	60	80
	Pumping	MW	135	190	145	200
Discharge	Generating	m^3 s^{-1}	60	100	23	36
	Pumping	m^3 s^{-1}	51	82	55	87

of the second-stage pump–turbine units allowed for an increase in size and capacity within the constraints of the original cavern and powerhouse excavation limits.

Although the lower reservoir is interconnected with the Ambleve River, the system has many aspects of a closed system.

References

[1] Federal Energy Regulatory Commission, (FERC) http://www.ferc.gov/industries/hydropower/gen-info/regulation/pump.asp
[2] Price A (2000) Recent developments in the design and applications of a utility-scale energy storage plant. *Proceedings of the Electrical Energy Storage Systems Applications and Technologies International Conference 2000.*
[3] Schoenung S (2001) Characteristics and technologies for long- vs. short-term energy storage. *Sandia National Laboratories.* Report SAND2001-0765. March 2001.
[4] Herr M (2002) *Economics of Integrated Renewables and Hydrogen Storage Systems in Distributed Generation.* London, UK: Imperial College of Science, Technology and Medicine, University of London.
[5] Butler P, Miller J, and Taylor P (2002) Energy storage opportunities analysis. Phase II final report: A study for the DOE energy storage systems program. Sandia Report SAND2002-1314. Unlimited Release. May 2002.
[6] Boyes J (2000) Energy storage systems program report for FY99. Sandia Report, SAND2000-1317, 2000.
[7] Baxter R and Makansi J (2002) *Energy Storage. The Sixth Dimension of the Electricity Value Chain.* Pearl Street Inc.
[8] Gandy S (2000) A Guide to the Range and Suitability of Electrical Energy Storage Systems for Various Applications, and an Assessment of Possible Policy Effects. Master Thesis, Imperial College of Science, Technology and Medicine.
[9] Schoenung S (2002) Characteristics of energy storage technologies for short- and long-duration applications. *Proceedings of the Electrical Energy Storage Systems Applications and Technologies International Conference 2002 (EESAT2002),* San Francisco, CA, 15–17 April.
[10] Swaminathan S and Sen R (1997) Electric utility applications of hydrogen energy storage systems. DOE/GO/10/70-T19, US Department of Energy.
[11] Ter-Gazarian A (1994) *Energy Storage for Power Systems.* Peter Peregrinus Ltd.
[12] Schoenung S (2001) *Characteristics and Technologies for Long vs. Short-Term Energy Storage.* Report SAND2001-0765, March 2001. Sandia National Laboratories.
[13] Gordon S and Falcone P (1995) The emerging roles of energy storage in a competitive power market: Summary of a DOE workshop. Sandia Report. SAND95-8247.
[14] Tanaka T and Kurihara I (1998) Market potential of utility-purpose energy storage in Japan up to the year 2050. *Proceedings of the Electrical Energy Storage Systems Applications and Technologies International Conference 1998 (EESAT98),* Chester, UK, 16–18 June.
[15] www.voithhydro.com/.../VSHP090009_Pumped_Storage_72dpi.pdf
[16] Houdeline JB, Lavigne S, Mora P, *et al.* (2005) *Single Stage Reversible Pump/Turbine – From Design to Experience.* Austin, TX: Waterpower XIV.
[17] Houdeline JB, Verzeroli JM, Kunz T, and Schwer A (2010) *Reversible Pump/Turbine and Generator-Motor Design The Last PSP Achievements Experience in Asia.* IWPC.
[18] Electric Power Development Company (EPDC) www.jpower.co.jp/english/index.html
[19] MITI Ministry of Economy, Trade and Industry http://www.miti.gov.my/cms/index.jsp/
[20] Beyer Th. http://www.hydroworld.com/index/display/article-display/351208/articles/hydro-review-worldwide/volume-15/issue-1/articles/goldisthal-pumped-storage-plant-more-than-power-production.html
[21] Chinese National Committee on Large Dams (2010) *Current Activities. Dams Construction in China.* IWHR.
[22] Energy Division of the American Society of Civil Engineers (1995) *Hydroelectric Pumped Storage Technology: International Experience.*
[23] Lejeune A and Pirotton M *Exploitation de la Centrale de Coo.* LHM, University of Liege.

6.16 Simplified Generic Axial-Flow Microhydro Turbines

A Fuller and K Alexander, University of Canterbury, Christchurch, New Zealand

© 2012 Elsevier Ltd. All rights reserved.

6.16.1	Introduction and Context	436
6.16.1.1	What Is a Simplified Generic Axial-Flow Microhydro Turbine?	436
6.16.1.2	Who Would Use Such a Turbine?	438
6.16.1.3	Representative Designs	438
6.16.1.4	Energy Alternatives and Unconventional Economics	442
6.16.1.5	What Is Specific Speed?	442
6.16.2	Component-Level Design Methods	443
6.16.2.1	Water Supply	444
6.16.2.1.1	Penstock	445
6.16.2.1.2	Open flume	449
6.16.2.2	Volute	449
6.16.2.2.1	Runners need swirl	450
6.16.2.2.2	Volute characterization	451
6.16.2.2.3	Swirl predictability	451
6.16.2.2.4	Limits of the tangential inlet volute	452
6.16.2.3	Runner	453
6.16.2.3.1	Specifying runner geometry	454
6.16.2.3.2	An example runner	455
6.16.2.3.3	How specific speed influences blade shape	455
6.16.2.4	Draft Tube	456
6.16.2.4.1	The effect of diffuser inlet swirl	460
6.16.3	Turbine Selection from an Existing Range	460
6.16.4	Direct Sizing	463
6.16.5	Conclusions	465
Further Reading		465
References		466

Nomenclature

A area
A_R diffuser area ratio
C penstock cost
C_p pressure coefficient
d diameter, penstock inner diameter
F_{AM} angular momentum flux factor
g acceleration due to gravity
h head loss
H total head
$K_{1,2,3}$ penstock optimization constants
K_d diffuser loss coefficient
L penstock length
L_R diffuser length ratio
$\dot{m}$ mass flowrate
N runner speed (rev min^{-1}), diffuser length
N_R Reynolds number
N_S specific speed
p pressure
P power
Q turbine volumetric flowrate

Q_a available volumetric flowrate
r radial position
R diffuser wall radius
T torque
U blade velocity
V velocity
β local blade angle
δ blade trailing edge deviation
Δ change of indicated quantity
η efficiency
θ azimuth position in runner frame, diffuser half-angle
ρ density
ψ blade setup angle
ω runner angular velocity

Subscripts–Stations

0 headwater
1 between intake and penstock
2 between penstock and volute

3 between volute and runner
4 between runner and draft tube
5 tailwater
i component inlet
o component outlet

Subscripts–Turbine components
d draft tube
g generator
i intake
m transmission (mechanical)
p penstock
r runner
t turbine composite
T site total
v volute

Subscripts–Velocities
a axial
$_{abs}$ absolute
n normal
r radial
$_{rel}$ relative
t tangential

Subscripts–Pressures
d dynamic
s static
t total

Subscripts–Runner blade location
h hub
LE leading edge
t tip
TE trailing edge

Subscripts–Heads
g gross
n net

6.16.1 Introduction and Context

6.16.1.1 What Is a Simplified Generic Axial-Flow Microhydro Turbine?

A simplified generic axial-flow microhydro turbine describes a class of devices for extracting energy from an elevated source of water, whose main design points are communicated in this chapter.

These devices are 'simplified' relative to large-scale turbines in terms of their geometry, construction, and operation. They are designed to operate at peak efficiency at a single speed and flowrate with minimal maintenance. Their geometry is fully fixed, the transmission is direct, and guide vanes are omitted.

The devices are 'generic' in that while each individual turbine is tailored to its site, the process for designing it is flexible enough to accommodate a range of inputs. This is especially relevant for microscale turbines, which, being smaller, may be mounted entirely above ground on a foundation, reducing civil works complexity. Best practices can be applied consistently, and consultation required for each turbine is reduced or eliminated.

'Axial flow' describes the path the water takes through the turbine runner. The runner is the device that extracts work by rotating under a torque. From Euler's turbomachinery equation, runner torque is equivalent to the change in angular momentum of the water. For an incompressible fluid, this is proportional to the fluid's change in rV_t. In an effort to reduce energy lost to residual swirl, most turbine designs use a stator to convert some of the static pressure to a tangential velocity, after which the runner reduces the swirl back to zero. Turbine head is also proportional to this change in angular momentum, such that in general low-head turbines, which cope with less swirl, are of the axial-flow variety. Axial flow corresponds to a high specific speed.

'Microhydro' describes the size of a turbine in terms of its power output. The product of head and flowrate is relatively small. It does not mean microscopic, although relative to existing gigawatt-scale installations, microhydro turbines are roughly a million times smaller. The term itself is a bit vague, and has been variously described as turbines producing from 1 to 10 kW up to 35 to 100 kW. In many cases, the term 'microhydro' could be replaced with 'community scale' or 'household scale' due to its frequent applications in those duties. For this chapter, about 2–35 kW is the range of interest.

Starting from the descriptions above, the focus can be narrow further, beginning with the aim to design simple turbines. Classically, one of the most complicated and expensive components of a small turbine has been the governor. When hydraulic turbines are employed to generate alternating current (AC) electricity, the frequency of the generator's electrical output is typically proportional to runner speed. In this case, the runner speed must be strictly controlled to meet electrical supply quality standards under varying flow conditions. Fortunately, the maturity of digital power electronics in recent decades has provided an economical answer to the challenge of governing small turbines: electronic ballast governors. In a typical simple installation, the electronic controller senses the frequency output of the generator directly and if the frequency is above the target it switches in more resistance from a bank of resistors to slow it down. If the frequency is too low, it reduces the generator's electrical load and the runner speeds up. This loop is performed several times per second so that the effects of discrete switching are not noticeable. This balancing of real user loads with ballast loads ensures that the electrical quality is high at all times, ready for a user load to be switched on. The requirement for the ballast resistors is to be able to absorb the turbine's full output, although there is no reason the ballast load cannot contribute to a useful purpose such as water or space heating. While the electrical design of a microhydro installation is of great importance to the operators, it is outside the scope of this chapter, which focuses on the hydraulic aspect of turbine design. As turbine designers, we will simply assume that the generator is a directly driven four-pole induction motor and as such requires a nominal 1500 rev min^{-1} to produce standard 50 Hz AC output.

Most typically, microhydro turbines are considered to function as a run-of-river system, that is, they have no appreciable storage capacity. Depending on the size of the parent stream, the turbine flow may be a significant portion of the total stream flow. While an ideal, simplified stream would have a constant flowrate all year-round, real streams exhibit significant variation with the seasons and from year to year. This fact is best illustrated by a flow duration curve (FDC), which is essentially a histogram of flowrate and shows what proportion of the total sample period the flowrate exceeded a particular value. **Figure 1** is an example of a typical New Zealand mountain stream, where the flood flows are many times the dry season flow.

Fully fixed turbines in particular require a careful study of at least 1 year's hydrological data, and are generally given a design flowrate of the minimum expected stream flow less any required flows reserved for other purposes, which we will call the available flowrate, Q_a, shown in **Figure 1**.

This does have the downside of not capturing available energy all of the time, but avoids the complications of automated governing and varying output, which is especially important in situations where the water turbine is the sole source of electricity for its users.

In the absence of adjustable turbine geometry, the turbine flowrate will be a unique function of turbine head. As the speed is fixed by the governor, and the site head will not likely vary appreciably, the scale of the turbine needs to be such that the design flowrate is passed under these precise conditions, otherwise the relatively narrow peak of efficiency of the fully fixed turbine will never be utilized. From the perspective of the turbine manufacturer, a fully fixed turbine greatly reduces mechanical complexity and simplifies construction. However, the same decision works against the designer, who must ensure that the geometry specified will result in a turbine operating at peak efficiency. Once built, there will be no way to recover from bad design.

In an effort to produce not a single efficient axial-flow turbine design, but a method for producing a uniformly efficient range of turbines, each step of design of each component must be left as flexible as possible. For the

'intake' this means being able to cope with different sites' requirements for a weir, penstock, or open flume without breaking compatibility with turbines unless absolutely necessary;

'volute' this means being scalable to different sizes and adaptable to different swirl requirements with high confidence in the resultant exit flow from a minimum number of template designs;

'runner' this means choosing leading edge blade angles by a method of zero torque from a valid upstream surface, choosing trailing edge angles for purely axial flow, and choosing a blade shape that lies on a plane, allowing the entire blade – traditionally one

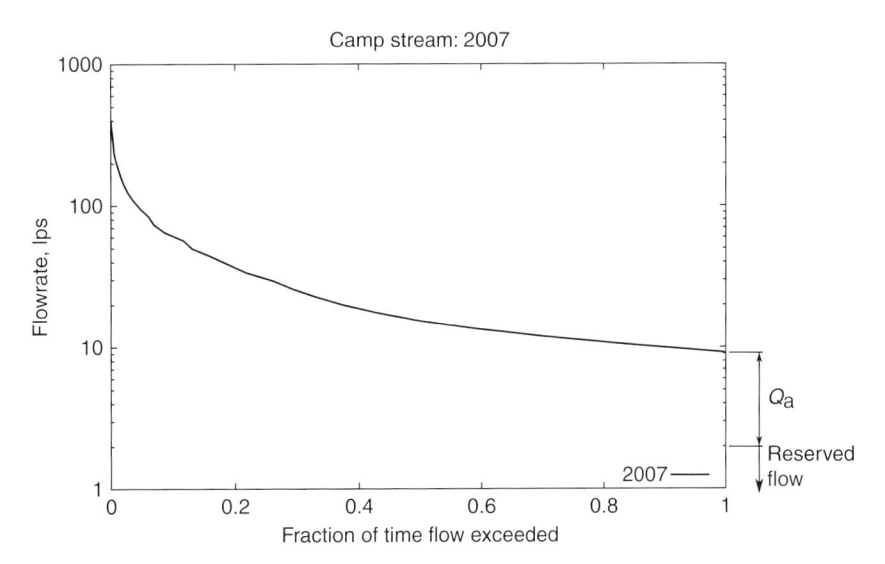

Figure 1 A typical mountain stream flow duration curve.

of the most geometrically complicated turbine components – to be cut from flat sheet steel and welded directly to a cylindrical hub; and

'draft tube' this means avoiding the pitfalls of overly optimistic diffusers, understanding how the runner exit flow impacts the draft tube's performance, and giving the draft tube appropriately weighted attention due to the amount of dynamic head it is responsible for recovering.

As are all geometrically related groups of turbine designs, axial-flow turbines are employed over a limited range of specific speeds where they perform more efficiently than competitive designs such as Pelton or Francis turbines. Turbines that are exactly similar geometrically and dynamically, but differ only in scale, will have the exact same specific speed. Specific speed will be defined later in detail, but it is helpful to think of it as a function of runner speed, turbine head, and turbine power which quantifies the general shape of the turbine with a single number. To illustrate the two ends of the spectrum, turbines typified by

'high head and small flow', such as Pelton wheels, have a low specific speed, and

'low head and large flow', such as unshrouded free-stream turbines, examples of which are marine current turbines and wind turbines, have a high specific speed.

6.16.1.2 Who Would Use Such a Turbine?

Users of axial-flow microhydro turbines are a varied bunch, including

- communities looking for a first or more dependable or economical source of electricity;
- farmers, tourist lodges, camping huts, ski lodges, or other remotely located off-grid users with suitable streams;
- developed nations attempting to reduce environmental impact and utilize existing renewable resources; and
- wastewater treatment and irrigation works where low-head water is an inevitable by-product of the primary activity.

The sites may be relatively modest in absolute power terms, but are more than sufficient to be valuable to an individual or small group, where a typical Western lifestyle requires an average of about 2 kW per person, although peaks are generally several times that. Their low-head and run-of-river nature means a good site will not require extensive damming or flooding. Next, let us take a look at some modern examples of this class of turbines.

6.16.1.3 Representative Designs

To illustrate the variety of designs under the heading of small, low-head hydro, some examples of recent commercial and academic work, including the authors', are listed below. Not all designs shown are strictly intended for microscale installations.

StraflowMatrix/HydroMatrix (546–700 kW, 5.5–30.5 m; VA Tech HYDRO GmbH)
Square, self-contained turbine units designed for a fixed flowrate are stackable in rectangular arrays to meet various flowrate requirements. The StraflowMatrix turbine unit is characterized by a unique generator configuration where the rotor coils are arranged around the periphery of the runner shroud, while the earlier HydroMatrix uses a more conventional bulb configuration [1]. Flow is from left to right.

StraflowMatrix/HydroMatrix
VA Tech HYDRO GmbH

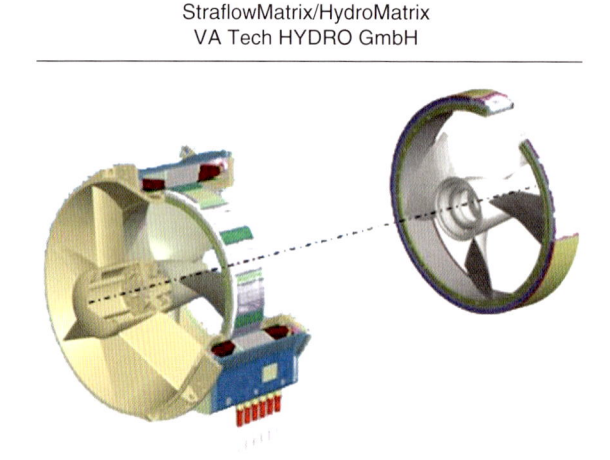

eKIDS (1–200 kW, 2–15 m; Toshiba)
Japanese manufacturer Toshiba markets a four-model microhydro lineup generating from 1 kW up. The intended audience is mainly city utilities and waste management facilities where residual head exists in an industrial environment and the recovered

energy can easily be injected back into the grid. The limited English product literature states the turbines' ability to operate off-grid [2]. Flow is from right to left.

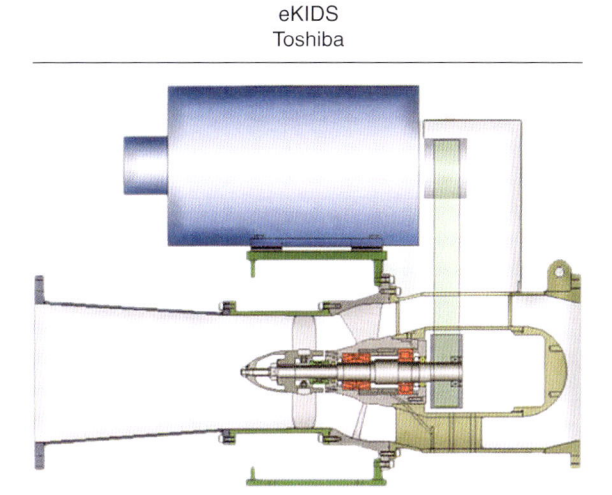

eKIDS
Toshiba

Siphon propeller turbine (10 kW, ~3 m; IT Power)
The UK firm IT Power has developed a belt-driven propeller turbine that operates fully above the headwater to reduce civil works complexity and protect the turbine from flooding damage, but requires priming by external means for start-up [3]. Flow is from a headwater to the right, through the conical draft tube to the tailwater channel on the left.

Siphon propeller turbine
IT Power

VLH turbine (100–500 kW, 1.4–2.8 m; MJ2 Technologies S.A.R.L.)
The Canadian-developed very low-head turbine is different from most other low-head designs. Rather than attempting a smaller size and higher speed turbine to achieve a more compact machine, the designers have opted for a large through-flow area and low peak efficiency speed which keep runner exit velocities to a minimum, which avoids the need for a draft tube. Water-to-wire efficiencies are about 79% for the smaller turbines. Testing took place at the University of Laval [4]. Flow is from upper right to lower left.

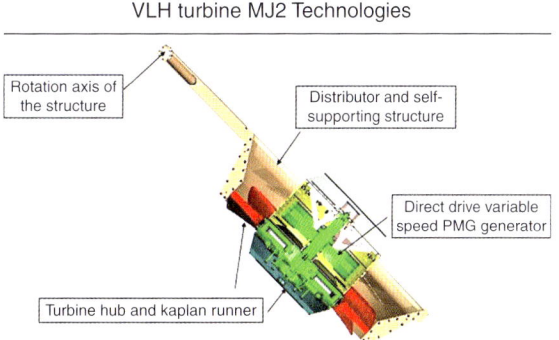

VLH turbine MJ2 Technologies

'Vaneless' turbine (– kW, – m; Swiderski Engineering/Rapid-Eau)
Turbine design has patent protection
Designed with a low fish mortality rate in mind, the omission of flow-spanning vanes evidenced in this design should also translate to reduced susceptibility to clogging by entrained leaves and other trash, which is an important requirement for microscale turbines. No test results are available. Testing took place at the University of Laval. Flow enters the volute through the rectangular opening and leaves for the runner to the left.

'Vaneless' turbine
Swiderski Engineering / Rapid-Eau

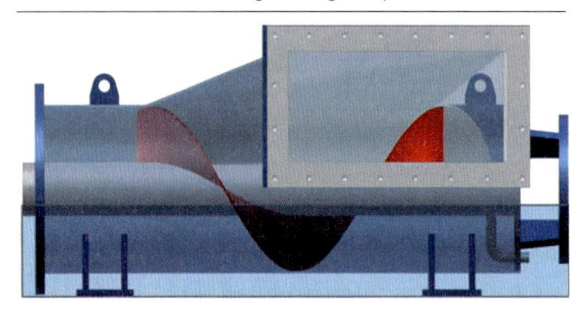

Sub-kilowatt (0.3–1 kW, 1.5–4 m; Exmork/Energy Systems & Design/Yueniao)
These four companies manufacture or supply a similar class of turbines: a one-piece turbine unit consisting of a runner and guide vanes connected to a directly driven generator by a rigid frame, which is meant to be directly fitted to a circular orifice in the bottom of an open race. They are generally unregulated and produce either standard AC directly or DC power for charging battery banks. Exmork (http://www.exmork.com) is a Chinese manufacturer whose products are distributed in Canada by PowerPal and in Europe by Kleinstwasserkraft Klopp. PowerPal (http://www.powerpal.com) is a Canadian company that manufactures in Vietnam and imports. Energy Systems & Design (http://www.microhydropower.com) is a Canadian company that produces a 1 kW turbine similar to PowerPal's. Yueniao (http://www.yueniao.com/) is a Chinese manufacturer that produces a similar 1 kW turbine. Flow is top to bottom in the picture shown.

Sub-kilowatt
Exmork/Energy Systems and
Design/Yueniao

PAT (– kW, – m; various)
Using pumps as turbines (PAT, or BUTU in Spanish) provides an alternative to purpose-built turbines where pumps are available at competitive prices, and can cover essentially the same head and flow range as the pumps themselves. High specific speed axial-flow pumps, which also operate at relatively high specific speeds as turbines, are more expensive per kilowatt than their lower specific speed centrifugal counterparts. If the efficiency handicap of operating in reverse is overcome, the simplicity of a single unit containing turbine and generator is certainly attractive [5, 6]. Flow enters the volute by the tangential pump outlet and leaves through the pump eye.

PAT
various

Archamedean screw (5.5–63 kW, 1.6–8 m; Western Renewable Energy)
The UK firm Western Renewable Energy has developed the Archamedean screw as a relatively novel method of making electrical energy from a low-head source. It is comparable to traditional waterwheels in that it requires considerable civil works and a significant speed increase in order to generate 50 Hz AC power, which came about when shaft output was directly used for mechanical work, which could utilize the slow rotational speeds directly. Flow is top to bottom along the channel by parcels which are separated by the helical blade of the runner.

Archamedean screw
Western Renewable Energy

Giddens propeller turbines (1.4–4.1 kW, 3.4–10.7 m; University of Canterbury)
The Mechanical Engineering Department of University of Canterbury has developed a range of radial-flow, mixed-flow, and, more recently, axial-flow propeller turbines called Giddens propeller turbines, named after late civil engineering professor Peter Giddens, who spearheaded the project, designed for forgiving construction and reliable use in remote locations. As with all typical tangential inlet turbines, the flow enters tangentially and swirls around while gaining a radial component before exiting axially.

Giddens propeller turbines
University of Canterbury

6.16.1.4 Energy Alternatives and Unconventional Economics

Users may be characterized by their access to alternative energy options, which may be superior to microhydro in terms of cost or available output. The alternatives for those considering microhydro as a source of electricity are a mix of renewable and nonrenewable sources, and may be a subset of grid connection, diesel or gasoline generators, solar photovoltaic (PV) cells, and wind turbines. The energy users may see

- microhydro as one of several energy supply options, or
- microhydro as the only energy supply option.

The distinction becomes particularly important in off-grid installations, where the retail cost of electricity is not available as a benchmark for project cost effectiveness. While options such as grid connection or generators have a well-defined capital and ongoing cost, microhydro sites are less forgiving in some ways when it comes to developing a site to the needs of the users. The total output is hard-limited by the site, not by available cash, and the capital and labor external cash cost of developing a microhydro site to the point that it is generating useful power will vary significantly depending on the resourcefulness of the developers and the local availability of materials, which cannot necessarily be stated for alternative sources. This phenomenon is what is commonly referred to as unconventional economics.

6.16.1.5 What Is Specific Speed?

As mentioned earlier, when comparing different turbine designs, turbine-specific speed is a useful parameter for quantifying families of turbines, that is, turbines of similar shape but different size. The rigorous nondimensional form of specific speed is shown in eqn [1]:

$$n_S = \frac{\omega \sqrt{P}}{\sqrt{\rho}(gH)^{5/4}} \quad [1]$$

However, as a parameter used extensively across the breadth of the topic of turbomachinery, it appears in various convenient forms. The form adopted by the authors is shown in eqn [2]:

$$N_S = \frac{N\sqrt{P_r}}{H^{5/4}} \quad [2]$$

where P_r denotes the shaft power out of the runner, before the transmission.

Specific speed has been shown to be the best dimensionless parameter for characterizing the general shape of a hydroturbine from only rotational speed, head, and flow, and therefore its suitability to a given flow regime.

From an efficiency standpoint, the efficiency of a turbomachine peaks at some finite specific speed and then decreases monotonically from there as a consequence of handling larger flows, meaning a larger proportion of the total inlet head is dynamic head to which hydraulic losses are proportional. The result is that the practical peak efficiency attainable for a given scale of output is lower at very high specific speeds [7].

From a compactness standpoint, efficient high specific speed machines are physically larger per unit power and incur higher construction and transport costs in general than their higher head cousins. This is a direct result of the site's lower head, or power density, its unit of power per unit weight of water, from the formulation of potential energy for a parcel of water,

$$H_g = \frac{d}{dt}\left(\frac{dE_p}{g\rho\, dV}\right) = \frac{P}{g\rho Q} \qquad [3]$$

where E_p is the potential energy and V is the volume.

6.16.2 Component-Level Design Methods

A turbine is simply a combination of individual components acting in concert; therefore, a design methodology can be presented for each individual component. First, heavily used definitions are introduced. Then, Sections 6.16.2.1–6.16.2.4 present the authors' methods for designing or specifying each of the turbine's main hydraulic components: intake and supply, volute, runner, and draft tube. Although not treated here, the transmission and generator efficiencies can significantly reduce the final output power if not handled properly. Base SI units are used unless noted.

The main components of a penstock-based scheme are introduced in **Figure 2**, along with the numbered stations which define the boundaries of each component for energy accounting purposes.

Several terms will be used more than others to refer to heads, and need defining here. Site gross head is the vertical distance between the headwater and the tailwater

$$H_g \equiv H_0 - H_5 \qquad [4]$$

net head is the total head across the turbine

$$H_n \equiv H_2 - H_5 \qquad [5]$$

and runner head is the total head across the runner

$$H_r \equiv H_3 - H_4 = \eta_i \eta_p \eta_v H_g \qquad [6]$$

Unless mentioned, stated heads are total, that is, static plus dynamic.

A head loss is denoted by an h, with a subscript indicating either the particular component or the stations between which the loss is incurred. For example, h_p represents the head loss in the penstock, and h_{03} represents all total head losses upstream of the runner.

The concept of efficiency is integral to the development of a system such as a water turbine. Its definition varies with the component or system it describes, but it is generally a ratio of work out to work in. Components that have no work output, for example, the penstock, volute, and draft tube, do not possess an efficiency in this strict sense but are given an efficiency symbol to show their effect on net head. These components' 'efficiencies' are defined as the total head loss they incur divided by the site gross head, except for the draft tube, whose exit dynamic head is also a loss.

'Intake' efficiency is defined as 1 minus the head loss incurred by the intake, including any grates or trash racks, divided by the site gross head, and is neglected in this treatment.

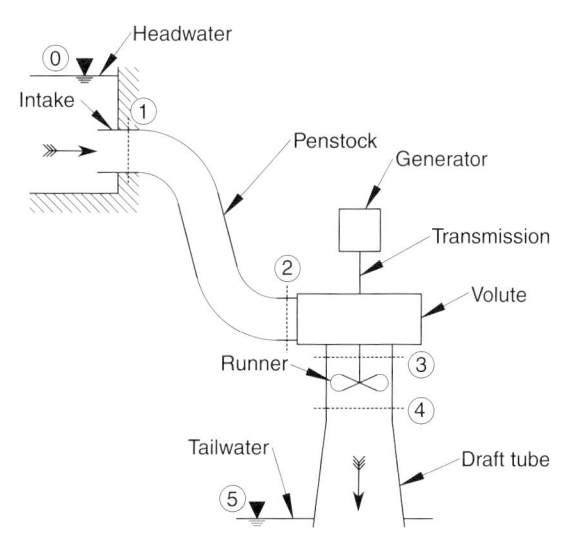

Figure 2 Penstock-supplied microhydro scheme components and stations.

'Penstock' efficiency is defined as 1 minus the head loss incurred in the penstock divided by the site gross head. Minor losses are excluded.

$$\eta_p \equiv 1 - \frac{h_p}{H_g} \quad [7]$$

'Volute' efficiency is defined as 1 minus the head loss incurred in the volute divided by the site gross head.

$$\eta_v \equiv 1 - \frac{C_{p_v p_{d_i}}/\rho g}{H_g} \quad [8]$$

'Runner' efficiency is defined as the work output at the runner shaft divided by available hydraulic power. It does not include the effect of bearing or seal losses.

$$\eta_r \equiv \frac{P_r}{\rho g Q H_n} \quad [9]$$

'Draft tube' efficiency is defined as 1 minus the head loss incurred in the volute divided by the site gross head.

$$\eta_d \equiv 1 - \frac{K_{dp_{di}}/\rho g}{H_g} \quad [10]$$

'Transmission' or mechanical efficiency is defined as the work done on the generator divided by the work received from the runner.

$$\eta_m \equiv \frac{P_s}{P_r} \quad [11]$$

'Generator' efficiency is defined as the electrical power at the generator terminals divided by the mechanical power delivered by the transmission.

$$\eta_g \equiv \frac{P_e}{P_s} \quad [12]$$

Runner head can now be written as site head reduced by the plumbing component efficiencies

$$H_r = \eta_i \eta_p \eta_v \eta_d H_g \quad [13]$$

Mechanical power extracted by the runner can be written as its hydraulic power input reduced by the runner efficiency

$$P_r = \eta_r \rho g Q H_r \quad [14]$$

And finally, the available electrical power can be written as the runner mechanical output reduced by the transmission and generator efficiencies

$$P_e = \eta_m \eta_g P_r \quad [15]$$

Turbine efficiency is an oft-mentioned term that bundles the efficiency terms of the volute, runner, and draft tube.

$$\eta_t = \eta_v \eta_r \eta_d \quad [16]$$

A catch-all efficiency term may be defined as the product of all component efficiency terms,

$$\eta_T = \eta_i \eta_p \eta_v \eta_r \eta_d \eta_m \eta_g \quad [17]$$

where the site's electrical output is defined by

$$P_e = \eta_T g Q H_g \quad [18]$$

6.16.2.1 Water Supply

If beginning upstream of the turbine, the first component encountered is the intake and supply. The intake and supply are the hydraulic and structural interface between the turbine and its source of water. Especially for low-head turbines, this construction component of a microhydro project has the potential to determine the project's economic sensibility. It is for this reason that the hydrology and site should be carefully examined, and the intake chosen to suit.

Let us begin by stating several important definitions and assumptions integral to the intake and supply design method presented here. As the turbines of concern are fully fixed and will operate at single condition, the variability of a real site and hydrology need simplifying to this level of detail. The site gross head, H_g, may be defined as simply the vertical distance between headwater and tailwater, since stream depth is typically small by comparison and will not vary much with flowrate. Available volumetric flowrate, Q_a, may be taken as constant due to the fixed-output design criterion, and is defined as the minimum expected stream flow minus any minimum-flow requirements, ecological or otherwise, imposed by regulation or common sense. In this case, the minimum stream flow can be taken from the full-year hydrology record if the turbine will work year-round, or some subset of the full year if the turbine will operate only seasonally. If the turbine will be relied upon only during the wet season, this will result in a higher design flowrate. The consequence of this simplification is clearly that the turbine, even if operating efficiently, will not be able to 'follow the peaks' of flowrate, and hence power, throughout the seasons. The upshot is that the users are able to rely on a constant output. After a site is selected for penstock development, the penstock will have a known length, L. The penstock material cost, C, is assumed to be proportional to material volume. The head loss incurred in the penstock, h, is a variable to consider while optimizing penstock flowrate, Q, and penstock inner diameter, d. The average slope of the penstock, S, is estimated from H_g/L. For the purposes of optimizing the intake and supply, the turbine is idealized by a fixed-efficiency machine, which allows the product QH to represent the turbine power being optimized.

At the risk of oversimplifying matters, sites where axial-flow microhydro turbines are applicable will fall into one of two categories depending on the most convenient and economical method for transmitting the stream's flow and head to the turbine:

'lower head sites' may make use of an open-flume arrangement where the turbine is mounted near headwater level and supplied directly by an open channel, exhausting to the tailwater through a draft tube which is entirely in suction, and

'higher head sites' may make use of a weir-and-penstock arrangement where a low dam forms a small reservoir upstream to ensure continuity of flow, while the backwater is allowed to escape through a pipe running down to the turbine, which sits near tailwater level, so that the draft tube is less critical.

These two categories are purposefully vague, in that intake selection is very site-specific, meaning that H_g and S give an incomplete description of the site, and idiosyncrasies of the site may become important. Having said that, when designing the intake, the way in which the site head is developed is of fundamental importance. The slope between intake and tailwater level is the key parameter besides head that determines the appropriate intake type. When $S < 0.25$, the water supply cost as a fraction of the total project cost will probably be the main factor of the project's return on investment. For either type of supply, open flume or penstock, for a given head a gradually sloping section will be more expensive to develop than a steeply sloping one due to penstock length. For example, if a useful head is developed over only a relatively flat length of streambed, the length of penstock, or alternatively the length of channel that must be extended along a contour, will be longer than if the entire head is developed at a compact site, the ideal case being a conveniently located waterfall. This being said, open channels are generally more economical for conveying a large flowrate with a small total head loss for long stretches, while a penstock allows large pressure heads to be transmitted down steep slopes.

In microhydro installations that are not grid-connected, the retail cost of electricity is of little importance. In this case, where energy is needed and not optional, projects may be further divided into two categories of relative stream size, depending on whether

'available stream power is more than desired' and the turbine may be sized to provide sufficient power, bypassing much of the total flow, that is, $Q < Q_a$, or

'available stream power is less than required' and the turbine design must be carefully optimized to produce the most energy possible, that is, $Q = Q_a$.

This dichotomy highlights whether the project developers are in the luxurious position of having surplus stream power at their disposal. A site whose total power is on the verge of being insufficient may increase its turbine head slightly by further investment in larger penstocks, although this will be less cost effective compared to the case of a surplus site where the entire turbine could be scaled up slightly to accommodate more flow. While a site offers a fixed potential, developers determine how efficiently and economically power is extracted primarily by considering their cost constraints. Depending on the site, the intake and supply may be more or less important in the total project cost, but regardless of cost, the hydraulic design of either the supply penstock or open channel will be crucial for the hydraulic performance of the turbine downstream.

6.16.2.1.1 Penstock

Assuming that a weir and penstock supply makes sense for a given site (H_g, Q_a, and S), d and h_p/H_g are the key controlling parameters of penstock cost and available hydraulic power. The following analysis is an adaptation of Alexander and Giddens' work [8] on penstock optimization. Importantly, it can be shown that for a given d, the product QH is maximized when h_p/H_g is 1/3. From the definition of available turbine head shown in eqn [19]

$$H = H_g - h_p \qquad [19]$$

and penstock head loss due to wall friction shown in eqn [20]

$$h_p = K_1 L Q^2 \qquad [20]$$

where K_1 characterizes the pipe and is defined in eqn [21]

$$K_1 = \frac{f}{2g\left(\frac{\pi}{4}\right)^2 d^5} \qquad [21]$$

turbine power shown in eqn [22]

$$P = K_2 Q H_n \qquad [22]$$

may be rewritten as a function of the penstock losses as shown in eqn [23]

$$P = K_2 Q (H_g - K_1 L Q^2) \qquad [23]$$

where K_2, defined in eqn [24], characterizes proportionality of turbine power to the product QH_n and turbine efficiency, η_t, is assumed constant

$$K_2 = \eta_t \rho g \qquad [24]$$

Differentiating the polynomial in eqn [22] with respect to Q gives the second-order polynomial equation [25]

$$\frac{dP}{dQ} = K_2 H_g - 3 K_1 K_2 L Q^2 \qquad [25]$$

into which eqn [20] rearranged to give K_1 may be substituted back, giving eqn [26]

$$\frac{dP}{dQ} = K_2 (H_g - 3 h_p) \qquad [26]$$

Setting this expression equal to zero reveals that, by assuming a turbine efficiency independent of flow conditions, maximum turbine power for a given diameter penstock simply occurs when Q is such that $h_p/H_g = 1/3$. This is a somewhat counterintuitive message that it is optimum to lose a third of the available head considering that commercial hydro schemes are designed to have penstock losses on the order of 5%. This result is useful to select the most cost-effective penstock. It may not be the developer's choice, but it simply shows that it will cost more to use more of the site's head.

Of more general use is to consider the site fixed as before, but choose d in the interest of controlling cost or maximizing power. A methodology will be presented that accomplishes both, but a key point to make is that if the available stream power is near the turbine power being developed, the method should handle this discontinuity realistically. To begin with, and before the analysis gets too bogged down in mathematical details, it is helpful to develop a computer tool that allows the developer to calculate results over a range of diameters for varying sets of inputs. As far as the authors are aware, there is no information in the literature concerning precise optimization of hydro scheme intake and supply, the likely reason being that as has been pointed out, even on a large scale, the natural character of sites and hydrology escape neat characterization by one or two parameters. Even less regular are the characteristics of the developers' material supply and possible economies achievable locally and their labor force.

Building on the power maximization developed over eqns [19]–[26], let us highlight the key points of penstock selection. In an effort to maintain simplicity and transparency, only losses due to wall friction are considered and any turbulent development length near the inlet is assumed to be small, both of which result in optimistic estimates of power and efficiency. The penstock should be kept free of any unnecessary bends or lossy fittings to avoid additional losses. For estimation purposes, penstock cost is assumed to be proportional to material volume and wall thickness is assumed to be proportional to inner diameter, such that penstock material cost can be defined as in eqn [27]

$$C = K_3 L d^2 \qquad [27]$$

where K_3, which has the unit of cost per volume, is defined as in eqn [28]

$$K_3 = C_V \frac{\pi}{4} \left(\left(\frac{d_o}{d}\right)^2 - 1 \right) \qquad [28]$$

with C_V the cost per unit volume of penstock material and d_o the penstock outer diameter.

The general method for each value of d is as follows:

1. Let H_g and Q_a define the site.
2. Define a target value for h_p/H_g of 1/3 from the above analysis.
3. Calculate Q required to achieve the target penstock loss using Moody diagram or equivalent.
4. However, if $Q > Q_a$, set $Q = Q_a$ and calculate resultant h_p and update H_n.
5. Regardless, calculate turbine power from eqn [23].
6. Calculate penstock cost from eqn [27].
7. Repeat for a range of d.

Point 4 refers to the discontinuity mentioned earlier, the essence being that if the penstock is made large enough that the full available stream flow Q_a does not incur the target head loss, then Q is limited to Q_a and the head loss will simply be lower than the optimum value. The implications on cost of reducing the value of h_p/H_g will be apparent in the results.

To illustrate the use of the results in decision making, the sample calculation from Reference 8, has been reworked using this method. The important parameters are $H_g = 12$ m, $Q_a = 0.090$ m^3 s^{-1}, $L = 72$ m, $K_3 = 1764$ \$ m^{-3}, and surface roughness $\varepsilon = 0.1$ mm. Furthermore, to reflect the fact that only a range of discrete diameters are available, the commonly available diameters given in Reference 8 are represented by circles along a continuous line in each of the figures that follow.

Figure 3 illustrates the point that when d is large enough that Q is limited to Q_a while attempting to maintain optimum h_p/H_g, at still larger diameters it will be less than the target value. Let us call the diameter where $Q = Q_a$ and h_p/H_g is the target value as the critical diameter, d'. In **Figures 3–8**, $d' = 0.184$ m.

Figure 4 shows the same information, only directly in terms of available turbine head. At diameters larger than d', the reduced head loss means that $H_n \to H_g$.

Figure 5 shows how Q increases as the square of diameter to maintain optimum h_p/H_g up until $d = d'$ at which point Q is limited to Q_a for any larger diameter.

Maximizing the hydraulic power delivered to the turbine downstream is of critical importance to the project developer. For a real penstock, the power delivered, which can be extracted by a turbine, is proportional to QH_n, but we have already shown that the losses that detract from H_n are proportional to Q^2. Balancing the increased turbine power, which is $\propto Q$, with penstock loss, which is $\propto Q^2$, is the essence of penstock optimization. **Figure 6** shows how the hydraulic power delivered grows rapidly with d up to d', at which point growth slows as Q is limited to Q_a and $\lim_{d \to \infty} H_n = H_g$. **Figure 6** also shows that while eqns [19]–[26] do indeed maximize power for given diameter assuming that $Q_a \gg Q$, in the case where flowrate is limited and a range of penstock diameters are available, a larger penstock will inevitably produce less losses and a larger potential turbine power.

Just as important as power, however, is the matter of penstock cost. The cost-per-volume relationship is sufficient for showing trends, and **Figure 7** shows the result of eqn [27] evaluated over the range of d shown, which shows that cost is $\propto d^2$.

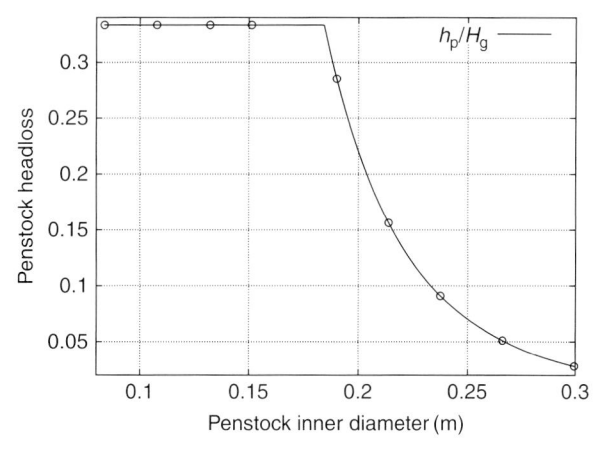

Figure 3 Head loss in penstock due to wall friction.

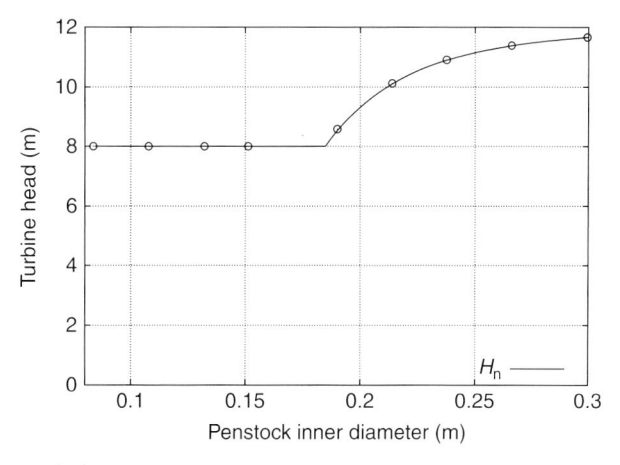

Figure 4 Available turbine head from penstock.

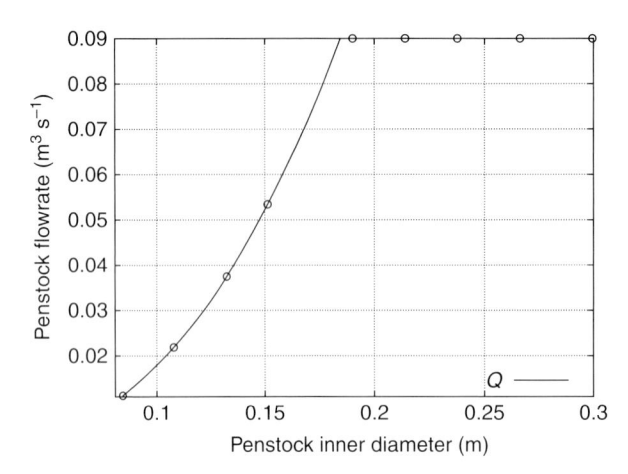

Figure 5 Penstock volumetric flowrate.

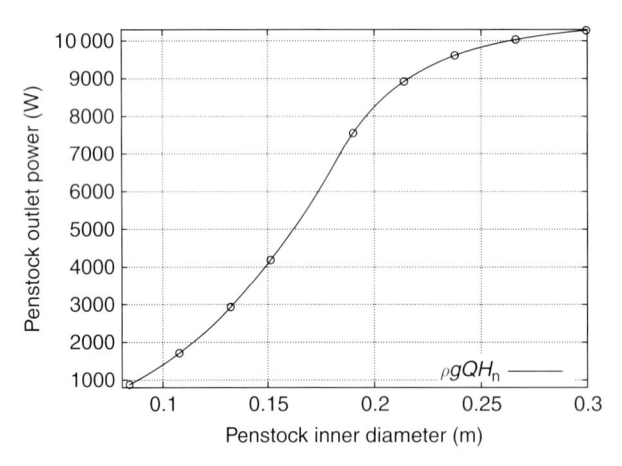

Figure 6 Penstock outlet hydraulic power.

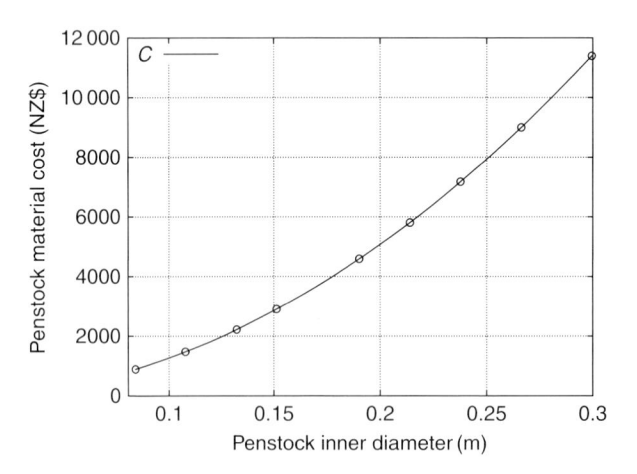

Figure 7 Penstock material cost.

Possibly more important than total cost is some indication of cost effectiveness: what benefit does one purchase with a larger penstock? Since power only asymptotically approaches a limit as $d \to \infty$, but cost climbs $\propto d^2$, delivered hydraulic power per penstock cost is a clear variable to be maximized in order to make a reasoned economic decision. **Figure 8** shows the clear result that d', by the method presented, is in fact the most economical choice of penstock diameter. The reader will no doubt recall the importance of developer constraints, however, and recognize that if the power developed using penstock of diameter d' is

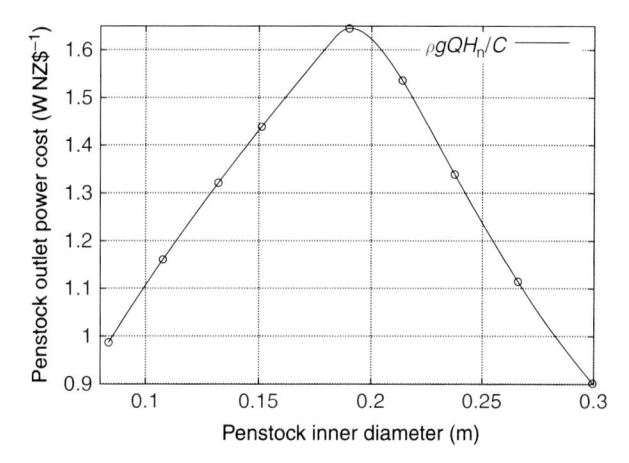

Figure 8 Penstock hydraulic power delivered per material cost.

insufficient, and 1/3 more power would make the difference, larger penstocks may be utilized to further reduce penstock head loss, although it will be expensive.

6.16.2.1.2 Open flume

In cases where an open flume is appropriate, this type of intake can be made much simpler and more direct than an equivalent penstock installation. Particularly, where a reasonable length of draft tube is approximately the same as the site head, the turbine may be oriented vertically with the runner slightly below headwater level in an open flume, with an aligned draft tube directly underneath diffusing the turbine's exit flow down to tailwater level. As the authors are currently in the process of expanding their turbine range to lower head, higher specific speed models, open flumes will likely play a role in their efficient development; however, we have little personal experience at this time. The reader is referred to the section on open-channel design in Reference 9 for the channel cross-section and slope requirements, and Reference 10 for construction and general open-flume site layout. Similarly, the equivalent of the volute in an open-flume design takes a completely different form than penstock-supplied turbines, being an extension of open-channel design, and will be left out from further discussion.

6.16.2.2 Volute

Continuing downstream from the intake and supply, and assuming the supply type chosen is a penstock, the next component encountered is the volute, which is the hydraulic interface between the penstock and runner. While axial-flow turbines need not have a volute (see Section 6.16.1.3 for examples), the designs presented here do have volutes. The task of the volute is to efficiently transform the inlet pipe flow into circumferentially uniform swirling flow with a velocity distribution matched to the runner's leading edge angles. Whereas the penstock design method was a balance of economic and hydraulic optimization, volute design is a purely hydraulic matter.

Let us again start the section by stating several important definitions and assumptions integral to the volute design method presented here. Where a turbine is supplied by a single penstock, as here, the connection to the penstock is achieved with a single tangential inlet, due to the penstock's outermost face being tangential to the volute body so that the outermost streamline travels smoothly into the volute body. Since an axial-flow runner has an annular ring inlet duct section, the volute takes the form of a closed vessel connecting the circular inlet section with the annular outlet section. The axes of the inlet and outlet are commonly perpendicular, so that if the runner axis is vertical, the penstock axis is horizontal. **Figure 9** shows the $N_S 176$ turbine in Reference 11 with key dimensions labeled, which conforms to this description.

The width of the passage between the volute wall and the runner outlet pipe is defined in eqn [29].

$$W = \frac{1}{2}(\varnothing C - \varnothing D') \qquad [29]$$

Three orthogonal velocities are useful when describing the flow within a volute. Tangential velocity, V_t, is the velocity about the volute exit centerline, and is positive in the direction of penstock flow. Radial velocity, V_r, is the velocity perpendicular to the volute exit centerline, and is positive inward. Axial velocity, V_a, is the velocity parallel to the volute exit centerline, and is positive in the direction of volute exit flow. Where convenient, X, Y, and Z directions may be referred to. The X-axis is parallel to the penstock axis and positive in the direction of flow. The Y-axis is parallel to the shortest line connecting the penstock axis and the volute exit axis and positive in that direction. The Z-axis is parallel to the volute exit axis and positive opposite the direction of exit flow. Azimuth angle θ is defined as positive about the Z-axis in the direction of flow.

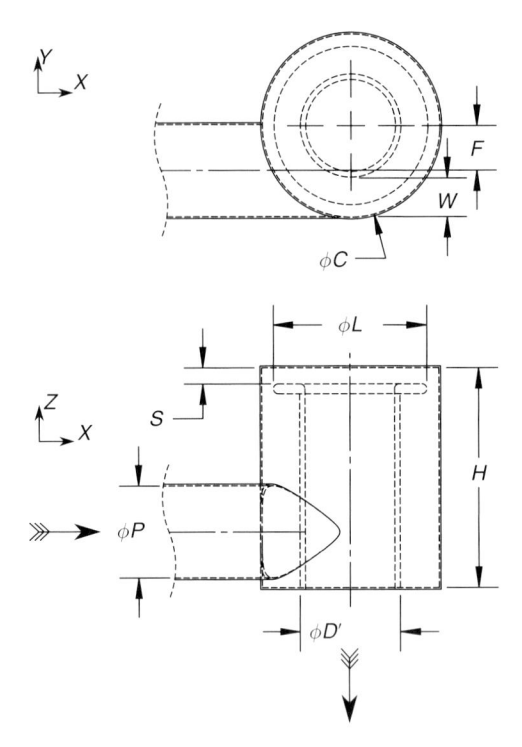

Figure 9 Single tangential inlet volute.

6.16.2.2.1 Runners need swirl

In the event that guide vanes are omitted to prevent them from being clogging with entrained debris, the volute must be designed to provide the required flow. Swirl, or more specifically, rV_t, is what the runner changes in the passing flow in order to extract work. This is succinctly presented in the Euler turbomachinery equation shown in eqn [30], which states that the runner torque is proportional to the flowrate and to the change in the value of rV_t.

$$T = \dot{m}\,\Delta(rV_t) \qquad [30]$$

The relationship between torque and turbine head is revealed if eqn [30] is multiplied by runner angular velocity, ω, resulting in several representations of runner power, shown in eqn [31],

$$P = \eta_t g \rho Q H_r = \omega T = \omega \rho Q ((rV_t)_3 - (rV_t)_4) \qquad [31]$$

where η_r is the runner efficiency, and then divided through by $\dot{m}$, giving eqn [32],

$$\eta_r H_r = \frac{\omega}{g}\left((rV_t)_3 - (rV_t)_4\right) \qquad [32]$$

which shows that both torque and head alike are proportional to the change in rV_t across the runner.

With these relations in mind, consider a 'rubber turbine' of fixed P and ω, or N, where the up and down arrows in the following equations represent an increase or decrease in the adjacent quantity, and the initial perturbation is to slightly increase the turbine's flowrate, while all subsequent adjustments are those required to fit the stated constraints. The list is meant to be read in order from top to bottom, and shows why, all other things being equal, high specific speed turbines need less swirl.

1. $P_r = \eta_t \rho g Q^\uparrow H_r{\downarrow}$, Q is increased and the change in H_r is inversely proportional.
2. $N_{S\downarrow} = \dfrac{N\sqrt{\eta_t \rho g Q^\uparrow}}{H_r^{3/4}\downarrow}$, N_S increases.
3. $P = T\omega$, P and ω are constants, so T is unchanged.
4. $T = \rho Q^\uparrow \Delta(rV_t)\downarrow$, $\Delta(rV_t)$ changes inversely proportional to Q, as shown in eqns [30]–[32].

While the Euler equation is suitable for radial-flow turbines where the runner leading and trailing edge stations are completely described by a single r and V_t, axial-flow runners have velocity distributions that vary along the blade span, and the integral of the product of this quantity and mass flux must be integrated over a surface that passes the turbine's full flow. This integral will be introduced in eqn [35]. The volute's task is to convert static pressure to a uniform swirl velocity V_t while passing the design flowrate, such that the flow angle relative to the runner is very close to the blade angle spinning at its design speed, which allows smooth and efficient operation of the runner.

It is generally assumed as a reasonable goal that the swirl leaving the runner should be minimized, as it represents lost energy, which leads us to the conclusion that runner torque will equal $\oint \rho r V_t \, dQ$ evaluated across the runner leading edge swept surface. The common practice in volute design is based on a simple principle: rV_t is constant throughout the volute. For an incompressible fluid, this leads to eqn [33],

$$(rV_t)_3 = (rV_t)_2 \quad [33]$$

where the right-hand side represents conditions at the runner leading edge and the left-hand side represents the penstock conditions just prior to entering the volute. It is a simple exercise to solve for tangential velocity along the leading edge.

The authors' research has shown that this does not necessarily hold for simplified single tangential inlet volutes designed for high specific speed turbines [12]. The single tangential inlet design, with inlet offset decreased and diameter increased to attain the weak exit swirl required by high specific speed turbines, deviates from such a shape, and the volute's internal geometry is no longer torque-free.

6.16.2.2.2 Volute characterization

Two key parameters that largely describe a volute's performance are the total head loss coefficient, C_p, and the angular momentum flux factor, F_{AM}. Because of the turning that is required to happen within the volute, losses may be expected, which not unlike the penstock head loss, h, reduce the head available to the runner for useful work, H. C_p is the total head loss incurred by the volute divided by inlet dynamic head, and is defined in eqn [34].

$$C_p = \overline{pt}_2 - \overline{pt}_3 \overline{pd}_2 \quad [34]$$

where the overbars indicate area-weighted averages of the variables underneath. It must be determined experimentally, but should be expected to be only a weak function of Reynolds number at the high Reynolds numbers typical of hydro turbines so that it may be applied to geometrically similar models.

The angular momentum flux factor, F_{AM}, is defined in eqn [35], where the subscripts indicate that the surface integrals are to be taken over representative inlet and outlet surfaces of the volute.

$$F_{AM} = \frac{\oint_o \rho r V_t V_n \, dA}{\oint_i \rho r V_t V_n \, dA} \quad [35]$$

The parameter F_{AM} is likewise only weakly dependent on Reynolds number when well beyond the turbulent transition point.

6.16.2.2.3 Swirl predictability

After questioning the swirl behavior of two model volutes, the $N_S 176$ and $N_S 544$ models in Reference 11, measurements were performed, which confirmed that not all volutes can be expected to obey the usual zero-torque, or free-vortex, assumption, which proposes that the volute exerts no torque on the fluid passing through it. The parameter F_{AM} was the logical ratio to quantify a volute's relative deviation from zero torque. In fact, measurements to be published on this topic [12] revealed that a volute similar to that shown in **Figure 9** possessed an F_{AM} of 1.78, meaning that the value of rV_t leaving the volute and entering the runner is 1.78 times that which enters the volute from the penstock. The volute obviously does exert a torque on the flow in this case. Although not a direct loss, a volute that provides an exit flow different from that which the runner has been designed for will reduce the runner performance due to larger angles of attack at the runner leading edge and causes it to run at a different point of its efficiency curve.

Upon closer examination of the $N_S 176$ volute, it is quite easy to imagine the streamlines leaving the penstock and being turned by the internal geometry of the volute. Furthermore, postprocessing of numerical results on that geometry, shown in **Figure 10**, clarifies the matter, showing the variation of rV_t along streamlines integrated over a cut plane through the middle of the penstock.

From **Figure 10**, it seems quite clear that for certain designs, the internal geometry of the volute may exert a significant torque on the flow. A good design will not present this uncertainty of flow behavior, but if the design is in question, a simple potential streamline sketch on a cut surface through the inlet midplane will approximate how the flow will adjust once inside the volute. In the case of the $N_S 176$ volute, the two main features of the inlet geometry of the volute seem to be

1. the relatively large tongue angle and
2. the fact that the streamtube is turned from the penstock offset $\overline{r}_i$ to the larger offset $\overline{r}_i'$, while the passage width is reduced.

These factors combined will no doubt increase the value of rV_t from the value slightly upstream in the penstock, as predicted in **Figure 10**, which shows the value of rV_t along streamlines emanating from the penstock. The streamlines are created from numerical results [12] confirming the measured value of F_{AM}.

This turning is not necessarily present in all volutes; in fact, commercial turbine volutes have traditionally been designed with the assumption of zero torque in the absence of measurements. An example is given by Malak et al. [13], where velocity measurements were made throughout the volute interior, and the value of F_{AM} can be calculated as approximately unity. The geometry also conforms very closely to the classical snail-shell volute shape. This shape is not merely an aesthetic coincidence, but rather the consequence of shaping the volute walls to fall along theoretical streamlines of potential

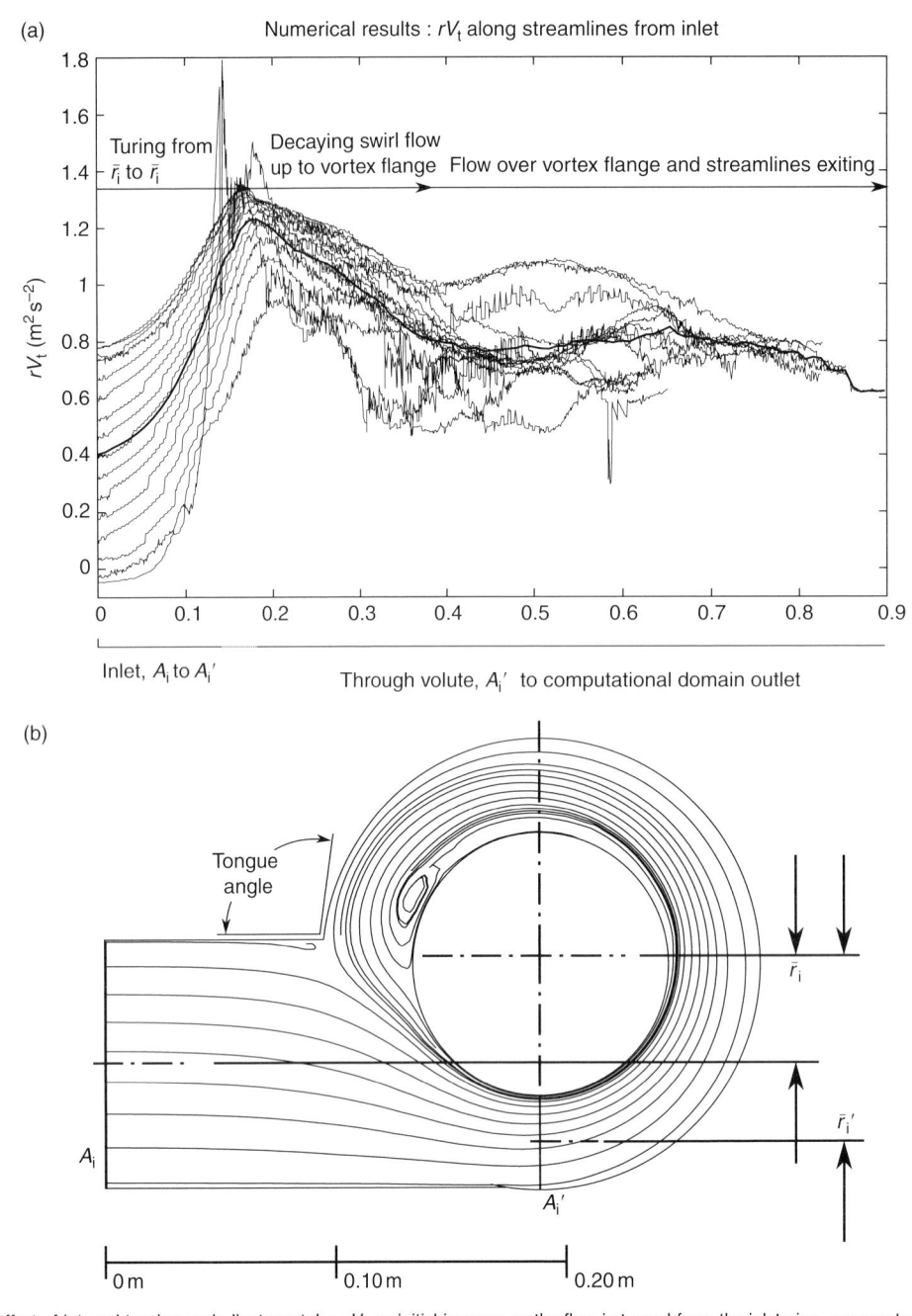

Figure 10 The effect of internal turning on bulk streamtube rV_t: an initial increase as the flow is turned from the inlet pipe, represented by its mean radial offset, $\bar{r}_i$, to the space between the main volute wall and the outlet pipe, represented by a more meaningful streamtube offset inside the volute, $\bar{r}_i'$, which is coincident with the peak of mean rV_t in (a) near 0.2 m along the streamlines, followed by a steady decay due to the effect of wall shear stress. (a) rV_t variations along streamlines emanating from cross-inlet rake, with their mean shown a heavier line. (b) In-plane streamlines in a cut plane passing through the inlet at mid-Z.

vortex-plux-sink flow. Bhinder [14] presents a volute design method intended to produce circumferentially uniform flow that is based on this assumption.

6.16.2.2.4 Limits of the tangential inlet volute

After revealing that some volute designs in use behave quite differently from what simple analytical methods would predict, it is useful to take stock of these designs, to see if they are worth their lack of predictability, and if not, what alternatives are present. Although eqn [33] presents a simplistic and flawed approach for volutes in general, it may be updated in light of the parameter F_{AM}, as shown in eqn [36], although the convenience of the analytical method is somewhat negated by the need for an experimentally determined value.

$$(rV_t)_3 = F_{AM}(rV_t)_2 \quad [36]$$

As a commonsense look at the inlet geometry of the simple volute in **Figure 10(a)** would lead the observer to believe, and research [12] has indicated, the value of F_{AM} will tend to be greater than unity for the reasons listed in Section 6.16.2.2.3. To prevent the flow from being torqued upon entering the volute, all streamlines entering the volute should be able to proceed straight ahead and begin swirling. Referring to **Figure 9**, $\varnothing P$ should not be more than W, or **Figure 10(b)**, the width of section A_i should not be greater than the width of section A_i'. The naming of the three possible cases of relative width of the volute inlet,

'pure swirl', where $\varnothing P = W$,
'expansion swirl', where $\varnothing P < W$, and
'contraction swirl', where $\varnothing P > W$

was introduced in Reference 15 in reference to swirl chemical reactors, although the descriptions remain apt in discussion of turbine volutes. The $N_S 176$ volute is clearly a case of contraction swirl, and the geometry is seen to produce $F_{AM} \gg 1$, whereas the Malak volute [13] was a case of pure swirl and $F_{AM} \approx 1$. Referring to eqn [36], this leads us to conclude that the minimum value of $(rV_t)_3$ from a single tangential inlet volute is limited by the increase of F_{AM} as $(rV_t)_3$ is decreased from expansion, to pure, to contraction swirl. Available data are insufficient to correlate the relationship of F_{AM} with $\varnothing P/W$, although this geometric property would appear to be a key factor. From **Figure 10(b)**, another likely possibility is that as the inlet moves to more extreme contraction swirl, A_i ceases to be an inlet to a reasonably torque-free region, whereas A_i', which is downstream of the torque-affected inlet and tongue region, becomes a better representative inlet. If the penstock streamtube is assumed to maintain the penstock's cross-sectional area, and therefore its velocity is unchanged, the section A_i' can be used to determine $(rV_t)_2$.

6.16.2.3 Runner

Continuing downstream from the volute, the next component encountered is the runner, a rotating blade cascade. The blades deflect the flow as they rotate, and through the resulting pressure differential, the flow exerts a torque on the cascade. Some of the material presented here was first presented in Reference 11.

Let us begin by stating several important definitions and assumptions integral to the runner design method presented here. A large part of maintaining simplicity of design in a turbine is keeping the runner as easy to build as possible. Complicated designs, while they certainly can be more efficient, are difficult to manufacture accurately, and it has been decided to use flat blades which are less efficient in absolute terms, but have a geometry that is easier to communicate clearly. The leading and trailing edge stations are denoted by subscripts 3 and 4, respectively. Similarly, blade hub and tip are represented by subscripts 'h' and 't'. **Figure 11** shows a cylindrical hub with a single blade attached, with relative and absolute velocity vectors attached to both the leading and trailing edge at the blade tip, where $U = \omega r$ is the local blade velocity, β is the blade angle, where 90° is facing directly upstream, and δ is the trailing edge deviation, the angle between β_4 and exit velocity relative to the runner, $V_{4\text{rel}}$. All other symbols are as defined in previous sections. Other important runner dimensions not shown are the tip diameter, d_t, and the hub diameter, d_h. The principle action of the runner of decreasing swirl is shown as the absolute velocity vector $V_{3\text{abs}}$ enters the runner with some swirl in the direction of the runner's rotation, and $V_{4\text{abs}}$ is purely axial.

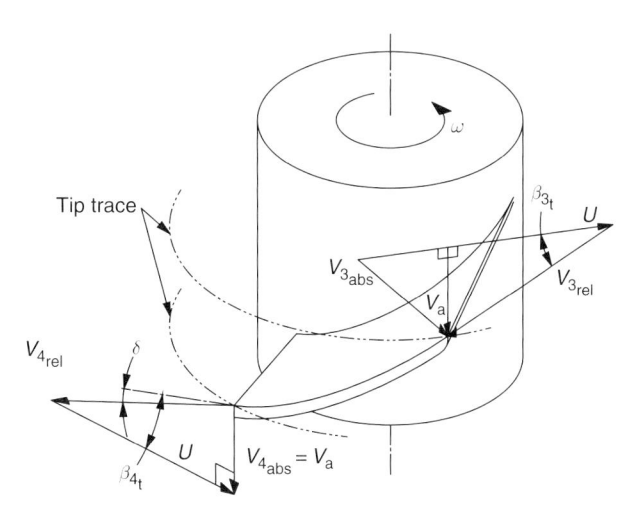

Figure 11 Runner velocity diagrams for the blade tip.

6.16.2.3.1 Specifying runner geometry

Blade construction is achieved by cutting each blade to the desired profile from a flat plate, positioning against the hub at the specified setup angle ψ, and welding in place. Figure 12 illustrates this setup.

Runner speed, N, is fixed when the turbine is directly driving an induction motor as a generator, and a four-pole motor will deliver 50 Hz at a nominal speed of 1500 rev min^{-1} plus roughly 3% slip, for a design N of 1550 rev min^{-1}. Hub-to-tip ratio, d_h/d_t, has been fixed at 0.6, as it fits with common practice and provides a balance between the excessive spanwise variations in flow angle needing to be accommodated when d_h/d_t is much smaller, and the sudden-expansion losses incurred downstream of the hub with it are much higher, although there is no reason not to consider d_h/d_t as a flexible variable used to optimize runner performance.

The runner blades described here are cut from a flat steel plate and left unbent or rolled. The consequence of this geometry is that blade curvature and dihedral are a function of azimuth angle and nominal blade angle, rather than being specified explicitly. The local blade angle, β, as defined in Figure 13, at any point on the blade plane can be determined using eqn [37].

$$\beta = \tan^{-1}(\tan\psi \cos\theta) \qquad [37]$$

This is the angle between the blade plane and the runner plane, on a plane tangent to the idealized cylindrical streamtube passing through the point of interest. Notice that the local blade angle is only a function of azimuth angle, θ, and ψ, which is a constant for a given runner.

Equation [37] may be solved for θ to provide the azimuth angle where the local blade angle occurs, and this result is presented in eqn [38].

$$\theta = \cos^{-1}\left(\frac{\tan\beta}{\tan\psi}\right) \qquad [38]$$

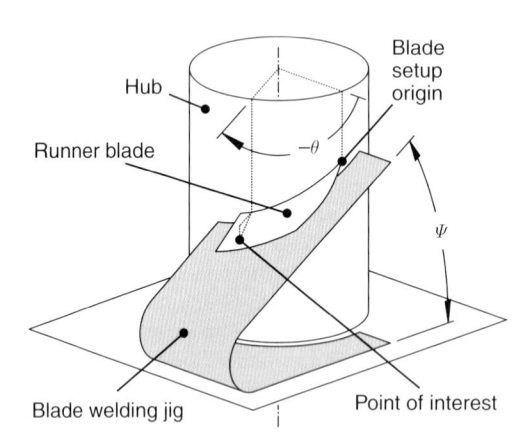

Figure 12 Runner blade assembly setup, showing that when the blade is welded in place, the blade setup origin corresponds to $\theta = 0$, where dihedral is zero.

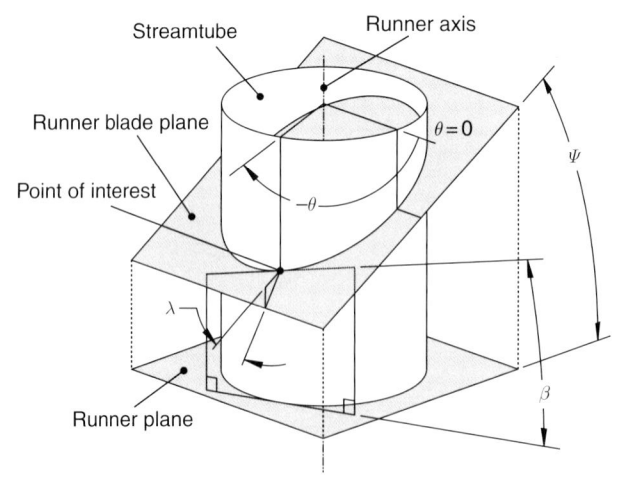

Figure 13 Intersection of cylinder and runner blade plane showing definition of local blade angle β on the assumed cylindrical streamtube, local dihedral angle λ, azimuth angle θ, and blade setup angle ψ.

Now, a velocity triangle can be drawn at any point on the leading edge of a runner blade based on the tangential velocity, V_t, the runner velocity, ωr, and the axial velocity, V_a, as shown in the velocity triangle diagram (Figure 11). V_a is assumed uniform over the runner leading edge surface. The tangential velocity V_{t3} is found using eqn [36] from a trusted upstream surface based on either a valid zero-torque region or knowledge of the volute's value of F_{AM}. From this, the leading edge blade angle for zero angle of attack may be determined. The required blade angle β_3 along the leading edge is given in eqn [39].

$$\beta_3 = \tan^{-1}\left(\frac{V_a}{\omega r - V_t}\right) \quad [39]$$

Similarly, the trailing edge blade angles are found by constructing a velocity diagram at the trailing edge to achieve $V_{t4} = 0$, taking into account a reasonable value for δ, which is on the order of 5° and varies with blade loading. The trailing edge blade angle is given in eqn [40].

$$\beta_4 = \tan^{-1}\left(\frac{V_a}{\omega r - 0}\right) - \delta \quad [40]$$

Although this method is inherently limited in accuracy since V_{a3} and V_{t3} are obtained on the backs of fairly sweeping assumptions about the velocity distribution, it is adaptable in the sense that new blade profiles could be constructed to accommodate measured velocity profiles leaving the volute.

6.16.2.3.2 An example runner

Now that the runner construction method has been presented, let us consider an example in order to illustrate the character of the resulting design. We will specify the operating conditions and runner and hub diameters outright as those of the N_S544 presented in Reference 11, including an expected trailing edge deviation of 5°, repeated here in Table 1 to get on with the runner design, and delay discussion of turbine sizing, which has been deliberately avoided so far.

One major downside to the use of flat blades is that camber and chord length, and therefore blade loading, are fully defined after ψ, β_3, and β_4 are chosen. Figure 14 shows the resultant camberline, and how, unfortunately, the resulting maximum camber point is nearer the trailing edge, increasing blade loading near the point where separation is most likely to occur even on a properly shaped turbine.

6.16.2.3.3 How specific speed influences blade shape

To illustrate how specific speed influences runner design, Figure 15 shows how the fast runner speed and low-head characteristic of high specific speed turbines combine to effect the velocity triangles and the required flow deflection. Note that in Figure 15(b) the leading edge is always placed at $\theta = 0$ since ψ has been set equal to β_3, which gives the largest blade area using this method. The method described in Reference 11 is a compromise variant of the strict adherence to eqns [39] and [40] presented here, where ψ is set equal to β_{4t} putting the leading edge tip at $\theta = 0$, rather than β_{4h}, which gives leading edge angles with a negative angle of attack, but a blade with considerably more area. It is debatable whether this added area reduces real blade loading, as the camberline is then of the reflexed trailing edge type. The results were obtained by varying specific speed as described in the numbered list in Section 6.16.2.2.1, which means that P and ω are constant. The key points to note are that the required flow deflection decreases with increasing N_S and ω, and that in this one-dimensional (1D) approximation, the deflection in the blade frame excluding deviation, $\beta_3-\beta_4$, is only a couple of degrees.

This highlights a practical difficulty in achieving high efficiency in high-speed runners in general. Assume for example that a 1D swirling flow exists, with a constant flowrate and some angle χ off of axial, and we wish to design a runner to remove the swirl. From the values in Table 1, χ would be approximately 12°. For a moment, assume that we also have the freedom to choose any reasonable runner speed. How will this effect the required blade angles and, more importantly, their difference from the leading to trailing edge, $\beta_3-\beta_4$? Following from the logical conclusion that when $N = 0$, $\beta_3-\beta_4 = \chi$, we can show that $\lim_{N \to \infty}(\beta_3-\beta_4) = 0$. Even when $N = 1500 \,\text{rev min}^{-1}$, the blade deflection required has been reduced to several degrees, which is on the order of manufacturing tolerances. Figure 16 shows the blade angles required for the 1D example case.

Figure 17 shows how the difference required decreases with speed.

Table 1 Runner design example starting values

Parameter	Value
Q	0.123 m³ s⁻¹
N	1491 rev min⁻¹
d_t	0.225 m
d_h	0.136 m
$(rV_t)_3$	0.193 m² s⁻¹
δ	5 deg

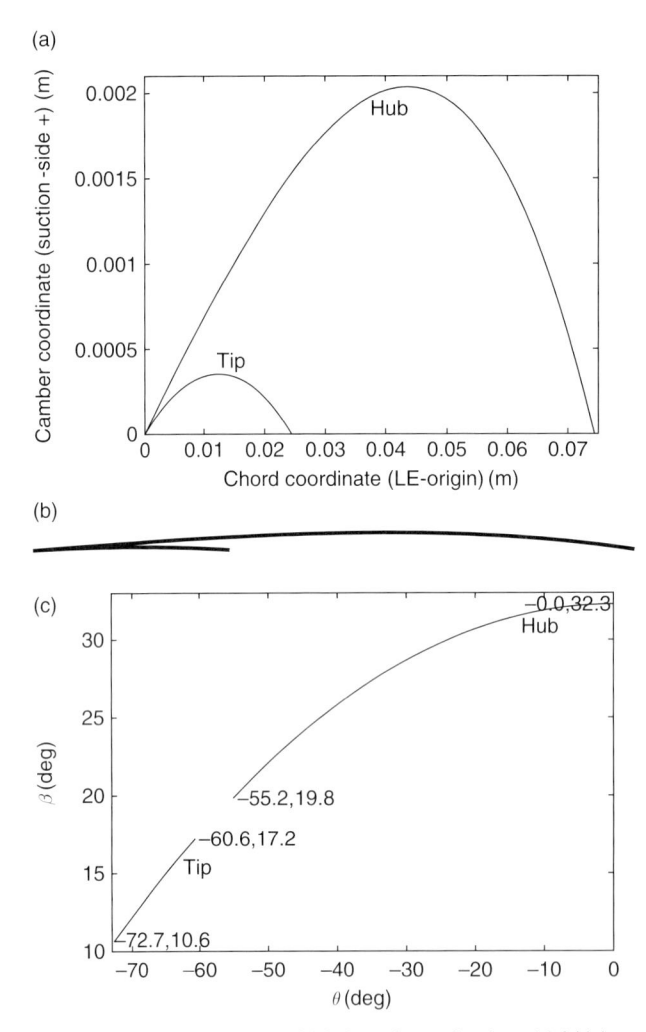

Figure 14 Various representations of the resulting camberline when a blade is cut from a flat sheet. (a) A high specifc speed axial runner blade hub and tip camberline, scaled to show details of the hub and tip camberlines. (b) Camberline unscaled to show the true shape, where the common left-hand point is the leading edge for both. (c) Blade angle β plotted versus azimuth angle θ.

6.16.2.4 Draft Tube

After leaving the runner trailing edge, the next component encountered is draft tube, a duct that carries water away from the turbine and exhausts to the stream below. The draft tube serves two main purposes, the importance of each varying with specific speed:

'recover runner exit static head' by providing an airtight duct to connect the runner to the tailwater, such that the suction due to the weight of the water column in the duct will add to the net head acting across the runner, and

'recover runner exit dynamic head' by choosing a divergent duct to decelerate the flow as it passes through the draft tube, converting as much as possible the velocity head $V_{a4}^2/2g$ to static head acting across the runner.

Static recovery is naturally only important when the turbine elevation above the tailwater is a significant proportion of H_g. It is not uncommon for very high-head turbines such as Pelton wheels, whose total head may be more than 100 m, and a turbine setting above tailwater only 1 or 2 m to omit the draft tube altogether.

Dynamic recovery is naturally only important when the runner exit dynamic head $V_{a4}^2/2g$ is a significant proportion of H_g. The reason that V_{a4} is used and not simply V_4 is that the swirl velocity is not significantly affected by the draft tube as is the axial velocity, and passes through, meaning $V_{t4}^2/2g$ should be considered a loss, which is a major reason to design runners for zero exit swirl. The combination of low head and large flowrate, that is, a high specific speed, can produce conditions where the runner exit dynamic head is a third or more of the total head. In these cases, draft tubes are critical in recovering that portion of the total head [7, Table 5.1; 16, Figure 6.11].

Let us begin by stating several important definitions and assumptions integral to the draft tube design method presented here. Due to its simplicity, economy, and performance, the conical diffuser is often used. It may be rolled from sheet steel and seam-welded. Its inlet and outlet faces are denoted by subscripts 4 and 5, respectively. The draft tube exhausts directly to tailwater.

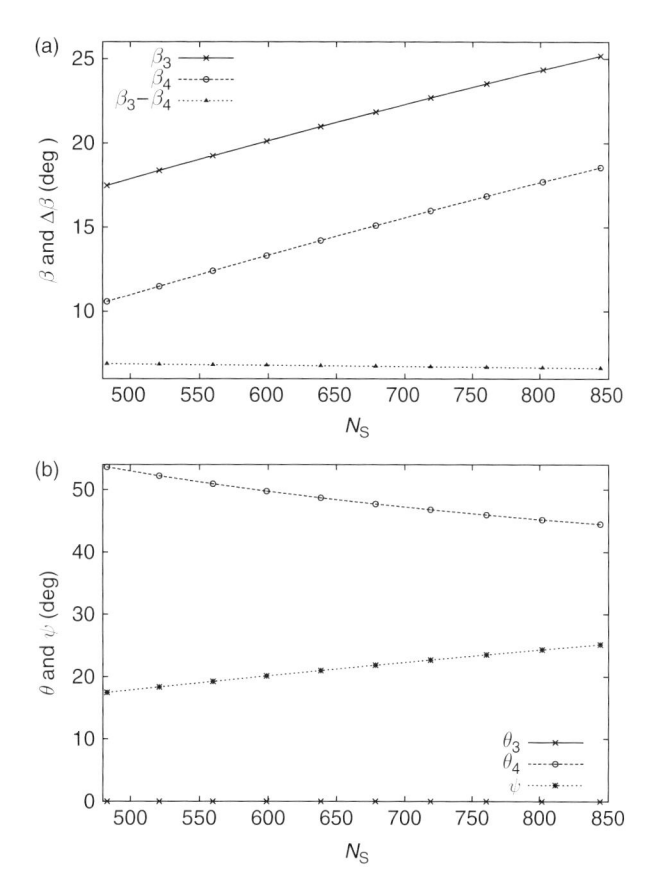

Figure 15 One-dimensional flow deflection as a function of specific speed, where reported β and θ are for the mean radius. (a) Leading and trailing edge blade angle and their difference vs specifc speed. (b) Leading and trailing edge azimuthal location and blade setup angle.

Also, in microhydro schemes, it may be possible to orient the turbine so that the draft tube and turbine axis are aligned, which will eliminate losses due to bends in the high-velocity region following the runner. It has a circular inlet and outlet, and its diameter increases linearly with length. Key conical diffuser dimensions are given in **Figure 18**.

The length and area ratios defined in eqns [41] and [42] nondimensionalize the diffuser's shape.

$$L_R = \frac{N}{R_1} \quad [41]$$

$$A_R = \left(\frac{R_2}{R_1}\right)^2 \quad [42]$$

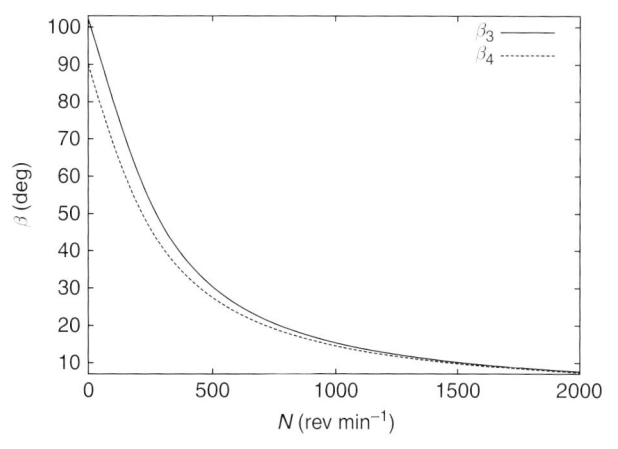

Figure 16 Leading and trailing edge β required for a constant incoming swirl angle and flowrate.

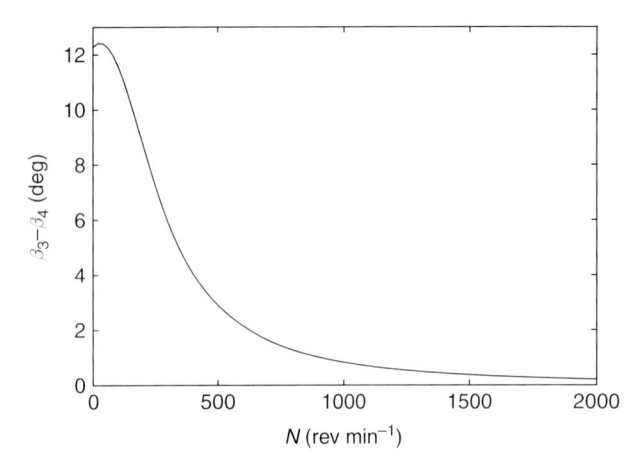

Figure 17 Flow deflection in the runner frame required for a constant incoming swirl angle and flowrate.

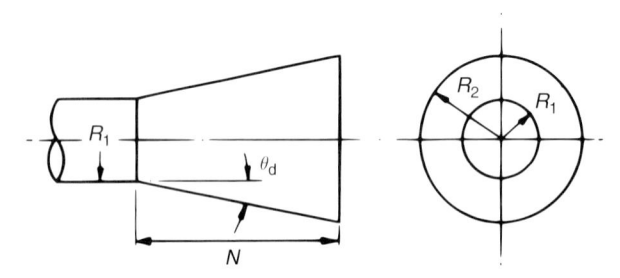

Figure 18 Conical diffuser dimensions.

Note the use of different subscripts here.

The most important performance measure of the draft tube is the amount of inlet dynamic pressure it successfully converts to static pressure. C_p, which is defined in eqn [43], is the ratio of inlet dynamic pressure converted to a pressure increase

$$C_p = \frac{\overline{p_{s4}} - \overline{p_{s5}}}{\overline{p_{d5}}} \quad [43]$$

where the overbars indicate area-weighted averages of the variables underneath. The two extreme cases of a diffuser are

$C_p = 0$ which is throttling valve, where total pressure decreases by the dynamic pressure in the valve throat and static pressure is unchanged,

$C_p = 1$ and an ideal diffuser which decelerates the inlet flow to a uniform outlet velocity distribution of infinite diameter such that total pressure is unchanged and static pressure increases by the dynamic pressure at the diffuser entrance.

An ideal diffuser can be considered a 1D control volume with one inlet and one outlet, where the ideal pressure recovery coefficient, C_{p_i}, is only a function of A_R, as defined in eqn [44].

$$C_{p_i} = 1 - \frac{1}{A_R^2} \quad [44]$$

Any static pressure losses due to wall friction are also contained within C_p, so real conical diffusers cannot be expected to have a C_p much greater than about 0.8 due to space and cost limitations and wall friction losses. When the diffuser has a free discharge, that is, there is no pipework following the diffusing section, a loss coefficient may be defined, as in eqn [45].

$$K_d = 1 - C_p \quad [45]$$

Length ratio, L_R, area ratio, A_R, Reynolds number, N_{R1}, and an indication of boundary layer thickness are sufficient to describe a diffuser's geometry and operating conditions for most purposes. When well above the turbulent transition N_R, inlet velocity distribution, particularly near the walls, becomes much more important than gross Reynolds number because separation in diffusers occurs near the wall when the wall-adjacent streamlines are decelerated to the point of stagnating and then reversing. Regions of sluggish flow effectively change the shape of the diffuser and limit C_p.

Contour maps showing C_p as a function of L_R and A_R are typically created for a given set of inlet conditions. Results presented in **Figures 19** and **20** from Reference 17 are for an inlet Reynolds number of 10^6 with no swirl, but the axial velocity distributions are different, showing the potential for increasing C_p by thinning the inlet boundary layer.

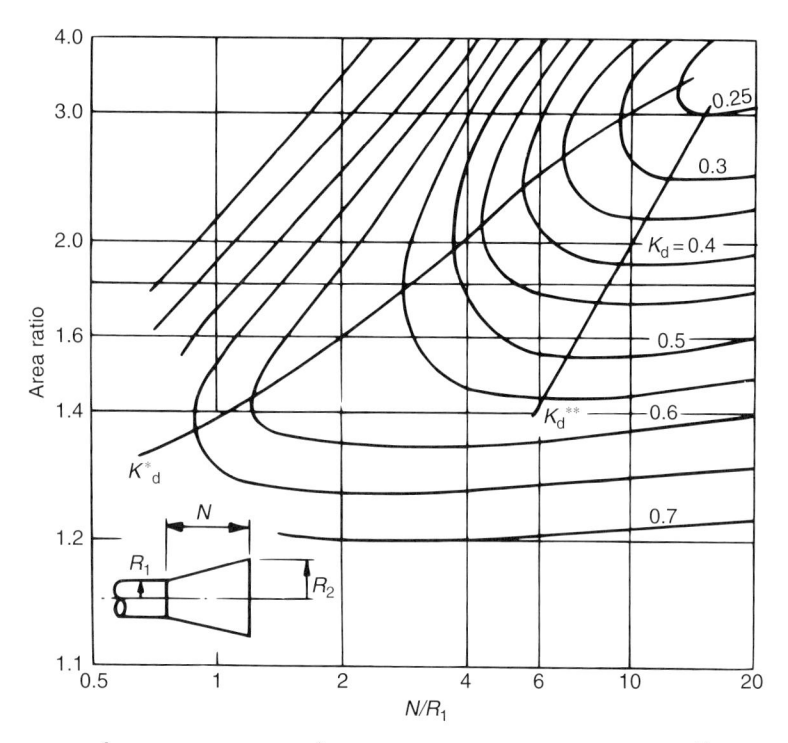

Figure 19 Conical diffuser K_d, $N_R = 10^6$, 'thick' boundary layer; K_d^* line denotes minimum K_d for a given L_R and K_d^{**} line denotes minimum K_d for a given A_R. Reproduced with permission from Miller DS (1978) *Internal Flow Systems*. British Hydromechanics Research Association [17, Figure 11.4].

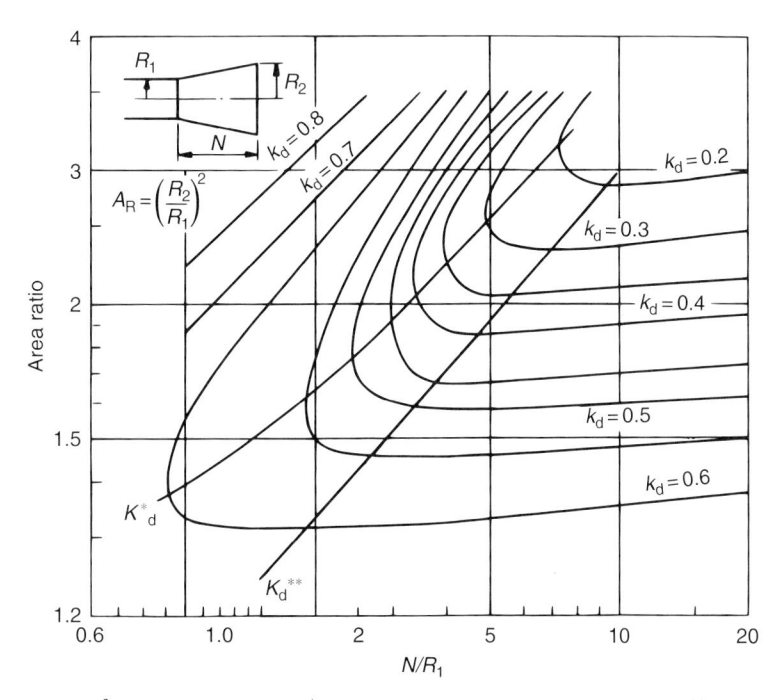

Figure 20 Conical diffuser K_d, $N_R = 10^6$, 'thin' boundary layer; K_d^* line denotes minimum K_d for a given L_R and K_d^{**} line denotes minimum K_d for a given A_R. Reproduced with permission from Miller DS (1978) *Internal Flow Systems*. British Hydromechanics Research Association [17, Figure 11.5].

The accepted optimum conical diffuser full angle is fairly small at 7°, which means that efficient diffusers of a useful area ratio will be relatively long in terms of L_R, but for microscale turbines this may not be unacceptably long in real terms. One should not naively assume that even a slight increase in angle beyond the optimum will lead to an increase; once the threshold angle for separation on the diffuser's wall is crossed, performance will be greatly reduced.

6.16.2.4.1 The effect of diffuser inlet swirl

Some attention has been paid over the years to the effect of swirling inlet flow on the performance of axial diffusers, due to the fact that the potential for designing swirling flow following rotating blades is clearly present, and also because diffusion is a notoriously inefficient process, which turbomachinery designers are in the business of working around. The beneficial effects of swirl on the performance of diffusers are often cited [18, 19], and the swirl itself has been characterized with respect to its intensity [20]. Swirl tends to create radial and axial pressure gradients. This is intuitive if you consider a case of steady swirling flow in a pipe, where the centripetal force causing circular motion to any given fluid particle is necessarily provided by an increasing pressure on the particle's outermost face. This increase in pressure from the centerline to the wall in comparison to the purely axial case reduces the unfavorable pressure gradient along the wall, which is where boundary layer growth and flow reversal are most likely to occur and reduce the effect area ratio of the diffuser. However, the negative effect of swirl, especially that of the free-vortex type which is characterized by large axial and swirl velocity gradients near the centerline, is to cause a positive axial pressure gradient along the centerline and thereby encourage the very behavior that is being attenuated along the wall, namely, flow reversal and an effective reduction in cross-sectional area. As can be seen from the experiment, the maximum pressure recovery for a given diffuser occurs at some intermediate swirl strength, presumably due to the optimum balance between these two effects, which act to distort the exit velocity profile from the optimum uniform distribution [19]. Although it is possible to describe a scenario where swirling flow, through manipulation of pressure gradients and velocity distribution, reduces separation by keeping streamlines pushed out to the wall, there is no hard evidence to support the claim. Testing of the effects of swirl alone is difficult to control for since imparting swirl to an axial flow alters the axial velocity distribution. Since the axial velocity distribution is known to be of importance, it is difficult to separate the effects of swirl when the axial component is dependent.

6.16.3 Turbine Selection from an Existing Range

Microhydro projects all generally strive to achieve maximum economy. This means locally sourcing materials and labor, ideally volunteered, whenever possible. But before any work can be done or supplies purchased, a design must be drawn up specific to the project at hand. If the project's developers wish to simply pay for a turnkey system, engineering consultation to describe the site works and turbine design will become a significant fraction of the entire project's cost. It is the authors' goal to reduce this cost by providing an open-source range of efficient turbines which developers may choose from to fit their particular site. Specific speed is used to describe a family of geometrically similar turbines, and this section describes a method whereby an appropriate existing turbine design may be scaled to suit a particular site. **Figure 21** graphically presents the fundamental array of available turbine forms and sizes, which turbines at the University of Canterbury have been designed to fill.

Although the methods behind producing **Figure 21** were covered in Reference 11, where the similarity rules for scaling a single turbine to cover a range of power outputs are based on BS 60995 [21], the method presented here is the authors' recommendation for how to use such a selection matrix, where one already exists. The basic procedure, introduced in Reference 22, is

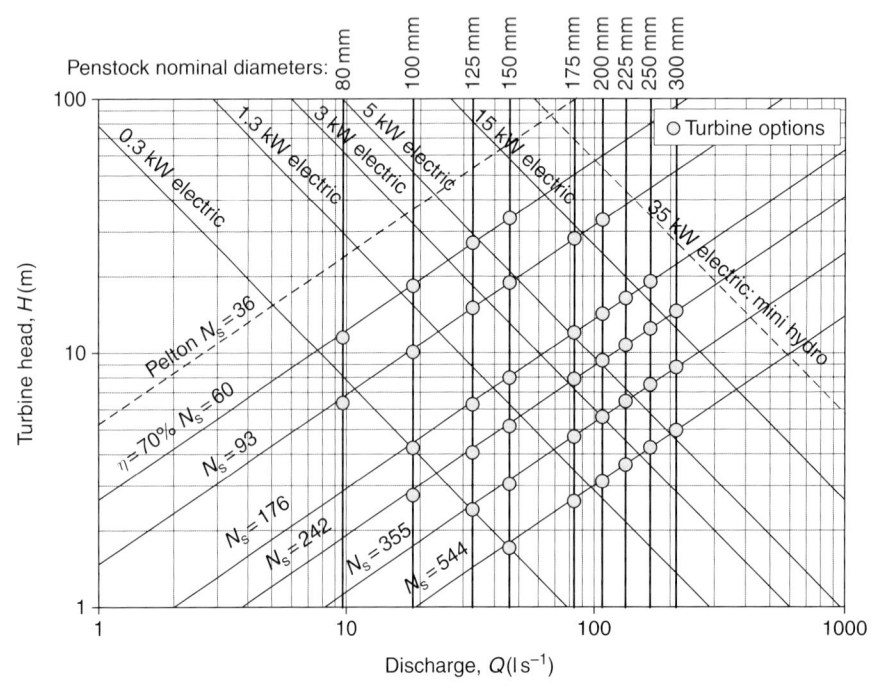

Figure 21 University of Canterbury turbine matrix.

1. measure Q_a and H_g,
2. select the penstock diameter,
3. calculate h_p and update H,
4. select appropriate turbine form from **Figure 21**, and
5. use the ratio of penstock diameter to scale from known turbine geometry to site geometry.

Site measurement: Step one is simple but is the basis of all work to follow. H_g and Q_a must be determined as accurately as possible and with a slight conservative bias to avoid efficiency reduction and power quality issues during operation. Refer Reference [10] for details on site measurement and preparation.

Penstock diameter selection: Select the penstock diameter corresponding to the nearest flowrate less than Q_a in **Figure 21**.

Calculating H: From Q_a and penstock diameter, calculate h_p and update turbine head, H.

Turbine form selection: Locate the appropriate turbine nearest the intersection of Q_a and H in **Figure 21**. In the general case where the intersection of H and Q_a does not lie exactly on a distinct turbine form, Reference 22 offers some suggestions for accommodating the mismatch.

Scale reference turbine to site: Arriving at a final scaled turbine requires the dimensions of the chosen specific speed turbine to be known. The full dimensions of the four Canterbury propeller turbines normalized to $N = 1500 \, \text{rev min}^{-1}$ and $P_r = 1$ kW are tabulated in **Table 2**, which correspond to the dimensions labeled in **Figures 22** and **23**. The turbines' operating conditions are tabulated in **Table 3** [11].

For a given specific speed, the geometrical form of the turbine remains the same, but the size may be scaled. This allows the design for each specific speed to be scaled to match the discrete optimum penstock flows and to deliver, from the appropriate heads and discharges, the power bands shown in **Figure 21** to cover the microhydro range. When scaling from a reference machine, the resultant hydraulic efficiency may be estimated from an empirical function of the physical size (or discharge, depending on the model used) of so-called majoration effects [9]. Moving along lines of constant N_S in **Figure 21** represents the process of scaling to a geometrically similar machine, that is, a machine with the same value of N_S, but the calculations to produce such a graphical tool are as follows:

1. calculate H_R, Q_R, and N_R directly,
2. calculate P_R to see what you will get out compared to the reference turbine, and
3. calculate L_R to scale the reference turbine.

To find the ratio of the size of the site and reference turbine, it is necessary to calculate the linear scaling ratio of the real to the prototype machine, L_R. Since the turbine forms represented in **Figure 21** need to remain geometrically similar when scaled, the scaling ratio between the two is equal to the ratio of the site and reference penstock diameter. This ratio may then be used to scale the complete set of dimensions from either **Table 2**.

Table 2 Reference propeller turbine performance and dimensions

N_S	176	242	355	544
d_t (mm)	129	131	143	182
d_h (mm)	83	78	86	110
Draft tube included angle, A (deg)	16	6	16	7
Vortex flange-to-outlet throat, B (mm)	201	383	169	271
Main casing radius, R_C (mm)	126	157	144	204
Scrolled casing radius exponential factor, $R_C p$			0.0024	0.0020
Runner casing inner diameter, $\varnothing D$ (mm)	129	132	144	184
External draft tube length, E (mm)	329	481		
Inlet-to-runner centerline offset, F (mm)	61	73	198	324
Outlet cone-to-draft tube gap, G (mm)	5	13	7	0
Main casing internal length, H (mm)	301	381	240	810
Inlet centerline-to-vortex flange offset, K (mm)	196	225	62	555
Vortex flange outer diameter, $\varnothing L$ (mm)	208	228	268	207
Inlet inner diameter, $\varnothing P$ (mm)	129	168	240	243
Vortex flange inner diameter radius, RR (mm)	4	5	7	4
Vortex flange-to-main casing clearance gap, S (mm)	26	46	85	65
Outlet throat inner diameter, $\varnothing T$ (mm)	86	100	111	184
Vortex flange outer diameter radius, RV (mm)	4	5	7	6
Inlet diffuser plan width, W (mm)			250	322
Inlet diffuser full length, X (mm)			478	1179
Outlet cone included angle, Z (deg)	16	11	16	0

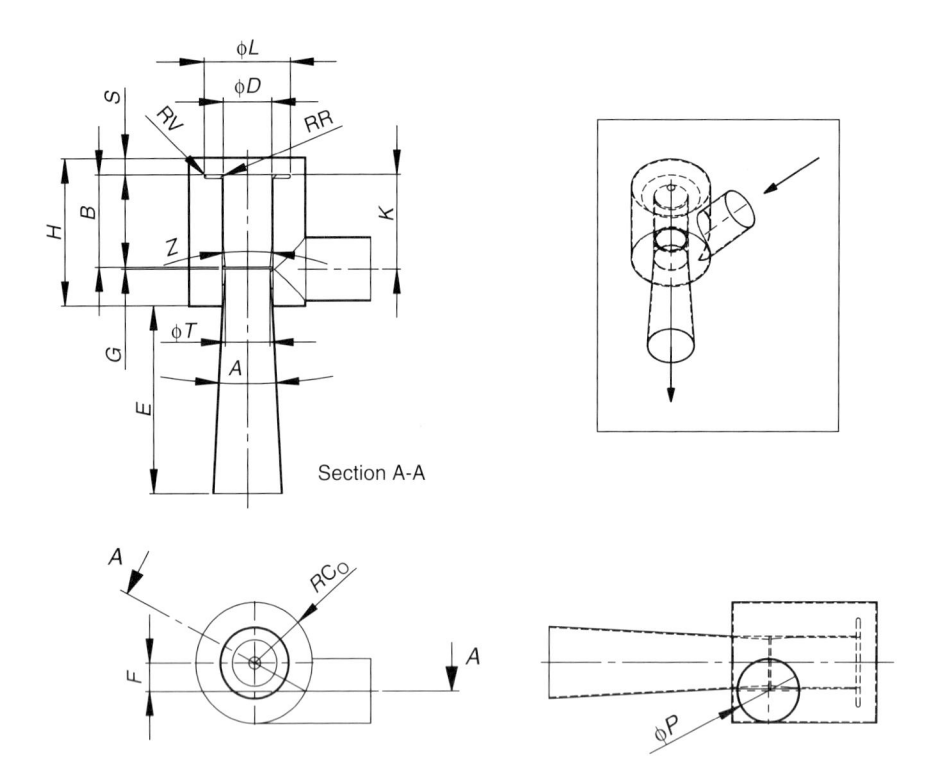

Figure 22 N_S176 and N_S242 dimensions.

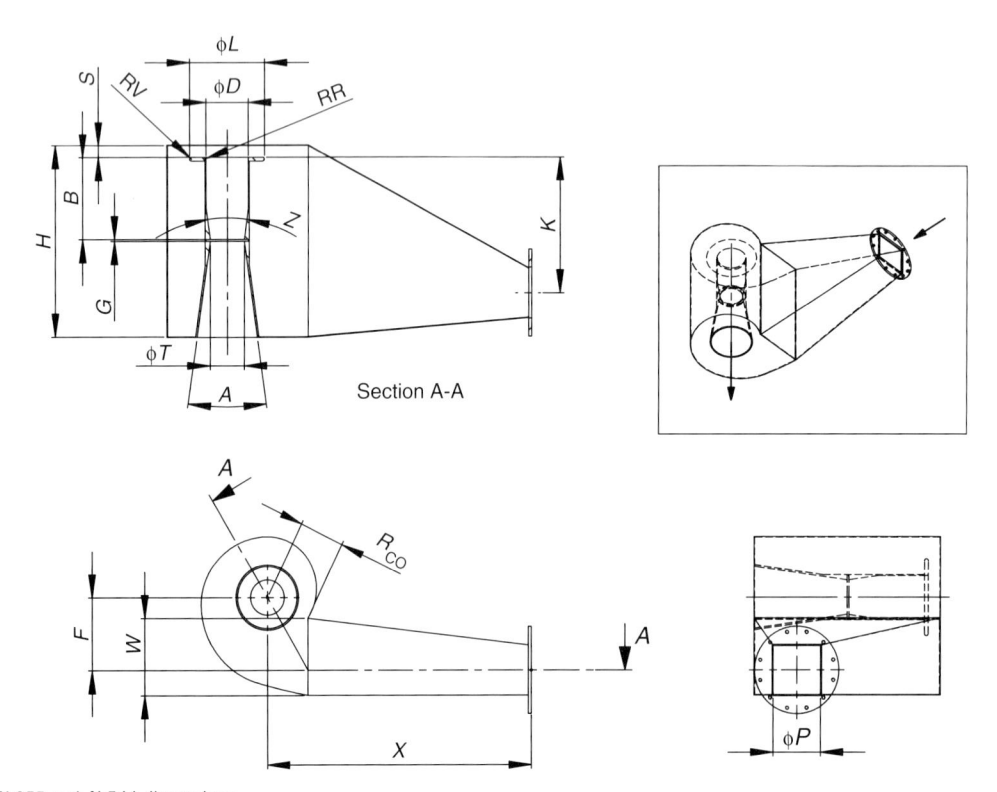

Figure 23 N_S355 and N_S544 dimensions.

Table 3 Reference propeller turbine performance

N_S	176	242	355	544
P_r (kW)	1	1	1	1
N (rev min^{-1})	1500	1500	1500	1500
H_t (m)	5.56	4.31	3.17	2.25
Q (l s^{-1})	25	32	46	66
η_t	0.727	0.747	0.727	0.665

6.16.4 Direct Sizing

By comparison with the previous section, a more direct method to turbine design will now be presented. The method is direct in that it does not involve scaling of an existing turbine to fit a given site. Because its result is only dependent on a collection of sets of rules – a set for each component – it might be called a generic or modular design method. If validated, it is highly amenable to automation:

1. measure site,
2. prioritize supply,
3. size turbine (analytical flowrate prediction), and
4. design individual components

Although the process involves designing each component individually, a considerable amount of automation could significantly reduce the amount of human interaction required.

Site measurement: As with any turbine development, the site's topography, in particular H_g and L, and hydrology, Q_a, must be determined first. As before, although the hydrology will in general have a significant range between dry and wet season mean flows, not to mention the occasional flood, Q_a, the flowrate available for the turbine is taken as slightly less than the minimum expected flowrate for the season of usage.

Prioritize supply design: Using eqn [22] assuming $K_2 = g\rho$, calculate the maximum site potential. Assuming a penstock supply is suitable, use the penstock analysis and optimization methods presented in Section 6.16.2.1.1 to prioritize either cost effectiveness or power, and then determine penstock diameter. Calculate h_p/H_g and update H_n.

Size turbine: Instead of using turbomachinery similarity laws to scale existing turbines of known size and flowrate to a new size so that the flowrate of the new turbine is known, several methods are presented here, which attempt to bridge the gap between the dictated boundary conditions of each component and the actual flowrate. For the simplest possible hydraulic component – a length of straight pipe with fully developed steady flow – the Moody diagram can be used. Although more complex, turbine sizing involves quantifying the functional relationship between flowrate and turbine shape and size in a similar manner. There are several ways to approach flowrate prediction.

'Scaling a geometrically similar turbine' of known flowrate, head, and speed is the method presented in Section 6.16.3, and has the least uncertainty, but also the least flexibility of the sizing methods presented. Alternatively, the unit flow coefficient could be used if known for the geometry being used [7]. These parameters are turbine flowrate nondimensionalized by size and head. If it is known for a similar design, the flowrate for the design in question is only a matter of evaluating this expression [7, Figure 6.4] in a graphical representation of such data.

'Fully analytical flowrate prediction' is the estimation of flowrate through a turbine from basic fluid and thermodynamic principles as a function of the turbine's size, shape, and head. This approach will be elaborated below.

'Computational fluid dynamics tools' may be used to predict flowrates through arbitrary complete turbines, from which a flow coefficient may also be determined for the turbine in question, although the computational model may require validation.

To clarify the meaning of fully analytical flowrate prediction, the principles involved will now be introduced. The following 1D methods can be derived using the same principles for a different geometry. They have the same limitations as most 1D models, which is that unless they are validated and adjusted with empirical data, they may useful for only predicting trends or producing order-of-magnitude estimates. Runner efficiency, in particular, is assumed constant.

Single tangential inlet volute: Begin by equating the two expressions for turbine power, based on energy and momentum.

$$P_r = \eta_r \rho g Q H_r = T\omega = \rho Q \, \Delta(rV_t)\omega \qquad [46]$$

Next, divide through by ρQ, which gives an alternate form of the Euler equation, similar to eqn [32]:

$$\eta_r g H = \Delta(rV_t)\omega \qquad [47]$$

With the stipulation that swirl leaving the runner is negligible, the change in angular momentum across the runner is then simply the angular momentum reaching its leading edge.

$$\eta_r g H = r_3 V_{t3} \omega \qquad [48]$$

Equation [48] may then be rearranged such that runner inlet tangential velocity is written as primarily a function of ω and H_r.

$$V_{t3} = \frac{\eta_r g H_r}{r_3 \omega} \qquad [49]$$

Although tangential inlet volutes are typically designed not to change the fluid's angular momentum, if experiments are carried out to measure the volute's ratio of exit to inlet angular momentum, this effect may be included using the parameter F_{AM}. Swirl reaching the runner leading edge can then be written in terms of the volute inlet velocity and offset and the volute's swirl modification behavior.

$$V_{t3} = F_{AM} \frac{r_2 V_{t2}}{r_2} \qquad [50]$$

The two expressions for V_{t3} may now be equated giving

$$\frac{\eta_r g H_r}{r_3 \omega} = F_{AM} \frac{r_2 V_{t2}}{r_2} \qquad [51]$$

Since the flow in a tangential inlet is, clearly, tangential, eqn [51] can be rearranged to give volute inlet velocity as a function of primarily H_r, N, and volute inlet offset, r_2.

$$V_2 = V_{t2} = \frac{\eta g H_r}{\omega F_{AM} r_2} \qquad [52]$$

Combining penstock velocity its cross-sectional area in a statement of continuity gives the final result,

$$Q = \frac{A_2 \eta_r g H_r}{\omega F_{AM} r_2} \qquad [53]$$

Axial flow with guide vanes: The tangential-volute method just presented is of limited use for sizing in that it does not relate flowrate to the size of the runner section, but rather the volute inlet section. This method is developed from the same principles, but because it is for an axial-flow runner with an axial-flow stator just upstream, the runner and guide vanes share the same section, so the resulting area can be used for both, essentially sizing the runner. **Figure 24** shows the geometry described.
In this case, the link between turbine geometry and runner inlet swirl is simply

$$V_{t3} = \frac{V_a}{\tan \beta_{gv}} \qquad [54]$$

where V_a is the uniform axial velocity throughout both the guide vanes and the runner. Next, just as eqn [49] was equated to the momentum-derived expression of V_{t3} for a tangential inlet volute, eqn [50], eqn [54] is also set equal to eqn [49], except that this time the runner radius does not cancel out, giving the analytical expression for flowrate

$$Q = \frac{\eta_r g H_r A_3 \tan \beta_{gv}}{r_3 \omega} \qquad [55]$$

where A_3 and r_3 are the runner cross-sectional area and representative radius in this 1D model.
Radial-inflow guide vanes: Another example of a fully analytical flowrate prediction model is the guide vane function introduced in Reference 7, which is reproduced in eqn [56] using this chapter's notation.

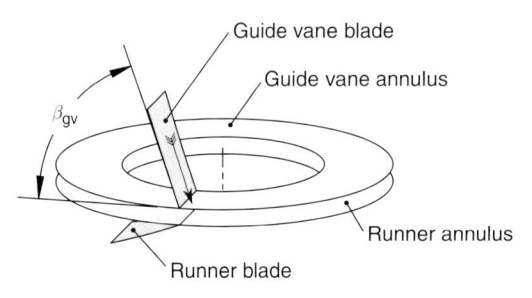

Figure 24 Fully axial turbine showing stator (guide vane) and runner annuli.

$$Q = \frac{g\eta_{\mathrm{h}} H_{\mathrm{n}} + \omega^2 r_4^2}{\dfrac{\omega \tan^{-1} \alpha_{\mathrm{gv}}}{2\pi B} + \dfrac{4\omega r_4 \tan^{-1} \beta_4}{d_{\mathrm{t\,TE}}^2}} \quad [56]$$

where B is the guide vane span, η_{h} accounts for turbine leakage, gv denotes the guide vane cascade, and all other notations are is as defined in this chapter.

Although any of these methods may be used to scale a turbine's key hydraulic dimensions, a single set of continuous expressions cannot be trusted to provide reasonable turbine designs over the entire range of specific speeds. Some piecewise logic is needed to accommodate unavoidable discreteness inherent in real turbines such as available penstock sizes, material thicknesses, and fastener sizes, to name a few.

Component design: Once the scale of the turbine has been determined, each component may be designed. This process is guided by several constraints:

- Turbine speed is usually fixed in order to cooperate with a generator.
- Volute design is guided by a possibly modified zero-torque assumption, whereas flow leaving properly loaded guide vanes may be assumed tangential to the vane camberline.
- Runner leading edge angles are fixed by ω and the local values of r and V_{t}, which come from the torque calculated in eqn [30].
- Runner trailing edge angles are fixed by ω and the target exit swirl distribution, typically none.
- If the runner blades are constrained to lie on a plane, as described in Section 6.16.2.3, the only variable left to adjust is the setup angle, ψ, which is typically set to the largest blade angle required, which is generally where the leading edge meets the hub. More complex geometry allows more flexibility in designing for a particular blade loading if separation proves an issue.
- The cost invested in the draft tube will be proportionate to the amount of dynamic head leaving the runner as a proportion of runner head, $(V_3^2/2g)/H_{\mathrm{r}}$.

6.16.5 Conclusions

This chapter has endeavored to present an intentionally self-centered view of the design process of microhydro turbines, one that encapsulates the current state of design practice at the University of Canterbury. This work to produce freely available designs for a range of simple, effective, appropriate turbines began in 1981 with the efforts of the late professor Peter Giddens and is continued today by the authors, with numerous student projects spanning the nearly three decades in between.

Moving forward from the $N_{\mathrm{S}}544$ design, current research is focused on understanding the interconnectedness of volute geometry and turbine performance, with a view to increasing the range's maximum specific speed while keeping the essence of simplicity alive.

As the true measure of the worth of these turbines is their uptake in the real world, our long-term goal is to incorporate the full range of validated turbine designs, from radial to axial flow, into a free, automatic design tool, such that the user can input their site data and particular constraints, and be presented with a full set of drawings, tailored to their particular site. It is hoped that the axial-flow turbines presented here, and their cousins to come, will provide a valuable contribution.

Further Reading

While this chapter attempts to be fully informative within the fairly narrow scope of the authors' axial-flow microhydro turbines, many complementary sources of information exist and bear mentioning here.

Hydrodynamic Design Guide for Small Francis and Propeller Turbines [7] is a slim, straightforward, but fairly comprehensive and detailed manual of turbine design, focused more on larger, more complex turbines, with references to more detailed sources and Francis and propeller volute and runner design examples. This text gives a detailed step-by-step method of designing the essential turbine components of volute, runner, and draft tube, although some external references are needed for details.

Microhydro Design Manual [10] is a different type of comprehensive text with discussion weighted more toward the specifics of hydrology, site measurement, site layout, and little instruction on actual turbine design. This text is more appropriate for those looking to purchase an appropriate turbine and oversee its correct installation and operation.

Guide on How to Develop a Small Hydropower Plant [16] is the European Small Hydropower Association's take on a comprehensive small hydropower guide. It is similar in scope to Reference 10, in that it does not cover detailed turbine hydraulic design but gives a clear enough overview for a developer looking to select equipment, rather than design it.

Kempe's Engineers Year-Book [23] is a vast single volume of practical engineering knowledge and contains roughly 30 pages of information on turbine design, although it is aimed more at the design of large turbines.

The Design of High-Efficiency Turbomachinery and Gas Turbines [24] provides comprehensive background reading on basic thermodynamic and hydrodynamic principles used throughout turbine design.

Fluid Mechanics [9] is an excellent basic fluids text for fundamental reference.

Motors as Generators for Micro-Hydro Power [25] is a useful introduction to the most commonly used electrical hardware in microhydro installations.

References

[1] Schlemmer E, Ramsauer F, Cui X, and Binder A (2007) HYDROMATRIX and StrafloMatrix, electric energy from low head hydro potential. *International Conference on Clean Electrical Power 2007*, pp. 329–334, May.

[2] Hiroyoshi T (2006) Feature of micro hydraulic power unit (Hydro-eKIDS), and its application. *Energy* 56(266): 53–59.

[3] IT Power http://www.itpower.co.uk/Technologies/Hydro.

[4] MJ2 Technologies http://www.vlh-turbine.com/EN/html/The_VLH_Range.htm.

[5] Whitfield A and Mohd Noor AB (1994) Design and performance of vaneless volutes for radial inflow turbines. Part 1: Non-dimensional conceptual design considerations. *Proceedings of the Institution of Mechanical Engineers* 208: 199–211.

[6] Williams A (2003) *Pumps as Turbines: A User's Guide*, 2nd edn. Rugby, UK: ITDG Publishing.

[7] Hothersall R (2004) *Hydrodynamic Design Guide for Small Francis and Propeller Turbines*, pp. 61, 63. Vienna, Austria: United Nations Industrial Development Organization.

[8] Alexander K and Giddens P (2008) Optimum penstocks for low head microhydro schemes. *Renewable Energy* 33(6): 1379–1391, Section 16.

[9] White FM (2003) *Fluid Mechanics*, 5th edn, p. 765. New York: McGraw-Hill.

[10] Harvey A (1993) *Microhydro Design Manual*. Exeter, UK: Intermediate Technology Publications.

[11] Alexander K, Giddens P, and Fuller A (2009) Axial-flow turbines for low head microhydro systems. *Renewable Energy* 34(1): 35–47.

[12] Fuller A and Alexander K (in press) Single tangential inlet vaneless volute swirl measurement.

[13] Malak MF, Hamed A, and Tabakoff W (1987) Three-dimensional flow field measurements in a radial inflow turbine scroll using LDV. *Journal of Turbomachinery* 109: 163–169.

[14] Bhinder FS (1969–1970) Investigation of flow in the nozzle-less spiral casing of a radial inward-flow gas turbine. *Proceedings of the Institution of Mechanical Engineers* 184(Pt. 3G(11)): 66–71.

[15] Legentilhomme P and Legrand J (1991) The effects of inlet conditions on mass transfer in annular swirling decaying flow. *International Journal of Heat and Mass Transfer* 34(4/5): 1281–1291.

[16] European Small Hydropower Association (2004) *Guide on How to Develop a Small Hydropower Plant*. Technical report, ESHA.

[17] Miller DS (1978) *Internal Flow Systems*. British Hydromechanics Research Association.

[18] Morel T and Arndt REA (1984) Potential for reducing kinetic energy losses in low-head hydropower. In: ASME (ed.) *Project Report No. 221*, pp. 57–65. New Orleans, LA: ASME, 9–14 December.

[19] Yasutoshi S, Nobumasa K, and Tetsuzou N (1978) Swirl flow in conical diffusers. *Bulletin of the Japanese Society of Mechanical Engineers* 21(151): 112–119.

[20] Armfield SW and Fletcher CAJ (1989) Comparison of κ–ϵ and algebraic Reynolds stress models for swirling diffuser flow. *International Journal for Numerical Methods in Fluids* 9: 987–1009.

[21] International Organization for Standardization (1995) BS EN 60995:1995 determination of the prototype performance from model acceptance tests of hydraulic machines with consideration of scale effects. Standard BS EN 60995:1995, International Organization for Standardization.

[22] Alexander K and Giddens EP (2007) Microhydro: Cost-effective, modular systems for low heads. *Renewable Energy* 33(6): 1379–1391.

[23] Taylor EQ (1984) *Kempe's Engineers Year-Book*, ch. F5. London, UK: Morgan-Grampian.

[24] Wilson DG (1984) *The Design of High-Efficiency Turbomachinery and Gas Turbines*. Cambridge, MA: MIT.

[25] Smith N (1994) *Motors as Generators for Micro-Hydro Power*. Southampton Row, UK: Intermediate Technology Development Group.

6.17 Development of a Small Hydroelectric Scheme at Horseshoe Bend, Teviot River, Central Otago, New Zealand

P Mulvihill, Pioneer Generation Ltd., Alexandra, New Zealand
I Walsh, Opus International Consultants Ltd., New Zealand

© 2012 Elsevier Ltd. All rights reserved.

6.17.1	Introduction	467
6.17.2	Background	468
6.17.3	Scheme Layout and Specifications	468
6.17.4	Project Development and Processes	470
6.17.5	Land Tenure	470
6.17.6	Resource Consents	470
6.17.7	Project Management	472
6.17.8	Contract Framework	472
6.17.9	Interesting Features of Design and Construction	473
6.17.9.1	Control Valve Positioned at the Tunnel Outlet	473
6.17.10	RCC Dam Design and Construction	473
6.17.10.1	Geological and Hydrological Setting	473
6.17.10.2	Site Layout	474
6.17.10.3	RCC Mix Design and Handling Characteristics	475
6.17.10.4	GIN Foundation Grouting Method	477
6.17.10.5	Commissioning/Performance Monitoring	480
6.17.11	Conclusions	482
References		483

6.17.1 Introduction

Horseshoe Bend Hydroelectric Scheme is a small 4 MW project owned by Pioneer Generation Ltd. and located on the Teviot river 15 km east of Roxburgh in the lower South Island of New Zealand.

The scheme consists of a 13 m-high roller compacted concrete (RCC) dam, a 180 m-long tunnel, 800 m-long steel pipeline and penstock, and a powerhouse housing a 4.3 MW horizontal Francis turbine and a 5 MVA generator. The gross head of the scheme is 93 m. Although small in scale, the project presented risk management challenges in gaining land access and workable resource consents, achieving financial control in an area renowned for cost overruns, and establishing suitable contracting frameworks. The challenges faced in obtaining consents and in design and construction were also compounded by regulatory processes, resistance from environmental groups, tight time frames, and a difficult and remote location. This chapter outlines these challenges and some of the processes and methods used, to achieve a successful project outcome, coming in ahead of time and on budget.

In New Zealand today embarking on the development of any new project, especially one using natural resources such as land and water, has significant associated risks. Any significant construction project such as a hydroelectric scheme still involves tackling the age-old technical problems of gaining sufficient confidence in the hydrology and foundation conditions, refining mechanical and electrical hardware and control, and finding the most cost-effective transmission option. Once the conceptual design work is completed, further issues arise such as creating a design team and choosing suitable project management and contractual frameworks. The overall objective of these investigations is to finally achieve a product fit for the purpose at a reasonable cost. However, in the initial stages of progression of a hydroelectric development project, even a small one, environmental issues, along with the issue of securing some form of tenure over the required land, can govern the overall viability of the project.

In our environmentally conscious society there exist many challenges such as sustainable management of resources, ensuring that the adverse environmental impacts and risks associated with a project are kept to a minimum, and convincing the regulating authorities and the public at large that the negative impacts of any proposal are outweighed by the long-term benefits. To add to these problems, projects involving hydroelectric schemes have a checkered history and are generally seen by environmental groups and the general public as having significant negative environmental impacts.

With the transition from public-sponsored projects, and the associated enabling legislation, to private development and the 'user-pays' environment over the past 20 years in New Zealand, the land required for hydroelectric projects must be obtained through negotiation. When there are a number of landowners involved, this can lead to significant problems.

The Horseshoe Bend hydroelectric project, although small in scale, encountered during the development and construction phases many of the issues and risks associated with any 'greenfield' hydropower development project in New Zealand in recent times.

6.17.2 Background

The Teviot river has a history of water resource development for mining, irrigation, and hydroelectric power, which dates back to the latter part of the nineteenth century. Being one of the four small hydroelectric stations on the river, Horseshoe Bend is situated 15 km east of Roxburgh at approximately 600 m asl. The site location is shown in **Figure 1**. The scheme is located on a 2.5 km long gorge section of the river.

Pioneer Generation Ltd. (PGL) and its predecessor Central Electric Ltd., and Otago Central Electric Power Board have a history of water resource development for hydroelectricity on the Teviot river dating back to 1924. The most recent stations were constructed in the early 1980s, which included a 1.6 MW and an 8 MW station, and the associated head works.

The conceptual design of the Horseshoe Bend scheme was carried out during the mid 1980s with the proposed scheme layout being very similar to that which was finally constructed.

Serious investigations into the viability of the project began in 1992. Problems were encountered in acquiring easements and purchasing land at a reasonable cost, and gaining some form of tenure from the Department of Conservation (DOC) for the areas of the marginal strip (Queens Chain) required for the dam abutments and other structures associated with the scheme. This issue was resolved through an amendment to the Conservation Act in 1996.

During the investigation phase in the 1980s the foundations of possible dam sites were exposed and a report on surface investigation was completed. The conceptual design assumed that an arch structure would be built on the site following on from the successful experience at other sites on the river. A review of the dam concept was undertaken and further geotechnical investigations carried out by Opus International Consultants in late 1996. The site investigations revealed that rock relaxation, weathering processes, and the presence of foliation shears were significant factors at the dam site, which could adversely impact the design of the arch structure. During the process of reviewing alternatives to the arch concept, the physical limitations of the site to provide adequate diversion capacity during construction were highlighted. This coincided with the overtopping and failure of Opuha Dam during construction, in February 1997, which raised awareness of diversion issues in the Regional Council. After consideration of all these issues, it was concluded that the RCC dam concept was the most appropriate one for the site.

After significant input into consultations with the affected parties and environmental impact assessment, applications for resource consents under the Resource Management Act 1991 were lodged in March 1997. Resource consents for the project were gained by late 1997, with many of the construction and equipment supply contracts, including a design–build contract for the dam, being signed in mid 1998. The scheme was constructed over the following summer and completed on budget and 3 weeks ahead of schedule in April 1999.

6.17.3 Scheme Layout and Specifications

The scheme is a run-of-the-river type project with limited daily storage. The main storage for the river is located at Lake Onslow some 7 km upstream from Horseshoe Bend.

The site is characterized by a section of the river flowing through a deeply incised gorge. The area is surrounded by farmland, and the land occupied by the scheme includes the riverbed, a marginal strip, and some privately owned

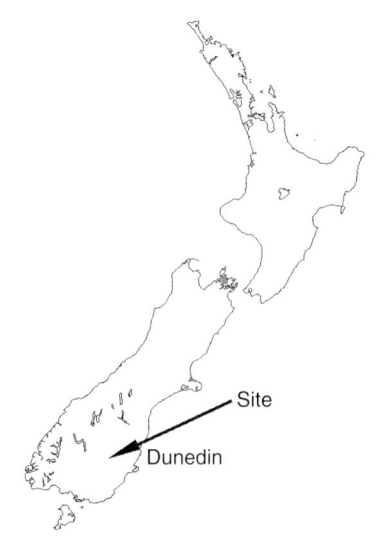

Figure 1 Location of the Horseshoe Bend hydroelectric project.

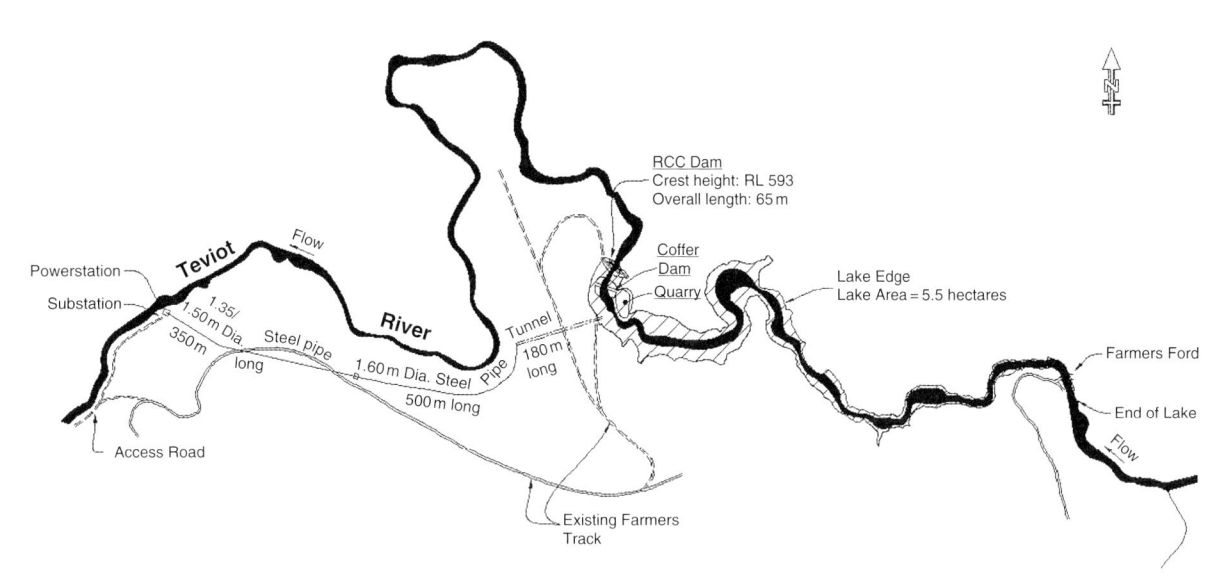

Figure 2 Layout of the scheme.

farmland. The potential for electricity generation at this location is a product of the physical characteristics present. The river level at the dam site is approximately 82 m above the river level at the powerhouse site. With the addition of the dam the overall gross head of the scheme is approximately 93 m. The average flow through the Horseshoe Bend site is $3.52\,\mathrm{m^3\,s^{-1}}$. The maximum peak output from the scheme is 4.3 MW with an annual production of 18–20 GWh. The layout for the scheme is shown in **Figure 2**. This form of development is similar to that used successfully for the existing schemes on the lower river.

The dam or intake weir consists of an RCC structure with the spillway 10.5 m above the river level. The overall height of the dam is approximately 13 m. The spillway crest is 32 m long with the overall length of the dam being approximately 65 m. Other ancillary items forming part of the dam include sludge and residual flow valves and water level measuring equipment. The dam passes a residual flow of $312\,\mathrm{l\,s^{-1}}$ and has no fish pass.

The reservoir behind the dam has a maximum storage level at spillway crest of 593 m asl. The lake behind the dam is small, covering an area of approximately 5.5 ha, and contains approximately 260 000 m³ of water. This reservoir forms a long narrow lake extending 1.3 km upstream from the dam structure. The daily variation in reservoir level during the winter months is approximately 1.0 m although, under the resource consents, the level can be varied up to 1.5 m.

The tunnel inlet is located upstream of the dam. The control valve for the penstock is located at the downstream end of the tunnel. The tunnel itself is 180 m long with a D cross section 2.5 m high. Due to its low overburden ratio the tunnel is concrete-lined along its full length. The pipeline is made up of a 500 m long, 1.6 m diameter buried steel section and a 350 m section of exposed steel penstock.

The powerhouse consists of a 13 × 10 m color steel structure with limited crane capacity. It houses a horizontal Francis turbine manufactured by Turab of Sweden and a 5 MVA synchronous generator manufactured by ABB, South Africa. The station also houses a $1\,\mathrm{m^3\,s^{-1}}$ bypass disperser valve to sustain river flow in an emergency shutdown. The control system designed and constructed by Marlborough Lines Ltd. enables the station to run unmanned with infrequent operator visits.

The output voltage of the station is 6.6 kV which is stepped up to 33 kV for transmission. The construction proposal of the scheme included a small substation and approximately 16 km of 33 kV transmission line, from the Horseshoe Bend power station site to link with the existing network at the Michelle power station on the lower river.

Interesting features in the design and construction of the scheme to cut costs and improve the scheme's environmental image includes the following:

- Use of weak onsite schist aggregates for dam construction.
- Construction of the RCC dam with an unformed upstream face.
- Installation of the penstock control valve at the outlet of the tunnel in preference to the inlet.
- Providing only limited crane capacity in the powerhouse despite the installation of a 42 ton generator. The generator was installed using jacks and load skates.
- Although not a requirement of the resource consents, a $1\,\mathrm{m^3\,s^{-1}}$ bypass valve was installed in the station to augment the river flow downstream of the station during emergency flow shutdown. It was considered that in case of an emergency shutdown the time delay of 20 min for the water released from the dam to travel down the gorge would result in a significant visual impact on the river.

6.17.4 Project Development and Processes

The risks to any development project involving the use of natural resources in the present economic environment are in general common to all. Viability is dependent upon obtaining suitable resources such as land and resource use consents, and a suitable rate of return, which is a function of upfront capital cost, operating cost and long-term projection of the value of the product produced.

Early in the project development phase the areas of risk associated with the project were identified as follows:

- Obtaining tenure for the areas of land required to construct and operate the scheme.
- Obtaining resource consents with sufficient scope and conditions to ensure viability of the project.
- Project management frameworks and control of financial risks.
- Control of the risks associated with the contract frameworks, given the limited geological data, tight time frames, new technology, and requirement for design development of the dam during construction.
- Risk mitigation and contingencies for unforeseen events during construction (e.g., floods).

The project management of land and resource consent procurement was carried out in-house, with relevant expertise and legal advice being bought in wherever required.

6.17.5 Land Tenure

The Horseshoe Bend project was surrounded by privately owned farmland. The Teviot river is bordered on both sides by a 20 m wide marginal strip owned by the Crown and administered by DOC. The riverbed is also Crown-owned and administered by the Commissioner of Crown Lands. The route for the power line crossed land owned by eight different landowners. Negotiating and obtaining the required tenure for structures over these different properties presented significant challenges, as there was no 'For Sale' sign on the gate and the standard 'willing buyer, willing seller' did not exist.

The final outcomes included the following:

- Adjacent to the area involving the majority of construction activities and structures a 760 ha block of private land was purchased. The land was then leased back to the vendor during the development phase of the project. This land has since been subdivided with 70 ha being retained for the scheme and the remainder sold.
- Option agreements for other parcels of land required for the reservoir and for laying of roads were obtained under various commercial arrangements including agreed purchase prices and 'disturbance' payments.
- Option agreements for easements for the power line route were negotiated, including a nominal annual lease payment based on energy production and the average annual spot price in the New Zealand energy market.
- Tenure for structures on the marginal strip was obtained on the basis of a lease with and annual payment linked to energy production and the average annual spot price in the New Zealand energy market. Negotiating this agreement with DOC was a difficult process as there was little precedent.
- Gaining tenure for the riverbed, which is owned by the Crown, was also difficult, as few procedures existed to enable this process. Despite the dam and reservoir having been constructed, a formal agreement is yet to be signed.

6.17.6 Resource Consents

Early in the development, obtaining resource consents in a climate of opposition from environmental groups and the public was seen as one of the major risks to the project. After significant consultations it was also apparent that even if resource consents were granted the attached conditions could also make the project unviable.

In the initial stages prior to firming up the detailed project design, affected parties were identified and discussions held to identify the major environmental issues. Further studies that were carried out then focused on these issues. This approach proved to be cost-effective and less time-consuming than trying to identify and study the possible issues prior to the consultation process.

Once the detailed studies were completed, further consultation was initiated and, wherever possible, solutions were proposed to address ongoing concerns. The consultation process was time-consuming and costly, but in the main produced sustainable outcomes both from an environmental and from an economic point of view.

The main issues included the following:

- The perception that the cumulative environmental effects of development of many small generating schemes by individual power companies are greater than those of one large scheme supplying the same total amount of power.
- Protection of Iwi and heritage values, including the possibilities for a future eel fishery. This included the arguments for and against inclusion of a fish pass on the dam structure.
- Protection of the habitat of native fish and native falcon that occasionally nested in the area.

- Landscape effects.
- Effects on the fishery and ecological values in the area. The effects included the impacts of the dam and the associated impoundment blocking fish passage, effects of the residual flow in the gorge section of the river between the dam and the powerhouse, the effects on diurnal variations in flow, and ramping rates on the river downstream of the powerhouse.
- Farmers' concerns regarding the effects of flow variation on the natural stock boundary afforded by the river, impacts on development of local roads, and the introduction of noxious weeds by construction equipment.
- Management of dam and reservoir safety both during construction and during operation.

Positive impacts of the scheme were seen as follows:

- Provision of further energy generation in the region.
- Enhanced road access to the river and improvements to road laying infrastructure in the area.
- Financial contribution to the local community in the form of a development levy.
- Provision of a lake fishery afforded by the reservoir. Some members of the fishing fraternity saw this as a negative impact as a portion of the river fishery would be lost and replaced by a lake fishery, the type of which there are many in Otago.

Mitigation measures to address some of the environmental concerns included the following:

- Limiting the daily operating range of the reservoir to 1.5 m.
- Maintaining a residual flow of $0.315 \, m^3 \, s^{-1}$ between the powerhouse and the weir, which in hydrological terms is equivalent to the natural 7-day 10-year return period low flow for that stretch of the river.
- Inclusion of a bypass valve on the dam for emergency station shutdown situations to limit the impacts of rapid flow changes in the residual river.
- It was considered unnecessary to put a fish pass on the dam, as the exotic fish populations above and below the dam were self-sustaining, and in the case of native fish the migration of the koaro into the upper catchment could endanger the resident rare galaxid population.
- Limiting the rate of change, and the upper and lower bounds of flow ramping downstream of the powerhouse, to reduce the impact on the fishery and natural stock boundary afforded by the river.
- Ongoing monitoring of the aquatic environment downstream of the powerhouse to assess the impacts of ramping.
- The land use and construction issues were covered by design and color selection to soften the structures, sedimentation control, additional fencing, upgrading of roads, and ongoing control of weed infestation.

The processes and methods for successfully gaining resource consents are well documented and have been talked through at some length in New Zealand over the past 10 years. Many of the processes followed for Horseshoe Bend emulated the examples of good practice carried out in the past. However, it is worth highlighting some aspects that did ease the process.

To overcome or at least soften the negative public perception of the project, the help of a public relations consultant was enlisted and a plan was formulated early in the consultation process. This involved the following:

- Setting up contact with the media at an early stage. Representatives of each of the major newspapers were taken to the site and given a full briefing, and throughout the project this was updated regularly. This had many advantages including building a proactive relationship and avoiding any 'myths' being created in the media by groups opposed to the project. Once a myth or negative perception has been established with the general public, it is hard, if not impossible, to erase it; cell phone towers is an example of this.
- Being open and frank about the negative environmental aspects as well as about the benefits of the project. This ensured that no surprises were unearthed at later stages of the process that could impact the credibility of the company.
- Keeping the issues and benefits of the project as local (Central Otago) to avoid being dragged into nationwide arguments.
- Having the technical people involved in the project front the media. This involved a certain amount of training for these individuals. This added significant credibility to the information passed to the media and the perception of the public at large.
- Themes and messages passed through the media were kept simple and were repeated often in different ways.

The public relations exercise was only part of the consultation process but it was felt that this is a powerful tool that can work for a hydroelectric project, and surely the investment reaped significant benefits in creating a positive image of the project in the eyes of the public, which in turn assisted the process of gaining resource consents.

During the risk management review of construction methods for the dam, a concrete-faced rockfill structure was estimated to be a financially competitive, if not the cheapest, option. However, the RCC option was chosen as it was considered the lowest-risk alternative because of the following reasons:

- Under the conditions of any resource consent the physical location of the dam structure is generally fixed to what is nominated. Any variation to this location resulting from, for example, uncovering unforeseen foundation conditions would require revisiting

the consultation and hearing process resulting in significant time delays. It was considered that the RCC option was less sensitive to these problems than other dam models.
- The issue of flood diversion capacity during construction of the dam was accentuated during the resource consent application process by the events at Opuha. The options for a large diversion work were limited by the physical characteristics of the site and the cost impacts on the project. The size of the river diversion chosen had a reasonably high probability of being overtopped, but this risk was considered acceptable, as the consequences of overtopping were significantly reduced by using the RCC option. In several cases overseas, RCC dams had been overtopped by floods during construction with only minor damage to the works, limited disruption to the construction program, and no risk to the downstream inhabitants. Later in the project, the insurers who covered the public liability and contract works insurance for the dam also viewed these characteristics favorably.

To conclude, despite significant opposition to the scheme especially from the fishing fraternity, the resource consents were gained without full reference to the Environment Court.

6.17.7 Project Management

Hydroelectric projects, and especially those that involve significant foundation construction such as dam projects, are renowned for cost overruns. Smaller projects such as Horseshoe Bend are even more sensitive to overruns than the larger ones. For example, seven of the eleven small hydroelectric projects built in the early 1980s experienced significant cost overruns [1].

As mentioned above, PGL had significant experience in developing and managing small hydro projects in Central Otago. The company had developed an independent culture over 70 years of operation and had taken on many previous projects using in-house staff and resources. Design and construction of a significant proportion of the project were outside the scope and expertise of its existing resources, but the company was still keen to have an involvement in all phases of the project along with accepting some of the risks of maintaining this involvement. The reasons for this were that PGL had a known standard of quality it wished to achieve and felt that it was the best judge of fit-for-the-purpose criteria for the project, and that the involvement would also enhance the existing expertise within the company. The management and directors of the company were obviously keen to increase certainty and reduce the financial risks of the project wherever possible, but realized that to shed the risk completely would result in adding a significant premium to the project costs. Some of the key principles PGL took into the project were as follows:

- To maintain an in-house involvement with the project to ensure that the standard of design and construction was fit for the purpose, while keeping the costs to a minimum and enhancing expertise within the company.
- Wherever possible, use contractors based in Central Otago. It was considered that the scheme and dam construction had potential to inject significant revenue into the Roxburgh and Central Otago economy. Using local resources was seen as having long-term benefits in gaining local support for future projects and providing opportunities and employment for PGL customers.
- When forming and administering any contractual relationships during the project, the overall objective was for all parties to benefit financially and avoid situations where one party could derive a significant financial windfall at another's expense. It was considered that the project would have a greater chance of a successful outcome if this objective could be achieved.

PGL considered many project management options and, taking into account the above criteria and time frames involved, decided upon a partnering option with a local contractor, Fulton Hogan Central (FH), who could call upon nationwide resources as required. A design–build contract for the dam was negotiated with FH and, to avoid duplication, FH resources were also used for onsite management of health and safety and the role of Engineers Representative for other civil contracts. Opus International Consultants were chosen as consultants for the project. PGL used in-house resources to fulfill the engineer-to-contract role.

Some features worth noting with regard to the project management includes the following:

- Given the short time frame for design and construction (~ 10 months), decisions were required quickly and the team of people involved had to work well together from the outset. PGL worked hard on choosing a small team of people with the required attributes to achieve this.
- The Opus offer included allowance for importing expertise from overseas in a peer review capacity for the RCC dam design and construction. This proved invaluable for all parties concerned including the contractor.
- Throughout the project, significant time was spent reviewing the sensitivities of the project cost to design changes and acceptable risks that could be taken while still achieving the design objectives.

6.17.8 Contract Framework

In the past, civil contracting has been an area of significant financial risk to hydroelectric projects especially those involving dams. In the case of Horseshoe Bend this risk was compounded by the following factors:

- The small scale of the project resulted in limited initial subsurface investigations.
- The technology for using RCC on a significant scale was new to New Zealand contractors.

- The short time frames available and use of local low-strength aggregates meant that a significant portion of the dam design needed to be developed during the contract period.

All of these factors influenced the choice of the design–build contract form for the dam. The contractor chosen had the advantages of being local and having proven quality systems that were seen as essential when using new technology.

Other civil construction contacts for the project including the tunnel, penstock, and road laying were let out as separate contracts and were of a scale that local contractors could provide competitive bids.

The mechanical, electrical, control, and transmission components of the project were procured under a design–build contract, and the powerhouse was constructed using in-house resources.

Some features worth noting with regard to the contracting framework includes the following:

- The design–build contract for the dam was based on an amended form of NZS 3910. This standard was used in preference to other proprietary international design–build contract standards, as all parties were reasonably familiar with the document.
- PGL made significant input into the design and specification process as part of cost control management, and they had definitive ideas on the standard of product and final outcomes they wished to achieve.
- PGL was conscious of possible conflicts of interest in a design–build contract for a dam project. To overcome this, a clause was included in the special conditions requiring the contractor, their designer, and the client to sign off at significant milestones (e.g., completion of foundation excavation, diversion works, etc.) during the construction. At these milestones the standard achieved was required to be to the satisfaction of all parties concerned. If any party was not satisfied the construction could not proceed to the next stage until the concerns were resolved. PGL also employed an independent peer reviewer, and part of his brief was to inspect and report on the works at these significant milestones.

6.17.9 Interesting Features of Design and Construction

6.17.9.1 Control Valve Positioned at the Tunnel Outlet

Most hydro project designs include some form of control valve located at the upstream end of the water conveyance system. This is to allow for emergency shutdown and dewatering for routine maintenance of the pressurized conveyance system.

In the case of Horseshoe Bend, this would have required the installation of a penstock gate at the upstream end of the 180 m long 2.5 m high tunnel. During the concept-and-design phase of the project the option of installing a butterfly control valve was chosen. This option was chosen after consideration of the following factors:

- There was significant cost benefits in purchasing a 1.6 m diameter butterfly valve versus a penstock valve to seal off a 2.5 m high D cross section tunnel.
- The tunnel was constructed using very competent schist rock and fully lined. Therefore the probability of the tunnel requiring regular long-term maintenance was considered low.
- Presence of a storage dam relatively close upstream of the intake afforded significant control of inflows into the scheme. In addition, the intake reservoir storage was relatively small and there was sufficient valve capacity on the dam to drain the reservoir over a short period to dewater the tunnel.

A 1.6 m diameter butterfly control valve was installed in the pipeline at the downstream end of the tunnel with a fail-safe battery-powered backup shutdown system.

6.17.10 RCC Dam Design and Construction

6.17.10.1 Geological and Hydrological Setting

Geological setting. The Teviot river at the dam site is incised into the terrain some 20–30 m. Isolated rock outcrops are present along the river banks, but the side slopes are typically 2H:1V. The river gorge is cut into relaxed quartzofeldspathic schist with flat-lying foliation. A thin mantle of loess and colluvium is present, and the degree of weathering of the schist rock is reflected in the variable side slopes of the gorge. The rock is moderately weathered to around RL 588 m, and slightly weathered to around RL 581 m at the river level.. Horizontal foliation shears are present in the abutments, although no wide shears were identified immediately below the river channel. Steeply dipping orthogonal joint sets are present throughout the site, some with silt infilling following relaxation of the rock mass. Very high water flows (>100 lugeons) were measured in the shallow relaxed abutment zones during packer testing, but low permeability conditions (generally 0–5 lugeons) were measured in the rock below the relaxed zone.

Hydrology. The dam site is situated 7 km downstream of the controlled outlet of Lake Onslow. The catchment area above the dam site is 209 km^2, and the probable maximum flood peak (PMF) has been assessed at 335 m^3 s^{-1}. The spillway operational design capacity has been set at 0.6 PMF (=200 m^3 s^{-1}), with provision to also pass the full PMF flow.

Reservoir storage is provided in Lake Onslow, and the volume impounded by the low dam will not contribute significantly beyond daily flow balancing. Lake Onslow was deepened during the construction period to reduce the peak flow rates in the river and in the diversion works.

6.17.10.2 Site Layout

The location of the dam was dictated primarily by the need to be within 200 m downstream of the tunnel portal that supplies the low-pressure pipeline and penstock, to make effective use of the topography. Detailed dam layout decisions were made on the basis of quarry development considerations near the right abutment, suitability for temporary diversion layout, and local foundation rock conditions. A layout plan of the site is presented as **Figure 2**. The scale of the development was not large enough to justify the establishment costs normally associated with RCC production, but by keeping the setup and production costs tightly under control, it was possible to economically produce the relatively small quantity of RCC required.

Diversion and outlet works. Diversion capacity of $19 \, m^3 \, s^{-1}$ was provided using twin 1600 mm diameter steel pipes encased in conventional concrete to provide for diversion flow and subsequent operational discharges to the river channel. This diversion capacity (at overtopping of coffer dam level) was selected at an 11% assessed probability of exceedance over the critical 4-month RCC construction period. As the small reservoir can be effectively dewatered for major maintenance activities, only simple outlet control valve gear is provided for residual flow, compensating flow, and dewatering purposes. The adoption of twin conduits allowed minimum flows to be maintained during the transition from the diversion to operational mode and during later maintenance activity.

Spillway. The 28 m wide spillway incorporates an ogee crest profile and energy dissipating steps transitioning to an effective 0.8H:1V slope. The stage discharge rating for a 28 m long crest at RL 593 m is shown in **Figure 3**. A length of reinforced concrete training wall on top of the RCC at each abutment directs the spillway flow into the main river channel downstream. The stilling basin is formed by a natural lateral contraction immediately downstream of the dam, and a conventional concrete apron was constructed in the original river channel.

Cross section. Larger RCC dams generally incorporate a formed vertical upstream face [2] to make maximum effective use of the RCC volume used. However, the small size of this dam and the desire to achieve maximum seepage path lengths along open joints in the abutment rock resulted in the adoption of an unformed face profile as shown in **Figure 4**. The structure is approximately 16.5 m high and 65 m long and has a concrete volume of approximately 7000 m³. The dam was constructed in continuous 300 mm lifts (two lifts placed per day) with contraction joints cut by vibrating plate at 14 m centers. Conventional concrete is incorporated into the river channel infill, and the crest has been detailed in conventional concrete to provide increased mechanical strength and frost resistance over the RCC. Grout enrichment of RCC [3] was used at abutment contact zones and around water stops to improve watertightness. A 3 m wide upstream cement mortar bedding strip was used between lifts to control leakage. Further RCC issues are discussed in References 4–6.

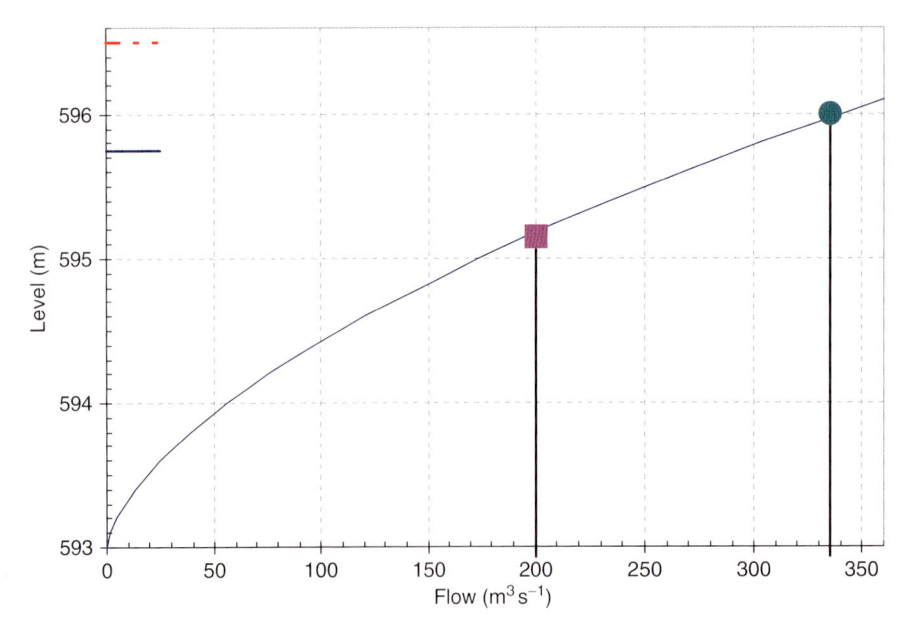

Figure 3 Spillway discharge rating.

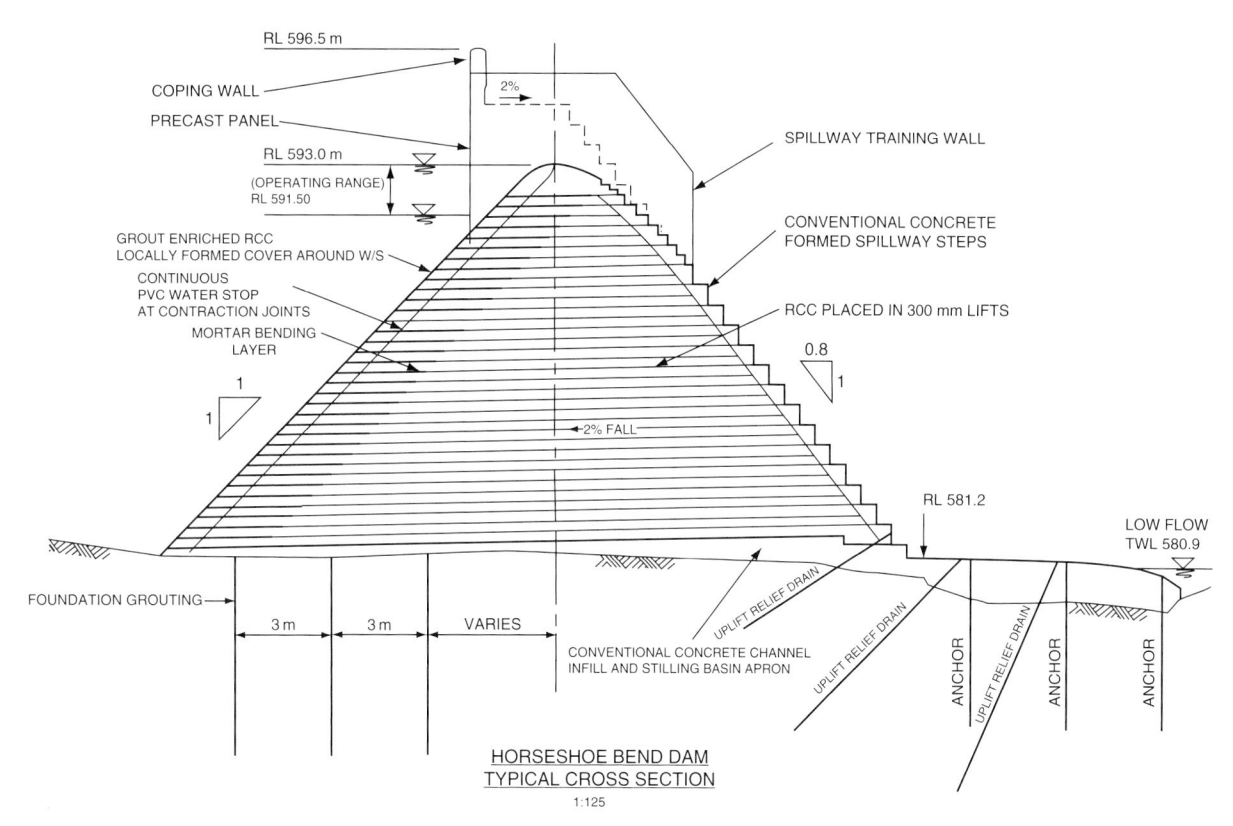

Figure 4 Typical cross section.

6.17.10.3 RCC Mix Design and Handling Characteristics

The design called for four zones of RCC to be incorporated in the dam. The bulk placement was unmodified RCC as delivered from the plant. The RCC was modified by the *in situ* introduction of cement–water grout (grout enriched RCC or GE-RCC) to improve the shear strength, durability, adhesion, and waterproofing in selected areas of the dam such as at abutment contact zones and around water stops, and cement mortar was used between lifts at strategic areas to seal possible seepage paths. Additional *in situ* modification of RCC that would potentially be exposed to frost and spillway discharge was allowed for in the design in lieu of conventionally batched structural concrete. The final decision to use this air-entrained mix was subject to the results of field trials. The RCC was manufactured predominantly from crushed schist aggregate quarried onsite. The schist rock obtained from the quarry had an unconfined compression strength across the foliation in the range 20–40 MPa, and a tensile splitting strength across the foliation of 0.7–1.0 MPa. Aggregate absorption (< 1% limit) was a convenient measure of the degree of weathering in the quarry. The schist product tends to produce excessive silty fines in relation to the sand fraction obtained, so imported Roxburgh sand was added to the blend to achieve the required particle size grading. There was no source of fly ash or other cementitious substitute, so low-heat cement alone was used with a water-reducing agent. A long-term compressive strength of 15 MPa (average) was initially established for the RCC mix based upon the cement content expected to be used, but this figure was higher than necessary for the structural demands in the internal zones of the dam. Conventional concrete and/or air-entrained GE-RCC for use in exposed zones had a specified 28-day compressive strength of 25 MPa.

Lab trials. Initial laboratory testing of schist aggregates obtained from the diversion excavation was used to establish the specified acceptance criteria for the aggregates to be won from the production quarry onsite. The degree of weathering of the rock samples was assessed to establish the weathering, crushing, and absorption criteria for the production quarry. Laboratory trial mixes [7] commenced with an aggregate and sand blended grading curve at 30–38% passing 4.75 mm, then progressively increasing up to 52% passing 4.75 mm. The most suitable trial mix was established with 50% passing 4.75 mm and including 18% screened Roxburgh East Sand. Cement contents of 135, 143, and 150 kg m^{-3} were examined, with the 150 kg m^{-3} mix being adopted for the field trial. Water/cement ratios from 0.8:1 to 1.0:1 (w/w) were examined and 0.9:1 w/c was adopted with a high-range water reducer to produce a Vebe consistency of around 25 s. The Vebe apparatus was based upon the USAC CRD C53-96a test method modified to suit a 50 Hz vibrating table. The 91-day compressive strength of this adopted mix was tested at 15.5 MPa. Grout enrichment of the adopted lab mix was examined at total cement contents in the range of 215–285 kg m^{-3}, at total w/c ratios from 0.70:1 to 0.80:1 (slump 40–180 mm).

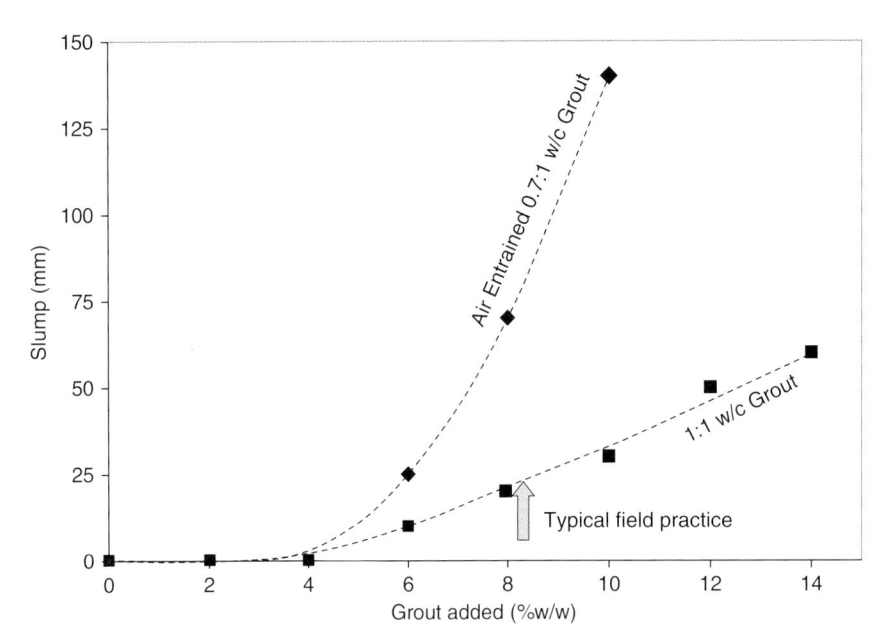

Figure 5 Grout enrichment.

Air entrainment was achieved by agitating the enrichment grout, but the final air content in the mix was found to be inconsistent. Grout-mixing was found to require considerable energy input, and the most effective grout distribution was achieved by placing the grout at the bottom of the lift and allowing the heavy aggregate to displace the aerated grout under vibration. The very high air content required in the grout lowered the density to such an extent that it would not readily work down into the underlying RCC mix. The transition from zero-slump to low-slump properties is shown in **Figure 5**.

The 91-day compressive strength of GE-RCC lab trial specimens was found to range from 17.0 to 21.5 MPa, well below the target 25 MPa value. The decision on the use of air-entrained GE-RCC on the downstream face was reserved pending results from the field trial pad.

Field trials. Following the production of aggregates from the onsite quarry and commissioning of the pug mill plant, a trial pad was constructed on 7 January 1999 which included a formed-step face. Compaction with a Dynapac CA151 7.5 ton 1.67 m wide self-propelled single-drum vibrating roller was evaluated to confirm that this unit, which was narrower and lighter than the specified plant, was suited to the application. The compaction target was 98% of the theoretical air free (TAF) density, that is, 2% air voids maximum. Both low- and high-frequency modes were found to be suitable with up to 8–10 passes on 300 mm lifts. The RCC mix at this time was still somewhat sandy (50% passing 4.75 mm) and dry (Vebe 25 s). The twin-probe nuclear density meter (NDM) as specified was not available in New Zealand, so a single-probe Troxler 3440 unit, normally used for soil testing, was used at 100 mm and 250 mm direct transmission depths. The aggregate grading of the trial pad RCC was found to be on the fine side of the specified envelope with 8–10% passing 75 μm and 52% passing 4.75 mm. Additional water was found to be necessary to achieve satisfactory workability of the mix. The w/c ratio needed to be raised to around 1.15:1, and there was concern regarding the effect of this on strength. The 7-day compressive strength results for the pad were 7.5–8.0 MPa, although some test results were as low as 5 MPa. Production commenced with the cement content increased to 162 kg m^{-3} while the strength was established by further testing.

Enrichment of the mix placed in the trial pad proved to be impractical in other than very small quantities owing to the degree of vibration required to achieve effective mixing. Immersion vibrators (electric 50 mm) were found to be not powerful enough, which was contradictory to the laboratory experience that had indicated that the risk of overvibration was a real possibility. The decision was made to not progress to full air-entrained GE-RCC production, and conventional concrete was adopted for the downstream face zone.

RCC production. Quality control measures included monitoring the crushing and weathering resistance of the aggregate, and absorption and soundness of the aggregate; wash-grading of the wet mix; and accelerated curing of test cylinders to give daily feedback on performance. Workability was measured with a Vebe apparatus. Compaction effectiveness was monitored using an NDM to confirm that voids were below the 2% limit. Water cooling of aggregates was needed to keep the mixing temperature below 20 °C. As the pug mill mixer operates on a continuous feed basis rather than as a batch process, there was a need to continuously obtain feedback on the output. Intensive monitoring of the initial six lifts resulted in further changes to the mix design as shown in **Figure 6**. The grading was modified to reduce the sand content outside of the specified envelope and to increase the water content. A Vebe consistency of 16 s was targeted, and the wet mix showed much improved resistance to segregation in the feed-out bin. A cement content of 162 kg m^{-3} was retained, and a 0.96:1 w/c ratio was adopted.

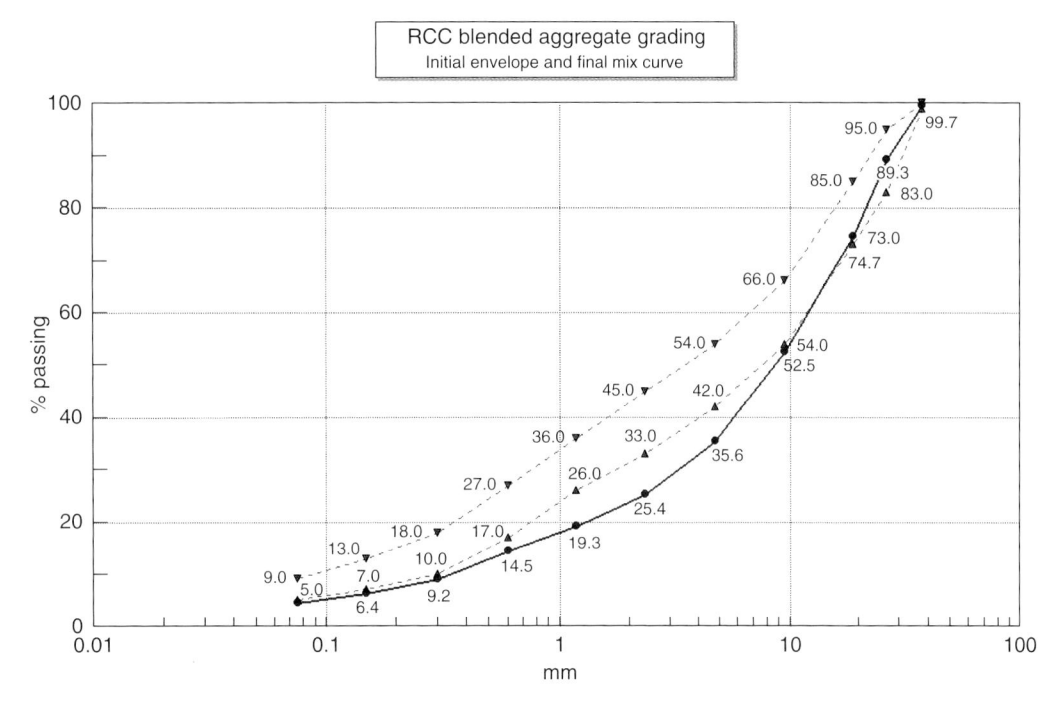

Figure 6 RCC aggregate grading.

Compacted density results were close to the 98% TAF threshold, but measurements in the 97–98% range were not uncommon. The results are shown in **Figure 7**, with the TAF density results above 100% indicating a slight variability in the mix and/or the NDM test method.

The 7-day compressive strength results were initially inconsistent, varying from 5 MPa to 10 MPa and higher. Variation in the aggregate stockpiles and difficulty in maintaining plant calibration were thought to be the key influences on consistency of performance. The mix adopted for the bulk of the production (lift 8 and above) was not varied, but mixing plant control was improved from lift 24, as illustrated in **Figure 8**. A summary of the design mix is tabulated below.

Unmodified RCC		GE-RCC	
Fines volume	11.2%	Grout w/c	1.00
Paste/mortar	51%	Application rate	200 kg m^{-3}
Cement	162 kg m^{-3}	Effective cement	231 kg m^{-3}
Water/cement	1.08	Effective w/c	1.05

Compressive strength gain for the RCC test cylinders taken from lift 24 onward is shown in **Figure 9**. The average, 10 percentile, and 90 percentile compressive strength of the 150 mm diameter test cylinders is shown for the accelerated 65 °C 18 h tests, together with the laboratory-cured 7-, 28-, and 90-day tests. The accelerated-cure test with its 24 h turnaround gave a reasonable degree of correlation with lab-cured cylinder strengths as shown in **Figure 10**.

Conventional concrete has been retained for the downstream face and high-level upstream face zones. Grout enrichment has been restricted to abutment contact and water stop zones, which do not require higher compressive strength. Higher water content grout (1:1 w/c) has been used to achieve the required field mixing efficiency.

Grout enrichment (non-air entrained) was found to be most effective at a total cement content close to 230 kg m^{-3}, a slump of less than 40 mm, and a compressive strength equivalent to that of the base RCC mix.

6.17.10.4 GIN Foundation Grouting Method

Although the dam is only a low-head structure, a moderate amount of foundation grouting has been undertaken to seal the larger rock defects within the remaining relaxed rock mass to control the tendency for erosion of the joint infill material. Potential hydraulic displacement of rock blocks during grouting was a concern, given the orientation of the defect planes, so the grout intensity number (GIN) technique [8–10] was adopted. This grouting technique provides for improved injection control in these circumstances, through the continuous adjustment of injection pressure subject to the rate of grout take experienced. The method is

478 Design Concepts

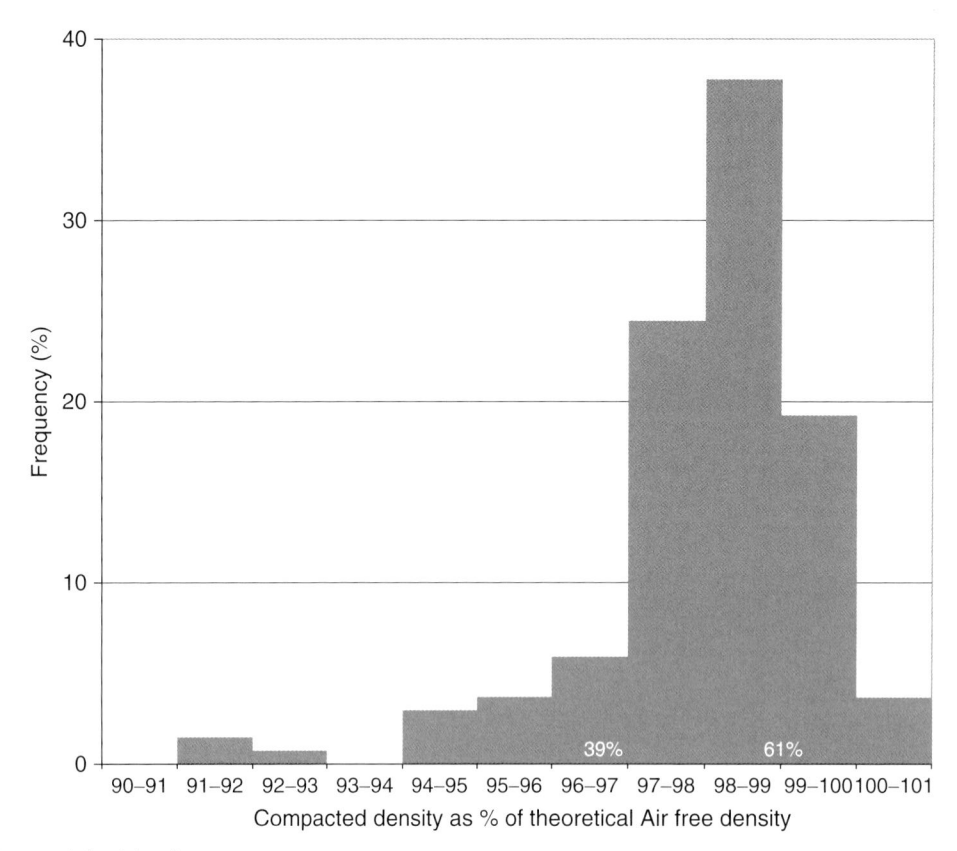

Figure 7 RCC compacted wet density.

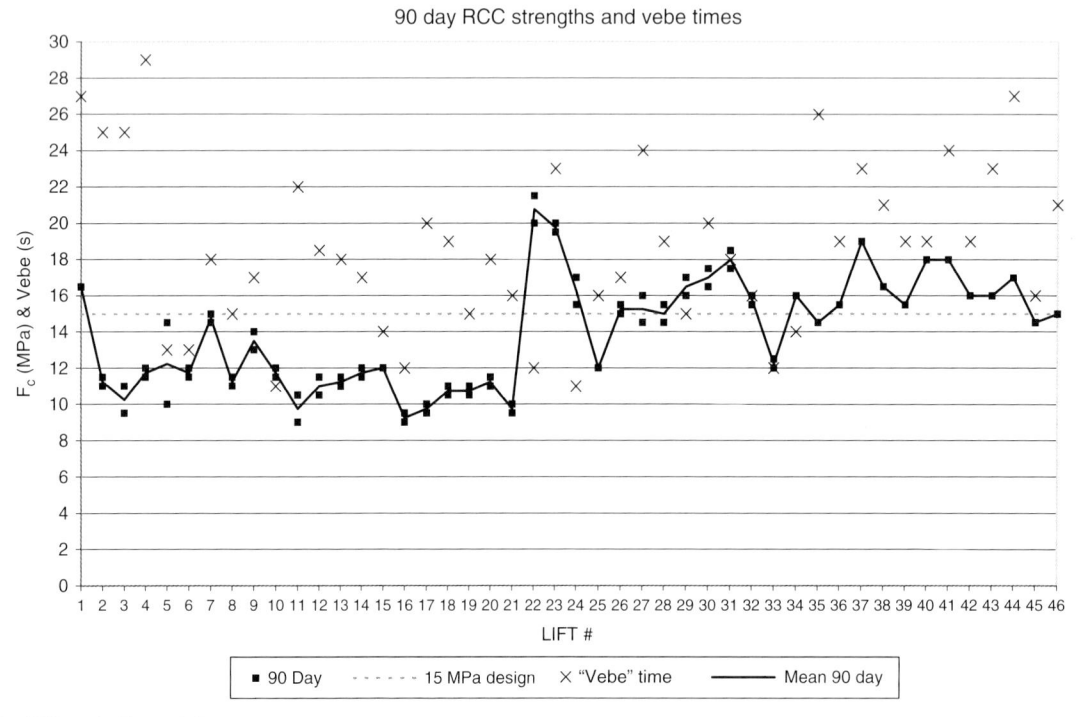

Figure 8 RCC production variation.

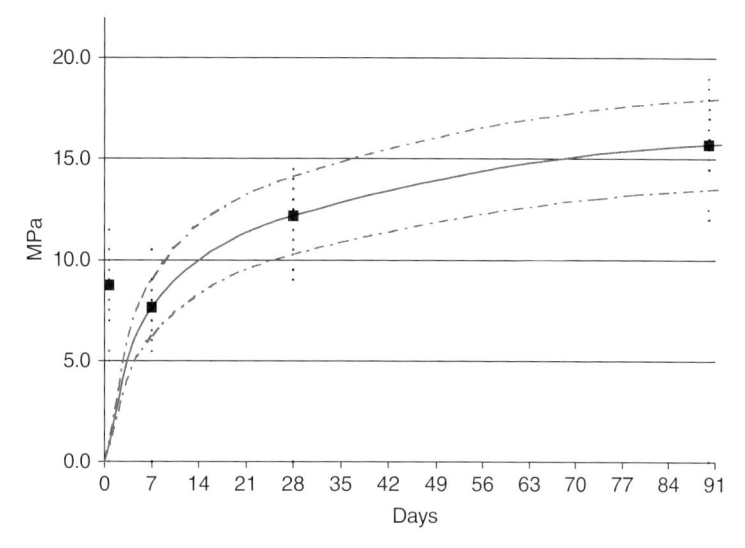

Figure 9 RCC strength gain.

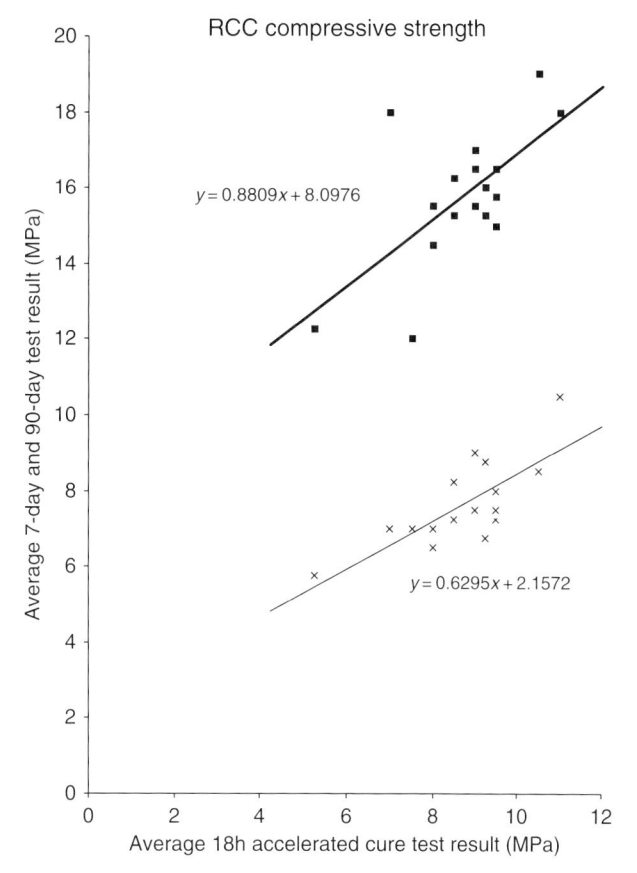

Figure 10 Accelerated test correlation.

based upon the use of a single grout mix, which simplifies field operations. As there was no need to penetrate very fine joints, a 0.6:1 w/c mix (w/w) with a water-reducing agent was adopted. The GIN curves used for the foundation and abutment zones are shown in **Figure 11**.

The depth of initial grouting was limited to 6 m to match the highly relaxed zone identified in the packer tests. It was also recognized that further strategic grouting could readily be carried out during commissioning if required.

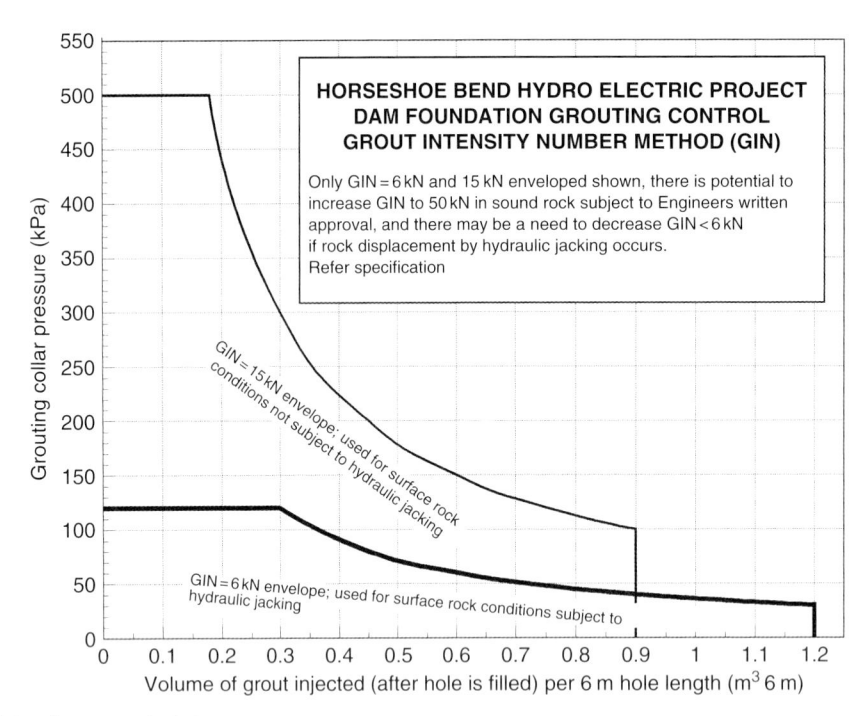

Figure 11 The GIN intensity curves adopted.

6.17.10.5 Commissioning/Performance Monitoring

During commissioning in April–May 1999, a significant seepage flow of some $100 \, l \, min^{-1}$ developed within the right abutment below the shallow grouted zone. The reservoir level was lowered, and a remedial grouting program was carried out. The water path was effectively intercepted, and the seepage was reduced to a few liters per minute. **Figure 12** illustrates the grouting pattern designed to intercept the foliation shears and the steeply inclined joint passing under the dam blocks.

Standpipe piezometers have been installed within the dam and foundation to monitor uplift pressures. Seepage at RCC lift joints and along the vertical contraction joints was evident at commissioning, and **Figures 13–15** illustrate the uplift that has been measured on the lift joints at three locations within the RCC dam. Possible uplift profiles have been included, although, given the lack of instruments, these are somewhat speculative.

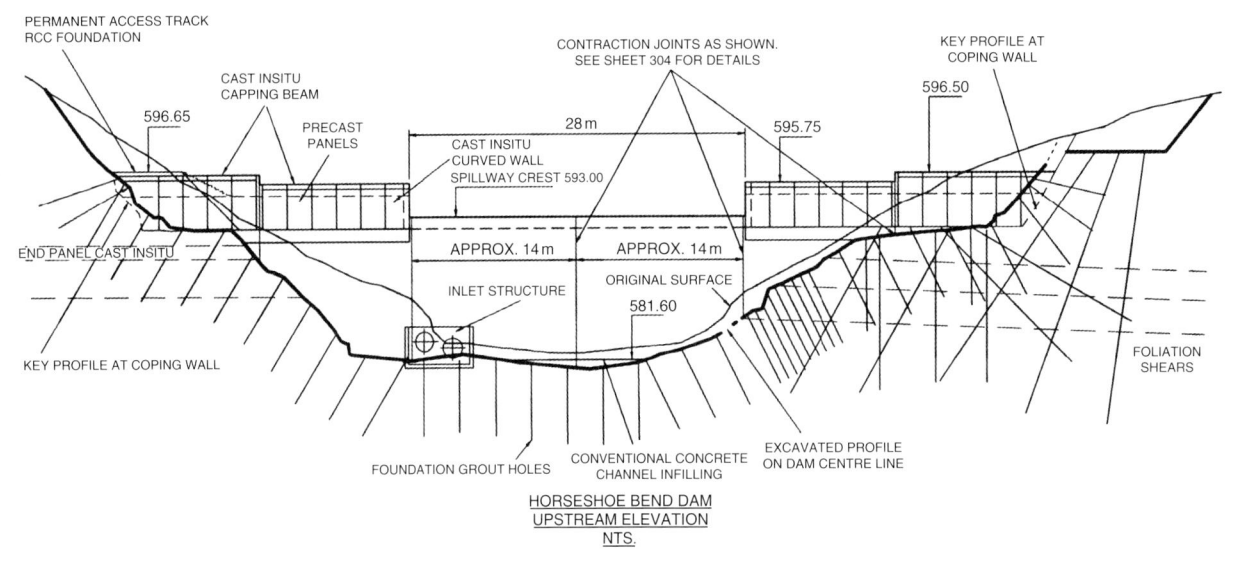

Figure 12 Upstream sectional elevation.

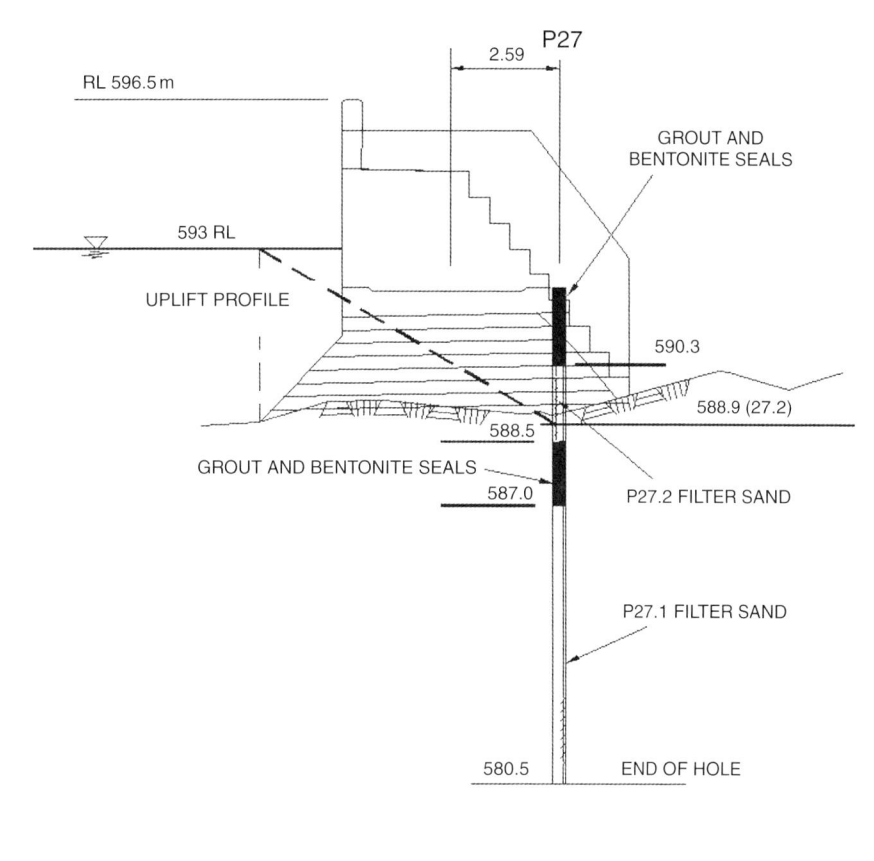

Figure 13 Cross section at right abutment.

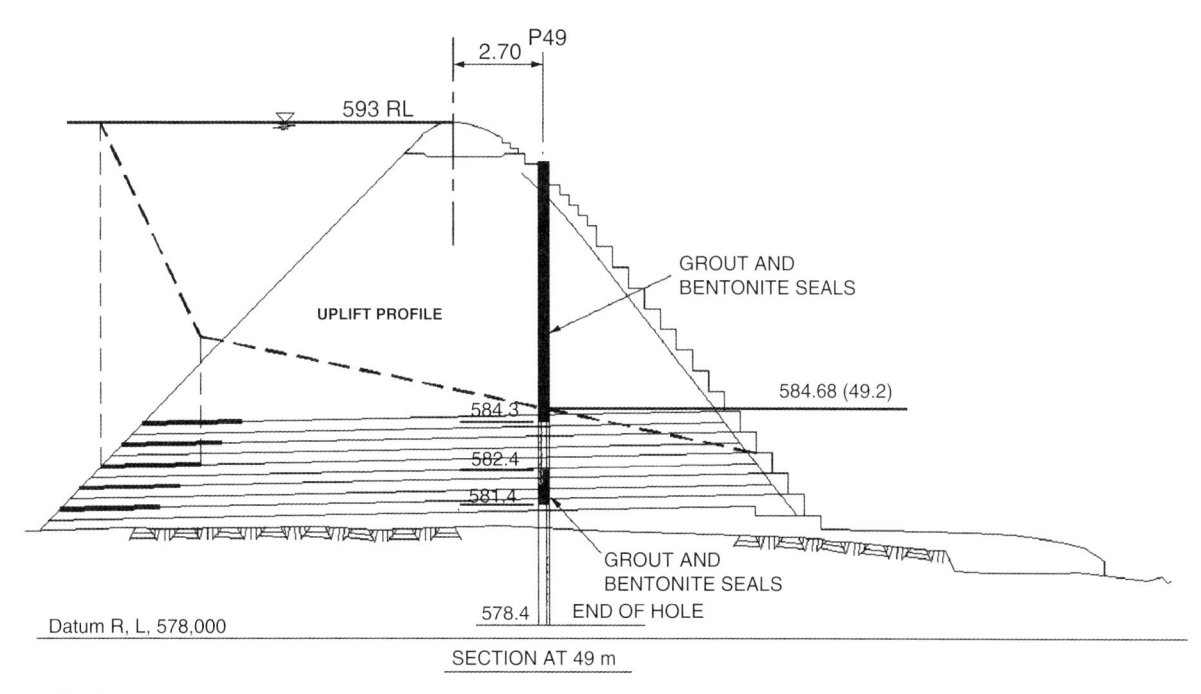

Figure 14 Central cross section.

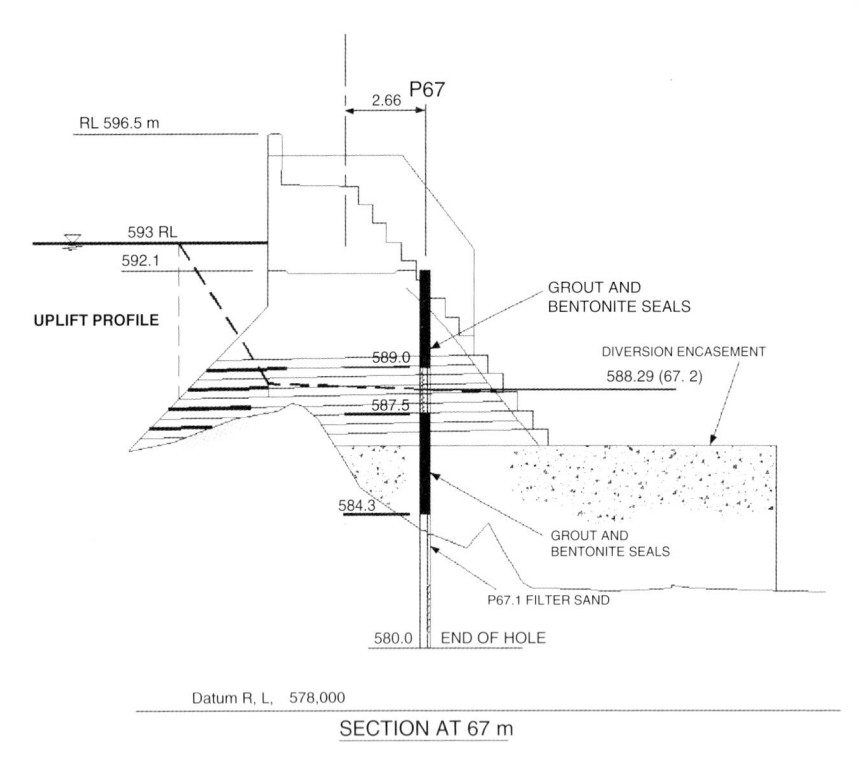

Figure 15 Cross section at left abutment.

6.17.11 Conclusions

The Horseshoe Bend hydroelectric project, although small in scale, encountered during the development and construction phases many of the issues and risks associated with any 'greenfield' hydropower development project in New Zealand in recent times. In the initial stages of progression of a hydroelectric development project today, it is not necessarily technical issues but environmental issues, and securing some form of tenure over the required land, that govern the overall viability of the project.

Gaining tenure over the required land was time-consuming and resulted in the negotiation of agreements in many different forms. The process of gaining resource consents also required significant consultation with the affected parties. Advising the media early on in the process and communicating balanced information was a key element in avoiding misinformation gaining credence and influencing public perception. The consultation and environmental assessment process was time-consuming and costly but in the main produced sustainable outcomes both from an environmental and from an economic point of view.

Although design and construction of a significant proportion of the project was outside the scope and expertise of PGL's existing resources, maintaining a significant involvement in the project management of the design and construction phases of this development proved successful. It is the view of PGL that with small hydropower development projects the developer needs to actively drive the risk management processes, as this role cannot be effectively delegated.

For critical path items such as the dam construction, and mechanical and electrical component supply and installation, the design–build contract form was the only means of achieving the tight project construction time frame. Potential conflicts of interest that can arise in this contractual environment need to be recognized and dealt with through the implementation of a robust review process with adequate intermediate steps (hold points) incorporated to avoid possible program delays. Tendering of other elements of the project as a series of smaller-scale contracts proved successful, enabling local contractors to successfully bid for the work and thus providing a boost to the local economy.

Although there was a lack of RCC construction experience in New Zealand, the adoption of RCC technology was driven by the lower risk profile that this approach offered. Low sensitivity to both diversion flood risk and unforeseen foundation conditions was the major factor in favor of RCC. Notwithstanding the small scale of this development, the application of RCC construction has proven to be a practical and economical alternative to traditional dam-building techniques at this site. The schist aggregate has proven to be suitable for the moderate strength being sought.

In situ grout enrichment of the RCC has been shown to be effective for modifying the properties of the material in strategic areas. Mixing of grout *in situ* is very labor-intensive, so the extent of GE-RCC must be limited if the efficiency benefits of the RCC process are not to be compromised. Enrichment using air-entrained grout in place of conventional concrete did not prove to be practical and was abandoned.

The GIN grouting technique has also proved to be an improvement over the traditional limiting-pressure method at this site.

To conclude, PGL initially set out to construct a small hydropower development project that was environmentally sustainable and economically viable, and provided an example of what could be achieved with similar developments in the future; the end result exceeded expectations.

References

[1] Electrical Supply Authorities (1987) Local hydro-electric power schemes. Report of the Audit Office. New Zealand; February 1987.
[2] Forbes BA (1995) Australian RCC practice, nine dams each different. International Symposium on Roller Compacted Concrete Dams, October 1995.
[3] Forbes BA and Williams JT (1998) Thermal stress modelling, high sand RCC mixes and *in-situ* modification of RCC used for construction of the Cadiagullong dam NSW. ANCOLD Conference and Proceedings, Sydney, Australia.
[4] Forbes BA (2000) Solving some long-standing RCC concerns. *The International Journal on Hydropower and Dams* 7(3).
[5] Forbes BA (1999) Grout enriched RCC: A history and future. *International Water Power & Dam Construction*.
[6] Forbes BA, Lichen Y, Guojin T and Kangning Y (1999) Jiangya dam, some interesting techniques developed for high quality RCC construction. International Symposium on RCC Dams, Chengdu, China, April 1999.
[7] Roller compacted concrete. *Technical Engineering and Design Guides as Adapted from the US Army Corps of Engineers*, No. 5. New York: ASCE Press.
[8] Lombardi G and Deere D (1993) Grouting design and control using the GIN principle. *International Water Power & Dam Construction*.
[9] Lombardi G (1996) Selecting the grouting intensity. *HydroPower and Dams* 4.
[10] Ewert FK (1996) The GIN principle, Parts 1 & 2. *International Water Power & Dam Construction*.
[11] Horseshoe Bend Hydro-electric Scheme (1997) Assessment of Effects on the Environment; Central Electric Ltd.; March 1997.

6.18 Recent Achievements in Hydraulic Research in China

J Guo, China Institute of Water Resources and Hydropower Research (IWHR), Beijing, China

© 2012 Elsevier Ltd. All rights reserved.

6.18.1	Introduction	485
6.18.2	Energy Dissipation	487
6.18.2.1	Slit Bucket	487
6.18.2.2	Flaring Pier Gate	489
6.18.2.3	Jet Flows Collision with Plunge Pool in High-Arch Dams	492
6.18.2.4	Orifice Spillway Tunnel	494
6.18.2.5	Vortex Shaft Spillway Tunnel	495
6.18.3	Aeration and Cavitation Mitigation Measures	497
6.18.4	Flow-Induced Vibration	499
6.18.5	Discharge Spraying by Jet Flow	499
6.18.6	Hydraulic Field Observations	502
References		504

Glossary

Aerator A special device used to tract the air into the bottom floor of a spillway tunnel or chute spillway. It consists of air vent, offset of ramp.
Flaring pier gate One type of energy dissipater. The pier on the downstream part is expanded and the width between the piers is reduced. It is applied on the surface spillway to form a 3D flow.
Jet flow collision Collision of jet flows from the surface spillway and the middle outlet before impinging into the plunge pool to increase the ratio of energy dissipation.
Orifice spillway tunnel One type of energy dissipater. It consists of one or several orifices installed inside a spillway tunnel.

Plunge pool A water body formed by a secondary dam built just downstream of the dam for dissipation of energy.
Slit bucket One type of energy dissipater. The width of the flip bucket is contracted symmetrically or asymmetrically. It can be applied in the outlet of the spillway tunnel, chute spillway, surface spillway, and middle outlet. The flow through the slit bucket is contracted latitudinally and dispersed longitudinally.
Spraying Rainfall is formed by splashed jet flow with a high intensity during the discharging.
Vortex shaft spillway tunnel One type of energy dissipater. A vortex chamber is connected to a vertical shaft and then a spillway tunnel. The vortex chamber can form a rotating flow.

6.18.1 Introduction

The hydraulic research has achieved noticeable improvements as the hydropower projects have been developing at a faster rate in China since the 1980s, mainly on the new energy dissipaters, aeration and cavitation mitigation, pressure fluctuation and flow-induced vibration, flow discharging spraying, and prototype observations [1]. Table 1 gives the typical characteristics of Chinese hydropower projects, which are high dams in narrow valleys with large discharge flows.

General Report of the 13th Congress of ICOLD [2] gives statistics of discharge facility applications worldwide with the physical parameters of L/H and P and their combinations (see Figure 1). The author has put the parameters of some selected projects from China and the United States into the same figure for comparison.

Table 1 and Figure 1 show that (1) most dams are over 200 m high and some are nearly 300 m high; the highest dam under operation is Ertan Arch Dam with a maximum height of 240 m and the highest dam under construction is Jinping Arch Dam with a maximum height of 305 m; (2) the discharge flow is over $20\,000\,m^3\,s^{-1}$ and the largest one is $102\,500\,m^3\,s^{-1}$ in Three Gorges Project; this indicates that the unit width discharge flow is usually over $200\,m^3\,(s\text{-}m)^{-1}$; (3) more than one type of discharge facilities are found in different types of dams, such as the surface spillways combined with middle outlet, chute spillway, or tunnel spillway; (4) some new types of energy dissipaters are involved, such as flaring pier gates with stilling pool or with roller compacted concrete (RCC) stepped spillway, flip buckets with plunge pool, orifice spillway tunnel, or vortex spillway tunnel; (5) high head and large gates are used.

As the complicated hydraulics is the key issue in the design and operation, and the characteristics of energy dissipaters of dams in China are difficult to determine, efforts have been made during the designing stage based on the physical model experiments. To verify the scientific research and designing solutions, several hydraulic field observations on large projects have been undertaken when they are in operation.

Table 1 Typical hydraulic characteristics of Chinese hydropower projects

No.	Name of project	Type of dam	H (m)	Q (m³ s⁻¹)	Surface spillway b × h (m²)	Middle outlet b × h (m²)	Bottom outlet b × h (m²)	Chute spillway b × h (m²)	Spillway tunnel b × h (m²)	Energy dissipater	Note
1	Three Gorges	PG	183	102 500	22–8 × 17	2–8 × 11	23–7 × 9			Surface spillway and middle outlet	u/o
2	Xiangjiaba	PG	161	48 680	5–19 × 26	7–7 × 11				Surface spillway and middle outlet and stilling basin	u/c
3	Ankang	PG	128	37 000	5–15 × 17	5–11 × 12	4–5 × 8			Flaring pier gate and still basin	u/o
4	Wuqiangxi	PG	84.5	55 962	9–19 × 23	1–9 × 13	5–3.5 × 7.0			Flaring pier gate and still basin	u/o
5	Longtan	RCC	216	35 500	7–15 × 20		2–5 × 8			Flip bucket	u/c
6	Guangzhao	RCC	195.9	9 857	3–16 × 20		2–4 × 6			Flip bucket	u/d
7	Dachaoshan	RCC	115	23 800	5–14 × 17.8		3–7.5 × 10			Flaring pier gate and roll bucket	u/o
8	Longyangxia	PG/VA	178	6 000		1–8 × 9	1–5 × 7 1–5 × 7	2–12 × 17		Flip bucket	u/o
9	Wujiangdu	PG/VA	165	21 350	4–13 × 18.5	2–4 × 4.4		2–13 × 18.5	2–9 × 10.44	Flow over powerhouse	u/o
10	Jinping I	VA	305	15 400	5–11.5 × 10	5–5 × 6			1–14 × 12	Flip bucket and plunge pool	u/c
11	Xiaowan	VA	292	20 683	5–11 × 15	6–6 × 5			2–10 × 12	Flip bucket and plunge pool	u/c
12	Xiluodu	VA	273	50 311	8–12.5 × 18		7–5 × 6		4–14 × 12	Flip bucket and plunge pool	u/c
13	Baihetan	VA	277	44 151	6–12.5 × 18	7–5 × 8			4–14 × 11.3	Flip bucket and plunge pool	u/d
14	Ertan	VA	240	23 900	7–11 × 11.5	6–6 × 5	4–3 × 5		2–13 × 13	Flip bucket and plunge pool	u/c
15	Goupitan	VA	225	26 950	6–16 × 15	7–6 × 7	2–6 × 7			Flip bucket and plunge pool	u/c
16	Dongjiang	VA	157	7 830					1–D10 1–D8.5		u/o
17	Shuibuya	CFRD	233	15 243			2–4 × 5	1–10 × 7.5 2–10 × 7.5		Chute spillway and slit bucket	u/o
18	Tiansheng-qiao I	CFRD	178	21 750				5–15 × 18	1–6.4 × 7.5	Chute spillway and flip bucket	u/o
19	Gongboxia	CFRD	127	7 500			1–7.5 × 6	5–13 × 20	1–7 × 10	Chute spillway and vortex shaft tunnel	u/o
20	Nuozhadu	ER	258	35 300				2–14 × 16 10–15 × 20	2–5 × 8.5	Chute spillway and plunge pool	u/d
21	Pubugou	ER	186	9 780				3–12 × 16	1–9 × 9 1–12 × 7.5	Chute spillway and spillway tunnel	u/c
22	Hongjiadu	ER	182	6 996					12 × 7.5	Chute spillway slit bucket	u/o
23	Xiaolangdi	TE	154	17 063				3–11.5 × 17	3–D14.5 3–D6.5 1–10 × 12 1–10 × 11.5 1–10.5 × 13	Chute spillway, tunnel spillway, orifice tunnel and stilling basing	u/o

PG, gravity dam; RCC, roller compacted concrete dam; VA, arch dam; CFRD, concrete faced rockfill dam; ER, rock-filled dam; TE, earth-filled dam; u/o, project under operation; u/d, project under design; u/c, project under construction.

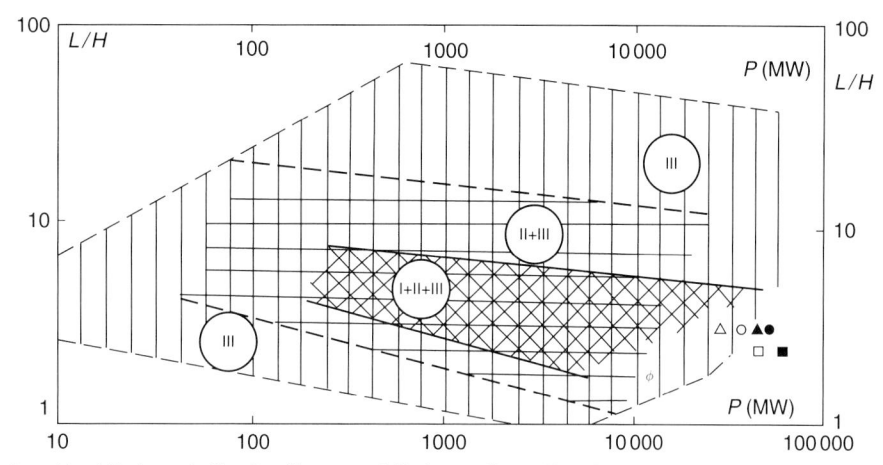

Figure 1 Statistics of combined discharge facilites I, spillway tunnel; II, chute spillway; III, surface spillway and middle outlet; L, length of dam crest (m); H, dam height (m), P, 0.0098AZ (MW); Q, discharge flow (m³s⁻¹); Z, head difference between design reservoir water level and original river bed (m); •○, Xiaowan Dam; ▲ △, Eartan; ■ □, Goupitan; Φ, Mossyrock Dam. Original figure is taken from GR 50 of the 13th Congress of ICOLD [2]; the marked points are made by author for comparison.

6.18.2 Energy Dissipation

6.18.2.1 Slit Bucket

As the valley is usually narrow in the west and there is a large discharge flow during the flood season, the normal energy dissipaters are not suitable. The slit bucket is specially developed for such kind of situations and it can make the flow contracted at the end of the bucket and project it dispersing in the sky longitudinally. The advantages are high efficiency of energy dissipation and less scours in the riverbed. The systematic physical model studies are conducted to understand its hydraulic characteristics. The model tests have found out that (1) the Froude number in front of the slit bucket should be larger than 3.5, (2) the angle of the bucket can be changed between −10° and +45°, and (3) the scour in the riverbed can be reduced by 1/3 to 2/3 compared to the normal bucket with an angle of 30° [3]. This kind of dissipater was first applied in the sky-jump spillway of Dongjiang Project in the early 1990s with the unit width discharge flow reaching 600 m³ (s-m)⁻¹. The prototype observations performed in 1992 (see **Figure 2**) show a good relationship between model and prototype on jet flow and scour patterns although the discharged flow does not reach the design value [4].

This new technique has been widely applied to more than 10 projects in China and also included in the 'Design Specification for River-Bank Spillway'.

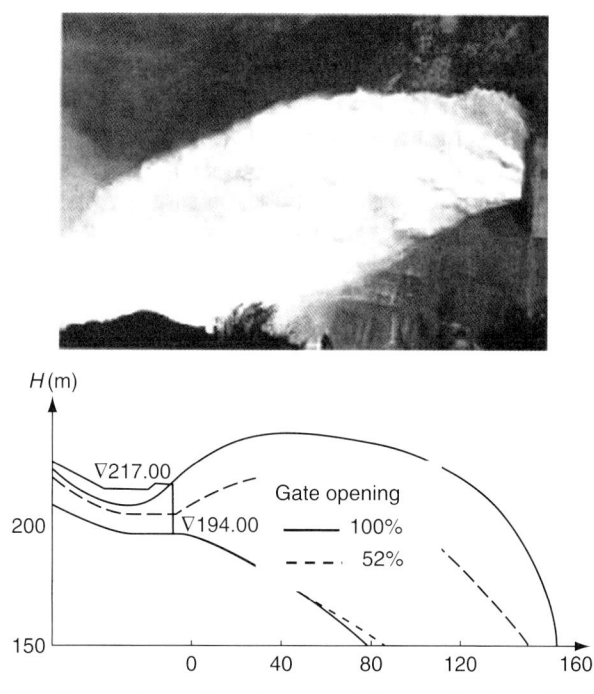

Figure 2 Flow pattern of slit bucket in Dongjiang sky-jump spillway (upper one in the case of real operation, bottom one in the case of model test).

In recent years, the slit bucket has been studied in three large spillway tunnels in China. They have common characteristics in which dam height is about 300 m, discharge capacity of each tunnel is over 3500 m³ s⁻¹, and discharge head is over 200 m. The details of the tunnels are given in **Table 1**.

The spillway tunnel in Xiaowan Project is located on the left bank. Reducing the riverbed and bank erosion is one of the tasks during the design as the river valley is very narrow and the rock on the right bank further downstream of the energy dissipation zone is not strong enough to resist erosion.

Four types of flip buckets have been studied (see **Figure 3**) [5]. As the injection angle of flow in type (a) is too small, it results in jet flow to close the right and erodes the right bank in the original design. A large backflow appears along the left bank with a maximum return

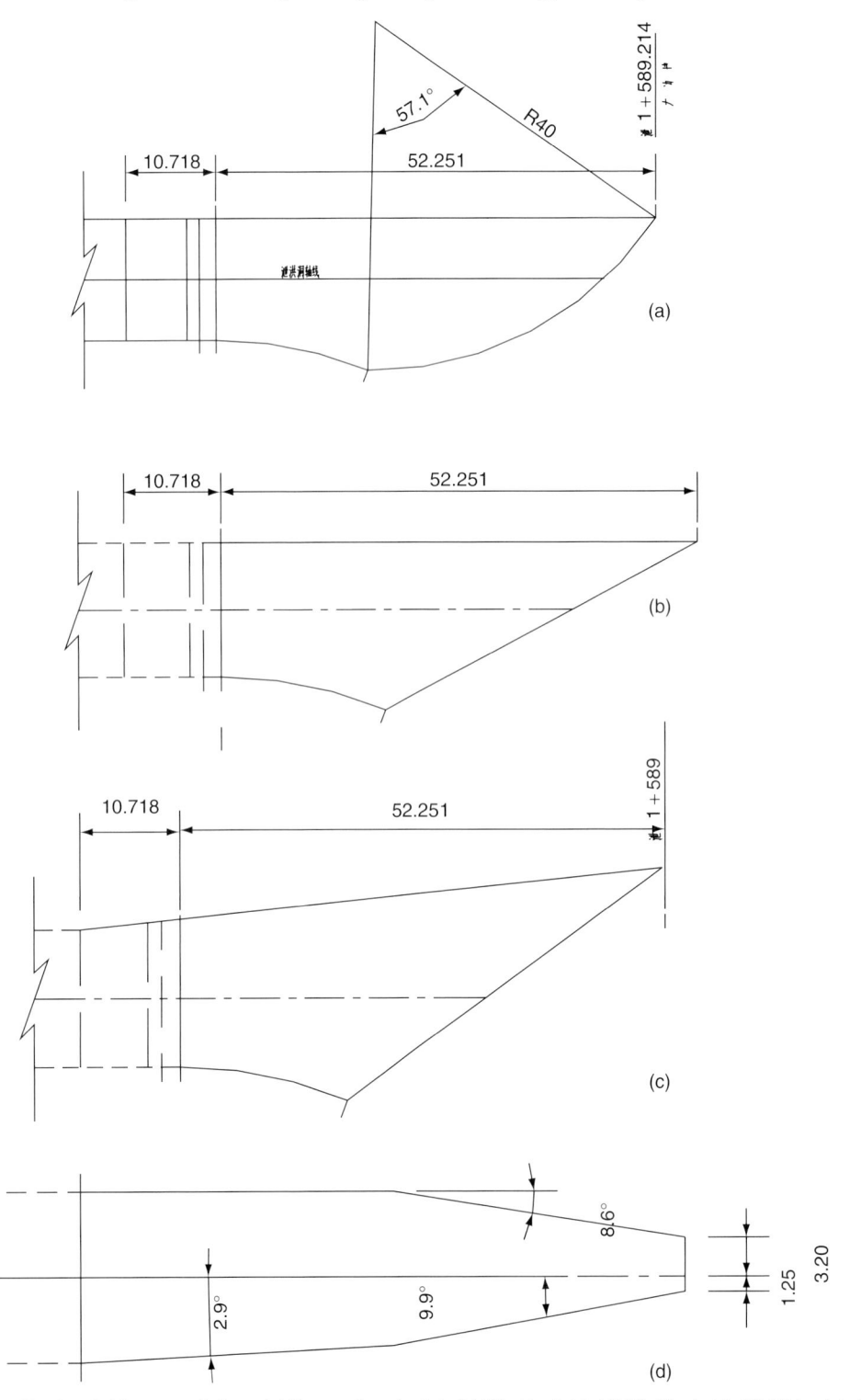

Figure 3 Four types of buckets in Xiaowan arch dams. (a) Tongue shape bucket. (b) Tilted bucket I. (c) Tilted bucket II. (d) Slit bucket. (H_{max} = 292 m, Q_{tunnel} = 3535 m³ s⁻¹).

flow velocity of 14 m s^{-1}. The maximum depth of scour pit is 15.6 m and close to the right bank. Another two types of flip buckets, type (b) and (c), have been proposed and tested. The scours in riverbed have not been improved ideally. The scours are still close to the right bank.

Type (d) slit bucket is finally adopted by the design through optimization. The direction of jet flow is adjusted and dispersed along the river channel. The riverbed and bank erosion has been reduced greatly. The configurations of the final design are that (1) the width of edge is reduced from 14.0 to 4.45 m with the contraction ratio of 0.3178. The axis of slit bucket is asymmetrical with the tunnel axis. The left-side wall is 3.2 m from the axis of the central line and the right-side wall is 1.25 m from the central line. (2) Two steps of contraction are selected on the right-side wall in which the first contraction is 27.251 m long with a contraction angle of 2.86° and the second contraction is 25.0 m long and a further contraction angle of 7.08° is applied. (3) One step of contraction on the left-side wall is 25.0 m long with an angle of 8.64°.

The physical model tests with a scale of 1:45 show that flow surface is raised suddenly through the slit bucket, and flow is dispersed longitudinally in the range of 200 m downstream of river reach slightly close to the left bank without a return flow (Table 2). The maximum flow velocity along the right bank is less than 8 m s^{-1}, which is reduced by 40% compared with the original design, and the maximum scour depth is 8 m in the case of low downstream river level. In most cases, the flow velocities along both banks are less than 5 m s^{-1}, which reduce the protection work greatly. A slight scour is measured in the design and under check flood operation modes because the water depth downstream is much larger. Similar physical model tests have been performed on the Xiluodu 3# spillway tunnel and Jinping spillway tunnel. Expected results have been obtained which reduced the scours downstream riverbed greatly. Figure 4 gives the scours on riverbed by Jinping spillway tunnel under the designed reservoir water level [6]. The maximum scoured depth on the proposed plan is 6.3 m.

6.18.2.2 Flaring Pier Gate

This new type of energy dissipater, state of the art, is specially developed for Ankang Hydropower Project [7, 8]. The stilling basin of the project is located on a curved river reach and the riverbed is with a low ability of anti-scourging. The other reason is that the construction has been proceeding and the length of the stilling basin cannot be further lengthened. A new concept of energy dissipation has been proposed for this project, that is, combining the flaring pier gates on surface spillway with stilling basin, to make the flow out of pier gates contracted laterally and dispersed longitudinally, which changes a two-dimensional (2D) flow into a 3D flow and increases the energy dissipation ratio (see Figure 5). The strong 3D turbulent flow can create aeration in the flow through lateral space. High ratio of energy dissipation makes the length of the stilling basin to be shortened and the construction work reduced.

This new energy dissipater was applied to the Ankong Hydropower Project in the middle 1970s with the maximum unit width discharge flow of 254 m^3 (s-m)$^{-1}$. Finally, the length of the stilling basin is reduced by one-third. In fact, the Panjiakou surface spillway is the first one that adopted this kind of energy dissipater in the world. Further inventions have been made by combining with bottom outlets in Wuqiangxi and Baise Hydropower Projects, or RCC stepped spillway in Shuidong and Dachaoshan Hydropower Projects.

Dachaoshan Hydropower Project is an RCC gravity dam with a maximum height of 111 m and a unit width discharge flow of 193.6 m^3 (s-m)$^{-1}$. The energy dissipater is a flaring pier gate with stepped spillway, and the roll bucket is adopted in the downstream. Special measure has been taken in the design that the first step is two times higher than normal ones, so that it will make the flow project over several steps and a large cavity is formed under the jet flow; thus, more air enters the bottom of the flow. The hydraulic field observation was carried out under normal water level in 2002 when the reservoir was filled for the first time. The observed results [9] show that (1) the pressure variations on steps have been changed a lot, and the pulsation pressure is as high as 10 kPa (see Table 3); and (2) the air concentration on steps is over 30%, which is much higher than the chute spillway. The analysis indicates that the first high step plays an important role in cavitation mitigation on steps (see Figure 6). Slit bucket can also be applied to surface spillway to reduce the scouring downstream. Guangzhao RCC Dam is a good example. The dam height in Guangzhao is 195.5 m with a maximum discharge flow of 9857 m^3 s^{-1}. Three surface spillways and two bottom outlets are adopted in the design.

Traditional flip bucket is used in the beginning of the design. As a 30° bucket angle is taken in the middle one and 22° in the side one, the elevation difference between the middle one and the side one is 2.75 m. The buckets on both sides are slightly contracted from the width of 16 to 13 m.

Table 2 Scour depth and location by slit bucket in Xiaowan Project

No.	Operation mode	Reservoir water level (m)	Downstream water level (m)	Location of maximum depth of scoured pit (m) Change	Distance from axis	Maximum scoured pit (m) Elevation	Depth
1	Start operation	1236.50	998.92	0+325	5 m to the right	972.3	7.7
2	Start operation	1236.50	1000.60	0+330	5 m to the left	972.8	7.2
3	Start operation	1236.50	1002.69	0+310	5 m to the left	973.8	6.8
4	P = 1%	1236.90	1010.18	0+320	0	976.8	3.2
5	Design flood	1238.30	1012.73	0+340	5 m to the left	079.3	0.7
6	Check flood	1242.51	1016.70	Not scoured			

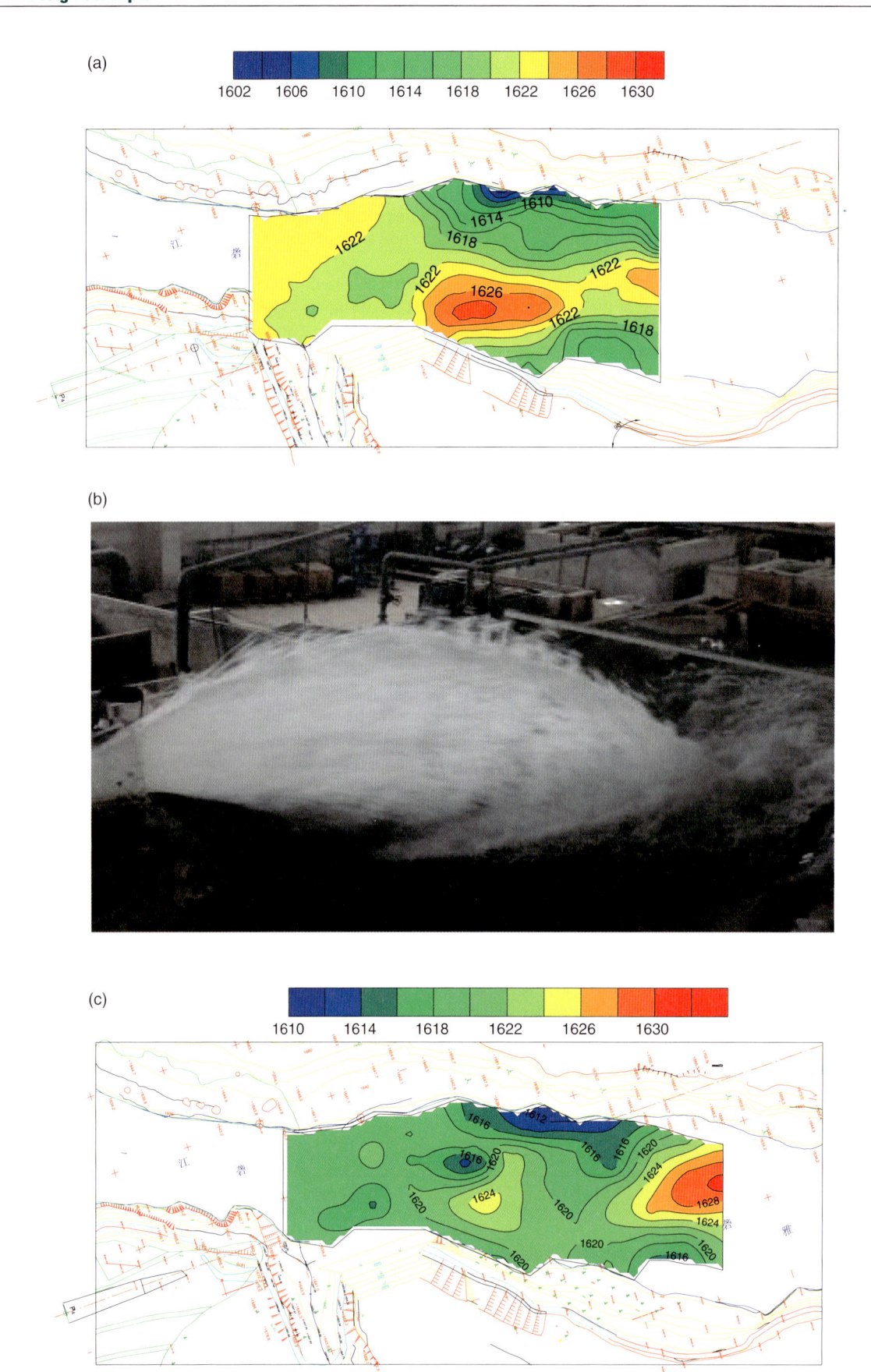

Figure 4 Scours by two types of buckets and flow pattern of slit bucket under the designed reservoir water level in Jinping spillway tunnel: (a) scour by oblique bucket; (b) flow pattern of slit bucket; and (c) scour by slit buctet (model scale 1:30), ($H = 278$ m, $Q_{tunnel} = 3535$ m^3 s^{-1}).

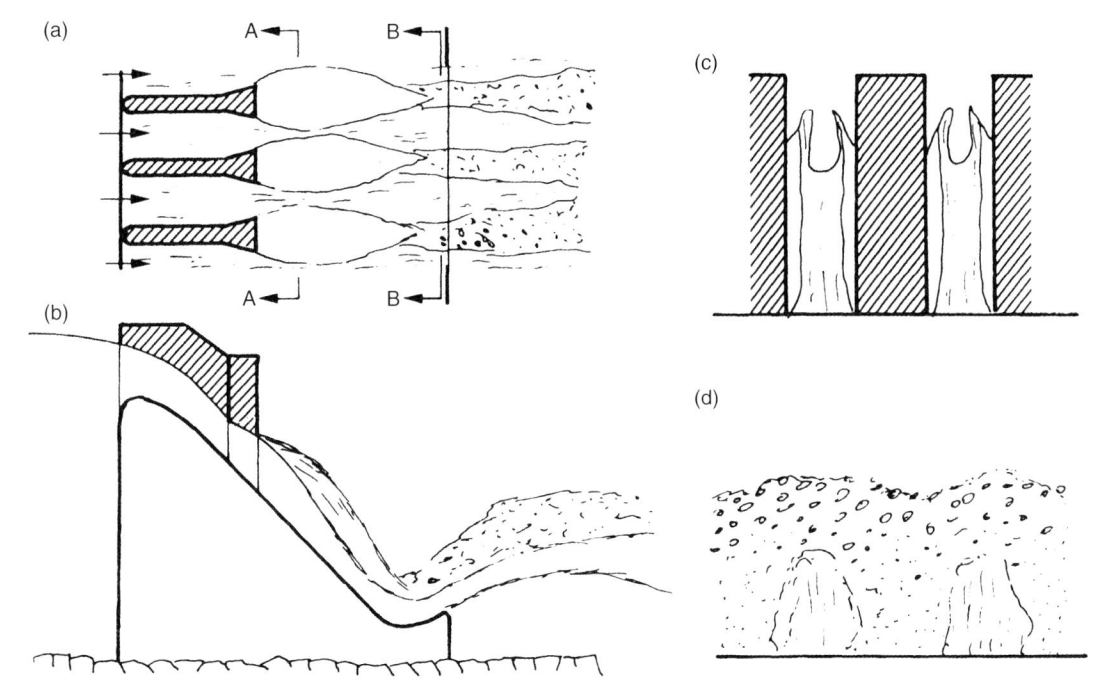

Figure 5 3D energy dissipation by flaring pier gate. (a) Plan view, (b) side view, (c) section A–A, (d) section B–B.

Table 3 Pressure and air concentration on steps in Dachaoshan Hydropower Project

Step no.	P_{ave}/σ (kPa)	P_{max}/P_{min} (kPa)	V (m s^{-1}) on the height of 3 cm/ 8 cm/15 cm Case I	Case II	Air concentration (C%) Case I	Case II
15#	3.8/3.2	44.3/−3.3	21.7/26.2/27.5	21.3/26.8/27.9	32.0	35.8
21#	6.5/6.4	62.0/−7.8	21.3/24.2/26.6	23.0/26.2/25.9	39.1	45.5
26#	2.9/2.7	28.7/−10.2	−/23.5/26.9	−/26.0/28.9	49.5	51.3
30#	5.5/6.8	78.8/−12.9	21.0/24.7/27.1		39.5	

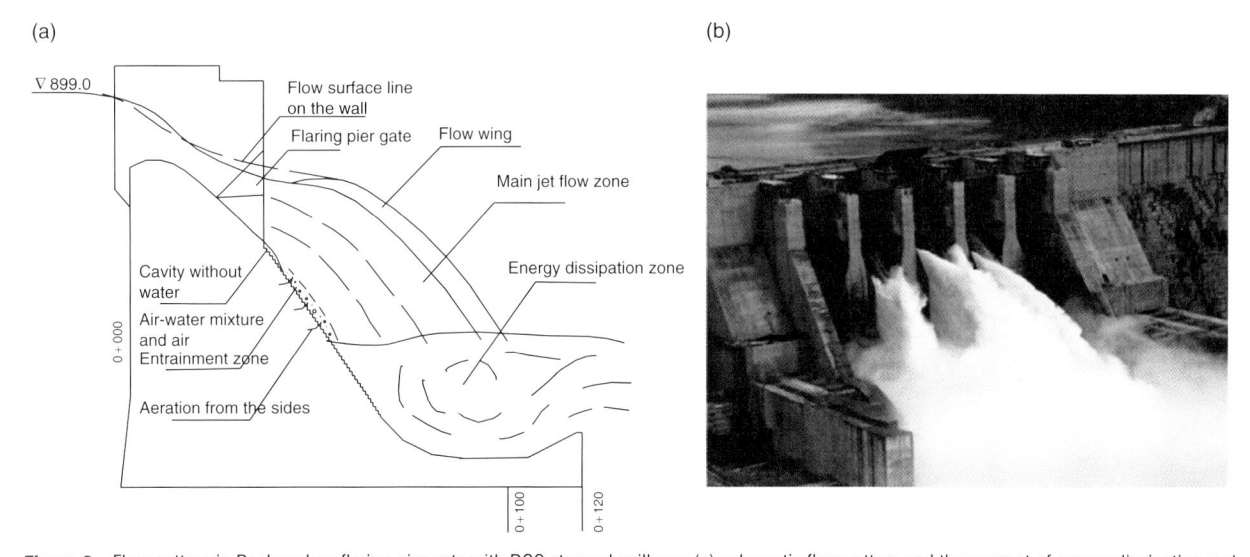

Figure 6 Flow pattern in Dachaoshan flaring pier gate with RCC stepped spillway: (a) schematic flow pattern and the concept of energy dissipation and (b) operation in case I.

The physical model tests show that (1) the overburden in the energy dissipation zone is almost scoured to the downstream with a maximum scouring depth of 26 m, and solid rock on the riverbed is scoured by 5 m; (2) toes of side banks are also scoured; and (3) the scoured materials are accumulated around the tailrace with a maximum height of 18–20 m, which will severely affect the operation of power plant.

The proposed slit bucket [10] is designed with (1) bucket angle of −10° applied for all three with a contraction ratio of 0.3; and (2) unsymmetrical contraction on side buckets and symmetrical contraction on the middle one with the width of edge of 4.8 m. **Figure 7** gives the comparison of scouring pattern by two types of flip buckets under the check flood operation mode. It indicates less scouring by slit bucket.

6.18.2.3 Jet Flows Collision with Plunge Pool in High-Arch Dams

As some high-arch dams are constructed in narrow valleys, the collision of energy dissipation by jet flows of surface spillways and middle outlets and a large plunge pool downstream is often chosen. The very successful project is the Ertan high-arch dam.

The design criteria on the slab of plunge pool are that the maximum impinging pressure must be less than 15×9.81 kPa. Commendable efforts on the arrangements of the spillways, middle outlets, and plunge pool have been made and measured by the physical models during the design stage, such as the impinging angle between surface spillway and middle outlet, the shape of flip bucket of surface spillway, the length of plunge pool and the elevation of the floor considering the excavation, and the height of secondary dam. The final solution on the arrangement of discharge facilities in Ertan Dam are seven surface spillways and six middle outlets, and the length of plunge pool is 330 m with a 32 m high secondary dam (see **Figure 8**). Different flip buckets are adopted in every opening of the surface spillway. The maximum discharge flow through the surface spillways and middle outlets is $16\,300\,\text{m}^3\,\text{s}^{-1}$

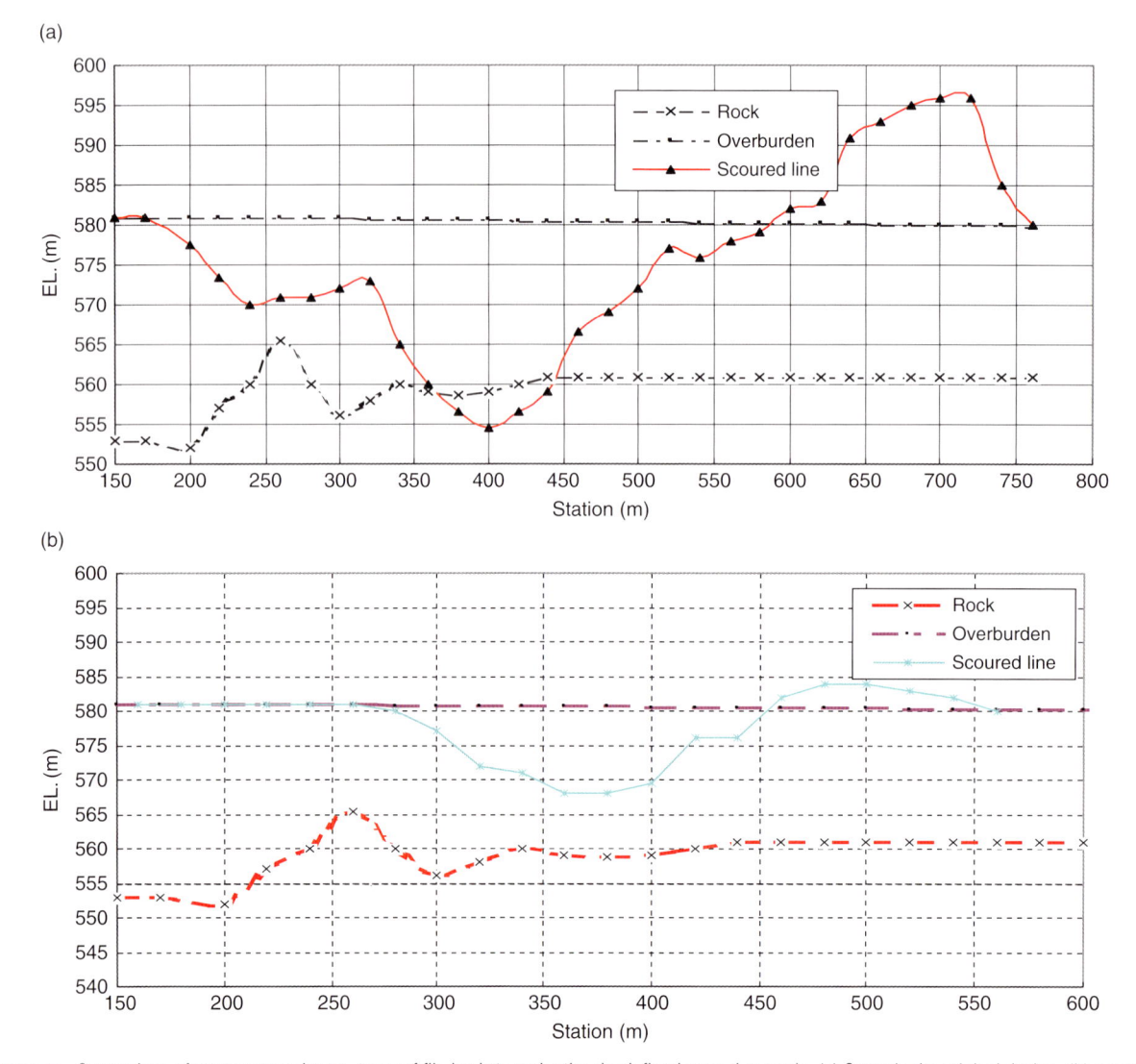

Figure 7 Comparison of scour pattern by two types of flip buckets under the check flood operation mode. (a) Scour in the original design, (b) scour in the proposed slit bucket.

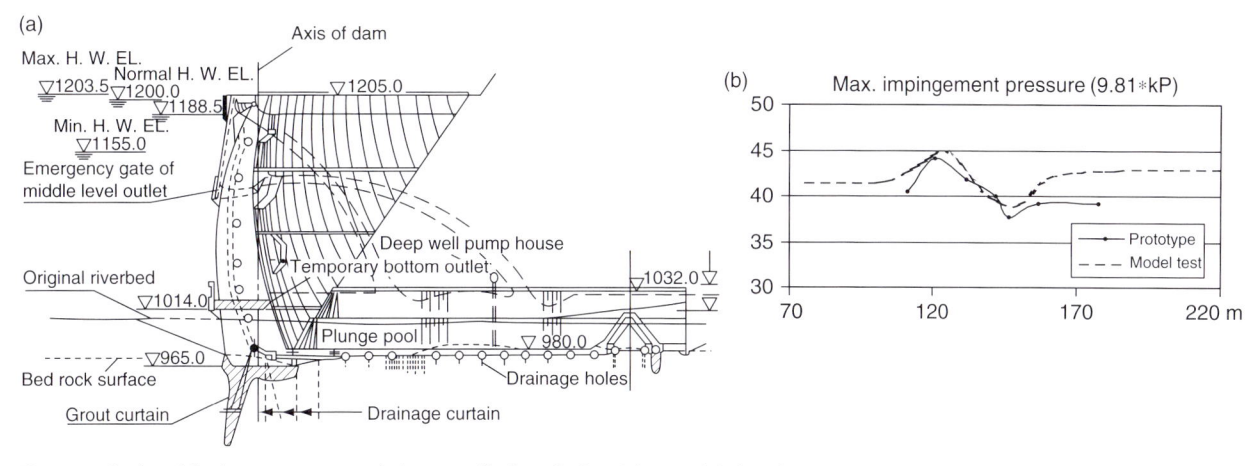

Figure 8 Design of discharge structures and plunge pool in Ertan Project: (a) general design of energy discharge structures and (b) comparison between the prototype measurements and model tests in Ertan plunge pool.

(68.2% of total discharge) and the critical situation is the independent operation of surface spillway with the maximum impingent pressure of 14.0×9.81 kPa under the check flood reservoir water level.

The Ertan Arch Dam was completed in 1999 and hydraulic field observation was carried out in the same year. The field observation results are in good agreement with the model's results [11], shown in **Figure 8**. The field observations are carried out under the design reservoir water level with a discharge flow of 8000 m^3 s^{-1} (four surface spillways and four middle outlets).

The design concept of energy dissipater in Ertan Dam is accepted by other high-arch dams, such as Jinping (305 m), Xiaowan (292 m), Xiluodu (278 m), Baihetan (277 m), Goupitan (232 m), and Laxiwa (250 m), which all have large plunge pools with a length of about 400 m and secondary dams with a height of about 40 m.

As the pressures on the vertical wall of the differential buckets in Ertan are quite low, even negative, the differential flip buckets between surface spillways are recommended and studied on Xiaowan, Goupitan, Xiluodu, and Baihetan arch dams. The angles of buckets change from $-35°$ to $10°$, which makes the jet flows separated along the plunging pool and the impinging pressures reduced greatly. For example, the maximum discharge flow through seven surface spillways and eight middle outlets in Xiluodu Project has increased from 30 000 to 33 800 m^3 s^{-1} with the bucket angles of surface spillways from $-30°$ to $10°$ and the maximum impinging pressure being controlled under 13.0×9.81 kPa. The angles in Xiaowan Arch Dam are from $-20°$ to $10°$ and in the Baihetan from $-35°$ to $20°$; the maximum discharge flow can be increased by about 10%. The bucket shape is also an important factor to spread the flow to lateral directions and reduce the impinging pressure. **Figure 9** gives the flow

Figure 9 Flow pattern by surface spillways of Baihetan Arch Dam.

pattern through surface spillway in Baihetan Arch Dam from the physical model [12]. The surface spillways and middle outlets are all optimized.

6.18.2.4 Orifice Spillway Tunnel

The principle of energy dissipation of orifice spillway tunnel is sudden contraction and then sudden expansion through the orifices. It was first applied in the Mica Dam in the 1980s but the discharge capacity was less than $1000\,m^3\,s^{-1}$. The first large-scale orifice spillway tunnel was adopted in Xiaolangdi Project by reconstruction of diversion tunnel in the 1990s.

Xiaolangdi Project has a rockfill dam with a maximum height of 154 m and a total discharge capacity of $17\,063\,m^3\,s^{-1}$. All discharge structures are located on the left bank, including one chute spillway, three spillway tunnels, three orifice tunnels, and three silt flushing tunnels. The powerhouse is also located on the left bank. The main consideration on the orifice spillway tunnel is cavitation. The objectives of studies include optimization of the number, interval, orifice plate shape, adoption of abrasion-resistant concrete, and inclined ratio on the top of the chamber to increase the pressure of the tunnel. The final design of the orifice tunnel is that three orifice plates are installed in the horizontal pressurized tunnel with an interval of $3D$ (D is the diameter of the tunnel, $D = 14.5$ m). The contract ratios of these are 0.690, 0.724, and 0.724, respectively, which result in a strong rotation, shear and turbulent flow, dramatic energy dissipation, and reduction in velocity to about $10\,m\,s^{-1}$ (see **Figure 10**). More details of the research had been considered during the design, including the different scales of conventional model tests, depressurized model tests, and intermediate prototype observation in the Baozhusi silt tunnel.

The orifice spillway tunnel was first operated in April 2000 and hydraulic field observations have been carried out with the working heads of 70 and 100 m on 1# tunnel and 100 m on 2# tunnel [13, 14]. The parameters observed are pressure and flow noise in the pressurized tunnel; pressure, cavitation noise, air entrainment, and air concentration in the open flow tunnel; and strength and stress on the radial gate.

The model test results and field observations show that they are in consistency with the energy dissipation ratio and pressure distribution (see **Figure 11**). A slight cavitation noise is still observed at the gate opening ratio from 0.96 to 0.99 (see **Figure 12**). Sound increment of spectrum level at 11.6 to 27.0 dB in a high-frequency band is observed. But no cavitation damage is found during inspection after several rounds of operation.

The scale effect on cavitation has been a cause for concern during the design. Several physical model experiments, under the normal atmosphere condition and depressurized condition, are carried out with the model scale of 1:40 to 1:30 [15]. An intermediate test on the silt flushing tunnel in Pikou Project was performed for further analysis of scale effect. **Table 4** shows that

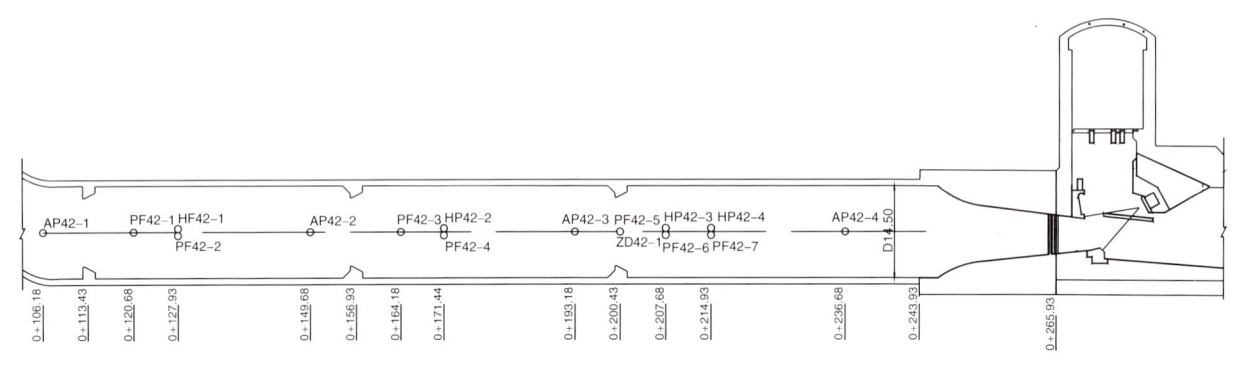

Figure 10 Pressure and hydrophone sensors arrangement in Xiaolangdi 2# orifice tunnel.

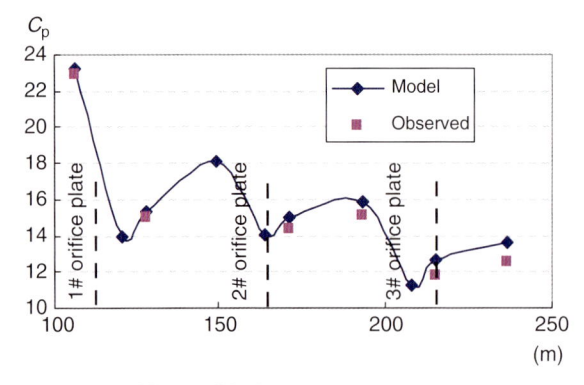

Figure 11 Pressure coefficients of 2# orifice tunnel in Xiaolangdi Project.

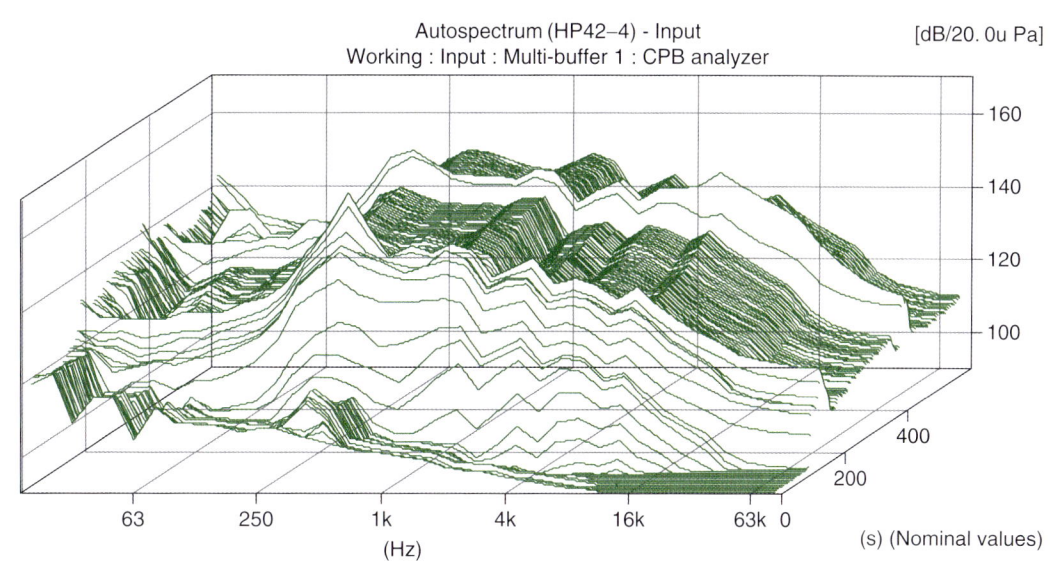

Figure 12 Spectrum of flow noise downstream of the third orifice plate during the gate opening.

Table 4 Flow cavitation numbers of three orifice plates at the full gate opening (σ)

	Model test			Observation
Water head (m)	130.0	105.0	85.0	103.0
1st orifice plate	5.23	5.18	5.19	5.29
2nd orifice plate	5.04	5.01	5.04	5.14
3rd orifice plate	4.40	4.41	4.42	4.37

the flow cavitation numbers based on the observed data under the working head of 103 m are very close to the ones calculated based on the physical model test results under the working head of 105 m. It indicates that the previous studies and methods are able to predict the cavitation characteristics of orifice tunnel at the design working head – the flow cavitation intensity at the full reservoir water level will be greatly changed.

6.18.2.5 Vortex Shaft Spillway Tunnel

Rebuilding the diversion tunnel into the spillway tunnel is another way to reuse the diversion tunnel. It can also solve the difficulty in the arrangement of connection tunnel by flexible arrangement of the intake. The vortex spillway tunnel is one way to reuse the diversion tunnel. There are two types of vortex shaft spillway tunnels: one is vertical and the other is horizontal, both making the flow run in rotation to dissipate the energy.

Shapai Project is the first one to adopt a vertical vortex shaft spillway tunnel in China which is reconstructed from diversion tunnel [1]. The discharge capacity is about 250 m^3 s^{-1} with a head of 100 m. The first operation began just after the '5-12' Wenchuan earthquake in 2008 in China to control reservoir water level from overtopping.

The hydraulic studies on large-scale vortex shaft spillway tunnel are much more challenging and have been carried out in Xiluodu and Gongboxia Projects.

The Xiluodu Hydropower Project has an arch dam with a maximum height of 278 m and a maximum discharge flow of about 50 000 m^3 s^{-1}. There are four large spillway tunnels with a maximum discharge flow of about 4000 m^3 s^{-1} each. The vertical vortex shaft spillway tunnel is an alternative to discharge the extra flood; otherwise, an additional long tunnel must be built. The design is a conventional intake with a head of 60 m and a short connecting tunnel of about 100 m long, a one-fourth of elliptical curve at the end of the horizontal tunnel connecting to a chamber with a diameter of 22 m, and a vertical shaft with a diameter of 16 m connecting to the original diversion tunnel. As the energy dissipation head is about 220 m and the maximum discharge flow is 2700 m^3 s^{-1}, the cavitation must be carefully considered. To increase the wall pressure especially on the lower part of the shaft and to increase the flow cavitation number, an orifice plate on the lower part of the shaft can be considered and it is effective. A plunge pool in the diversion tunnel is used as it is easy to be built and has the same function as the orifice. The energy dissipation ratio of such arrangement can reach up to 85% [16] (see **Figure 13**).

Figure 13 Xiluodu vortex shaft spillway tunnel model.

The horizontal vortex spillway tunnel is studied for the Gongboxia Hydropower Project [17]. The original spillway tunnel is rebuilt by a diversion tunnel in a conventional way with an inclined tunnel. During the excavation of the intake, it is found that the geological condition is not favorable, and hence another solution must be considered. By analysis and comparison, a horizontal vortex spillway tunnel is selected. The discharge head is 100 m and the maximum discharge flow is 1100 m^3 s^{-1}. The diameter of the vertical shaft is 9 m and a one-fourth of elliptical curve is connected to the diversion tunnel. The diameter of the horizontal vortex tunnel is 11 m and it is 50 m long (see **Figure 14**). A 40 m long plunge pool and a special energy dissipater are adopted for energy dissipation. A physical model with a scale of 1:40 is built for pressure, velocity, and aeration measurements. As the vertical shaft is a pressurized flow, a circular orifice plate is adopted for air entrainment and cavitation mitigation.

The real operation of the horizontal vortex tunnel and field observation was carried out in August 2006. The reservoir water level during the observation was close to the normal water level, which means that the discharge water head and capacity were about 104 m and 1130 m^3 s^{-1}, respectively. The main parameters observed are water levels in reservoir and river channel downstream, pressure distribution, flow pattern in the horizontal tunnel, airflow and its velocity distribution in the air vent, air concentration, structure vibration, and structure dynamic response such as displacement, stress, and strain [18].

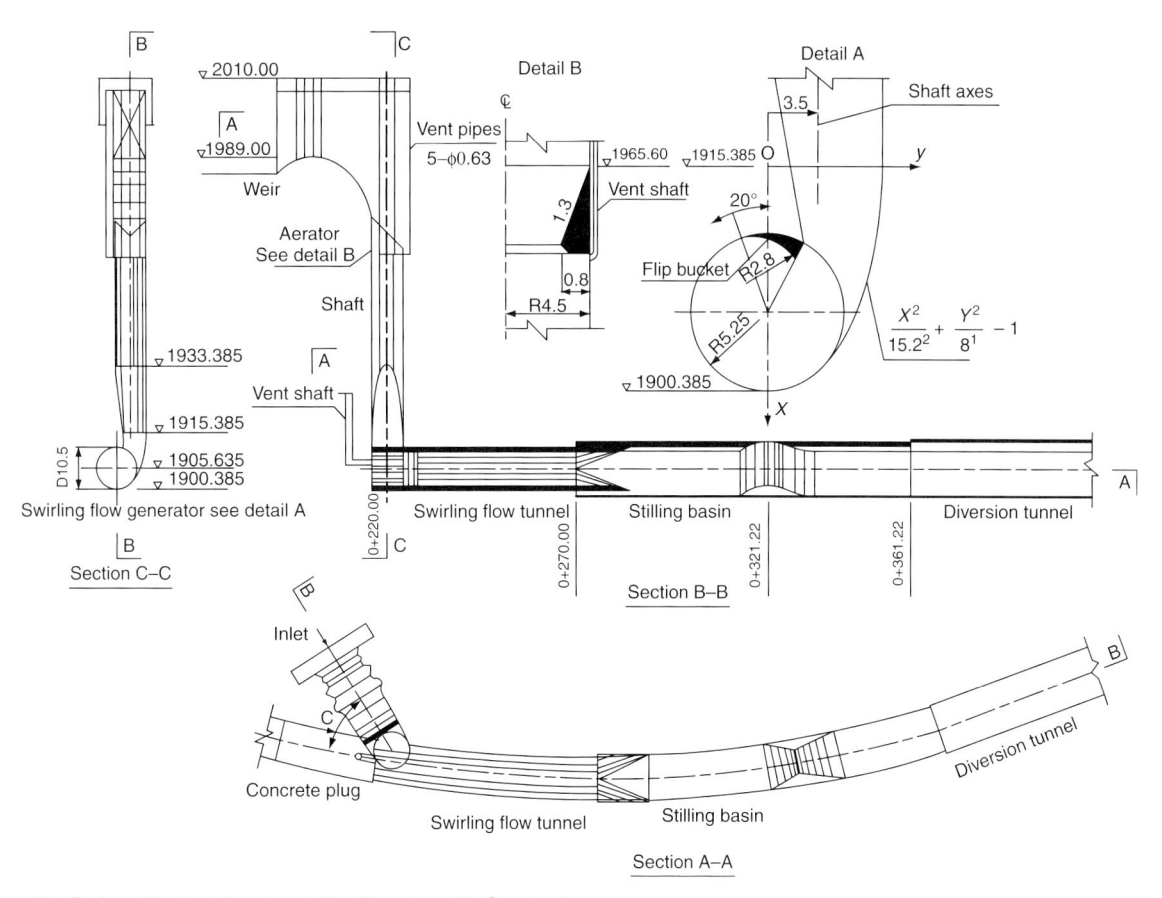

Figure 14 Design of horizontal vortex shaft spillway tunnel in Gongboxia.

The main observed results are (1) the air vent works well after the gate opening and the maximum airflow velocity is about 120 m s^{-1} with a maximum airflow of 403.4 m^3 s^{-1}, which is about 36% of the flow rate in the spillway tunnel. The cavity length downstream of the aerator is about 17 m, which is about 3 times the model result. The near-wall air concentration in the shaft is more than 8%, which is much higher than the one obtained on the floor from the other projects. Enough airflow is not only good for cavitation mitigation, but it can also increase the energy dissipation ratio. (2) The pressure distribution has a good agreement with the model tests. (3) The energy dissipation ratio is up to 84.5% and the velocity in the horizontal part of the tunnel is less than 15 m s^{-1}. (4) Flow is very smooth during the gate opening and closing as well as the full opening operations observed by a video camera installed in the crown of the tunnel. (5) The dynamic responses of the structure are all under the design conditions. All measured results have been applied to the safety assessment of the project and provided useful information for further research and design of similar vortex shaft spillway tunnels.

The diameters of the vortex chamber and shaft, connection tunnel and elliptical curve, energy dissipater, and aerator can be determined from the research on Shapai, Xiluodu, and Gongboxia vortex shaft spillway tunnels.

6.18.3 Aeration and Cavitation Mitigation Measures

The main characteristics of Chinese dams are high head and large discharge flow. Therefore cavitation damage must be paid much more attention. Since the first aerator was successfully applied in the Fengjiashan spillway tunnel in 1979, much more studies have been carried out on hydraulic characteristics. Systematic studies on the aeration and cavitation mitigation measures have been carried out [19, 20] and have been accepted by the Design Code of Spillway, such as the determination of offsets and offset combined with ramp, and calculation of length of cavity, airflow, pressure in cavity, air concentration in cavity, and protection length. Design Code of Spillway indicates that (1) the aerator must be adopted with a velocity over 35 m s^{-1}, (2) the air concentration along the floor should be more than 4–5% to mitigate the cavitation damage, and (3) the length between two aerators has to be about 120–150 m. The studies also show that the negative pressure downstream of offset should be kept around −10 kPa. As the physical model test results are usually smaller than prototypes by the scale effect of similarity, further studies should be conducted on the proper prediction from the physical test results.

The offset, which is perpendicular to the spillway floor, combined with a small ramp is commonly applied. The height of the offset is about 1 m and it depends on the discharge flow and the slope of the chute. The vertical offset combined with the lateral offset is also considered by the reason of a round water stop arranged in the case of high-head radial gate while the water head is as high as 60–70 m. The ramp is not recommended when the aerator is placed on the steep inclined part of the spillway tunnel because the flow could be projected to the ceilings of the tunnel, which might cause a temporary pressurized flow.

The critical point of cavitation protection in the spillway tunnel is the end of inversed part after the inclined tunnel where the tunnel is changed to be horizontal and the pressure along the floor is changed dramatically. Several projects have incurred serious damages downstream from this point. A differential type of aerator is specially studied for the Ertan spillway tunnel when the Froude number, which makes a good aeration on the floor (see **Figure 15**), is not big enough.

The water head between the reservoir and the end of the inversed tunnel in Ertan 1# spillway tunnel is 102 m and the discharge capacity is 3700 m^3 s^{-1}, which results in high-speed flow and aeration. Therefore the measure of cavitation mitigation should be of more concern. The recent model tests show that the air concentration by different aerators is not appreciated for the side wall, and the air concentration is nearly zero. Therefore, a 3D aerator for both bottom and lateral aeration is proposed and systematic studies have been carried out [21]. **Figure 16(a)** shows the details of its configuration. The minimum air concentration on the side wall is larger than 1%. The field observation between the second and third aerators in Ertan 1# spillway tunnel is performed after the rebuilding of the aerators [22]. Much air is entrained through the air vents with the air concentration in the air cavity over 83%. The minimum measured air concentration on the farthest point, about 200 m from the second aerator, is 4.2% after the 3D aerator is applied but the previously observed minimum air concentration on the same point was only 2.8%. The higher the air concentration on the floor, the better the effect of cavitation mitigation. Inspection on the tunnel after 190 h of operation has not found any cavitation damages.

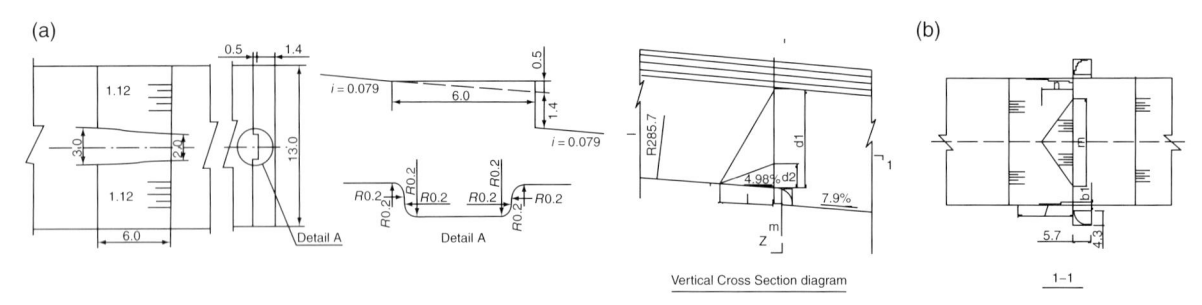

Figure 15 Aerators in Ertan 1# spillway tunnel: (a) differential aerators and (b) modified 3D aerator.

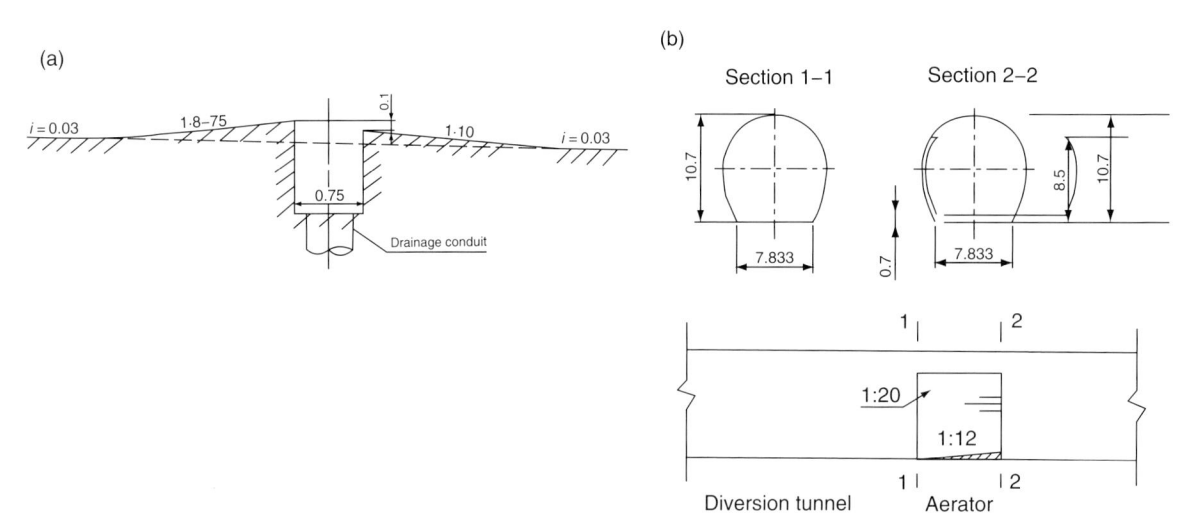

Figure 16 Special aerators: (a) aerator in Longyangxia Project and (b) aerator in Zipingpu Project.

Some special types of aerators or ramps (see **Figure 16(b)**) are applied when the tunnel slope is too small and special configurations have been determined; for example, the combination of upstream and downstream ramps and groove is applied to the spillway tunnel in Longyangxia [1]. A circular ramp is applied to the Zipingpu tunnel spillway [23].

The research has found that the conventional step of offset is not satisfied in a large slope of tunnel; for example, the slope in Xiaowan spillway tunnel is over 10%, as insufficient air is trapped and air cavity is sometimes filled by water.

A two-step aerator has been developed based on the physical model experiments [5] (see **Figure 17**). Type (a) is used only in the first aerator and type (b) is used for the remaining six aerators. H1 and L1 in the second aerator are larger than the others as it is located at the end of the inversed part of the inclined tunnel. Air concentration on each aerator is satisfied (see **Figure 18**).

6.18.4 Flow-Induced Vibration

In recent years, special attention has been paid to the problem of high-velocity-flow-induced vibration because lots of large dams are constructed in China. Vibration problems often arise at hydraulic gates, trash racks, pipes, and dams of discharging flow. The mechanism of vibration is very diverse and complex. Two general types of flow-induced vibration may be distinguished: (1) extraneously induced vibration, such as turbulence vortex-excited vibration; and (2) instability-induced vibration, caused by flow instability or movement instability.

According to the research of some engineering examples, the vibration problem of sluice structure can be forecasted or limited in two ways. (1) During the design stage, the physical or mathematical model should be used to predict the dynamic response for the hydraulic structure in complex flow situation, such as high-head and large dimension gates, and high-discharging arch dams. Great progress has been made in the mathematical simulation of fluid and dam, sluice gate structure coupling vibration by using a finite element method. There is a mature experience in the physical modeling of flow and concrete structure coupling vibration by using tailor-made latex. During the past 10 years, many efforts have been made in developing a special hydraulic-elastic material for the modeling of fluid-induced steel structure vibration, and now it is also becoming a mature modeling technique which is widely used in the flow-induced gate vibration research. (2) During the operation stage, the prototype research should be done to evaluate the degree of vibration or to avoid harmful vibration. Field observations are also used to verify whether the projects or gates are in safe operation conditions or to calibrate the research. Many observations of large-sized radial gates of surface spillways and some radial gates with high heads have been carried out by research engineers (see **Figures 19** and **20**).

6.18.5 Discharge Spraying by Jet Flow

The impinging of jet flows or free flows may create spraying rainfall especially in the case of high dam operation with large discharge flow that may damage some structures. This causes the switch yard to break down in the initial operations

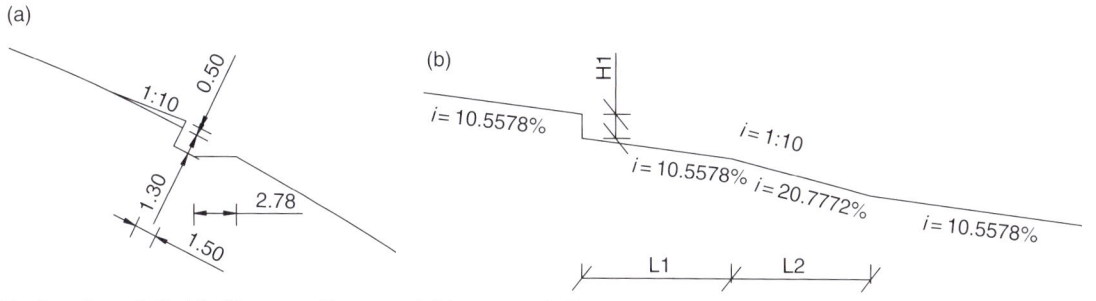

Figure 17 Two steps of offset for Xiaowan spillway tunnel: (a) one step of offset in the original design and (b) two steps of offset after modification.

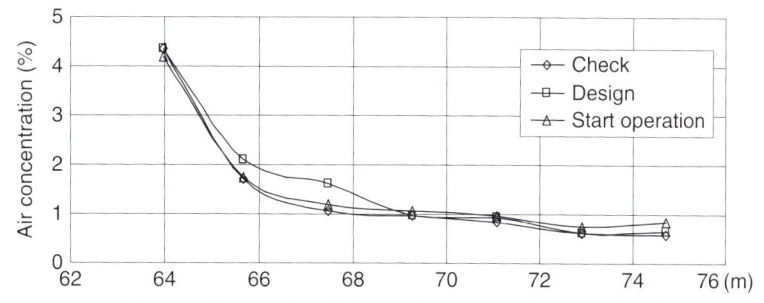

Figure 18 Air concentration measurement between the second and third aerators along the floor at different water levels.

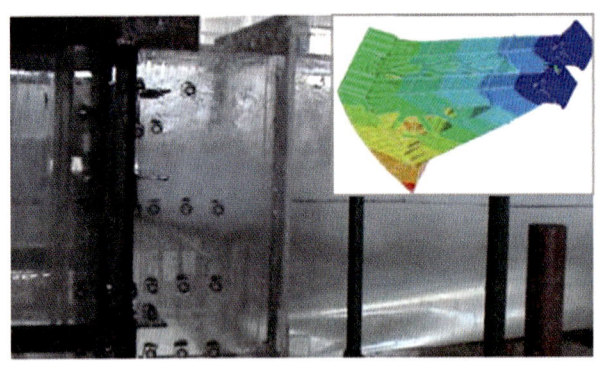

Figure 19 Gate vibration test of bottom outlet in Xiaowan Hydropower Project.

Figure 20 Prototype research of gate vibration in Xiaolangdi Project.

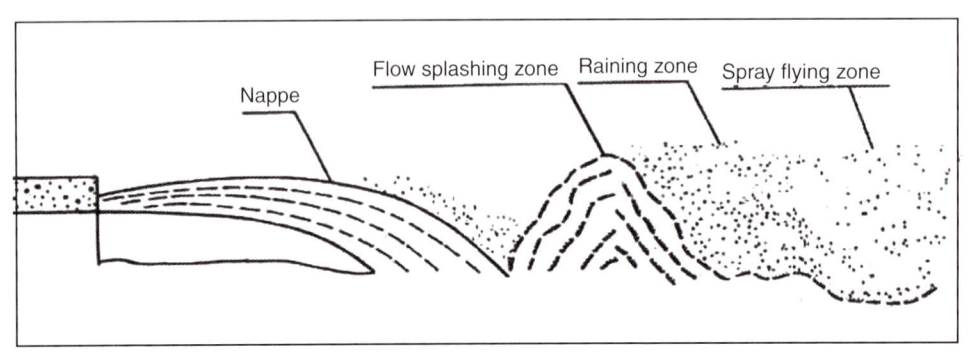

Figure 21 Mechanism of discharging spray.

in Liujiaxia and Xin'anjiang Hydropower Projects, blocks the access to the powerhouse in Dongfeng Hydropower Project, and brings about landslides in Lijiaxia Hydropower Project. The discharging spraying appeared since the 1970s and a wide range of research has been carried out in China mainly through model tests, numerical simulations, and field observations.

As there is a large-scale effect between the physical model test and prototype, and the numerical model is very difficult to be verified, the field measurement on the spraying intensity is quite reasonable. The measurement technique was first applied to Dongjiang Project in 1992. The project is a 157 m high-arch dam built in a very narrow valley. Two spillways were built on the right embankment and one on the left. The total discharge capacity is $5610 \, m^3 \, s^{-1}$. For the purpose of spraying intensity measurement, a special digital rain gauge is developed and the data are acquired by an SG20 hydraulic parameter system controlled by computer. **Figure 21** shows the mechanism of discharging spray that the flow splashes and becomes a main source of spraying and makes a very strong and intensive rainfall. The rainfall distributions and the effects have been analyzed.

The designers of the later projects have learnt much more about discharging spray and make a proper arrangement of the structures and bank protections. There are several other projects that have been measured in the same way, such as Lijiaxia for the purpose of making proper protection on both banks, and Manwan and Dachaoshan Projects for the purpose of making proper design of access gallery to the powerhouse [1, 24].

A field observation of discharging spray was carried out in Ertan Project in the case of surface spillway operation, middle outlet operation and their combinations, and spillway tunnel operations when the project was first put into operation in 1998 and 1999 (see **Figures 22** and **23** and **Table 5**).

Some results have been achieved. (1) There are high spray intensities on the two banks in the case of I–III under the elevation of 1115 m. (2) The spray intensity in case III is higher than that in case I or II. This shows that the collision by surface

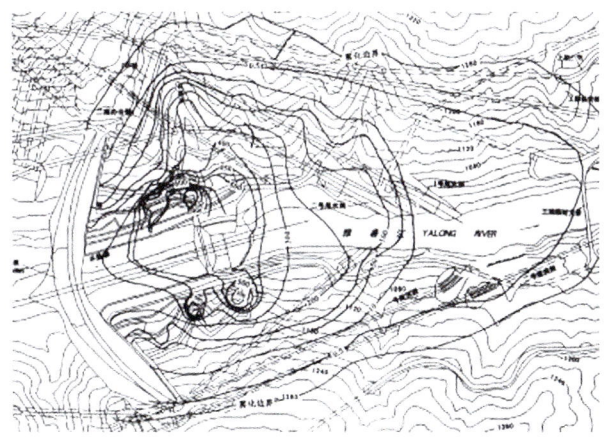

Figure 22 Intensity distribution of discharge spray by four surface spillway and four middle outlets in Ertan Project in 1999.

Figure 23 Discharge in Ertan Project in 1999.

Table 5 Main cases and the results of spray measurement in Ertan Project

Case	Gate openings	Reservoir level (m)	Discharge ($m^3 s^{-1}$)	Spray intensity ($mm\,h^{-1}$) 2# tailrace platform	Left bank	Right bank
I	6 middle outlets	1199.69	6856	7.1	833	491
II	7 surface spillways	1199.71	6024	1.8	850	305
III	4 surface spillways + 4 middle outlets (1, 2, 6, 7 surface spillways and 1, 2, 5, 6 middle outlets)	1199.33	7757	104	1180	750
IV	1# and 2# spillway tunnels	1199.78	7378		1000	422

spillways and middle outlets would cause much stronger spray intensity than one by surface spillways or middle outlets although such type of discharge arrangement could have a high efficiency of energy dissipation. (3) There is a strong spray close to the 2# tailrace platform, which indicates that it could be very difficult to have access to the gallery for checking the plunge pool during the discharging [11].

The spray rainfall intensity in Ertan Project is very valuable to the further understanding of the mechanism of discharging spray of high dams. The research based on this measurement and other several field observations make a prediction for new projects. The main parameters of influence range L and ξ are determined by Rayleigh method analysis [25]. The experimental formulas are brought up to estimate the influence range and the rainfall distribution of spray in the practice. This method has been used to predict spraying rainfall distribution in Xiaowan Project and allows the designer to make a proper protection especially on the right bank downstream of the tunnel spillway.

6.18.6 Hydraulic Field Observations

Hydraulic field observation is a valuable tool to analyze and explain the scale effects and make a proper prediction according to physical model studies. It is also one of the measures to evaluate the safety of the projects and the potential damages. More than 80 field observations on hydraulics have been carried out in China during the past 50 years and techniques have been developed especially on the instrumentations. The field observation on Foziling Dam was carried out on the flow pattern, the pressures, and aeration on chute spillway in Moshikou in the early 1950s. The cavitation damage was measured in Xin'anjiang in the 1960s.

The modern hydraulic field observations began from 1979 in Fengjiashan concentrating on air concentration as the aerator was applied in China for the first time. Then many researches on the pressures and air concentrations were carried out for Wujingdu, Dongjiang, Dongfeng, Ertan Projects, and so on. As there was some vibration on the radial gate in Liujiaxia, systematic studies together with field observations were performed on the aspect of flow-induced vibration both for gates and dams. Many large and high-head radial gates have been observed, such as the $19 \times 26\,m$ radial gate of surface spillway in Wuqiangxi in 1997 and the $13 \times 13.5\,m$ radial gate of spillway tunnel in Ertan. Similar to the discharge spray field measurement performed in Dongjiang Arch Dam in 1992, studies on Dongfeng, Manwan, Lijiaxia, Dachaoshan, and Ertan have been done for different purposes. The mechanism of cavitation damage is studied together with the field observations in many projects. Some relative parameters are also measured at the same time, such as the pressures, air concentrations, and airflow rate. More than 15 parameters have been measured up to now.

The comprehensive hydraulic field observations have been carried out in Ertan Project and ship lock of Three Gorges Project. The observations in Ertan include plunge pool, surface spillways, spillway tunnels, middle outlet gate and spillway tunnel gate, dam vibrations, and discharge sprays. **Figure 24** gives a flow chart of hydraulic field measurement method and data acquisition system which integrates advanced modern techniques, such as computer control, instrumentation, and diagnosing and processing [26]. This technique has been widely applied to large hydraulic structure observations. The measurement results contribute a great deal to the improvement of the design of hydraulic structures, such as large plunge pool, spillway tunnel, and bank protection on discharging spray, which are also important data for the evaluation of safety of hydraulic structures, mainly introduced in sections above.

The five-stage ship lock in Three Gorges Project is the largest one in the world. Many concerns have been concentrated on the filling and emptying systems, such as the pressures and cavitation behaviors in valve chamber and diversion systems, aeration and airflow during the valve opening, vibrations of valves and lock gates, stress on the lift poles of valve and AB connect pole, operation of valves, wave in the lock chamber and in the channel during the gate opening and closing, and

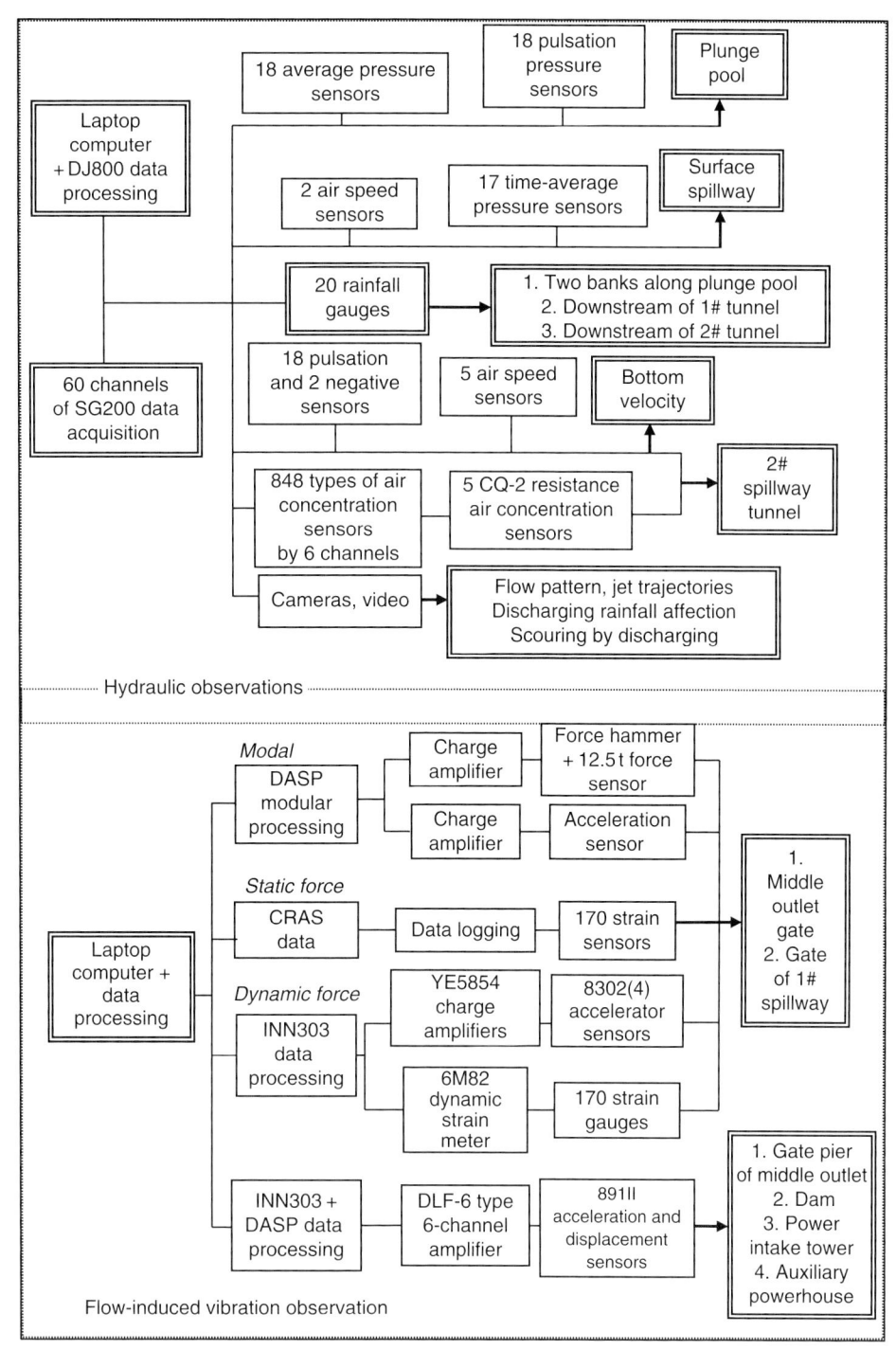

Figure 24 Flow chart of hydraulic field measurement method and data acquisition system.

tensile stress of ship rope during filling and emptying. The measurements of the valve and filling systems on try-operation were carried out in late 2002 with 10 cases [27]. The measurements on ship-lock operation were carried out in the summer of 2003. The measured results prove that the hydraulic behaviors of filling and emptying system are satisfied and the ship lock is in good operation and has achieved the design requirements. Adjustment on valve operation mode reduces the time of filling and emptying and increases the efficiency of ship transportation. **Figure 25** gives a measuring arrangement in filling system and **Figure 26** gives some measured results.

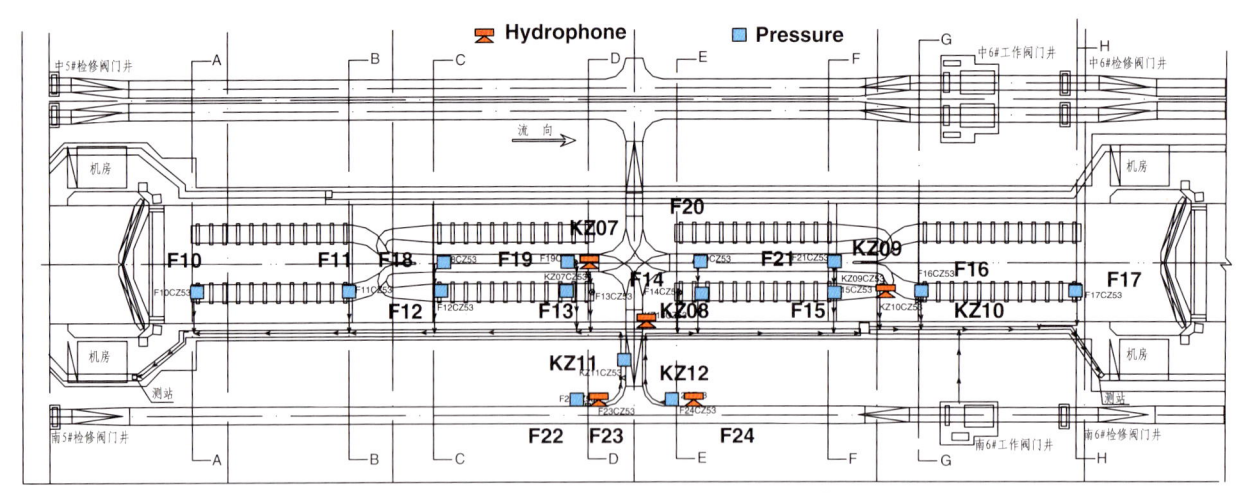

Figure 25 Measurement arrangement in the filling and emptying system in the ship lock of Three Gorges Project.

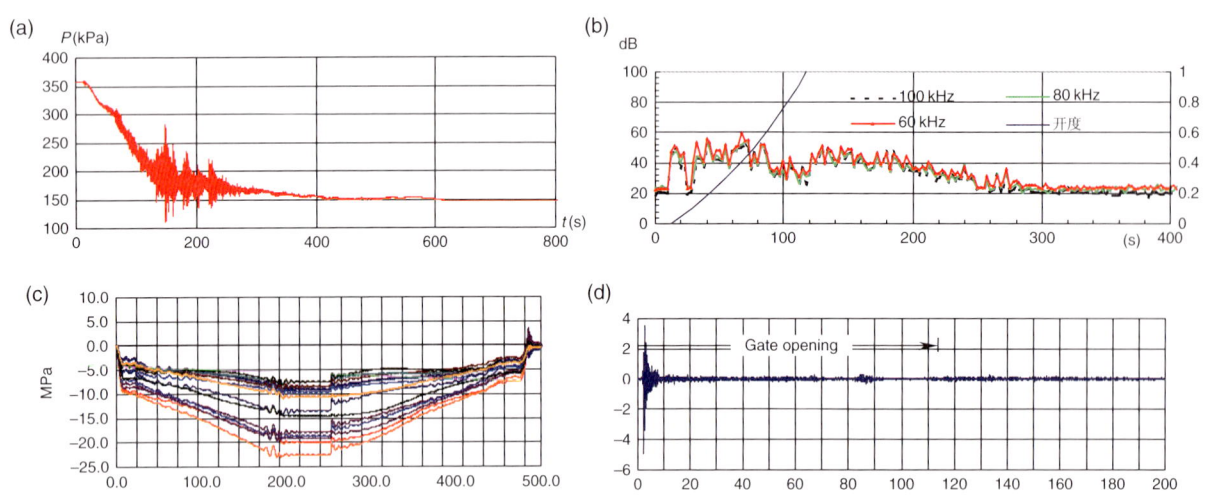

Figure 26 Measurement results in the ship lock of Three Gorges Project: (a) pressure vs. time in T pipe; (b) flow noises in valve chamber; (c) stress in A pole of gate; and (d) acceleration speed of valve.

References

[1] Pan JZ and He J (2000) *The Fifty Years of Chinese Dams*. China: China HydroPower Publication.
[2] Semenkov VM (1979) Large-capacity outlet and Spillways, G.R.50. *Proceedings of IV*. 13th ICOLD, Paris. pp. 1–110.
[3] Gao JZ and Li GF (1983) Studies on the application of Slit Bucket energy dissipators. *Journal of Water Resources and Hydropower Engineering* 14(3): 1983.
[4] Tong XW, Li GF, Xie SZ, et al. (2000) *The Contracted Types of Energy Dissipaters with High Water Head and Large Discharge Flow*. China: China Agriculture Publication.
[5] Sun SK, et al. (2006) Energy dissipation research of spillway tunnel by physical model (scale 1:45) on Xiaowan Project (in Chinese). *HY-2006-3-27*. China: China Institute of Water Resources and Hydropower Research (IWHR).
[6] Zhang D, et al. (2008) Energy dissipation research of Spillway tunnel by physical model (scale 1:30) on Jinping I Project (in Chinese). *HY-2008-065*. China: China Institute of Water Resources and Hydropower Research (IWHR).
[7] Lin BN and Gong ZY (2001) Contracted and flaring pier gates energy dissipaters (in Chinese). *Proceedings of Lin Binnan's Works*. China: China HydroPower Publication.
[8] Xie SZ and Lin BN (1992) Mechanism of energy dissipation by flaring pier gates and its hydraulic calculations (in Chinese). *Journal of Water Power* Volume 1.
[9] Guo J and Liu ZP (2003) Field observations on the RCC stepped spillways with the flaring pier gate on the Dachaoshan Project. *Proceedings of the IAHR XXX International Congress*. Theme D, pp. 473–478. August 2003.
[10] Sun SK, et al. (2005) Energy dissipation research of surface spillway by physical model on Guangzhao Project (in Chinese). *HY-2005-3-12*. China: China Institute of Water Resources and Hydropower Research (IWHR).
[11] Gao JZ and Liu ZP (2001) Prototype observation of hydraulic and flow-induced vibration for Ertan Project. *Proceedings of the Special Seminar, IAHR XXIX Congress*. Beijing, China, 2001.
[12] Sun SK, et al. (2006) Energy dissipation research of surface spillways and middle outlets by physical model on Baihetan Project (in Chinese). *HY-2006-3-55*. China: China Institute of Water Resources and Hydropower Research (IWHR).
[13] Guo J and Liu JG (2001) Pressure observation for the 1# Orifice tunnel in Xiaolangdi Project. *Proceedings of the Special Seminar, IAHR XXIX Congress*. Theme D, pp. 663–668, September 16–21, Beijing, China.

[14] Wu YH, Guo J, Zhang D, and Liu JG (2004) *The Prototype Observation of Hydraulic and Flow-Induced Vibration on Xiaolangdi 2# Orifice Tunnel*. China: China Institute of Water Resources and Hydropower Research (IWHR).
[15] Li ZY (1997) The hydraulic research on orifice tunnel (in Chinese). *Journal of Hydraulic Engineering* Volume 2.
[16] Dong XL and Guo J (2000) The study on a vortex shaft spillway tunnel with high water head and large discharge flow (in Chinese). *Journal of Hydraulic Engineering* 11: 27–33.
[17] Dong XL and Guo J (2003) The characteristics and operation reliability analysis on vortex spillway tunnels (in Chinese). *Journal of Water Power* 29(4): 33–35.
[18] Chen WX, Liu JG, Guo J, *et al.* (2007) Prototype observation of shaft horizontal vortex spillway of Gongboxia Hydropower Project. *Proceedings of the IAHR XXXII Congress*. Venice, Italy. pp. 529–536.
[19] Shi QS, *et al.* (1983) Hydraulic experimental studies on aeration mitigation (in Chinese). *Journal of Hydraulic Engineering*, volume 3.
[20] Pan SB and Shao YY (1999) Design and Application of Aeration (in Chinese), Special Thesis on the Design Code of Spillway, edited by China South Design Institute. China: China Water Resources and Hydropower Publication.
[21] Zhang D and Liu ZP (2005) Research on 3D aeration infrastructure shape of high-head, large-discharge spillway tunnel. *Proceedings of the IAHR XXXI Congress*. Theme D, September 11–16, Seoul, Korea.
[22] Zhang D, *et al.* (2006) *Hydraulic Field Observation of 1# Spillway Tunnel on Ertan Project (in Chinese)*. China: China Institute of Water Resources and Hydropower Research (IWHR).
[23] Chen WX, Li GF, Xie SZ, and Yang KL (2007) Study on aerators of high head spillway tunnels. *Proceedings of the IAHR XXXII Congress*. Theme D, July 1–6, Venice, Italy, pp 748–755.
[24] Guo J and Liu ZP (2005) Field study on the spray by discharging. *Proceedings of the IAHR XXXI Congress*. Theme D, September 11–16, Seoul, Korea.
[25] Sun SK and Liu HT (2005) The rainfall distribution of atomized flow in large dams. *Proceedings of the IAHR XXXI Congress*. Theme D, September 11–16, Seoul, Korea.
[26] Guo J (2000) Hydraulic prototype observations for Ertan Project. *Proceedings of the ICOLD XX International Congress*. Beijing, China. vol. 1, pp. 484–488.
[27] Liu JG and Wu YH (2004) The hydraulic field observation of the ship locks in three gorges project (in Chinese). *Journal of China Three Gorges Construction* (1): 17–22.